ISBN 978-1-5282-9468-3
PIBN 11007037

1 MONTH OF
FREE
READING

at
www.ForgottenBooks.com

By purchasing this book you are eligible for one month membership to ForgottenBooks.com, giving you unlimited access to our entire collection of over 1,000,000 titles via our web site and mobile apps.

To claim your free month visit:
www.forgottenbooks.com/free1007037

English
Français
Deutsche
Italiano
Español
Português

www.forgottenbooks.com

Mythology Photography **Fiction**
Fishing Christianity **Art** Cooking
Essays Buddhism Freemasonry
Medicine **Biology** Music **Ancient
Egypt** Evolution Carpentry Physics
Dance Geology **Mathematics** Fitness
Shakespeare **Folklore** Yoga Marketing
Confidence Immortality Biographies
Poetry **Psychology** Witchcraft
Electronics Chemistry History **Law**
Accounting **Philosophy** Anthropology
Alchemy Drama Quantum Mechanics
Atheism Sexual Health **Ancient History**
Entrepreneurship Languages Sport
Paleontology Needlework Islam
Metaphysics Investment Archaeology
Parenting Statistics Criminology
Motivational

Transactions
of the
Illuminating Engineering
Society

VOL. XVIII
JANUARY–DECEMBER
1923

Contents
Papers, Reports and Discussions

Subject Index
Personnel Index

ILLUMINATING ENGINEERING SOCIETY
29 WEST THIRTY-NINTH STREET
NEW YORK, N. Y.

CONTENTS

PAPERS, REPORTS AND DISCUSSIONS

This list gives the subjects and authors in sequence as published in the TRANSACTIONS. The issue number (1 to 10) of the TRANSACTIONS precedes the page number.

SUBJECT INDEX

FOREWORD—The various topics and sub-topics are listed alphabetically. It is suggested that related or functional topics should be reviewed when one desires all references on a given subject. The issue number (1 to 10) of the TRANSACTIONS precedes the page number.

PERSONNEL INDEX

The letter "d" indicates matter given in the discussion of certain papers. The issue number (1 to 10) of the TRANSACTIONS precedes the page number.

TRANSACTIONS
OF THE
ILLUMINATING ENGINEERING SOCIETY

Vol. XVIII   January, 1923   No. 1

Suggestions for Improving the TRANSACTIONS

B EGINNING its second year's work the Committee on Editing and Publication wishes to set forth plans for the development of the TRANSACTIONS in the hope that through the continued interest on the part of each individual member and his
greater participation in the supplying of interesting news of the industry, we may enlarge the various sections which were added
during the past year and introduce new departments when the
demand becomes apparent. The aim of the committee on assuming office last year was to increase the general interest in the
TRANSACTIONS. The scientific value has always been maintained
by the very excellent papers which have been available for publication. A demand was evident for news of the industry, and of
the activities of the Society's committees and sections in order
that the non-technical member and the member at a distance from
the influence of the Sections might be placed more closely in
touch with the very notable work accomplished from time to time
by the Society's representatives. For this purpose, during the
past year, a rearrangement was made of the TRANSACTIONS and
several new sections are now open to the members for their contributions.

The editorial department which was a new feature inaugurated
last year will be continued and plans are under way to present
editorials of more value and interest to the membership.

In "Reflections" we have a news gathering section for the
dissemination of information interesting to the illuminating engineering profession. In it will be presented reprints from maga-
zines and papers, new items about notable lighting installations,
new instruments, etc. Members are urged to send in clippings
or to write news items for publication.

The "Papers" section can be improved if those members who have papers or discussions to present will aid in decreasing time between the presentation of the paper and its publication. Delays are occasioned by a number of causes, not any one of which is responsible for a large amount of lost time but in the aggregate interfering with printing schedules. Papers pass through at least eight different channels before printing so that a delay on the part of any one, or the cumulative small delays on the part of several may prevent prompt publication in the TRANSACTIONS.

Section officers and managers, and the various committee chairmen are especially requested to send more information about the activities of their respective sections or committees for insertion in "Society Affairs." The place of the meeting, the time and name of the speaker and his subject are not the most interesting features of the section meeting. It would, however, be of greater value to the membership if, in addition, a short abstract were given of the paper presented and the important points emphasized in its discussion. An exchange of ideas regarding new methods or suggestions for stimulating the interest of the members in the meetings of their respective sections or chapters, could be fostered by the various secretaries in this department.

In general, the committee bespeaks from the membership a little more personal interest and aid in developing the news bearing portion of the TRANSACTIONS to the point where it will be of equal value with the material in the "Papers" section, which has been a characteristic feature of the TRANSACTIONS in the past.

The committee also solicits communications which should be of general interest to the membership, including not only information of important activities, but should if possible express the writer's opinions, recommend courses of action, criticise present practices or anticipate future endeavors.

COMMITTEE ON EDITING AND PUBLICATION

REFLECTIONS

Highway Lighting Financed

THERE has recently been passed by New York State a measure which allows the counties of the State to appropriate money for the illumination of highways. This bill is an important and logical step toward providing means of financing highway illumination, placing authority in the hands of the board of supervisors of any county to provide for lighting public highways, or portions thereof, or bridges, located in such county outside of cities and villages. The board must submit proposals to the State Commissioner of Highways when it is desired to provide lighting for some highway or bridge. The Commissioner either approves the proposals or suggests such modification as he sees fit, returning them to the board for final adoption. The expenses for installing and maintaining the lighting system are provided and appropriated in the same manner as other county expenses.

The importance of highway lighting is becoming more seriously recognized in all parts of the country and the passage of this bill will prove a big help in the rapid development of New York's highways to meet the requirements of modern traffic. Other states will doubtless consider the inauguration of similar legislation.

The Light Cure

TREATMENT of disease by light rays is reviewed in an editorial contribution to *The Journal of the American Medical Association* (Chicago). The writer is somewhat cautious in commending it, but thinks that it has been undoubtedly beneficial in some maladies, and that it has possibilities. That sunlight is beneficial to green plants, he notes, is an every-day observation. Through the intermediation of chlorophyl (green vegetable coloring matter) light energy is stored under conditions most advantageous to mankind. There is a wide-spread conviction that sunlight is health giving to man as well as to vegetation; but proof of this traditional belief, he thinks, is not so easily secured.

"Sunlight is only one of numerous environmental factors to which the human organism is subject, and they can not readily be dissociated so that each can be charged with its specific responsibility for well-being or the reverse, as the case may be. It would be manifestly unfair to say that heliotherapy (sunlight treatment) is an entirely unexplored field; but it will scarcely be denied that its present claims and its accomplishment are essentially based on empiricism. To admit this is not derogatory to the possibilities of medical treatment through the agency of light rays, but rather a challenge to promote the scientific aspect of the subject. The latest studies of the sunlight treatment of rickets, notably those of A. F. Hess and others in this country, should give a marked impetus to the investigation of the physical factors as well as the clinical results.

The instance cited is a notable one for heliotherapy. Clark has suggested other fields in which the already evident possibilities of light therapy deserve experimental consideration from a clinical standpoint. The pioneer work of Finsen emphasizes the importance of considering a diversity of forms of radiant energy in skin affections. The relative safety of treatment with ultraviolet rays over roentgen-ray exposure should add to the desirability of careful consideration of the respective merits. In tuberculosis, especially surgical tuberculosis, heliotherapy has long had advocates. The observations made have warranted the suspicion that light of short wave-lengths, which is known to have marked bactericidal effects, may not be without salutary influence in the treatment of wounds. Other suggestions might be cited, while the familiar inflammatory reaction known as sunburn is an omnipresent reminder of the potency of absorbed ultraviolet light. It should be remembered that the potent rays of shorter wave-length do not penetrate glass. The shorter the wave-length, the smaller the layer of skin that will absorb the rays. Artificial lights, if glass-covered, are therefore harmless and therapeutically weak. Sunlight rarely contains enough for ultraviolet rays to produce injury. Consequently, heliotherapy that demands highly potent effects must look to artificial sources of radiation. The quartz mercury arc and bare metallic arcs are known to belong in the potent class, and, as has often been warned, may be extremely injurious so that the eyes should be protected from them."

PAPERS

THE RELATIVE PERFORMANCE OF TUNGSTEN FILAMENT LAMPS UNDER TEST UPON ALTERNATING AND DIRECT-CURRENT CIRCUITS*

BY JOHN W. LIEB**

Under service conditions some incandescent lamps are operated on alternating-current circuits and others are operated on direct-current circuits. In laboratory practice, however, because of the ease and flexibility of alternating-current transformation, lamps are usually life-tested with 60-cycle alternating current, and only rarely are they tested with direct current. In the lamp testing work of the Association of Edison Illuminating Companies, however, it has been the practice for some years to test a small percentage of samples upon direct current in order to ascertain differences, if any, in performance under the two methods of operation. The purpose of this paper is to make available to illuminating engineers facts as to differences in alternating current and direct-current performance which have been disclosed by these tests.

HISTORICAL

At the time of the appearance of the tantalum lamp, there was no evidence or data available indicating that the carbon and Gem lamps of that day were in any way differently affected in

Note: In order to ascertain the performance and quality of incandescent lamps and thus stabilize the purchase or sale of such lamps, the Association of Edison Illuminating Companies in 1896 agreed with certain manufacturers of lamps upon a procedure for conducting laboratory tests. Such tests have been made continuously ever since. The results of these tests are not published but are available to the companies for whom they are made and furnish an index of the quality of the product and in formulating such specifications for the product as may be agreed upon

*A paper presented at the Annual Convention of the Illuminating Engineering Society, Swampscott, Mass., Sept. 25-28, 1922.

**Vice-President, The New York Edison Co., New York City.

The Illuminating Engineering Society is not responsible for the statements or opinions advanced by contributors.

from time to time between purchasers and manufacturers of lamps. The data in this paper are taken from tests made in the course of this work by Electrical Testing Laboratories under the supervision of the Lamp Committee of the Association of Edison Illuminating Companies. These. are laboratory tests in which to a certain extent the more trying conditions of actual service are modified in order to secure reproducibility and comparability of test results. Thus the voltage regulation is superior to that of service conditions, and, unless otherwise stated, the lamps while under test are not subjected to appreciable vibration and are protected from shock and jar, except upon occasions when they are removed from test racks for examination.

their performance, depending upon whether they were burned on alternating or direct current. Even at the present time, in spite of systematic tests on carbon and Gem lamps calculated to bring out such a difference, if it existed, none has been found. When the tantalum lamp first made its appearance, a number of the earliest lamps were submitted to test under the direction of the A. E. I. C. Lamp Committee, and it was found that when they were tested in the ordinary way on alternating-current circuits, the life was very much shorter than had been claimed for them. It was at that time a new idea to seek for differences in performance of incandescent lamps depending upon the character of the current employed but in order to seek an explanation for the differences found in the case of the new lamps some of them were tested on direct current, and the life figures thus obtained were found to be in general agreement with those that had been published abroad. On 60-cycle alternating current a life of only about one-fourth of that observed with direct current was found on test. The first published announcement of this difference was made in a paper before the American Institute of Electrical Engineers[1] using as a basis the tests made for the Lamp Committee. In this paper also was given the physical reason accounting for this difference. It was shown that microscopic examination of the filaments revealed that they were subject to crystallization and so-called "offsetting" a breaking up of the filaments which went on much more rapidly with alternating current than with direct current, and figures illustrating that effect were published.

[1] C. H. Sharp, TRANSACTIONS, A. I. E. E., Vol. 25, Page 828, 1906.

The following statement appeared in the A. I. E. E. paper:

> "The conclusion is unavoidable that this lamp (namely the tantalum lamp) at the present time is essentially a direct-current lamp."

and again:

> "No such effect is observable with the tungsten lamps. Tests of Electrical Testing Laboratories show quite definitely that their life on direct current and on alternating current is the same. This has also been proved by elaborate experiments to be true of carbon filaments."

The above statement with regard to tungsten filament lamps refers of course to the squirted and sintered tungsten filament of that time. A draw wire filament, the desirability of which was early recognized, was scarcely looked forward to as being a possibility, because of the well known difficult characteristics of the metal tungsten. However, it is very interesting to note that in the discussion of this paper Dr. C. P. Steinmetz made the following statement:

> "The manufacture of tungsten lamps would possibly be simplified if the metal could be made ductile. I must draw your attention to one feature, however, of all the incandescent filaments, the only one not suited for alternating current is the ductile tantalum filament, which shows a feature similar to the crystallization of wrought iron under rapidly oscillating stress, while the osmium and carbon filaments, which have no fibrous structure, show no inferiority with alternating current. It appears to me possible, therefore, that a ductile tungsten filament may be unsuitable for alternating current."

When through the labors of Dr. W. D. Coolidge at the Research Laboratory of the General Electric Company the "impossible" was achieved, and ductile tungsten became a reality, the prophetic surmise of Dr. Steinmetz was justified. It was found in reality that the early ductile filament tungsten lamps showed a decided tendency to offsetting which, due to the difference in the character of the crystallization was greater with alternating than with direct current. However, the research scientists who were working on the incandescent lamp were not at the end of their resources, and means have been found whereby this effect has been very greatly reduced. The method most commonly employed is understood to be that devised by Coolidge, which consists in adding a small quantity of thoria to the tungsten. The result of such a procedure has been to produce lamps which are commercially satisfactory when operated either on alternating or on direct-current circuits. Indeed there is no general recognition of any difference in performance even by the lamp industry itself. It has remained for the more refined and searching tests of the laboratory to detect and measure these differences.

TEST PERFORMANCE OF TUNGSTEN FILAMENT LAMPS

In the tests forming the basis of this paper, the average voltage is carefully established and is maintained within about one-fourth per cent, plus or minus, for the alternating-current tests, and about one-half per cent, plus or minus, for the direct-current tests. The wave form of the 60-cycle current employed is substantially sinusoidal. Except as otherwise stated, the vacuum type lamps have been operated with axes either horizontal or vertical and gas-filled lamps have been operated tip downward. Also, except as otherwise stated, the operation has been upon racks practically free from vibration and at voltages corresponding closely to rated initial efficiencies. Four or five times during the life of lamps of normal longevity they have been removed from the test racks for photometry and examination, but otherwise they have not been handled.

Tests have been made of both the straight filament vacuum type lamp and of the coil wound gas-filled lamp. They have included a considerable number of different brands which come under investigation by the Association of Edison Illuminating Companies. In both types and among these different brands performance differences have been found in varying degrees. Apparently these appear to be to some extent within the control of the manufacturer, although possibly improvement in certain particulars may be had only at the expense of inferiority in some other respects.

In the case of the vacuum type lamp, the following characteristic differences have been definitely established as a result of the tests:

1. The filament resistance increases more rapidly with age on a-c. than on d-c.
2. The fragility of the filament increases more rapidly with age on a-c. than on d-c.

3. The mortality *rate* increases more rapidly during life on a-c. than on d-c.

4. The efficiency during life declines less rapidly on a-c. than on d-c.

5. The total life is less on a-c. than on d-c.

Similar differences in performance are encountered also to a greater or less extent among gas-filled tungsten lamps, although as a general proposition the differences are not so well established and, in the case of some gas-filled lamps, a reversal of results is found in certain particulars, as will be shown later.

Crystallization on Alternating and Direct Current.—Something of the difference in crystallization characteristics of filaments when operated on alternating and direct current is conveyed by Fig. 1, in which the cuts are photomicrographs (magnified in the original about 100 diameters) of sections of 300-watt new tungsten lamp filaments. The cuts at the left show new filaments; the cuts in the middle show crystallization of corresponding filaments which have been operated upon direct current until the lamps have lost about 20 per cent of their initial lumens; the cuts on the right show crystallization resulting from corresponding operation upon 60-cycle alternating current. Fig. 2 illustrates crystallization in the case of 60-watt vacuum type lamps, the photomicrographs being similarly arranged. The filaments operated upon direct current are characterized, generally, by greater surface disorder, while those operated upon alternating current are broken up into larger crystals, apparently with deeper seated lines of cleavage. These changes appear to be characteristic effects of the operation of both vacuum and gas-filled types of Mazda lamps upon direct current and alternating current respectively.

Extent of Comparative Life Tests upon Alternating and Direct Current.—Beginning with 1910, small groups of fairly comparable tungsten filament lamps have been tested each year with 60-cycle alternating current and with direct current. The samples tested have ranged in number from ten to forty lamps per group. The total number of lamps involved in these comparisons throughout the years covered by the tests are:

Alternating current	2,039 lamps
Direct current	865 lamps

From time to time these samples have been of the 25, 40, 50, 60 and 100-watt Mazda B type lamps and 75, 100, 200 and 300-watt Mazda C type lamps. Prior to 1920 the tests were made with a view to arriving at the "useful" life and therefore were discontinued shortly after the lamps had declined to 80 per cent of initial lumens. Subsequent to 1920 the tests have been continued until the lamps failed.

Lamp Efficiencies.—The efficiencies at which these lamps have been operated have varied somewhat throughout the years. To avoid the complication thus introduced into the comparison of performances, the data shown in this paper have been adjusted by applying the recognized exponents connecting efficiency and life to the performance at the efficiencies at which they were actually tested so as to reduce the performance in each case to the equivalent performance at such efficiency as would result in an average total life upon alternating current of 1,000 hours, thus making the life results comparable even though the tests were actually made at different efficiencies.

USEFUL LIFE TESTS

In the accompanying table all tests made on the mean useful life basis are summarized with respect to those particulars treated of in this paper. The useful life(hours life to 80 per cent of initial lumens or to earlier failure) is shown, because most of the test results are available in that form. This measure of lamp life performance served a useful purpose in emphasizing the importance of good candlepower maintenance, but is no longer relied . upon to evaluate lamp performance.

In considering the practical value of the lamps during their useful life a tendency was found for the superior lumen maintenance to compensate more or less for the inferior mortality record of the Mazda B lamps operated upon alternating current. Among the vacuum lamps of the smaller sizes the difference in mortality rate was most notable. Among the Mazda C or gas-filled lamps there were two types of filaments appearing in the tests shown in the table as respectively I and II. Among these lamps in general the superior lumen maintenance upon alternating current was not offset, as in the case of the vacuum lamps, by a marked betterment of mortality rate upon direct current. The

SUMMARY OF RELATIVE TUNGSTEN LAMP LIFE PERFORMANCE ON
ALTERNATING AND DIRECT CURRENT

		Alternating current		Direct current		
Year	Lamp rating in watts	No. of lamps tested	Per cent burn-out above 80 % of initial lumens	No. of lamps tested	Per cent burn-out above 80 % of initial lumens	Ratio d-c. "useful" life to a-c "useful" life
			Mazda B Lamps			
1911-12	25	33	66	23	0	100.7
1912-13		214	52*	42	5*	110.5
1914-15		215	42*	40	0*	91.4
1916-17		30	80	29	7	115.8
1918-19		30	50	26	15	87.3
1910-11	40	44	4.5	50	0	87.5
1911-12		29	34	23	9	89.3
1912-13		150	39*	38	0	93.9
1914-15		231	23*	41	0	86.2
1916-17		28	25	28	7	88.7
1916-17	50	23	48	30	0	113.0
1918-19		24	21	27	0	83.2
1919-20		9	22	10	0	88.0
1920-21		30	8	30	8	84.8
1911-12	60	24	29	24	0	114.8
1912-13		133	18*	28	18*	80.8
1914-15		289	18*	37	8*	84.2
1916-17		19	0	20	0	87.0
1911-12	100	23	0	19	0	80.7
1912-13		116	7*	33	6*	103.1
1914-15		68	13*	24	17*	87.9
1916-17		20	0	19	0	66.4

*Per cent of failures throughout a comparable period, but independent of actual decline in lumens.

		Alternating current		Direct current		
Year	Lamp rating in watts	No. of lamps tested	Per cent burn-out above 80 % of initial lumens	No. of lamps tested	Per cent burn-out above 80 % of initial lumens	Ratio d-c. "useful" life to a-c "useful" life
			Mazda C Lamps			
1916-17	75 (II)	7	100	6	17	56.6
1918-19	(I)	22	50	23	39	84.7
1916-17	100 (I)	10	40	10	0	64.0
1918-19	(I)	20	50	25	12	74.5
1919-20 A	(I)	4	50	5	0	100.4
B	(II)	10	100	4	75	79.8
C	(I)	5	40	5	20	61.2
1920-21	(I)	20	40	20	15	60.7
	(II)	10	100	10	67	64.6
1916-17	200 (I)	9	11	10	0	86.8
1918-19	(I)	10	30	8	13	67.0
1919-20 A	300 (II)	10	80	5	20	76.3
B	(II)	9	78	5	0	108.1
C	(I)	4	50	5	20	45.8
1918-19	500	8	13	8	13	109.4

result was a rather general shortage of *useful* life for the direct-current lamps, even though as shown elsewhere, the total life has been found during the last two years to be slightly longer upon direct current than upon alternating current.

TOTAL LIFE TESTS

In the following paragraphs are summarized the relative test data for the last two years, during which period the total life—to failure in each case—was determined.

Change in Light Production and Efficiency throughout Life— Vacuum Type Lamps.—In the accompanying Fig. 3, performance curves are shown for 50-watt, S-19 bulb Mazda B lamps of Edison, National and Westinghouse makes combined. These have been adjusted in each case to an efficiency which would produce an average total life of 1,000 hours when operated in the laboratory upon 60-cycle current. Curves of performance with direct current at the same initial efficiency are included in the same diagram for comparison. The lamps were of 1921 production. The average lives when operated at the same initial efficiency are:

50-WATT MAZDA B LAMPS

	60-Cycle A-c.	D-c.
Average hours total life	1,000	1,600
Relative mean lumens per watt throughout life in per cent of that at initial	100%	91%

The better maintenance of lumens in alternating-current operation combined with a slightly increased rate of decline in watts contributes to a much better maintenance of efficiency throughout life. When operated at the same *initial efficiency,* corresponding lamps in direct-current service decline in lumens somewhat more rapidly, consume relatively more watts, and therefore operate throughout at a lower efficiency. The average total life of the d-c. lamps is longer by about 60 per cent than when operated on alternating current.

Adjustment to equivalent *mean efficiency throughout total life* for purposes of comparison, such as that called for by the latest standard specifications for incandescent lamps, establishes the

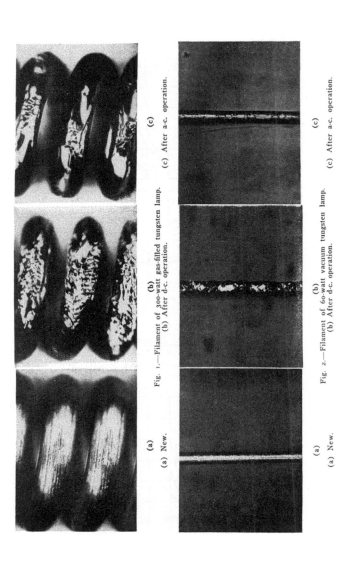

(a)   **(b)**   **(c)**

Fig. 1.—Filament of 300-watt gas-filled tungsten lamp.

(a) New.   (b) After d-c. operation.   (c) After a-c. operation.

(a)   **(b)**   **(c)**

Fig. 2.—Filament of 60-watt vacuum tungsten lamp.

(a) New.   (b) After d-c. operation.   (c) After a-c. operation.

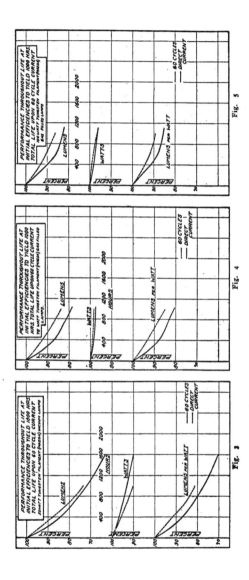

Fig. 3

Fig. 4

Fig. 5

fact that for a given mean efficiency throughout life the lamps will actually operate for a longer period upon alternating current than upon direct current. When, however, as is the custom, they are operated at labeled voltage, a longer life with a somewhat lower mean efficiency is had in direct-current operation than in alternating-current service.

Gas-filled Lamps.—The differences in performance with alternating and direct current are smaller among gas-filled tungsten lamps than among vacuum tungsten lamps. Fig. 4 shows average performance curves for 75-watt lamps of 1921-22 manufacture, the figures for Edison, National and Westinghouse makes being consolidated. The same characteristic differences in lumens which were noted in the vacuum type lamp are here manifested. The watts, however, are slightly lower upon direct current and the resultant life is but little longer than that upon alternating current. Adjusted to equivalent mean efficiency throughout life for purposes of comparison, the life on alternating current is, as in the case of vacuum lamps, longer than that on direct current.

Fig. 5 shows typical performance curves for 100-watt lamps of Edison and National makes combined. The average relative lives of these gas-filled lamps are:

	75-Watt		100-Watt	
	60-cycle a.-c.	d.-c.	60-cycle a.-c.	d.-c.
Average hours total life	1,000	1,043	1,000	1,108
Relative mean lumens per watt throughout total life in per cent of that at initial	100%	97%	100%	95%

In General.—The general indication is that by the time the lamps have reached three-quarters of their total life, the lumens of those operated upon direct current are 3 to 5 per cent lower and the watts, in the case of the vacuum lamps, are about 1 per cent higher and, in the case of the gas-filled lamps, about 1 per cent lower when the lamps are operated upon direct current.

Relative Increase in Filament Fragility Throughout Life.— There is but little definite information on fragility available to the writer, and this data is entirely of laboratory origin. It is known that all tungsten filament lamps are relatively sturdy when new, but become fragile after some hours of operation and increase in fragility thereafter as they are burned. This effect is accelerated if the lamps are subjected to vibration or jar. Perhaps the most reliable indication of fragility is that deduced from the operation upon stationary racks supplied with respectively 60-cycle current and direct current, and afforded by filament breakage in incidental handling for photometric testing of lamps which are being life-tested. This handling is substantially equivalent for the two kinds of operation. Such filament breakage summed up in the following table, is the total breakage encountered in the entire life-testing process in the course of handling, photo-metering and inspecting of lamps, during equal hours of operation.

HANDLING BREAKAGE DURING LAMP LIFE TESTS

	A-c.	D-c.
	%	%
Vacuum type lamps		
25-watt lamps	12.7	6.0
40-watt lamps	3.4	1.6
50-watt lamps	9.6	1.1
60-watt lamps	3.9	6.8
100-watt lamps	5.9	3.0
Gas-filled lamps		
75 and 100-watt lamps	13.0	18.4
200-watt lamps	9.5	5.3

Mortality.—In studying the laboratory performance of incandescent lamps, it is our practice to consider as natural failures lamps which actually "burn out" while on the test racks, as well as those having filaments which are broken in the ordinary course of careful handling incidental to the testing work. In the statistics presented in this paper, all filament ruptures are therefore assumed to be natural failures provided they occur at times when the filament is believed not to have received any unusual jar or blow which might have broken it.

The mortality characteristics of 50-watt Mazda B lamps and 100-watt Mazda C lamps upon alternating and direct-current circuits respectively are shown in Figures 6 and 7. Among the Mazda B lamps the mortality rates are quite similar for the first 800 hours, after which failures among lamps on alternating current occur with increasing frequency, while the mortality rate does not change much throughout on direct current. Among the 100-watt Mazda C lamps, those operated on direct current are characterized by more failures during the first 900 hours and subsequently by a lower rate of failure.

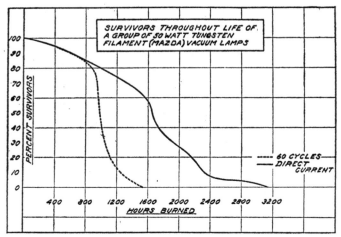

Fig. 6

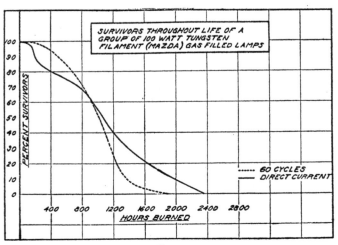

Fig. 7

The statistics of these failures, classified in 200-hour periods, with data adjusted to efficiencies necessary to yield a total life of 1,000 hours upon 60-cycle current, are summarized in the following table, which shows a greater concentration of failures between 600 and 1,200 hours among lamps operated upon alternating current with fewer relatively long lived lamps. It is difficult to dissociate this result from that previously shown of increased breakage in alternating-current operation particularly among vacuum lamps, since they may have a common origin in the greater offsetting of filaments encountered in alternating-current operation, due to the peculiar crystallization.

FAILURES IN 200-HOUR PERIODS

Hours	50-Watt vacum		Gas-filled	
	60-cycle a-c. %	d-c. %	60-cycle a-c. %	d-c. %
0-200	0.0	0.0	0.0	10.0
200-400	0.0	3.5	5.0	7.5
400-600	6.9	3.5	10.0	5.0
600-800	7.0	6.8	10.0	10.0
800-1000	41.0	7.0	20.0	10.0
1000-1200	34.8	6.8	27.5	20.0
1200-1400	0.0	3.5	17.5	5.0
1400-1600	10.3	7.0	5.0	10.0
1600-1800	0.0	24.0	2.5	7.5
1800-2000	0.0	6.8	2.5	5.0
2000-2200	0.0	10.3	0.0	5.0
2200-2400	0.0	10.3	0.0	5.0
2400-2600	0.0	3.5	0.0	0.0
2600-2800	0.0	0.0	0.0	0.0
2800-3000	0.0	3.5	0.0	0.0
3000-3200	0.0	3.5	0.0	0.0
	100	100	100	100

CONCLUSION

This paper sets forth only the relative and comparable laboratory performances of tungsten lamps on 60-cycle and direct current. It does not show absolute life values.

Commercial tungsten lamps give satisfactory performance in service on either alternating or direct-current circuits. It is only through careful laboratory tests directed toward the particular end of discussing differences however minute, that any characteristic differences in performance have been established. The data from these tests are not conclusive as indicating clearly any superiority of one type of electric service as far as the performance of the lamps may be concerned, but they do supply interesting information which may be of importance in research work and for the investigation of filaments directed toward further improving the lamp product.

In the testing of incandescent electric lamps in the laboratory, the tendency is to seek accurate reproducibility of results through the elimination or standardization of variables. However strong the arguments in favor of this course, it is important to remember always that there are many variables in the use and performance of lamps in actual service which must be taken into account in evaluating lamps and in practical illuminating engineering. In order to attain reproducibility, laboratory tests must be conducted under ideal conditions, with uncertain and disturbing service conditions eliminated as far as possible so as to make the test results comparable. These laboratory tests must be supplemented, however, from time to time by special investigations to ascertain the influence of each of the important service variables, to the end that our knowledge of the actual performance of lamps in service may be comprehensive and of practical value. The difference in performance of tungsten filament lamps upon alternating and direct current is an illustration of this point, presenting interesting information from a scientific standpoint, but numerically not of sufficient significance to be of themselves a compelling factor in the practical application of lamps to commercial light circuits.

PARTIAL BIBLIOGRAPHY OF PERFORMANCE OF TUNGSTON FILAMENT LAMPS

Amrine, T. H. and Guell, A.—Tests of Tungsten Lamps, *Elec. World*, Vol. 55, March 17, 1910, p. 698.

Coolidge, W. D.—Ductile Tungsten, *Trans. A. I. E. E.*, 1910.

Howell, J. W.—Metal Filament Lamps, *Trans. A. I. E. E.*, Vol. 29, 1910, p. 927.

Langmuir, I. and Orange, J. A.—Tungsten Lamps of High Efficiencies, I and II, *Trans. A. I. E. E.*, Vol. 32, pp. 1,913 and 1,935.

Lavender, F. H. R.—Tests on Metallic Filament Lamps, *Journal, Inst. E. E.*, Vol. 44, Feb., 1910, p. 181.

Merrill, G. S.—Tungsten Lamps, *Trans. A. I. E. E.*, Vol. 29, 1910, p. 1,709.

Middlekauff, G. W., Mulligan, B. and Skogland, J. F.—Life Testing of Incandescent Lamps at the Bureau of Standards— U. S. Bureau Standards Scientific Paper No. 265, March 16, 1916.

Patterson, C. C. and Rayner, E. H.—Life Tests on Glow Lamps, *Electrician* (London), Vol. 65, Sept. 9, 1910, p. 893.

Sharp, C. H.—New Types of Incandescent Lamps, *Trans. A. I. E. E.*, Vol. 25, 1906, p. 815.

Sharp, C. H. and Millar, P. S.—Comparative Tests of Glow Lamps, *Electrical World*, Vol. 56, Sept. 22, 1910, p. 665.

DISCUSSION

W. H. ROLINSON: It is not my wish to discuss in detail Mr. Lieb's paper, as the value of the technical work contained, therein is sufficient unto itself. I wish, however, to take the opportunity before this Society of expressing my appreciation of the valuable work which has been done by the Lamp Committee of the Association of Edison Illuminating Companies in the matter of making better illuminants. I think as an illuminating engineer I may safely say that one of the principal reasons you have these wonderful incandescent lamps to-day, uniform in quality, uniform in light output or nearly so, and the reason you have satisfactory performance on any kind of current is in no small measure due to the research conducted through many years by this Lamp Committee. I believe the whole lighting industry owes them a deep debt of gratitude.

J. B. TAYLOR: There are many ways in which incandescent lamps can be tested and compared. The user of a comparatively small number of lamps,—for example, the average householder,— naturally wishes to keep the sockets filled with serviceable lamps, but often worries along with partial replacements because of the special effort in time and planning to select and purchase a few renewals. It is probably procrastination rather than expense which delays replacements with the majority of customers.

Now a bare statement of one of the conclusions of this paper, that a lamp operated on alternating current will fail sooner than the same lamp would if operated at the same voltage on direct current, is a disturbing fact to lodge in the backs of the heads of a large number of customers of a-c. service. The fact, also found in the paper, that for the shorter life there are other characteristics which compensate, or perhaps more than compensate, is not easy to demonstrate and is beyond the comprehension of many.

From the data presented it appears that if further tests are made, or new curves plotted on a different basis of comparison from data already available, other conclusions just as true but giving a different impression to the casual lamp user may be presented. Such a comparison would be not on the basis of operating the lamps on a-c. and d-c. at a voltage giving equal light efficiency when the lamps are new, but at one voltage for a-c. and another voltage for d-c. which will give the same average efficiency figured throughout the life of the lamp. In order to obtain equal average efficiency and also equal life from two identical lamps, one on a-c., the other on d-c., the terminal voltage d-c. must be kept higher than the a-c. voltage. More concretely, though I am jumping at the figures to illustrate the point: 115-volt lamp on a-c., maintain 113 volts; 115-volt lamp on d-c., maintain 117 volts.

By the time these lamps have lived their life, each will have given approximately the same number of lumen-hours and each will have caused the meter to register about the same number of kilowatt-hours. But the a-c. lamp will have worked more evenly, and the d-c. lamp will have worked hard when young and then taken things easier after middle life.

WARD HARRISON: There are several matters pertaining to lamp performance which have long been puzzling, but which are cleared up at once by the admirable presentation of the facts to which we have just listened. For example, it has been our experience that the life of incandescent lamps in industrial plants is frequently longer than on our own test racks even though operated at the same average voltage. Inasmuch as the lamps out in service are subject to more severe vibration than on our racks and also since for the same average voltage the extremes are much greater it has always been hard to believe the figures. However, when we couple with Mr. Lieb's data the fact that perhaps a majority of the industrial plants use direct current the answer is obvious. It likewise becomes clear why we sometimes find lamps operating at considerably less than rated candle-power in such plants.

Louis Bell: I think Mr. Lieb is vastly to be felicitated on the successful outcome of this long and laborious investigation.

I remember very long ago now, when the tantalum lamp first turned up, conducting some experiments leading to results such as Mr. Lieb showed, and it then became clearly evident that the tantalum was not in it as an a-c. proposition.

With the advent of the tungsten lamp the same question in a very greatly modified form began to come up, to which the answer has just been given.

The interesting thing from the practical standpoint, aside from noting the really small difference in the performance of the lamps as a whole, is that now for the first time we seem to have, at least for a-c. service, the kind of lamp that we have been looking for for a good many years, the kind of lamp which goes on cheerfully at a good efficiency up to a certain point, and then presto, goes out. Dr. Osler some years ago made strenuous remarks about the desirability of chloroforming some of his dear colleagues at a certain point in their senescence. His theory seems to have been rather neatly carried out on this a-c. use of the tungsten lamp: it goes to about a thousand hours and then it drops rapidly, going down with increasing speed, so that nobody is encouraged to use it; after about the proper length of life, it is down-and-out. It is just what we have been looking for in a general way. The remarkable thing about it is that a lamp which was not produced with any conception of this particular property should have automatically shown it.

G. H. Stickney: Question has been raised as to the method of plotting the curves. I prefer that used in the paper, for the reason that it is simple and definite, and does not suggest an over emphasis on the slight difference. The data is of interest to those who study lamp performance with exactness. I can see how lamp engineers, testing authorities, and some central station experts would desire this information. They would know how to use it, and could shift the curves to meet the requirements of a specific problem.

The tests show so little difference between the performance on d-c. and a-c., that so far as the vast majority of light users are concerned, no distinction should be made. They will be benefited by regarding the lamps as interchangeable.

Some ten years or more ago, there was quite a little correspondence published in the electrical press over the effect of various a-c. wave forms on lamp performance. I assume this would effect an even smaller difference than that between a-c. and d-c. I would like to inquire if anything regarding wave shape was brought out in these tests.

A. H. TAYLOR: I wanted to ask whether there was any difference found in initial efficiencies of lamps measured on d-c. and a-c. For instance, if you measure the lamp at 110 volts, a-c., would the candlepower and efficiency be the same at 110 volts, d-c.? I would like to know in photometering these life test lamps, whether lamps which were life tested on a-c. were photometered on a-c. for the purposes of this investigation?

L. C. PORTER: Mr. Lieb touched on one point in his paper which seems to me to suggest some further valuable research. Where we are studying the details of lamp performance and getting down to such minute differences as are brought about between the use on a-c. and d-c., it certainly would be interesting to get more accurate data than is now available for the use of lamps, for example, under vibration and for the use of flashing lamps. The applications are coming now of lamps in many cases where they are subject to different kinds of vibrations, particularly in railway signal work. We have been up against quite a lot of difficulty, due to the lamps that are put on the signals being subject to vibration each time a train passes, and it seems to make quite a little difference in the performance of the lamps as compared with their performance when they are not subject to such vibration.

Also, in the application of lamps in flashing signals, such as these highway crossing beacons, and signals for lighthouse service; those are fields in which lamps are going to be used in increasing numbers and further research on these conditions would certainly be a very valuable contribution.

P. S. MILLAR: Dr. Taylor's suggestion about purchasing lamps of one or another voltage, depending upon whether they are to be used upon a-c. or d-c. seems to me to involve a needless complication. As Mr. Lieb has pointed out, even in the case of the vacuum type tungsten lamps, where the performance differences are larger, the somewhat longer life is balanced by a lower mean efficiency throughout life, so that the user receives service from the lamp of substantially the same value in either case, and in either event the lamp life is so adequate that the difference is not felt by the user of the lamp.

Dr. Taylor asks one or two questions which might in part be answered by saying that at Electrical Testing Laboratories the photometry is carried out on storage battery circuits. As to the measurement of any possible slight differences in light output on respectively d-c. and a-c., it may be said that in our commercial lamp testing work we have not made a study of that matter. In commercial lamp testing we have quite difficulties enough in determining with necessary accuracy the efficiency of lamps on direct current without worrying about any possible slight difference in initial efficiency there may be as between a-c. and d-c. operation.

E. C. CRITTENDEN: If you refer to the articles published, which Mr. Stickney mentioned, you will find these later questions answered. As I recall, the difference in efficiency between a-c. and d-c. of the same equivalent voltages is of the order of one-sixth of one per cent.

C. H. SHARP: I just want to point out that that depends upon the diameter of the filament. If the filament is fine enough so that the thermal lag is very small, then you are going to get the effect of a power factor in an incandescent lamp, that is a leading current. But with commercial lamps, of 25 watts or above, the difference of efficiency at given voltages on d-c and a-c. is so small that we have never been able to discover it—and we have tried.

JOHN W. LIEB: Questions that have been asked have already been answered more or less satisfactorily in the discussion as it has developed. On the question of the service satisfaction afforded by lamps having somewhat longer or shorter lives, I want to say that the problem is psychological as well as economical.

A few years ago in the course of the work of the Lamp Committee of the Association of Edison Illuminating Companies, an attempt was made to ascertain what life characteristics are most desired by the average user of lamps. Within certain limits longer lamp life may be had for slightly increased cost of light, when this cost is taken to include both the cost of the current and the cost of the lamp. Conversely, within certain limits, light may be had at a somewhat less cost through incurring more frequent outages and lamp replacements due to shorter life of lamps. Seven or eight alternative propositions were presented to a variety of people who pay lighting bills. Thus, under certain conditions, in a moderate sized residence, a lighting service may be had at a cost of $3.00 per month with a failure of one lamp per month on the average. Or, by paying $3.06 per month for light, the failures might be reduced on the average to one every six weeks, or by paying $3.27 per month for light, the frequency of failures might be reduced to one in three months.

The conclusion of this canvass was that there is no general inclination either in the direction of obtaining light at the lowest possible cost irrespective of frequency of lamp failures or in the direction of reducing the frequency of lamp failures to a very low figure at some added expense for light. The choice of individuals ran the whole gamut of the seven or eight alternatives, there being no consensus indicated.

Out of this experience emerged the curious indication that 1,000 hours, adopted in a more or less haphazard way as a general figure to represent desirable life for incandescent lamps, probably meets the requirements of commercial and domestic service as well as any other figure that can be adopted.

R. E. MYERS* (Communicated): Mr. Lieb's most interesting paper brings out the fact that alternating current burning has a rather marked effect on the life and filament crystallization of some types of lamps in comparison with direct current.

Figs. 1 and 2 show that the filament crystallization or grain growth is considerably increased by burning on alternating current. It is evident, from a consideration of the tables given in this paper, that this crystallization has a considerable effect on the

*Chief Engineer, Westinghouse Lamp Company, Bloomfield, N. J.

life and strength of the lamps. To make this more obvious, data in slightly different form, which was obtained from the tables on life test, has been arranged in the following comparison:

	Per cent burnout above 80 per cent initial lumens		Handling breakage during the lamp life tests	
Mazda B	A.-C. Per cent	D.-C. Per cent	A.-C. Per cent	D.-C. Per cent
25-watt	50.0	5.0	12.7	6.0
40-watt	27.2	2.25	3.4	1.6
50-watt	23.2	2.1	9.6	1.1
60-watt	17.8	7.3	3.9	6.8
100-watt	7.5	6.3	5.9	3.0
Mazda C				
75-watt	62.0	34.4 75 and 100-watt 13.0		18.4
100-watt	58.0	21.6	—	—
200-watt	21.0	5.5	9.5	5.3

This data shows that the per cent failures above 80 per cent initial lumens are considerably greated on a-c. than on d-c., particularly in the smaller wattages. This is natural, since burnouts are more or less closely related to deleterious filament crystallization.

A corresponding difference apparently exists in the handling breakage during life test. As before, the breakage on a-c. is considerably greater than on d-c., particularly in the smaller sizes of filament, where filament crystallization is more of a factor in the strength.

For some reason there is a reversal in the 75 and 100-watt gas-filled lamps but this need not affect the general conclusion, since slight discrepancies can easily be explained by differences in filaments, no special effort, of course, having been made to have filament from the same spool in the lamps which were submitted to the alternating current and direct current burning.

The exact cause of the increased crystallization on a-c. burning is perhaps unknown. However, a filament operating on a-c. must be subjected to internal thermo-mechanical strains due to the temperature fluctuations which occur twice on each cycle. The thinner the filament, the greater is this temperature fluctuation. We can readily imagine that these thermo-mechanical strains, due to expansion and contraction, would bring about slippages at the crystal surfaces, which might result in the development of weak spots, which might cause burnouts or general lack of strength in the lamp.

Though the above seems likely to be the most important action of the a-c., we may note that the swinging or vibration of a filament section may also affect its life. In the commercial use of a lamp such vibration is more likely to occur with a-c. than with d-c. operation. A section of filament that is not under tension between the supports has no natural frequency of vibration but any section that is under tension has such a natural frequency.

If an outside rhythmic disturbance of this frequency is communicated to the filament, the latter may respond and a slight disturbance may cause a violent vibration. Lamp filaments burning on either a-c. or d-c. would vibrate in response to mechanical vibrations of this kind but with alternating current, the possibility of such a rhythmic disturbance is always present, since we have on the one hand, the force of the earth's magnetic field pulling upon the current-carrying filament and on the other hand, we have the alternating increase and decrease of length of the filament section caused by expansion and contraction. This may set the section into transverse vibration, just as a string can be made to vibrate transversely by having one end fixed and the other attached to a tuning fork in such a way that the tension is alternately increased and decreased as the fork vibrates. Since ordinarily, however, it is only occasionally that filaments vibrate, we cannot attach much importance to the effect of vibration upon lamp life. At least we can eliminate with fair safety this item from our consideration of the life tests given by the author.

One would say, therefore, that it is probable that the crystal slippages produced by alternate expansion and contraction are the main cause of the difference between a-c. and d-c. burning.

In Mr. Lieb's paper, reference is made to a paper by Dr. C. H. Sharp, in the *Transactions of the A. I. E. E.*, Vol. 25, page 828, and the discussion of same by Dr. C. P. Steinmetz. In Dr. Sharp's paper the statement is made or implied that the pressed tungsten filament does not crystallize. This statement was made six months before the first low wattage, pressed filament tungsten lamps were put on the market in this country, which occurred in the summer of 1907. Moreover, the conclusion reached by Dr. Sharp was based on the testing of a very limited number of foreign made lamps and these were of a size in which the filament does not offset readily.

In Dr. Steinmetz's discussion of Dr. Sharp's paper, he predicted that a ductile tungsten filament, if obtained, would be unsuitable for a-c. In other words, from his knowledge of ductile tantalum and other pure metals such as wrought iron, he concluded that pure tungsten would also offset under a-c. Since it was not common knowledge in 1906 that the crystallization could be delayed or prevented from the practical standpoint, it was perfectly natural for Dr. Steinmetz to conclude that ductile tungsten would crystallize, where the pressed tungsten filament apparently would **not.**

In Mr. Lieb's paper, credit is given to Dr. W. D. Coolidge for the addition of thoria to drawn wire for the purpose of preventing tungsten filament crystallization. The facts are, however, that tungsten filament has a natural inclination, like most metals, to crystallize whether the filament is pressed or drawn. This tendency to crystallize can be hastened in certain ways and it can also be retarded or prevented in so far as the practical life of the lamp is concerned. The prevention of crystallization in tungsten filament has been the subject of special investigation in the Westinghouse Lamp Company for many years. Dr. Anton Lederer of the Westinghouse Austrian factory, in the year 1906, or perhaps earlier, first conceived the idea of preventing or delaying the injurious crystallization of tungsten, which occurs during the burning life of the filament. This was at least four years before the idea of using offset-preventing material was introduced into the Coolidge drawn wire application.* Dr. Lederer applied for a patent in the United States in December, 1906, and this, with basic claims, was granted to him in April, 1916. His English patent was issued in 1907.

Dr. Lederer's invention covers the introduction of thoria and other oxides into tungsten for the prevention or delay of injurious crystallization; that is, crystallization which interferes materially with the normal life of the lamp.

Since, as will be shown, pure tungsten offsets in both pressed and drawn wire filaments, the inference drawn by Mr. Lieb from Dr. Sharp's work is, in the opinion of the writer, incorrect. However, it is rather interesting to note that the Patent Office placed the same interpretation on Dr. Sharp's statement by citing same as a reference against Dr. Lederer's application. Nevertheless

*See History of Coolidge Patent.

the Patent Office was convinced, by evidence submitted by the Westinghouse Lamp Company covering certain phases of their work along this line, that Dr. Sharp's statement was not a proper reference as first given, and it was withdrawn.

Two years before the Coolidge drawn wire patent was granted, we discovered in our Laboratories at Bloomfield, that oxides other than those mentioned in the Lederer application prevented the early crystallization of fine tungsten filament. The introduction of lime and magnesia into the filament gave about as good results as are obtained by the use of thoria. Casein was used as an organic binder with the metal tungsten powder to make the so-called paste from which the filament was pressed. This casein was put through a very careful purification process in order to free it from ash. The metallic tungsten used was also extremely pure. To this mixture, oxides of lime and magnesia were added. Filaments made from this mixture would burn throughout their normal life without showing offsetting at all, or at least not before the desired life was attained. Filaments made up from the same tungsten metal and the same casein without the oxides were practically worthless on account of filament crystallization or offsetting. Such filaments on the average did not give much more than 30 per cent of the life obtained from the filaments containing the lime and magnesia.

I would like to mention also in this connection that offsetting in vacuum is a function of the size of the filament. Although offsetting occurs sometimes in 0.6 ampere filament and larger filament, it is not generally injurious to the life of the filament on sizes carrying 0.4 ampere. Lamps of 25 watts of the 100-volt grade and smaller would be practically worthless if offset resisting materials were not introduced into the filament.

Naturally, when the manufacture of drawn wire was started at the Westinghouse Lamp Company about two years before the issue of the Coolidge drawn wire patent, we looked for offsetting in pure drawn wire and we found it. The introduction of lime and magnesia or thoria into the drawn wire, as with the pressed filament, prevented objectionable offsetting. As before stated, presentation of the above facts overcame the examiner's misinterpretation of the reference to Dr. Sharp's statement that tungsten filament lamps did not offset.

Microphotograhs are shown below, illustrating offsetting in pressed and drawn tungsten filaments.

The similarity in the crystallization of pure tungsten, whether it is pressed or drawn, is illustrated by Figs. 8 and 10. Very definite offsets are shown in both microphotographs.

Figs. 9 and 11 show the crystallization of the pressed and drawn filament respectively, with thoria present. Fig. 9 shows a somewhat rougher condition than Fig. 11 for two reasons; first, pressed filament is not as smooth initially as drawn wire and second, the crystallization is further developed due to the greater burning period.

In Figs. 12 and 13 the surface crystallization is emphasized by reflected light. These are not the same filaments as those pictured in Figs. 10 and 11.

In Figs. 14 and 15 cross sections are shown of pure drawn wire and thoriated drawn wire filaments, which have been burned for a short time to develop crystallization. It is quite evident that the thoria influences crystal growth to a marked degree. As the filament shown in Fig. 14 continues to burn, the larger crystals will more or less rapidly absorb the smaller ones and in a comparatively short time they will extend across the diameter of the wire. When this condition is attained, the development of offsets is the natural result.

Fig. 15 shows that the presence of thoria has a marked influence on crystal growth. The thoria, which can be seen in the photographs in the form of fine dots, gradually works in between the crystal surfaces during subsequent burning and greatly retards the formation of the larger crystals which, under certain conditions, produce offsetting.

Fig. 8.—Pressed filament of pure tungsten. Burned 839 hours.

Fig. 9.—Pressed filament, 0.8 per cent thoria. Burned 1760 hours.

Fig. 10.—Drawn filament of pure tungsten. Burned 823 hours.

Fig. 11.—Drawn filament, 0.8 per cent thoria. Burned 876 hours.

Fig. 12.—Drawn filament of pure tungsten showing surface crystallization.

Fig. 13.—Drawn filament of 0.8 per cent thoria showing surface crystallization.

Fig. 14.—Section of drawn filament of pure tungsten.

Fig. 15.—Section of drawn filament of 0.8 per cent thoria

REPORT OF COMMITTEE ON MOTOR VEHICLE
LIGHTING 1921-1922*

Revised Specifications. The principal work of the Committee during the past year has been the revision of the "Rules Governing the Approval of Headlighting Devices for Motor Vehicles." This revision was approved by the Council of the Society at its meeting on January 12th. The rules so revised, together with a statement regarding the necessity for the revision have been included in the Interim Report of this Committee which was published in the TRANSACTIONS of the Society for February, 1922.

The revised rules were at once adopted by the states of Massachusetts and Connecticut. At the July meeting of the Conference on Motor Vehicle Administrators a resolution was unanimously adopted that this Conference accept and approve these revised rules. This conference is composed of Motor Vehicle Administrators of the states of Maine, New Hampshire, Vermont, Massachusetts, Rhode Island, Connecticut, New York, New Jersey, Pennsylvania and Maryland The approval of the I. E. S. rules by these states should go a long way toward establishing them as the recognized standard. Furthermore, the Society of Automotive Engineers have adopted as Recommended Practice for Laboratory Tests for Regulatory Purposes Part 1 of the rules, which covers actual tests and numerical values, the only portion of the rules which they deem comes within their scope. For Laboratory Tests for Desirable Illumination they have adopted as Recommended Practice values providing a higher road illumination.

American Engineering Standards Committee. As noted in last year's report, the rules of this Committee were laid before the American Engineering Standards Committee for adoption as a tentative American standard. No action had been taken by that

*A report presented at the Annual Convention of the Illuminating Engineering Society, Swampscott, Mass., Sept. 25-28, 1922.

time, and when it was apparent that a revision of the rules would be required, the A. E. S. C. was requested to reserve action until the revision had been completed. When this had been done, a special committee of the A. E. S. C. met to consider the matter. This Committee consisted of the following persons:

C. F. Clarkson, *Chairman*, Society of Automotive Engineers
C. W. Alling, Underwriters Laboratories
M. G. Lloyd, Bureau of Standards
Bert Lord, Conference of Motor Vehicle Administrators
R. W. E. Moore, Electrical Manufacturers Council
A. C. Morrison, Gas Group, Compressed Gas Mfrs. Assn.
J. W. O'Connor, International Traffic Officers' Assn.
David Van Schaak, Safety Group, Aetna Life Insurance Co.
C. H. Sharp, Illuminating Engineering Society.

The Committee concluded, after considerable deliberation, that while there was abundant evidence to show that the old specification represented the standard in general use, yet the revised rules having been adopted at that time by only two states hardly came under the same category. It was decided therefore to suspend action with a view to seeing if the revised rules should not be adopted by a sufficient number of states so as to make it perfectly clear that they were a generally adopted American standard. Evidently the adoption of the rules by the Interstate Conference of Motor Vehicle Administrators has now changed this whole situation.

Rear Lamp Specifications. Mr. A. W. Devine of the State of Massachusetts, called the Committee's attention to the fact that a recently passed Massachusetts law required the use of approved tail lamps in that state. Under this law the registrar may require such tests as he may deem sufficient and may issue regulations governing the use of approved tail lamps. Mr. Devine further suggested that it would be desirable if this Committee were to draw up a set of rules and specifications for such tests. Inasmuch as this was a matter which involved various structural questions and was of very wide interest, the Committee invited the Lighting Division of the Society of Automotive Engineers to join with it in its consideration. A joint meeting of the two committees was accordingly held and the matter discussed at great length.

Finally a joint sub-committee consisting of

A. W. Devine, *Chairman*
E. C. Crittenden
C. E. Godley
W. F. Little
H. H. Magdsick
C. A. Michel
G. H. Stickney

was appointed to make a draft of specifications and to submit them to the main Committee at an early date. This meeting was held late in the year and the joint sub-committee has not yet reported its findings.

Standard Colors of Traffic Signals. At the request of the Council of the Society this Committee addressed a letter to the American Engineering Standards Committee, calling attention to the importance of the standardization of colors for traffic signals, a matter which with the increased number of traffic signals on the street is becoming very urgent, and suggesting that a conference of interested parties be called by the American Engineering Standards Committee on this subject. A similar letter requesting a conference was sent in by the President of the International Traffic Officers Association. Acting upon this suggestion the A. E. S. C. called a conference which was held on April 10, 1922 and which was very largely attended by representatives of a great variety of interests. The matter was discussed and the conference resolved that the standardization of colors of traffic signals should be carried out under the auspices of the American Engineering Standards Committee, and in accordance with the regular procedure of that Committee. The matter of the appointment of sponsor bodies for putting this into effect was referred to a sub-committee. The Illuminating Engineering Society was prominently mentioned as a sponsor body for this work, but the representative of this Society on the Committee did not urge this, deeming the matter under consideration to be one which is to only a secondary degree a matter of illuminating engineering. The sponsor bodies finally chosen were the National Safety Council and the Bureau of Standards. Subsequently the American Association of State Highway Officials was named as a third sponsor body. The Illuminating Engineering Society therefore, after having proposed the matter, is in a position where it can participate actively in the deliberations of the Sectional Committee which will have the matter in charge without assuming the heavy burden of sponsorship.

It will appear from the foregoing that the year has been one of considerable activity on the part of the Committee, and it is to be hoped that its efforts will result in a further contribution to public welfare which can be accredited to the Illuminating Engineering Society.

Respectfully submitted,

COMMITTEE ON MOTOR VEHICLE LIGHTING 1921-1922

F. C. CALDWELL	W. A. McKAY
G. N. CHAMBERLAIN	A. L. McMURTRY
P. W. COBB	H. H. MAGDSICK
E. C. CRITTENDEN	L. C. PORTER
A. W. DEVINE	C. H. SHARP, *Chairman*
J. A. HOEVELER	G. H. STICKNEY
W. F. LITTLE, *Secretary*	L. E. VOYER

DISCUSSION

E. C. CRITTENDEN: As Mr. Porter has already said, a campaign of education is the thing that is most needed in order to accomplish an improvement in lighting conditions on our highways. There is one aspect of this matter that I would like to emphasize; that is that the requirements or specifications for which the Society's Committee is largely responsible are especially fitted for such a campaign of education. They have this advantage because they are intended primarily not to restrict the driver, but to give him a good driving light. That is, the specifications are twofold: they not merely limit the amount of light in the direction of oncoming vehicles, to avoid dazzling approaching drivers; but they also provide for a good light on the road, and it is this aspect of the specifications which has been strengthened in the recent revision.

If this feature of the specifications is emphasized, so as to secure the support of automobilists, much better results can be obtained than by stirring up increased police activity.

WARD HARRISON: The proposed standardization of traffic signals discussed in Dr. Sharp's report interested me particularly. Colored signals are, of course, used very extensively on railroads but likewise assurance is had by periodic examination that those who are required to see the signals are reasonably free from color blindness. It would seem impractical to make a similar rule applying to automobile drivers, but the fact must not be lost sight of that a considerable proportion of the population have seriously defective color vision. I believe that it was stated before this Society some years ago that the railroads were seriously considering the use of light signals distinctive in shape or position rather than in color so that engineers otherwise competent would not need to be disqualified simply because of their color vision. It was also thought that there would be less chance of mistaking signals under adverse circumstances. The same arguments appear to have equal or greater force in the case of traffic signals.

A. W. DEVINE: The new tail-light law in Massachusetts went into effect on July 27th, and a temporary approval was given covering all tail-lamps which complied with the old law; that is, the law which required that the number plate be so illuminated as to be visible at 60 feet. Of course, that approval really covered no tail-lamps, or possibly one in a thousand. Massachusetts, however, has gone ahead on this proposition in a more or less indefinite way, and the matter is still open, as far as specifications are concerned. We have no specifications. The Registrar has, however, issued a letter in which he states it has been decided that all tail-lamps, to have approval, must have a slot opening or window covered with glass, regardless of illumination of the plate. The reason for that is the ordinary deterioration and coloring of celluloid.

A further requirement is a light source of two candlepower or more, and with acetylene lamps, a quarter foot burner. However, no specifications have been adopted, nor have any regulations been made, and it is quite possible that there may be changes made before any devices are approved.

Massachusetts is now open for applications for approval of tail-lamps, and I imagine that possibly about the first of November an approved list will be issued. If it is possible, we shall attempt to approve some of the lamps that are now in use; that is, we will let the owners adapt them to the new law. Before the first of January, we will get the approved list out, and it ought to serve as a basis, or as additional information for, the Sub-Committee which we have, and assist them in drawing up a set of specifications for approval of tail-lamps—particularly with regard to the illumination of the rear number-plate—which will be a real standard.

ELIHU THOMSON: I am very much interested in this discussion, because I have to comply with these laws, of course. Many times I have gone to Boston and have found several vehicles without lights, fore or aft. They pass the traffic officer, and he does not do anything about it.

The other night, we counted at least twenty-four machines that had no lights, in a run of about two or three miles, and if lighted ahead there was no tail-light; some were parked along sidewalks without any lights. We saw one machine without any lights, which we followed into Boston, and we saw it again an hour after on the boulevard going to Lynn. It had run in and out from Boston, passing traffic officers a number of times, and had not been held up or stopped.

It seems to me the psychological end of this matter is just as important as the making of rules and regulations, rules that are made to be followed. The innocent or slight offender will often be stopped when the real criminal—the speeder—as you know, escapes. That is the real problem. We can make all the regulations we please, but if the people are not going to be reached in some way and absolutely forced to comply with them, they are almost useless.

F. C. CALDWELL: I would like to call attention however to the fact that we have got to entirely reverse popular opinion with regard to automobile headlighting regulations. I have found that the almost universal idea with regard to such regulations is that they are solely to stop glare. The idea that regulations are made to help the driver, by insisting on his having adequate illumination, has not reached the public yet to any appreciable degree, and the universal ignorance with regard to the principles of light holds here as strongly as anywhere. For instance, we had one headlight submitted in Ohio in which the light was expected to start from the filament, make a turn or two inside the lamp and come out through a hole in the front, headed straight for the road.

L. C. PORTER: I was very much interested in what Dr. Thomson said about the machines running without lights. On the other hand, I certainly want to extend my congratulations to Mr. Devine and his Department for the control which they have over motor vehicle lighting here in Massachusetts. During the past summer, I have done quite a little night driving and I have driven through some states where practically no attention was paid to the headlighting. I have driven through some very mountainous country and over some very crowded boulevards, and I find that the conditions in Massachusetts and also in Connecticut are very much better than they are in some of the other states. It was surprising to me to find such a noticeable difference, and I think the State of Massachusetts is to be congratulated upon having gone as far as it has.

I believe, as I said before, the next big point is the education of the public. I came over from Boston last night and I do not know how many cars we passed on the way over. On most of them, the headlights were good. Some of them were not good, but it was plainly evident in those instances that it was simply a case of the headlights not being focused. And I am sure, from the many people that I have talked with, the headlights are not focused because they do not know how to do it. We have good equipment now, and the point should be to get across to the driver the necessity of focusing, and show him how to do it.

INFLUENCE OF DAYLIGHT ILLUMINATION INTENSITY ON ELECTRIC CURRENT USED FOR LIGHTING PURPOSES IN THE DISTRICT OF COLUMBIA*

BY A. SMIRNOFF**

SYNOPSIS: This paper represents the result of an investigation to find the "psychological darkness" outdoors, when generally artificial light is switched on in the business district, causing a sudden increase in the consumption of electrical current. This "psychological darkness" is determined at approximately 1,500 foot-candles on a horizontal surface of daylight intensity. The advance determination of this darkness may be utilized by central-stations for economical reasons and for effecting more efficient service to the customers.

It is apparent that electric current used by a city for lighting purposes will show an increase on cloudy days over clear days. The diagrams for the load on clear days do not show much difference between themselves, excepting on Saturdays and Sundays for obvious reasons. Therefore these days are not taken in consideration in this paper. Cloudy days, however, vary considerably. The solution of the why, when, and how is the aim of this discussion.

The District of Columbia has a double system of electrical supply, alternating current for the greater part of the residential section and direct current for the greater part of the business section; thus an opportunity is given to study the two parts of the city separately. To make it still more favorable to study the influence of daylight, it must be mentioned that there is a comparatively small industrial load in Washington, D. C.

Comparing the alternating current diagrams for the residential part of the city on clear and very cloudy days, Curves A

*A paper presented at the Annual Convention of the Illuminating Engineering Society, Swampscott, Mass., Sept. 25-28, 1922.

**Engineering Dept., Potomac Electric Power Co., Washington, D. C.

The Illuminating Engineering Society is not responsible for the statements or opinions advanced by contributors.

and B in Figure 1, it is seen that though there is an increase in the
current used, this increase is comparatively small. This is explained by two facts: (1) The current is paid for by the users
themselves and consequently they are inclined to be economical;
(2) Most of the residents are not at home during the day. The influence of the second fact is shown by Curve C in Figure 1, for
a cloudy Saturday, when business people had already returned
home.

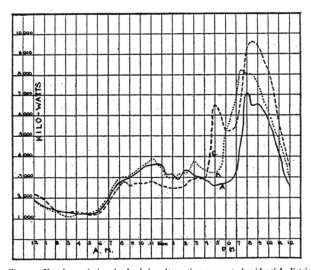

Fig. 1.—Hourly variation in load in alternating current (residential district).
   A—Typical clear day July 27, 1921.
   B—Typical cloudy day August 3, 1921.
   C—Typical cloudy Saturday, May 28, 1921.

The intensity of daylight shows quite a different influence on
the direct current, serving most of the business section. Here
are less economical restraining forces and the electric light is
turned on as soon as an uncomfortable darkness is felt and the
user, the employee, does not pay for the current used. Therefore
this field is more open to study.

Aside from calling a day clear or cloudy it is essential how cloudy the day is. As the current diagrams show, there may be a cloudy day and yet there is no marked difference between the diagram of this cloudy day and a clear day. Again it may be cloudy during the greater part of the day without change in the general use of current up to a certain point when the current

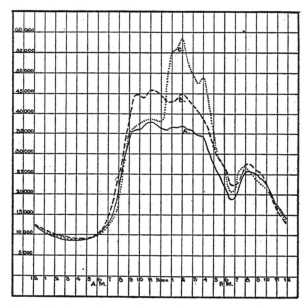

Fig. 2.—Hourly variation in load, direct current (business district).
A—Typical clear day July 7, 1921.
B—Typical cloudy day July 1, 1921.
C—Typical day with sudden dark cloud July 29, 1921.

curve shows a sudden jump, see Figure 2. The darkness at which this sudden rise occurs may be called psychological darkness, that is the point when people feel the need of artificial lighting in the business section of a city. The determination of this darkness is now considered.

It was clear that it was necessary to determine the intensity of daylight illumination prevailing over the city computed in foot-candles. This was done with a Sharp-Millar Photometer and it was found that the sudden rise in current consumption occurred regularly when the illumination intensity went below 1,500 foot-candles approximately in the open air. This minimum incidentally corresponded to 360 foot-candles at the window looking west at four o'clock P. M. and about 9 foot-candles six feet away from the window. The lower the intensity went the higher the current consumption, but all variations over 1,800 foot-candles had only a negligible influence.

The knowledge of this psychological darkness may have a varied practical application and therefore it is necessary to show how it is practicable to have a record of daylight without the necessity of having special observers fitted with photometers.

Through the courtesy of the U. S. Weather Bureau and Mr. Irving F. Hand of the Solar Radiation Investigation Section, permission was obtained to go over the records of this section. In this section there had been used for several years a Callendar's Receiver in connection with a Callendar Wheatstone Bridge Recorder; by these instruments a graphic record had been produced of the solar radiation in gram-calories.[1] There also had been

TABLE I.—ILLUMINATION EQUIVALENT OF ONE GRAM-CALORY PER MINUTE PER SQUARE CENTIMETER OF SOLAR HEAT ENERGY, WITH THE SUN AT DIFFERENT ALTITUDES

Air mass	1.1	1.5	2.0	2.5	3.0	3.5	4.0	4.5	5.0	5.5
Solar altitude	65°.0	42°.7	30°.0	23°.5	19°.3	16°.4	14°.3	12°.6	11°.3	10°.2
	Foot-Candles									
All observations	6,320	5,770	5,490	5,170	4,820	4,780	4,670	4,610	4,600	4,480
Observations on good days only	6,720	6,600	5,580	5,310	5,120	4,780	4,670	4,610	4,600	4,480

calculated by the Chief of this Section, Dr. Herbert H. Kimball, the illumination equivalent of one gram-calory per minute per square centimeter of solar heat energy with the sun at different altitudes[2] making it possible to read the graphic record directly in foot-candles.

Now, during the business hours between 8:00 A. M and 4:30 P. M., in the summer, where a sudden thunderstorm causes

[1] For description see *Monthly Weather Review*, August, 1914.
[2] *Monthly Weather Review*, November, 1919.

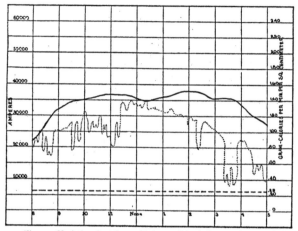

Fig. 3.—Variation in solar radiation and load (amperes) on typical
variable summer day. Load ————, radiation ....

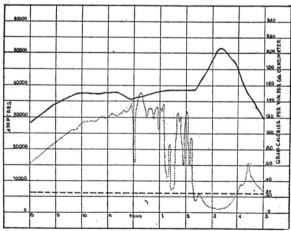

Fig. 4.—Variation in solar radiation and load (amperes) on typical
extremely variable summer day. Load ————, radiation. ....

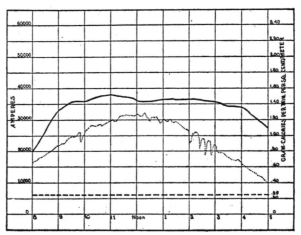

Fig. 5.—Variation in solar radiation and load (amperes) on typical clear summer day. Load ——————, radiation ....

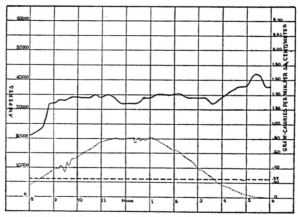

Fig. 6.—Variation in solar radiation and load (amperes) on typical clear winter day. Load ——————, radiation ....

a sudden increase in the use of current, the altitude of the sun varies between 60 and 40 degrees for which altitude at Washington, D. C. one gram-calory per minute per square centimeter corresponds roughly to 6,000 foot-candles.

All calculations are intentionally made only approximate as generally people also act only approximately in the same way under the same conditions.

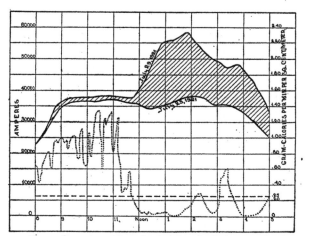

Fig. 7.—Comparison of load of typical clear day (July 25, 1921) and cloudy day with difference in load due to effect of cloud. Load ————
radiation July 29, 1921. . . . .

In other words, as soon as the solar radiation falls below 0.25 gram-calory energy equivalent to 1,500 foot-candles artificial light is generally used. This theoretical conclusion is splendidly proved by the comparison of current diagrams and solar radiation records as shown in Figures 3 to 7 inclusive. By reversing the current readings on these diagrams the interdependence will be still clearer.

The farther and longer the solar radiation falls below 0.25 gram-calory, the higher the consumption of current and by comparison it is possible to determine the approximate equivalent of current used for a certain darkness. Such a tentative equivalent table is shown in Figure 8, derived from Figures 3 to 7. Naturally this equivalent is not the same for all cities or for all hours of the day and can be determined by comparison in each case individually, but once determined it is characteristic that this equivalent is approximately constant for the same city, the same hour of the day regardless of the different seasons. Compare the four o'clock equivalent derived from summer readings,

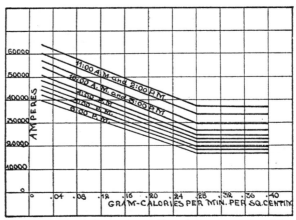

Fig. 8.—Chart indicating the probable equivalent current demand for a given degree of darkness at different hours.

Figure 8, with the solar radiation at four o'clock P. M. in January and the corresponding current curve, Figure 7. Naturally there must be made an allowance for the development of the city, increased number of customers, etc.

A few tentative suggestions are given here where the application of the psychological darkness may result in some economy and improvement of the lighting systems.

(1) During the summer months the current used in business hours, with the exception of the noon hour, is usually constant and also lower than in the winter, but the summer is characteristic for its sudden thunderstorms accompanied very often by

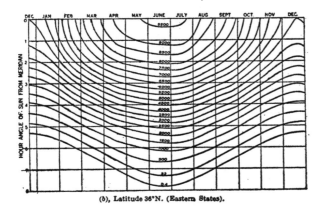

(b), Latitude 36°N. (Eastern States).

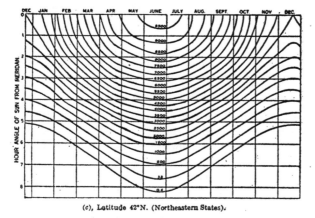

(c), Latitude 42°N. (Northeastern States).

Fig. 9.—Total illumination on horizontal surface (foot-candles) calculated for different latitudes at normal incidents for all hours and months of the year.

heavy clouds, that cause the usage of current to rise suddenly. In view of this under certain circumstances and in certain localities there are kept in reserve expensive storage batteries or additional generators are maintained to provide a certain reserve. Both systems represent a certain loss, especially where electric current is produced by coal. To eliminate this a warning could be given ten or fifteen minutes in advance not only of an approaching storm, but also of the degree of darkness caused by the clouds by the use of either a Callendar's Receiver or a receiver using the principle of the thermopile being developed at present by the Solar Radiation Investigation Section of the U. S. Weather Bureau. Receivers could be placed on the roofs of the outlying sub-stations about ten miles away from the business section of the city in several directions, which by means of a milli-ammeter and a relay could be made to ring a bell as soon as the darkness, causing a fall in solar radiation, reaches the energy requirement of 1,500 foot-candles. Warning could be then transmitted by the operator with the indication of the direction of the storm to the central station.

(2) By using on these sub-stations specially calibrated ammeters or several relays for different intensities of illumination it would be possible to know at the same time in advance the approximate height of the current that will be used by having prepared an equivalent table similar to the one in Figure 8, specially adapted for the system and different hours. Thus unnecessary over-increase of power will be avoided.

(3) A recording instrument would be especially useful for the system operator who would be able to read directly from the instrument the approximate amount of current used at any instant of a storm without the necessity of calling up the sub-stations.

(4) By using a Solar Radiation Receiver and Recorder similar to the one used by the Weather Bureau and the tables[3] prepared by Prof. H. H. Kimball of the total illumination on horizontal surface (foot-candles) calculated for different latitudes at normal incidents for all hours and months of the year as shown in Figure 9, it will help to make advance calculations for the

[3] *Monthly Weather Review*, November, 1919, page 789.

load curves months ahead. In these tables the curve showing 1,500 foot-candles and below commands special interest. Thus probably it will be possible to calculate the winter peaks, coal consumption, etc., allowing a certain percentage for steady growth of the city, local characteristics, special affairs, exhibitions, election years, etc.

REPORT OF THE COMMITTEE ON NOMENCLATURE AND STANDARDS—1922*

Last year the Committe had the pleasure of reporting that the International Commission on Illumination at its Paris meeting in July, 1921, had adopted definitions of fundamental quantities conforming very closely with the corresponding definitions in the Committee's earlier Reports. This year it is able to record another definite step in the fact that the Society's reports on "Illuminating Engineering Nomenclature and Photometric Standards" have been formally approved as an American Standard by the American Engineering Standards Committee.

In submitting these Reports for such approval the Committee thought it best to adhere exactly to the international definitions already agreed upon, in the hope that this course would facilitate the attainment of a still greater degree of international uniformity in nomenclature and practice. The official text of these definitions is in French, and an authoritative English translation has yet to be agreed upon by the British and American delegates to the International Commission. The Committee has therefore used the provisional translation published in the 1921 Report.

The definitions approved, with the exception of the six which have been adopted by international action, are identical with those given in the 1918 Report. There has been so much demand for copies of that Report that the supply is exhausted and it is considered desirable to reprint the standard form of the definitions in full. In order to make clear that these Reports do not cover standards of illumination such as fall within the province of other committees of the Society, the word "photometric" has been inserted in the title.

*A report presented at the Annual Convention of the Illuminating Engineering Society, Swampscott, Mass., Sept. 25-28, 1922.

The Illuminating Engineering Society is not responsible for the statements or opinions advanced by contributors.

FURTHER REVISION OF I. E. S. NOMENCLATURE AND STANDARDS

While the general acceptance of the Committee's recommendations has been gratifying, it is recognized that these standards are by no means perfect or complete. The Committee believes that its work should be continued particularly with a view to clarifying the definitions and making them as useful as possible for all who have occasion to discuss illumination problems. The Society has been designated by the American Engineering Standards Committee as sole sponsor for this field of work, and a "Sectional Committee" composed of representatives from various engineering and commercial organizations is being formed to work with our own Committee in the revision of the nomenclature and standards.

In the meantime the work of revision has been actively taken up and has progressed so far that the Committee wishes to present for discussion the revised draft which follows. It should be understood that this draft is tentative, being subject both to the decisions of the Sectional Committee mentioned above and to discussion with other countries in so far as it affects international practice. The order of definitions is considerably changed, and for convenience in comparing the revised wording with the approved Standards the numbers of corresponding sections in the latter are hereby given in parenthesis.

It may be of interest to note that the term "Luminaire" which was recommended by the Committee last year, and is defined in Section 57 below, has been formally approved by a resolution of the Standards Committee of the American Institute of Electrical Engineers.

PROPOSED DRAFT FOR DISCUSSION

1. (1) **Light:** The term light is used in various ways:

(1) To express the visual sensation produced normally when radiant flux (*q. v.*) within the proper limits of wave-length, of sufficient intensity and of sufficient duration, impinges on the retina.

(2) To denote the luminous flux (*q. v.*) which produces the visual sensation.

(3) By extension, even to denote radiant flux of wave-lengths outside of the visible spectrum (*e. g.*, ultra-violet light).

2. (2) **Radiant flux,** is the rate of energy radiation, and is expressed in ergs per second or in watts.

3. (3) **Luminous flux** is the rate of energy radiation evaluated with reference to visual sensation.

> Although luminous flux must strictly be defined as a rate, in practice it is often referred to as if it were an entity or quantity.

4. (12) **Lumen:** The unit of luminous flux is the Lumen. It is equal to the flux through a unit solid angle (steradian) from a uniform source of one international candle.

5. (8) **The Luminous intensity** of a source in any direction is the flux per unit solid angle (steradian) from the source in that direction.

> The flux from any source of dimensions which are negligibly small by comparison with the distance at which it is observed, may be treated as if it were emitted from a point. For apparent candle-power see Section 70.

6. (10) **International Candle:** The unit of luminous intensity is the International Candle, such as has resulted from international agreement between the three national standardizing laboratories[1] of France, Great Britian and the U. S. A. in 1909.

This unit is conserved by means of incandescent electric lamps in the laboratories which remain charged with its conservation.

7. (11) **Candlepower** is luminous intensity expressed in candles.

8. (9) **Illumination** at any point of a surface is the luminous flux density at that point, or, when the illumination is uniform, the flux per unit of intercepting area.

9. (13) **Lux:** The practical international unit of illumination is the Lux. It is equal to one lumen per square meter, or it is the direct illumination on a surface which is everywhere one meter distant from a uniform point source of one international candle.

> As a consequence of certain recognized usuages, the illumination can also be expressed by means of the units, phot and foot-candle.

[1] These laboratories are the Laboratoire Central d'Electricité, Paris, the National Physical Laboratory, Teddington, and the Bureau of Standards, Washington.

10. (13) **Phot:** Using the centimeter as the unit of length the unit of illumination is one lumen per square centimeter and is called the Phot.

11. (13) **Foot-Candle:** Using the foot as the unit of length, the unit of illumination is one lumen per square foot, and is called the Foot-Candle.

12. (new) **Lighting:** The time integral of luminous flux is designated by the term lighting.

13. (new) The **Lumen-hour** is the unit of lighting. It is the lighting represented by a rate of one lumen continued for one hour.

14. (14) **Exposure** is the product of an illumination by the time.

The microphot-second (0.000,001 phot-second) is a convenient unit for a photographic plate exposure.

15. (15) **Brightness** of a surface is the luminous intensity per unit of projected area.

16. (15) **Units of Brightness:** The C. G. S. unit of brightness is one candle per square centimeter of projected area. In the English system the unit of brightness is one candle per square inch of projected area.

> A surface of unit brightness emits one lumen per steradian per unit of projected area.

17. (16) The **Lambert** is a practical unit of brightness. It is equal to a brightness of $1/\pi$ candle per square centimeter of projected area. It is the average brightness of a surface emitting or reflecting one lumen per square centimeter, or the uniform brightness of a perfectly diffusing surface emitting or reflecting one lumen per square centimeter.

For most purposes the millilambert, 0.001 lambert is the preferable practical unit.

> Brightness expressed in candles per square inch may be reduced to lamberts by multiplying by $\pi/6.45 = 0.487$
>
> A perfectly diffusing surface emitting one lumen per square foot will have a uniform brightness of 1.076 millilamberts.
>
> In practice no surface obeys exactly the cosine law of emission or reflection; hence the brightness of a surface generally is not uniform but varies with the angle at which it is viewed.

18. (4) **Visibility** of radiation of a particular wave-length is the ratio of the luminous flux at that wave-length to the corresponding radiant flux.

19. (70) **Visibility Values:** The following values for visibility and for relative visibility (maximum visibility being taken for this purpose as unity) are recommended.

(For table of values see Section 70, of Illuminating Engineering Nomenclature and Photometric Standards).

20. (5) **The Mechanical equivalent of light** is the ratio of radiant flux to luminous flux for the wave-length of maximum visibility, and is expressed in ergs per second per lumen, or in watts per lumen. It is the reciprocal of the maximum absolute visibility.

As a standard value for the mechanical equivalent of light, the figure 0.0015 watt per lumen is recommended.

> This term has been used in a variety of senses. As here defined it refers only to the minimum mechanical equivalent of light and corresponds to monochromatic light of maximum visibility. The reciprocal of this quantity is sometimes called the luminous equivalent of radiation.

21. (6) A **Luminosity curve** of a source of light is a curve showing for each wave-length the luminous flux per element of wave-length. Therefore it gives, wave-length by wave-length, the product of the radiant flux and the visibility.

22. (7) The **Luminous efficiency** of any source is the ratio of the luminous flux to the radiant flux from the source. For practical purposes it is usually expressed in lumens per watt radiated.

23-56. These sections would be the same as Nos. 17-50 of the approved Standards.

57. (new) A **Luminaire** is a complete lighting unit consisting of a light source, together with its direct appurtenances, such as globe, reflector, refractor, housing and support. The term is used to designate lighting fixtures, wall brackets, portable lamps or so-called removable units.

58. (51) A **Primary luminous standard** is one by which the unit of light is established and from which the values of other standards are derived. A satisfactory primary standard must be reproducible from specifications.

59. (52 and 53) A **Secondary standard** is one calibrated by comparison with a primary standard. The use of the term may also be extended to include standards which have not been directly measured against the primary standards, but derive their assigned values indirectly from the primary standards.

> Because of the lack of a satisfactory primary standard of light the unit is actually maintained in most laboratories by electric incandescent lamps serving as reference standards. The values assigned to these standards were originally agreed upon as representing the average value of the accepted primary standard as nearly as this could be determined. This procedure is formally recognized in France and the United States.

60-70. Same as present 54-64.

71. (65) **Spherical reduction factor** of a lamp is the ratio of the mean spherical to the mean horizontal candlepower of the lamp.

> In the case of a uniform point source, this factor would be unity, and for a straight cylindrical filament obeying the cosine law it would be $\pi/4$.

72. (66) **Photometric Tests:** The results of photometric tests should not be stated in candlepower unless the measurements are made at such a distance from the source of light that the latter may be regarded as practically a point. When measurements of lamps with reflectors, or other accessories, are made at distances such that the inverse square law does not apply, the results should always be given as "apparent candlepower" at the distance employed, which distance should always be specifically stated. For ordinary illuminants with shades or reflectors a distance of 3 meters (10 feet) is recommended.

73. (new) **Apparent candlepower** of an extended source of light is the candlepower of a point source of light which would produce the same illumination at the distance employed.

74, (67) Photometric Units, Symbols and Abbreviations.

Photometric quantity	Name of unit	Symbols and defining equations	Abbreviation for name of unit
1. Radiant flux	{ Erg per second { Watt		
2. Luminous flux	Lumen	F, Ψ	
3. Luminous intensity	{ Candle	$I = \dfrac{dF}{d\omega}, \Gamma = \dfrac{d\Psi}{d\omega}$	cp.
4. Illumination	{ Lux, phot { foot-candle	$E = \dfrac{Fd}{dS}$	ph., fc.
5. Exposure	Phot-second Micro-phot-second	Et	phs., μphs.
6. Brightness	Candle per sq. cm. Candle per sq. in. Lambert Millilambert	$b_i = \dfrac{dI'}{dS\cos\theta}$	— — L mL
7. Visibility	—	$K = \dfrac{F_\lambda}{\Phi_\lambda}$	
8. Reflection factor		ρ	
9. Absorption factor	—	α	—
10. Transmission factor	—	τ	
11. Mean spherical candlepower		scp.	
12. Mean lower hemispherical candlepower		lcp.	
13. Mean upper hemispherical candlepower		ucp.	
14. Mean zonal candlepower		zcp.	
15. Mean horizontal candlepower		mhc.	

16. A source of unit spherical candlepower emits 12.57 lumens.

17. 1 lumen is emitted by a source whose spherical candlepower is 0.07958.

18. 1 lux = 1 lumen incident per square meter = 0.0001 phot = 0.1 milliphot.

19. 1 phot = 1 lumen incident per square centimeter = 10,000 lux = 1,000 milliphots = 1,000,000 microphots.

20. 1 milliphot = 0.001 phot = 0.929 foot-candle.

21. 1 foot-candle = 1 lumen incident per square foot = 1.076 milliphots = 10.76 lux.

22. 1 lambert = 0.3183 candle per square centimeter = 2.054 candles per square inch.

23. 1 candle per square centimeter = 3.1416 lamberts.

24. 1 candle per square inch = 0.487 lambert = 487 millilamberts.

25. A perfectly diffusing surface of 1 millilambert brightness emits 0.929 lumen per square foot. (See also Section 17).

75. (68) **Alternative Symbols:** In view of the fact that the symbols heretofore proposed by this committee conflict in some cases with symbols adopted for electric units by the International Electrotechnical Commission, it is proposed that where the possibility of any confusion exists in the use of electrical and photometric symbols, an alternative system of symbols for photometric quantities should be employed. These should be derived exclusively from the Greek alphabet, for instance:

    Luminous intensity....................................Γ
    Luminous flux..Ѱ
    Illumination...β

                         Respectfully submitted,

COMMITTEE ON NOMENCLATURE AND STANDARDS

LOUIS BELL,   HOWARD LYON,
E. C. CRITTENDEN, SECRETARY,   C. O. MAILLOUX,
W. A. DOREY,   A. S. McALLISTER,
W. J. DRISKO,   W. E. SAUNDERS,
E. J. EDWARDS,   C. H. SHARP, CHAIRMAN,
G. A. HOADLEY,   C. P. STEINMETZ,
E. P. HYDE,   G. H. STICKNEY.
A. E. KENNELLY,

DISCUSSION

C. H. SHARP: The changes proposed are in the way of simplifying the definitions and of making the arrangement of them more systematic and logical.

I wish to reiterate what the Secretary said: that the Committee would like the benefit of the criticisms and suggestions of the members of the Society, particularly of those who are really practicing illuminating engineering and who are using these things.

F. A. BENFORD: In listing these units, Item 4, which is the lumen, a derived unit, is placed before Item 6, which is the International Candle, the fundamental unit of photometry.

I think the order of these two items should be inverted because the simpler and fundamental units should come first and the complex and derived units should follow. This order of precedence might well be followed throughout the entire list, if possible, so arranging matters that each complex unit follows the units from which it is derived.

F. E. CADY: Mr. President, having had occasion to present the subject of photometric units and standards to classes in college for the last four or five years, I have found it very difficult to present to the students in a form which they can grasp clearly and distinctly, a number of the definitions as given by the Committee on Nomenclature and Standards. This is particularly true of the term "light."

The Committee gives three definitions of the term "light" yet there is omitted what I feel—and Dr. Worthing of our Laboratory feels the same way— is a very common usage, namely: that term which would have the same relation to luminous flux which energy has to radiant flux, or the time integral of luminous flux.

This proposal was sent to the Committee and resulted in the use of the new term "lighting," which is under Section 12, and while I realize this is probably the result of considerable deliberation on the part of the Committee, yet both Dr. A. G. Worthing and I felt that it was not a satisfactory solution.

The term "lighting" is used in so many ways that to attempt to confine it to cover this idea of the time flux, time luminous flux, would create, it seems to us, more confusion than clarity.

Dr. Worthing has asked me to present the following written discussion: "This year's report of the Nomenclature and Standards Committee corrects certain rather glaring errors and brings the list of standardized terms and definitions much nearer the form which finally represent approved usage.

"To my mind there are still a few outstanding points needing consideration. There is a common usage of the word 'light' in which the term bears the same relation to 'luminous flux' as energy does to 'radiant flux.' It is illustrated by the sentence 'The sun radiates light.' In response to a suggestion by Mr. Cady and myself that this usage be recognized and that the 'lumen-hour' be named the unit of its measurement, the Committee has replied by giving the name 'lighting' to our 'light.' Their 'lighting' is also to be measured in 'lumen-hours.'

"One of the tests for terms of this kind is that of usage. Consider them such statements as these"—and here he is using the term "lighting" as it has been defined in this new section—" 'The lighting emitted by an ordinary 25-watt, 110-volt vacuum incandescent lamp during its life amounts to about 250,000 lumen-hours.'

"According to the quantum theory of radiation, lighting is emitted in small bundles. One is tempted to ask: 'Where is the pun?'"

While lighting as a term may well describe the result produced by light and is certain to continue to be so used, it seems peculiarly unsatisfactory when substituted for 'light'."

M. LUCKIESH: The last time I remember hearing this report read, I brought up a question as to the terminology of "light" and the Committee is still dodging it. They have here three definitions, two of which are entirely superfluous, becouse they are also defined as radiant flux and luminous flux.

Here we are an Illuminating Engineering Society and we have not nerve enough to tell what light is. It is the forty-sixth word in the Bible, and you have all heard that quotation: "Let there be light, and there was light." Certainly God Almighty did not mean ultra-violet, infra-red, radiant flux, or anything else but visual sensation.

I feel this matter is of great importance. We should adopt that first definition and put the other two in a foot-note. We should try to get that used universally, because we have radiant flux, we have visible radiation, infra-red radiation, ultra-violet radiation and terms for everything else but light as a visual sensation.

I have not heard heretofore anything about this definition "lighting," but it does not bring to my mind what I think of when we speak of lighting. To me, lighting involves in some way distribution. This is a good definition, I think, if you are concerned with the lighting of a single point. I can not quite analyze why I do not like it, but I certainly would never use that term in that manner. And, as Dr. Worthing has said, I think usage is so established in other ways than this that that could never be adopted.

LOUIS BELL: I am very much in accordance with what Mr. Luckiesh has said regarding the desirability of pinning ourselves down to a simple, sensation definition of light. To do anything else is raising the old question as to whether there can be a sound where there is no ear to hear it, and all that sort of thing, which is metaphysics, and not common sense. Carlyle once said of metaphysics that: "High air castles are cunningly built out of words, well laid, moreover, in good logic mortar, wherein, however, no knowledge comes to lodge."

I think when we have a simple definition of light like the primary one, it is best to put the other ones in foot-notes, recognizing the fact that they are used.

Regarding the written communication which Mr. Cady read, I think that the classes which the gentleman teaches had better be instructed in simple, plain English. The phrase "the sun radiates light" certainly does not connote to me continuity, or the integration of that light over minutes, hours, days, or eons. I think it simply expresses the fact that it radiates that which gives us a sensation of light, and that is all there is to it.

We should stop splitting hairs on matters of that kind and try to eliminate and throw out of usage loose expressions like that on which this particular objection is based.

G. A. Hoadley: The difficulty in undertaking to define something that everybody understands is very great indeed. That first definition has shown itself to be satisfactory to a number of the members present. It is not satisfactory to me at all, and the reason is that it is the definition from the physiological point of view.

If you take the physical point of view, the term light may be defined as the thing that produces a sensation, and not the sensation itself. I believe that is a more accurate definition; it is, at any rate, from the physical standpoint.

T. W. Rolph: I would like to call attention particularly to Item 57, the definition of "luminaire." The Committee has supplied us here with a word which is extremely useful and fills a need which has long been felt. The old term of "light-unit" or "lighting-unit" to describe a lamp with its various accessories is crude; it is a very poor term to use. Furthermore, we never got it across to the public; they always confused it with some kind of unit of measurement, and it never came into general use anywhere except among ourselves.

Now we have a new word, which is a coined word, and which definitely describes the thing that we want to describe. I would advocate that we all take great pains to use the new word on every occasion, and that we drop completely from our terminology the words "light-unit" and "lighting-unit."

C. H. SHARP: We have been accused of dodging the definition of light for years. The Committee has said that the term light is used in various ways, and here are three of them. I presume it is used in some other ways, and I do not think there is anything in that statement there that can be disputed. It is so: "light" is used in different ways.

Some of us want to use the term light to express the sensation, and some of us want to use the term light to denote luminous flux, and, we are going to do it, just the way we wish individually to do it!

But where the Committee has been perfectly definite has been in the definition of luminous flux. Luminous flux is defined as "the rate of energy radiation evaluated with reference to visual sensation." There is no dodging there, and the point of view of the Committee is that luminous flux is the thing we have to deal with. We are not dealing with this indefinite term "light." The only marked feature of the discussion of the proposed term "lighting" was that nobody seemed to offer any other term. Really, a name is needed. Perhaps Professor Thomson has something in his mind that would fit in. He has been thinking about these things for more years than most of us. We need a term for the integral of light, for the light-hours. We used to use candle-hours, and now we speak of lumen-hours. We ought to have a term to designate that integral quantity, and this is one suggestion for it. I hope a better one may be found, but until then there is no use finding fault with this one. It is not perfect, but let us hold on to it until we have a better one.

M. LUCKIESH: As I understand it, this is a Committee on Nomenclature and Standards, whose job it is to give us definitions. None of us are going to use in every day writing and speaking the words "luminous flux" for "light," and it seems to me that we require a standardized term or definition for the word "light."

Of course there is no blunder here at all; the term "light" is used in various ways, and this report gives three ways. But if the Committee is going to assume that its job is to record the various ways in which all of these terms are used, we are going to have a book which could be classified as humor. Certainly it would be humorous but valueless as terminology.

E. C. CRITTENDEN: This is a good sample of what happens at all of our Committee meetings, where the chief difficulty is to close the discussion. It usually closes because we have to catch our trains.

There are several important points which have been raised here. Mr. Luckiesh wishes to give the Committee the duty of censoring the language used by members in dealing with light. If this Society had the authority and would delegate it to the Committee to prescribe in what sense we should use the term "light," we might decide on one of these definitions, but we have not that power. We can not say light means this or that, and limit you or anybody else to using the term in that sense. I think the Committee through its years of activity has been useful partly because it has kept in view practical and attainable ends. It has not tried to do things that it could not possibly do. It has tried to give some guidance to natural development which would meet with approval and would come into actual use.

Personally, I am inclined to think the Committee itself has wasted many hours in discussion of this first definition, some of which might better have been given to more practical terms, such as "luminaire." I do not see that we gain much by arguing over these general definitions. If Dr. Bell will pardon me, I will bring out one reason by reference to his remarks; Dr. Bell argued that light ought to mean *sensation* only. Immediately thereafter he used the sentence "The sun radiates light" to illustrate his argument on another point, saying that this means it radiates that which gives us a sensation of light. He has argued for one definition and then used the other. That is what we all do.

Therefore, there is not much use of our trying to hold Dr. Bell and Mr. Luckiesh down to one definition, as long as they admit that they use both within five minutes. The Committee has thought it best to recognize actual established practice, rather than to go counter to it.

As to Mr. Cady's proposal that "light" ought to be extended to cover still a fourth meaning, the time integral of luminous flux, the opinion of the Committee is definite. We can not agree to that, because it is absurd to say that the same word ought to mean the integral and the thing you integrate. They are two different things, and you do not want the same name for both. "Lighting" may not be the right name, but we certainly feel that the use of "light" for the integral and also for the rate or flux would lead to still further confusion.

I might point out that quite independently our confrères in Germany, who are quite logical in these matters, have during the year arrived at a corresponding set of terms. They have adopted a term corresponding to "lighting," that is, the integral or quantity of light; starting with that they have proceeded to give as successive derivatives, light flux, intensity, and illumination. The fact that the need of such a term is being recognized thus independently is rather significant. We will leave the question open for suggestions during the year as to a good term for that quantity.

The point Mr. Benford brought up as to the order of two of these definitions, Numbers 4 and 6, has in it an idea which is at the basis of many of our difficulties in discussing these terms. In a few years we have changed the very basis of our thinking. A few years ago we used to think in terms of the source of light, that is, of the candlepower; we now deal with lumens, or flux, and this is a very fundamental change.

When we speak of lumens, we are thinking primarily of light as the thing which falls on objects and makes it possible to see them, or we are thinking of the flux through a space. But in our measurements, we still have to start from a concrete standard of candlepower. The logical unit, to fit our way of thinking of light nowadays would be a certain amount of flux of a given quality. That is a thing which we can not yet measure with sufficient precision to base our measurements upon it. We must still go back to somewhat archaic standards of candlepower, and we have as a result a certain confusion of thinking. We talk about lumens, we calculate in terms of flux, and then in all measurements we start out from a source of specified candlepower.

That is one reason why these sections appear to be in the wrong order. Logically, however, the definitions start with luminous flux as the fundamental thing; then naturally the unit of luminous flux follows. The definition of candlepower from this point of view is secondary, and therefore it is put second.

I may say that one of the difficulties in reaching an international agreement on definitions is the fact that this practice of thinking in terms of light flux is not so common abroad as here. We hope that our own point of view, which we consider more advanced, will soon become more widely accepted, and that a greater degree of uniformity in the definitions may be obtained between the different countries concerned.

ELIHU THOMSON: Really, I do not think I am competent to discuss this question, which seems to give rise to such difficulties. It is the same old question of course that we have been familiar with for so many years in sound, where there is sound, there is this, that or the other thing, whether sound is the vibration of the air, or the effect on the auditory nerve collected and passed on to the brain.

It seems to me it is the old story of having one term to mean many things. We speak of a "light" color, using it as an adjective, and a dark color, and there again we have another use of the term "light," meaning the effect of the light sent back from a colored surface.

A DIRECT READING AND COMPUTING ATTACHMENT FOR SPHERE PHOTOMETERS*

BY BEN S. WILLIS**

SYNOPSIS: Computing scales for use in connection with the photometering of vacuum tungsten lamps on life test have been in use for quite a number of years. The increasing use of the integrating sphere has made it desirable to adapt this arrangement to the sphere photometer.

The scientists at the Bureau of Standards have developed a very ingenious mechanical arrangement for this purpose which could be adapted to any photometer with a movable comparison lamp and fixed head and test source.

The paper with illustrations describes in detail the construction of the device which converts into logarithmic motion the inverse square motion of the comparison lamp of a sphere photometer. The arrangement of computing scales which has been in use for a number of years at the Bureau of Standards on a horizontal bar photometer has been adapted to this device with a consequent saving of time and labor on life test measurements made in the integrating sphere. Preliminary tests show a high degree of precision.

A set of computing scales which have proved of great value in convenience and the saving of time, has been in use for a number of years at the Bureau of Standards in connection with the photometering of vacuum tungsten lamps on life test. Since the description of the watts per candle computer was published[1] Mr. J. F. Skogland has added a scale for reading the per cent of initial candlepower and a computer based on the lamp characteristics which enables efficiencies at the life-test voltage to be read directly from the computer on the photometer while the lamps are photometered at a lower voltage which gives a color match with the standards used. These additional scales will be

---

*A paper presented at the Annual Convention of the Illuminating Engineering Society, Swampscott, Mass , September 25-28, 1922. Published by permission of the Director of the Bureau of Standards of the U. S. Department of Commerce.

**Assistant Physicist, Bureau of Standards, U. S. Department of Commerce, Washington, D. C.

[1] *Bull. Bur. Stds.*, Vol. 12, p. 605, 1915-16; *Bur. of Stds., Sci. Paper No.* 265; TRANS. I. E. S., Vol. X, No. 8, p. 814, Nov. 20, 1915.

Fig. 1.—Attachment assembled on sphere photometer.

described by Mr. Skogland in a forthcoming paper. The increasing use of the integrating sphere in the photometry of both vacuum and gas-filled lamps has made it desirable to adapt this arrangement of scales to the sphere photometer and the present paper describes a device which has been designed for this purpose.

The operation of the scales in use on the bar photometer depends on the fact that in the case of a photometer with movable head and fixed test and comparison lamps the candlepower scale very closely approximates a logarithmic scale ruled to the proper base. This makes it possible to use auxiliary logarithmic scales in connection with the candlepower scale like an ordinary slide rule for computing efficiencies, per cent candlepower and characteristic relations, the photometer supplying the required motion of the scales. But in the case of a photometer with movable comparison lamp and fixed head and test source, in which class a sphere photometer must necessarily fall, the candlepower scale deviates widely from a logarithmic scale. Therefore some method of compensation must be employed before computing scales can be used. This has been acomplished in the attachment to be described. Incidentally the motion is also reduced to one-half of the comparison lamp motion, giving scales of a more convenient size, and is brought around 90° from the line of motion of the comparison lamp, thus placing the scales in a position for easy reading.

The principal feature, and the one upon which the operation of this device depends, is a cam so constructed that for any given motion of the comparison lamp, and consequently of the scales attached to it, a motion of the reference plate will be produced equal to the difference between the motion of the comparison lamp and the corresponding distance on the logarithmic scale. This brings the index and computing scales, which are attached to the reference plate, to their proper position relative to the candlepower scale. Reference to Figures 1 and 2 will show how this is accomplished.

Attached to the comparison lamp carriage, L, is a rack, F, which moves in guides attached to the photometer bar and engages a large gear, G, at the end of the rack. Gear, G, rotates on an axis set in the bed, B, and on this axis are also the cam, C, and the small gear, g, fixed to move with G. The diameter of g is

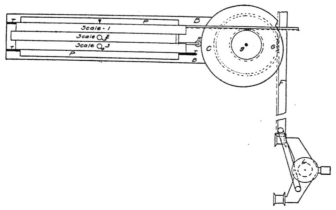

Fig. 2.—Diagrammatic sketch of attachment.

one-half that of G so that Scale 1, which carries a rack engaging g, moves always one-half as far as the comparison lamp. A weight attached to Scale 1, by a cord running over a pulley serves to take up the backlash in the gears and keep them always in contact on the same side of the teeth. Thus the motion of Scale 1 with reference to the *bed* is proportional to the motion of the comparison lamp, and if the index were on the bed, the scale of candlepowers would be one of inverse squares. Such a scale would deviate too much from a logarithmic scale to permit its use as a slide rule, being too open at the lower end and too close at the upper end. In order to compensate for these differences a movable reference plate, P, is introduced which moves on the tracks, T, cut in the bed, and which is kept in contact with the cam, C, at the roller, R, by a spring not shown in the diagram. The cam deviates from a circle at any angle of rotation by the same amount that the logarithmic scale deviates from an inverse square scale at the point corresponding to this angle. Thus the index and Scales 2 and 3 are moved the amount necessary to bring them opposite the proper point on Scale 1.

The scales, ruled to the proper base carry the following values:

1.   Lumens and per cent of initial lumens on upper side, and amperes on lower side.

2.   Efficiencies.

3.   Scale from lamp characteristics.

These scales are the same, except for size, as those now in use on the bar photometer for horizontal candlepower measurements.

Scale 1 is free to move on the reference plate, deriving its motion directly from the comparison lamp, while Scales 2 and 3 are set to the proper place and then locked to the reference plate by means of the set screws, S, so that their position with reference to the plate does not change when the comparison lamp is moved.   Adjustments are provided so that the scale can be set to read correctly when the comparison lamp is at the distance corresponding to the lumen reading at the center of the scale.

When not in use, *i. e.,* whenever runs other than life test are being made, the bed, carrying everything except the rack, F, can be lifted off from the photometer and stored out of the way.   This is a convenient feature since at the present time the same sphere is being used for both life test and precision photometry.

In its present form the lumen range is from 700 to 2,300 which corresponds to a motion of the comparison lamp of about 56 cm. or a motion of the scale itself of about 28 cm.   This range corresponds to about one-half the length of a slide rule so that the scale has divisions the same size as those of a 20-inch slide rule.   The scales are easily removed and it is intended that additional ones be ruled to take care of intensities above and below this range.

Table I shows the magnitude of the errors resulting from the use of this device.   In the first column are given various distances of the comparison lamp, in the second the relative lumen values corresponding to these distances, and in the third the readings taken from the lumen scale of the device.   The largest difference is of the order of 0.2 per cent which is within the range of errors in reading the scale.

TABLE I.—MAGNITUDE OF THE ERRORS RESULTING FROM USE
OF ATTACHMENT.

Comparison lamp distance—cm.	Relative intensity computed	Relative intensity observed
66	229.6	229.6
70	204.1	204.0
75	177.8	177.9
80	156.2	156.2
85	138.4	138.4
90	123.5	123.6
95	110.8	110.8
100	.100.0	.100.2
105	90.7	90.8
110	82.6	82.5
115	75.6	75.5
119	70.6	70.6

The writer is indebted to Mr. J. F. Skogland for advice and assistance in adapting his computing scales to this device, and to Mr. R. Hayes, whose careful work in cutting the cam is largely responsible for the high degree of precision attained.

PLOTTING OF SPECTROPHOTOMETRIC DATA*

BY FRANK A. BENFORD**

SYNOPSIS: The plotting of spectrophotometric data is looked upon as a means of making the data more easily read and understood, and to gain these ends it is considered essential to standardize certain features such as the limits of wave length covered by the web, the color boundaries, and some common unit for measuring wave lengths. Some of the distinctions between a curve showing energy and one showing light are pointed out and a form of curve combining the advantages of both is introduced. The real purpose of the "unity" point is considered and note is made of how this custom alters the position of curves and reduces them from relative values to comparative values. The form of a spectrophotometric curve is influenced by the kind of spectrum, either continuous or discontinuous, and by the type of instrument used in testing. The use of a suppressed zero on the vertical scale of the web leads to a certain loss in the purely graphic value of a curve, and several instances are given of the errors that may be introduced by having a curve plotted to a wrong scale although the reading at various wave lengths may be apparently in the correct relation.

INTRODUCTION

The data collected during a spectrophotometric test are best presented in the form of a curve. During recent years the value of presenting data by a plotted line has become recognized in all branches of industry but in no case is the graphical presentation more useful than in this type of photometry. As the art becomes more highly specialized, greater care is needed in preparing the curve because not everyone has time to keep fully informed on all the increasing details that enter into its plotting. When this stage is reached the next logical step is to have a general agreement among all concerned about many of the details so that standards of practice can be established. This renders it easier for the occasional user of such data to absorb such information as the data may contain with the least expenditure of time and effort, which is after all the main purpose of all graphical presentation.

*A paper presented at the Annual Convention of the Illuminating Engineering Society, Swampscott, Mass., Sept. 25-28, 1922.

**Physicist, Illuminating Engineering Laboratory, General Electric Co., Schenectady, N. Y.

The Illuminating Engineering Society is not responsible for the statements or opinions advanced by contributors.

There are several distinct types of data that require different treatments in plotting. It is seldom that the emissivity of a light source is given in other than relative units, but on the other hand, data of transmission and reflection must be plotted to an absolute scale. In giving examples the writer has confined himself to data drawn from the files of the laboratory as these were all at one time the subject of considerable thought and the practical problems that had to be solved would hardly have suggested themselves from a purely theoretical study of the subject.

This society has always paid considerable attention to the color of light sources and to the reflectivity of surfaces, but in the presentation of the data each author has laid out the web and plotted the data as seemed best to himself. The resultant lack of uniformity in methods and terms might possibly lead to confusion and this paper is an attempt to present in an orderly manner some of the details of plotting spectrophotometric curves that have formarly been taken as a matter of course, and it is offered as a contribution toward a general uniformity of practice.

THE EXTREME AND USEFUL LIMITS OF VISION

The eye is sensitive to radiations of wave length 0.39μ to 0.76μ or 0.77μ, but these are extreme limits that can be attained only under certain rare conditions. For all ordinary purposes of illumination the limits are better placed at 0.42μ to 0.68μ, because beyond these values the amount of light is so small as to be of small practical value and accurate testing beyond these limits is difficult. In making up a web to plot the data it is convenient to extend the limits to 0.40μ and 0.70μ. This web has several important features, one of which will be discussed later. For the sake of simplicity and uniformity it has been the practice of the writer to make the web extend from 0.40μ to 0.70μ even if the data points covered only the central part of the web.

COLOR BOUNDARIES

There does not seem to be any generally accepted definition of the spectrum colors in terms of wave length. The boundaries given below are by Listing and they have been used by a number of writers.

>Below 0.397 μ, ultra-violet;
>0.397 μ to 0.424 μ, violet;
>0.424 μ to 0.455 μ, indigo;
>0.455 μ to 0.492 μ, blue;
>0.492 μ to 0.575 μ, green;
>0.575 μ to 0.585 μ, yellow;
>0.585 μ to 0.647 μ, orange;
>0.647 μ to 0.723 μ, red;
>above 0.723 μ, infra red.

Two changes might be made in the above list. One would be to limit the range to 0.40 μ and 0.70 μ, and the other would be to drop indigo. There seem to be a large number of people (the writer included) who do not see any color between 0.424 and 0.455 sufficiently distinct from blue and violet to be called a separate color. The list, as amended for use in this laboratory is given below.

>Below 0.400 μ, ultra-violet;
>0.400 μ to 0.424 μ, violet;
>0.424 μ to 0.492 μ, blue;
>0.492 μ to 0.575 μ, green;
>0.575 μ to 0.585 μ, yellow;
>0.585 μ to 0.647 μ, orange;
>0.647 μ to 0.700 μ, red;
>above 0.700 μ, infra-red.

It is a great convenience to have these color boundaries marked in some manner on all spectrophotometric webs. Most people think in terms of color rather than wave lengths and even if the addition of color boundaries to the web makes it appear a little more complicated the net result is a pronounced gain in clarity.

CHOICE OF UNITS FOR DESIGNATING WAVE LENGTHS

There are, rather unfortunately, three units of length used to designate wave lengths of light. There is the micron, or 10^{-6} meter, designated by μ. On this scale green light is indicated by 0.55 μ. A second scale is the ninth meter, or 10^{-9} meter, designated by $m\mu$, on which scale green is indicated by 500 $m\mu$.

Then there is the Ångstrom scale or tenth meter (10^{-10} meter) on which green is 5,500 A. The last scale is universally used in precision measurements of wave lengths where there may be as high as seven or eight significant figures. The illuminating engineer when dealing in colors is seldom interested beyond the third figure of the wave length and the Ångstrom scale contains one or more useless figures. The micron scale, or 10^{-6} meter, is much better in this respect and a web marked in microns can be marked clearly and with sufficient accuracy with no more than two significant figures at the foot of each main ordinate. The third figure can be estimated and this gives a wave length accuracy sufficient for all ordinary purposes.

ENERGY vs. LIGHT

There are at least two common ways of plotting spectrophotometric data. One way is to give the relative energy at each wave length throughout the visible region, paying no attention to the visibility of radiation except in the matter of limits. This curve is, therefore, purely an energy curve and a great deal of interpretation is needed to get it into terms of light. The greatest error this type of curve may lead to is assigning too much importance to radiation at the ends of the visible spectrum.

One way out of this dilemma is to reduce the energy curve to a curve of light by multiplying the various energy values by the visibility at the different wave lengths. The area under the reduced curve now represents light and each portion of the spectrum is given to a proper scale. The great shortcoming of this curve is that it has lost much of its graphic value, due to the strong characteristic form of the visibility curve covering and concealing the variations that were easily seen in the original curve.

A little more than a year ago the writer had occasion to plot the transmission data of twenty-one dyes and filters used in the motion picture industry. An attempt was made to plot the data so that the curve would show:

(a) transmission of each wave length;
(b) average transmission for entire visible spectrum;
(c) relative luminosity of the different parts of the transmitted spectrum.

The web adopted was constructed as follows: A visibility curve was plotted to a large scale and the ordinates measured at close intervals. Beginning at 0.40 μ the ordinates were added together one at a time and the summation curve so obtained was plotted. This curve thus represented the summation of the light from a source of uniform intensity from end to end of the spectrum.

The areas under the visibility curve from 0.40 μ up to various wave lengths were read from this summation curve and used to get the spacing of the ordinates of Figure 1. As a result, the space between, say 0.55 μ and 0.56 μ on the web is proportional to the light between these limits from a source of uniform intensity of radiation.

A curve of transmissivity, or reflectivity, or radiation, plotted on this web fulfills conditions (a), (b) and (c) above and at the same time all parts of the curve are readable and retain an individuality of form midway between the energy curve and the reduced luminousity curve. The area under such a curve of transmissivity divided by the whole area of the web gives the net transmissivity which may be obtained with a planimeter or by measuring ordinates.

To make the difference more apparent the same data have been plotted in Figure 2 to a scale of evenly spaced ordinates. The area under 59 A is 29 per cent of the total area of the web, while the area under 30 A is 34 per cent of the total. Filter 30 A thus appears to have the higher transmission of the two, whereas, the other filter 59 A has a higher transmission in the ratio of 0.487 to 0.265.

At the extreme ends of the web of Figure 1 the ordinates are crowded together, but with this exception all parts of the curve are readable and the purely graphic value of data so plotted is much higher than when on the customary web.

POINT OF EQUALITY

In studying the light from any given source we consider the quantity and quality as two distinct characteristics and in nearly all cases two distinct tests, photometric and spectrophotometric, are made. The spectrophotometric data need not then be plotted to an absolute scale as the curve is supposed to

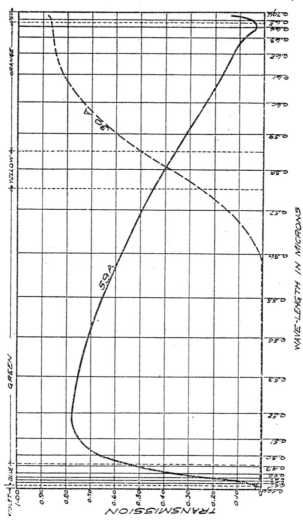

Fig. 1.—Transmission curves of Wratten filters 59A and 30A.

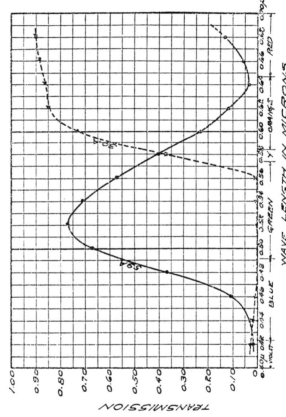

Fig. 2.—Transmission curves of Wratten filters 59A and 30A, evenly spaced ordinates.

show only relative intensities throughout the spectrum. Under this condition any scale of ordinates may be used, but there has grown up a certain convention in the use of the vertical scale that makes for ease in comparing the data of various curves.

A number of years ago it was customary to make all emission curves pass through unity at 0.59 μ. It was then the general opinion that the visibility was greatest in the yellow region, and this opinion persisted until accurate tests were made of both visibility and the emission characteristics of light sources. Later work has shown that under normal illumination the peak of the visibility curve lies at about 0.556 μ and this thus becomes a natural fixed point in plotting curves. In place of using 0.556 μ it is customary to use 0.550 μ as the common crossing point for all emission curves. This is exactly in the center of the 0.4 μ— 0.7 μ web. Another factor that has helped to bring about the change from 0.59 μ to 0.55 μ is the gradual change in color of the common illuminants. The early illuminants, such as candles, oil lamps, and gas flame fed with sodium, were predominately yellow in color, and the center in a vertical direction of the emission curve was at about 0.59 μ. The light sources used in the arts and sciences to-day are much hotter and the center of the curve has shifted over to, or beyond, 0.55 μ.

It might be noted that the peak of the visibility curve moves to shorter wave lengths as the illumination is decreased and there is no particular theoretical virtue in adopting exactly 0.556 μ or any other point. The choice is more logically made for considerations of convenience and uniformity.

Some special problems arise in connection with the unity point when we are dealing with discontinuous spectra. Here, any curve whether smooth or broken, does not represent at all accurately the finer details of the distribution of energy, although the general nature of the spectrum may be shown by either type of curve with all desirable accuracy. Of the two the smooth curve may be simpler to read and understand. In Figure 3 is shown a series of test points that are connected by a broken line that crosses the 0.55 μ ordinate at 0.92. A smooth curve that was drawn in order to compare with another spectrum that is actually smooth is shown passing through 0.55 μ at 1.00 on the scale of relative energy. This illustrates very well the difficulty

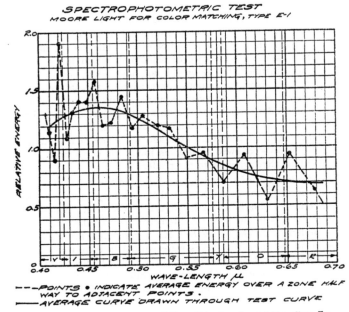

Fig. 3.—Relative energy curve of Moore light for color matching, Type E-1.

that may arise between two curves drawn from the same data. As the entire level of the curves is determined by the value at 0.55 μ this is sometimes a serious question. If the two curves in Figure 3 are made to pass through a common point at 0.55 μ it is readily seen that the general level of the broken curve will no longer agree with that of the smooth curve. This difficulty in a slightly different form arises when two or more tests are made on a discontinuous spectrum of varying intensity, such as that from an arc lamp. Any lack of agreement at 0.55 μ influences all the other points if we insist upon an exact agreement between curves at this point and it seems better practice to get a general agreement throughout the spectrum rather than at some selected point.

Every spectrophotometric curve should have shown along with it the actual data points. It often happens when dealing with continuous spectra that a smooth curve can be drawn through all points, but even here they should be given. The smooth curve of Figure 3 is an extreme case of smoothing which the writer believes was justified, but it would not have been fair to the reader to have given the smooth curve without the data points and claimed it was a good average curve. With all data points given, each reader is in a position to accept, reject or alter the curve to suit himself. This condition arises continually in all of the more difficult tests with arcs and while the man plotting the curve must be allowed to use his judgment in all cases, he should extend that privilege to the reader.

The greatest use of the common point is in the assembling of several curves on the same sheet so that they may be compared. An example of this is given in Figure 4 where an attempt is made to so present the spectrophotometric data of daylight and tungsten light that they may be readily compared for quality. Thus, if all the tungsten curves were plotted in their true relation for any given lamp, the 2,335° curve would be lowest of all and the 3,015° curve would be above the other three and none of the curves would cross. But as we are ordinarily more interested in comparisons than in relations, we resort to the device of the common point.

PRISM SPECTRUM AND GRATING SPECTRUM

In testing with a continuous spectrum the operator has considerable freedom in the choice of test points, and usually they will be selected at equal wave length spacings, regardless of whether the spectrum is formed by a prism or a grating. But when a discontinuous spectrum is being measured for light certain precautions must be rigidly observed. One of these is that no section of the spectrum must be omitted, otherwise an important line or band might be missed, and no section of the spectrum should be included in the readings on two adjacent test sections.

The most reliable way of fulfilling these conditions is to move the telescope of the spectrophotometer through the width of the eye slit for each test point. With a prism spectrum this gives test points spaced about eight times as far apart in the red region

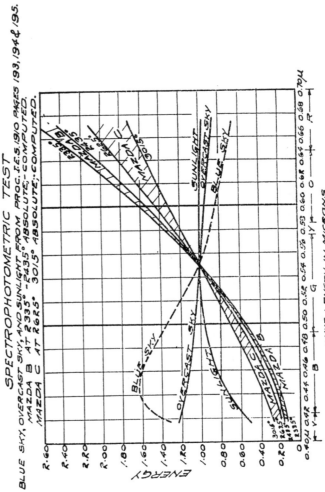

Fig. 4.—Energy curves of blue sky, overcast sky, sunlight, Mazda B and Mazda C tungsten-filament lamps.

as in the violet region. This in turn influences the form of the plotted curve and may make one end seem more uneven than the other, as in Figure 5, full line curve. A normal spectrum, as from a grating, will give under the above test conditions a series of steps of uniform width and the real unevenness of the spec-

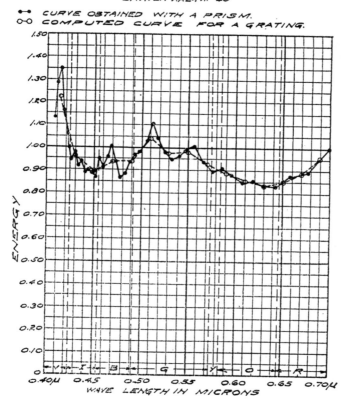

Fig. 5.—Energy curve of Beck searchlight.

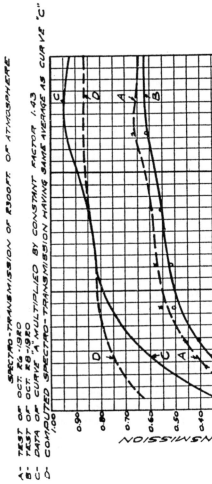

SPECTRORADIOMETRIC TEST

SPECTRO-TRANSMISSION OF 2,300 FT. OF ATMOSPHERE

A- TEST OF OCT. 26-1920
B- TEST OF OCT. 28-1920
C- DATA OF CURVE "A" MULTIPLIED BY CONSTANT FACTOR 1.43
D- COMPUTED SPECTRO-TRANSMISSION HAVING SAME AVERAGE AS CURVE "C"

WAVE LENGTH IN MICRONS

Fig. 6.—Transmission curves of 2,300 feet of atmosphere.

trum will be more fairly shown. The two curves of Figure 5 illustrate this point. The full line curve is from a test with a prism and dotted line curve is computed from the same data using a uniform spacing of 0.02 μ such as might be employed with a normal or grating spectrum.

If a smooth curve is drawn among the points the differences arising from the different natures of the two instruments will largely disappear.

ABSOLUTE AND COMPARATIVE VALUES

There are at least two kinds of spectrophotometric curves that should always be plotted to an absolute scale. These are curves of reflectivity and transmissivity. There is no justification for not having these curves to exact scale, for if they are displaced upward or downward by a constant factor, they cease to truly represent the facts. To illustrate, consider the curve A of Figure 6 giving the spectro-transmission of 2,300 feet of air on a certain night, when the average transmission was found by an independent method to be 0.58 for the entire luminous spectrum. If through some error the average transmission had been determined as 0.83 the spectro-transmission data would have been plotted as curve C, while for an actual transmission of 0.83 the spectro-transmission curve would be curve D, which differs greatly from C. It is thus evident that the curve of transmission must be plotted to correct scale to represent the facts.

Another example of the necessity of having absolute rather than relative values is given in Figures 7, 8 and 9. In Figure 7 two curves, A and B, were plotted so that they could be compared. In this first test it was necessary to separate, by computation the front surface losses from the transmission and silver loss occuring in a certain glass mirror. In the second part of the test, Figure 8, it was necessary to separate the double surface loss from the transmission loss, and the data again had to be to absolute scale. The data of these two tests combined to give the reflectivity of the silver as shown in Figure 9. The crosses are computed from curves A in Figures 7 and 8 while the circles are computed from curves B in the same figures.

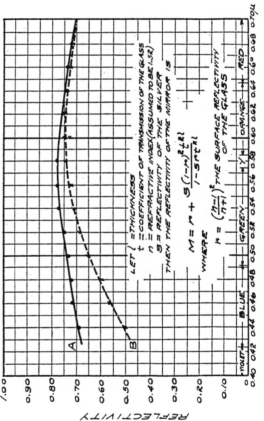

Fig. 7.—Reflectivity curves of a silvered mirror.

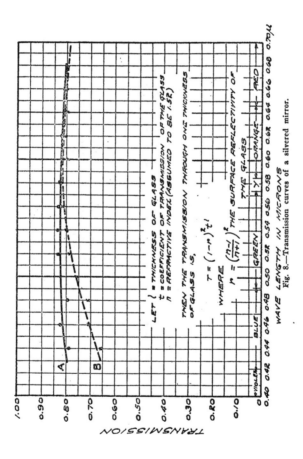

Fig. 8.—Transmission curves of a silvered mirror.

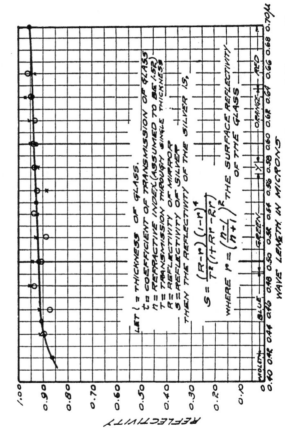

Fig. 9.—Reflectivity curves of a silvered mirror. Data computed from curves in Figures 7 and 8.

It will be observed that the data points are badly scattered and it is apparently impossible to get a smooth curve such as would be expected for the reflectivity of silver. The computation of each point included six different photometric determinations and the experimental errors added to produce the large variation shown. There were several severe handicaps, other than physical, accompanying this test, but in spite of these the data were kept to absolute scale as far as possible and the final results were satisfactory. In this particular case, keeping to correct scale was more important than the precision of the photometric values, and this is very often the case.

There is one practice connected with the use of an absolute scale that is a common source of annoyance, or even of error. This is the "suppressed zero" or web beginning with some figure other than zero on the vertical scale. The practice is particularly deceiving when only a few of the lower divisions are dropped so that in place of reading from 0 to 100 it reads from, say 20 to 100. To get a correct idea of the data plotted on such a web the individual points must be read and there is little gain over a tabulation of the same data.

There are, of course, cases where a suppressed zero is useful, particularly where the data is restricted to a small section of web, say between 80 and 100 on the vertical scale and the work has been done with precision, so that the smaller details of the curve have some meaning other than experimental errors. If, however, there is evidence of experimental errors of sufficient magnitude to prevent a smooth curve from passing through all points, it is then better to return to a full web. In Figure 9 the data points are all between 0.85 and 1.00 but to have used an enlarged web covering only this range would have been an absurdity.

In making tests for the transmissivity of glass there is always a correction for the surface losses to be considered. If the refractive index is 1.52 this loss is 0.0427 for each face, or a net transmission of 0.916 for perfectly clear glass. Samples of glass often come in such shape that they can be tested only through a small thickness and the absorption is small. If one sample shows a net transmission of 0.900 and another shows

0.884, both having an index of 1.52, the second absorbs twice the light of the first and not 11.6 per cent against 10 per cent as would be indicated if we neglected to consider the influence of the surface loss. These data, if plotted too low on the web by some constant factor, say 0.95, will indicate entirely different absorbing and heating qualities, and for some types of service the glass would certainly be rejected.

It is very apparent that for this type of curve the values must not only be relatively correct, but they must also be absolutely correct in the sense that they be placed as nearly as possible in the correct position in the web between 0 and 1 on the transmission scale.

DISCUSSION

NORMAN MACBETH: It may be that the points covered by Mr. Benford have been generally agreed to in other laboratories, but I think that we have a very considerable need for a recognition of standards in this kind of work.

I have recently had a great many spectrophotometric curves to compare with previous published data and it is almost impossible to get a graphic idea of just what the differences are, that is, to find a curve that a clear graphic representation of what it stands for.

The fixing of the color boundaries is an exceedingly important item, and this new plot for visibility data appears to me to be of more than unusual value. In our radiant energy distribution of various sources, undoubtedly the visibility curve is the curve for consideration, which as it is ordinarily given is practically without graphic significance. Mr. Benford's plot permits a more graphic representation of the distribution.

There is one other point on which agreement should be recorded, that is in comparing various energy distribution curves, whether the equality point shall be at 0.55, 0.56 or 0.59 μ. I have been interested in checking up the radiant energy distribution of near-daylight sources, and I recently had occasion to replot a curve from equality at 0.59 μ to equality at 0.55 μ. At 0.59 μ it looked like a north skylight equivalent but when replotted with equality at 0.55 μ, it practically dropped below the 100 per cent line and became what you would call a sunlight equivalent.

Another point of importance is the suppressed zero, and it was suggested that possibly the Society suffered in the report of the Secretary. A good many members, I understand, in the rear of the hall only saw the curve* and not the scale values. The curve showed a peak of 1,500 members in 1910 and in 1919 dropped half way down towards the base, apparently showing a 50 per cent decrease. The base line was at 700, however, and the drop from 1,500 to 1,100 was more nearly 25 per cent. That was due to the suppressed zero, and I agree with Mr. Benford that curves should be plotted to show at a glance the result secured when the actual figures are noted and weighed.

F. E. CADY: In addition to the points mentioned in the paper, it might be well to point out for the benefit of those not versed in this method of visualization that in choosing coordinates, values should be chosen wherever possible so that the resultant curve will have a general direction approximating 45° to the axes. If the coordinate scales are such that the curve is almost vertical, it is obvious that any slight variations in the ordinate values will tend to be concealed, just as in the case of horizontal values referred to by Mr. Macbeth.

I do not wish to criticize this paper, which I consider a very valuable summarizing of the desirable characteristics of curves. However, while the author has laid considerable emphasis on the desirability of using this special web in which luminosity is taken into consideration, rather than mere energy, as far as I have been able to make out, the only curves in which he follows his recommendation are in the first two. All of the other curves seem to be plotted on a basis of equal energy ordinates.

A. L. POWELL: We are certainly particularly indebted to Mr. Benford for this exposition of the characteristics of spectrophotometric curves. I frequently have occasion to explain to students a little bit regarding such curves and it is indeed difficult to tell the story. I do not know of any text book or any place in technical literature where we can find a clear explanation of the various characteristics of the curves. I would like to see Mr. Benford add an appendix to his paper illustrating the four or five curves which are fundamental indicating what these show and the relation between them.

* See page 753, I. E. S. Transactions, Dec., 1922.

A hasty survey indicates these to be as follows: (1) Transmission curve of absorbing or color modifying medium. (2) Relative energy curves of standard illuminants, which are pretty well established. (3) Relative energy curve of illuminant fitted with modifying media (1 x 2). (4) Sensibility curve for the average eye (more or less standard). (5) Luminosity curve of the equipment (1 x 2 x 4). (6) Curve (5) plotted to the new web proposed by Mr. Benford.

It is always desirable to have curve (1) when presenting data on color filters and the like. Unfortunately, many of the laboratories neglect to present this curve giving their results in curve (3). The latter being in relative values alone, we cannot go back to the fundamental, absolute data which is necessary in transforming results to some other of the curves. With curve (1) available and curves (2) and (4) standardized, we can then transform data at will from one form of curve to another, depending on the requirements of the occasion. We can also plot curves at any particular point of equality say 0.55-0.57 or 0.59 μ that may be standardized.

E. P. HYDE: Not only is my auditional memory bad, but I find that my auditional intelligence is not good; I have to see a thing to really understand it, and unfortunately since my arrival I have not had time to read this paper of Mr. Benford's, which I wanted very much to read because it is a subject in which I am very much interested.

But from hearing him present it, I either failed to get a reference to a point which I think of considerable importance, or else it was not there, and I will leave it to him afterwards to say which was the case.

I refer to the importance in many cases of plotting curves on the basis of equal luminosity. By relative luminosity, we mean a curve drawn to arbitrary ordinates, which represents the product of the visibility and the energy.

In comparing several light sources, the energy curves of which we know, I think we are likely to be misled if we restrict ourselves in all cases to plotting them to agreement at some definite wave length. It seems to me it is far preferable in many cases to plot those curves on the basis of equal luminosity area, choosing as the visibility curve some standard adopted by the Society. There is one illustration of how important this may become and how it may be overlooked.

In determining the visibility curve of the eye, of course it is necessary to get data on a great many individuals. The problem is how to average those data after you get them. At one time they were simply all plotted to unity at some chosen wave length and averaged that way. That is obviously unfair, because if you take a man who has a rather exaggerated sensibility in the blue end or in the red end, with the rest of it normal, you are giving a good deal more weight to his whole curve than you should give.

The next step, historically was to average these visibility curves on the basis of equal visibility area, but in my judgment that is just an erroneous or almost as erroneous as the other method because we are not dealing with a uniform energy distribution,— that is an energy distribution in which the energy intensity ordinate is the same at all wave lengths.

The real basis for comparison, according to my notion, is this: that each of these observers should be assumed to give the same flux values as each of the other observers, for the distribution of energy which we will accept as the standard distribution; and in the absence of any other, I would suggest that at the present time we take the energy distribution of a four-watt per candle carbon lamp, since that is yet, so far as I know, the lamp to which we ultimately refer for our unit of candlepower. If A, B and C are reading visibilities with an energy curve of that character, their visibility curves may come, as I said before, very high in the red or high in the blue, but if you multiply their visibility curves by the energy curve and reduce them all to the same area, you are starting on the assumption that A, B, C and D all read a 16 candlepower lamp as 16 candlepower. Then you want to know what relative value they give to the red and the green and the blue, and it seems to me that it is necessary in that particular case to equate the luminosities of the different individuals at this particular energy distribution corresponding to our standard lamp, whatever it may be, and average them.

That is an illustration of the necessity of doing this in one particular case, and I am inclinded to think that there are many cases where we will get a much more intelligent understanding of the facts if we plot our spectrophometric data, I am not referring now to reflection or transmission but to emission curves, on the basis of equal luminosities rather than on the basis of equal ordinate for energy at some arbitrary wave length.

LOUIS BELL: I want to reinforce what Dr. Hyde has just been saying regarding the combination of the curves of various eyes, when averaged, to obtain the average eye.

One thing which we very frequently forget in making estimates of this kind, as Dr. Hyde was just saying, is that different individuals not only have different color sensibilities but different total sensibilities in a very material degree. We do not often realize that there are probably as many people who are plus sensibility in one sensation as people who are minus. There are minus red folk, so-called color blind; there are also a good many plus red, and in the same way there are plus greens and plus blues as well as minus greens and minus blues in their color sensibilities. It also has turned out through some of the investigations that one man may have all three of his sensations slightly plus and another all three minus, or respectively one man have two high sensations and one low, and the other two low and one high. Consequently, in order to obtain the average performance of the eye for the purposes which we have just been indicating, it is necessary to take account of the total sensibility as well as the color difference sensibility, and to make our average on the basis of that, and not of variations from the normal energy distribution spectrum as regards our sensibilities. It is perhaps rather a fine point, but one finds in investigating eyes for color that it is a much more considerable factor than would naturally be thought.

F. A. BENFORD: I see Mr. Cady rather suspects I do not like to take my own medicine. As a matter of fact, I believe I mentioned in the paper that these curves were taken from the files of the Department, and while I am not quite certain of the dates, I believe that that special curve is, in date of issue almost the latest one in the paper, and while we have not adopted it as a regular thing I hope that I can get enough people to agree with me to adopt it and use it.

I am perfectly willing to take Mr. Powell's suggestion and write an appendix to this, if he thinks it is worth while.

Dr. Hyde had a happy word to say when he said he was picture-minded. I think that most engineers are picture-minded. They can remember what they see better perhaps than what they can hear, and if you present a curve to them and be sure and get the frame around the curve right, they will remember it properly. A great deal depends on that frame.

The matter of equal luminosity is rather important but there is this point about it: that very often when plotting curves both to a unity point in the middle of the web, the curves will have approximately equal luminosity. In fact, I have often used that method in hurried work. If you choose point 0.59 μ for your unity point, that would not be so.

H. P. GAGE (Communicated): A spectroscopic study of color filters has been undertaken by Mr. Benford and by me independently; in his case the study of light distribution from different sources and the light transmission of dyed color filters for motion picture projection, in my case the study of colored glasses for railway signaling and the fitting of a colored glass to high temperature tungsten light sources to produce the same spectrophotometric light distribution as natural daylight. Our problems of illustrating and comparing the energy distribution and the color transmissions have been the same but quite naturally some of the details have been worked out differently. My remarks are not offered as criticism of Mr. Benford's work but as a caution that before standardizing on any web for graphic representation more complete agreement should be secured among the different observers as to the best average values to assign to variable physiological functions of the eye. I refer to the exact form to be assigned to the luminosity curve and to the wave lengths to be chosen as the division points between the colors.

For the extreme useful limits of vision 0.42 μ to 0.68 μ are hardly great enough to include all color effects, as for example the peculiarities of cobalt glass and many green dyes are due to the transmission band in the extreme red beyond 0.68 μ. In my work I have generally used the scale and made measurements from 0.41 μ to 0.71 μ, although I sometimes make measurements as far as 0.73 μ.

Color Boundaries.—In illustrative curves it is useful to designate the colors of each part of the spectrum. In 1909 Dr. P. G.

Nutting published a "Method for Constructing the Natural Scale of Pure Color," (Reprint No. 118 from the Bulletin of the Bureau of Standards, Vol. 6). Dr. Nutting divides the spectrum into twenty-two colors equally spaced from its neighbor. In my work I have chosen color numbers 2, 7, 15, 18 and 20 as the division points in the spectrum. Other observers in my laboratory agree that there is no very serious error in using these points but individual differences of opinion occur. The Bureau of Standards show on one of their curve sheets division points between the colors the same as Mr. Benford uses except the division between violet and blue is given as 0.430 μ instead of 0.424 μ. The division points are tabulated below.

Benford	Bureau of Standards	H.P. Gage
0.424 μ	0.430 μ	0.435 μ
0.492 μ	0.492 μ	0.490 μ
0.575 μ	0.574 μ	0.574 μ
0.585 μ	0.585 μ	0.595 μ
0.647 μ	0.647 μ	0.626 μ

Before rigidly adopting any set of division points numerous observers should be consulted.

Plotting transmission curves to illustrate the luminosity of different portions of the spectrum strikingly reveals the reasons for many color effects which are not apparent from the usual transmission plot. It is not, however, sufficient to plot the curve in which each ordinate is multiplied by the luminosity of that portion of the spectrum or in which the distances between the abscissae are divided by this luminosity unless the luminosity is calculated for the particular light source used. The luminosity curves published are for light sources of equal energy distribution and differ from actual sources. The appearance of the usual signal purple glass is vastly different with an electric light source than with the kerosene flame for which it was designed. A special "Electric Purple" is required for the electric lamp. Plotting the resulting luminosity curve of light transmitted by these two glasses with sources of 1,900° K. and 2,375° K. readily reveals the reason. A similar method of plotting curves in which the luminosity curves of the individual color sensations, red, green and blue, might be developed but I fear that somewhat prolonged study would be required before the results could be presented as clearly and revealingly as by the method of plotting resultant luminosity curves on which are designated the colors of each portion.

Points of equality for plotting light sources of different colors while usually chosen as 0.59 μ or 0.55 μ may for special purposes be more advantageously taken at some other point, as for example at 0.41 μ when calculating absorbing glasses to change the color of a tungsten lamp to that of natural daylight.

The usual numerical value of transmission is required in calculating areas for determining luminosity of transmitted light and similar factors but is not so useful as the logarithm of transmission variously designated as density, coefficient of extinction, transmissive index, etc. This is the one value which is doubled by doubling the thickness of the absorbing medium and involves no calculating of odd exponents for different thickness as does the usual numerical value.

The plotting according to frequencies or reciprocal wave lengths is useful in many cases especially where relations between "black body" light sources at different temperatures are involved.

Mr. Benford's paper is a valuable contribution to the art of graphically illustrating some aspects of color which are clear to students of the subject but rather obscure to those unfamiliar with it.

ABSTRACTS

In this section of the TRANSACTIONS there will be used (1) ABSTRACTS of papers of general interest pertaining to the field of illumination appearing in technical journals, (2)ABSTRACTS of papers presented before the Illuminating Engineering Society, and (3) NOTES on research problems now in progress.

HYGIENIC VALUE OF LIGHT IN OFFICES*

BY A. B. EMMONS, 2ND. **

An office worker must *see* in order to do her work. If she sees perfectly she will do the work without strain. Other conditions being equal, she will do more work in a given time and she will make less errors than if she sees with difficulty or dimly.

As Director of the Harvard Mercantile Health Work it has been my duty to personally survey twenty-five stores in six cities including Cleveland, Washington, New York, Newark, Pittsburgh and Boston. Only two of these stores make routine eye examinations of applicants. To explain this extensive lack of intelligent health supervision it may be said that stores have lagged behind industry in changing their conception that health, bad or good, was a personel matter not an industrial concern. They are beginning to find that the wage earner unorganized and unaided can neither afford nor obtain first class medical care and health supervision, and that this failure is costly to the worker and the store.

One of the chief objects of the Harvard Mercantile Health Work is to prove that a store, which from a health standpoint is a group of workers, can obtain through proper organization and administration good health work, to the great advantage of the workers and the work, and at reasonable rates, resulting in profit to the enterprise.

*Abstracted from a paper presented at the Annual Convention of the Illuminating Engineering Society, Swampscott, Mass., Sept. 25-28, 1922.

**M. D., Director Harvard Mercantile Health Work, Boston, Mass.

The best vision possible for each individual is the goal. How near is this goal to-day? One store with excellent health work reports the examination of 983 individuals. Five hundred and seventy-one or 58 per cent had normal vision without glasses. One hundred and eighty or 18 per cent had subnormal vision, who already wore glasses. Two hundred and thirty-two or 23.6 per cent with subnormal vision not wearing glasses.

The frequency of substandard vision is indicated by The Eyesight Conservation Council, (Times Building, New York City) Bulletin I, a part of the Waste in Industry study. A number of carefully recorded investigations covering 15,906 workers show from 58 per cent to 83 per cent had visual errors needing correction. Rejection in the National Army for eye trouble is quoted as 21.7 per cent. Such reliable figures make it obvious that it is safest to presume that a worker has visual error needing correction until proved otherwise.

The best eye examinations to-day not only test the vision at 20 feet, but determine the vision at and furnish glasses suited to the *actual working distance* of each worker.

When it is determined that our workers have eyes, which can see their work, the next question is how best to light that work.

What is the condition of lighting in store offices? Careful footcandle readings have been made only in a few stores where improvements were planned. These will be continued in other stores. I give here a personal estimate of the lighting of the twenty-five store main offices. Seven offices had poor artificial light, eleven had fair artificial light, and seven had good day light.

Mr. Julius Daniels[1] has presented to this society a study of department store illumination. He does not specify office readings in these stores. If we assume that the offices were no better lighted than the public selling spaces, eight of the twenty-seven stores Mr. Daniels visited had average readings of at least four foot-candles, while nineteen had readings averaging below four foot-candles.

It is also fair to assume that few other stores have equally good lighting with these two groups of selected stores. It is quite likely, therefore, that most store offices have artificial lighting considerably below four foot-candles, the minimum standard set

[1] Notes on Department Store Illumination—TRANS. I. E. S., Vol. XV, page 709.

by this society which recommends the reasonable intensities of from four to eight foot-candles for office lighting.

. We find, then, that approximately three in every four workers have substandard vision and that about three in every four selected stores have insufficient light for good vision.

There are three simple requirements for efficient illumination, light in sufficient quantities, diffusion, and elimination of glare.

Experience leads to a few practical suggestions for improving office lighting.

Natural or *day light*, if rightly used is the best illumination, and the cheapest. Studios, operating rooms and laboratories are always planned to use the northern light, preferably from a single source. The modern industrial building is planned to utilize this principle by means of the saw-toothed roof with windows to the north. If store offices could be placed on the top floor of the building and the saw-toothed roof used, the best possible light would be obtained, as well as excellent natural ventilation, easily controlled.

When the office has windows with considerable direct sunlight, this may be utilized by installing French blinds, the slats of which are light colored and highly polished. The sunlight can then be reflected upwards and inwards to the ceiling from which it will be diffusely reflected downward to the work. Sunlight cut off by shades is light wasted.

Obstructions to light such as file cases, partitions, high desks and the workers' bodies themselves were often found to reduce both natural and artificial light. The proper arrangement of desks is such that the light source is over and behind the left shoulder. Diffuse light from above is also good. Light from the right or from in front either casts a shadow or glares in the eyes. This principle is well illustrated in the Code of Lighting issued by the Illuminating Engineering Society.

The plane on which the work rests should be at or nearly at a right angle to the source of light. When the light source is directly over head and the work is filing, and therefore near the vertical, the light is parallel to the work instead of at a right angle. In such a case the light on the file card may be less than half that on the top of this same file.

The best angle for a desk top is a slope of about three inches in twenty-four, similar to the old ledger desk top. This brings the working surface nearly at a right angle to a light source over the left shoulder. It also prevents the necessity of stooping over the desk.

A proper work chair[2] supporting the small of the back while in the working position helps much to maintain the correct working posture.

Reflected light or glare from the working surface of the desk may be much reduced by a dull green linoleum desk top.

Natural lighting must be considered in relation to ventilation. In cold weather a window transom opening inward or a ventilator at the inner side of the sill sloping at an angle of about thirty degrees from the perpendicular gives the best mechanism for well controlled natural ventilation. The mechanism controlling the light must be co-ordinated with that of ventilation.

When we have learned to control the humidity to the desirable 35 per cent to 40 per cent we shall have added one more important factor to the comfort of the office worker.

One more factor in clear seeing has frequently been found lacking. The *work* is not distinct. Old faded cards and too transparent paper and ink do not have sufficient contrast to be easily legible. Too small type or too narrow spacing make difficult reading.

Artificial Lighting

When light must be added to or entirely replace daylight we enter the field of endeavor in which this Society is primarily interested, namely artificial lighting.

Eye comfort at work and good seeing conditions rather than appearance are desirable in office lighting. Ferree and Rand are giving needed facts in regard to this important element, eye comfort.

Elimination of glare, experience shows, is *the* factor above all others, to give eye comfort. To obtain sufficient light, a minimum of five foot-candles well diffused with the least possible glare is desirable. The totally indirect lighting, where conditions are suitable, best fulfil these requirements. Semi-indirect is often satisfactory.

[2] The Work Chair, by A. B. Emmons, 2nd, M. D. and Joel E. Goldthwait, M. D., *Journal of Industrial Hygiene*, Sept., 1921.

The foot-candle meter is, perhaps, the best educator in good lighting. It may be used by anyone quite satisfactorily and with reasonable accuracy. The lighting facts are evident at once. There is no difference of opinion.

The office manager who measures his light with the foot-candle meter will immediately seek a reasonable standard with which to compare his own lighting. This standard he will find in the "American Standard, Code of Lighting" and publications of reputable manufacturers of lighting equipment. Failure to maintain the equipment, especially to replace old and dim lamps is a common cause of poor lighting in store offices.

A record system checked by foot-candle readings at regular intervals is the only way to insure continued good lighting.

Summary

The office manager, responsible both for production and the producer, needs a simple practical statement of the factors involved in seeing. He should realize that approximately three in every four working people have substandard vision needing correction, and three in every four store offices have substandard lighting.

To get the facts about his own office he should arrange for eye examinations by an opthalmologist who should be informed of the type of work of each individual examined. One industrial firm figured the increased production more than paid the cost of these examinations. The copy should be easily read. He should take foot-candle readings of actual working conditions comparing these with the standard codes.

To improve lighting he should consider more daylight, removal of obstructions, proper placing of desks, files, and machines, and better artificial lighting.

Totally indirect light I believe gives the greatest eye comfort, diffusion and least shadows. Semi-indirect comes next in meeting the needs. His lighting will degenerate and his plans fail unless he establishes an effective maintenance system checked by foot-candle readings.

Such improvements in seeing not only result in improved and increased work of the office force but in a higher morale.

* * * * * * * * * * * * * *

In discussing this contribution the following facts were mentioned: Mr. L. B. Marks suggested that the problem should not be handled alone by the hygienist, the views and experience of the illuminating engineer should be carefully considered in its solution, indirect lighting is not always the best type for the office as sometimes its installation produces poor results.

The opinion was expressed by Mr. G. H. Stickney that co-operation should be encouraged between the physician and the illuminating engineer in securing wider use of good lighting. The difficulty in reading faded print or typed matter on thin paper can be avoided by the use of higher intensities. Glare can be minimized by using skill in planning various types of illumination. Indirect lighting is used to good advantage in offices as well as direct and semi-indirect, the choice depending upon local conditions. It is a difficult matter to define glare with any degree of exactness, and problems involving glare have to be handled rather empirically, on the basis of individual judgment and experience.

SOCIETY AFFAIRS

SECTION ACTIVITIES

NEW YORK Meeting—December 14, 1922

The New York Section met at the Engineering Societies Building on December 14 to hear a paper, "The Development of Period Styles in Luminaires" presented by Mr. J. W. Gosling of the Illuminating Engineering Laboratory of the General Electric Co. of Schenectady, N. Y. Mr. Gosling illustrated his talk by means of crayon sketches and his presentation proved to be very interesting. There were present about one hundred members and guests and owing to the length of the paper no discussion was held.

NEW ENGLAND Meeting—December 6, 1922

The first meeting of the year was held on December 6, at which time Professor A. G. Webster of Clark University, Worcester, Mass., gave a paper, "Let there be Light" which was accompanied by lantern slides. The section met at the Industrial Lighting Bureau, which is located in the Rogers Building. After the discussion of the paper, a buffet lunch was served; about fifty members and guests were present.

CHICAGO Meeting— December 15, 1922

At the meeting of the Chicago Section held in the rooms of the Western Society of Engineers on the evening of December 15, Professor A. M. Wilson of the University of Cincinnati, Cincinnati, O., spoke on the topic, "Correct School Illumination and its Relation to the Health and General Welfare of the American Youth." Members of various Boards of Education in Chicago and surrounding towns were invited to attend.

PHILADELPHIA Meeting—December 12, 1922

A joint meeting of the Philadelphia Section with the Fixture Manufacturers' Association and the Fixtures Dealers' Association was held on December 12 at the Engineers' Club. Mr. M. Luckiesh of the National Lamp Works of Cleveland, O., presented a paper, "The Design and Installation of Luminaires" which was based upon the "Tentative Code of Luminaire Design" which was printed in the December TRANSACTIONS. There was a general discussion of the paper. Thirty-two members and guests were present at the meeting.

TORONTO Meeting—November 20, 1922

The papers presented at the last I. E. S. Convention in Swampscott, Mass., were the topics of the meeting of the Toronto Chapter on the evening of November 20, at the University of Toronto. Mr. W. H. Woods gave an interesting resumé of the papers, and Mr. Watson Kintner, of Hamilton, Ont., led an interesting discussion of the Convention papers, in which many of those present participated. About twenty-one members and guests were in attendance.

Meeting—December 8, 1922

A joint meeting with the Toronto Chapter of the A. I. E. E. was held on the evening of December 8, at University of Toronto. Mr. Arthur J. Sweet, of Milwaukee, Wis., presented a paper, "Street Lighting." There were one hundred and fifty members and guests present.

COUNCIL NOTES

ITEMS OF INTEREST

At the meeting of the Council on December 14, 1922, the following were elected to membership:

Three Members

FIELD, OSCAR S., Hall Switch and Signal Co., Garwood, N. J.

GODLEY, CHARLES E., Edmunds & Jones Corp., 4440 Sawton Ave., Detroit, Mich.

O'MEARA, THOMAS J., Hall Switch and Signal Co., Garwood, N. J.

Sixteen Associate Members

BEALS, GILLSON W., Ivanhoe-Regent Works of G. E. Co., 5716 Euclid Ave., Cleveland, Ohio.

BIRGE, NATHAN R., General Electric Co., Schenectady, N. Y.

DUNCAN, FREDERICK B., Duplex Lighting Works of G. E. Co., 6 West 48th St., New York, N. Y.

FISKE, JOHN M., Montana State College, Bozeman, Mont.

HORNER, JR., MERRITT, Edison Lamp Works of G. E. Co., Harrison, N. J.

KELCEY, GEORGE G., American Gas Accumulator Co., 999 Newark Ave., Elizabeth, N. J.

KILLEEN, JOHN F., Mitchell-Vance Co., 505 West 24th St., New York, N. Y.

LEIGHLEY, HARRY M., Public Service Corp., 80 Park Place, Newark, N. J.

McGLYNN, JAMES J., Lit Brothers, 8th and Market Sts., Philadelphia, Pa.

McNAIR, GRAYSON B., Westinghouse Elec. and Mfg. Co., 1062 Gas and Elec. Bldg., Denver, Col.

MORRISON, WILLIAM G., Edison Lamp Works of G. E. Co., Harrison, N. J.

SMITH, ELMER A., Warren & Wetmore, 16 East 48th St., New York, N. Y.

TEBERG, ERNEST J., Public Service Co., of Northern Illinois, 72 W. Adams St., Chicago, Ill.

TILDEN, EARL A., Westinghouse Elec. and Mfg. Co., 40 Clinton St., Newark, N. J.

VAN FLEET, GEORGE LEE, Greene Electric Light and Power Co., Greene, Iowa.

VASSAR, HERVEY S., Public Service Electric Co., 102 River St., Newark, N. J.

Fourteen Sustaining Members

CONSOLIDATED ELECTRIC LAMP Co., Danvers, Mass.
Frank W. Marsh, Official Representative.

FITCHBURG GAS & ELECTRIC LIGHT Co., 537 Main St., Fitchburg, Mass.
Frank S. Clifford, Official Representative.

HARTFORD ELECTRIC LIGHT Co., 266 Pearl St., Hartford, Conn.
W. D. Gorman, Official Representative.

LYNN GAS & ELECTRIC Co., 90 Exchange St., Lynn, Mass.
H. K. Morrison, Official Representative.

MALDEN ELECTRIC Co., 139 Pleasant St., Malden, Mass.
J. V. Day, Official Representative.

MALDEN & MELROSE GAS LIGHT Co., 137 Pleasant St., Malden, Mass.
Harry Walton, Official Representative.

MONTPELIER & BARRE LIGHT & POWER Co., 20 Langdon St., Montpelier, Vermont.
C. J. Cookson, Official Representative.

NEW ENGLAND POWER Co., 35 Harvard St., Worcester, Mass.
Hugo Rocktaschel, Official Representative.

SALEM ELECTRIC LIGHTING Co., 205 Washington St., Salem, Mass.
S. Fred. Smith, Official Representative.

CHARLES H. TENNEY & Co., 201 Devonshire St., Boston, Mass.
Cyrus Barnes, Official Representative.

UNION LIGHT & POWER Co., Franklin, Mass.
E. S. Hamblen, Official Representative,

WETMORE-SAVAGE Co., 78 Pearl St., Boston, Mass.
R. M. Topham, Official Representative.

WORCESTER ELECTRIC LIGHT Co., 11 Foster St., Worcester, Mass.
F. H. Smith, Official Representative.

WORCESTER SUBURBAN ELECTRIC Co., Palmer, Mass.
H. M. Parsons, Official Representative.

The General Secretary reported the death of one associate member, C. W. WAGGONER, Univ. of West Virginia, Morgantown, W. Va.

CONFIRMATION OF APPOINTMENTS

The appointments of the following committee members and representatives were confirmed:

As Members of the Committee on Membership—W. E. Clement, H. G. Smith, and H. H. Smith.

As Member of the Committee on Nomenclature and Standards—M. Luckiesh.

As Member of the Committee on Lighting Legislation—E. Y. Davidson, Jr.

As Member of the Committee on Lecture Courses—G. H. Stickney.

As Member of the Committee on Research—Thomas Scofield.

As Members of the Committee to Co-operate with U. S. National Committee of the International Commission on Illumination to Arrange a Meeting in 1924—G. H. Stickney, Chairman; Louis Bell, Clarence L. Law, A. S. McAllister, W. J. Serrill.

As Representatives at 1922 Boston Meeting of A. A. A. S.—Louis Bell and Charles F. Scott.

As Members of the Committee on Sustaining Members and Advertising—H. F. Wallace, Chairman; R. B. Burton, H. Calvert, E. Y. Davidson, Jr., W. M. Goodrich, F. T. Groome, R. S. Prussia, J. L. Wolf.

As Members of the Committee on I. E. S. Exhibit at Fixture Market, January 15-20, 1923, Cleveland, Ohio—E. Y. Davidson, Jr., E. D. Story, W. Sturrock, A. S. Turner.

NEWS ITEMS

BULLETIN OF THE COMMITTEE ON LIGHTING LEGISLATION

Oklahoma Industrial Lighting Code

Oklahoma is the eighth state in the union to put into effect an industrial lighting code based upon the I. E. S. Code. The other states are Pennsylvania, New Jersey, New York, Wisconsin, Oregon, California, Ohio. The adoption of a code is now being considered in several other states.

The Oklahoma code became effective July 1, 1922. It is issued as Bulletin No. 3 of the Department of Labor and its title page bears the heading "Industrial Code Rules Relating to Lighting of Factories and Mercantile Establishments, Based on the Illuminating Engineering Society's Lighting Code (American Standard)." The Oklahoma code rules and notes and recommendations follow closely the lines laid down in the New York State Code, the details of which have been published in the TRANSACTIONS.

Enforcement of New York State Industrial Lighting Code

Mr. James L. Gernon, Director, Bureau of Inspection, Department of Labor, New York State, has submitted the following statistics relating to the violations of the lighting code rules and the compliances with orders issued by the Inspection Department.

The Inspection Department reports cordial co-operation of factory owners in most cases. During the past two years the attention of the Department has been directed principally to the correction of flagrant cases of glare from bare or from inadequately shaded lamps. Plans are now on foot looking to the enforcement of the rule relating to minimum permissible intensities for detailed industrial operations and processes.

N. Y. STATE INDUSTRIAL CODE

Fiscal Year, July 1, 1921-July 1, 1922.

The following orders were issued and compliances obtained:

Factories

Eyesight protection glare		To prevent accidents Proper placement Sufficient light
Orders issued	3,543	488
Compliances secured	3,416	440
Pending	127	48

Mercantile Establishments

Orders issued	94	85
Compliances secured	65	51
Pending	29	34

Total, 4,210 orders issued concerning lighting, with 3,972 compliances obtained or 94.3 per cent of the orders complied with, leaving but 238 uncomplied lighting orders.

For the preceding fiscal year, July 1, 1920, to July 1, 1921, the following is the record:

Fiscal Year, 1920-1921

Factories

Eyesight protection		To prevent accidents
Orders issued	9,969	1,009
Compliances obtained	9,302	933
Pending	667	76

Mercantile Establishments

Orders issued	170	144
Compliances obtained	170	108
Pending	0	36

Total, 11,292 orders issued with 10,513 compliances obtained, or 93.1 per cent of the orders complied, leaving 779 uncomplied orders at the end of this fiscal year, which were, however, complied with during the fiscal year succeeding.

Revision of Code of Lighting School Buildings

The School Lighting Code issued in 1918 is being revised by the Committee. The following Sub-Committee on revision was appointed: M. Luckiesh, Chairman; H. B. Dates, E. Y. Davidson, Jr., W. J. Serrill, G. H. Stickney. The Council has authorized the Committee to carry out the revision under the auspices of the American Engineering Standards Committee, with the Illuminating Engineering Society, the American Institute of Architects, and the National Education Association acting as joint sponsors. Arrangements for final revision have not yet been consummated.

Co-operating Members of the Committee

The following gentlemen, mostly non-members of the Society, have accepted the invitation of the Committee to co-operate in activities relating to the industrial lighting code and the school lighting code. The co-operating members will not be called upon to attend regular meetings but will give the Committee the benefit of their suggestions and criticisms chiefly by correspondence. The Co-operating Members are: S. K. Barrett, N. Y. University; Lewis T. Bryant, N. J. State Dept. of Labor; C. B. Connelley, Penna. State Dept. of Labor and Industries; Frank I. Cooper, Architect, Boston, Mass.; J. C. Cronin, Penna. State Dept. of Labor and Industries; Thomas C. Eipper, N. Y. State Dept. of Labor; Jas. L. Gernon, N. Y. State Dept. of Labor; C. H. Gram, Oregon Bureau of Labor; H. H. Higbie, University of Michigan; Wm. A. Howe, N. Y. State Dept. of Education; G. E. Kimball, Calif. Industrial Accident Commission; C. I. Shirley, Board of Education, Newark, N. J.; C. B. J. Snyder, Architect, Board of Education, N. Y. C.; Ethelbert Stewart, U. S. Commissioner of Labor Statistics; E. L. Sweetser, Mass. Dept. of Labor and Industries; F. H. Wood, N. Y. State Dept. of Education.

MEASURING LIGHT

The Bureau of Standards at Washington has issued Scientific Paper No. 447, describing a large sphere built at that Bureau and giving a complete explanation of the theory of spheres as light measurers and of the precautions which must be taken in order not to make mistakes when using them. The particular sphere described is 88 inches in diameter and weighs about a ton, but for different purposes spheres from 3 inches to 10 feet in diameter have been used. Some of these, for example, are made as attachments for pocket size "Illuminometers" and are carried about to measure the reflections from walls and ceilings and from samples of paper, paint, and other materials, while modified forms on a large scale have been used to catch the beams from full-sized army searchlights and measure the light in them.

CLEVELAND TO HAVE THE 1923 LIGHTING FIXTURE MARKET

The Fourth Annual Lighting Fixture Market of the Associated Lighting Industries, which consists of the National Council of Lighting Fixture Manufacturers, the Illuminating Glassware Guild and the Lighting Fixture Dealers Society of America, is to be held in Cleveland from January 15 to 20, 1923.

The program of the convention will be devoted exclusively to educational work which should prove extremely interesting and instructive. Among the special features will be a practical demonstration of selling to retail customers, which should give to the dealers and their salesmen many new and effective selling arguments. It has been decided to confine this meeting solely to lighting fixture dealers and manufacturers, because it is felt that by so doing each one present will secure a better understanding of how to promote his own business, as well as to gain a closer feeling of co-operation and unity in the industry.

I. E. S. EXHIBIT AT CLEVELAND

At the invitation of the Illuminating Glassware Guild, the I. E. S. will display an educational exhibit at the Winton Hotel. The Council at its last meeting appointed a committee consisting of Messrs. Davidson, Story, Sturrock and Turner to arrange the details.

ADVISORY BOARD ON ILLUMINATING ENGINEERING RESEARCH

Pursuant to discussion and presentation of a paper delivered by Dr. Edward P. Hyde at the last meeting of the Division of Engineering, National Research Council, held December 5, 1922, indicating the importance of a study in "Relation of Quality and Quantity of Illumination to Efficiency in the Industries," it was voted at that meeting that the Division appoint the Research Committee of the Illuminating Engineering Society its *Advisory Board on Illuminating Engineering Research.*

The Division has in each broad field of engineering an advisory board. Meetings are not held frequently. From time to time it is found convenient to call upon these members for advice in connection with certain problems.

STREET LIGHTING DEVELOPMENTS

Talking recently before the Convention of the American Society for Municipal Improvements on the subject "Modern Requirements in Street Lighting," Mr. Thomas W. Rolph managing engineer, Street Lighting Department, Holophane Glass Company, discussed the influences that necessitated better and higher standards of street lighting in our cities to-day. The two factors on which he dealt particularly were, first, the tremendously wide application of the automobile to modern civilization and, second, the influence of the moving picture industry.

Lighting equipment and incandescent lamp manufacturers have kept pace with the expansion that has constantly demanded better equipment both from the standpoint of utility and design. More satisfactory street lighting has been retarded, however, because municipalities have not possessed the funds necessary to undertake to any appreciable extent lighting improvement work.

Mr. Rolph also spoke on the comparative efficiencies and merits of porcelain enameled reflectors, diffusing glassware and prismatic refractors for street· lighting installations, and told of experiments conducted to ascertain proper mounting heights and spacings.

Mr. Earl A. Anderson, Engineering Department, Nela Park, told of the numerous developments, such as the new 2,500-candlepower gas-filled lamp; the gradual replacement of the overhead wooden-pole mast-arm type by poles and equipment of beauty and greater efficiency; the tendency toward underground circuits, as well as citing the progress made in street lighting in several American cities within the past year.

One interesting highlight touched upon in Mr. Anderson's talk was the relation of better lighting to accident and crime prevention. He pointed out that the annual loss in dollars and cents directly chargeable to inadequate city street lighting was more than the total annual expenditure made for street lighting in the country.

A MILLION PEOPLE HAVE VISITED ELECTRIC HOMES IN 1922

Recent returns indicate that over a million people have learned something more about electricity and good illumination in the home during the past year. Over sixty-one electric homes have been exhibited to the public, with attendance figures varying from 6,500 to 37,000 people per home, or an average of 20,000 visitors.

Recently in Cleveland, O., the fifth electric home of that city closed its doors after a most successful demonstration.

It is evident that the achievements of the Electric Home movement have been most noteworthy, and are deserving of the support of every one interested in better illumination of the home.

PROFESSIONAL OPPORTUNITIES

ASSISTANT SALES MANAGER: A man from twenty-five to thirty-five who is familiar with illuminating engineering and who feels qualified to act as assistant sales manager in a growing concern.   R-1

ELECTRICAL ENGINEER: Young graduate engineer to specialize in testing of incandescent lamps.   L-1

LABORATORY TEST AND RESEARCH MEN: Several positions open for young technical graduates. Men must be willing to travel. Write stating training, experience, etc.   T-1

ILLUMINATION INDEX

PREPARED BY THE COMMITTEE ON PROGRESS.

An INDEX OF REFERENCES to books, papers, editorials, news and abstracts on illuminating engineering and allied subjects. This index is arranged alphabetically according to the names of the reference publications. The references are then given in order of the date of publication. Important references not appearing herein should be called to the attention of the Illuminating Engineering Society, 29 W. 39th St., New York, N. Y.

*Not previously reviewed.

Psychological Review

The Effect of Variations of the Intensity of Illumination of the Perimeter Arm on the Determination of Color Fields—
C. E. Ferree and Gertrude Rand   Nov.   457

Public Works

Street Lighting Construction and Apparatus—   Nov.   242

Rept. Lab. Asahi Glass Co.

Protective Spectacles for Ultra-violet Rays—   Inagaki   1921

Revue Generale de L'Electricite

Lampe a incandescence a filament droit pour recherches de laboratoires et experiences de cours—   A. Cotton   Oct. 21   125D

Science Abstracts "A"

A New Method of Investigating Colour Blindness with a Description of Twenty-Three Cases—
(*Roy. Soc. of Edinburgh, Proc.*, 42, 1, pp. 75-88, 1921-22).   R. A. Houstoun   Sept.   712

Dispersion of Light in Fluorescent Solutions—
(*Accad. Lincei, Atti.*, 31, 1, pp. 157-160, Feb. 19, 1922).   A. Garrelli   Sept.   717

Effect of Impurities on Recrystallization and Grain Growth—
(*Inst. Metals, J.*, 27, p. 107; *Disc. and Corres*, 148, 1922. *Eng.*, 113, pp. 342-346, March 17, 1922).   Research Staff of the General Electric Co., London   Sept.   732

Behavior of Carbon at High Temperatures—
(*Zeit. Elektrochem.*, 28, pp. 183-185, April 1, 1922).   F. Sauerwald   Sept.   734

Textile World

Light Without Glare—   Nov   81

Transactions of the Faraday Society

High Temperature Phenomena of Tungsten Filaments*—   C. J. Smithells   Feb.   485

*Not previously reviewed.

OFFICERS & COUNCIL

1922-23

President

WARD HARRISON,
Nela Park,
Cleveland, Ohio
Term expires Sept. 30, 1923

Junior Past Presidents

GEO. H. HARRIES,
208 South LaSalle Street,
Chicago, Ill.
Term expires Sept. 30, 1923

GEO. S. CRAMPTON,
1700 Walnut Street,
Philadelphia, Pa.
Term expires Sept. 30, 1924

Vice-Presidents

F. M. FEIKER,
36th Street and 10th Avenue,
New York, N. Y.
Term expires Sept. 30, 1923

O. L. JOHNSON,
847 W. Jackson Blvd.,
Chicago, Ill.
Term expires Sept. 30, 1924

WM. J. DRISKO,
Mass. Inst. of Technology,
Cambridge, Mass.
Term expires Sept. 30, 1924

G. BERTRAM REGAR,
1000 Chestnut Street,
Philadelphia, Pa.
Term expires Sept. 30, 1924

General Secretary

SAMUEL G. HIBBEN,
165 Broadway,
New York, N. Y.
Term expires Sept. 30, 1923

Treasurer

L. B. MARKS,
103 Park Avenue,
New York, N. Y.
Term expires Sept. 30, 1923

Directors

Terms expire Sept. 30, 1923	*Terms expire Sept. 30, 1924*	*Terms expire Sept. 30, 1925*
WALTON FORSTALL, 1401 Arch Street, Philadelphia, Pa.	F. C. CALDWELL, Ohio State University, Columbus, Ohio	FRANK R. BARNITZ, 130 East 15th St., New York, N. Y.
ADOLPH HERTZ, Irving Place and 15th Street, New York, N. Y.	AUGUSTUS D. CURTIS, 235 W. Jackson Blvd., Chicago, Ill.	CLARENCE L. LAW, Irving Place and 15th Street New York, N. Y.
FRANK S. PRICE, 160 Pearl Street, Boston, Mass.	Preston S. MILLAR, 80th St. and East End Ave., New York, N. Y.	A. L. POWELL, Fifth and Sussex Sts., Harrison, N. J.

SECTION & CHAPTER OFFICERS

Chicago Section

CHAIRMAN . . . . Albert L. Arenberg, 316 S. Wells St., Chicago, Ill.
SECRETARY . . . . F. A. Rogers, Lewis Institute, Madison and Robey Sts., Chicago, Ill.

MANAGERS

J. H. Allen . . . . Monadnock Block, Chicago, Ill.
F. A. DeLay . . . . 72 W. Adams St., Chicago, Ill.
F. Lee Farmer . . . . 216 S. Jefferson St., Chicago, Ill.
W. M. Goodrich . . . . 500 South Clinton St., Chicago, Ill.
J. J. Kirk . . . . 72 W. Adams St., Chicago, Ill.

New England Section

CHAIRMAN . . . . R. R. Burnham, 119 Water St., Boston. Mass.
SECRETARY . . . . C. A. Strong, 60 India Street, Boston, Mass.

MANAGERS

W. V. Batson . . . . : . . . . . . 210 South Street, Boston, Mass.
J. Daniels . . . . 39 Boylston Street, Boston, Mass.
J. A. Toohey . . . . Box 3396, Boston, Mass.
L. T. Troland . . . . Harvard Univ., Cambridge, Mass.
J. L. Tudbury . . . . 247 Essex Street, Salem, Mass.

New York Section

CHAIRMAN . . . . A. L. Powell, Edison Lamp Works, Harrison, N. J.
SECRETARY . . . . J. E. Buckley Irving Place and 15th Street, New York, N. Y.

MANAGERS

R. B. Burton . . . . 6 W. 48th Street, New York, N. Y.
J. R. Fenniman . . . . 130 E. 15th Street, New York, N. Y.
A. M. Perry . . . . 10th Ave. and 36th Street, New York, N. Y.
H. A. Sinclair . . . . 526 W. 34th Street, New York N. Y.
Geo. Strahan . . . . 507 W. 24th Street, New York, N. Y.

Philadelphia Section

CHAIRMAN . . . . Howard Lyon, Welsbach Co., Gloucester, N. J.
SECRETARY . . . . H. Calvert, 2114 Sansom Street, Philadelphia, Pa.

MANAGERS

H. B. Andersen . . . . 20 Maplewood Ave., Philadelphia, Pa.
C. E. Clewell . . . . Univ. of Penn., Philadelphia, Pa.
M. C. Huse . . . . 1000 Chestnut Street, Philadelphia, Pa.
J. B. Kelley . . . . 235 Market Street, Philadelphia, Pa.
W. E. Saunders . . . . Broad and Arch Sts., Philadelphia, Pa.

Cleveland Chapter

CHAIRMAN . . . . C. L. Dows, Nela Park, Cleveland, Ohio
SECRETARY . . . . E. W. Commery, Nela Park, Cleveland, Ohio

MANAGERS

E. A. Anderson . . . . Nela Park, Cleveland, Ohio
G. A. Hausman . . . . 6th and Rockwell Sts., Cleveland, Ohio
Phil. C. Keller . . . . 5716 Euclid Ave. Cleveland, Ohio
A. J. Thompson . . . . 226 St. Clair Ave. N. E., Cleveland, Ohio
J. L. Wolf . . . . Builders Exchange, Cleveland, Ohio

San Francisco Bay Cities Chapter

CHAIRMAN . . . . Leonard E. Voyer, Rialto Bldg., San Francisco, Calif.
SECRETARY . . . . H. H. Millar, 1648 16th Street, Oakland, Calif.
TREASURER . . . . J. A. Vandegrift, 1648 16th Street, Oakland, Calif.

MANAGERS

W. W. Hanscom . . . . 848 Clayton Street, San Francisco, Calif.
R. A. Hudson . . . . 703 Rialto Bldg., San Francisco, Calif.
Romaine Myers . . . . 204 Bacon Building, Oakland, Calif.
R. S. Prussia . . . . 705 First National Bank Bldg., San Francisco, Calif.
Miles F. Steel . . . . 590 Howard Street, San Francisco, Calif.

Toronto Chapter

CHAIRMAN . . . . G. R. Anderson, Univ. of Toronto, Toronto, Canada
SECRETARY-TREASURER . . . . M. B. Hastings, 62 Front St. W., Toronto, Canada

Executive Committee

G. R. Anderson . . . . Univ. of Toronto. Toronto, Can.
F. T. Groome . . . . 119 Fairview Ave. West Toronto, Can.
M. B. Hastings . . . . 62 Front St. W., Toronto, Can.
J. T. Scott . . . . : . . . 221 Dufferin St., Toronto, Can.
W. H. Woods . . . . 507 Brunswick Ave. Toronto, Can.

CALIFORNIA:	San Francisco	Romaine W. Myers, 204 Bacon Building, Oakland.
	Los Angeles	F. S. Mills, 631 Pacific Finance Building.
COLORADO:	Denver	G. O. Hodgson, Edison Lamp Works.
CONNECTICUT:	New Haven	Charles F. Scott, Yale University. J. Arnold Norcross, 80 Crown Street.
	Hartford	R. E. Simpson, Travelers Insurance Co.
DISTRICT OF COLUMBIA:	Washington	R. B. Patterson, 213 14th Street. E. C. Crittenden Bureau of Standards.
GEORGIA:	Atlanta	Charles A. Collier, Electric and Gas Building.
IDAHO:	Boise	W. R. Putnam, Idaho Power Co.
KENTUCKY:	Louisville	H. B. Heyburn, Second and Washington Sts.
LOUISIANA:	New Orleans	Douglas Anderson, Tulane University of Louisiana.
MASSACHUSETTS:	Springfield	J. F. Murray, United Electric Light Co.
	Worcester	Carl D. Knight, Worcester Polytechnic Institute.
MICHIGAN:	Detroit	F. R. Mistersky, 129 Kenilworth Avenue. Charles Monroe, Detroit Edison Co.
MINNESOTA:	St. Paul	A. L. Abbott, 174 E. 6th Street.
	Minneapolis	G. D. Shepardson, University of Minnesota.
MISSOURI:	St. Louis	Louis D. Moore, 1130 Railway Exchange Bldg.
NEW YORK:	Rochester	Frank C. Taylor, 166 Shepard St.
	Schenectady	S. H. Blake, General Electric Co.
OHIO:	Cincinnati	A. M. Wilson, University of Cincinnati.
	Columbus	F. C. Caldwell, Ohio State University.
	Toledo	W. E. Richards, Toledo Railway and Light Co.
OREGON:	Portland	F. H. Murphy, Portland Railway Light and Power Co. Electric Building.
PENNSYLVANIA:	Pittsburgh	L. O. Grondahl, Carnegie Inst. of Technology
RHODE ISLAND:	Providence	F. A. Gallagher, Narragansett Elec. Lighting Co.
TEXAS:	Austin	J. M. Bryant, University of Texas.
UTAH:	Salt Lake City	R. H. Ashworth, Kearns Bldg.
VIRGINIA:	Charlottesville	W. S. Rodman, University of Virginia.
WASHINGTON:	Seattle	Fred. A. Osborn, University of Washington.
WISCONSIN:	Milwaukee	F. A. Vaughn, Metropolitan Block, Third and State Sts.
CANADA:	Montreal	L. V. Webber, 285 Beaver Hall Hill.
SOUTH AFRICA:	Johannesburg	H. A. Tinson, South African Gen. Elec. Co., Southern Life Buildings, Transvaal.

OFFICIAL REPRESENTATIVES TO OTHER ORGANIZATIONS

On the United States National Committee of the International
Commission on Illumination . . . . { L. B. Marks
{ Preston S. Millar

On American Institute of Electrical Engineers Standards Committee . Clayton H. Sharp

On Governing Board of the American Association for the Advancement
of Science . . . . { Ernest F. Nichols
{ Clayton H. Sharp

On the Advisory Committee, Engineering Division, National Research
Council . . . . Edward P. Hyde

On the Fixture Manufacturers Committee to prepare a Code
of Luminaire Design . . . . { E. W. Commery
{ J. W. Gosling
{ S. G. Hibben
{ M. Luckiesh
{ R. H. Maurer
{ A. L. Powell
{ Samuel Snyder

On A. E. S. C. Sectional Committee, Building Exits Code . . . . R. E. Simpson

COMMITTEES

1922-1923

Except as noted below, all committees are appointed by the President, subject to the approval of the Council, and terminate at the time of the first Council meeting of each new administration, in the month of October. The duties of each committee are indicated.

WARD HARRISON, President, Ex-officio member of all Committees.

(1) STANDING COMMITTEES AUTHORIZED BY THE CONSTITUTION AND BY-LAWS.

COUNCIL EXECUTIVE.—(*Consisting of the President, General Secretary, Treasurer and two members of the Council.*) *Act for the Council between sessions of the latter.*

Ward Harrison, Chairman.
Nela Park, Cleveland, Ohio.
Walton Forstall,   Clarence L. Law,
S. G. Hibben,   L. B. Marks.

FINANCE.—(*Of three members; to continue until successor is appointed.*) *Prepare a budget; approve expenditures; manage the finances; and keep the Council informed on the financial condition.*

Adolph Hertz, Chairman.
Irving Place & 15th St., New York, N. Y.

Walton Forstall,   D. McFarlan Moore.

PAPERS.—(*Of at least five members.*) *Provide the program for the annual convention; pass on papers and communications for publication; and provide papers and speakers for joint sessions with other societies.*

J. L. Stair, Chairman.
235 W. Jackson Blvd., Chicago, Ill.
A. L. Powell, Vice-Chairman.
Fifth and Sussex Sts., Harrison, N. J.

E. A. Anderson,   Howard Lyon,
Louis Bell,   Norman D. Macdonald,
Geo. G. Cousins,   F. H. Murphy,
H. H. Higbie,   J. J. Ryan.
Albert Scheible.

Chairman of Section and Chapter Papers Committees, Ex-officio Members.

EDITING AND PUBLICATION.—(*Of three members.*) *Edit papers and discussions; and publish the Transactions.*

Norman D. Macdonald, Chairman.
80th St. and E. End Ave., New York, N.Y.
Allen M. Perry,   Ralph C. Rodgers.

GENERAL BOARD OF EXAMINERS.—(*Appointed by the President.*) *Pass upon the eligibility of applicants for membership or for changes in grade of membership.*

Norman Macbeth, Chairman.
227 W. 17th St., New York, N. Y.
Clarence L. Law,   L. J. Lewinson.

(2) COMMITTEES THAT ARE CUSTOMARILY CONTINUED FROM YEAR TO YEAR.

LIGHTING LEGISLATION. — *Prepare a digest of laws on Illumination; coöperate with other bodies in promoting wise legislation on illumination; and prepare codes of lighting in certain special fields, to function also as a Technical Committee on Industrial Lighting.*

L. B. Marks, Chairman.
103 Park Avenue, New York, N. Y.

W. F. Little, Secretary.
80th St. and E. End Av., New York, N. Y

Louis Bell,   J. A. Hoeveler,
W. T. Blackwell,   Clarence L. Law,
F. C. Caldwell,   M. G. Lloyd,
C. E. Clewell,   M. Luckiesh,
Geo. S. Crampton,   R. H. Maurer,
H. B. Dates,   A. S. McAllister,
E. Y. Davidson, Jr., W. J. Serrill,
W. A. D. Evans,   R. E. Simpson,
S. G. Hibben,   G. H. Stickney.

MEMBERSHIP. — *To obtain additional individual memberships.*

G. Bertram Regar, Chairman.
1000 Chestnut St. Philadelphia, Pa.,

J. F. Anderson,   F. W. Bliss,
A. L. Arenberg,   R. I. Brown,
C. A. Atherton,   W. E. Clement,
W. T. Blackwell,   George G. Cousins,

IV

COMMITTEES -- 1922-1923.

Terrell Croft,
J. Daniels,
B. Y. Davidson, Jr.,
Robert B. Ely,
R. W. Everson,
H. W. Fuller,
H. B. Hobson,
J. A. Hoeveler,
Merritt C. Huse,

C. C. Munroe,
F. H. Murphy,
H. G. Smith,
H. H. Smith,
E. W. Spitz,
Clare N. Stannard,
O. R. Toman,
L. J. Wilhoite,
L. A. S. Wood.

MOTOR VEHICLE LIGHTING.

C. H. Sharp, Chairman.
8oth St. and E. End Ave., New York, N.Y.

G. H. Stickney, Vice-Chairman.
Fifth and Sussex Sts., Harrison, N. J.

F.C. Caldwell,
A. W. Devine,
C. E. Godley,
C. A. B. Halvorson,
J. A.Hoeveler,

W.F. Little,
W.A. McKay,
A. L. McMurtry,
H. H. Magdsick,
L. C. Porter,
U. M. Smith.

L. B. Voyer, Representative in State of Cal.

NOMENCLATURE AND STANDARDS—*Define the terms and standards of Illumination; and endeavor to obtain uniformity in nomenclature.*

E. C. Crittenden, Chairman.
Bureau of Standards, Washington, D. C.

Howard Lyon, Secretary.
Welsbach Co., Gloucester, N. J.

Louis Bell,
G. A. Hoadley,
B. P. Hyde,
A. E. Kennelly,

M. Luckiesh,
C. O. Mailloux,
A. S. McAllister,
·C. H. Sharp,
G. H. Stickney.

PROGRESS.—*Submit to the annual convention a report on the progress of the year in the science and art of Illumination.*

F. E. Cady, Chairman.
Nela Park, Cleveland, O.

G. S. Crampton,   W. E. Saunders.

RESEARCH.—*Stimulate research in the field of illumination; and keep informed of the progress of research.*

E. P. Hyde, Chairman.
Nela Park, Cleveland, O.

Louis Bell,
W. T. Bovie,*
Alex. Duane,
Knight Dunlap,
E. N. Harvey,*

M. Luckiesh,
Thomas Scofield,*
C. H. Sharp,
C. A. Skinner,*
L. T. Troland,
F. H. Verhoeff.*

(3) TEMPORARY COMMITTEES FOR SPECIAL PURPOSES.

SKY BRIGHTNESS.

H. H. Kimball, Chairman.
U. S. Weather Bureau, Washington, D. C.

E. C. Crittenden,
E. H. Hobbie,
Bassett Jones,

W. F. Little,
M. Luckiesh,
L. B. Marks.
I. G. Priest.*

COMMITTEE TO COÖPERATE WITH FIXTURE MANUFACTURERS

To Prepare Code of Luminaire Design

M. Luckiesh, Chairman,
Nela Park, Cleveland, O.
E. W. Commery, Secretary,
Nela Park, Cleveland, O.

S. G. Hibben,
J. W. Gosling,*

R. H. Maurer,
A. L. Powell,
Samuel Snyder.

COMMITTEE ON NEW SECTIONS AND CHAPTERS

D. McFarlan Moore, Chairman,
Edison Lamp Works, Harrison, N. J.

COMMITTEE TO PREPARE A BULLETIN ON RESIDENCE LIGHTING BY GAS

Howard Lyon,   R. H. Maurer.

COMMITTEE TO PREPARE A BULLETIN ON RESIDENCE LIGHTING BY ELECTRICITY

Samuel G. Hibben, Chairman.
165 Broadway, New York, N. Y

Geo. Ainsworth,
F. E. Cady,
O. H. Caldwell,
W. J. Drisko,
R. B. Ely,
Gertrude R. Ferree,*
R. S. Hale,
F. Y. Joannes,*
Clarence L. Law,

M. Luckiesh,
W. R. M'Coy,
Herman Plaut,
A. L. Powell,
F. B. Rae, Jr.,*
R. C. Rodgers,
J. W. Smith,*
J. L. Stair,
E. D. Tillson,
D. H. Tuck

* Not members of the Society but coöperating in the work of these committees.

SUSTAINING MEMBERS

Alexalite Co., New York, N. Y.
American Gas & Electric Co., New York, N. Y.
American Optical Co., Southbridge, Mass.
Bausch & Lomb Optical Co., Rochester, N. Y.
Beardslee Chandelier Mfg. Co., Chicago, Ill.
Benjamin Electric Manufacturing Co., Chicago, Ill.
Boston Consolidated Gas Co., Boston, Mass.
The Brascolite Co., St. Louis, Mo.
Brooklyn Edison Co., Inc., Brooklyn, N. Y.
Brooklyn Union Gas Co., Brooklyn, N. Y.
Cambridge Electric Light Co., Cambridge, Mass.
Central Electric Company, Chicago, Illinois.
Central Hudson Gas & Electric Co., Newburgh, N. Y.
Commonwealth Edison Co., Chicago, Ill.
Consolidated Electric Lamp Co., Danvers, Mass.
Consolidated Gas, Electric Light & Power Co., Baltimore, Md.
Consolidated Gas Co. of New York, New York, N. Y.
Consumers Power Co., New York, N. Y.
Cooper-Hewitt Electric Co., Hoboken, N. J.
Cox, Nostrand & Gunnison, Brooklyn, N. Y.
Cutler-Hammer Manufacturing Co., Milwaukee, Wi-
Dawes Brothers, Chicago, Ill.
Detroit Edison Co., Detroit, Mich.
Duquesne Light Co., Pittsburgh, Pa.
East Side Metal Spinning & Stamping Co., New York, N. Y.
Edison Electric Illuminating Co. of Boston, Boston, Mass.
Edison Lamp Works of General Electric Co., Harrison, N. J.
Electric Outlet Co., Inc., New York, N. Y.
Electrical Testing Laboratories, New York, N. Y.
Fitchburg Gas & Electric Light Co., Fitchburg, Mass.
Oscar O. Friedlaender, Inc., New York, N. Y.
Georgia Railway & Power Co., Atlanta, Ga.
Gillinder Brothers, Port Jervis, N. Y.
Gillinder & Sons, Inc., Philadelphia, Pa.
Gleason-Tiebout Glass Co., Brooklyn, N. Y.
Hartford Electric Light Co., Hartford, Conn.
Haverhill Electric Co., Haverhill, Mass.

Holophane Glass Co., Inc., New York, N. Y.

The Horn & Brannen Mfg. Co., Inc., Philadelphia, Pa.

Hydro-Electric Power Commission of Ontario, Toronto, Canada.

Hygrade Lamp Co., Salem, Mass.

Krich Light and Electric Co., Newark, New Jersey.

Kristiania Elektrecitetsvork, Kristiania, Norway.

E. B. Latham & Co., New York, N. Y.

Leeds & Northrup Co., Philadelphia, Pa.

Lightolier Co., New York, N. Y.

Lynn Gas & Electric Co., Lynn, Mass.

Macbeth-Evans Glass Co., Pittsburgh, Pa.

Malden Electric Co., Malden, Mass.

Malden & Melrose Gas Light Co., Malden, Mass.

McGraw-Hill Publishing Co., New York, N. Y.

Millville Electric Light Co., Millville, N. J.

Montpelier & Barre Light & Power Co., Montpelier, Vt.

Narragansett Electric Lighting Co., Providence, Rhode Island.

National Lamp Works of General Electric Co., Nela Park, Cleveland, Ohio.

National X-Ray Reflector Co., Chicago, Ill.

New Bedford Gas & Edison Light Co., New Bedford, Mass.

New England Power Co., Worcester, Mass.

New Haven Gas Light Co., New Haven, Conn.

New York Edison Co., New York, N. Y.

New York & Queens Electric Light & Power Co., Long Island City, N. Y.

Pennsylvania Power & Light Co., Allentown, Pa.

Peoples Gas Light & Coke Co., Chicago, Ill.

Pettingell-Andrews Co., Boston, Mass.

Philadelphia Electric Co., Philadelphia, Pa.

Philadelphia Electric Manufacturing Co., Philadelphia, Pa.

L. Plaut & Co., New York, N. Y.

Portland Railway, Light & Power Co., Portland, Ore.

Potomac Electric Power Co., Washington, D. C.

Providence Gas Co., Providence, R. I.

Public Service Corporation of New Jersey, Newark, N. J.

Public Service Company of Northern Illinois, Chicago, Ill.

Rochester Gas & Electric Corp., Rochester, N. Y.

Max Schaffer Company, New York, N. Y.

Salem Electric Lighting Co., Salem, Mass.

Shapiro & Aronson, Inc., New York, N. Y.
Societa Idroelettrica Piemonte, Turin, Italy.
Southern Electric Co., Baltimore, Md.
Sterling Bronze Co., New York, N. Y.
Stone & Webster, Boston, Mass.
Suburban Gas & Electric Co., Revere, Mass.
The United Gas Improvement Co., Philadelphia, Pa.
Charles H. Tenney & Co., Boston, Mass.
Tungsten Wire Works, A. B., Stockholm, Sweden.
Union Gas & Electric Co., Cincinnati, Ohio.
United Electric Light & Power Co., New York, N. Y.
Union Light & Power Co., Franklin, Mass.
Alfred Vester Sons, Providence, R. I.
Welsbach Co., Gloucester, N. J.
Welsbach Street Lighting Co., Philadelphia, Pa.
Westinghouse Electric & Manufacturing Co., East Pittsburgh, Pa.
Westinghouse Lamp Co., Bloomfield, N. J.
Wetmore-Savage Co., Boston, Mass.
Worcester Electric Light Co., Worcester, Mass.
Worcester Suburban Electric Co., Palmer, Mass.

TRANSACTIONS
OF THE
ILLUMINATING ENGINEERING SOCIETY

| VOL. XVIII | FEBRUARY, 1923 | NO. 2 |

Tail Lights and License Plates

FOR SEVERAL YEARS, the sizes of number plates for automobiles have been increasing to accommodate the larger serial numbers, but no one has seen to it that tail lights were so changed as effectively to illuminate this larger area. As a result, almost all the modern equipments are violating the legal requirement that rear number plates shall be so lighted as to be legible at a distance ordinarily specified as 50 or 60 feet.

As an illustration of the situation, the officials of an eastern state, recently stopped over four hundred automobiles and found but one which had satisfactory number plate lighting. It is obvious that such condition of affairs cannot continue, since it is fully as important to be able to read the identifying numbers after dark as in the daytime.

It has remained for Massachusetts to take the first step to correct this anomalous situation. By a recently enacted law, motor vehicles must be equipped with approved rear lamps. Under this law, which is just going into effect, a large number of tail lamps have been submitted for approval. It is so obviously important that improvement be made, that it is generally assumed that other states will fall in line and make the requirements more or less similar to those of Massachusetts. On the other hand, it is almost of equal importance that the action of the several states should be as uniform as possible. It would be very unfortunate if different equipments were required in different states, especially if visitors were required to make changes in their equipment.

The Massachusetts Motor Vehicle Department early recognized that it was not enough simply to have a well designed lantern, but that it was also necessary to so fix the relation between the lantern and the plate as to make sure that the illumination would be directed on the number plate. They have therefore required the submission of the complete equipment, including the plate support. They have handled the problem in a commonsense way, but in order that there may be uniformity and action throughout the country, it seemed desirable to have some general specifications covering the requirements.

The Society's Committee on Motor Vehicle Lighting was therefore asked to draw up such specifications. They in turn invited the Lighting Division of the Standards Committee of the Society of Automotive Engineers to collaborate. A joint sub-committee, under the chairmanship of Mr. A. W. Devine of the Massachusetts Registry of Motor Vehicles was appointed. The membership of this sub-committee spent several months studying the problem and making the tests. These data, together with the experience in Massachusetts, were brought together at a meeting held in December. A comparison with the data from various sources, showed a surprisingly close agreement, and the Committee had relatively little difficulty in arriving at an acceptable set of specifications.

These specifications were then submitted to a joint meeting of the main committees, to which state officials, lantern manufacturers, and all others known to be actively interested in the problem, were invited. It is a testimonial to the work of the sub-committee that only slight changes were suggested, and the specifications were approved without a single dissenting vote.

These specifications are printed in this issue of the TRANSACTIONS. They seem to be the best that can be drawn at the present time, and entirely practicable. It is recognized that the conditions to be met are quite variable, and that it is quite conceivable that experience in applying the specifications will suggest further improvements. The specifications apply, of course, to the present form of number plates, which are seen by reflected light.

The scheme of using stencil plates to be illuminated by transmitted light, was considered by the Committee, but it did not seem desirable to prepare specifications covering this type ·of· license plate until there was more assurance that such plates would be furnished by state authorities.

One of the problems faced by the Committee was the lack of uniformity in the plates issued by different states. To meet this condition, the Committee passed a resolution calling attention to the need of suitable standardization, and suggesting certain limiting dimensions, use of contrasting colors and the use of numerals of characteristic shape and specified widths of stroke and spacing.

The tail light problem is much simpler than that of the headlight, inasmuch as the equipment is much less sensitive to adjustment in operation, and there is good reason to believe that a really effective step has been taken toward the establishment of means of providing good illumination on the number plates.

In taking this action, the Society and those who co-operated with it, have contributed a real service to the public.

G. H. STICKNEY, *Vice-Chairman*,
Committee on Motor Vehicle Lighting.

REFLECTIONS

Candlesticks Found in Tutankhamen's Tomb

ACCORDING to a recent wireless dispatch to *The New York Times*, Mr. Howard Carter removed on January 18th, from Tutankhamen's Tomb, four ancient candlesticks, or supports for miniature torches. They are considered by Professor Breasted as among the most important objects in the tomb because archæologists did not know how the Pharaohs illuminated their palaces at night.

One of the candlesticks contained a wick, which was seemed to be made of twisted linen, which was probably soaked in oil on the same principle as modern oil lamps.

The experts do not agree as to whether these candlesticks were used for domestic lighting or for religious processions. They have a curious shape. The general outline is the key of life, supported by bronze wrists and hands and overlaid with gold in the parts where the light was unlikely to flare.

The expression, "The Key of Life," used in describing the candlesticks removed from the tomb refers to what is also known to Egyptian archæologists as the Angh, a sort of staff resembling a cross with a loop at the top, and usually held by one of the Egyptian dieties or kings. In their religious services it represented generation, or enduring life.

Reckoning the Cost of Daylight

THE Illuminating Engineering Society convention paper by Luckiesh and Holladay dealing with the cost of daylight stirs up a question which is of real importance in modern city planning. It extends in fact not only to large industrial or office buildings, but even to hotels and residences. In its larger aspect the problem is this: Given ground space for a substantial skyscraper, say 20,000 square feet, how much of such space does it pay to reserve for light wells or their equivalent, looking at the matter in all aspects and reckoning the full cost of daylight lighting and also of artificial lighting at current rates? Of course, it will readily be granted that to the degree in which outside lighting can conveniently be obtained without sacrificing of ground space it is always worth obtaining; but in the inner portions of the available space the loss of floor area due to a light well is serious, and the effectiveness of such a well in the case of very tall buildings becomes dubious unless considerable area be sacrificed. Therefore is it not wise to depend more upon artificial light than we have of late been in the habit of doing?

Few people realize that window space costs in dollars and cents more wall space. This is true not only in the case of an industrial building with highly specialized steel window frames, but as regards a common frame house. Moreover, the amount of heat required in our climate during a considerable part of the year is greatly increased when very liberal window space is provided. Heat is lost through the windows much more rapidly than through the walls, and consequently more must be furnished. Repairs, renewals and cleaning of windows also rise to a very perceptible figure, so that, including the interest on the investment, the authors find that, speaking in general terms, the total annual costs of lighting by window spaces and by artificial light are usually not widely different, while in not a few instances artificial lighting carried to the point indicated by good modern illuminating practice is actually cheaper than natural light.

This is not at all to offer the suggestion that we had better do without windows entirely unless they are needed for ventilating purposes, but to show that in cases where ground space is valuable and the uses of a building are such as permit of effective artificial lighting such light may be not only economical, but on the whole quite as desirable as natural lighting. If a building is to be used for industrial purposes, experience shows that without great sacrifice of ground area it is extremely difficult to get good natural lighting, and it is also a well-known fact that first-class artificial light keeps production up to at least the point reached by the average daylight conditions. It is a curious turn of affairs that has brought about a profit balance on the side of artificial light, a result perhaps of our insistence on crowding

together in cities and putting our buildings on end and in contact
with each other instead of spreading them out for the maximum
benefit of light and air. Society is penalized by this sort of
crowding in many ways, but it is a condition and not a theory
with which we have to deal, and from present data it looks very
much as though the future architect and contractor would have
to consider the economic factor of artificial lighting as one bearing somewhat heavily on the utilization of the space at hand.—
Electrical World, October 21, 1922.

Some of the Problems of Flicker Photometry

W E HAVE learned to look to Ferree and Rand for highly
ingenious experimenting and interesting results. The paper
prepared by them for the convention of the Illuminating Engineering Society deals with the often-discussed matter of flicker photometry. Of course, it is well known that in the comparison of
lights of different colors the flicker instrument gives results materially different from the equality-of-brightness photometer. It is
generally regarded, too, as more reliable, but where a discrepancy
exists and there is no available method to ascertain with absolute
precision the cause of the variation, it is always worth investigating. In the case of flicker photometry the research is fraught
with almost every conceivable kind of difficulty, because little is
definitely known about the relation of the sensation of flicker to
the color stimuli that reach the retina and the speed at which
they are developed. It is this last-named factor with which the
paper before us deals; that is, the growth of color sensation as
a function of time in the transient sensation obtained in the use
of a flicker photometer. The study made in the paper under
notice bears definitely on the rate of growth with time of the sensations of color. This rate of growth proves to vary greatly,
like all similar functions, in different persons and different states
of adaptation in the same individual.

So far as Ferree and Rand's investigation goes, it does not
deal yet with a sufficient group of observers to allow a safe
average to be taken, nor does it make altogether clear the particular state of adaptation reached during the experiments. It does,
however, bring out very clearly that there is a somewhat large
factor of uncertainty in flicker due to the variations in the rela-

tively slow rise of sensation for the several colors. What is perhaps the most striking result of this very interesting investigation is that it was found that, using the flicker photometer on the one hand and the equality-of-brightness photometer as the "blink" instrument on the other, the results dropped into coincidences. In other words, taking the variations of sensation in the flicker method as they were found, glimpsing the equality-of-brightness field for the same observing time led to the same result as in the case of the flicker. A large element of uncertainty appears, then, to be introduced by the use of the brief exposures of the flicker disk as against the steady stimulus of the ordinary equality-of-brightness method.

It should here be noted that a flicker photometer of the Simmance-Abady type gives a very different character of exposure. as regards its time relations from the instrument used in the experiments here described. This may indicate the reason why the photometer performs a good deal more steadily than these very experiments appear to show. The improved relation between the two general methods obtained with very short exposures in the equality-of-brightness comparison suggests an interesting side line of research on the effect of such brief exposures on simultaneous contrast of colors, which is known to be one of the troublesome factor in photometry of colored lights. We trust that by next season's convention the authors of this painstaking study will have had opportunity to try out this and many other collateral experiments which naturally suggest themselves. It is good team work to co-operate in this kind of research.—*Electrical World*, November 18, 1922.

Artificial Lighting Competes with Daylight

A NOVEL lighting installation was recently made under a bridge of the New York, New Haven and Hartford Railroad in Providence, R. I. A street extends a considerable distance under this bridge with the result that it appears very dark to the automobilists in going from the brightly sunlighted street to the darkened street beneath the bridge; then again on leaving the darkened area under the bridge and coming out into sunlight, the drivers were blinded, creating extreme discomfort and endangering the safety of the public.

For ordinary illumination at night, lamps of small size were sufficient to light the roadway, but in order to relieve the contrast between the sunlight and the dark area under the bridge, a higher level of illumination was necessary. This fact resulted in the installation of a system of units with 1,000-watt lamps, which brought the illumination to a level which greatly relieved the effect of contrast. The new lighting system is now in operation during the entire day providing greater comfort and safety to the drivers and pedestrians who travel on this street.

Several other installations of this sort, but on a smaller scale have proved successful. A similar scheme has been planned for the New York-New Jersey-Hudson River vehicular tunnel, by which the transition from daylight to the tunnel lighting is made gradual.

PAPERS

THE COST OF DAYLIGHT*

BY M. LUCKIESH AND L. L. HOLLADAY **

SYNOPSIS: Detailed estimates have been made of the initial and annual costs of natural and of artificial lighting.

The initial net cost of equipment for natural lighting includes, the difference in the cost of the building with and without windows and skylights, the cost of ground area occupied by light-courts, and the cost of extra heating system to supply the difference in heat losses from windows and skylights and from the wall replaced by the glass areas. The annual cost of natural lighting includes interest on the initial net cost, depreciation, cost of repairs, washing and extra fuel. Wall space occupied by windows has not been charged to natural lighting although this is an appreciable item in many cases. The cost of artificial lighting used in the daytime to reinforce daylight has not been charged to natural lighting in making the cost estimates. Deterioration of interiors due to natural lighting and various items of minor importance have been omitted. In fact computations were confined to the major factors which are quite tangible.

The cost of natural lighting is compared in each of nine representative cases with the cost of electric lighting. The initial cost of equipment for electric lighting consists of the cost of wiring, fixtures, and portables, and its annual cost consists of interest upon the initial investment, depreciation, repairs, lamp renewals, and electric energy. Adequate and proper artificial lighting was assumed in all cases excepting the first dwelling.

According to the estimates the initial net cost of natural lighting for dwellings is from 110 to 225 per cent of the initial net cost of electric lighting; for apartments and hotels from 100 to 140 per cent; for offices about 160 per cent; and for art galleries from 80 to 180 per cent. The annual cost of natural lighting was found for dwellings to be from 95 to 125 per cent of the annual cost of electric lighting; for apartments, hotels, and offices where artificial lighting is freely used during the daytime, from 50 to 100 per cent; and for art galleries about 135 per cent.

INTRODUCTION

In the study of lighting it must often occur to some of us that natural lighting costs something, still relatively few persons appear to be conscious of its cost. How often does the consumer of artificial light bring up the question of cost without even hinting of applying this unwelcome characteristic to natural lighting!

* A paper presented at the Annual Convention of the Illuminating Engineering Society, Swampscott, Mass., September 25-28, 1922.

** Laboratory of Applied Science, Nela Research Laboratories, National Lamp Works of G. E. Co., Cleveland, Ohio.

The Illuminating Engineering Society is not responsible for the statements or opinions advanced by contributors.

In fact many times we have been met with the statement that natural light "costs nothing." This is true quite generally outdoors but is rarely true indoors. Many items contribute to the cost of natural lighting such as the greater cost of constructing windows and roof-skylights than plain walls and roofs respectively; the cleaning and replacement of glass; the greater loss of heat from glass areas than from ordinary walls and ceilings; the valuable wall-space and ground-area consumed by windows and light-courts respectively; and various other factors. Furthermore natural lighting fails at the end of each day and is often inadequate during the day and its distribution in interiors is not often ideal.

The cost of natural lighting is of interest to everyone if for no other reason than that its recognition tends to make us more lenient toward our artificial-lighting bills. But the cost of natural lighting and certain undesirable characteristics such as undependability, difficulty of control, and its inadequacy in congested districts afford questions of real economic interest. Where land-values are extremely high, light-courts and even windows place a burden upon natural lighting worthy of consideration from an economic standpoint. Of course, we must have ventilation in buildings but the present areas of windows and of light-courts are usually more than adequate.. In fact artificial ventilation is practised to some extent and can serve us adequately and oft-times better without depending upon natural-lighting openings.

But it is not the aim of this paper to discuss the advantages and disadvantages of natural lighting. Such a discussion would lead far afield and would finally become entangled in opinions and speculations. The aim is to compare the cost of natural lighting with that of artificial lighting.

One of the authors[1] has discussed this subject in various places during the past ten years chiefly with the hope that architects might be aroused to give greater consideration to this aspect of natural lighting and to publish discussions of the subject. Not having seen any data published by architects we have set ourselves the task of making certain computations and estimates assuming what we believe to be representative cases and avoiding

[1] M. Luckiesh, Light. Jour., Vol. 4, Oct., 1916, p. 229; Illum. Engr., London, Nov., 1916; The Lighting Art, McGraw-Hill, 1917; Artificial Light—Its Influence on Civilization, 1920.

extreme cases. The results certainly indicate approximately the costs of artificial and of natural lighting. From them it is seen that the cost of natural lighting is appreciable. The results show that there are real economic questions to consider in providing natural lighting. As our cities become more congested it appears probable that natural lighting must give way more and more to artificial lighting. Furthermore in certain buildings we are bold enough to state that natural lighting should be excluded in favor of artificial lighting.

In making these computations, estimates are unavoidable and conditions are extremely variable. Therefore the results may not indicate more than a general approximation of the costs. However, the computations were made in great detail, with the intention of conservative judgment, and the data employed in each case are representative of actual existing buildings. We present only a few of the many details of computation but enough, we hope, to enable anyone to criticize or to modify the results in accordance with his judgment. To recapitulate, we make no greater claim for the results than to show the approximate costs of artificial and of natural lighting in a number of representative types of modern buildings, such as residences, apartment buildings, office buildings, hotels, and other public buildings.

In computing the cost of artificial illumination attention is confined only to electric lighting with incandescent filament lamps.

COST OF NATURAL LIGHT

In estimating the initial cost of natural lighting obtained by means of windows, we include the cost of window-frames, sash, glass, accessories, trimmings, installation and painting, less the cost of wall and plaster replaced by the window. To this initial net cost of window is added the cost of extra radiation and heating system required to supply the increased heat loss of window surface over wall-area. Overhead skylights are treated in a similar manner.

The annual cost of natural lighting includes the interest at 6 per cent on the net initial cost of windows and heating system, net annual repair and depreciation of windows and additional heating system, the cost of washing of windows, and the cost of fuel to supply the additional heat loss.

Strictly the cost of artificial lighting used to reinforce natural lighting in the *daytime* should be charged to natural lighting thus reducing the costs of artficial lighting and increasing those of natural lighting. This applies to some extent to all interiors but this cost has not been charged to natural lighting. However it is included in the cost of artificial lighting.

We have not included in the cost of natural lighting such items as window-shades, the spoilage and cleaning of interiors and furnishings due to dust and to leaks and various related items. We have confined the maintenance charge solely to the window and sky-light areas.

COST OF ARTIFICIAL ELECTRIC LIGHTING

The initial investment of electric lighting includes the cost of electric wiring and cabinets, wall and ceiling fixtures, table and floor lamps, shades and reflectors, and other similar equipment.

The annual cost of electric lighting includes the interest on the initial outlay, depreciation and repairs, lamp renewals, and the cost of electric energy.

The cost of artificial lighting includes even those which could be justly charged to natural lighting when the former is used to reinforce inadequate natural lighting in the *daytime*. This applies to some extent in all interiors, for example in basements and closets of residences, in hallways, in offices, in hotel rooms, etc., natural lighting in the daytime is often reinforced by artificial lighting.

I. SEVEN-ROOM FRAME HOUSE WITH HOT-WATER HEAT

SLIGHTLY ABOVE AVERAGE ARTIFICIAL-LIGHTING EQUIPMENT

	Natural lighting	Artificial lighting
Initial cost of installation	$601	$265
Total annual cost of lighting	93	75

Natural Lighting.—In estimating the cost of natural lighting in residences let us consider first a modern two-story frame building, about 27 feet x 30 feet plan, which includes a living-room, dining-room, kitchen, three bedrooms, a bath, hallways, basement, and front and back porches. In this house there are 14 windows 36 inches x 57 inches, six windows 30 inches x 42 inches, and 7 windows 33 inches x 24 inches, or a total of about 288 square feet of window area. The cost per window for these three sizes are

$27.75, $17.00, and $10.75 respectively, or about $1.95 per square foot of window. Deducting $0.37 per square foot for weatherboarding and plastering replaced by window, the net costs respectively for each size of window are $22.50, $13.75, and $8.70 or about $1.58 per square foot of window. Since there are 288 square feet of window surface, the initial net cost of windows is $455. According to our estimates and assumptions the interest at 6 per cent upon this initial investment of $455 is $27.30; the depreciation at 2.83 per cent is $12.85; the repairs and painting of these 27 windows per year, less the cost of painting the weatherboarding is about $6.25; and the cost of washing them four times a year is about $11.50. Therefore the total cost of maintaining them is $57.90 per annum.

According to our assumption of temperature differences and radiation our estimates indicate that 0.283 square feet of additional radiation is required for each square foot of window area to supply the net heat loss chargeable to windows. Assuming the cost of a hot-water system to be $1.80 per square foot of radiation the initial cost of heating system would be $0.51 per square foot of window or a total of $146 for the 288 square foot of windows in the house to supply the net heat loss due to windows.

The annual charge for heating these 288 square feet of window area would include interest upon the $146 at 6 per cent or $8.76, depreciation at 3.4 per cent or $4.97, repairs of about $0.82, and the cost of coal at $10.50 per ton when the furnace is operated at half capacity for seven months of the year or about $20.70; or a total annual cost of supplying the heat lost through windows of about $35.25.

On summarizing we obtain the costs of natural lighting presented at the beginning of this section.

Artificial Lighting.—In estimating the cost of lighting this house by electricity let us assume first a meagerly equipped home but somewhat above the average as shown by a survey which is now in progress. We assume for this home a ceiling-fixture, two portable lamps, and two convenience outlets for the living room; one ceiling-fixture, one portable and one convenience outlet for the dining room; one ceiling-fixture and one convenience outlet for the kitchen; one ceiling-fixture, one portable lamp, and one convenience outlet for each of the bed rooms; two wall-brackets in the bathroom, one ceiling-fixture and one convenience outlet in the hallway; one ceiling-fixture on the porch; three drop cords in the basement. According to our assumption these represent an initial cost at $5 per outlet of $100 for wiring; $100 for fixtures; and $65 for portables; or a total initial cost of $265.

The annual charge against electric lighting includes interest at 6 per cent on the investment equal to $15.90; depreciation on the wiring at 3 per cent, depreciation on the portables at 5 per cent, depreciation on the fixtures at 4 per cent or a total depreciation of $10.25 per annum; an annual repair of $7.90; lamp renewals of $4.35; and the cost of 460 kw-hr of energy at $0.08 per kw-hr or $36.80. Therefore the annual cost of electric lighting is $75.20 according to these assumptions. A summary of total costs is presented at the beginning of this section.

II. SEVEN-ROOM BRICK vs. FRAME HOUSE

SAME LIGHTING EQUIPMENT AS IN I

	Natural lighting	
	Brick	Frame
Initial cost of installation	$505	$601
Total annual cost of lighting	86	93

For a comparison between the costs of lighting a frame and brick residence let us consider a two-story brick house of the same size as the frame building, with the same number and area of windows, and an identical heating plant.

The initial cost per square foot of window is assumed to be $1.95 and the cost per square foot of brick wall replaced by window to be about $0.65, therefore the initial net cost per square foot of window is $1.30 and the initial total cost for 288 square feet of windows is $374.

The interest at 6 per cent upon this initial net cost is $22.45; depreciation at 2.83 per cent is $10.60; net repairs and painting is $9.60; and washing is, as before $11.50. Therefore the annual charge for the 27 windows is $54.15.

The heat loss per square foot of window, 72.1 B. t. u. per hour, is the same for the brick house as the frame house. However the net heat loss per square foot of window is computed to be 44.8 B. t. u. per hour as compared with 49.7 B. t. u. per hour in the case of the frame house. The initial cost of the hot-water heating required for supplying the net heat losses from the windows is found to be $131.40 and the annual cost of this additional heating including interest, depreciation, repairs, and fuel is found to be $31.75.

On summarizing we obtain the comparative costs of natural lighting presented at the beginning of this section, for the brick and the frame house.

III. SEVEN-ROOM BRICK HOUSE

ADEQUATE LIGHTING EQUIPMENT

	Natural lighting	Artificial lighting
Initial cost of installation	$658	$610
Total annual cost of lighting	111	120

In order to obtain a further comparison of the natural and artificial lighting of residences we have considered a more adequately lighted house. A total window area of 374 square feet is assumed and a more elaborate electric lighting equipment was provided.

Assuming the same conditions as in the previous cases we find the net initial cost of windows to be $486.20. The annual cost consists of the following items: interest, $29.17; depreciation at 2.83 per cent or $13.75; repairs at 2.5 per cent or $12.15; and washing four times per year or $14.96. The total annual cost of windows is found to be $70.03.

Assuming the same heat losses per square foot the initial cost of heating system to supply the net heat loss is $171.60. The annual cost of supplying the heat loss consists of the following items: interest, $10.30; depreciation $5.84; repairs $0.96; fuel $24.30. The total annual cost of heating system to supply the net heat loss is $41.40.

This seven-room house equipped for adequate lighting is assumed to have the following equipment. The living-room contains a ceiling-fixture, one table lamp, one floor-lamp, two candle-lamps and four convenience outlets; the dining-room contains a ceiling-fixture, two wall-brackets and one convenience outlet; the kitchen contains a ceiling-fixture, a drop-cord or wall-bracket, and two convenience outlets; each of the three bedrooms contains a ceiling-fixture, two wall-brackets, a table-lamp and two convenience outlets; the bath-room contains two wall-brackets and a convenience outlet; each of the two hallways contains a ceiling-fixture and a convenience outlet; each of the two porches contains a ceiling-fixture and a convenience outlet; and the basement contains four drop-cords and a convenience outlet.

The initial investment in this more adequately equipped seven-room house, therefore, includes the expense of fifteen ceiling-outlets, ten wall-bracket outlets and 19 convenience outlets, all at $5 per outlet or a total of $220 for electric wiring; it includes $245 for fixtures; and $145 for table and floor lamps; or a total of $610 for electric wiring, fixtures and portables.

The annual cost of the artificial lighting consists of an interest charge of $36.60; depreciation of $23.65; repairs of $6.85; lamp renewals of $5.06, and the cost of 598 kw-hr. of energy at $0.08 per kw-hr. or $47.84 for electric energy. This is a total for electric lighting of $120 per annum.

The summary of costs is presented at the beginning of this section.

IV. FIVE-ROOM BRICK APARTMENT
STEAM-HEATED

	Natural lighting	Artificial lighting
Initial cost of installation	$225	$230
Total annual cost of lighting	39	65

It is assumed that this steam-heated five-room apartment has only one-half the window area as the first house considered or 144 square feet and that the electric lighting is meager but representing the average condition.

Using the same assumptions and conditions as in previous cases the initial net cost of the windows is found to be $187.20 and the net annual cost of these windows to be $27.22.

The initial cost of the steam-heating system is assumed to be $1.40 per square foot of radiation. The annual cost per square foot of radiation is computed to be $0.146 including interest, depreciation, and repairs; and the cost of fuel is computed to be $0.28 per square foot of radiation per annum. The initial cost of the heating system to supply the net heat loss of the windows is found to be $37.80. The total annual cost to supply the heat losses due to windows is found to be $11.50.

The summary of costs chargeable to natural lighting are presented at the beginning of this section.

In this apartment there are assumed to be five ceiling-outlets, two wall-brackets, and five convenience outlets which at $7.50 per outlet represent an initial cost of $90.00 for wiring.

An allowance of $85 is made for fixtures and $55 for brackets making a total initial cost of $230.00.

The annual cost consists of the following items: interest, $13.80; depreciation, $9.13; repairs, $2.60; lamp renewals, $3.83; 443 kw-hrs. at $0.08 per kw-hr. or $35.44 for electric energy. This totals $64.80 per year for electric lighting.

V. MODERN HOTEL; ROOM AND BATH

	Natural lighting	Artificial lighting
Initial cost of installation	$161	$115
Total annual cost of lighting	14	14

We assume a hotel in the down-town section of Cleveland, for instance where real estate is worth $60 per square foot; we assume that the building is 14 stories high; that 13 per cent of the possible floor space is sacrificed for light-courts; that the building is one-story high above ground on light-courts and covered with sky-lights over one-fourth this area. We further assume that a system of heating and ventilation is installed for washing and heating the air.

Assuming the initial cost of a high-class pressed-steel frame window is $2.50 per square foot, and the brick wall replaced is $0.75 per square foot or a net cost per square foot of $1.75, with interest at 6 per cent, depreciation at 3 per cent, repairs 0.6 per cent, and washing four times per year at $0.01 per square foot the annual cost per square foot becomes $0.208. Assuming a room and bath to have but one 4 foot x 6 foot window the initial net cost of window would be $42.00, and the annual cost of interest, depreciation, repair and washing would be $5.

Assuming an initial cost for the heating and ventilating system of $5.50 per square foot of radiation, and an interest of 6 per cent, depreciation of 3.4 per cent, repairs of 0.6 per cent, and fuel at $7 per ton to operate the heating system at half capacity for seven months of the year, and electric energy for the ventilating fans at one-fourth the cost of fuel, then the annual cost of operating one square foot of radiation becomes $0.80. Since one square foot of window is computed to lose a net amount of heat of 44.8

B. t. u. per hour, one square foot of window requires 0.187 square feet of radiation. Therefore the initial cost of heating and ventilating systems required per square foot of window is $1.03; and the annual cost for heating one square foot of window is $0.15. Therefore the initial net cost of heating and ventilating system for the 24 square feet of window surface for a room and bath is $24.72; and the annual cost is $3.60.

Assuming real estate worth $60 per square foot the building 14-stories high and 13 per cent of possible floor space taken up in light-courts, the net cost of real estate per square foot of floor is $0.56. If the room and bath be assumed to contain 168 square feet of floor space, the initial cost of real estate chargeable against light-courts is $94.08, and the annual charge for interest at 6 per cent is $5.65.

The large portion of the cost of natural lighting which is due to the area occupied by light-courts makes this a matter of concern. Where land-values are still higher the cost of light-courts becomes even more important.

Summarizing we obtain the costs presented at the beginning of this section.

Assuming that we have one ceiling-outlet, two wall-bracket outlets and one convenience outlet in each room, and two wall-bracket outlets and a convenience outlet in each bath and allowing $10 per outlet we find the initial cost of wiring to be $70. If in addition to the fixtures there is a table-lamp we assume an initial cost of fixtures and portables to amount to $45 or a total initial cost of electric wiring, fixtures and portables of $115.

The total annual charge will consist of a charge of $6.90 for interest, of $2.27 for depreciation, of $1.18 for repairs, of $0.99 for lamp renewals, and a charge for 65 kw-hr. of electric energy at $0.04 or $2.60. These represent an annual cost of $13.94 for electric lighting for a room and bath.

VI. A SUITE OF OFFICE ROOMS

| | Natural lighting | | Artificial lighting |
	without light-court	with light-court	
Initial cost of installation	$667	$1,584	$976
Total annual cost of lighting	86	141	290

In such a case as the hallway assumed in this suite, artificial light must be used a considerable portion of the day. This is also true of artificial light in the offices. Strictly, the costs of artificial light to reinforce the natural lighting should be charged against natural lighting. This has not been done but if it had been done the costs of artificial lighting would have been reduced and the costs of natural lighting would have been correspondingly increased.

For a comparison of natural and artificial illumination we have assumed a suite of rooms in an office building consisting of six 14 feet x 16 feet offices with three offices on a side and a 7-foot hallway between. We assume further that each room is illuminated by two 4 feet x 5 feet windows and by four 200-watt semi-indirect ceiling-fixtures, and that the air for ventilation is washed and heated before being circulated.

Assuming a net initial cost per square foot of window to be $1.75 and its annual cost to be $0.208, we find the initial net cost for windows in the whole suite to be $420; and the annual cost to be $49.92.

The heating system necessary to supply the heat losses due to windows assume as in the hotel, to have an initial net cost per square foot of window of $1.03; and the annual cost to be $0.15; therefore, the initial cost of heating system to supply the net loss of heat through windows is $247.20 and the annual cost is $36.

Summarizing we find the costs as presented at the beginning of this section.

Assuming four semi-indirect ceiling-fixtures and three convenience outlets in each office and one direct ceiling-fixture in the hall, we have at $10 per outlet, a total initial cost of $430 for electric wiring. Assuming, $22.50 for each office fixture and $6 for the hall fixture we obtain a total initial cost of $546 for fixtures or a total initial cost of wiring and fixtures of $976.

The annual cost of electric lighting according to the assumptions and computations consists of $58.56 for interest, of $21.84 for depreciation, of $9.76 for repairs, of $36.90 for lamp renewals, and of the cost of 4,080 kw-hr at $0.04 or $163.20 for electric energy. These represent a total annual cost for a suite of six rooms and hall of $290.26 for artificial lighting.

If we assumed, as in the case of the hotel that there be an initial cost of $0.56 per square foot of floor area for light-courts, the initial cost of natural lighting would be increased by $917.28 and would become $1,584.48; and the annual cost of natural lighting would become $140.96.

The costs are summarized at the beginning of this section.

ART MUSEUM

VII. TOP-LIGHTED PAINTINGS' GALLERY, 33 FEET x 115 FEET

WITH LOUVERS FOR CONTROLLING DAYLIGHT AND ARTIFICIAL
LIGHTING UNITS ABOVE THE SUB-SKYLIGHT

	Natural lighting	Artificial lighting
Initial cost of installation	$8,017	$4,550
Total annual cost of lighting	942	720

We have assumed an art museum, built of marble, in which some of the rooms are top-lighted by means of skylights and others are side-lighted by means of windows. We have estimated the cost of natural lighting with and without louvers for controlling the light.

First let us assume a group of top-lighted paintings' galleries with southern exposure, having a roof-skylight and a sub-skylight and a system of louvers for controlling the daylight. The artificial lighting units are above the sub-skylight.

The group of rooms have a total floor space 33 feet x 115 feet; the outer skylight is 28 feet x 115 feet, and the sub-skylight is 23 feet x 115 feet. Assuming a net cost of $0.20 per square foot for the outer skylight and $0.80 for the sub-skylight, we have an initial cost of outer and inner skylights of $2,770. Assuming an annual cost per square foot of $0.044 for the outer skylight and $0.146 for the sub-skylight, the total cost for interest, depreciation, repair and washing of skylights becomes $472 per annum.

The initial cost of adequate louvers to control the daylight is estimated to be $4,500; and assuming an interest, depreciation, repair and maintenance charge of 7.5 per cent, the annual cost becomes $337.50 for the controlled system of natural lighting.

Estimating a net loss of 8.7 B. t. u. per hour per square foot of sub-skylight and estimating the initial and annual cost per 100 B. t. u. per hour of capacity to be $3.25 and $0.575 respectively we obtain an initial cost of additional heating and ventilating system of $747; and an annual cost of interest, depreciation, repair, and fuel cost of $132 for supplying the net heat losses.

Summarizing we obtain the costs of natural lighting as presented at the beginning of this section.

For artificial illumination let us assume that above the sub-skylight which is of "pebbled" wire glass there are 130 150-watt Mazda daylight lamps fitted with silvered glass projector units. Assuming an initial cost of $20 per outlet for wiring and $15 per lighting unit we obtain an initial investment cost of $2,600 for wiring and $1,950 for lighting units or a total initial cost of $4,550. Allowing an interest, depreciation and repair charge of 7.25 per cent, and an energy consumption of 7,120 kw-hr per annum at $0.04, we obtain an anual cost of $329.88 for interest, depreciation and repair, a cost of $105 for lamp renewals, and a cost of $284.70 for electric energy or a total annual cost of $719.58 for electric lighting.

VIII. TOP-LIGHTED PAINTINGS' GALLERY, 52 x 88 FEET
WITHOUT LOUVERS BUT WITH LIGHTING UNITS ABOVE SUB-SKYLIGHT

	Natural lighting	Artificial lighting
Initial cost of installation	$3,254	$4,088
Total annual cost of lighting	577	479

This gallery is assumed not to have louvers to control natural lighting and is lighted by means of a large number of units above the sub-skylight the light being reflected generally downward. In this case we have assumed a floor-area 52 feet x 88 feet which is lighted by a roof-skylight covering the entire room and by a sub-skylight 30 feet x 72 feet.

Using the same estimates as in the previous case we obtain an initial cost of skylights of $2,643 and an annual cost of skylights of $469. Using the same estimates for cost of heat loss we obtain an initial cost of $611 and an annual cost of $108 to supply the net heat loss. Summarizing we obtain the costs of natural lighting presented at the beginning of this section.

For artificial illumination we have assumed that the sub-skylight is etched wired glass and that the lighting units are placed between the skylight and the sub-skylight. There are 32 outlets equipped with 150-watt Mazda daylight lamps and mirrored glass reflectors; and 188 outlets equipped with 50-watt Mazda daylight lamps and metal reflectors. Assuming the initial cost of the electric wiring to be $12 per outlet, the total initial cost of wiring would be $2,640. Assuming the initial cost of fixtures to be $10 per mirrored glass reflector, and $6 per metal reflector we find the total initial cost of fixtures to be $1,448.

With interest at 6 per cent, depreciation on wiring and fixtures 1 per cent, and repairs 0.25 per cent we find the annual costs to be $245.40 for interest, $40.90 for depreciation, $10.22 for repairs, $46.20 for lamp renewals, and the cost of 3,410 kw-hr. of energy at $0.04 or $136.40 for electric energy; a total annual cost of $479.12 for artificial lighting.

IX. SIDE-LIGHTED GALLERIES, 33 FEET x 88 FEET

	Natural lighting	Artificial lighting
Initial cost of installation	$3,210	$1,800
Total annual cost of lighting	308	214

Now let us consider a group of rooms with a northern exposure which are side-lighted by six 6 feet x 13 feet double glazed windows and artificially lighted by twelve 300-watt Mazda daylight lamps in enclosed prismatic glass units.

The net initial cost of the bronzed frame double window in white marble walls is estimated to be $6 per square foot therefore the initial net cost of the 468 square feet of windows is $2,808.

Assuming interest at 6 per cent, depreciation at 0.8 per cent, repairs at 0.3 per cent, and washing four times per year at $0.02 per square foot of window, the annual cost is $0.506 per square foot of window; or for the group of rooms $236.81.

Estimating 26.4 B. t. u. net loss of heat per square foot of double window, the initial net cost of additional heating and ventilating system per square foot of window is $0.858; and the annual cost is $0.152 per square foot of window. Therefore the initial cost of heating and ventilating system to supply the net heat loss from the windows of the whole group of rooms is $401.60 and the annual cost of the additional heating and ventilating system including fuel is $71.20.

Sumarizing we obtain the costs presented at the beginning of this section.

Assuming the initial cost of wiring per outlet to be $75 and the cost per enclosed prismatic unit to be $75 we obtain for the 12 outlets an initial cost of $900 for wiring, and $900 for fixtures or a total investment of $1,800 for electric lighting.

Assuming 6 per cent interest, 1 per cent depreciation, 0.25 per cent for repairs, we obtain an annual cost of $130.50 for investment, depreciation and repairs. Also assuming 1,530 kw-hr of electric energy consumed per annum, the cost of lamp renewals is $21.85 and of electric energy at $0.04 is $61.20. The total annual cost of artificial lighting according to these estimates is $213.55.

CONCLUSION

It is the aim of this paper to present cost-data pertaining to natural and artificial (electric) lighting systems for a few representative interiors. In making these computations actual buildings have been used as a basis in each case although in some cases certain minor modifications were made which, in our judgment, made them more generally representative of their class. In the interest of space conservation a vast number of details of computations and estimates have been omitted in the preparation of this paper; however the chief assumptions and estimates are presented. It should be recognized that conditions upon which costs are based vary considerably but having chosen actual buildings it is thought that the costs presented in the foregoing paragraphs are at least approximate.

It is seen that the cost of natural lighting is of the same order of magnitude as that of artificial electric lighting in all cases.

It is not suggested that we eliminate windows from residences but is felt that cognizance of the fact that daylight costs are appreciable in residences should make the householder more lenient toward the costs of adequate and proper artificial lighting.

In the case of office buildings, hotels, museums, and other public buildings, especially in congested districts where land-values are excessive, it is seriously suggested as one of the authors has suggested on other occasions, that the cost of natural lighting be given greater consideration in general than in the past. Light-courts in many cases contribute a large burden of cost to natural lighting and in many of our offices and hotel rooms they do not provide adequate natural lighting. In other words during many hours of a year artificial lighting must reinforce natural lighting even in the daytime. But in the foregoing sections this has not been changed to natural lighting, but has been borne by artificial lighting.

In many cases windows also contribute much to the cost of natural lighting by utilizing valuable space.

The psychological effect of the absence of windows is well recognized but in our congested cities many persons must work under inadequate, or total absence of daylight. Many night-workers gain little or no advantage from windows excepting the psychological effect of "natural" ventilation. If congestion increases, economic considerations will surely compel the reduction of areas of light-courts and of windows and will perhaps eliminate them entirely in some cases.

Artificial ventilation and adequate artificial lighting can supply the actual utilitarian demands. It is easy to prove that many interiors existing to-day would be improved as work-places if the outlay for daylight could be applied to the rehabilitation of the artificial lighting.

In some cases, such as museums, there appear to be no defensible reasons for not eliminating the cost of daylight in some of the rooms while the building is being construced. Adequate and proper means of artificial lighting can be installed, operated, and maintained at costs comparable with and in some cases even less than adequate natural lighting systems. This plan applies particularly to spaces in the central portion of large buildings, and substitutes for more or less undependable natural lighting, readily controlled modern artificial lighting.

The questions involved in substituting artificial lighting for natural lighting are numerous and complex and a discussion of them would lead us far from the objective of this paper. Some of them have been briefly touched upon to show why we set ourselves to the task of estimating the cost of daylight. This phase may form the basis of another paper.

DISCUSSION

L. B. MARKS: The work of Messrs. Luckiesh and Holladay as evidenced in this paper, is worthy of most serious consideration.

About six or seven years ago, Mr. Macbeth and I appeared before the Heights of Buildings Commission in the City of New York, on the question of the cost of daylight, the very question that Mr. Luckiesh is taking up, and we both came to the same conclusion that Mr. Luckiesh has come to in his paper, although our work related to office buildings very largely. The Heights of Buildings Commission was taking up the question of securing better daylighting facilities in the new buildings that were going up in the congested districts of the City of New York, and was laying the foundation for a zoning law to restrict the height of buildings. The Commission wanted to find out where the line should be drawn, and in that connection the question of the cost of daylight and the cost of artificial light arose. Mr. Macbeth's figures I thought, were very convincing. He showed that in some cases it would not pay to restrict the height of the buildings, and that in these cases artificial light would be actually cheaper than daylight.

The only question left for consideration at that time was: "Can you supply an artificial lighting system that is as good as natural lighting?" Of course our answer was, "Yes, beyond question."

NORMAN MACBETH: I want to compliment Mr. Luckiesh on digging into this particular field and putting it on record; there has been a tremendous amount of wrong assumption on the apparent low cost of daylight and tremendously high cost of artificial light. If more of this kind of work can be done, it will help us in rendering the right kind of service, in terms of what that service is worth.

In that cost connection, when people bring up the point of how much more can be saved by this kind of lighting as against that other kind of lighting, they should be shown how few these dollars actually are, and instead of saving on light use more of it, more effectively and turn to some other channel to save. Mr. Luckiesh has referred to the cost of cleaning windows. There are very few offices indeed where the cost of towel service and ice water does not exceed the cost of adequate lighting in that office. There are few places, as we have brought out in school discussions for instance, where an extra piece of underwear on the children and a little less heat would not supply more than enough money to double the lighting intensities and make the lighting at least adequate.

This case of the Heights of Buildings Commission Mr. Marks referred to was a rather odd situation. I was chairman of the Executive Committee of the I. E. S. at that time and asked Mr. Marks to accompany me. We were asked to appear as experts for the plaintiff and turned out to be helpful to the defense. The question, as I recall it, put by the Chairman was: "We want you gentlemen to tell us how necessary daylight, natural daylight, is on the lower floors of office buildings and how harmful artificial light is on the eyes of the workers."

As I recall the incident, one comparison I used was: "Let us agree that artificial light costs money. We believe also that daylight costs money. Let us assume that you cannot have a downtown office building higher than four stories, with the present width of streets, and have adequate daylight throughout those floors. That means you are going to have to build office buildings in the marshes of New Jersey to take care of the extra stories you are going to pull off of these downtown buildings. It means that a building to-day of twelve stories, that has a rental rate of $3.00 per square foot per annum, has got to come down to four stories and have a rate of upwards to $12.00 per square foot—a 300 per cent increase."

It happened that I had gone to this meeting a little early and I found it was lighted by two semi-indirect fixtures with a 200-watt lamp in each. To bring about a comparison between daylight and artificial light—and this office, incidentally, was on about the twelfth floor, and right across from where the old Equitable Building used to be, the time was shortly after the fire—I said, "Gentlemen, we can come to a decision on this. Let us pull down the shades and put on the artificial light. You have been sitting here in daylight and you know how satisfactory and absolutely necessary it is; just look around now and size it up, appraise it. Then we will pull down the shades and put on the artificial light." We did, and they took a vote, and as I recall it the vote favored artificial light. The cost for that kind of lighting did not exceed 3 per cent of the present annual rental.

While it has been assumed that daylight is free, it has also been assumed that it is pure, and daylight near our street surfaces is far from it. I am referring now of the use of daylight for color. It has been the custom to use daylight in the naming of colors. I do not believe daylight should be used exclusively in the observation of colors. I had this matter up with a big publishing company a short time ago and I asked them this question: "Have you any idea of the reader-hours under artificial light, as compared with daylight?" and added, "I believe if you check it up, if you can secure the information, and are considering your reader and what you want your reader to see, you will conclude that all this important passing upon color ought to be done under artificial light and not exclusively under daylight."

To go back to an ordinary illustration: you are supposed to use distilled water in a storage battery, but if you take a porcelain vessel and put it out in the rain and catch that rain water, you will get pure water for your battery. Yet you dare not take it from the leaders, nor off the roof, because that water is contaminated, nor would you consider dipping it up out of the gutter. I have used that illustration, comparing the possibility of purity of light with water, by stating, if a man leaves his position behind the counter in a store and is on the way over to the water cooler for a drink, if when passing the front door on his way over he notes that it is raining, he will not say, "Well, I'll just take some of that water out of the gutter and drink it." He would not think of it. Yet why should he use, in accurate color work, the kind of daylight that he gets on any of the lower floors in any of our city buildings.

SAMUEL G. HIBBEN: In venturing a word about this paper, I realize I am sitting at the feet of learning, with deference to the Chairman, but I know of a problem in connection with the utilization of daylight which was interesting.

It is well known that one of the biggest costs of daylight arises from the radiation of heat through a glass window surface. This led to the development or the attempt to develop a window glass opaque to heat but translucent to light. In some factories that I investigated, the radiation of heat in the winter time through the windows was about ten times that lost through equivalent solid concrete walls.

A glass was developed of slightly amber color, and was tried first in the skylights of large shops, in an attempt to keep out the excess heat of the sun in the summer time and conversely to retain the heat in the building in the winter. That development has not been entirely satisfactory, and I mention it here because work is now being done by some of the glass manufacturers to use prisms, instead of color in the body of the glass, and to deflect the heat by those prisms, but not to so influence the light. That is something I hope to see reported before the Society in a year or two.

A survey of several large factory buildings gave us some interesting facts, not the least of which was that illumination on the lower floors of a factory building, due to daylight, was quite frequently greater than on the upper floors, and the tests seemed to indicate that the illumination coming into the building was largely from the diffusion of light from the neighboring grounds, the light being diffused into the building and toward the ceiling. A light colored factory ceiling under those conditions greatly aided the illumination toward the center of the room.

Ordinarily a person is surprised at the large variation in daylight when retreating from the windows. The curve of intensities would not be symmetrical with regard to the center of the symmetrical building, but offset to the south, giving us more daylight illumination along that side. In other words, we might have, for ten feet in from the south wall, a higher average value, than ten feet in from the north. And if a building was constructed in such a way as to have the long axis north and south then as the sun passed from morning to evening, we would have high daylight values along the east half up till noon. Toward the evening hours, the curve would reverse with the minimum toward the east, natural sunlight coming from the west. That led of course, to an attempt to have the factory owner or manager turn on his rows of artificial lights as this daylight shifted. In other words, we would have artificial lighting from two or three rows along the west ceilings in the morning. I mention this because the outcome of these investigations led to an attempt to utilize an automatic artificial light switch, which is another subject for consideration and further discussion. The Government for similar use has developed a "sun valve," used for isolated navigation lights, but which I believe may be made successful in buildings. It is a switch, which will as daylight fails, automatically turn on some rows of artificial lights, to equalize that changing conditions of daylight.

F. C. CALDWELL: Closely connected with what Mr. Hibben has said, there is not only the matter of the superior distribution of light that you can get by using artificial light in the case of deep rooms, but there is also the point that in some cases, if you do not have to provide daylight at the center of the room, the possibility of making the room deeper might mean an important reduction in the cost of the building.

H. W. DESAIX: I believe there is another factor which enters into the cost of daylight that has not been mentioned, and that is the cost of spoilage through sunlight. I am interested in the textile field, and principally in silk. Unless silk in a loom is kept moving, it will spoil through fading, and will also sag. They have to keep certain tension on the warp in order to keep the stretch for the shuttle to travel across. I have noticed invariably that architects in laying out a building for silk purposes will put on a saw-tooth roof and steel sash, and the minute they do that they add to the cost of special shades, and make it necessary to have covers for every loom. That is not only magnified over the week end, for instance Saturday afternoon and Sunday, but also during the day; if a machine has to be shut down for repairs, they will very often forget to cover the warp, which is the silk, and not work on the machine for several hours. When the weaver comes back he will find he has lost his tension, and when the goods are finished you will find a streak in it, which increases the number of seconds.

GEO. A. HOADLEY: I hope when you have succeeded in eliminating daylight from our city buildings that you will permit us to retain just a little daylight for those who have to live in the country. (Laughter). If the only purpose of a window was to introduce daylight into a home, then we could close up all our walls and have walls like that of a certain morgue-like building that is near my home, that belongs to a college fraternity, into which they do not want outside people to look. But I have a little satisfaction in looking through a window and treating it as a part of the decoration of the home, for the purpose of giving a connection between Mother Nature outside and Father inside. (Laughter).

When we come to reduce everything to a matter of dollars and cents, which is the idea brought forward by one of the speakers here, we seem to carry out the idea of that advertisement, "Buy a Ford, and spend the difference." Of course, it depends somewhat on what you want to buy—and I would like a little daylight in my house.

E. D. TILLSON: In one of the former conventions a speaker called attention to the fact that in the country districts the people had keener eyes or longer sight than residents of cities, and I have speculated as to whether there was not much value in windows in that respect. When we coop people up in an inside office, where they must face a wall at close range, and cannot look out of windows at distant objects, my opinion is that they will have a tendency to be short-sighted, and perhaps suffer impairment of the eyes.

Perhaps I have experienced a little of that myself. I work in a laboratory or gallery where there are no windows, but only skylights; and face a blank wall a good part of the time. I am quite short-sighted. That may or may not be in consequence of having to stare at the wall, but it does seem more comfortable to be in an office where one can look out of the window at something far away.

This may be considered as aside from the subject, but I note that the paper has been turned in part to a discussion of the relative cost of artificial, and natural light; and in such a balance, the cost of impaired eyesight, if such enters at all, should no doubt be considered.

NORMAN MACBETH: It is not proper to discuss the discussion, but I would like to say something on this matter that Mr. Tillson has brought up. The point has frequently been made that Eastern men on going West have been amazed at the apparent long-sightedness of the Western men on the plains. They see a little speck way off on the plains and they say, "Hello, here is Bill

Jones coming; I wonder what he wants." All the Easterner sees is the speck. The plainsman knows Bill Jones has a horse with a peculiar swing, and he gets the faintest impression of that motion way off in the distance. Similar stories have been told about sailors.

I can stand on 42nd Street, New York, and pick out a Broadway or Crosstown car from many blocks' distance, merely because the Broadway and 42nd Street Crosstown are both red cars; the designation of both cars is up on the top in white letters on a black ground, and "42nd Street Crosstown" covers the whole width of the sign and "Broadway" is considerably shorter. Westerners want to know how I can read it at such a distance.

That explanation has been proven to my satisfaction and I believe is the most reasonable one to account for the apparent long-sightedness of sailors and plainsmen particularly.

H. B. DATES: On the effect of environment upon eyesight, I am inclined to differ with Mr. Tillson, because the surveys that were made in the rural districts and in the city schools show there is fully as large a percentage, if not more, country children who have defective eyesight as city children. I do not think therefore you can lay it up to that fact.

E. D. TILLSON: I am just quoting what was brought out under test.

M. LUCKIESH: Mr. Hibben brought up a point about the transmission of radiant energy by glass. Unfortunately, glass can never be so constituted as to eliminate conduction losses. The content of a glass will do much toward eliminating or reducing the transmission of infra-red, but our experiments show that the content makes little difference in actual conduction, and most of the heat loss is due to conduction. Glass does not transmit the radiation of long wave-lengths emitted by radiators and other objects at ordinary temperatures.

One gentleman mentioned spoilage due to daylight. We eliminated consideration of actual cost of spoilage, of cleaning of materials, deterioration of window shades, etc., because these introduce the element of judgment. The case is plain enough, without heaping any more upon daylight.

I was very sorry to note that the discussion of two of the gentlemen indicated that they seemed to have the idea that we were trying to abolish daylight. I thought we made it very plain in this paper that we are merely considering the economies of daylight just as we consider artificial light. I assure Dr. Hoadley we are not aiming to interfere with daylight outdoors at all. (Laughter). He can always have all he wants as far as we are concerned. I certainly like the outdoors and daylight, and there is no finer picture than a sunset or clouds and blue sky viewed through a window. I am always going to have windows in my house. (Laughter). When it comes to ventilation, I do not need the fifty-six windows that I have in my house.

We should consider conditions as they are and it is foolish to put in skylights when artificial light is better and less costly. Many of the windows I have seen in offices look out upon a 10-foot x 10-foot court, which sometimes is a most filthy, foul outlook. I would much rather hang paintings where those windows are and have artificial light.

I want to reiterate that in no sense do we want to eliminate daylight. We think a recognition of the cost of daylight will be a very influential psychological factor in overcoming indifference and prejudice toward artificial lighting. We do say there are many interiors to-day where the use of artificial lighting and the exclusion of daylight would greatly improve the situation and would be economically advisable. Furthermore there is plenty of evidence that people enjoy themselves under artificial light as well as under daylight. Many seem to be willing to take the chances on staying up all night and missing the daylight. (Laughter).

LIGHTING FOR MOTION PICTURE STUDIOS*

BY F. S. MILLS**

SYNOPSIS: The author sets forth the use of various illuminants in the field of the moving picture industry; the high intensity arc lamp for special effects, the Cooper-Hewitt mercury vapor lamp for general actinic lighting, and the high wattage Mazda C lamp for diffused illumination.

Mention is made of the wiring and distribution system on stages which entails great flexibility in order to carry the variable load. On "location" the power is produced by means of very large motor generator sets mounted on portable trucks.

Numerous illustrations of interior and exterior views of the studio in action and night lighting on "location" are shown, as well as various types of lamps used.

The modern magic of the motion picture studio presents many angles of romance, but probably the most facinating romance of all can be found in the electric lights. When the cameraman cries; "Light 'em up," and shafts of light come from the "spots," floods of light from the huge sun arcs, and the pale blue rays are sent out by the long Cooper-Hewitt tubes, it is no wonder that the casual visitor to the studio is amazed.

In the early days of the motion picture, very little artificial lighting was used. With mere sunlight no special effects such as are achieved in motion pictures to-day could be obtained; also it must be understood that on a clear day light values do not remain the same, thus it is necessary to use artificial light to standardize the light.

Old Sol has been cheated of part of his glory in the modern lighting systems now in use in the large motion picture producing plants. In the early days of the film industry, artficial lights

---

* A paper presented at the Annual Convention of the Illuminating Engineering Society, Swampscott, Mass., Sept. 25-28, 1922.

The illustrations for this paper are used through the courtesy of Paramount Pictures.

** Resident Engineer, National X-Ray Reflector Co., Hollywood, Calif.; formerly Engineer for the West Coast Studios of the Famous Players—Lasky Corporation.

The Illuminating Engineering Society is not responsible for the statements or opinions advanced by contributors.

were unheard of and the sun was the source of all photographic illumination. When cloudy weather came along, however, there had to be a suspension of activities. It was also impossible to get anything but flat, plain photography by the use of sunlight, hence special effects were out of the question.

As the art became daily more complex, photographic novelties in the form of lighting effects were introduced. It also became possible, with the aid of lights, to defy the sun—to "shoot" in rainy or cloudy weather as well as on the brightest days; and at the present time, so practical and so thorough have become the systems of electrical illumination now in vogue, that hardly an interior scene is made without the aid of these powerful illuminating agencies.

At first the motion picture industry used natural daylight in the taking of pictures with the motion picture camera. This was followed by the use of a reflector of silver paper pasted on sections of wood about 3 feet x 4 feet, thus catching the sun and reflecting it back into the faces of the actors and actresses. The next advance in lighting was the use of a carbon arc. At present there are two forms of lighting used, one is the carbon arc and the other the mercury vapor unit. From the carbon arc there is developed what is termed the "high intensity arc," the latest development is the adoption of a 36-inch navy searchlight. In this, diffusing glass is used to get a deflection of 20° and 40° divergence. This form of lighting from the searchlight is the last step that has been taken in the motion picture industry.

We have been greatly handicapped in the use of the type "C" lamp owing to its lack of actinic ray. However voltage manipulation, such as a 10 per cent rise in voltage, greatly increases its actinic value. Type "C" lamps have been used successfully for diffused illumination on gold picture frames and from practically one-half voltage up to and over 10 per cent normal voltage was registered by the camera. This was done by building a cove lighting system around three life size picture frames and showing the gradual increase of intensity on models within the frame, from practically no intensity to full intensity, and then fading out through the control of the voltage by rheostats on motor generator sets. This method was used in a picture of Miss Bebe Daniels directed by Mr. Franklin. Then under another condition, the 1,000-watt flood-lighting projectors have been successfully used on some scenes in the picture "The Fighting Chance" directed by Chas. Maine.

Cooper-Hewitt tubes are used to secure average daylight lighting in studios which are completely darkened, but out in the West, nearly all the stages are open with glass roofs, using white muslin diffusing cloth to cut down the bright intensity. They are now using in addition, a black diffuser cloth to maintain certain average intensities throughout the scenes. This "darkening-down" with black diffusing cloth is done owing to the fact that a sequence may be started in the forenoon and run past daylight value about four o'clock in the afternoon during summer months, or light values may change through cloudy condition of the sky which may change the character of the scene.

A development which has recently been used, is an overhead dome using carbon arc in a metal lined trough in a cove. This has been used in semi close-ups and close-ups for getting perfect diffusion overhead and down the face for elimination of certain shadows. These domes and other direct lighting domes are also used where wall sets are built and no provision is made for the placing of carbon arcs along the side of the room.

In the use of table lamps, stand lamps, etc., these are always equipped with a carbon arc. See Figure 1. Formerly they have been using the plain hand feed arc, but an automatic arc was developed to make and break by opening and closing of switch of the "set." Considerable trouble was experienced with this carbon as it often smoked during the taking of a scene. To overcome this objection, a special carbon was used. In the filming of any scenes where an actor lights a cigarette or cigar, it is always done by a little baby arc to get effect of the light on his face. The wires run down his sleeve to the floor, and is always "struck" by the actor himself, using what is called the pencil carbon with a consumption of 1,000 watts or 10 amperes. In the filming of fire places, in addition to the gas logs with the gas burning, carbon arcs are placed in the back so as to get the fire effect into the face or into the room. Ordinarily the flame would not "pick up" or photograph. In addition to this development, there have also been designed what are called "suspended floor lighting

units" for close-ups and semi close-ups to overcome certain "neck-
lace" lines on the faces of people taking part in the picture.

The intensity of illumination used on sets varies considerably
with the action, the character of the set itself, wall papers used
and furniture. There are seldom two sets of any size lighted the
same, special study being given to where the natural source
of light would be, and then the units placed to follow the natural
shadows which would be secured from a room lighted in various
manners, such as with brackets, overhead fixtures or other
methods of lighting. In camera work, particular care is taken
to prevent light or shadow across the nose. In lighting of actors
and actresses, consideration is always given to tests of the "make
up" which on some complexions acts different than on others.
There are many grades of powder used. For example, some
people of a rugged complexion have blood vessels close to the
surface of the skin and these must be eliminated by heavier pow-
ders. Red is never used in the make up except on the lips.
Eye brows and eye lashes are darkened with Mascaro. It has
always been attempted to bead the eye lash with little balls of
Mascaro at the end which directs shadows over and under the eye.

Spots of 70 amperes over sets are always used for back lighting
of the characters such as shown in Figure 2. This is done to
make the character stand out from the wall, otherwise what is
termed "flat light" would be secured on the screen and in this
flat lighting there is no division made between the scene and the
people themselves in action. This is a very difficult way of lighting owing to the fact that these spots which are used in large
numbers in sets, are hand fed and quite often are used to spot
the actors and actresses in a set during action. When spots are
used in sets of this character, one must be very careful that the
camera is not picking up the foreground or floor of the room. If
so, a frost paper is used in combination with the spot for breaking
up the circles caused by the spot on the floor. This is also used
to soften one side of the lighting of the room by the use of the
spot in combination with frosted paper in frost frames. Considerable study is also given to the use of these spots on top of
sets for "raking" of walls and also are placed in such a position
as to concentrate the full value of their light on brackets, silk
shades, to have them stand out from the wall and make them look
as though they were lighted.

Fig. 1.—Carbon arc portable table lamp

Fig. 2.—Seventy-ampere spots used for back lighting of characters.

Fig. 3. Carbon arcs in reflecting hoods.

Fig. 4. Close up of carbon arcs in reflecting hoods.

Fig. 5.—Large transformers used for stage lighting.

Fig. 6.—View of studio interior showing wiring.

· Owing to the fact that the round frosted lamps used in brackets on walls will not photograph and "spots" are used to light the silk shade and to direct a ring of light directly over them on a statue, vase or table, as though the lighting were secured from the bracket directly over such article. This form of illumination makes the composition more complete and stops any possibility of flat lighting.

In all sets usually employed, the lighting is secured from a bank of carbon arcs used in reflecting hoods mounted on a stand, as shown in Figure 3. These arcs are equipped with a diffusing glass ribbed, having a very fine stippled finish, as it is very dangerous to use an open arc, because of its effect upon the eyes of the players. This glass is used on lighting units to secure the hard side of the illumination. A method has been devised of building a box frame around both of the lighting hoods and using a tracing paper. See Figure 4. This arrangement softens down the quantity of intensity and gives more character to two sides of a set. This lighting is very technical and is generally followed through with the lighting specialists on a set in combination with the camera man and director to secure the correct direction of action and placing of shadows.

In the use of both the 70 ampere spot and the broad side unit as shown in Figure 3, there were designed a series of louvres, like shutters, made of aluminum. These louvres or shutters are placed on approximately two-inch centers down the length of the box frame as shown in Figure 4, inside of the curtain and so controlled on tracks that by the operation of one lever all of these fins will open or close simultaneously. In combination with the shutter, a diffusing cloth is used to stop any shadows caused by the irregular opening or closing of these shutters. A similar type of unit has been designed for the use of the 36-inch navy searchlights from which are secured lighting effects.

Great use is found for this diffusing cloth and shutter unit where the lighting is from alternating current. Most motion picture cameras are equipped with an automatic "fadeout," but owing to the extreme danger of synchronizing the cranking with the cycles of the alternating current, the automatic "fadeout" is seldom used on the camera when alternating current service is employed. Therefore, by the use of these special shutters, a complete set can be faded in and out of the picture by opening and closing these shutters which accomplishes the same thing as the automatic "fadeout" on the camera, and without the danger of synchronization with the camera.

Trouble has been frequently experienced due to the fact that on long scenes the camera will start normal cranking, which is 14-16, and increase or decrease the speed from this normal value, and by so doing cause flickers to appear in the lighting of the picture itself. The camera has synchronized with the alternating current, which is a problem that has been faced by the motion picture producers in the use of alternating current.

One of the largest motion picture producers to-day is using very successfully, alternating current, 50 cycles, having no investment in power apparatus other than one three-unit motor generator set, and three units for supplying direct current to the high intensity arcs. In this case, the power company delivers service at 2,300 volts and installs the necessary capacity in transformers, Figure 5, 250 kw. for stage lighting. A very heavy load is built up quickly in this form of illumination. Every unit used averages 70 amperes and when a carbon arc first strikes, the current is excessive but quickly settles back to its rated value.

Great study has been given to the wiring and distribution system on stages owing to the heavy load. Where a number of stages are used in one plant for the making of motion pictures, the total capacity may be on any one stage at any one time. This entails great flexibility and a system where provision must be made for wiring, distribution and carrying of a load at one point of the total capacity of the plant. See Figure 6. There have been sets in one picture alone where total connected load has been over 14,000 amperes of carbon arcs alone. There is no consideration given the overloading of any apparatus up to the danger point, and it is not unusual at times to see the insulating compound running out of cables and oil bubbling out of transformers. Many directors are not posted technically as to the dangers of a continuous load and are at times liable to burn the lights, where unless carefully watched, may cause serious break-downs and tie up a complete studio.

The most difficult problems faced in the matter of lighting is on "locations." See Figure 7. Very often it is necessary to build

Fig. 7.—Searchlight used in "Night shooting" for George Melford's Paramount picture, "Burning Sands," on location at Oxnard, Calif.

Fig. 8.—Portable truck carrying motor generator sets.

high tension transmission lines, taking service from a power company at 15,000 volts, put up a complete line for a number of miles, setting a bank of transformers and then stepping down to the voltage of the lighting equipment. It has been found where these large sets are built on "location" and impractical to build transmission lines, the studios have been forced to construct very large motor generator sets mounted on portable trucks as shown in Figure 8. The expense alone of the truck was in excess of $30,000. In this case a marine type of gas engine is used, carrying two submarine type generators. All of this apparatus is mounted on a 7½ ton White truck especially geared for mountains, as locations are picked by directors without any consideration for transportation or how power can be secured. When necessary to build sets and locate them in remote regions, these plants must be sent.

In a recent production it was not feasible to take the snow scenes in or near Southern California due to the fact that the picture was written around the Canadian Northwest where snow and blizzards were necessary. All these scenes and sets were constructed at Truckee, Calif., where the complete company stayed for six weeks. In this case the settings were placed three and one-half miles from the nearest power line, and it was necessary to build a 2,300 volt line, and set transformers. It was found necessary to shut down some of the buildings in Truckee, Calif., in order to get the necessary power load to take the pictures. This transmission line was built along a roadway without a pole line. The insulators and pins were therefore installed on trees heavily covered with snow. During this picture and their night work, much time was lost owing to heavy snow fall where it was necessary to stop work and dig the snow from around the sets.

One of the most notable lighting achievements accomplished during the past year, was the night lighting of the big Siamese exterior setting which was built at Balboa, Calif., for some of the spectacular scenes in Cecil B. DeMille's production, "Fools Paradise." Without a doubt this was one of the biggest motion picture illuminating propositions ever undertaken. The lighting equipment required to properly illuminate that set, if out in a continous line, would extend for a quarter of a mile.

The cost of lighting scenes varies according to the size of the set and the consequent number of lights used. For unusually large settings, requiring a great deal of light, the cost might range from twelve to fifteen dollars an hour for lights.

Of course, the motion picture industry is still comparatively new, and we are still in the experimental stage so far as lighting is concerned. Not only in the studios, but in the movie theatres, there is considerable room for improvement. A spectator entering the largest of our cinema palaces is generally plunged from bright daylight into murky darkness, he passes through the reverse experience when he leaves the theatre, and the eyes suffer a consequence strain, as the pupil does not have time to adjust itself to the sudden change.

DISCUSSION

L. C. PORTER: In the past most of the motion picture work has been done on black and white film, but the time is unquestionably near when there will be a great deal of color work, and in color work the incandescent lamp is proving very effective. It is only recently that we have known how to handle panchromatic film and by the use of panchromatic film and incandescent lamps we can get some very beautiful effects. I think we will see the incandescent lamp come into considerably wider use in studio work. As a matter of fact, this large lamp you saw out here the other night was developed for motion picture studio work.

One of the problems in connection with the use of such units is to get rid of the heat. There is a tremendous amount of energy concentrated on the actors and actresses, and they are attempting to overcome that by using large mirrors and by using, you might say, indirect lighting from the lamps.

FLICKER PHOTOMETRY*
I. THE THEORY OF FLICKER PHOTOMETRY

BY C. E. FERREE AND GERTRUDE RAND**

SYNOPSIS: The subject of the photometry of colored light is not yet fully understood. Each year new investigators bring forth additional data leading to a clearer conception of the intricacies of the problem.

There are two principle means of comparing light intensities, one known as the "equality of brightness" method, and the other as the "flicker" method.

In the equality of brightness method, two objects in juxtaposition are illuminated in such manner that they appear equally bright to the eye. In the flicker method, a device is used which, in rotating, is exposed, first to the light from one light source and then to that from the other. When the intensities of light from the two sources are unequal, a noticeable flicker occurs. When this flicker disappears, a theoretical balance is obtained.

Discrepancies have always been known to exist between the results obtained from these two methods. In the first half of this paper, the investigators point out the causes for these discrepancies. They show that difficulties are encountered with the flicker method, because the intensity of the sensation (in the eye) does not vary directly with the intensity of the light, but is dependent upon the time this light has to act on the eye and also on the color. They have found that the flicker and equality of brightness methods agree if the eye is exposed the same length of time with both methods.

They further show that with the time of exposure ordinarily employed in flicker photometry, the eye does not reach the condition of maximum sensation, or in other words, get the full effect of the light, which, of course, obtains when the equality of brightness method is used, for there a prolonged exposure or sighting of the photometric field takes place.

In the second half of the paper, which is entitled "Comparative Studies of Equality of Brightness and Flicker Photometry, with Special Reference to the Lag of Visual Sensation" the authors first describe the ingenious apparatus with which they made these tests.

Figures and curves from the data seem to justify the conclusions mentioned above. They indicate that for various wave-lengths or colors, there is a rise of sensation to a maximum value which is different for each color and for each intensity. In other words, it will take a longer or shorter time for the eye to receive its maximum sensation from a given color at a given intensity, in comparison with the time for the eye to reach its maximum sensation from the same color at a different intensity or from a different color at either of these intensities. Furthermore the curves representing the rise of sensation for the different colors and white light intersect and stand generally in irregular relationships to each other throughout their course, which irregularity of relationship varies with the intensity of lights.

*A paper presented at the annual convention of the Illuminating Engineering Society, Swampscott, Mass., September 25-28, 1922.

**Department of Psychology, Bryn Mawr College, Bryn Mawr, Pa.

The Illuminating Engineering Society is not responsible for the statements or opinions advanced by contributors.

This naturally explains why the effect of varying the speed of rotation or length of · exposure with the flicker method shows varying results and also why the effect of varying the intensity of the lights compared produces a variation both of the amount and the type of disagreement between the two methods of photometry. Because of this variation of type of disagreement with intensity it is possible to find a single intensity for each pair of lights at which agreement occurs for the two methods with the speed of rotation ordinarily employed with the flicker method. Above and below this intensity comes underestimation or overestimation by the method of flicker depending upon the color selected for comparison with the white light. No single intensity could be found at which agreement occurred for more than one pair of lights. These most favorable intensities were quite widely separated for the four pairs of lights used and the percentages of disagreement at each of these intensities for the remaining pairs were quite large.

The possibility of the use of the eye in the measurement and comparison of light intensities has been a subject of great interest for more than a hundred years. The lack of proportionality in its response to intensity and the variation in this selectiveness of response with wave-length have thus far limited its use to a balancing or equalizing of light intensities. Without an undue amount and detail of calibration it could not be used to measure light intensities directly.

The eye as a balancing instrument.—Lights may be balanced either with regard to their energy content or to their power to arouse sensation. While only a limited use can be made of the eye for the former purpose, an important field is found for the latter,—the rating of lights by the eye for its own use, or photometry. However, the employment of the eye for rating lights for its own use is by no means free from difficulty. Here again a selectiveness of response to the different wave-lengths is, depending upon the method used, a serious source of trouble. Important points to be considered in this regard are variations in the intensity of the response with time of exposure and their differences for the different wave-lengths of light. The sensation rises to its maximum through an interval of time and then dies away because of a progressive loss of sensitivity, or adaptation of the eye. Moreover, the rate of rise and fall varies both with the wave-length and the intensity of light. That is, there is a lag and a decay in the response of the balancing instrument both of which are variable functions of the composition and intensity of the lights to be balanced. In short the eye as a balancing instrument may be likened roughly to weighing scales *which never quite attain to stability or constancy of balance* when the objects or commodities to be balanced are not of the same kind.

The result obtained depends upon how long the objects to be balanced remain on the scales, the stability increasing, however, with the increase of time beyond a certain value, the time required for the instrument to give its maximum response. Moreover to make the situation still more complicated, the effect on the value of the balance varies with the amount of the commodity present.

From the above considerations it is quite obvious that the length of exposure of the eye to the lights to be balanced is an important factor in the result that will be obtained when these lights differ in composition. The effect of time or length of exposure of the eye as a factor should not be ignored, therefore, in the comparative study and evaluation of the fitness of photometric methods for the purpose for which they are to be used, namely, the rating of lights for the work which the eye is called upon to do. Length of exposure is, we believe, an important, if not the chief point of difference between the two methods of arriving at a balance which are now the most widely used: the methods of flicker and equality of brightness. In case of the former a slow succession of exposures is given of a value much less than the time required for the eye to give its maximum response; and in the latter, a single exposure many times greater than the time required for the eye to give its maximum response. When the difference in the way the eye is used in the two methods is considered, it is not strange that a poor agreement in result should obtain for lights of different composition. Indeed it is somewhat surprising that agreement should ever have been expected. The reason for the disagreement will become clearer perhaps on a more detailed consideration of the effect of length of exposure on the eye's response.

For exposures less than the time required for the eye to give its maximum response, such for example as are used in the method of flicker, the following facts may be noted. (a) The rate of rise differs greatly for the different wave-lengths of light. Also the time required for the sensation to reach its maximum value is short, of the order of tenths or hundreds of a second; and the rate of deviation from equality of response for equal exposures as the sensations rise towards their maxima is rapid. Thus when it is sought to establish a photometric balance between lights differing in composition by means of exposures of the small values used in the method of flicker, not only will the ratings of the exposed lights sustain a direct relation to the values which the sensations have reached in the individual exposures and a balance be established which is not a true one for the longer exposures which the eye receives in the greater part of the work which it is called upon to do; but a considerable difference in result should be expected for small differences in the length of exposure. That is, not only may the balance established by the method of flicker be considered an inapt one for the purpose for which the results are to be used, but it is subject to no small amount of change by small changes in the speed of rotation of the flicker disc or other apparatus used to present the two lights successively to the eye. The first of these points is of course of the greater importance; the second is of consequence, however, because of the difficulty of standardizing the speed of rotation of the disc. The sensitivity of the method requires that the lowest speed be used that gives no flicker. In case the lights differ in color value, this lower limit of speed varies with the different wave-lengths for the same observer and for the same wave-length for different observers, the net result of which is to affect both the amount of the disagreement with the equality of brightness method and the precision of the determination. Obviously too a difference in result should be expected from any inequality of length of exposure to the two lights as happens in case of the method of flicker when the rotating sectors exposing the two lights are given unequal values. For a demonstration of this point see "A Preliminary Study of the Deficiencies of the Method of Flicker for the Photometry of Lights of Different Color, *Psychol. Rev.*, 1915, XXII, pp. 110-162. (b) The rapidity of the rate of rise is strongly affected by the intensity of light. The effect of intensity too is different for the different wave-lengths. Even the order of ranking of the different wave-lengths as to rate of rise is changed with change of intensity. That is, for a given observer the rate of rise at a low intensity may be faster for blue and green than for red and yellow; and at a higher intensity this order may be reversed. Intensity of light may thus be expected to be an important factor in any balance between lights of different composition with exposures shorter than the time required for the eye to give its maximum response. (c) The state of adaptation of the eye is doubtless a factor in the rate of rise, perhaps also field size, although as yet the extent of these influences has not been systematically investigated. (d) Individuals differ apparently not only as to the rate of rise, but as to the relative rates of rise for the different wave-lengths.[1] For example, it may be expected that for some observers at a given intensity of light the sensations aroused by the short wave-lengths rise more rapidly than those aroused by the long wave-lengths, and for others the reverse of this is true; also that there are individual differences in the effect of intensity of light both on the rate of rise and on the relative rates of rise for the different wave-lengths.

For exposure longer than the time required for the sensation to reach its maximum value, such for example as are used in the equality of brightness method, the effect of length of exposure is of comparatively little consequence in the use of the eye as a balancing instrument. (a) The rate of decay of sensation is slow. The drop is comparatively rapid just after the maximum is reached, but the curve soon straightens out, becoming less and less steep as the exposure is continued. For exposures of the order of value used in the equality of brightness method, the course of sensation is quite stable as compared with the rapidity of change that comes before the maximum is reached. Moreover, the order of magnitude of exposure is that which the eye ordinarily receives in viewing its objects: thus the balance obtained is better suited to the purpose for which it is intended. Also the rate of deviation in the course of sensation for the different wave-lengths is slow, therefore small differences in length of exposure have comparatively little effect on the value and precision of the balance obtained. (b) Intensity of light has an effect on the rate of decay of sensation, but there is, so far as we can discover, no difference of any very great practical

[1] This comment is not based upon a direct or systematic investigation of individual and group differences in the rise of sensation, but seems to be a reasonable inference from the collective results of several investigators of this and related functions. It is strongly indicated also by the individual and group results in flicker photometry, interpreted in the light of the results we have obtained with regard to the relation of the disagreement of the two photometric ratings and the changes and reversals of this disagreement to the rise of sensation curves.

significance in the relative effect in case of different wave-lengths. The reversal, for example, in the relative ranking with reference to rate of change for the short and long wave-lengths, which occurs as an effect of intensity in the rate of rise, never takes place so far as we know in the rate of decay. (c) There are some individual differences in the relative rates of decay for the different wave-lengths but again these differences may be expected to be small as compared with the differences in the rate of rise.

If the choice were between single exposures of the length ordinarily used in the method of flicker and those of the length used in the equality of brightness method, there is, aside from the question of ease of making the determination, little doubt that it would be uniformly in favor of the longer exposures, for there would be just as little reason for rating lights for working purposes with an under-exposed eye as there would be for using a false balance as a weighing instrument. There seems to be a belief among some, however, that in the succession of exposures the impressions summate to such an extent as to overcome the effect of the under-exposure. To obliterate the effect of the under-exposure either the sensations must rise to their full value or the sensation which lags the most must also summate the faster by an amount which just compensates for its slower rate of rise in the single impression. The former position is out of the question; the latter, as we have shown in a former paper, can not be assumed without violating well known laws relating to the factors which influence the persistence of vision. It could not be held, for example, unless it were assumed that the sensation which shows the greater lag and is therefore of lesser intensity, should also show the greater tendency to persist or to carry over to the succeeding impression, which is contrary to fact.

It still may be considered a point for discussion, however, with regard to what degree it should be held that the difference in lag between the sensations aroused by the single exposures used in the method of flicker is obliterated in the succession of exposures. Broadly considered three positions are possible with regard to this point for the rates of succession employed in the method of flicker. (1) The difference is not obliterated at all. In this case the flicker balance should deviate from the equality of bright-ness balance in direct proportion to the difference in lag for the single exposures. (2) The difference is in part obliterated, but is still present to a degree which renders the method questionable in principle for the rating of lights for the use of the eye when there is a considerable difference in color or composition. And (3) the difference is entirely obliterated or so nearly so as to be of no practical consequence for the validity of the method. The first is the position indicated by the results to be given in this and the following paper. If there is a summating action due to the succession of exposures, there is no detectable *difference* in this action for the two lights compared. That is, the results given by the method of flicker have been found by us to agree within the limits of the sensitivity of the judgment with the results given by the equality of brightness method when the time of exposure by the latter method is made equal in value to the individual exposures in the method of flicker. In other words the equality of brightness method within the limits men-tioned gives the same results as the method of flicker when the eye is given the same length of exposure by both methods. This would not be the case (a) were the difference in the length of exposure in the two methods not the cause of their disagreement in result; and (b) were the effect of the under-exposure on the results obtained by the method of flicker obliterated by any de-tectable amount in the succession of exposures.

The behavior of the eye in flicker photometry.—The utilization of the absence of flicker to detect the equality of brightness in two differently colored impressions is based on the following prin-ciples. (1) Brightness flicker is due to a succession at certain rates of speed of impressions differing in luminosity or bright-ness. No flicker occurs then when the impressions are of the same brightness. (2) The phenomenon is very sensitive to changes in the luminosity of the successive impressions. (3) Color fusion takes place at a much lower rate of succession than brightness fusion. This leaves the brightness flicker outstanding in a field uniform as to color quality.

All of these principles are sound as applied to detecting small differences in the luminosity or brightness of the successive sensations. For this purpose the phenomenon of flicker is excellent; but trouble arises when we try to say that lights which. arouse sensations of equal brightness for one length of exposure to the eye will arouse sensations of equal brightness for any length of exposure. For, change the length of exposure either for the equality of brightness or the flicker method and a different rating is given to the luminosity or sensation arousing power of the two lights, inconsiderable it is true for the equality of brightness method unless an exposure is used shorter than is required for the sensation to reach its maximum. That is, change the rate of speed of the flicker disc and a different rating is made of the two lights. Make an equal and corresponding change for the equality of brightness method, starting from the same length of exposure as was given by the flicker disc, and a corresponding and approximately equal change in the result is obtained. For example select a certain rate of speed for the flicker disc, the rate at which the method is most sensitive. Determine the value of the individual exposures at this rate of speed. Set a rotary tachisto-scope to give an exposure equal to this value and judge the two lights by the equality of brightness method. An agreement of result is obtained within the limits of the sensitivity of the judg-ment. Increase or decrease the rate of speed of the flicker disc and again determine the value of the individual exposures. Set the rotary tachistoscope to give this value of exposure and again make the rating by the equality of brightness method. Again agreement in result is obtained but the rating by both methods is different from the previous rating. The equality of brightness and flicker methods agree in result if the eye is given the same lengths of exposure to the lights compared in the two methods; they do not agree if the eye is given one length of exposure by the flicker method and another by the equality of brightness method. Nor will the equality of brightness method agree with itself in any two determinations, if in one the eye is under-exposed to the lights compared and in the other it is fully exposed as it is in the accustomed use of the equality of bright-ness method. Moreover, there is no discoverable reason why agreement should be expected in either of the above cases, once the facts about the difference in the variation of the in-tensity of sensation with the time of exposure of the eye are known for lights of different composition.

The method of flicker would present no difficulty as a means of rating lights with regard to their power to arouse the brightness sensation if the intensity of the brightness sensation were a regular function of the intensity of light. However, the intensity of sensation is an irregular function of three interacting variables: wave-length and intensity of light and time of action on the eye. It is safe to predict that when the relation of the intensity of sensation to intensity of stimulus is finally and correctly written for vision, intensity of sensation will not be expressed as a regular function of one variable (intensity of stimulus) as it is in the Fechner formulation; but as an *irregular function of three variables: wave-length, intensity and time of exposure.*[2] No one would think of attempting to express the response of the selenium cell, the photoelectric cell or the photographic plate to the different wave-lengths of light as a function of one variable and yet the eye is admittedly more complex in its responses than any of these reactors.

Factors which affect the results in flicker photometry.—Starting from the above premise as to the influence of length of exposure on the eye as a balancing instrument, it might be instructive to inquire into the number of factors or changes in the working conditions that might be expected to affect the results of the method of flicker. These factors so far as we have been able to discover, are: (1) Wave-length or composition of light. Lights of different wave-length or composition rise to their maximum value in sensation at different rates, *i.e.*, they show different amounts of lag. Lights of different composition which arouse sensations of equal brightness at the optimum value of exposure will arouse sensations of unequal brightness for the short exposures given by the sectors of the flicker disc run at the speeds ordinarily employed.

(2) Intensity of light. The lag changes with change of intensity of light and the change is different in amount for the different wave-lengths. The effect of change of intensity of light on the sensations aroused by exposures of the length used in the method of flicker is to change the relative levels to which

---

[2] Even when the wave-length and exposure are held constant, the intensity of sensation is not a regular function of intensity of light. The deviations from regularity are quite large when the eye is exposed for a time shorter than is required for the sensation to reach its maximum.

the sensations aroused by the lights compared are allowed to rise, and thus the value of the balance which is established by the method of flicker.

(3) The rate of speed at which the flicker disc is rotated. A change in the rate of speed changes the value of the individual exposures and this changes the relative levels to which the sensations compared are allowed to rise during the individual exposures. Also this change might be expected to have and does have a different effect for different intensities of the same pair of lights.

(4) The relative lengths of exposure given by the lights compared or the relative values of the open and closed sectors of the disc. In practice the open and closed sectors are given the same value. However, if these values are made unequal, the levels to which the sensations are allowed to rise change correspondingly and the relative rating of the two lights as to luminosity should be changed. For example, if an observer is used who underestimates red and yellow and overestimates blue and green at a given intensity of light by the flicker method, a change in the value of the sectors to give a shorter exposure to red or yellow and a longer exposure to blue or green should cause a greater underestimation of red or yellow and a greater overestimation of blue or green.

(5) Field size may be and probably is a factor insofar as changes in the field size may effect the relative rates of rise of the sensations aroused by lights of different composition. The effect of field size on the rate of rise of sensation presents an interesting subject for further investigation and one which, so far as we know, has not been undertaken.

(6) State of adaptation of the retina. This may also be a factor through a possible influence on the rate of rise of the sensations aroused by lights of different compositions.

Results will be given in Part II showing the effect of the first three of the above factors. Accompanying these results curves will also be given showing the rate of rise of sensation plotted in just noticeably different steps (the time j. n. d.) for the same observer at each of the intensities and compositions of light employed in the photometric work.

Difference in length of exposure the cause of the disagreement between the flicker and equality of brightness ratings.—The evidence in support of this point may be summarized as follows.

(1) Our first and more fundamental evidence is the comparison of the type and amount of the disagreement between the two methods with the brightness to which the sensations have been allowed to attain for an exposure time equal to that of the individual exposures used in the method of flicker. As a basis for this comparison the brightness was measured in terms of just noticeable differences from zero. A close agreement was found. The rise of sensation curves were determined with the same apparatus, the same observer, the same composition and intensities of light, and the same state of adaptation as were used in the flicker and equality of brightness comparison.

(2) The effect of intensity of light on the type and amount of the disagreement was determined and the result was checked up on the rise of sensation curves for three of the seven intensities used. Again the disagreement was found to sustain a close relation to the difference in height to which the sensations are allowed to rise for these intensities with the lengths of exposure used in the method of flicker.[3] Furthermore an intensity was found for each pair of lights at which agreement occurs for the two methods of photometry with speed of rotation of the disc ordinarily employed. Above and below this intensity came underestimation or overestimation of the colored light depending upon the color selected for comparison with the white light. No single intensity could be found at which agreement occurred for more than one pair of lights. These most favorable intensities were quite widely separated for the four pairs of lights and the percentages of disagreement at each of these intensities for the three remaining pairs were quite large.

(3) The effect of speed of rotation of the flicker disc on the type and amount of disagreement was determined for each of the intensities of light employed. A number of speeds were used. A very considerable effect on the disagreement was found, varying

[3] A comparison of the correspondence between the two types of photometric rating and the differences in height to which the sensations have been allowed to rise for the given exposures can be made with exactness only with regard to the type of disagreement and the order of ranking as to amount. Comparisons of actual amounts of disagreement with differences in sensation level must necessarily be rough.

with the intensity of light employed. At the intensity at which
agreement occurred for the two methods at the most sensitive
flicker speed, variation in speed produced its least effect, as might
be expected from the small angle of deviation of the rise of sensation curves for the lights compared at their point of intersection.
In some cases the closest agreement came when the speed was
low, in others when it was high. The result would obviously
depend upon the relation of the rates of rise of sensation for
the two lights above and below the points from which the change
of speed was made. In much the greater number of cases, however, the closest agreement came with the slower speeds and the
longer exposures. Considering the trend of results very broadly,
it may be said that as the speed is decreased and the length of
exposure increased, agreement is approached irregularly as a limit,
the limiting value of exposure being that used in the equality of
brightness method.

(4) The effect of varying the relative values of the exposure
to the lights compared was determined. The effect of this unbalancing of the flicker photometer, if a disturbance in the adjustment of an instrument already out of balance can be called unbalancing, varies both with the intensity of the light and the speed
of rotation of the disc. In general the effect of a relative
decrease of the exposure to a light already underestimated and
an increase in the exposure of the light with which it is compared
tends to increase the amount of the underestimation of that light
by the method of flicker, and conversely. It would seen that a
degree of physical unbalance or ratio of open to closed sector
might be found for any given intensity of light and speed of
rotation of the disc that would give agreement between the two
methods, or would even change an underestimation to an overestimation. This latter point has not yet been verified by us.

(5) Each pair of lights at three intensities was rated by the
equality of brightness method with a value of exposure equal
to that of the individual exposure used in the method of flicker.
These short exposures were made by means of a rotary tachistoscope. Agreement of result was obtained within the limits of
sensitivity of the judgment. In every case the flicker result
fell within the small range of values that was judged equal by
the equality of brightness method. This might well be considered

the most important step in the argument. It was added as a final confirmation not only to show that difference in time of exposure is the cause of the disagreement between the two methods, but to ascertain whether there was any detectable effect of summation on the disagreement, due to the succession of impressions given in the method of flicker. If agreement in result is obtained by the two methods under this condition it would seem reasonable to conclude that no considerable effect could be attributed to summation. The judgment by the equality of brightness method for the short exposures was not as difficult as might be supposed because the color difference was very much reduced by the brevity of the exposure; nor was the sensitivity of the method reduced to any considerable extent by the shortness of the exposure.

The lights used in these studies differed greatly in composition. The colored lights were selected from the spectrum of a Nernst filament by a slit 0.5 mm. wide and all trace of impurities was absorbed out by means of thin gelatines placed over the objective slit. They were red 675 $m\mu$; yellow 579 $m\mu$; green 515 $m\mu$; and blue 466 $m\mu$. Seven intensities were used, ranging from 12.5 to 50 meter-candles; the field size was 1.9 degrees. The standard lamp was a 32 cp., 4.85 watts per mean spherical candle, carbon lamp. When lights so different in composition as these are used, there can be no doubt that there is a disagreement in result by the two methods of photometry of a type and amount that can not be ascribed to uncertainty of the judgment. The consistent throw of the disagreement in one direction or the other for a given observer for a given intensity and difference of composition of light by an amount greatly in excess of the mean variation by either method indicates clearly that there is a physiological basis for the disagreement. For example, if the disagreement were due merely to an error of uncertainty in the judgment, the throw for a given observer would be as apt to be in one direction as the other. Or if the disagreement were due to a difference in the pattern or concept of what is equality of brightness in the presence of hue difference, as some have suggested, there should not be a complete and consistent reversal in type of disagreement for the same observer for the same pair of lights and hue difference with no other change of conditions than change of intensity. A change in the subjective pattern could account only for such phenomena as inconsistency of rating by the equality of brightness method for the same observer at different times, and for differences in rating by different observers. A consistent throw in any one direction by the same observer at widely separated intervals and an equally consistent reversal in the direction of the throw with change of intensity could scarcely be accounted for on the ground of unreliability or inconsistency of the judgment due to an instable or fluctuating pattern. The disagreement is unquestionably physiologic, not mental,—grounded in the nature of the response of the balancing instrument not in the evaluation of that response by the judgment. It would seem just as plausible to ascribe the Purkinje shift and other phenomena well recognized and established as optical, to an error of judgment as this equally consistent disagreement of the flicker and equality of brightness methods. The facts are that we have one type of effect of change of intensity on the eye as a balancing instrument when exposed for a half second or more, and another when exposed from 0.015 to 0.025 second. The two effects do not run parallel. They would run parallel, or approximately so, if for a given difference of wave-length and of time of exposure the response of the eye were a regular function of intensity instead of a complicated function of three interacting variables.. That is, even with wave-length and time of exposure held constant, the response of the eye is an irregular function of intensity, quite strikingly so when the time of exposure is less than that required for the sensation to rise to its maximum. Whatever may be said for the Weber-Fechner law as expressing a characteristic of response of the eye with exposures equal to or greater than that required for the sensation to reach its maximum, surely little or nothing can be said in its support for exposures shorter than this.

Individual and group differences in flicker photometry.—That there is a great deal of difference in the behavior of the flicker photometer with different observers is well known. The most that can be hoped is that observers will ultimately be found to fall into rather large groups or types. This tendency to group according to type was brought out clearly in the work of Ives and Kingsbury[4] and of Crittenden and Richtmyer.[5] It has also come

[4] Ives, H. E. and Kingsbury, E. F. Trans. I. E. S., 1915, X, p. 203.
[5] Crittenden, E. C. and Richtmyer, F. K. Trans. I. E. S., 1916, XI, p. 331.

out strongly in the drill work among our own students during the course of several years. For the observer used in these experiments, red was overestimated at 50 m. c., underestimated at 25 and 12.5 m. c., and agreement with the equality of brightness method at 29.39 m. c. But there is no guarantee whatever that this result will be true for another observer selected at random. A group of observers, however, may be found which shows roughly the characteristics of this observer. A fair percentage of such observers has been found among the students in our laboratory. As yet, however, we have not had the opportunity to compare their flicker results with those for the rise of sensation. In relation to explanation, obviously a study should be made of individual and group differences in the rise of sensation. The work on rise of sensation so far as it has been carried seems to indicate that great individual differences may be expected.

The rating of lights in relation to their service to the eye.—The question may be raised as to how much lights must differ in composition before the disagreement between the flicker and equality of brightness ratings comes to be a serious matter in practical work. Is the discrepancy great enough to be of considerable consequence for the range of difference of composition found among artificial illuminants? Comparisons of artificial illuminants have been made by Wilde,[6] Bell,[7] Crittenden and Richtmyer[8] and others. It has not been our purpose in this series of studies to go into that question at all. Our object has been merely to make an analytical and explanatory study of the phenomeon of flicker under various conditions pertinent to its application as an indicator in making a photometric balance.

While it may be said that the equality of brightness method gives a rating for exposures more nearly in accord with those ordinarily used in the act of seeing than does the method of flicker, and should therefore, be regarded as the standard photometric method, still as was shown in the paper presented by us to this Society last year, even the equality of brightness method does not rate lights differing widely in composition according to

[6] Wilde, L. W. "The Photometry of Differently Colored Lights," *The Electrician* 1909, LXIII, pp. 540-541.

[7] Bell, L. "Chromatic Aberration and Visual Acuity," *Electrical World*, 1911 LVII, pp. 1163-1166.

[8] Crittenden, E. C. and Richtmyer, F. K. *loc. cit.*

the power they give the eye for clear seeing. That is, it was demonstrated in that paper that spectrum lights made photometrically equal by the equality of brightness method give rather widely different acuities, speeds of discrimination, and powers to sustain acuity. Furthermore a continuation of that study this year has shown that spectrum lights made equal both as to saturation and brightness also give differences in acuity, speed and power to sustain acuity, though not nearly so great. Obviously a method based on some aspect of acuity would rate lights differing in composition more nearly in accord with their serviceability to the working eye than any method in which the ratings are made in terms of brightness. Acuity, however, is an insensitive indicator of changes in light intensity at medium and high illumination. Perhaps the needed sensitivity could be added by using speed of discrimination as an indicator, instead of acuity as determined in the conventional way. While acuity changes slowly with intensity at medium and high illuminations, speed of discrimination changes rapidly with increase of illumination at these intensities and very rapidly indeed at low illuminations. The rating of lights in terms of their ability to give equal speeds of discrimination of a 1 min. opening or less in a broken circle, or some other suitable detail, would not be a difficult or infeasible task. Nothing more would be involved in that task than indicating which way the opening was turned and the judgment, moreover, would present no more difficulty for lights differing in color than for lights of the same color. The method too would have the very great advantage of an objective check on the correctness of the judgment.[9] Also acuity itself changes rapidly at low illuminations. That is, a method of rating lights based on acuity at low illuminations would have a great deal of sensitivity. If, as stated by some writers, the Purkinje phenomenon is absent in the fovea,

[9] If at high intensities the speed of discrimination for the size of test-object used should be less than the time required for the sensation to reach its maximum, the objectionable effect of under-exposure could be eliminated by having the surface covering the test-object of the same coefficient of reflection as the test surface. This pre-exposure which could be made of any length desired, would add to the time the light acted on the eye. The eye would thus not be under-exposed to the light which is being rated. That is, the short exposure would apply only to the discrimination of detail at a given level of brightness, not to the brightness which would be aroused. Again the smaller the test-object, the longer would be the exposure needed to discriminate the required detail. This principle also could be used, if desired, to guard against under-exposure. The decrease in the size of the test-object, by increasing the difficulty of the task set to the eye would, moreover, add to the sensitivity of the method for picking up differences in intensity.

advantage could be taken of that fact always to make the ratings at low illumination. The detail to be discriminated would always fall well within the fovea. It might be well to revive the acuity method and see what can be done with it in terms of facts that have been brought out in recent work. Where the task set to the eye is merely the discrimination of brightness, it is clear that the rating of the lights to be used should also be in terms of luminosity or brightness; but where the task is the discrimination of the form or detail of illuminated objects, the rating of the lights in terms of some aspect of acuity would be more appropriate to the use to which they are to subserve.

Theories as to the cause of the disagreement between the flicker and equality of brightness methods.—Three factors have in the main been assigned as the cause of the disagreement between the flicker and equality of brightness ratings: the effect of intensity of light, the size of the photometric field, and the time of exposure of the eye to the lights compared. Lauriol,[10] Dow,[11] Millar,[12] Ives,[13] and Luckiesh[14] have investigated the effect of intensity of light; Schenck,[15] Dow,[16] and Ives the effect of size of field. With regard to intensity of light as a factor, Lauriol and Dow claim that the relative shift in the brightness of the different colors at low illuminations is shown by both methods. The shift for Dow, however, is more pronounced in the equality of brightness than in the flicker determinations. For Lauriol the shift for the different colors varies in magnitude by the two methods and in some cases in direction. Millar, on the other hand, photometering a mercury against a carbon lamp, claims that the Purkinje phenomenon is not shown at all by the flicker method at

[10] Lauriol. "Le photomètre à papillotement et la photométrie heterochrome." *Bull. Soc. Intern. des Electriciens,* 1904, pp. 647-652.

[11] Dow, J. S. "Color Phenomena in Photometry." *Philos. Mag.,* 1906, XII, Ser. 6, p. 131.

[12] Millar, P. S. "The Problem of Heterochromatic Photometry." Trans. I. E. S., 1909, IV, p. 769.

[13] Ives, H. E. "Studies in the Photometry of Lights of Different Colors." *Philos. Mag.,* 1912, XXIV, Ser. 6, pp. 149-188.

[14] Luckiesh, M. "Purkinje Effect and Comparison of Flicker and Equality of Brightness Photometers." *Electrical World,* March 22, 1913, p. 620.

[15] Schenck, F. "Ueber die Bestimmung der Helligkeit grauer and farbiger Pigmentpapiere mittels intermittirende Netzhautreizung." *Pflüger' Archiv,* 1896, LXIV, pp. 607-628.

[16] Dow, J. S. *op. cit.* pp. 130-134; "Physiological Principles Underlying the Flicker Photometer," *Philos. Mag.,* 1910, XIX, Ser. 6, pp. 58-77.

low illuminations; while Ives and Luckiesh go to the other
extreme and declare that a reverse Purkinje effect is obtained
by the flicker method.

Dow, a follower of the *duplicitäts* theory of color vision,
believes that there are two kinds of flicker, a coarse flicker (rod
flicker) and a fine flicker (cone flicker). The final adjustments
for the photometric balance by the method of flicker are made
by the fine or cone flicker. The photometric balance then by the
method of flicker is always in terms of cone functioning. The
equality of brightness balance, however, is in terms of both rod
and cone functioning, unless a field size is chosen which falls
within the fovea.[17] If a field size is chosen which falls within
the fovea the balance is established by both methods in terms
of cone functioning and a substantial agreement in result should
be expected. These explanatory principles proposed by Dow.
although not completely worked out and elaborated by him, may
be fairly singled out, we think, as one type of theory as to the
cause of the disagreement of the results by the two methods.

Time will not be taken to develop all of the arguments
against Dow's conception of rod and cone functioning as the
cause of the disagreement. However, since the importance of
intensity as a factor has been noted by him and other writers, it
might be well to consider whether its effect can be explained in
terms of the rod and cone hypothesis. If the sole cause of the
disagreement is in terms of a difference in rod and cone functioning, it is not readily seen how intensity could affect the disagreement by the two methods at all when a field size is chosen
which falls within the fovea; for with fields of this size the
balance is established in terms of cone functioning by both
methods; and intensity, if it affects the results by one method,
should correspondingly affect the results by the other. The
effect of intensity, however, is not only readily explained as an
effect of lag on the balance established by the method of flicker,
but it is just what would be expected from a study of the curves
for the rise of sensation for different intensities of light. Even
the disagreement as to the effect of the change from high to low
intensities found by Lauriol, Dow, Millar, Ives and Luckiesh
is what might be expected in terms of this explanatory principle.

[17] It may be pointed out that in terms of the theory, the proportion of cone
functioning would increase with increase of intensity of light for extrafoveal field sizes.

That is, these men were probably working with different types of observer,—Ives and Luckiesh with a type for which it might be inferred red and yellow rise to a higher value than blue and green for the length of exposure used in the method of flicker for the lower intensities,[18] and the converse for the higher intensities. There seems to be no explanation at all for such differences in effect in terms of the Dow adaptation of the *duplicitäts* theory of color vision.

With regard to size of field as a factor Schenck found that a decrease in size lowered the mean variation for the flicker method and decreased the luminosity value for all colors. Dow found that as the size of field was decreased red and yellow lightened relatively to blue and green. Ives[19] admitting the disagreement and accepting size of field and intensity of stimulus as important causes, sought to determine whether a field size and an intensity could be found for which the methods agree. He photometered different portions of the spectrum against carbon lamps at a number of intensities and with different field sizes, and found in general that the luminosity curves obtained by the two methods differed. The difference, however, was less for high intensities and small field sizes than for low intensities and larger field sizes. He concludes that high illuminations (25 m. c. and above) are the most favorable to agreement.[20] Our results would seem to indicate that perhaps this generalization was based on the use of too limited a range both of the higher intensities and of wave-length. Agreement was found by us for the spectrum red (675 $m\mu$) and the carbon lamp at 29.39 m. c.; for the spectrum yellow (579 $m\mu$) at 26.12 m. c.; for the green (515 $m\mu$) 11.91 m. c.; and for the blue (466 $m\mu$) at 14.91 m. c. Twenty-five meter-candles, the value which he estimated his most favorable intensity approximately to be,[21] was found by us to be compara-

[18] For an example of this type of observer, see M. A. Bills, "The Lag of Visual Sensation in its Relation to Wave-length and Intensity of Light." *Psychological Monographs*, No. 127, p. 93.

[19] Ives, H. E. *loc. cit.*

[20] "Of considerable practical interest is the fact that the flicker method most nearly agrees with the method of equality of brightness at high illuminations." [Ives, *Philos. Mag.*, 1912, XXIV, (6), p. 182.] "An illumination of 25 m. c. was chosen since by the means described below this was found to correspond closely to the high illumination at which the results of the flicker and equality of brightness methods became the same." (*op. cit.*, p. 854.)

[21] Ives, H. E. "The Spectral Luminosity Curve of the Average Eye," *Philos. Mag.*, 1912, XXIV, (6), pp. 853-863.

tively favorable to agreement for red and yellow (the range of spectrum used by Ives was 653 to 517 $m\mu$) ; but the disagreement for these colors become large again as the intensity was increased up to 50 m. c. Moreover 25 m. c. was not at all favorable for the blue and green. The most favorable intensity is the one which allows the sensations aroused by the lights compared to rise to the same level or nearly so for the short exposures used in the method of flicker. Just what this intensity is can not be predicted in advance for any observer even for one pair of lights.

A table is appended, Table I, in which is shown in percentages the difference in result obtained by the five observers used by Ives at the intensity of light (250 Illumination Units) which of those used he calls most favorable to agreement by the two methods. It will be seen from this table that the disagreement is by no means eliminated. In the table percentage overestimation is designated by $+$ and underestimation by $-$.

TABLE I.[22]—Showing in Percentage the Difference in Result Between the Methods of Flicker and Equality of Brightness for the Five Observers Used by Ives at "250 Illumination Units," His Most Favorable Intensity.

λ	H. E. I.	M. L.	P. W. C.	C. F. L.	F. E. C.
	Per cent	Per cent	Per cent	Per cent	Per cent
0.653 μ	—12.0	+29.0	—18.0	—51.0	—50.0
0.643 μ	— 3.6	+56.0	— 7.0	—31.0	—23.7
0.632 μ	— 4.3	+20.0	—15.5	—45.5	—12.9
0.622 μ	— 7.3	+12.5	— 4.2	—42.0	—15.9
0.612 μ	—10.0	+ 8.3	— 6.5	— 7.5	+ 0.3
0.594 μ	— 1.0	— 0.5	— 0.5	+ 7.5	+ 5.8
0.574 μ	+ 0.5	— 2.4	— 2.5	+27.0	— 5.9
0.555 μ	— 0.4	— 8.0	—11.9	— 4.8	+ 8.9
0.545 μ	— 3.1	— 8.4	—13.8	+ 3.0	+ 8.0
0.535 μ	— 1.9	— 4.0	—12.6	—12.0	+14.3
0.526 μ	0	— 8.7	—21.4	— 3.0	+35.5
0.517 μ	+ 0.6	—10.8	—13.8	—33.0	+30.0

It seems quite clear that field size and intensity can not play a very important rôle as a foundation cause of the disagreement in the results of the two methods, and that agreement can not come for lights of all compositions through a search for one most favorable field size and intensity. We have found only one way in which agreement can be brought about for lights of all compositions, namely, by giving the eye the same length of exposure by both methods. When an exposure is selected for the

[22] These percentages were computed by use from Table II, p. 186, *Philos. Mag.*, 1912, XXIV, (6).

equality of brightness method equal in value to that given by the individual exposures in the method of flicker, the desired agreement alone is obtained. When however, agreement is wanted only for a given pair of lights the case is different. An intensity can be found that will satisfy this condition, namely, that intensity for which the sensations aroused by the two lights have risen to the same level for the lengths of exposure given in the method of flicker. Such an intensity has been found by us for each of the four spectrum lights when matched against the carbon lamp. This intensity, however, is different for each of the spectrum lights.

That time of exposure is the cause of disagreement was, so far as we know, first suggested by Morris-Airey. It would seem a priori that time of exposure might affect the results of the method of flicker in one or both of two ways,—through a difference in the rise of sensation for the different colors during the individual exposure; and through a difference in their persistence between exposures. Time of exposure might affect the results by the equality of brightness method through a difference in the decrease or decay of sensation when the eye is exposed through a longer period than is required for the sensation to reach its maximum, although the chances of this effect being of any considerable importance are slight owing to the comparatively slow rate of decrease of sensation after the maximum has been reached. A brief discussion of how time of exposure might be expected to be effective is given by Morris-Airey, but no experimental work was done. In an article appearing in *The Electrician*, Dec. 17, 1909, LXIV, p. 401, he says: "In recent years photometers of the flicker type have obtained a certain degree of prominence, but from time to time suspicions have been raised as to whether the principle on which flicker photometers are based does not involve physiological phenomena which disturb the conditions of illumination so that the numbers obtained are not a true representation of the illuminating powers of the sources to be compared. Physiologists are at variance with regard to color theory. * * * * * With every new change of color theory the physicist would have to change his starting point in the explanation of a phenomenon like the flicker effect and I suggest that a more stable foundation for a physical theory is

to be found in the experimental examination of the growth and
decay of the retinal stimulus due to differently colored lights
without any attempt to connect the form of the curves obtained
with physiological theories which are admittedly of uncertain
stability." Morris-Airey then cites a phenomenon described by
G. N. Stewart (Proc. Roy. Soc. Edin., 1888) which is merely a
re-discovery by Stewart of a phonomenon described by Fechner[23]
many years before and now called Fechner's colors. Stewart
observed that when a mirror reflecting white light was rotated at
different rates of speed, the image of the source of light appeared
successively in different colors. This selective unbalancing of
the retina's action to give the component colors instead of white,
obviously could be due either to difference in rate of rise of the
component color excitations or to a difference in their rate of
persistence through the dark interval between the flashes or to
both. This phenomenon can be more conveniently demonstrated
by the rotation of either white or black sectors (180° each), a
black spiral on a white disc, or a specially constructed disc
(Benham's disc) made up of sectors of 180° of white and 180°
of black. On the white sector are painted narrow 45° arcs at
different distances from the center of the disc and differently
spaced with relation to the two edges of the white sector. These
black arcs are seen in different colors depending upon their spacing in relation to the edges of the white sector, and the colors
change as the speed of rotation of the disc is changed.

The rate of growth and decay as an explanatory principle was
considered by Ives[24] who seemed to believe, however, that Dow's
rod and cone theory offered greater possibilities of explanation.
As an objection to the explanation suggested by Morris-Airey
it is interesting to note that Ives says: "It is not established that
at the same illumination the rates of rise of sensation are actually
different for different colors * * * * *   * * it is clear that these
phenomena of rise of sensation must be studied with the same
observer and apparatus by which are made flicker measurements
in which they may play a part." In Jan., 1914, the theory advanced
by us in this paper was first proposed before the Philadelphia

[23] Fechner, G. T. "Ueber eine Scheibe zur Erzeugung Subjectiver Farbon," *Poggen,
Annal.*, 1838, XLV, pp. 227-232.

[24] Ives, H. E. *Philos. Mag.*, 1912, XXIV, (6) pp. 178-181.

Section of this Society and experimental data offered in its support. This paper was reviewed in an editorial in *The Electrical Review and Electrician* Mar. 7, 1914, pp. 478-479, and was published in full in *The Psychol. Rev.*, 1915, XXII, pp. 110-162. Later in 1914 Luckiesh[25] explains the discrepancy between the flicker and equality of brightness ratings in terms of the different rates of growth and decay of the color sensations; but seems inclined to reject this explanation in an appendix to the same paper.

The theory proposed by us differs from that suggested by Morris-Airey and Luckiesh in that we attribute the cause to differences in the rate of rise of sensation, not to both rate of rise and decay. Difference in rate of decay as a factor is, we grant, an *a priori* possibility; but we feel reasonably sure that it is not a factor of any considerable consequence because if it were, its effect would be manifested through a *differential* summation effect. That there is no detectable differential summation effect seems pretty well proven by the fact that within the limits of sensitivity of the equality of brightness judgment the same results are given by both methods of rating when the same lengths of exposure are used,—*that is the difference in level to which the two sensations rise in a single exposure by the flicker method is not decreased by a detectable amount in the succession of exposures.*

[25] Luckiesh, M. *Phys. Rev.*, 1914, (2), IV, pp. 1-11.

FLICKER PHOTOMETRY*
II. COMPARATIVE STUDIES OF EQUALITY OF BRIGHTNESS AND FLICKER PHOTOMETRY WITH SPECIAL REFERENCE TO THE LAG OF VISUAL SENSATION

BY C. E. FERREE AND GERTRUDE RAND**

The Lag of Visual Sensation in Relation to the Disagreement of Flicker and Equality of Brightness Photometry for Lights of Different Compositions.

The experimental work under this heading covers two points: (1) A comparison of the rating of four spectrum lights at three intensities against a carbon lamp by the methods of flicker and equality brightness; and (2) a determination of the curves of the rise of sensation for these five lights at each of the three intensities using the same apparatus, the same observer and the same state of adaptation of the eye. The results obtained enable us to make a direct comparison of the type and roughly the amount of the disagreement between the two methods of photometry with the actual luminosities or brightnesses to which the two sensations had been allowed to rise for values of exposure equal to the individual exposures used in the method of flicker. As a basis for this comparison the brightnesses were measured in terms of just noticeably different steps from zero. No experimental evidence more direct as to type than this can be offered, we believe, on the relation of lag to the disagreement in question. If it be found, for example, that for a given intensity of light, red and yellow are underestimated by the method of flicker and blue and green overestimated, then it should be shown by the curves that the sensations aroused by the red and yellow lights had risen to a lower and by the blue and green to a higher value than the sensation aroused by the light of the carbon lamp. The value which the sensation would attain can be read off directly from the curve and comparisons made with the type and amount of disagreement between the two methods of photometry. In the remaining chapters the evidence offered by this comparison is supplemented and rounded out by results when the eye is given the same length

*A paper presented at the annual convention of the Illuminating Engineering Society Swampscott, Mass., September 25-28, 1922.
**Department of Psychology, Bryn Mawr College, Bryn Mawr, Pa.

of exposure by both methods of photometry and by showing the effect of varying the speed of rotation of the disc and the intensity of light on the amount and type of disagreement on the results in question.

The colored light was taken from the spectrum of a Nernst filament operated at 0.6 amp.; a narrow band in the red at 675 $m\mu$, of yellow at 579 $m\mu$, of green at 515 $m\mu$, and of blue at 466 $m\mu$. The breadth of slit used in isolating these bands was 0.5 mm. In every case the light was examined for impurities at the objective slit by means of a small Hilger direct vision spectrometer provided with an illuminated scale. When found impurities were absorbed out by thin gelatine filters placed over the objective slit, so selected as to absorb as little as possible of the useful light. The spectrum lights were photometered against a 32 cp. carbon lamp, 4.85 watts per mean spherical candle. The photometric comparisons were made and the rise of sensation curves determined at 12.5, 25, and 50 m. c. as measured by the equality of brightness method.

The plan of the apparatus used is indicated in Fig. 1 and a photograph is shown in Fig. 2. The light from the objective slit was collimated by the lens A, Fig. 1, and focussed on the eye by the lens B. The image formed fell within the pupil. By substituting this lens system for the eye-piece of the spectroscope the need and inconvenience of the use of an artificial pupil was eliminated, and a surface of uniform color was presented to the eye. Lens B was diaphragmed by a circular opening E of such a size and such a distance from the eye as to furnish a photometric field of 1.9 degrees. Between the lenses A and B was interposed the 90 degree sectored disc D mounted on an electric rotator; and between lens B and the opening E, the disc F of the rotary tachistoscope used in making the exposures for determining the rise of sensation curves. Discs D and F were both of light sheet aluminum. Disc D was coated with magnesium oxide deposited from the burning metal; Disc F was painted mat black.

The method of making the photometric comparisons was as follows. For the equality of brightness determinations, disc D, reflecting the light from the carbon lamp L* was turned so that its edge bisected vertically the photometric field at E, and the luminosity of the colored light was increased or decreased by adjusting the collimator slit until it equalled by the equality of brightness method the 12.5, 25 or 50 m. c. reflected from the disc D after the correction had been made for the absorption of lens B. The sensitivity of the determination by the equality of brightness method was approximately 2 per cent, varying around this value with the color and intensity of the light. For the flicker determinations the disc D was rotated and the position of the carbon lamp adjusted on the bar until no flicker was obtained at the lowest possible speed that would give no flicker. At this speed the sensitivity of the determination was around 0.4 per cent varying with the color and intensity of the light.

In making the determinations for the rise of sensation curves for color, the disc D was turned completely out of the beam of light so that the entire field at E was seen as colored; and for the curves for white it was turned so as to cut off the colored light completely and reflect the light from the carbon lamp so that the entire field at E was seen as white (see footnote below). The exposure disc F of the rotary tachistoscope was then turned into position. This disc was compounded from two pairs of discs having radii respectively of 17 and 24 cm. These discs were mounted on the axle of the tachistoscope the position of which was so adjusted that the circular edge of the smaller pair of discs bisected vertically the field seen at E. These halves, right and left, of the photometric field formed the standard and comparison fields for the determination of the rise of sensation curves. This determination was made as follows. The smaller set of discs was closed completely and the larger was opened by a series of adjustments until, with a given speed of rotation of the discs, a just noticeable sensation was aroused. For the next observation the smaller pair of discs was given this value of opening and the opening of the larger pair was increased until a just noticeable difference in brightness was obtained. The series was continued, the comparison opening for one observation being made the standard for the next, until a value of exposure was obtained, no further increase of which caused a noticeable increase of brightness of the comparison field. This was accepted as the optimum value of the sensation and the sensations produced by

* For convenience this light will be designated as white in the rest of the paper and in the figures.

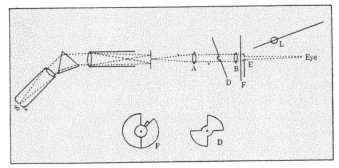

Fig. 1.—Plan of apparatus.

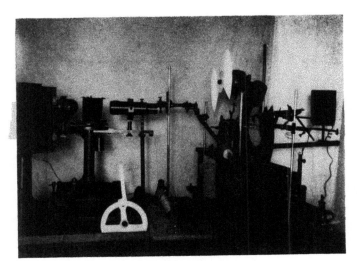

Fig. 2.—Photograph of apparatus.

other exposures were rated in terms of their number of just noticeable differences from zero. Further increase of length of exposure caused a decrease in the brightness of the comparison field and a continuation of the series beyond this point gave the just noticeable difference values needed to plot the other half of the curve which represents the temporal course of sensation, rise, and fall.

The exposure apparatus or rotary tachistoscope was made up of the rotation apparatus and two pairs of specially constructed sectored discs. The rotation apparatus consisted of a heavy Y-shaped stem with carefully turned bearings in which rotated a spindle fitted with an arbor and chuck to hold the exposure discs. Since only a single exposure was wanted for any observation, the discs were driven by a gravity or fall device. That is, on the other end of the spindle was the larger of a two cog-wheel gear system. Attached at its middle to the larger cog-wheel was a rod 1 meter long designed to carry on its two arms equal weights. For most of this work in order to give short exposures and high speeds, a heavy weight was carried on one arm alone. These weights which slip on the rod were of lead and were held in position by means of set screws. The rod was graduated in mm. so that at any time the position of the weights could be read directly from the axle. Before an exposure the rod was held in position by a catch, the height of which was adjustable. When the catch was released the weight fell rotating the discs with a velocity depending upon the height of the catch and the position of the two weights on the rod. After the fall the other end of the rod was caught automatically by a second catch.

Another factor in determining the length of the exposure is the displacement of the open sector from the path of the stimulus light or its position in the circle of rotation; for obviously upon this will depend, with a given height of the catch and a given position of the weights on the driving rod, the amount of acceleration which is given to the disc before the open sector reaches the path of the stimulus light. This value for the initial distance of the open sector from the path of the stimulus light was fixed by pinning permanently to the axle that disc of each set, the edge of which made the beginning of the exposure, so that these edges were in the same radial line.

The values of the open sectors were measured with a Vernier protractor graduated to seconds of arc. These measurements after any number of sets of observations could be converted into units of time by a simple process of calibration. Smoked paper was clipped to the disc across the open sector; the pendulum was released with the weights, the starting point, etc., just as they were in the original observation; and a time line was run across the open sector by an electric tuning fork the vibration frequency of which was known. The paper was removed, shellaced and the records counted at leisure.

Prior to the use of this method of determining the rate of rise of sensation, a preliminary study was conducted in our laboratory in which a comparison of results was made using all of the older methods and three new ones devised for the purpose.[26] The work was done with spectrum lights with the same observer and state of adaptation of the eye and as nearly as possible the same apparatus for each of the methods. As the result of this inter-comparison and interchecking of methods, the one used in the present study was selected as the most feasible and perhaps the surest in principle. The new methods, all of which were devised by the present writers, were designed not only to ascertain the time in which a given stimulus produces its maximum effect on sensation, but to make possible a comparison of the sensation values at any point in the rise or fall. There are two possibilities of making accurate and reproducible judgments of sensation, namely, the judgments of equality and just noticeable difference. Upon these judgments all quantitative work on sensory responses ultimately rests. Both of these were utilized in developing the new methods mentioned above. The development of these methods have, in fact, been just the systematic application of these two judgments to the problem in question in accord with the purpose mentioned earlier in the paragraph. The just noticeable difference obviously may be determined by holding the time of exposure constant and varying the intensity of light or conversely by holding the intensity of light constant and varying the time of exposure. The latter principle was employed in the method used in this work. The sensation value obtained by this method may thus be called a time just noticeable difference.

[26] For the details of this study, See M. A. Bills, "The Lag of Visual Sensation in its Rlation to Wave-length and Intensity of Light." *Psych. Monog.*, No. 127, p. 181.

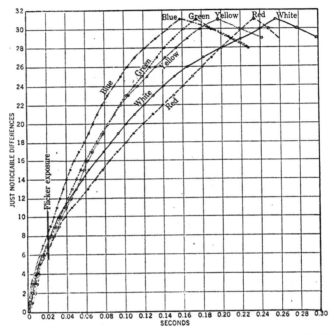

Fig. 3.—Curves showing the rise of sensation to the maximum for white, red, yellow, green and blue at 12.5 meter-candles.

Curves representing the results obtained for the rise of sensation are shown in Figs. 3-10. In these charts just noticeable differences in luminosity are plotted along the vertical and time of exposure along the horizontal coordinate. Fig. 3 shows the curves for the five lights at 12.5 m. c. At this intensity the curves were completed to the maximum and slightly beyond. As would be expected from the fact that all the lights were of the same photometric value, the same number of just noticeable differences (31) was obtained for each to the maximum. The optimum time of exposure for the blue was 0.157 second; for the green 0.174 second; for the yellow 0.195 second; for the red 0.233 second; and for the carbon lamp 0.254 second.

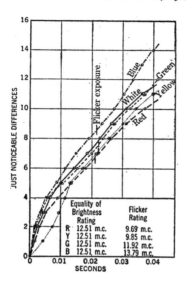

Fig. 4.—Curves showing the rise of sensation up to 0.04 sec. exposure for white, red, yellow, green and blue at 12.5 meter-candles.

In Fig. 4 the lower part of the curves of Fig. 3 is plotted on a larger scale. On these curves the height to which the sensation rises for a length of exposure equal to that of the individual exposures used in the method of flicker is indicated by a short vertical line. In the use of the flicker method we worked at the lowest possible speed that would give no flicker. Since this speed is slightly different for the different colors the values of the individual exposures were slightly different. On this chart are shown also the photometric values given to the colored lights both by the equality of brightness and the flicker methods. It will be noted that red, yellow and green are all underestimated by the method of flicker, red the most; and that blue is overestimated. An inspection of the curves shows that this is as it should be according to the relative heights to which the sensations rise for the time of exposure used in the method of flicker.

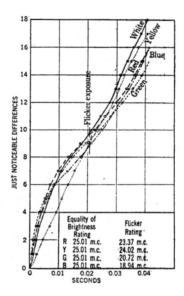

Fig. 5.—Curves showing the rise of sensation up to 0.04 sec. exposure for white, red, yellow, green and blue at 25 meter-candles.

In Fig. 5 curves are given for the rise at 25 m. c. These curves were not continued to the maximum because of the prohibitive amount of time and work required to make the complete series for lights of this intensity. The series was continued up to an exposure time of 0.04 second (15 to 18 just noticeable differences depending upon the color), which is well in excess of the flicker exposure. The photometric results show that all the colors were underestimated in the order from least to greatest of yellow, red, green and blue. The curves for rise show that for the value of exposure used in the method of flicker, the carbon light has risen to the highest value in sensation,—then yellow, red, green and blue.

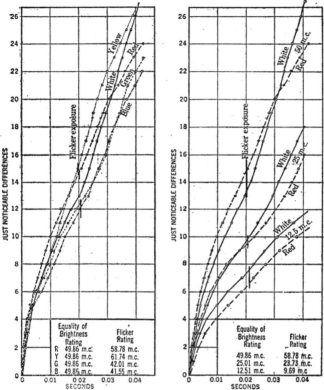

Fig. 6.—Curves showing the rise of sensation up to 0.04 sec. exposure for white, red, yellow, green and blue for 50 meter-candles.

Fig. 7.—Curves showing the rise of sensation up to 0.04 sec. exposure for red and white, 12.5, 25 and 50 meter-candles.

Fig. 6 shows the curves at 50 m. c. These curves likewise were continued only up to 0.04 second (20 to 27 just noticeable differences). The photometric data show an overestimation of red and yellow by the method of flicker at this intensity and an underestimation of green and blue. Yellow was overestimated quite a great deal more than red, and blue was underestimated slightly more than green. Several intersections in the curves will

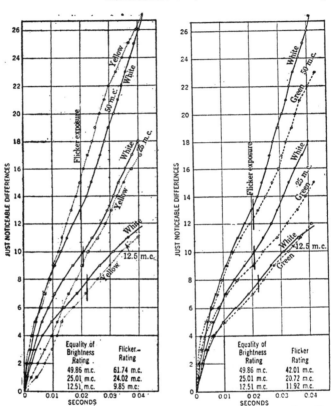

Fig. 8.—Curves showing the rise of sensation up to 0.04 sec. exposure for yellow and white, 12.5, 25 and 50 meter-candles.

Fig. 9.—Curves showing the rise of sensation for 0.04 sec. exposure for green and white, 12.5, 25 and 50 meter-candles.

be noted,—one for red and yellow at an exposure slightly less than that used in the method of flicker, and one for green and blue at an exposure slightly longer than the flicker exposure.

In Fig. 7 are shown for comparison the curves for red and white at the three intensities. An exact correspondence in the type of deviation of the photometric results and of their ranking as to amount with the respective differences in the heights to which the red and white sensations have risen, will be noted.

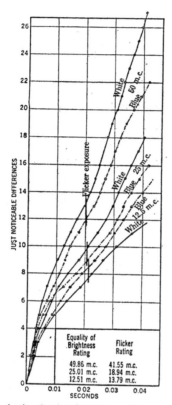

Fig. 10.—Curves showing the rise of sensation up to 0.04 sec. exposure for blue at 12.5, 25 and 50 meter-candles.

The greatest deviation is the overestimation of red at 50 m. c.; the next greatest is the underestimation at 12.5 m. c.; and the least the underestimation at 25 m. c. An inspection of the rise of sensation curves shows that red has risen to a higher level than white for the given exposure at 50 m. c. and to lower levels at 12.5 and 25 m. c. Also the difference in level is greatest at 50 m. c., next greatest 12.5, and least at 25 m. c. That is, the correspondence of the two sets of results is as it should be.

Similar comparisons are made of the curves for yellow and white in Fig. 8; for green and white in Fig. 9; and for blue and white in Fig. 10.

It may be stated that the photometric and rise of sensation work were done in quite different series. The chance for bias by pre-knowledge of the results of one type of work before the other was done was negligible. For example, while the work was scattered through the years 1917-1922, the final conversion of the settings of the exposure apparatus into units of time was not made until late in the spring of 1922.

The Effect of Varying Intensity on the Disagreement of the Results by the Flicker and Equality of Brightness Methods for Lights of Different Composition.

In the series of experiments planned to carry out the comparison between the disagreement in the results by the methods of flicker and equality of brightness on the one hand and the rise of sensation on the other, three intensities were used, 12.5, 25 and 50 m. c. Several intensities were added during the course of the work in order to investigate certain points that came out in the study of the relation of the two photometric methods.

It was found that the type of disagreement between the two methods was different for each of these three intensities. For example, at 12.5 m. c. blue was overestimated by 10.23 per cent; and red, yellow and green were underestimated by 22.54, 21.26 and 4.72 per cent respectively. At 25 m. c. red, yellow, green and blue were all underestimated,—red by 6.56 per cent, yellow by 3.96 per cent, green by 17.15 per cent, and blue by 24.28 per cent. At 50 m. c. red was overestimated by 17.89 per cent, yellow by 23.83 per cent; and blue and green were underestimated,—blue by 16.67 per cent and green by 15.74 per cent. At intensities somewhat lower than 12.5 m. c., *e. g.*, at 10.20 m. c., blue and green were overestimated and red and yellow were underestimated. That is, between these lower intensities and intensities in the neighborhood of 50 m. c. there was a complete reversal of the type of disagreement for the red and yellow and for the blue and green. Figs. 7-10 should be consulted for a comparison of the rise of sensation curves with the amount and direction of these deviations from those obtained by the equality of brightness method.

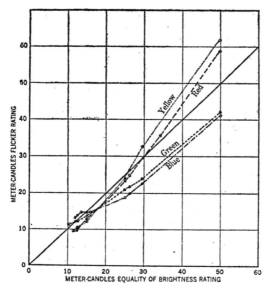

Fig. 11.—Curves showing the effect of varying intensity on the disagreement between the equality of brightness and flicker ratings for lights of different composition.

Since all of the colors showed the characteristic shift from underestimation to overestimation or the converse, it seemed reasonable to expect that an intensity might be found for each pair of lights at which agreement with the equality of brightness rating would occur. The search for this intensity was somewhat of a trial and error character controlled. however, by a fairly regular decrease in the disagreement as the intensity in question was approached. For example, red was overestimated at 50 m. c.; underestimated at 25 m. c.; and still more underestimated at 12.5 m. c. Agreement was found at 29.39 m. c. A similar indication of the region of intensity in which agreement might be expected to occur was had in case of each of the other colors. Agreement was found for yellow at 26.12 m. c.; for green at 11.91 m. c.; and for blue at 14.91 m. c. The region over which agreement could be obtained was narrow as is shown by the large angle at which the curve expressing percentage disagree-

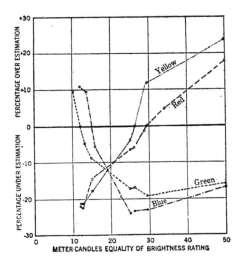

Fig. 12.—Showing in per cent the overestimation and underestimation of the flicker as compared with the equality of brightness ratings for lights of different compositions at different intensities.

ment of the results by the method of flicker crosses the line representing agreement for the different intensities as rated by the equality of brightness method (Fig. 12). The three remaining colors were photometered at each of the intensities at which agreement was found for one of the colors in order to get some idea of the deviation that might be expected for the other parts of the spectrum if one of these intensities were selected as standard. At 29.39 m. c. (agreement for red) yellow was overestimated by the method of flicker by 11.84 per cent; green was underestimated by 19.02 per cent; and blue was underestimated by 23.27 per cent. At 26.12 m. c. (agreement for yellow) red was underestimated by 6.36 per cent; green by 16.89 per cent; and blue by 23.43 per cent. At 14.91 m. c. (agreement for blue) red was underestimated by 14.69 per cent; yellow by 17.71 per cent; and green by 8.79 per cent. At 11.91 m. c. (agreement for green) red was underestimated by 21.92 per cent; and yellow by 21.25 per cent; blue was overestimated by 10.83 per cent.

These determinations were made at the speed of rotation of the disc ordinarily used in the method of flicker, *i. e.*, the speed which gives the greatest sensitivity. When the speed was changed agreement no longer occurred at these intensities. The amounts of disagreement for the other colors were also shifted by the change of speed. It is quite possible of course that another intensity may be found for each pair of lights above 50 m. c. or below 12.5 m. c. at which agreement would occur.

In searching for the intensity at which agreement occurs in case

TABLE I.—Showing the Effect of Intensity of Light on the Amount and Type of Disagreement of the Results by the Equality of Brightness and Flicker Methods.

Color	Photometric rating Equality of brightness method m. c.	Flicker method m. c.	Difference between the two methods m. c.	per cent	Revolutions per sec. flicker disc.	Length of individual exposures sec.
Red	49.86	58.78	+ 8.92	+17.89	12.5	0.020
(675 *mμ*)	34.03	35.56	+ 1.53	+ 4.50	12.4	0.02016
	29.39	29.39	0.0	0.0	12.2	0.02049
	26.12	24.46	— 1.66	— 6.36	12.2	0.02049
	25.01	23.37	— 1.64	— 6.56	12.15	0.02058
	14.91	12.72	— 2.19	—14.69	11.8	0.02119
	12.51	9.69	— 2.82	—22.54	12.0	0.02083
	11.91	9.30	— 2.61	—21.92	11.75	0.02128
Yellow	49.86	61.74	+11.88	+23.83	12.6	0.01984
(579 *mμ*)	29.39	32.87	+ 3.48	+11.84	12.2	0.02049
	26.12	26.12	0.0	0.0	11.9	0.02101
	25.01	24.02	— 0.99	— 3.96	12.2	0.02049
	14.91	12.27	— 2.64	—17.71	11.5	0.02174
	12.51	9.85	— 2.66	—21.26	11.35	0.02203
	11.91	9.38	— 2.53	—21.25	11.1	0.02252
Green	49.86	42.01	— 7.85	—15.74	12.3	0.02033
(515 *mμ*)	29.39	23.80	— 5.59	—19.02	12.4	0.02016
	26.12	21.71	— 4.41	—16.89	11.9	0.02101
	25.01	20.72	— 4.29	—17.15	11.8	0.02119
	14.91	13.69	— 1.31	— 8.79	11.5	0.02174
	12.51	11.92	— 0.59	— 4.72	11.3	0.02212
	11.91	11.91	0.0	0.0	11.2	0.02232
	10.20	11.18	+ 0.98	+ 9.61	11.0	0.02273
Blue	49.86	41.55	— 8.31	—16.67	12.3	0.02033
(466 *mμ*)	29.39	22.55	— 6.84	—23.27	12.0	0.02083
	26.12	20.00	— 6.12	—23.43	12.2	0.02049
	25.01	18.94	— 6.07	—24.28	12.0	0.02083
	15.80	14.91	— 0.89	— 5.63	12.0	0.02083
	14.91	14.91	0.0	0.0	12.0	0.02083
	13.60	14.91	+ 1.31	+ 9.63	12.0	0.02083
	12.51	13.79	+ 1.28	+10.23	11.6	0.02155
	11.91	13.20	+ 1.29	+10.83	11.8	0.02119

of blue, an interesting and very baffling phenomenon of low flicker sensitivity was found over a range on either side of the point at which agreement occurred with the equality of brightness method. A change from 13.6 to 15.8 m. c. as rated by the equality of brightness method, gave no difference in rating by the method of flicker. This low sensitivity by the method of flicker was very hard to understand until the rise of sensation work for 12.5 and 25 m. c. was completed and the curves plotted. A comparison of these curves showed very little difference in the sensation level for the lengths of exposure used in the method of flicker even for a change as great as from 12.5 to 25 m. c. That is, the rise of sensation curves for these two intensities of blue lie very close together for quite a little distance above and below the point representing the flicker exposure. We were confronted here then by the somewhat anomalous result that when the carbon standard was set to give 14.91 m. c. on the photometric field, the colored light could be changed through a range of 13.6 to 15.8 m. c., as rated by the equality of brightness method, without producing any change of result by the flicker method; yet the result of the comparison with the equality of brightness method had to be pronounced an overestimation of 9.63 per cent by the method of flicker at the lowest point of this range; agreement at 14.91 m. c.; and an underestimation of 5.63 per cent at the highest point in the range. As the sensitivity of the equality of brightness method was of the order of 1.5 to 2.5 per cent, we have here the unusual phenomenon of the equality of brightness method showing much greater sensitivity than the method of flicker.

The results of this work on the effect of variation of intensity are shown in Table I and Figs. 11 and 12. In Fig. 11 equality of brightness rating is plotted along the horizontal coordinate against the flicker rating along the vertical coordinate. Agreement between the two methods is represented by a solid line bisecting the angle formed by the two coordinates. Points on the curve for each of the colors falling above the line show a higher rating by the flicker than by the equality of brightness method or overestimation by the method of flicker; and points falling below the line a lower rating by the flicker than by the equality of brightness method or underestimation by the method of flicker.

Fig. 12 represents percentage overestimation and underestimation by the method of flicker. The rating in m. c. by the equality of brightness method is plotted along the horizontal coordinate and percentage disagreement along the vertical. The zero line parallel to the horizontal coordinate represents agreement between the two methods. Points above and below this line represent respectively percentage overestimation and underestimation by the method of flicker at the different meter-candle values. Agreement between the two methods occurs at the points where the flicker curves cross the zero line.

The Effect of Speed of Rotation of the Disc on the Disagreement of the Results by the Flicker and Equality of Brightness Methods for Lights of Different Composition and Intensity.

If time of exposure is a factor in the disagreement between the results of the flicker and equality of brightness methods, it would seem altogether probable that speed of rotation of the disc would affect the amount and perhaps the type of disagreement. The effect had been noticed in the work of our laboratory long before it was made the subject of a systematic investigation. In fact in the present series of studies this was our first point of attack on the problem.

In making an investigation of this point a means of producing very small changes of speed of rotation of the disc is required. For this purpose we were provided with a motor specially wound for speed control (500 to 6000 revolutions per minute) and a rheostat of the ordinary sliding contact type connected in series with a rotating drum wound with many turns of low resistance wire. The speed of rotation of the disc was automatically recorded by a small speed counter attached to the axle of the motor.

For each of the seven intensities of light mentioned in the preceding section, the following speeds were used: the speed giving the greatest sensitivity to the method of flicker, one speed lower and a number of speeds higher than this. For the most sensitive speed and the speeds higher than this, a position or a small range of positions could be found for the standard lamp which gave no flicker; for the lower speed, however, the balance had to be made in terms of minimum flicker. The sensitivity of the determinations varied considerably of course with change

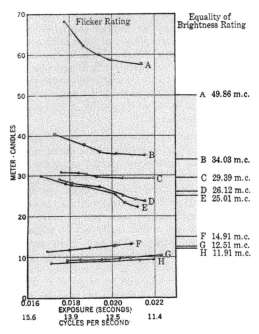

Fig. 13.—Showing the effect of varying the speed of rotation of the disc on the disagreement of the flicker and equality of brightness ratings at different intensities for red.

of speed. For the most sensitive speed it ranged from 0.20 to 0.68 per cent; for the lower speed, from 0.53 to 2.59 per cent; and for the higher speeds from 0.64 to 25.58 per cent.

The results showed consistently: (1) That the rating by the method of flicker varies with the speed of rotation of the disc. (2) That the flicker and equality of brightness ratings were in some cases in closest agreement when the speed of rotation of the disc was slow; in others when it was high. The result would obviously depend upon the relation of the rate of rise of sensation for the two lights above and below the point from which the change of speed was made. In much the greater number of

cases studied, it will be noted that the closest agreement came
with the slowest speed and the longest exposure. Considering
the trend of results very broadly it may be said that as the speed
is decreased and the length of exposure is increased, agreement
is approached irregularly as a limit, the limiting value of exposure
being that used in the equality of brightness method. And (3)
that the speed which gives the method of flicker its greatest sensitivity does not give the closest agreement in result with the
equality of brightness method.

The changes in result are quite striking. The details are given
in Tables II-IV and Figs. 13-16. In these charts meter-candles
as rated by the flicker method are plotted along the vertical coordinate and the length of exposures used, also cycles of the
flicker disc per second, along the horizontal coördinate. For
comparison the equality of brightness ratings for the different
intensities are indicated on the same meter-candle scale at the
right of the chart. A few of the results may be singled out here.
A change in the speed from 11.8 to 14.1 cycles per second (a
change in the length of the individual exposures from 0.02119
to 0.01773 second) for red at 50 m. c. changed an overestimation
by the method of flicker from 8.08 to 18.31 m. c., a percentage
change from 16.21 to 36.72. A change in speed from 12 to 16.2
cycles per second (a change in the exposure from 0.02083 to
0.01543 second) for blue at 50 m. c. changed an underestimation
of 7.34 to 12.67 m. c., a percentage change of from 14.72 to 25.41.
A change in speed from 11.9 to 14.8 cycles per second (a change
in the exposure from 0.2101 to 0.01689 second) for green at
50 m. c., changed an underestimation from 6.50 to 12.67 m. c.,
a percentage change of from 13.04 to 25.41. And a change of
speed from 11.9 to 14.4 cycles per second (a change in the exposure from 0.02101 to 0.01736 second) for yellow at 50 m. c.,
changed an overestimation of 9.76 to 21.56 m. c., a percentage
change of 19.57 to 43.24.

The results of these tables for 50, 25 and 12.5 m. c. should be
studied in connection with the rise of sensation curves for these
intensities. However it is a somewhat difficult task in this case
to make even roughly the comparisons needed. That is, while
the colors were set to give 50, 25 and 12.5 m. c. by the equality
of brightness rating. the white light which matches this by the

TABLE II.—Showing the Effect of Speed of Rotation of the Disc on the Type and Amount of Disagreement of the Results by the Equality of Brightness and Flicker Methods.
Color: Red (675 $m\mu$)

Photometric rating Equality of brightness method	Revolutions per sec. Flicker disc	Length of individual exposures	Photometric rating Flicker method	Difference between the two methods	Sensitivity Flicker method	Change produced dy variatiou of speed	
m. c.		sec.	m. c.	per cent	per cent	m. c.	per cent
49.86	11.8	0.02119	57.94	+16.21	2.23	10.23	17.66
	12.5	0.020	58.78	+17.89	0.58		
	12.9	0.01938	60.06	+20.46	2.13		
	13.4	0.01866	62.28	+24.91	4.28		
	14.1	0.01773	68.17	+36.72	11.29		
34.03	11.6	0.02155	35.16	+ 3.32	2.16	5.17	14.70
	12.4	0.02016	35.56	+ 4.50	0.67		
	12.9	0.01938	35.95	+ 5.64	2.20		
	13.4	0.01866	37.62	+10.55	5.48		
	14.5	0.01724	40.33	+18.51	16.62		
29.39	11.4	0.02193	29.39	0.0	2.01	1.54	5.24
	12.2	0.02049	29.39	0.0	0.61		
	13.0	0.01923	29.68	+ 0.98	2.97		
	13.3	0.01880	30.30	+ 3.09	8.65		
	13.6	0.01838	30.61	+ 4.15	11.31		
	14.2	0.01761	30.93	+ 5.24	13.89		
26.12	11.6	0.02155	23.80	— 8.89	1.81	5.62	23.61
	11.9	0.02101	24.24	— 7.20	1.44		
	12.2	0.02049	24.46	— 6.36	0.53		
	12.9	0.01938	27.12	— 3.83	2.39		
	13.8	0.01812	28.12	+ 7.66	6.37		
	14.3	0.01748	29.42	+12.63	7.38		
25.01	11.8	0.02119	22.36	—10.60	2.59	7.63	34.12
	12.15	0.02058	23.37	— 6.56	0.68		
	12.4	0.02016	25.39	+ 1.52	2.76		
	12.9	0.01938	26.63	+ 6.48	2.82		
	13.3	0.01880	26.89	+ 7.52	4.69		
	13.8	0.01812	27.68	+10.68	8.27		
	14.0	0.01786	28.01	+·11.99	8.46		
	15.0	0.01667	29.99	+19.92	9.46		
14.91	11.4	0.02193	13.15	—11.81	0.53	1.61	13.95
	11.8	0.02119	12.72	—14.69	0.39		
	12.4	0.02016	12.55	—15.84	0.64		
	13.2	0.01894	12.22	—18.05	4.42		
	13.9	0.01799	11.76	—21.14	4.84		
	14.8	0.01689	11.54	—22.61	6.07		
12.51	11.2	0.02232	10.40	—16.86	1.83	1.21	13.12
	12.0	0.02083	9.69	—22.54	0.20		
	12.7	0.01969	9.46	—24.37	4.45		
	13.5	0.01852	9.30	—25.65	6.55		
	14.0	0.01786	9.19	—26.53	20.06		
11.91	11.4	0.02193	9.34	—21.59	0.64	0.83	9.72
	11.75	0.02128	9.30	—21.92	0.22		
	12.3	0.02033	9.11	—23.52	0.88		
	12.8	0.01953	8.97	—24.70	2.22		
	14.0	0.01786	8.79	—26.21	5.36		
	14.6	0.01712	8.51	— 28.56	7.28		

TABLE. III.—Showing the Effect of Speed of Rotation of the Disc on the Type and Amount of Disagreement of the Results by the Equality of Brightness and Flicker Methods.
Color: Yellow (579 $m\mu$)

Photometric rating Equality of brightness method	Revolutions per sec. Flicker disc	Length of individual exposures	Photometric rating Flicker method	Difference between the two methods	Sensitivity Flicker method	Charge produced by variaton of speed	
m. c.		sec.	m. c.	per cent	per cent	m. c.	per cent
49.86	11.9	0.02101	59.62	+19.57	2.27	11.80	19.92
	12.6	0.01984	61.74	+23.83	0.58		
	13.2	0.01894	63.21	+26.77	2.04		
	13.8	0.01812	65.13	+30.62	7.10		
	14.4	0.01736	71.42	+43.24	10.16		
29.39	11.9	0.02101	32.59	+10.88	1.66	3.97	12.19
	12.2	0.02049	32.87	+11.84	0.64		
	12.7	0.01969	34.03	+15.78	1.73		
	13.2	0.01894	34.97	+18.98	3.75		
	14.0	0.01786	36.56	+24.39	6.93		
26.12	11.3	0.02212	25.88	− 0.92	1.51	3.75	14.48
	11.9	0.02101	26.12	0.0	0.57		
	12.2	0.02049	26.53	+ 1.57	3.39		
	12.7	0.01969	27.68	+ 5.97	3.79		
	13.0	0.01923	28.80	+10.26	5.73		
	13.5	0.01852	29.53	+13.06	8.04		
25.01	11.9	0.02101	24.69	− 1.28	1.42	5.96	31.83
	12.2	0.02049	24.02	− 3.96	0.46		
	12.9	0.01938	23.80	− 4.84	2.65		
	13.2	0.01894	21.40	−14.44	6.54		
	13.8	0.01812	19.67	−21.36	9.91		
	14.1	0.01773	18.73	−25.12	13.08		
14.91	10.9	0.02294	12.47	−16.37	1.36	1.29	11.53
	11.5	0.02174	12.27	−17.71	0.33		
	12.0	0.02083	12.14	−18.58	1.24		
	12.8	0.01953	11.83	−20.67	1.83		
	13.4	0.01866	11.46	−23.15	5.41		
	14.3	0.01748	11.18	−25.03	10.37		
12.51	11.2	0.02232	10.15	−18.86	1.87	1.93	20.84
	11.35	0.02203	9.85	−21.26	0.41		
	11.45	0.02183	9.69	−22.53	0.62		
	11.6	0.02155	9.46	−24.37	1.17		
	11.8	0.02119	9.40	−24.85	8.48		
	12.2	0.02049	9.22	−26.29	16.09		
11.91	10.9	0.02294	9.51	−20.16	0.84	0.52	5.78
	11.1	0.02252	9.38	−21.25	0.32		
	11.6	0.02155	9.30	−21.92	3.32		
	11.7	0.02137	9.09	−23.69	4.40		
	12.4	0.02016	8.99	−24.53	11.21		

TABLE IV.—Showing the Effect of Speed of Rotation of the Disc on the Type and Amount of Disagreement of the Results by the Equality of Brightness and Flicker Methods.

Color: Green (515 $m\mu$)

Photometric rating Equality of brightness method	Revolutions per sec. Flicker disc	Length of individual exposures	Photometric rating Flicker method	Difference between the two methods	Sensitivity Flicker method	Change produced by variation of speed	
m. c.		sec.	m. c.	per cent	per cent	m. c.	per cent
49.86	11.9	0.02101	43.36	—13.04	1.45	6.17	16.60
	12.3	0.02033	42.01	—15.74	0.47		
	13.4	0.01866	39.86	—20.06	2.31		
	14.3	0.01748	37.62	—24.55	7.55		
	14.8	0.01689	37.19	—25.41	14.20		
29.39	12.0	0.02083	24.69	—15.99	1.82	2.91	13.36
	12.4	0.02016	23.80	—19.02	0.55		
	12.7	0.01969	23.48	—20.19	3.07		
	13.2	0.01894	22.76	—22.56	6.76		
	13.5	0.01852	21.78	—25.89	14.00		
26.12	11.5	0.02174	22.20	—15.01	1.76	3.41	17.63
	11.9	0.02101	21.71	—16.89	0.45		
	12.6	0.01984	21.22	—18.77	2.54		
	13.3	0.01880	21.04	—19.46	4.99		
	13.7	0.01825	19.92	—23.75	11.04		
	14.4	0.01736	19.35	—25.92	11.94		
25.01	10.5	0.02381	21.40	—14.44	2.52	5.71	36.37
	11.8	0.02119	20.72	—17.16	0.34		
	12.35	0.02024	20.51	—18.00	2.49		
	12.75	0.01961	18.46	—26.20	6.23		
	13.5	0.01852	17.86	—28.60	7.45		
	14.1	0.01773	15.69	—37.28	7.89		
14.91	11.3	0.02212	13.70	— 8.10	1.10	1.23	9.86
	11.5	0.02174	13.60	— 8.79	0.37		
	11.9	0.02101	13.51	— 9.39	1.11		
	12.5	0.020	13.33	—10.60	3.30		
	13.0	0.01923	13.06	—12.41	3.90		
	13.6	0.01838	12.72	—14.69	8.17		
	14.0	0.01786	12.47	—16.37	11.95		
12.51	11.1	0.02252	11.96	— 4.40	0.68	1.12	10.33
	11.3	0.02212	11.92	— 4.72	0.34		
	11.6	0.02155	11.72	— 6.31	0.69		
	12.2	0.02049	11.39	— 8.95	3.07		
	12.6	0.01984	11.11	— 9.11	9.27		
	13.4	0.01866	10.84	—13.34	13.29		
11.91	10.8	0.02315	11.99	+ 0.67	1.08	1.48	14.07
	11.2	0.02232	11.91	0.0	0.34		
	11.4	0.02193	11.79	— 1.01	0.42		
	11.6	0.02155	11.52	— 3.28	1.74		
	11.8	0.02119	11.43	— 4.03	3.50		
	11.9	0.02101	11.25	— 5.54	5.96		
	12.4	0.02016	10.51	—11.76	12.55		
10.20	10.6	0.02358	11.39	+11.66	0.97	1.19	11.66
	11.0	0.02273	11.18	+ 9.60	0.36		
	11.5	0.02174	10.77	+ 5.59	1.21		
	11.8	0.02119	10.51	+ 3.01	2.28		
	12.4	0.02016	10.20	0.0	6.76		

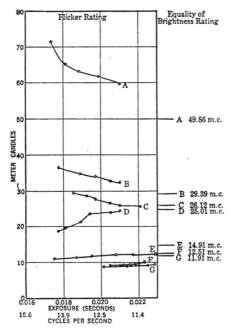

Fig. 14.—Effect of, speed of rotation of the disc for yellow.

method of flicker at the different speeds would not have been given these ratings. The intensities used then for the white light in this study of the effect of speed are only approximately represented by the curves for the rise of sensation for white light at 50, 25 and 12.5 m. c.

The Agreement of the Results by the Flicker and Equality of Brightness Methods when the same Lengths of Exposure are used in both Methods.

As a final proof that the discrepancy in the results between the flicker and equality of brightness methods is due to the difference in the time the eye is exposed to the lights compared by the two methods, it was decided to make the comparison using

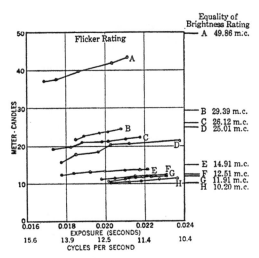

Fig. 15.—Effect of speed of rotation of the disc for green.

equal lengths of exposure by both methods. That is, the exposure of the eye in the equality of brightness method was cut down to equal that used in the method of flicker. These short exposures were given by means of the rotary tachistoscope described in a former section, provided, however, with only one pair of sectored discs. The photometric field was the same as was used in the previous equality of brightness comparisons. Some apprehension might be had as to the sensitivity of the equality of brightness comparison with these short exposures. The judgment, however, was not difficult because the saturation of the colors was greatly reduced by the brevity of the exposure and the brightness comparison was rendered much less uncertain thereby. The sensitivity of the determination ranged from 1.3 to 2.7 per cent.

The determinations were made for these values of exposure at 50, 25 and 12.5 m. c. In every case the flicker rating fell within the small range of values that were called equal in the short exposure equality of brightness judgment. The results of this work are given in Table VI.

TABLE V.—Showing the Effect of Speed of Rotation of the Disc on the Type and Amount of Disagreement of the Results by the Equality of Brightness and Flicker Methods.

Color: Blue (466 mμ)

Photometric rating equality of brightness method m. c.	Revolutions per sec. flicker disc.	Length of individual exposures sec.	Photometric rating flicker method m. c.	Difference between the two methods per cent	Sensitivity flicker method per cent	Change produced by variation of speed m. c.	per cent
49.86	12.0	0.02083	42.52	—14.72	2.28	5.33	14.34
	12.3	0.02033	41.55	—16.67	0.58		
	12.9	0.01938	41.07	—17.63	1.80		
	13.2	0.01894	40.33	—19.11	3.45		
	13.9	0.01799	39.40	—20.98	4.52		
	14.4	0.01736	38.49	—22.80	9.65		
	16.2	0.01543	37.19	—25.41	22.57		
29.39	11.7	0.02137	22.96	—21.87	1.78	3.87	20.28
	12.0	0.02083	22.55	—23.27	0.53		
	12.4	0.02016	21.87	—25.58	1.28		
	12.8	0.01953	21.48	—26.91	6.10		
	13.3	0.01880	21.22	—27.79	7.30		
	13.9	0.01799	21.13	—28.10	10.64		
	14.3	0.01748	19.09	—35.04	17.82		
26.12	11.85	0.02110	20.24	—22.52	1.19	3.20	18.78
	12.2	0.02049	20.00	—23.43	0.35		
	12.8	0.01953	19.75	—24.40	1.21		
	13.4	0.01866	18.88	—27.73	6.89		
	13.9	0.01799	18.14	—30.56	8.87		
	14.4	0.01736	17.04	—34.78	18.49		
25.01	11.7	0.02137	19.10	—23.64	0.84	3.98	26.31
	12.0	0.02083	18.94	—24.28	0.32		
	12.2	0.02049	18.70	—25.24	0.80		
	12.45	0.02008	18.58	—25.72	2.37		
	13.0	0.01923	17.58	—29.72	3.81		
	13.4	0.01866	17.17	—31.36	4.72		
	13.7	0.01825	16.53	—33.92	9.80		
	14.0	0.01786	15.12	—39.56	25.58		
14.91	11.7	0.02137	15.01	+ 0.67	1.13	1.35	12.39
	12.0	0.02083	14.91	0.0	0.34		
	12.4	0.02016	14.66	— 1.67	0.69		
	12.9	0.01938	14.66	— 1.67	0.98		
	13.6	0.01838	14.66	— 1.67	4.05		
	14.1	0.01773	14.66	— 1.67	5.25		
12.51	11.4	0.02193	13.60	+ 8.71	1.03	1.20	8.82
	11.6	0.02155	13.79	+10.23	0.36		
	12.2	0.02049	14.18	+13.34	1.27		
	12.8	0.01953	14.45	+15.50	3.11		
	13.4	0.01866	14.66	+17.18	5.25		
	14.2	0.01761	14.66	+17.18	7.44		
11.91	11.6	0.02155	12.89	+ 8.23	1.09	1.39	10.79
	11.8	0.02119	13.20	+10.83	0.23		
	12.35	0.02024	13.42	+12.68	2.01		
	13.0	0.01923	13.53	+13.61	5.76		
	13.6	0.01838	13.89	+16.63	11.45		
	14.7	0.01701	14.28	+19.91	17.15		

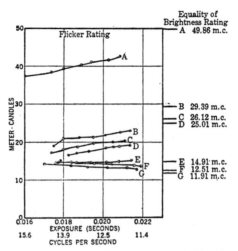

Fig. 16.—Effect of speed of rotation of the disc for blue.

TABLE VI.—Showing Agreement of the Results by the Flicker and
Equality of Brightness Methods When the Same Length
of Exposure is Used in Both Methods.

Color	Photometric rating equality of brightness method m. c.	Photometric rating flicker method m. c.	Length exposure flicker method sec.	Range of values judged equal by equality of brightness method with exposures equal to those used in flicker method m. c.
Red (675 $m\mu$)	49.86	58.78	0.020	57.48—60.05
	25.01	23.37	0.02058	23.20—23.80
	12.51	9.69	0.02083	9.61—10.02
Yellow (579 $m\mu$)	49.86	61.74	0.01984	60.05—62.79
	25.01	24.02	0.02049	23.80—24.43
	12.51	9.85	0.02203	9.66—10.19
Green (515 $m\mu$)	49.86	42.01	0.02033	40.70—42.85
	25.01	20.72	0.02110	20.28—21.18
	12.51	11.92	0.02212	11.68—12.20
Blue (466 $m\mu$)	49.86	41.55	0.02033	40.60—42.83
	25.01	18.94	0.02083	18.65—19.44
	12.51	13.79	0.02155	13.07—14.17

DISCUSSION

LOUIS BELL: I think we owe a considerable debt of gratitude to the authors of this paper for a really laborious and careful investigation of these discrepancies in photometry which have often bothered us, and it is extremely interesting to me to learn of the very great importance of the length of exposure under the equality of brightness method, as leading to the one easiest way of bringing it into agreement with the flicker method. In other words, the glimpse, the brief glimpse of the equality of brightness gives the same balance perception that the flicker photometry does. That is correct, is it not?

C. E. FERREE: Yes.

LOUIS BELL: Why this should be perhaps may be difficult to see. I have long had an impression that the phenomenon of simultaneous contrast was the principal thing that stood in our way in equality of brightness work, and I was very much interested to hear Dr. Ferree say that the color difference was very much less and tended toward equality in this glimpse work.

I think the elegance and the ingenuity of the apparatus is as interesting as the novelty of the results.

EUGENE C. CRITTENDEN: Mr. Chairman, I think it is fair to say that we have had more theorizing on the matter of flicker and equality of brightness, in proportion to the amount of experiments done, than almost any other subject, and the Society therefore is especially indebted to the authors for the amount of data they have put before us, based on definite measurements. It is therefore especially to be regretted that the printed paper was not available sooner so that we might have had an opportunity to study and partially digest this material, in order to discuss it more intelligently. I am sure we can all join in congratulating the authors on the clear way in which it has been presented, but there is so much presented here, that, we can hardly absorb it, no matter how well it is stated. I have therefore only a few questions to ask and these may possibly be covered in the paper.

Most of us who have dealt with comparisons of lights of different colors have come to put great faith in having a large number of observers to base our ratings on, rather than taking a small number, and I would like to ask, in the first place, about the basis for the comparison of flicker values with those which are given for the equality of brightness: whether those comparisons are based on several observers or whether as I understood, it is simply the judgment of one observer. There is such a mass of material presented here and the amount of work involved is so great that I hesitate to make criticisms that would involve a larger amount of work to meet them. I trust the authors will have no thought that we do not appreciate the amount of work involved in connection with this paper. I wish merely to emphasize the danger of basing conclusions on measurements by a single observer.

Another question is as to the definite significance of the details of the curves given; that is, as I understand it, these curves are built up by successive steps in such a way that necessarily the uncertainty is cumulative. That is, one step is based on another, as I get the significance of the curves, and a certain amount of wandering back and forth of the values is therefore to be expected, as for instance, in Fig. 9 on page 183, showing the green and white curves; the results as printed in the summary are so consistent that I suppose those crossings may be significant. The question I wish to ask is whether the authors consider those details are definitely established, or whether a repetition of the work by the same observer, or similar curves made by other observers, would show the same peculiarities.

I have one other question, with regard to the significance of the method of exposure: If I understand the operation of the tachistoscope, one steps from a practically dark field to the field in which the lights are shown; and my query is whether stepping from one color to another—for instance, from a red to a green— would give an effect which could be calculated from these results obtained in stepping from a dark field to a light field; in other words, whether the state of adaptation which results when stepping from the dark field to a light one is a basis for judgment as to what will happen when one goes from one light field to another light field, where the difference is not in intensity but in color.

B. E. SHACKELFORD: I would like to ask one question, although it is perhaps, as Mr. Crittenden remarks, covered in the paper. The question refers to the curves shown in Fig. 5 where I notice that the rise in response for white is at the extreme of all of the colors. I am not sufficiently familiar with the subject to be able to interpret this in my own mind, but I had rather expected some sort of a mean. Perhaps it is a summation of the lower intensities of the individual colors which make up the total of the white. I would like some information on the point. Have you any physiological explanation for that fact?

If, for instance, blue sensation rose more rapidly than red sensation, if I may use a short terminology, would you not at first hand expect that with a light which was composed of both of those colors the sensation would rise at a rate somewhere between those original rates? At first glance it appears that the effect may be explained by variations in the slope of the curves since the white of the given intensity is a summation of the lower intensities of the component colors.

C. E. FERREE and GERTRUDE RAND: We reply with some hesitation to Dr. Bell's discussion because we are not at all sure that we have interpreted his meaning correctly as to the role color difference and simultaneous color contrast may play in the disagreement of the flicker and equality of brightness methods. It is not our belief that the short exposure used in connection with the equality of brightness method produces agreement with the flicker method because it reduces the color difference between the two fields. The agreement comes because the two sensations compared have the chance to rise to the same brightness level by the two methods when the same length of exposure is used in both methods. Color difference as a subjective factor renders the judgment of brightness equality difficult to make by the equality of brightness method. This affects the sensitivity and precision of the judgment of brightness equality but it can not, we believe, affect the underlying physiological balance. To use a physical analogy it affects the ease and precision of the reading of the balancing instrument but in no way does it affect the balance given by the instrument. The reduction of the color difference by the short exposure then, so far as color difference is a disturbing factor, makes the instrument easier to read. The instrument gives the same balance when used by the two methods only when the lights are allowed to act upon it the same length of time. This principal has its foundation in the nature of the eye as a balancing instrument, namely its selectiveness of response to time of exposure when the lights to be balanced differ in wavelength or composition. The phenomenon of color contrast enters into the procedure only insofar as it increases the color difference and makes the index point of the balance: equality of brightness, more difficult to detect.

Dr. Shackelford states that in Fig. 5 "the rise for white is at the extreme of all of the colors" and asks for a physiological explanation. We would point out in reply that the phenomenon noted by Dr. Shackelford, namely a faster rate of rise for white than the colors, occurred only for the upper part of the segment of the curve shown in Fig. 5; and that it does not occur for any of the other intensities the results of which are charted in the paper. We would point out also that 25 m. c., the intensity represented in Fig. 5, was the only intensity at which all of the colors were underestimated by the method of flicker as compared with the equality of brightness method. That phenomenon is just as peculiar as the one for which Dr. Shackelford asks an explanation. The factual explanation for it, and at present we can go no farther than this, is found in the phenomenon noted by him, namely, that for the length of exposure occurring in the use of the flicker method white is allowed to rise to a higher level than any of the colors. An inspection of the curves for the relative rates of rise at 25 m. c. shows that from the threshold up to 0.009 sec. white rises more slowly than red, green and blue and faster than yellow; from 0.009 to 0.0185 sec. it rises faster than green, blue and yellow and more slowly than red; and from this point up to 0.04 sec. it rises faster than any of the colors. We have no explanation to offer for the variations in the relative rates of rise for the different lights at this intensity or for any of the other intensities represented in the paper. The problem of explaining differential lag and its peculiarities is a subject for further inquiry and investigation. It does not come within the scope of the present paper. The relative rates of rise are accepted by us as facts and the facts are compared with the disagreement of the two types of photometric rating. The significant thing in this regard is that at the intensity 25 m. c. for the length of exposure used in the method of flicker (*ca.* 0.021 sec.) white has risen to a higher level than any of the colors and at this intensity all of the colors are underestimated as compared with white by the flicker method.

In reply to Mr. Crittenden's first question as to the number of observers used in the various parts of the work, we would state that one experienced observer was selected for the comparison of the amount and type of disagreement of the flicker and equality of brightness ratings with the difference in level to which the sensations are allowed to rise with the length of exposure used in the method of flicker. The comparison of these two sets of data obviously had to be made for the same observer if it were to have any direct bearing on the question whether the difference in lag for the sensations aroused by the two lights is the cause of the disagreement in the two types of photometric rating. Nothing could have been accomplished towards this end, for example, by attempting to compare the rise of sensation curves for one observer with an average of the photometric results of a group of observers. If a group of observers were to be used, the comparison between the rise of sensation and the photometric results would have to be made separately for each observer of the group if the results are to have any bearing whatever on the point in question. We would have been glad to have made such a study on a number of observers if the amount of work involved had not been prohibitive. It would seem to us that Mr. Crittenden's question implies a confusion between the analytical and explanatory type of problem which we were attacking and such problems as the determination of a result for the average eye, etc. For the other parts of the work it was possible to use a plurality of observers. Up to the present time the results obtained for the effect of intensity have been verified as to their general features on six observers; for change of speed on 20 observers; and for the agreement of the flicker and equality of brightness ratings when the same length of exposure is used in both, on two observers.

Mr. Crittenden's second question pertains to safeguards in determining the rise of sensation curves. The determinations for only a small part of the curve could be made at a single sitting. How then could one put together in a single curve determinations made on different days and say that the differences found were due to changes in the length of exposure and not to some variations in the condition or sensitivity of the eye? Our reply is that in cases of this kind where a number of results taken at different times have to be compared with reference to the effect of a given variable, the same type of precaution has to be applied to the use of the eye as is exercised when a sensitivity tester is used in connection with a galvanometer or other physical instrument whose sensitivity is liable to vary from time to time. That is, on resuming the work at any particular sitting, it had first to be determined whether the curve made at the preceding sitting could be picked up or duplicated for a few points back. When this could be done we felt justified in assuming that the eye had not changed its characteristics of response with reference to the function investigated, and the work was allowed to continue on that day. When the curve could not be picked up, determinations were not made on that day. In no other way than this could curves of the regularity of those given in our paper have been obtained. Furthermore, each value that went into the curve as the increment of time required to produce a just noticeable change in sensation, was carefully checked up at each sitting by a number of repetitions. Done in this way, not more than five or six points of the curve were determined in a single day. Few pieces of work in physiological optics require more care than this work has required and few have received, or will receive, more care than it has received. In direct answer to Mr. Crittenden's question then we would say that we *do* consider these curves throughout definitely established for the observer in question. However, as we have stated in the paper, another observer or group of observers picked at random may be found who show a different type of result. But in case a different type of observer were found no conclusion could be drawn pertinent to the problem in hand until his curves for the rise of sensation were compared with his photometric results by the two methods. Curves very similar to ours in general characteristics were determined by M. A. Bills[27] in the Bryn Mawr laboratory in 1917 for lights of a lower range of intensities.

We interpret Mr. Crittenden's third comment, which has been somewhat difficult for us to understand, as follows: The rise of sensation curves show the relative levels attained by the two sensations only during the first exposure given by the flicker disc. Can it be assumed that these relations of level persist after the second, third, etc., exposure? In short, are the relative levels to which the two sensations rise in the single exposure modified by the succession of exposure occurring in the use of the flicker method? It was just to answer this question that we planned our last experiment,—a comparison of the results of the flicker with the equality of brightness method when using for the equality of brightness method a single exposure of the length of the individual exposures occurring in the method of flicker. This, it seems to us, should give the answer direct to that question. That is, one of the terms compared was the photometric balance obtained by a succession of exposures; the other, that obtained by a single exposure of the length of the individual exposure occurring in flicker, as per conditions of the question. If agreement were obtained it seems fair to assume that the relation of levels attained by the two sensations in a single exposure could not have been materially altered from any cause *whatsoever* in the succession of exposures. This does not mean, of course, that neither sensation has risen to a higher level in the succession of exposures than it attained in the single exposure. It means only that in rising, the relation of level has not changed by a detectable amount,—in other words, that the sensation which lags the more in a single exposure has not gained on the other sensation in the succession of exposures. Why should it? If an observer should be found for whom consistently no disagreement occurs in the two ratings,—and we have never heard that an observer exists for whom this is true over any considerable range of wavelengths at any intensity of light for lights of any considerable dif-

[27] "The Lag of Visual Sensation in its Relation to Wave-length and Intensity of Light," Pyschological Monographs, XXVIII, 101 pp.

ference in wave-length,—there are of course two possibilities of explanation:—either that there is no differential lag in sensation for the lights of different wave-length for this observer or there is a compensating, differentially modifying or summating effect on the rates of rise in the succession of exposures. Of these two possibilities our results indicate that the former is by far the more probable. That is, at the intensities at which agreement occurred separately for each of the pairs of lights used, the sensations had risen in the single exposure, so far as could be told, to the same level.

In brief, the comparison of the difference in level to which the two sensations are allowed to rise with the type and amount of disagreement of the photometric ratings for the same observer, state of adaptation of the eye, etc., for the different intensities of light employed, furnishes strong circumstantial evidence in support of our theory that differential lag for wave-length and intensity is the cause of the disagreement in the photometric ratings. The argument is supplemented and completed by the concordant evidence given by the effect of change in the relative lengths of exposure given to the two lights on the type and amount of the disagreement; by the effect of change of speed of rotation of the flicker disc; and particularly by the fact that agreement was found when the same lengths of exposure were used in the two methods.

SPECIFICATIONS GOVERNING THE ACCEPTABILITY OF ELECTRIC TAIL LAMPS FOR MOTOR VEHICLES*

COMMITTEE ON MOTOR VEHICLE LIGHTING
ILLUMINATING ENGINEERING SOCIETY

Adopted December 29, 1922.

INTRODUCTION

These specifications are drawn up to apply only to the Illumination by electric lamps, of opaque registration number plates for use on automobiles, trucks tractors and motorcycles.

DEFINITION

Tail Lamp.—A lighting unit used to indicate the rear end of a vehicle by means of a ruby light and illuminate the registration number plate.

Wherever reference is made in these Specifications to the registration number plate it designates a rectangle 16 inches long and 6½ inches wide (the top on level with center of bolt hole slot) designed to contain the registration numbers. This plane is one-sixteenth-inch in front of the plane of front face of the plate holder.

General Construction.

The opening in the lamp through which the light passes to the registration number plate shall be covered with colorless glass and shall be sufficiently large so that the light will cover the entire surface of the plate.

---

*These Specifications were unanimously adopted at a joint meeting of the I. E. S. Committee on Motor Vehicle Lighting with the Lighting Division of the Standards Committee of the Society of Automotive Engineers on December 29, 1922, and were approved by the I. E. S. Council on January 18, 1923.

The lamp and plate-holder shall bear such relation to each other that the major portion of the light incident at any point on the registration plate shall make an angle of not less than eight degrees with the plane of the plate.

The lamp shall be weather and dust proof and so constructed as to withstand the shock and vibration to which it is ordinarily subjected in use.

LABORATORY TEST

Samples for Test.

The samples submitted shall be representative of the lamp as manufactured and as marketed, and shall include the bracket to which the registration number plate is to be attached.

Marks of Identification.

Each lamp submitted must bear a distinctive designation prominently and permanently indicating the name or type of the lamp.

If the registration number plate holder is detachable from the lamp it shall bear the same distinctive name or make as the lamp.

Incandescent Lamps.

Incandescent lamps used in connection with the laboratory test shall be of 2 candlepower, 6-8 volt rating, of standard manufacture for automotive service. They shall be operated at 2 mean spherical candlepower during the test. In the case of tail lamps involving the use of special incandescent lamps, such incandescent lamps shall be submitted together with any necessary accessories.

Illumination Test.

The lamp and registration number plate holder which is included with the sample submitted shall be tested, using an incandescent lamp as provided above. The apparent illumination as measured on white blotting paper shall not be less than 0.5 foot-candle at any point on the registration number plate.[1] The ratio of maximum to minimum illumination shall not exceed 30.

---

[1] Since these specifications were adopted, it has been pointed out that blotting paper is subject to variations, therefore this sentence will probably be changed to read: "The illumination shall not be less than 0.5 foot-candle at any point on the registration number plate.

Requirements for Acceptance.

No lamp shall be considered acceptable unless it conforms with the requirements for measured illumination and unless the ruby light is visible for a distance of at least 500 feet.

No lamp shall be considered acceptable which is found unsatisfactory for any of the following reasons: Unstable or bad mechanical construction. Unduly dark or bright areas or excessive contrast in the illumination on the registration number plate. Cut-off of illumination within 1½ inches of the plate measured perpendicular to the plane of the plate at the edge furtherest from the lamp.

ABSTRACTS

In this section of the TRANSACTIONS there will be used (1) ABSTRACTS of papers of general interest pertaining to the field of illumination appearing in technical journals, (2) ABSTRACTS of papers presented before the Illuminating Engineering Society, and (3) NOTES on research problems now in progress.

THEORY, CONSTRUCTION, AND USE OF THE PHOTOMETRIC INTEGRATING SPHERE*

BY E. B. ROSA AND A. H. TAYLOR

During the last fifteen years great progress has been made in the production and utilization of light. With this progress has come the necessity for greater precision in the measurement of light output, and the Ulbricht photometric integrating sphere has been extensively utilized for this purpose. It is the only available instrument by which the total light output of sources can be determined in one measurement. There is no complete treatment of the theory, construction, and use of the sphere, available in English. This paper has been written to meet that need, and should be of interest to all who are concerned with photometry.

The majority of the spheres in use to-day are built of shaped sheet metal segments fastened to structural steel. Such spheres can be made of fairly light but substantial construction, but they are expensive to build. In the smaller sizes a cheaper construction may be realized by the use of papier-maché globes such as are used in schools.

In 1915, the authors designed and constructed an 88-inch reinforced concrete sphere at the Bureau of Standards. Steel T-rails were shaped into circular arcs and fastened to metal rings at each end, thus forming ribs of a spherical framework. These ribs were fastened together with metal bars and expanded metal lath, and the whole was plastered with concrete in the proportion

* Scientific Papers, No. 447., U. S. Bureau of Standards, Washington, D. C.

of one part Portland cement to two and one-half parts sand. The interior was swept out to a true spherical shape by means of a special sweep hinged at the centre of a pipe placed on the vertical diameter of the sphere. Later both interior and exterior surfaces were finished with white cement, giving a wall about one to one and one-half inches thick. A concrete sphere such as this, especially in the larger sizes, can probably be constructed more economically than a metal sphere of the same size.

This sphere has been thoroughly tested and found to give accurate results for many types of light sources. The sources tested included vacuum lamps both bare and with several types of glass reflectors and globes and a metal reflector. In all cases the agreement of the results with those obtained by point-by-point integration was practically within the range of ordinary errors of photometric measurements. Considerable care is necessary in the use of the sphere for measuring light sources which differ appreciably in their absorption of light, and in order to obtain reliable results on such sources proper corrections must be determined and applied.

The requirements for a satisfactory paint for photometric spheres are rather severe, and since no commercial paints fulfilling these requirements could be found, it was necessary to develop one. A good paint for this purpose is composed of alcohol, camphor, celluloid, and zinc oxide. It has a very high reflection factor, and does not discolor with age. The photometric equipment consists of a bar photometer, Lummer-Brodhun contrast photometric head, sectored disks, recording drum, and magnet, etc. The "substitution" method of photometry is used.

The theory of the sphere has been investigated, and tests have been made to verify certain phases of it, with satisfactory results. The best size and position of the opaque screen have been theoretically determined. It is found that as the lamp approaches the sphere wall at a point screen from the observation window, the brightness of the latter is reduced. This is due to the fact that as the lamp approaches the wall more of the direct light flux falls on the zone screened from the observation window, and hence it must be reflected from the sphere wall at least twice before it can reach the window.

One source of error in the use of such spheres is often overlooked. If the lamp being tested is somewhat blackened, or has parts which absorbed a considerable portion of light, the amount of light which it absorbs in the sphere will increase as it approaches the sphere wall. The amount of light absorbed by the lamp is also greater when the reflection factor of the sphere walls is high. These facts are of great importance if life-test lamps, appreciably blackened, are being photometered. It can be shown theoretically that a sphere to be used for life-test measurements of incandescent lamps should be painted with a paint having a reflection factor of about 80 to 85 per cent in order to minimize the errors of measurement.

The theory of the sphere assumes that its surface is a perfect diffuser, that is, that the specific brightness of any area is independent of the angle of view. Measurements of the diffusion characteristics of the sphere paint and the milk-glass observation window have been made. These show that the error arising from the measurement of a narrow beam of light in terms of a perfectly diffused light, or of light uniformly distributed in the sphere, is practically zero when the beam is directed at the sphere wall at a point about 80° on the circumference from the window.

The original paper also contains a bibliography of the subject.

AUTOMOBILE TAIL-LIGHTS

BY M. LUCKIESH*

It seems to me that the matter of tail-lights for automobiles should be given serious consideration as to color, brightness, and provision for illuminating the rear license number. I see no reason for a colored tail-light at all and I am in favor of a dense diffusing glass for the tail-light with adequate provision for lighting the license number.

*Director of Applied Science, Nela Research Laboratories, Nela Park, Cleveland, Ohio.

As we look down any of our city streets at night we are likely to see many colored lights which cause confusion and decrease the effectiveness of the colored lights which are actually necessary. Anything that can be done to lessen the number of colored lights encountered by the autoist, and particularly to retain red as a real danger signal, appears to me to be a great step in the right direction. The rear end of an automobile is not the dangerous end and therefore the use of a red tail-light is almost ridiculous. However, there are many danger points along our highways and especially along city streets where a red light must be used.

Some might favor the use of another color such as canary if the red tail-light is eliminated. However I see no reason at all for a colored tail-light. Red has the practical advantage of being very easily produced in glass and owing to certain characteristics of color vision that we encounter various red glasses appear approximately the same in hue. This is not so generally the case with any other color, and therefore if we would adopt yellow or green for example for the tail-light we would encounter many variations of these. I can see no reason for a colored tail-light, and many reasons for using an ordinary white light.

DEVELOPMENT, PRESENT STATUS AND PROBLEMS OF ELECTRICAL ILLUMINANTS

BY H. LUX

The author outlines the historic development of arc and incandescent lamps. It is shown that it is economically wrong to try to increase the amount of light from a filament by an increase of its temperature, because a visual efficiency of only 14.5 per cent can be reached theoretically. With increasing temperature the maximum of radiation is gradually shifted from the long-wave end of the spectrum into the short-wave end. At 3,600 deg. the radiation enters into the just visible red end at 800 $m\mu$ and leaves the visible end at 7,200 deg. at the violet end, at 400 $m\mu$. The highest visible efficiency is reached at 5,200 deg. at a wave-length of 556 $m\mu$. If we should succeed in developing a radiator emitting only visible rays, the highest possible light energy would be reached at 4,250 deg. with 248 lumens per watt. A light source with 100 per cent efficiency could be obtained from a radiator giving off rays near the 556 $m\mu$ maximum. Such a light would give the highest possible output—about 650 lumens for each watt input. But we are far from having reached the ideal light emission of 247.5 lumens per watt. The flaming arc gives 40 lumens per watt, and the mercury vapor lamp in a quartz tube 54 lumens per watt, but their light is quite disagreeable. Another matter of great importance is the light density. It rises enormously with the temperature. With a black body radiating at 6,500 deg. a density of 73,000 cp. to 150,000 cp. per square centimeter is obtained for a bright body or other selective radiator. Such a terrific density would cause instant blinding. A flaming arc gives a density up to 1,000 cp. per square centimeter, which is so high that artificial means have to be employed to lessen the glare. New and more promising ways of obtaining the ideal source of light are shown by the modern electronic theories. If electrons are emitted from a hot cathode and allowed to hit the atoms of a vapor or gas at low pressure, they will be reflected until the reflection potential is reached, when the entire energy of the impulse is absorbed and the radiation of a spectral line is generated without ionization. If now the pressure is increased to the ionization value, all electrons will be driven out of the atom and will unite to form a new atom, forming a complete spectrum. Assuming, for example, a tungsten cathode heated to 2,800 deg., an electron stream of 8.4 amp. per square centimeter is available, producing in the ideal gas a light at 2.2 volts, resulting in 12,300 lumens. To heat the cathode 126 watts will be required, resulting in 2,300 lumens, which would give a total of 12,300 + 2,300 ÷ (18.5 + 126), or 101 lumens per watt.— *Elektrotechnische Zeitschrift*, Nov. 23 and Dec. 7, 1922.

SOCIETY AFFAIRS

SECTION ACTIVITIES

CHICAGO Meeting—January 9, 1923

A joint meeting of the Chicago Section and the Illinois Chapter of the American Institute of Architects was held on January 9, at the Chicago Art Institute. Mr. A. F. Dickerson of the Illuminating Engineering Laboratory of the G. E. Co., Schenectady, presented a paper, "Exterior Building Illumination and Spectacular Lighting Effects." Some eighty-five members and guests were present and an interesting discussion was held.

PHILADELPHIA Meeting—January 9, 1923

The Philadelphia Section met at the Engineers' Club on January 9, to hear a paper, "Central Station Activities in Illuminating Engineering," presented by Mr. W. T. Blackwell of the Public Service Electric Light Co., Newark, N. J.

Mr. Blackwell stated that only nine utility companies had special departments for disseminating lighting information for the benefit of their customers. By means of lantern slides he showed the extent to which poor illumination existed and the possibilities for its improvement. It was mentioned that 80 per cent of industrial installations were inadequate. The speaker also showed diagramatically the sources from which Central Station customers obtain information relative to illumination and where they obtain their lamps and supplies. The responsibility for poor illumination was partly laid to' the electrical contractors.

The discussion brought out the point that electrical contractors could not seek to improve poor existing illumination on account of the labor and time involved in missionary work and the small financial returns. This is especially true when with much smaller efforts they can obtain more lucrative contracts for new lighting and power installations. Thirty-five members and guests were present at the meeting.

NEW YORK Meeting—January 11, 1923

A very interesting meeting was held by the New York Section on the evening of January 11, at the auditorium of the Railroad Branch, Y. M. C. A., 309 Park Avenue. The general topic of interest was Theatre Lighting, and two extremely interesting papers were presented. "Artistic Color Lighting in Motion Pictures Theatres" was described in an original manner by Mr. Sam Rothafel, Managering Director of the Capitol

Theatre and a paper, "Lighting Effects on the Stage,' was read by Mr. Louis Hartmann, Lighting Engineer for the David Belasco Productions.

In discussing the papers, Mr. W. H. Hall of the New York Calcium Light Co., gave some reminiscences of the days prior to the use of the carbon arc and the incandescent lamp when the calcium lights played an important part in stage lighting.

There were in attendance some two hundred and twenty-five members and guests.

Section Plans—The plans for subsequent meetings of the year for the New York Section are quite definite, although not entirely arranged.

In February, it is expected that a meeting will be held at the new American Art Association Galleries, Madison Ave. and 57th St., with a paper by Mr. Champeau on "The Lighting of Art Galleries." The speaker planned and installed the novel and effective installation used in these galleries and the meeting will give the members an opportunity to observe the results. Dr. C. H. Sharp has just returned from a trip abroad and will probably address the Section, giving comments on his observations of foreign lighting practice.

In March, the general question of hospital lighting will be discussed. Considerable investigation work has been done on the effect of various colors of room finish on the sensibilities of the patients. It is hoped that a paper will be available by the physician who conducted this research. Artificial daylight is now being successfully applied for the illumination of the operating table. The methods used and results secured will be demonstrated. It is possible that a third talk will be given on the other phases of hospital illumination.

In April, the general subject of office building and school illumination is scheduled. The results of some very comprehensive tests on the choice of equipment and maintenance characteristics will be presented.

For May, it is planned to hold the meeting in Paterson, N. J., the silk manufacturing center of the country. The membership will have an opportunity of inspecting actual installations of mercury-vapor and incandescent lamps in some of the silk mills and papers will be given on the methods of illumination.

It can be seen from the above, that the applied engineering phase of the Society's activities is stressed and the meetings are arranged to be especially interesting to the "men in the field" and be of real practical value.

CLEVELAND Meeting

The first meeting of the year for the Cleveland Chapter was held on January 23, at the Electrical League Rooms in the Hotel Statler. The subject of "Better Lighting Fixtures" was presented by Mr. M. Luckiesh of the National Lamp Works.

COUNCIL NOTES

ITEMS OF INTEREST

At the meeting of the Council, Cleveland O., January 18, 1923, the following were elected to membership:

Six Members

ANDERSON, J. F., Southern California Edison Co., 3rd & Broadway, Los Angeles, Cal.

DONLEY, H. B., Jno. W. Brown Mfg. Co., Marion Road, South Columbus, Ohio.

GILLIAM, H. R., Lancaster Lens Co., Lancaster, Ohio.

HIGGINS, A. W., So Minn. Gas & Electric Co., 110 S. Dearborn St., Chicago, Ill.

IVES, JAMES E., DR., U. S. Public Health Service, 16 7th St., Washington, D. C.

PULLEN, M. W., Johns Hopkins University, Baltimore, Md.

Thirty-Six Associate Members

ANDREWS, EDGAR M., 15 North 12th St., Richmond, Va.

BOWERS, CORWIN J., Columbus Ry. Power & Light Co., 102 N. 3rd St., Columbus, Ohio.

BROWN, RICHARD B., JR., Edison Elec. Illuminating Co. of Boston, 39 Boylston St., Boston, Mass.

CAIRNIE, H. W., Western Elec. Co., 385 Sumner St., Boston, Mass.

CALAHAN, H. C., Tri-City Electric Co., 14 Mechanic St., Newark, N. J.

CAMPBELL, JAMES P., Illuminating Company, 75 Public Square, Cleveland, O.

COUNSELLOR, W. M., Columbus Ry. Power & Light Co., 102 N. 3rd St., Columbus, O.

CROFOOT, CLARENCE E., Utica Free Academy, Kemble St., Utica, N. Y.

DOWNS, W. S., Public Service Elec. Co., 695 Bloomfield Ave., Montclair, N. J.

EVANS, DEAN P., Columbus Ry. Power & Light Co., 102 N. 3rd St., Columbus, O.

FERRIS, C. H., Illuminating Glassware Guild, 19 W. 44th St., New York, N. Y.

FREI, FRED J., 351 Lexington Avenue, New York, N. Y.

FULTON, ROBERT A., Cleveland Elec. Illuminating Co., 75 Public Square, Cleveland O.

JOHNSTON, H. L., Westinghouse Elec. & Mfg. Co., East Pittsburgh, Pa.

LENT, A. L., Detroit Edison Co., 2000 Second Ave., Detroit, Mich.

MANTER, E. W., Westinghouse Elec. & Mfg. Co., 10 High St., Boston, Mass.

McDonough, Thomas F., Benjamin Elec. Mfg. Co., 1225 L. C. Smith Bldg., Seattle, Wash.

McParland, John J., Pittsburgh Reflector & Illuminating Co., 1452 Broadway, New York, N. Y.

Nast, Cyril, New York Edison Co., 130 E. 15th St., New York, N. Y.

Nieder, Edward J., The Illuminating Co., Illuminating Bldg., Public Square, Cleveland, O.

Norton, C. A., Westinghouse Lamp Co., 165 Broadway, New York, N. Y.

O'Rourke, John A., Columbus Ry. Power & Light Co., 102 N. 3rd St., Columbus, O.

Outley, E. J., Detroit Edison Co., 2000 Second Blvd., Detroit, Mich.

Piez, Karl Anton, Columbus Ry. Power & Light Co., 102 N. 3rd St., Columbus, O.

Plaisted, L. H., Columbus Ry. Power & Light Co., 102 N. 3rd St., Columbus, O.

Poey, Charles, D., N. Y. & Queens Elec. Light & Power Co., Electric Bldg, Bridge Plaza, Long Island City, N. Y.

Prichett, George P., National X-Ray Reflector Co., 628 Witherspoon Bldg., Philadelphia, Pa.

Scheel, Alfred A., Lee Electric Co., 605 Davis Ave., Corning, Iowa.

Shaw, Harold, Electrical Extension Bureau, 613 Lincoln Bldg., Detroit, Mich.

Stolpe, Ernest F., United Gas Improvement Co., 20 W. Maplewood St., Philadelphia, Pa.

Story, Ernest D., Westinghouse Lamp Co., 165 Broadway, New York, N. Y.

Sulzer, G. A., Ophthalmologist, 327 E. State St., Columbus, O.

Teller, J. Paul, Westinghouse Elec. & Mfg. Co., 814 Ellicott Square, Buffalo, N. Y.

Thomson, Geo. L. A., Public Service Electric Co., 102 River St., Newark, N. J.

White, Albert H., Public Service Elec. Co., 80 Park Place, Newark, N. J.

Young, Fred R., Sieck-Boyington Elec. C., 209 W. Galena St., Freeport, Ill.

Six Sustaining Members

Boss Electrical Supply Co., 11 Peck St., Providence, R. I.
F. A. Boss, Official Representative

Fall River Electric Light Co., 85 N. Main St., Fall River, Mass.
R. F. Whitney, Official Representative.

Greenfield Elec. Light & Power Co., 41 Federal St., Greenfield, Mass.
H. E. Duren, Official Representative.

Lawrence Gas Co., 370 Essex St., Lawrence, Mass.
Fred H. Sargent, Official Representative.

PITTSFIELD ELECTRIC Co., Pittsfield, Mass.

E. P. Dittman, Official Representative.

SLOCUM & KILBURN, 23 N. Water Street, New Bedford, Mass.

A. H. Smith, Official Representative.

Two Transfers to Full Membership

KELCEY, GEORGE G., American Gas Accumulator Co., 999 Newark Avenue, Elizabeth, N. J.

SMITH, ELMER A., Warren & Wetmore, 16 E. 47th St., New York, N. Y.

One Transfer to Associate Membership

DONAHUE, E. J., Public Service Gas Co., St. Pauls & James Aves., Jersey City, N. J.

The General Secretary reported the death of two associate members,

WALLACE P. HURLEY, Westinghouse Elec. & Mfg. Co., 165 Broadway, New York, N. Y. and FREDERICK W. PRINCE, Western Electric Co., 195 Broadway, New York, N. Y.

CONFIRMATION OF APPOINTMENTS

The appointment of the following committee chairman and members were confirmed :

As Members of the Committee on Time and Place—Samuel G. Hibben, Chairman; L. B. Marks, and G. S. Crampton.

NORTHERN NEW JERSEY CHAPTER

Upon recommendation of the Committee to investigate the petition for the Northern New Jersey Chapter the Council voted to establish this Chapter. The center of the organization will be in Newark and its efforts will be confined to educational work, disseminating information of an established character. The plan is to educate the lighting trades by joint meetings with their associations, and then the light-using public through joint meetings with clubs, boards of trade and other organizations. The attempt will be made to interest the public in good lighting practice by holding a series of meetings in Jersey City, Paterson, and other neighboring localities.

COLUMBUS CHAPTER

A petition to form a Chapter in central Ohio was presented by Mr. G. F. Evans, of Columbus, and was referred by the Council to the Executive Committee for action by letter ballot.

TAIL LAMPS SPECIFICATIONS APPROVED

The Council formally approved the Specifications governing the acceptability of Electric Tail Lamps for Motor Vehicles. These Specifications were unanimously adopted at a joint meeting of the I. E. S. Committee with the Lighting Division of the Standards Committee of the S. A. E. on December 29, 1922.

The Council also approved that the desired scope of the joint sponsorship with the S. A. E. should embrace Motor Vehicle Lighting and not be limited solely to Headlighting.

COMMITTEE ACTIVITIES

COMMITTEE ON PAPERS

The Committee on Papers invites communications from members wishing to offer papers for presentation at the 1923 Convention which will probably be held in September. It will be appreciated also if members will advise of other developments in the science or art of illumination which might with advantage be included in the program.

The character of material desired for the convention presentation as set forth in the Society's Manuel is as follows:

Material Desired.

1. Material not closely related to illuminating engineering.

2. Collections of material in new and useful forms.

3. Subjects and treatments which are peculiarly appropriate for presentation before the Illuminating Engineering Society.

Material Not Desired.

1. Material not closely related to illuminating engineering.

2. Material already available in satisfactory form.

3. Prolix treatment.

4. Papers or discussions having the manifest purpose of advertisement.

Communications should be addressed to Mr. J. L. Stair, Chairman, Committee on Papers, I. E. S., 235 West Jackson Blvd., Chicago, Ill.

COMMITTEE ON MOTOR VEHICLE LIGHTING

On December 29, 1922, the Committee met with the Lighting Division of the Standards Committee of the S. A. E., having invited state officials, lantern manufacturers, and others interested.

The following resolution, prepared by the sub-committee on Tail Lights, was unanimously adopted:

"Whereas, the various state laws require that the rear registration plates be illuminated so that they are legible at a distance of approximately sixty feet and

Whereas, it is impossible to accomplish the proper illumination of the registration plate with the present lighting equipment, and

Whereas, the various motor car manufacturers and manufacturers of lighting equipment need definite standardization of registration plates in order to comply, and

Whereas, the color combinations, size of plate, size of stroke and spacing are of vital importance,

Therefore, it is the sense of this joint meeting of the Lighting Committees of the I. E. S. and S. A. E. that no specification for lighting the registration plate can be practically applied unless there is a reasonable standardization of registration plates as regards:

1. Maximum dimensions of the area to be illuminated;
2. Design and spacing of digits;
3. Color contrast.

The joint committees further suggest that such standardization might be approximately as follows:

1. That the area to be illuminated be included in a rectangle not larger than 16 x 6.5 inches.

2. That the design and spacing of digits be such that:

(a) The stroke be not wider than 0.5 inch.
(b) The spacing between strokes be not less than 0.5 inch.
(c) That a space of not less than 0.75 inch be provided between each third digit.
(d) The opening in such figures as 3, 6, 9 and 5 be kept wide so as to avoid confusion with each other or with the figure 8.
(e) The colors be so selected as to secure high contrast between the numerals and backgrounds."

It is planned to co-operate with other organizations in calling this resolution to the attention of the motor vehicle administrators of the various states.

The committee also adopted, unanimously, the Specifications governing the Acceptability of Electric Tail Lamps for Motor Vehicles, which had been prepared by the sub-committee, Mr. A. W. Devine, Chairman.*

The committee considered an inquiry from the A. E. S. C. as to the desired scope of sponsorship established in connection with the recently accepted standard specifications for motor vehicle headlights. It was the sense of the Committee, the S. A. E. representatives concurring, that it should embrace motor vehicle lighting, and not be limited solely to headlighting.

COMMITTEE ON NOMENCLATURE AND STANDARDS

At the meeting of the committee held at the General Office on January 23, plans for the work of the committee for the year were considered. It was decided to call attention to the membership of the Society and others interested the status of the revision of the Illuminating Nomenclature and Photometric Standards.

* These specifications are printed on page 208 of this issue.

*Revision of Illuminating Engineering Nomenclature and Photometric
Standards.*

The work of the Nomenclature and Standards Committee of the
Illuminating Engineering Society has contributed very greatly to uniformity of practice in this field, and the result of this work in the form
of a report called "Illuminating Engineering Nomenclature and Photometric Standards" was approved as an American Standard by the American Engineering Standards Committee on July 11, 1922. Copies of this
standard are now obtainable on application to the office of the Illuminating
Engineering Society.

Further revisions of this standard will be made jointly by the I. E. S.
committee and a sectional committee organized under the auspices of
the American Engineering Standards Committee. A preliminary revision was presented as a part of the report of the 1922 I. E. S. committee at the Swampscott Convention last fall. This report has just
been issued in the January number of the TRANSACTIONS of the Society.
The committees engaged in this work wish to make the standard as
practical and as widely useful as possible, and in order to attain this end
they wish to obtain criticisms from all who are interested in such subjects. It is therefore requested that suggestions and criticisms be sent
either to the general offices of the Society, 29 West 39th Street, New
York City, or to the Secretary of the committee, Mr. Howard Lyon,
Welsbach Company, Gloucester, N. J.

COMMITTEE ON NEW SECTIONS AND CHAPTERS

Mr. D. McFarlan Moore, Chairman, reports that one of the most
hopeful signs of growth of the Society is the pending formation of
Chapters in the following cities: Detroit, Mich.; Columbus, Ohio, Los
Angeles, Cal.; Newark, N. J.; Pittsburgh, Pa.; Rochester, N. Y., and
South Bend, Ind.

NEWS ITEMS

COUNCIL MEETS AT CLEVELAND

The January meeting of the Council was held in Cleveland on January 18, during the week of the 1923 Lighting Fixture Market of the
Associated Lighting Industries. Prior to the meeting a Council luncheon
was served in the private dining-room of the Electrical League, Hotel
Statler.

THE CHICAGO LECTURES

A Course of Lectures on Approved Methods of Illumination was held
in the rooms of the Western Society of Engineers on January 22 to 26,
1923. The lectures were open to architects and architectural engineers
and were given under the joint auspices of Armor Institute of Technology
and the Illuminating Engineering Society. The following program was
arranged for presentation:

Monday, January 22, 1923, 1:30 to 4:30 P. M.

Opening Address ....*Dr. Howard M. Raymond, Chicago, Illinois*
President, Armour Institute of Technology
Response ....*Mr. Alfred H. Granger, Chicago, Illinois*
President, Illinois Chapter, American Institute of Architects
Response ....*Mr. Frank E. Davidson, Chicago, Illinois*
President, Illinois Society of Architects
The Relations of Illumination to Modern Architecture
Mr. Ward Harrison, Cleveland, Ohio
President, Illuminating Engineering Society
Demonstration of Fundamental Lighting Principles and their Application
Mr. Albert L. Arenberg, Chicago, Illinois
Chairman, Chicago Section, Illuminating Engineering Society

Tuesday, January 23, 1923, 1:30 to 4:30 P. M.

Fundamentals of the Science and the Art of Illumination
Mr. E. H. Freeman, Chicago, Illinois
Professor Electrical Engineering, Armour Institute of Technology
The Choice of Reflecting, Transmitting and Diffusing Equipment in Lighting Installations ....*Mr. Arthur J. Sweet, Milwaukee, Wisconsin*
Consulting Engineer

Tuesday, January 23, 1923, 5:30 to 8:30 P. M.

Dinner ....*Edison Building*
Library and School Lighting...*Mr. H. H. Higbie, Ann Arbor, Michigan*
Professor of Electrical Engineering, University of Michigan
Inspection of Office Buildings .... { *Illinois Bell Telephone Bldg.* { *Commonwealth Edison Bldg.*

Wednesday, January 24, 1923, 1:30 to 4:30 P. M.

Methods of Office Lighting and Illumination Problems
Mr. S. G. Hibben, New York City
General Secretary. Illuminating Engineering Society
Residence Lighting ....*Mr. A. L. Powell, New York City*
Chairman, New York Section, Illuminating Engineering Society

Thursday, January 25, 1923, 1:30 to 4:30 P. M.

Lighting of Interiors for Public Assembly
Mr. J. L. Stair, Chicago, Illinois
Past Chairman, Chicago Section, Illuminating Engineering Society
Modern Store and Show-Window Lighting
Mr. Norman Macbeth, New York City
Consulting Engineer

Thursday, January 25, 1923, 5:30 to 8:30 P. M.

Dinner ....*Engineers' Club*

Light and Color as Instruments of Expression

Mr. M. Luckiesh, Cleveland, Ohio

Specialist in Color and Light

Inspection Visits .... $\begin{cases} Hamilton\ Club \\ John\ Crerar\ Library \end{cases}$

Friday, January 26, 1923, 1:30 to 4:30 P. M.

Factory Lighting ....*Mr. Otis L. Johnson, Chicago, Illinois*

Vice-President, Illuminating Engineering Society

Design of Illuminating Units......*Mr. George Ainsworth, New York City*

Lighting Fixture Designer

DR. SHARP ATTENDS BRITISH I. E. S. MEETING

Through the courtesy of Mr. Leon Gaster, Honorary Secretary of The Illuminating Engineering Society of London, Dr. Clayton H. Sharp was privileged to attend a meeting of that Society which was held on December 12, 1922.

Two papers were presented; one on Street Lighting and Safety by Leon Gaster, and the other on Street Lighting Requirements by Haydn T. Harrison. The meeting was presided over by Mr. A. P. Trotter whose experience in matters of street lighting and in the measurement of street lighting probably reaches back as far as that of any man who is active in the field to-day, and whom it was a great pleasure to meet once more. The meeting was well attended, one characteristic of the audience being that the average age of those attending seemed to be considerably greater than in the case of our own I. E. S. meetings.

Mr. Gaster's able paper made a plea for a large increase in street illumination for the sake of the increased safety which would result therefrom, and stated, amongst other things, data which have been presented before our own Society in support of his thesis.

The paper by Mr. Harrison referred back to the investigations of the Joint Street Lighting Committee made in 1911 and 1912, as a result of which the method of classifying street illuminations on the basis of minimum illumination was advocated. Mr. Harrison pointed out the advantages of this system and presented a scheme for what he called "longitudinal system" of street lighting by which the required minimum could be obtained with the greatest efficiency. He also noted the work which has been carried on in this country in the investigation of street lighting, notably as reported by the Street Lighting Committee of the National Electric Light Association, and cited also the work of Messrs. Millar and Lacombe. The author was inclined to be critical of the National Electric Light Association because of its rejection of the minimum criterion and its advocacy of street lighting contracts based on lamps in service rather than on illumination results. The discussion of the paper was of a very able order indeed, being participated in largely by individuals who are responsible for the actual street lighting work in England and elsewhere in the United Kingdom. One gentleman had come all the way from Glasgow in order to participate in this discussion.

Dr. Sharp was cordially invited to join the discussion and was most kindly received although he was obliged to differ radically from the main point of Mr. Harrison's paper and to defend and explain the reason for the action of the National Electric Light Association.

The meeting was a most interesting one from all points of view, and most instructive as showing how efficiently the British Society is operating, and what great attention is being given to its work by the leaders of thought and of practice in this field. It is to be hoped that some time it may be possible to carry out a joint meeting of the British and American Illuminating Engineering Societies. This would be of the greatest interest and profit to both organizations.

I. E. S. EXHIBIT AT THE 1923 LIGHTING FIXTURE MARKET

Through the auspices of the Illuminating Glassware Guild, the Illuminating Engineering Society had an Exhibit at the 1923 Lighting Fixture Market, held at the Hotel Winton, Cleveland. Ohio, January 15th to 20th. The Exhibit Room was approximately 12 x 15 ft. in size, and sections of the display are shown on the accompanying photographs. An outstanding feature of the Exhibit, which was entirely of an educational nature, was that of attracting attention to the work of the Illuminating Engineering Society.

At the entrance of the room a changeable electric sign alternated between displaying the words ILLUMINATING ENGINEERING SOCIETY and BETTER LIGHTING EXHIBIT. At the end of the room opposite the entrance a large United States map, equipped with miniature lamps, indicated the location of I. E. S. Sections, Chapters, and Local Representatives. On one side of the room a display of wall brackets with reflectors and shades showed the ratings with regard to shading, diffusion, transmission, etc., of various lighting media, as given in the I. E. S. Tentative Code of Luminaire Design. On the other side of the room an exhibit showed the development in table lamps from the standpoint of shading the light source. On the wall directly over the table lamp exhibit a display was made of good and bad practice in the use of wall bracket equipment. A display was also made of some of the different shapes of glass reflectors as used in residences on pendent type fixtures. By means of an Attractoscope,

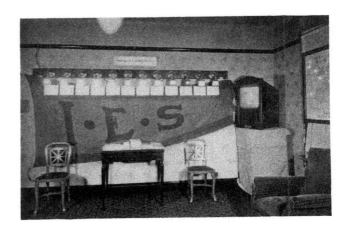

Views of the I. E. S. Educational Exhibit at the 1923 Lighting Fixture Market, Cleveland, Ohio.

pictures of good residence lighting were shown, and from a Glare Exhibit Box the uncomfortable lighting effect produced by a bare lamp located directly in the line of vision was demonstrated.

Three framed panels, each 2 x 3 ft., suspended from the wall mold·ing carried Lighting Legislation bulletins, Illuminating Engineering Society Codes, and other reports of the accomplishments of the I. E. S. committees. Copies of the Tentative Code of Luminaire Design, Illuminating Engineering Society TRANSACTIONS, and other literature pertaining to the work of the Society were available at the desk.

Mr. S. E. Doane, Chief Engineer of the National Lamp Works of General Electric Co., returned to this country on December 18, 1922, after a trip of three months in Europe. With headquarters in Paris he visited Milan, Budapest, Vienna, Geneva, Berlin and London.

A lecture to some three hundred members of the New York Training School for Teachers was given by Mr. S. G. Hibben on January 22, elaborating and explaining the I. E. S. Code of School Lighting.

A rather extensive portable industrial and commercial lighting demonstration which has been used by the Edison Lamp Works for the last two years has been in use this season in New England and at the various properties of the Consumer's Power Company in Michigan. This exhibition is under the direct charge of Messrs. W. H. Rademacher and R. E. Greiner who have been on such work ever since its inception and are experienced and forceful lecturers.

Mr. A. B. Oday addressed the class in Artificial Illumination under the direction of the Board of Education of New York City through Mr. F. J. McGuire on January 23 on "The Characteristics of Reflecting and Diffusing Media."

The Senior Electrical Engineers at New York University, in connection with their course on illumination by Prof. S. K. Barrett, visited the Harrison plant of the Edison Lamp Works on January 4. A trip was made through one of the lamp factories and the remainder of the afternoon was spent in demonstrations and talks by members of the Commercial Engineering and Lighting Service Departments.

Mr. A. L. Powell, addressed the Senior Electrical Engineers at Brooklyn Polytechnic Institute on "The Principles of Good Lighting" on January 9 and also delivered a talk at the Southern Association of Building Owners and Managers, Atlanta, Georgia, on "How to Make Office Lighting Satisfactory to Both Tenant and Owner" on January 16.

GENERAL OFFICE NOTES

DELINQUENT DUES

Some of our members are still in arrears for their dues for the present fiscal year. According to our constitution, these dues are payable in advance and become due on October first. One of our active members suggests the insertion of the following reminder:

"The Illuminating Engineering Society *absorbs* its income in the *diffusion* of *light* on *lighting*, so please duly *reflect* and then *transmit* your dues to the General Office, thus preventing *interference* of the propagation of *light* and *illumination.*"

ADDRESSES UNKNOWN

The General Office would appreciate any information from our members in regard to the present location of the following:

Latest Address Known

Maurice P. Brogan	Lumen Unit Company, Cleveland, Ohio.
Willis G. Gordon	Pacific States Electric Co., Seattle, Wash.
R. L. Sunstedt	Westinghouse Elec. & Mfg. Co., Seattle, Wash.

RECENT REPRINTS

There can be obtained from the General Office the following reprints of the Society's activities. Price per copy listed below:

"Code of Lighting School Buildings," as adopted in 1918. 25 cents.

"Code of Lighting Factories, Mills and Other Works Places." (American Standard). 25 cents.

"Illumination Engineering Nomenclature and Photometric Standards." (American Standard). 15 cents.

"Specifications of Laboratory Tests for Approval of Electric Head-lighting Devices for Motor Vehicles." (Tentative American Standard). 15 cents.

"Tentative Code of Luminaire Design." 25 cents.

OLD TRANSACTIONS FOR SALE

The committee on Editing and Publication is able to secure a limited number of various back issues of the TRANSACTIONS beginning with Vol. III, 1908. It is suggested to those members and others interested in completing former volumes to communicate immediately their wishes to the committee.

Unless a favorable demand is evidenced these former issues will not be acquired. Inquires should be addressed to Mr. N. D. Macdonald, Chairman, I. E. S. Committee on Editing and Publication. 29 West 39th Street, New York City.

ILLUMINATION INDEX

PREPARED BY THE COMMITTEE ON PROGRESS.

An INDEX OF REFERENCES to books, papers, editorials, news and abstracts on illuminating engineering and allied subjects. This index is arranged alphabetically according to the names of the reference publications. The references are then given in order of the date of publication. Important references not appearing herein should be called to the attention of the Illuminating Engineering Society, 29 W. 39th St., New York, N. Y.

National Safety News

Latest Practices in Factory Lighting—	G. H. Stickney	Sept.	36

Philosophical Magazine

The Measurement of Light—	John W. T. Walsh	Dec.	1165

Physical Review

A Study of the Exciting Power for Fluorescence of the Different Parts of the Ultraviolet Spectrum—	Leland J. Boardman	Dec.	552
Low Voltage Arcs in Diatomic Gases—	O. S. Duffendack	Dec.	665

Physikalische Berichte

Synet i Tusmorke (Color Physiology)	M. Tscherning	Nov. 14	1190

Popular Mechanics Magazine

Automobile Headlight Controller Dims Lights Gradually—		Dec.	879
Building Columns, Darkened, Illuminated by Strong Lights—		Dec.	848

Proceedings of the National Academy of Sciences

A Direct Method of Testing Color Vision in Lower Animals—	W. F. Hamilton	Dec.	350

Proceedings of the Royal Society

On the Fluorescence of Aesculin—	J. C. McLennan and F. M. Cale	Dec. 1	256
An Investigation of the Colour Vision of 527 Students by the Rayleigh Test—	R. A. Houstoun	Dec. 1	353

Railway Electrical Engineer

Power and Lighting Facilities on the Erie—		Nov.	373
Steel Car Lighting Fixtures—		Nov.	398
Principles of Car Lighting by Electricity—		Dec.	411

Revue Generale de L'Electricite

Le balisage lumineux—	E. Marcotte	Nov. 11	148D

Science

Colors for Traffic Systems—		Dec. 8	652

Science Abstracts

Phototropy and Photoelectric Effect (from *Soc. Chim. de France*, Bull. iv, 29, pp. 961-976, Nov. 20, 1921)—	P. H. Gallagher	Oct. 31	754
Photophoresis (from *Phys. Math. Soc. Japan*, Proc. 4, pp. 67-70, April-May, 1922)—	T. Terada	Oct. 31	755

*Not previously reviewed.

*Not previously reviewed.

TRANSACTIONS
OF THE
ILLUMINATING ENGINEERING SOCIETY

VOL. XVIII   MARCH, 1923   No. 3

Some European Impressions

IN ONE BRIEF period of three months abroad, which permits hardly more than a week in each country, one can obtain
very little real information on conditions excepting through the
courtesy of friends. It was my fortune to have good friends in
England, France, Italy, Austria, Hungary, and Germany, who
told me much of the chaos which has followed the war.

With the workings of our own ordered industry fresh in mind,
it was much as if time had reversed and I had stepped back
some years to visit these foreign countries. European industrial
leaders will all tell you that their countries are in process of recovery, but in some countries, notably in France and in England,
the recovery is far in advance of that in others, such as Germany.
In every country they have had word of our marvelous progress
in illumination and are anxious to establish contact with us. The
scientific and engineering men particularly are most anxious to
put the troubles of the past years behind and to get on a better
basis of understanding with America. America is recognized as
leading in the lighting field.

It is too soon after the wounds of the great war to expect to
find the various illuminating engineering societies well organized
and on an operating basis. It seems to me, however, that our
organization could well go out of its way to assist its European
friends of former years to get in step with conditions here once
more.

The informal meeting of the International Electrotechnical
Commission at Geneva brought together delegates from our allied
countries and our own and in addition, for the first time since
the war, delegates from Germany were present.

The American delegation comprised nine members, five of whom were members of the Illuminating Engineering Society; three of these five being Past Presidents. The subject which could be most properly mentioned in an I. E. S. publication was the conference on standardization of Edison base sockets of the medium and Mogul sizes. It is enough to say that the diplomatic course was well cleared and a procedure suggested which in all probability will result in an international standardization of the medium and Mogul sockets.

We sometimes refer to "standardization of bases" but it is obvious that it is rather standardization of the sockets which carries with it an ultimate standardization of bases. The converse of this would not be true.

Standardization is sometimes regarded as worth while in itself, whereas it is never so well worth while in itself as it is as a part of a larger program.

The manufacturers need the standardization much less than do the public. The state of heterogeneity which increases the cost of doing business merely increases the cost of goods to the ultimate consumer. These conditions of business must necessarily be the same for all manufacturers; hence these manufacturers are on an equal basis of competition, which basis, however, is on a higher level of cost than under the conditions of standardization, for many reasons.

Standardization, therefore, in this case is a factor in the cost of service to the public and it should be pursued with that in mind rather than as an end in itself.

The Committee confined itself to the medium and Mogul sizes of the Edison screw sockets, feeling that if they accomplished an agreement on these two sizes it would be possible at a later date to take up the more complicated questions relating to the less used sizes and types.

It would be fairly easy to find good lighting in any country of Europe, but, as one of my European friends who had been in this country remarked, "While you can always find good lighting in any foreign country if you know where to find it, you see it on every hand in America."

SAMUEL E. DOANE.

REFLECTIONS

Windowless Museums Forecast for Future

THE MUSEUMS of the future will have no windows and will be entirely dependent upon artificial lighting, says the New York State Committee on Public Utility Information, quoting Frederic A. Lucas, the Director of the American Museum of National History, as saying;

"Light is the great enemy of natural history collections. The lovely Luna moth fades after a few days' exposure; feathers of humming birds actually seem to disintegrate on long exposure to light; a few years ruin mammals, like deer and foxes. Some minerals even are affected by light, and rose quartz pales in the sunshine and must be kept in the dark."

Just what is it in plain ordinary daylight that does all this damage? The ultra-violet rays, says Mr. Lucas, and all attempts to filter out these rays through frosted or colored glass windows have not been successful. Electric light, he says, has many advantages. "It shines when and where it is wanted, and barring accidents, it shines at all times, with the same degree of intensity."

Mr. Lucas described a series of tests made by Sir Sidney Harmer, Director of the British Museum of Natural History, wherein objects were exposed to different kinds of light continuously for nearly three years. Of these tests, Mr. Lucas says:

"They show conclusively that electric light is much less harmful than daylight and Sir Sidney writes: 'A gallery without windows, lighted entirely by electricity, preferably not arc lights, would have great advantages.'"

The failure to produce a transparent, non-actinic glass capable of filtering out the destructive elements from sunlight, says Mr. Lucas, means that museum authorities must turn to electric light for the solution of the problem. It is estimated that the gain in wall space from the omission of windows would offset the cost of electric light. It also would eliminate dark corners in exhibitions halls.

The Coming of the Neon Lamp

FROM time to time a great deal has been heard about small neon lamps being put on the market and becoming fairly common in Europe. In this country they have so far been conspicuous mainly by their absence, but they are beginning to appear and may take their place among the regular lighting devices used for special rather than for general purposes. The brief report on some of these lamps, as tested in this country, printed last week is, therefore, a timely one.

Of course, the neon tube as a spectacular lighting "stunt" had, through the energy of M. Claude, been made known in Europe before the war, but these early tubes were of fairly high voltage and did not have a normal place in central-station activities. The lamps referred to by A. Palme in the last issue[1] are designed for 220-volt circuits, alternating or direct current, this voltage being a usual one in foreign distribution systems. The lamps are likely to be used mainly for advertising and similar purposes. it being possible to obtain a letter-shaped lamp working at an expenditure of not more than 2 watts or 3 watts per letter in the smaller sizes and giving the characteristic neon glow of vivid orange with the faint violet fringe about the electrodes due to a small amount of more refrangible light. The effect is characteristic and pleasing. These lamps may also find useful functions as fuse and switch indicators, polarity testers, and for similar purposes for which it is not now easy to get a lamp of really small size suited to the ordinary distribution voltages. In the United States a night lamp of this type has been adapted to work on 110 volts alternating current, giving a small fraction of a candle-power for an expenditure of energy of less than 2 watts.

Altogether, it would appear that the neon lamp is likely to have a real place in auxiliary service. The actual efficiency is low, the specific consumption of the small lamps here described being about 10 watts or 15 watts per candlepower, but where the only requirement is a very small amount of light for purposes that cannot really be considered within the scope of illumination high efficiency is not important, while the need of a lamp of very low power on circuits operated at the ordinary distributing voltages is a real one.—*Electrical World, February 3, 1923.*

[1] Palme, A., "The Neon Glim Lamp," *Electrical World*, 1923, Vol. 81, p. 216.

Street Lighting History

M ODERN electric street lighting, with the high efficiency tungsten light, had for its early prototype the horn-sided lantern hung out after dark by the keepers of inns and shops, with a view to attracting attention to their wares. It was entirely a private undertaking. The first recorded street-lighting city ordinance in London dates back to 1414, it being recorded that citizens having houses on certain streets were ordered to hang lamps before their doors at dark.

The New York State Committee on Public Utility Information, tracing the history of street lighting to date, says that the practice followed by so many thrifty citizens of using short candle-ends, that burned only for an hour or so, caused the regulation of this early system of street lighting to be turned over to the night watch and penalties were imposed on those householders whose lights went out.

In Paris, the first attempt at municipal street lighting was in 1558, when the city placed pitch-filled vessels on the street corners, as a measure of protection against night prowlers and marauders of that period. Gas began to be adopted in London in 1813, oil-burning street lamps being in general use before that time.

Electric street lighting is a development of the last forty years, beginning with the carbon arc, progressing with the carbon-filament lamps, the high-efficiency tungsten light now being used.— *New York Times*, February 25, 1923.

Manufacture of Lamps and Reflectors, 1921

A CCORDING to a recent announcement of the Department of Commerce reports made to the Bureau of the Census the value of products of establishments engaged primarily in the manufacture of lamps and reflectors amounted to $29,164,000 in 1921 as compared with $38,099,000 in 1919 and $16,638,000 in 1914, a decrease of 23 per cent from 1919 to 1921, but an increase of 75 per cent for the seven-year period 1914 to 1921. The industry includes the manufacture of automobile, carriage, wagon, and coach lamps, desk, table, and night lamps, portable lamps, street lamps, and lanterns, gasoline and kerosene lamps and fixtures, and reflectors, headlights, railroad track lights, etc. These

products do not include the arc, incandescent, and other electric
lamps for which statistics are given in connection with those
for electrical machinery, apparatus, and supplies.

In addition to the value of products of establishments assigned
to this classification, lamps and reflectors to the value of $1,609,-
000 in 1921, and $2,041,000 in 1919, and $735,000 in 1914, were reported as subsidiary products of establishments assigned to other
classifications.

Of the 157 establishments reporting products valued at $5,000
and over in 1921, 48 were located in New York; 35 in Illinois;
16 in Pennsylvania; 10 in Ohio; 9 each in California and Massachusetts; 6 each in Connecticut and Michigan; 4 each in Indiana
and New Jersey; 2 each in Maryland and Oregon; and 1 each in
Kansas, Missouri, Texas, Washington, West Virginia, and Wisconsin.

In November, the month of maximum employment, 7,233 wage
earners were reported, and in January, the month of minimum
employment, 5,685—the minimum representing 79 per cent of
the maximum. The average number employed during 1921 was
6,447, as compared with 8,360 in 1919.

The statistics for 1921, 1919, and 1914 are summarized in the
following statement. The figures for 1921 are preliminary and
subject to such change and correction as may be found necessary from a further examination of the original reports.

Items	1921	1919 [1]	1914 [1]
Number of establishments........	157	149	122
Persons engaged................	7,647	9,734	8,147
Proprietors and firm members...	132	116	87
Salaried employees.............	1,068	1,258	926
Wage earners (average number)	6,447	8,360	7,134
Salaries and wages..............	$9,930,000	$11,936,000	$5,518,000
Salaries...	2,467,000	2,644,000	1,345,000
Wages.......................	7,463,000	9,292,000	4,173,000
Paid for contract work..........	135,000	29,000	34,000
Cost of materials...............	14,014,000	18,429,000	8,012,000
Value of products...............	29,164,000	38,099,000	16,638,000
Value added by manufacture [2]....	15,150,000	19,670,000	8,626,000

[1] Statistics for establishments with products valued at less than $5,000 are not included
in the figures for 1921. There were 13 establishments of this class, reporting 4 wage
earners and products valued at $38,000. For 1919, however, data for 22 establishments of
this class, reporting 9 wage earners and products valued at $59,000, and for 1914, 29 such
establishments, with 36 wage earners and products to the value of $77,000, are included
in all items with the exception of "number of establishments."

[2] Value of products less cost of materials.

PAPERS

LIGHTHOUSES AND LIGHT VESSELS*

THEIR FEATURES OF ILLUMINATION

BY CAPT. SAMUEL G. HIBBEN**

SYNOPSIS: A survey of the general subject of important lighting features of the United States Government Lighthouses and allied lighted beacons used in navigation is presented. The author discusses the problems of geographical and optical range of the lights; the optical systems peculiar to the service and their classifications; the light sources; the supporting structures; and the historical features in brief that are necessary to set forth a complete and fundamental survey of this little-known subject.

The paper offers a record of progress to date, and indicates some of the recent developments that have come about by virtue of applications of incandescent lamps, replacing gas and oil burners.

Along the forty-seven thousand miles of the United States lake, river and sea coasts, are maintained over sixteen thousand aids to navigation. Here, and in other countries, the safety of navigation and the maritime commerce depends upon these day and night beacons, and because of their vital necessity to shipping and through the human interest that attaches to the picturesque,—often lonely, usually unfamiliar, always heroic—service of the lights, nearly everyone will find the subject engrossing.

In 1903 the Department of Commerce assumed charge of all navigation aids (formerly under the Treasury Department), and now this Department designs, installs and maintains all equipment under the Bureau of Lighthouses.

The scarcity of data and printed information bearing upon the Lighthouse Service leads to a brief presentation here of some engineering fundamentals, and the interjection of a few bits of history, in order to attain to a broader conception of the problems and the accomplishments.

*A paper presented at the annual convention of the Illuminating Engineering Society, Swampscott, Mass., September 25-28, 1922.

**Illuminating Engineer, Westinghouse Lamp Company; Capt. Engrs., R. C.

The Illuminating Engineering Society is not responsible for the statements or opinions advanced by contributors.

In 1921 there were in service in the United States

Lights, other than minor	1,860
Light vessel stations	49
Relief light vessels	14
Lighted buoys	629
Total lighted beacons	5,756
Fog signals	545
Bell and whistle buoys	362
Other buoys	7,193
Total aids	16,356

Travelers to Great Britian are familiar with the great Eddystone Lighthouse, one of the first of such towers built in the ocean, and which, curiously enough was lighted with twenty-four tallow candles as late as 1811. On our shores the first permanent lighthouse was erected near Boston, on Beacon Island, in 1716. The light was produced by oil lamps; later by the multiple-wick Argand burners and parabolic reflectors in 1811; finally with a 100,000 cp. revolving Fresnel lens in 1859.

When the United States Government was organized, in 1789, ten lights were taken over from the Colonies, and their improvement grew with the purchase, in 1812, of the patent rights to use "reflecting and magnifying lanterns," interpreted to mean Argand burners, parabolic reflectors and solid bulls-eye lenses. Now the engineer finds in the first order lights a lens mechanism that typifies the highest degree of applied illuminating engineering, and the recent applications of incandescent filament lamps to these lenses makes the following discussions of design and performance particularly timely.

THE GEOGRAPHICAL RANGE

In general two factors limit the distance at which a light at sea is visible; one is the luminous intensity and color of the source, or the condition of the atmosphere which governs the observer's judgment of this factor, and the other is the eclipsing of the light by the curvature of the earth.

It might seem that this latter distance would be easy to estimate and would be represented by the length of a right line from the source of light, tangent to the surface of the ocean. Referring to Figure 1, this theoretical geographical range would be the distance "Lr" or the very nearly equivalent distance "Pr," and would be represented by the expression, $\sqrt{2RH + H^2}$. In actual service the luminous rays from any light would traverse the atmosphere, coming in contact with strata of air whose density diminishes in proportion to their elevation, or changes in proportion to water vapor and dust content. The rays suffer successive refractions and consequently follow curved paths which are concave downward and have the effect of increasing the theoretical geographical range, which on this account becomes, as in Figure 1, "Pn." The beam of light for such a range would be directed along the line "LX" making an angle "a" with the horizontal plane through the light.

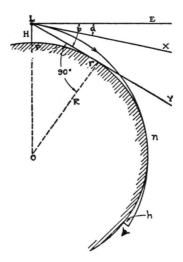

Fig. 1.—The geographical range of a light.

For any one position the atmospheric refraction will change with the seasons, the hours, and the meterological conditions, but for an approximate average the range "D" may be calculated from the formula, $D = \sqrt{\dfrac{RH}{0.42}}$ where "H" may be taken as the height of the source above mean low tide (In European Countries, height above high tide) and "R" the radius of earth's curvature. This formula enables the determination of the distance "Pn." If the eye of the observer is placed at a height "h" above the sea, the distance "Ph" or the maximum possible geographical range becomes $\sqrt{\dfrac{RH}{0.42}} + \sqrt{\dfrac{Rh}{0.42}}$. Obviously the beam should not be projected horizontally and neither should it be projected at an angle as large as "b." Table I gives the calculated geographical ranges using the above formula.[1]

Along the sea coasts and across large spaces of open water, the beam intensities of many of the larger lights, say of the first order, are of sufficient candlepower to penetrate ordinary atmosphere to a greater distance than the geographical range. It becomes therefore a question of deciding the height of the supporting structure in order to insure a geographical range sufficient to have the light visible at a maximum distance that would be useful to the navigator. It is easily conceivable that certain fixed lights would gain nothing by reason of elevation. In fact the geographical range increases very slightly with increments of height, and in cases like the original San Diego light, 422 feet above the sea, the obscuring of the light by fogs at higher levels may be seriously detrimental.

It is also interesting to note that some illumination of the areas close to the base of the lighthouse is desirable in numerous cases and one instance of this sort is found at the (157 feet high) Cape Henry Light Station. Where, as at this station, an important channel is relatively close to the base of the light and where a fairly good foreground illumination is desired at a distance of not less than 2,000 feet (neglecting atmospheric refraction at this range) it would seem desirable to have the bottom edge of the beam make an angle of about 4 degrees 5 minutes with the horizontal plane through the light. This will represent the angle "b" in Figure 2.

Mariners are concerned with the geographical range and with the height of the light because when they know their own elevations above the sea, likewise the height of the light that becomes

[1] Mariners often use a simplified form of this formula, where $D = 8/7 \sqrt{H,}$ "H" being in feet and "D" in nautical miles.

Fig. 2.—Conception of beam divergence.

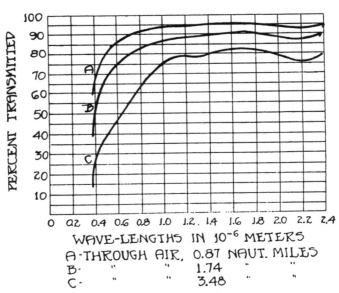

A·THROUGH AIR, 0.87 NAUT. MILES
B·   "   "   1.74   "   "
C·   "   "   3.48   "   "

Fig. 3.—Transmission of light through atmosphere.

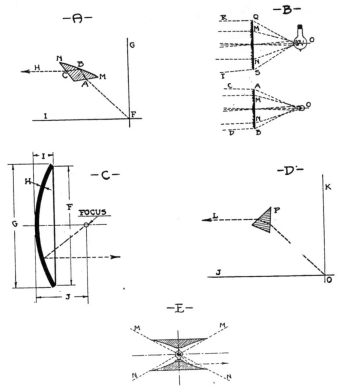

Fig. 4.—Illustrating the types of light control.

TABLE 1.

Geographical Ranges

For an observer of whose height "h," above the level of the sea, is equal to

Height of light above sea level		Distance of the point of contact of the luminous ray tangent to the surface of the sea $D = \sqrt{\dfrac{RH}{0.42}}$ $R = 6,366,953$ m.		3 m. 9.84 ft.		6 m. 19.68 ft.		9 m. 29.52 ft.		12 m. 39.36 ft.		15 m. 49.20 ft.	
Ft.	m.	In meters	In miles	In meters	In miles	In meters	In miles	In meters	In miles	In meters	In miles	In meters	In miles
32.5	10	12,312	6.65	19,056	10.29	21,849	11.80	23,993	12.96	25,800	13.93	27,391	14.79
36.1	11	12,913	6.97	19,657	10.61	22,450	12.12	24,594	13.28	26,401	14.26	27,992	15.12
39.4	12	13,455	7.25	20,232	10.93	23,025	12.43	25,169	13.59	26,976	14.57	28,567	15.43
42.0	13	14,053	7.56	20,752	11.22	23,575	12.73	25,719	13.89	27,526	14.86	29,117	15.72
45.9	14	14,503	7.57	21,312	11.51	24,105	13.02	26,249	14.17	28,056	15.15	29,647	16.01
49.2	15	15,079	8.14	21,823	11.78	24,616	13.29	26,760	14.45	28,567	15.43	30,158	16.29
65.6	20	17,412	9.40	24,156	13.04	26,949	14.55	29,093	15.71	30,900	16.69	32,491	17.55
82.0	25	19,465	10.51	26,212	14.15	29,005	15.66	31,149	16.82	32,956	17.80	34,547	18.65
98.4	30	21,325	11.53	28,070	15.16	30,863	16.67	33,007	17.82	34,814	18.80	36,405	19.66
114.5	35	23,034	12.44	29,778	16.08	32,571	17.59	34,715	18.75	36,522	19.72	38,113	20.58
131.3	40	24,625	13.30	31,369	16.94	34,162	18.45	36,306	19.61	38,113	20.58	39,704	21.44
147.5	45	26,116	14.10	32,862	17.75	35,655	19.25	37,799	20.41	39,606	21.39	41,197	22.25
164.0	50	27,531	14.87	34,275	18.51	37,068	20.02	39,212	21.17	41,019	22.15	42,610	23.01
180.5	55	28,875	15.59	35,619	19.23	38,412	20.74	40,556	21.90	42,363	22.88	43,954	23.74
196.7	60	30,159	16.28	36,903	19.93	39,696	21.44	41,840	22.59	43,647	23.59	45,238	24.43
213.1	65	31,590	16.65	38,134	20.59	40,927	22.10	43,071	23.26	44,878	24.23	46,469	25.09
229.5	70	32,575	17.59	39,319	21.21	42,112	22.74	44,256	23.90	46,063	24.87	47,654	25.73
246.0	75	33,819	18.21	40,463	21.85	43,256	23.36	45,400	24.52	47,207	25.49	48,798	26.35
262.5	80	34,825	18.81	41,579	22.45	44,362	23.96	46,506	25.11	48,313	26.09	49,904	26.95
279.0	85	35,776	19.33	42,640	23.03	45,433	24.53	47,577	25.69	49,394	26.67	50,975	27.53
295.5	90	36,637	19.95	43,681	23.59	46,474	25.10	48,618	26.25	50,425	27.23	52,016	28.09
311.5	95	37,949	20.49	44,693	24.13	47,486	25.64	49,630	26.80	51,437	27.78	53,028	28.64
328.0	100	38,935	21.02	45,679	24.67	48,472	26.17	50,616	27.33	52,423	28.31	54,014	29.17

visible when just rising out of the sea, they are able from Table
I, to estimate their distance off shore. Land-fall lights hence
should penetrate to the limits of the geographical range.

THE OPTICAL OR LUMINOUS RANGE

Were it not for the curvature of the earth as explained, the
maximum visible range of a light would depend on its beam
candlepower, but even more upon the opacity of the atmosphere.

Particularly along the sea coasts, one finds considerable variation in the air's water-vapor content, and many localities are subjected to severe fog conditions. Consequently the luminous
range for times of greatest need often becomes less than the
geographical range and less than a desirable or safe range.

Early designers of lighthouse apparatus estimated the theoretical maximum range of visibility upon the theory of rays transmitted through a vacuum, and having ranges proportional to
the square root of the source intensities. For example, a fixed
light having a beam intensity of 9.8 candlepower was seen on an
exceptionally clear night at a distance of ten kilometers and it was
therefore estimated that a neighboring light of 6,174 candlepower
intensity should be visible 251 kilometers. The above assumptions and calculations cannot be found true in practice, even
admitted that one observer's visual acuity would not differ materially from another's. In point of fact, long years of training
in maritime service increase the acuteness of vision and enable
mariners under fog conditions to estimate with fair accuracy
what intensities and what color changes they might expect in
any known white light.

Experiments made of atmospheric transmission of the sun's
rays from the zenith[2] indicate that the following approximate
figures hold:

Wave-length in microns	Percentage transmitted
0.40 Violet	47.5
0.45	55.3
0.50 Yellow	62.4
0.60	68.2
0.70	75.6
0.80 Red	80.1

[2] *Astrophysical Journal*, Vol. 19.

Other experiments with electric searchlight beams gave results as shown in Figure 3. In Table II are recorded the luminous ranges of oil vapor lights of varying candlepower but of constant color as reported by the French Lighthouse Service.

A theoretical way of arriving at a determination of the luminous range would be through the use of the formula $\dfrac{La^x}{x^2} = t$.

Here "a" $=$ that fraction of received light passing through an atmosphere of unit distance; "t" $=$ the smallest intensity of light at unit distance, which an observer can see through the clearest air; "x" $=$ the range; "L" $=$ candlepower of source. The coefficients "a" and "t" must be determined experimentally. Starting with full moonlight, one experimentor determined a value for "a" equivalent to 0.973 or in other words a certain kilometer of calm atmosphere on a clear night would absorb approximately 3 per cent of the initial light. A number of years ago the French Lighthouse Board determined the value of "t" to be equal to 0.01. The following Table II, is a summary of some of the above mentioned observations and are of value chiefly in indicating rough approximations.

TABLE II.—LUMINOUS RANGE—WHITE LIGHT.
From Observations of the French Lighthouse Service.

C. P. of source	Nautical miles		C. P. of source	Nautical miles	
	Clear air	Misty air		Clear air	Misty air
0.9	1.5	1.2	990	14.3	7.4
2.7	2.3	1.8	1,350	15.3	7.8
5.4	3.1	2.3	2,250	16.9	8.5
9.0	3.7	2.6	3,600	18.4	9.1
13.5	4.3	2.9	5,400	19.8	9.6
18.0	4.8	3.2	9,000	21.6	10.3
27.0	5.5	3.5	13,500	23.3	10.9
36.0	6.0	3.8	27,000	25.7	11.9
54.0	6.8	4.2	36,000	26.7	12.2
72.0	7.4	4.5	45,000	27.6	12.6
90.0	7.9	4.7	63,000	28.9	13.0
108.0	8.3	4.9	90,000	30.2	13.6
135.0	8.9	5.1	135,000	31.9	14.1
180.0	9.6	5.4	198,000	33.3	14.7
225.0	10.1	5.7	270,000	34.7	15.2
270.0	10.6	5.9	360,000	35.8	15.6
315.0	11.1	6.1	540,000	37.5	16.2
360.0	11.4	6.2	720,000	38.6	16.6
450.0	12.0	6.5	900,000	39.6	17.0
540.0	12.5	6.7	1,800,000	42.5	18.0
630.0	13.0	6.9	2,700,000	44.3	18.7

Evident from Table II, and as verified by practical experience, it is extremely difficult to materially increase the luminous range through fog by increasing the beam candlepower. Observers agree that the increase of visible range through heavy fog becomes very gradual when getting into higher candlepowers.

Along our Pacific coast the fogs are particularly troublesome. The Point Arena light, north of San Francisco, with its 380,000 cp. beam, is the maximum for that district.

Considering luminous range, or beam candlepower the revolving flashing lenses, by virtue of beam concentration, have the advantage over the fixed (360 degree) lights. For example the Seguin (Maine) fixed light, first order, is rated 22,000 cp. American Shoal (Florida) twenty-four panel first order, has 80,000 cp.; Hecta Head (Oregon), eight panel first order, 170,000 cp.; and Molokai (Hawaii) two panel second order is rated 620,000 beam candlepower.

Measuring the beam candlepower of a large lens is not an easy matter. There are nodes in the beam, and the inverse square law based upon foot-candle illumination of a target does not hold. One method of approximation is as follows :—:

Let L = beam candlepower, average
$\quad$ B = intrinsic brilliancy of source, in cp. per sq. cm.
$\quad$ W = effective width of source in the focal plane, in cm.
$\quad$ H = effective height of lens, in cm.
$\quad$ C = lens efficiency, roughly 66 per cent, without reflectors
Then $L = B \times W \times H \times C$ for fixed lenses.

For revolving (bulls-eye or bivalve lenses), substitute for $W \times H$ in this formula, the factor "A," = area in sq. cm. of the light projected on a plane perpendicular to the optical axis.

Vertical divergence, "a" of beam is calculated from the equation,

$$\operatorname{Tan} a/2 = \frac{h}{2F}$$, where "h" = height of source and "F" = focal distance.

Horizontal divergence "b," is found from the equation

$$\operatorname{Tan} b/2 = \frac{w}{2F}$$, where "w" = width of source.

Range lights, as used along a channel may blend on account of irradiation if they possess similar characteristics. Experiments indicate that first order lights must be separated by a visual angle of about 15 minutes in order to be seen as distinctly separate sources.

Associated with the optical range is the question of atmospheric changes in color of the original light and it is common knowledge that greens and blues are not satisfactory for lighthouse service. The ordinary action of water vapor is to redden the original white source, i. e., absorbing the short wave-lengths, and it seems advisable to have a light with initially as good a white color quality as possible,—much whiter in fact than the oil flame burners of the Argand type,—in order not to have these yellowish sources confused on foggy nights with the distinctly red beacons. It has been roughly estimated that the beam candlepower of a fixed light of the first order having an oil vapor incandescent mantle source, is reduced by 60 per cent on account of the red panes of glass in the outer lantern. In the case of the incandescent lamp source in the Cape Henry Light referred to elsewhere, it is estimated that the beam candlepower in the red sector will be 30 per cent of that in the white sector.

THE OPTICAL SYSTEMS

The light from the original source is re-directed or controlled by one of three general reflecting or refracting systems, or by a combination of them.

Where the light is reflected from a polished surface as in the case of Figure 4-C, the control is known as the Catoptric system. In the early lights the spherical metal mirrors were used which later developed into metal and then silvered glass paraboloids generated by revolving a parabola about its horizontal axis. These catoptric reflectors so obtained are designated as "photophores" and were used with the oil flame or the Argand burners. They developed into groups as illustrated by Figure 5. Obviously such combinations were serviceable for flashing or rotating lights but not generally satisfactory for fixed lights illuminating the entire horizon.

For the latter service there was employed the Catoptric system having the reflecting surface generated by revolving a parabola

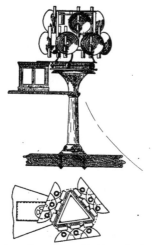

Fig. 5.—An early rotating catioptric mechanism.

around the vertical axis passing through its focus. This is known as the "sideral" type. Figure 4-E, illustrates the shape from which it will at once be seen that a large loss of light occurs in the wide angle "MON." Many of the Catoptric systems are still being used as for range lights, or in combinations with refracting lenses.

The second system known as the Dioptric, involves the refracting of the light rays by transmission through glass prisms and is represented by Figure 4-D. The cylindrical shaped lenses of the lower orders and the central bands of the higher order lenses as illustrated by Figure 4-B, operate on the dioptric principal.

Where refraction and reflection both occur, one has the Catadioptric system, Figure 4-A, which in practice is found in the upper and lower ring prisms of the lights of the higher orders.

Various systems of control may be better understood by reference to Figure 6, showing units that embody all three systems. These drawings form a valuable record of progress in light control apparatus.

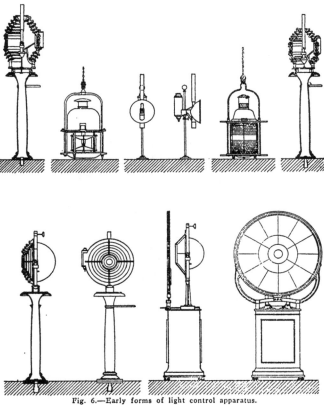

Fig. 6.—Early forms of light control apparatus.

Reference to Figure 7, shows a first order lens of the fixed light type with the central dioptric band and upper and lower catadioptric prisms. In designing lenses for the flashing lights as made up to this date, the section of the prism of Figure 4-A, is rotated around the axis "FI," to generate a ring prism, and the lenses built with similar concentric prisms are such as the Bivalve type.

Lens systems are classified according to the focal length, and as listed in Table III.

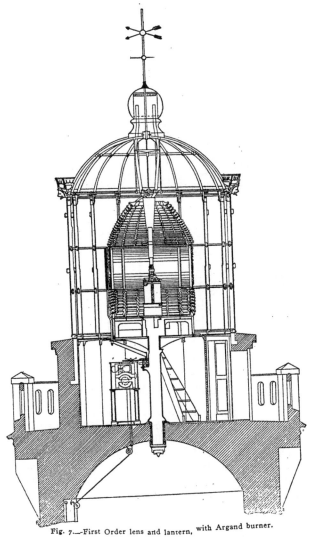

Fig. 7.—First Order lens and lantern, with Argand burner.

TABLE III.—CLASSIFICATION OF OPTICAL APPARATUS, DIOPTRIC
AND CATADIOPTRIC LENSES.

Order or diameter	Focal distance Millimeters	Focal distance Inches	Lens characteristics Height in mm. Diop.	Lens characteristics Height in mm. Catadiop	Lens characteristics Number of rings Diop.	Lens characteristics Number of rings Catadiop
Hyper-Radial	1,330	52.36	—	—	—	—
Meso-Radial	1,125	44.19	—	—	—	—
First Order	920	36.22	—	—	—	—
Second Order	700	27.56	—	—	—	—
Third Order, regular	500	19.68	—	—	—	—
Third Order, middle	400	15.75	795	1,140	17	4-8-9
Third Order, small	375	14.76	—	—	—	—
Fourth Order, large	300	11.81	—	—	—	—
Fourth Order, regular	250	9.84	509	700	11	0-6-8
Fifth Order	187.5	7.38	378	490	11	0-5-8
Sixth Order	150	5.90	245	370	9	0-4-7
200 Mm.	—	—	165	—	7	—
150 Mm.	—	—	156	—	11	—
140 Mm.	—	—	135	—	5	—
105 Mm.	—	—	105	—	5	—

The first catadioptric or Fresnel lens in the United States was placed at the Navesink Light on the New Jersey Highlands in 1841. In 1891 this was replaced by a first order Bivalve with an electric arc for the light source and up to this time this is the only operating first order light using electric current for the primary source, and having its own self-contained generating plant. The maximum beam intensity has been estimated at 25 million candlepower.

A typical fixed first order lens is in the familiar yellow and black Absecon lighthouse tower of Atlantic City, which was built in 1854 by L. Sautter et Cie.

The design and manufacture of the larger order lenses is a matter of extreme engineering accuracy and was not undertaken in this country until about 1910. Previous to that time the majority of lenses were manufactured in France and were so well made that many of them are still giving excellent service. The French glass was of excellent quality, known as the Saint Gobain glass with a refraction index of about 1.55, and of approximately the following formula:

Silica	72.0 per cent
Sodium carbonate	12.2 per cent
Chalk (lime)	15.7 per cent
Various alumina	0.1 per cent

TABLE IV.—PARABOLIC MIRRORS
(See Fig. 4C)

Nominal size		Dimensions in inches and millimeters									
		F		G		H		I		J	
Inches	Mm.	Inches	Mm.	Inches	Mm.	Inches	Mm.	Inches	Mm.	Inches	Mm.
11	279.4	10¾	273.1	11	279.4	¼	6.349	2 1/16	52.4	4	101.6
18	457.2	19 3/16	487.4	19 7/16	493.7	5/16	7.937	3 7/16	87.3	7⅞	200.0
24	609.6	25⅛	638.2	25¼	641.4	5/16	7.937	4 7/32	107.1	10	254.0
30	762.0	31 3/16	792.2	31 17/32	800.9	7/16	11.112	5½	139.7	12¼	311.1
36	914.4	37	939.8	37¼	946.2	7/16	11.112	6¼	158.8	14¾	374.6

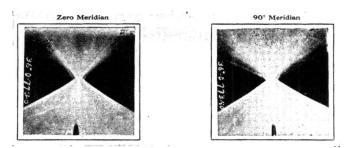

Fig. 8.—Focal point test of parabolic mirrors.

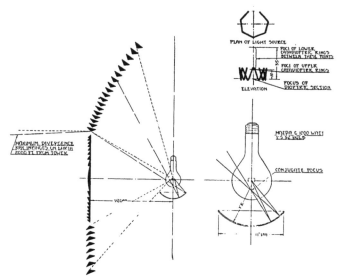

Fig. 9.—The adoption of an incandescent lamp to a First Order lens.

American lenses and mirrors are now manufactured of equal if not greater superiority to any others available. Some idea of mirror accuracy may be gained from the photograph of the focal point, Figure 8. The dimensions of typical parabolic silvered glass mirrors are given in Table IV.

It may be interesting also to note that most of the earlier imported French lenses were designed for a much larger light source than will be found in practice with the electric incandescent or arc lamp. Consequently a concentrated light source will not operate successfully in the larger order Fresnel lenses and some means must be provided with these old lenses of increasing the area of the source. Reference to Figure 9 will illustrate how in one case a spherical mirror is placed beneath the source, to reflect on to the upper catadioptric rings.

A clearer conception of the characteristics of fixed and rotating flashing beams may be obtained from Figure 10 where the bottom figures represent horizontal sections, and the upper figures an idealized or developed representation of the projected light.

THE LIGHT SOURCES

Three general types of light sources are now found in lighthouses and light beacon service. The oil flame wick burners remain in minor locations, which for the more intense lights, have the multiple wicks similar to Figure 11 and thus operated on the principle of the Argand burner. The incandescent mantle burners are quite numerous, using either acetylene gas or vaporized kerosene, but the most common is the open flame acetylene burner.

Many of our first order lights are equipped with the so-called I. O. V. Burners which have a special fabric mantle about 5 inches high and 1½ inches in diameter. The oil is mixed with air at a pressure of about 70 pounds and the carburetted vapor is conducted through a tube over the top of the mantle, thus supplying preheated fuel that allows of a very economical source. These I. O. V. burners have at least eight to ten times the brilliancy of the brightest Argand burner and in one common type are rated at 84 cp. per sq. cm. An average figure of fuel consumption may be taken from the records of the Sandy Hook Light which averages less than one gallon of kerosene per candlepower per year.

The incandescent mantles for the I. O. V. lights give good but short service and may have a life of almost any length, but seldom exceeding thirty days average service. Their renewal is a fair item of cost but the greatest objection to the I. O. V. source is the cleaning of the mechanism, the manual operation of the air compressor, the necessity when starting the light of generating by heating with a separate blow torch, and the nuisance of handling and supplying oil to a reservoir in the lantern.

As may be noted from the preceding section, the first order lenses were mostly designed for the oil flame burners and hence will not operate successfully if the original size of light source be greatly changed. Especially in the vertical dimension the size of the I. O. V. incandescent mantle is sufficiently great to secure the proper beam divergence.

A few light sources consume Pintsch gas in a small inverted mantle, but for rough service such as on a light buoy, the mantles will not survive and the most common source is the flat flame lava tip acetylene burner. These are found singly or in groups up to fifteen. A two-flame acetylene burner with pilot is rated at about fifty liters (1.75 cu. ft.) per hour and the most common single lava tip burners with pilot consume from eight

Fig. 10.—Idealized views of the beams.

Fig. 11.—The Argand type, multi-wick burner.

to thirty liters. An automatic valve forms a part of the acetylene burner which can be timed for any speed or duration of flashes that may be wanted.

The most recent application of electricity to the lighthouse service consists in the installation of the 1,000-watt, multiple, 110-115 volt Mazda lamp as shown by Figure 9. The advantage of this source lies chiefly in its better color and in the greater beam candlepower, as well as in the ease of control and supervision. The Cape Henry Light at this date is being so equipped, and the views of Figure 12 show at the left the I. O. V. burner, and at the right the electric lamp that replaces it. Indications are that the incandescent lamp will supply a satisfactory source of light for many of the first order lighthouses.

Acetylene will probably continue to supply the light for most of the buoys and for a number of the light vessels. However the 100-150-watt 110-volt G-25 bulb concentrated filament incandescent lamp is being successfully used for this service.

Where the large incandescent electric lights are to be given a flashing characteristic, an ordinary sign flasher driven by a one-eighth horsepower motor is being installed and may be easily set for any characteristic, such for example as the new Cape Henry Light, which will flash as follows:

one second on;   two seconds off;
one second on;   two seconds off;
seven seconds on;   seven seconds off.

The seven second flash period is useful in allowing mariners to take observations and to see foreground illumination. Tests on the 1,000-watt lamps show no unusual reduction in life on account of alternate heating and cooling. The filaments require about two seconds to cool to darkness.

LIGHT HOUSES—GENERAL CONSTRUCTION

The structural details of the support for any beacon,—its height, freedom from vibration, storage space for apparatus and materials, and day-time visibility,—are so closely connected with the performance and service of the light that a brief discussion of the supporting structures is justified.

The height of lighthouse towers in this country has been determined upon the basis of securing the necessary geographical range, though not exceeding a reasonable degree of solidity nor placing the light at a height where it might be obscured by high strata of changeable atmosphere. Cape Hatteras Light, Figure 13, 193 feet above the sea is the tallest lighthouse in this country. Cape Henlopen, the first tower built by the United States, in 1790, is 165 feet high and Tillamook Rock Light near the Columbia River mouth is 133 feet. At exposed stations the waves have been known to injure the lanterns, and glass panes in the latter lantern have been broken during severe storms.

Daytime visibility of a lighthouse tower is increased by a species of camouflage marking as illustrated by Figures 13 and 14. Wonderful engineering skill has been shown in developing lighthouse structures, and the illuminating engineer will be interested in examples as Figures 15 and 16 which are self-explanatory of the variety in construction. Examining such drawings as Figures 17 and 18 which illustrate respectively the French Lighthouse of Triagoz and of Cordouan, will afford better ideas

Fig. 12.—The old and the new light sources in Cape Henry Light.

Fig. 13.—Cape Hatteras, with I. O. V. 80,000 cp. light, visible 20 miles. Flash 1.4 sec., Eclipse 4.6 sec.

Fig. 14. Cape Henry, with electric 80,000 cp. lamp, visible 19 miles. Red Sector on Middle Ground Shoals.

Fig. 15.—Cobbs Point Bar, Potomac River, with oil burner, rated 490 cp., in white, and 150 cp. in red sectors.

Fig. 16.—Cape Charles, with I. O. V. 130,000 cp. light, visible 20 miles. "Old 45," flashing that cycle.

Fig. 17.—The Lighthouse of Triagoz.   Fig. 18.—The Lighthouse of Cordouan.

of the skill and engineering expended. This latter light was built originally in the year 1611 at the mouth of the Gironde river on the French sea coast, and had the first Fresnel lens, installed in 1823. These and similar structures embody grace and harmony, but from the engineering standpoint they also typify the results of centuries of lighthouse progress and are examples of the highest type of building that man has erected against destructive elements.

Excessive vibration of the towers must be guarded against in order to protect the clock-work machinery and especially the receptacles of mercury in which the larger flashing lenses are floated and rotated.

Lighthouse history is full of engrossing records of tower and lantern construction. One of the most severe problems, and one such example of hazardous building is the Minots Ledge Light six miles southeast of Boston Harbor, whose placement is considered to represent the most difficult engineering feat in the world.

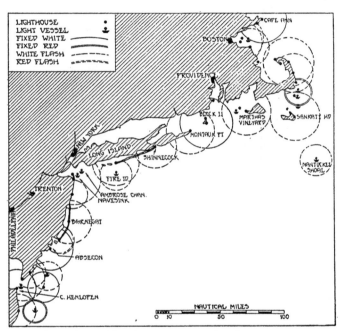

Fig. 19.—Major lights, Cape Ann to Cape Henlopen.

An exceptionally interesting item of lantern construction is found where the first order lenses are protected by cylindrical glass housings 10 to 15 feet in diameter and equally as high. The outer glass panes are sometimes of red glass, or the clear crystal panes are overlaid with colored panes to form colored sectors. In frequent cases like that of the familiar Absecon light at Atlantic City, a strong screen wire must be provided to protect these panes from birds and water fowl. Curtains must be drawn on the interior of the lantern to prevent the sun's rays from entering through the lens and destroying the light source when at sunrise and sunset the light would be focused within the lenses. The lanterns must be wind-tight to prevent drafts around the source, or windmill action on rotating lenses.

The ultimate endeavor on important sea coasts is to place the larger lights, or the landfall lights, on all important headlands. Along the most of the Atlantic Seaboard in fair weather, there is always in view a first order light, and the circles of visibility overlap as seen by Figure 19. The same thing applies to the Pacific Coast with the exception of three short intervals.

LIGHTSHIPS

Anchored on important shoals and in positions where fixed beacons cannot be maintained, the Lighthouse Service maintains the lightships such as illustrated in Figures 20 and 21.

The general illumination problems met here are quite similar to those of the fixed shore lights as regards geographical and optical range, but in addition possess further characteristics not unlike those of the lighted buoys. A lightship is built unlike any other supporting structure. Neglecting all of the interesting details of marine design, and from which are evolved a ship that can outride the roughest storms in the most exposed positions and which affords permanent living quarters for a crew of perhaps fifteen men, we may nevertheless note that from the standpoint of the stability of the light, these vessels are unexcelled for their particular function. For example their length (130 feet more or less) their displacement, and beam, are chosen to minimize the pitch and the roll, and the center of gravity is at just the proper distance below the metacenter to afford the most even keel and yet avoid the shocks and the whipping, thrashing action of the super-structure.

Cylindrical lenses, ordinarily of the 300, 375 or 500 mm. size, singly or in pairs, are carried at the mast heads, perhaps 40 feet above the deck. Naturally this weight (600-900 pounds) must be carefully accounted for, and furthermore the more severe the roll of the ship, the wider must be the vertical beam divergence in order to secure illumination of the horizon at all times.

Most of the light vessels are equipped with fixed lens systems, i. e., without rotating mechanism, and supply flashing or fixed white or red lights. The acetylene flame is the light source in the majority of cases, the incandescent mantles not having been found sturdy enough for this service. Flashing characteristics of the light may be provided by the automatic gas valves, and where two lights are operated simultaneously, a synchronising mechanism keeps them in step.

The electrification of the lightship units has not been carried on extensively, even though the ships themselves are electrically lighted and have self-contained generating systems. The new Diamond Shoals Light Vessel No. 105 for example has in addition to the ships generator a Delco lighting plant in duplicate and a 60-cell lead storage battery floating across either generator. This provides 110-115-volt service, operating usually off the storage battery during the night, and charging during daylight. As yet, however, this vessel carries a 480 cp. 1.25 cu. ft per hour acetylene flame burner in the 375 mm. lens.

The present status of the electrification of the mast-head lanterns consists in the use of a standard G-25 clear bulb Mazda C stereopticon lamp. This lamp with a filament approximately 6 mm. high gives a rather narrow beam and developments are under way to provide a filament source about 12 mm. in vertical dimension, thus increasing the divergence of the beam by perhaps 2¾°. The lightships have not so far been found to require an electric source greater than about 1,500 or 2,000 lumens output. However, the developments of an improved lamp filament arrangement to secure longer life and involving wider beam divergence may reduce beam candlepower to the extent of calling for wattages above 150.

The cylindrical or barrel lenses, limited to comparatively small diameters, are built up in segments or reinforced by astragals. The eclipsing action of these vertical or diagonal astragals would cause dark sectors on the horizon were it not for the fact that vertical triangular prisms are fitted on the inside of the astragals, to deflect the radiated light into the sectors of the adjacent astragal.

The familiar Nantucket lightship lying 200 miles east of New York City was the first anchored beacon on the Atlantic Coast and is perhaps in the most exposed position of any lightship. Further in the New York Harbor the Ambrose Channel Lightship is a familiar one to incoming travellers and is unique in that it was electrified in 1894 with a 60,000 candlepower arc lamp. Placing a light on Diamond Shoal, off Cape Hatteras, has long

Fig. 20.—Winter-Quarter Light Vessel, carrying 500 mm. acetylene flashing light, 600 cp. visible 12 miles.

Fig. 21.—Diamond Shoals Light Vessel No. 105, with electric incandescent lamps, 400 cp. flashing white, visible 11 miles.

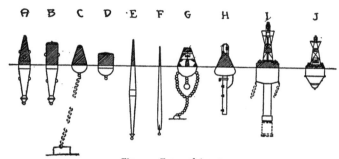

Fig. 22.—Types of buoys.

A—Tall nun buoy, B—Tall can buoy, C—Nun buoy, D—Can buoy, E—Iron spar buoy, F—Wooden spar buoy, G—Bell buoy, H—Large whistling buoy, I—Gas buoy, J—Gas buoy.

been a problem to navigation and it was not possible to secure a lightship there until in 1897 Light Vessel No. 69 was anchored with 900 feet of chain in about 30 fathoms (183 feet) of water.

LIGHT BUOYS AND MARKERS

Any traveller entering a port from seaward will notice red buoys carrying even numbers lying to starboard, and black buoys carrying odd numbers to port. Frequently these are the non-lighted "can," "nun," or "spar" buoys as shown in Figure 22, types A to F inclusive. A close observer will notice that shoals and divided channels will be marked by black and red buoys painted in horizontal stripes and that buoys in mid-channel will be marked black and white with perpendicular stripes. The problems of visibility of the non-lighted buoys are interesting and involve a more lengthy study of paints and exposed area than can be carried in the scope of this paper. The lighted buoys assume types shown by Figure 22, I and J, and Figure 23 illustrates the general design of the more recent and perhaps the largest light and bell buoy which will replace a light vessel. This buoy carries a 375 mm. acetylene lantern rated 480 candlepower and with an automatic valve giving a characteristic of one-half second on, and one-half second dark. Smaller buoys carry the 200 mm. lenses.

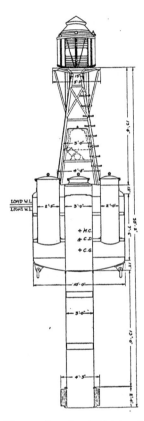

Fig. 23.—A modern lighted buoy.

In the body of the buoys of this type will be wedged one or more gas cylinders, carrying acetylene at about twelve atmospheres dissolved in acetone. As much as 1,000 cu. ft. of acetylene can be carried, which for such a typical cycle of burning as 0.3 second on and 2.7 seconds off, is sufficient to operate the light for some seven to eight months. The cylinders are charged with gas at a cost of approximately $22.00 per 1,000 cu. ft.

Some of the buoys carry Pintsch gas, and use a five-eighth cu. ft. burner with a mantle about one-half inch high and wide. Figure 23 illustrates also the tank compartments carrying carbon dioxide gas for the operation of the fog bell striker, that have a capacity of 1,200 pounds of compressed gas, which will operate the buoy continuously for a period of one hundred and twenty days, striking a 1,000 pound bell about 880,000 blows during this period.

In order to involve the least amount of roll and jump, the center of gravity must be carefully placed slightly below the metacenter and the buoy properly weighted so that it will not be necessary to provide too wide a beam divergence in order to illuminate the horizon when heavy swells are running. Carried at perhaps 15 feet above the water line a 375 mm. lantern might be expected to be visible at a geographical range of at least 10 miles.

The general types of beacons used mostly for the lighting of inland water ways, include oil or acetylene burners in lenses similar to the buoys, locomotive headlight lanterns, and Argand burners backed by mangin mirrors. In relatively few cases is electric service yet available.

The almost human sun-valve has been perfected to control the operation of unattended beacon lights, such as those operating on acetylene. It is sufficiently delicate to allow operation of the light only when the sun's heat is actually withdrawn from the lantern.

The lighting problems of buoys and markers are too numerous to be completely covered in this discussion, but they are the subject of earnest attention by Superintendents of the Lighthouse Service whose efforts are constantly directed towards adapting rather delicate mechanism to the most extreme conditions of storm, ice, and collision, and yet guarantee unchangeable and unfailing performance of the beacons as regards signals sent out by the light at night, by the color and shape of the structure by day, and by the fog bell or fog whistle at all times.

CONCLUSION

It is believed by the author that the successful application of the incandescent filament electric lamp will improve the performance of many navigation aids. Concurrent with the use of such sources, it may be expected that new designs of catadioptric and dioptric lenses, especially of larger orders, will be much smaller,—at least in diameter; less heavy or expensive; and that flashing by rotating mechanism will be replaced by current interruptors.

Grateful acknowledgement is made to the members of the United States Lighthouse Bureau, including Capt. King, Mr. Morse, and especially Mr. Hingsburg, First Asst. Superintendent of the 5th District, for their assistance and especially the latter's generous contributions to this paper. References have been taken from records of the Service, from reports of Mr. Putman, present Commissioner of Lighthouses, from memoirs of the French Commission des Phares, and from reminiscences of the faithful keepers of the lights who are trained in schools of loneliness but find their reward in the consciousness of their service to humanity, well done.

In this paper, "miles" refer to the knot of 6,086.44 feet, containing 1,000 fathoms or 1,852 meters. The liter is taken as 0.0353 cu. ft.

DISCUSSION

Louis Bell: I am sure we are greatly indebted to Captain Hibben for this very vivid explanation of an extremely beautiful branch of optical engineering in the devising of the lighthouse lenses themselves and of civil and mechanical engineering in the carrying out of the structures, which, resisting all attacks of the weather, enable the lighthouses to function properly and reliably through all sorts of storm and stress throughout the year even in the most severe conditions.

As he has just stated, there are possible great changes impending as regards future lights, but so splendid has been the work of the engineers who have already solved the problems of the lighthouse that it will be a long time perhaps before our chief lights are retired for depreciation. They come nearer to permanence than most structures with which we have to deal.

G. S. Crampton: May I ask why CO_2 is used to operate the bell?

S. G. Hibben: It is possible to carry a larger volume in a smaller space. Under extreme pressure, it is of course liquified, and it is non-explosive. It is also non-corrosive and cheap.

L. C. PORTER: I unfortunately did not hear the first part of Mr. Hibben's paper, and perhaps he has touched on the point which I have in mind. We have been doing a little work with the Lighthouse Department in connection with some of the smaller lights, particularly the range lights along rivers, for guiding the boats up and down the rivers. In those cases, we have been replacing some of the oil lanterns by small incandescent lamps operated from primary batteries. The advantage is that the oil lantern has to be visited very much more regularly than the incandescent lamp. The incandescent lamp, operated on an automatic flashing device and run from a primary battery, will last anywhere from six months to a year without renewals, which of course cuts down the labor expense considerably, not only on account of the infrequent times that the light has to be visited, but the renewals for batteries can be carried and handled much more easily than you can oil or bulky gas tanks.

There is another very interesting factor and that is in connection with the visibility of the electric light, as compared to the low power oil and the low power acetylene lights. The tests have been conducted, and I have seen the reports of the Lighthouse Department on them, stating that the incandescent lamps of equal candlepower to the gas could be seen at considerably greater range. Just why that is, I do not know, but it seems to have carried out also in railway signal practice, where we have been replacing oil signal lamps with electric lamps, really of lower candlepower than the oil flame, and yet we are getting greater range of visibility.

L. D. GIBBS: I would like to say from time to time articles will come out in the daily press that this person or that person or this engineering concern or that engineering concern has discovered a new lamp for penetrating fog, and that they are going to eliminate all dangers from fogs with the use of these lamps in lighthouses and along the coasts. I would like to ask Captain Hibben if there has been really any progress in that development. Also, I would like to ask Mr. Porter what the percentage of increase is in distance of visibility of the electric lamp over the oil lamp.

L. C. PORTER: That varied anywhere between 10 and 20 per cent.

S. G. HIBBEN: As to the ranges or the limits of visibility of the different colored sources, there are numerous data, which indicate that, for example, violet light or light of short wave-lengths, through the usual atmosphere, is transmitted slightly less readily than that of the red or longer wave-lengths. It is hard to get exact figures, because of the meteorological conditions varying so rapidly, but some tests made during the war of searchlight beams gave us curves which showed that for penetrations through about three and a half miles of atmosphere there was a much greater transmission in the yellow than in the violet.

We are still experimenting and we know that the yellowish-red light sources become distinctly still more red and still harder to see as the fog becomes denser. Water vapor seems to be the chief disturber. For example, a 500,000 candlepower source is found visible through about sixteen miles of misty air, and a 1,800,000 candlepower source of practically three times the brightness through only eighteen miles of air.

That touches on what Mr. Gibbs has mentioned, too: for the penetration of dense fog, I know of no lamp that is particularly good. Getting away from the yellowish sources of course is advisable and I presume the more blue white the source is, the less would be the probability of color mistakes, but nothing that I know of yet is a panacea for the trouble because we all know of fogs that would eclipse, at the range of only three or four miles, the brightest source of which we have any knowledge.

The important point is that the light color be not changed much. If the mariner does not see his light, he is ultra-cautious. If he sees a yellowish light and interprets it as being a red source, he is in immediate trouble. It is important therefore to have the original source as near white as possible, so that the effect of fog in making that source more yellow will still leave it distinguishable from red. If we start with an oil flame source and it becomes red, due to the absorption of the short wave-lengths in fog, then it may be interpreted to be a red source; so a white light source would lessen that difficulty.

F. W. BLISS: May I ask Captain Hibben why the fog causes a yellow tint in the light that is transmitted, when the blue light is transmitted better than the yellow?

S. G. HIBBEN: I would like to leave that discussion to the Committee on Automobile Headlights. (Laughter). There has been a lot of argument pro and con about that. My experience, and it is based on good theory and on such phenomenon as the sunset, is that the short wave-lengths are affected more by fog conditions than the longer wave-lengths. That seems natural and normal, and there is a tendency to decrease the short wave-length component of a beam. You do not increase the red component at all, you simply make that more prominent by absorbing the short wave-lengths.

F. W. BLISS: I do not see how the blue is transmitted any more than the yellow, if the yellow is what comes through.

S. G. HIBBEN: I do not know that it is, because the transmission of sunlight through atmosphere at 40 microns—that is, violet light shows in one case 47 per cent. So you have less obstruction of the long wave-lengths than of the short. Of course if you took any one individual section of the spectrum there would be a different story. Those light sources which are intentionally made yellow by a colored screen are usually so reduced in final or effective candlepower that their real penetration is less.

LOUIS BELL: Lord Raleigh long ago showed that the scattering of light, which of course dims the beam, was as the inverse fourth power of the wave-length, which fully accounts for the difficulty of the blue rays in getting through.

F. C. HINGSBURG* (Communicated): The paper presented by Capt. Samuel G. Hibben on "Lighthouses and Light Vessels" covers their features of illumination remarkably well, and makes available to the engineering profession valuable data heretofore obtainable only in the offices of the Lighthouse Service. Capt. Hibben is to be commended for the masterly and interesting manner in which he has presented the subject.

As one considers the progress of illuminating engineering as applied to lighthouses, beginning with wood fires, coal fires, candles, sperm oil, colza oil, lard oil, kerosene argand burners, kerosene oil vapor lamps and finally the electric incandescent

*First Ass't Supt., Fifth Lighthouse District, Baltimore, Md.

lamp, it will be seen that the light sources have become progressively smaller and the intrinsic brilliancy increased. The first order lens and hyper-radial lens were developed during a period when illuminating engineers could find no means to increase the intrinsic candlepower of the light source, but accomplished their aim by developing a combination of a large lens and large area of light source, resorting to 5-wick lamps, placed concentrically. With such a combination a beam candlepower of 6,000 was obtained at Cape Henry Light Station. With the development of the incandescent oil vapor light, reducing the area of light source from 25 square inches to 5 square inches, the beam candlepower through the first order lens was increased to 22,000. The necessity for a distinctive characteristic led to the development of the combination of Mazda C, 1,000-watt lamp with mirror, reducing the area of light source to about one square inch and increasing the beam candlepower through the first order lens to 80,000. The mirror was developed to create an image which when reflected into the upper catadioptrics, provided the necessary downward divergence of light needed to illuminate the foreground.

Capt. Hibben referred to the bivalve lens with electric arc installed at Navesink Light Station, New Jersey. The current for the electric arc was furnished by a 25 horsepower oil engine driven generator set, separately excited, and with four keepers in attendance, the operating expense was ridiculously high. The light source has been changed to incandescent oil vapor burned in a mantle. The establishment of light vessels and numerous lighted buoys at the entrance to New York has made it possible to change the illuminant in Navesink Light Station, and the high beam candlepower of this land fall light is no longer essential.

There is one function of the Lighthouse Service which Capt. Hibben has not covered and that is the maintenance of numerous minor aids in rivers, creeks and bays, for which flat wick lanterns with 200 mm. fresnel lenses are used. These lanterns are fitted with a reservoir holding about 3 gallons of kerosene which is fed through an atmospheric float feed valve to the lamp. The lamp burns eight days continuously without attention. Special refined kerosene is essential for satisfactory results. There has recently been developed an electric light operated on Edison primary cells for minor lights. The lamp is a special miniature 5-watt lamp with filament designed to give maximum candle-power through a 200 mm. fresnel lens with the necessary divergence of light for the purpose intended. The light is controlled by an electric flasher giving double flashes each of 1/10 second duration one second apart, with a time interval of about 4 seconds. Experiments have proved this characteristic to be very distinctive, the double quick flash giving as satisfactory results as a longer flash, but with maximum conservation of energy, the light burning 1/25 of the time. It is expected that the light can be operated a year without renewal of cells. The electric light was designed by A. W. Tupper of the Bureau of Light-houses.

As stated by Capt. Hibben, light vessels have been stationed on dangerous shoals where fixed aids cannot be maintained. Since the development of the A. G. A. acetylene lantern equipment, large buoys with 500 mm. and 375 mm. lanterns developing 600 and 480 candlepower respectively have been built, and so far as the light is concerned the buoy answers the requirements of aids to navigation as well as the more expensive light vessel. A modern light vessel will cost about $200,000, whereas the cost of the acetylene and CO_2 aerial bell buoy described by Capt. Hibben will cost $7,500 fully equipped. The distinct difference is the fog signal equipment.

When lighted buoys were first built they were designed in accordance with the principles of vessel design. A light vessel is dependent upon its shape for stability whereas a buoy is stabilized by virtue of its own weight. Referring to Figure 23, page 264, it will be noted that the buoy has a heavy counter weight at the bottom, lowering the center of gravity to a point a considerable distance below either the meta-center or the center of displacement, whereas in vessel design the c. g. lies between the center of displacement and meta-center. The buoy has a long pendulum period in comparison with the wave period, and has seaworthiness which does not permit the buoy to deviate more than a small angular distance from the vertical, and consequently each flash of light is directed towards the horizon, the angle of inclination being less than the divergence of light through the lens. While modern light vessels are designed with very small metacentric heights it is not feasible to limit the rolling angle so that each flash will be directed towards the horizon. Acetylene lights are found on most light vessels because of the greater height of light source and consequent greater angle of divergence of light, than obtained with electric lights. To overcome this objection, attempts have been made to keep the lanterns universally level by supporting the lantern in a universal joint and counterweighting the lantern. As the vessel rolls about the meta-center, the inertia of the counterweight defeats the purpose intended. A device is now under consideration for controlling the lantern by magnets, using a contrivance located at the meta-center to energize the control magnets, and as the lanterns are nicely counter balanced, there will be no inaccuracy due to inertia.

As stated heretofore light vessels are maintained because of their superior fog signal equipment. The aerial bell of 1,000-pound size struck by a mechanism operated on compressed carbonic acid gas is a recent development for fog signal equipment on buoys. Buoys so equipped will eventually replace minor light vessels in sheltered waters. The lightship stationed off the coast is considered the most necessary and efficient aid to navigation. Being equipped with characteristic fog whistle, submarine bell and radio fog signal, the light vessel safeguards navigation when the lights are rendered ineffective by fog. The aerial signals are usually a 12-inch chime steam whistle or a first class air siren. The submarine bell is lowered on a chain to a depth of 25-30 feet where the bell is struck by a clapper operated by compressed air and a code ringer. The characteristic radio signals are emitted on 1,000-meter wave and merchant vessels equipped with a direction finder or a radio compass can take bearing with very accurate results. With the radio fog signal it is feasible to proceed coastwise from light vessel to light vessel with absolute surety in fog, and in the event of meeting a passing steamer take bearings to avoid collision, and entering port masters can locate their position by bearings on two or more stations equipped with radio fog signals. The value of the light vessel as an aid to navigation has increased materially by the introduction of the radio fog signal.

A DISTRIBUTION PHOTOMETER OF NEW DESIGN*

BY C. C. COLBY, JR.,** AND C. M. DOOLITTLE***

SYNOPSIS: The field of usefulness of the distribution photometer is ex·
plained briefly. A review is made of the more common types of distribution photometers. The requirements which had to be met by the photometer described in
this paper are outlined. This is followed by a detailed description of the above instrument, including the methods and features of design which were employed to
meet the requirements. The ten foot integrating icosahedron is mentioned and an
outline given of the methods employed for transferring lighting units of different
types from icosahedron to distribution photometer, and vice versa, with a minimum
of time and labor. The electrical supply and control system for the entire room,
including the two above instruments, is described.

As the art of illumination advances, there is a constantly
growing demand on the part of those interested in the purchase
and installation of lamps and lighting units for more specific information regarding the performance of those units.   Running
parallel with this desire, there is naturally a demand for better
and more scientifically designed combinations of lamp and accessory; for light sources and systems of illumination which will be
more artistic, more restful to the eyes, and at the same time
more efficient in their distribution of light.   This development is
helped along to a considerable extent by educational work on
the part of the manufacturers of lamps and lighting equipment,
as well as by the natural desire for improvement upon the part
of all who are interested in the art.

It is, of course, in the development and measurement of sources
which more efficiently distribute their light that the distribution
photometer is found to be indispensable.   Instruments of this
sort have been made in a number of different forms.   The photometer which is described in this paper will be seen to be of the

*A paper presented at the Annual Convention of the Illuminating Engineer-
ing Society, Swampscott, Mass., Sept. 25-28, 1922.

**Physical Laboratory, Westinghouse Lamp Co., Bloomfield, N. J.

***Illuminating Engineer, Illumination Bureau, Westinghouse Lamp Co., New
York City.

The Illuminating Engineering Society is not responsible for the statements or
opinions advanced by contributors.

same general type as that of some others already in use, but it is believed that it differs sufficiently from its predecessors, to be of interest.

Before going into a detailed description of this instrument, however, a brief review will be made of the most common types of distribution photometers.

In the instrument shown in Figure 1, the arm supporting the unit under test rotates about a horizontal axis. Direct readings on the light source are made by means of the photometer head,

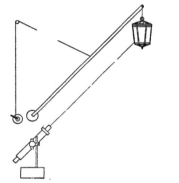

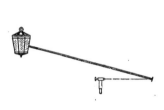

Fig. 1.—A simple photometer in which the unit is rotated about the measuring device.

Fig. 2.—A simple photometer utilizing a spot plate which is rotated about the unit.

which can be adjusted to point directly toward the light when the latter is in any position. The axes of rotation are so arranged that the photometer and light source are separated by a constant distance. This would in most cases be ten feet, as that is the standard distance which is used in the great majority of distribution photometer readings. This method is obviously unsuited to handle work on a large scale.

The arrangement shown in Figure 2, and the one just described, probably represent the simplest types of distribution photometers. Referring to Figure 2, the arm, or rod, rotates about a horizontal axis passing through the center of the light source. At the movable end of the rod is fixed a small diffusing plate or reflector, whose surface is perpendicular to the center line of the rod. The intensity of illumination on the diffusing plate is measured with a portable illuminometer, and from these readings the candlepower of the source is determined. We have used this method in the past, but it is, of course, quite slow and inconvenient.

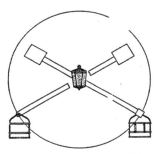

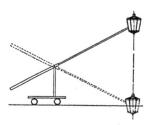

Fig. 3.—A "ferris wheel" type photometer.

Fig. 4.—One form of the Dibden photometer.

The apparatus shown in Figure 3 represents what might be termed the ferris wheel type of photometer. Two arms rotate in a vertical plane about opposite sides of the lighting unit. Each of these arms carries on one end a basket arrangement, or car, in which a photometer operator is seated. In each car is mounted a photometer capable of measuring the candlepower of the source when viewed from any direction. At the opposite end of each arm are mounted suitable counterweights. The arms are moved to form equal angles with the vertical, and simultaneous readings are taken from each position.

In Figure 4 is given a schematic drawing of one form of the Dibden photometer. The lighting unit can be moved vertically. From this unit an arm extends to the car, which travels on a horizontal track. The arm is pivoted where it is fastened to the car, so that, as the unit is raised and lowered the car travels forward and backward along the track. The photometer head is attached to the arm in such a manner that it always points toward the light. This apparatus can be set for any angle between the two extreme positions where the car is either directly above or directly below the unit. The Dibden photometer has been very satisfactory in certain cases, but was not suited for our needs because of the required height of more than 20 feet.

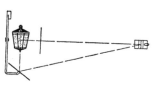 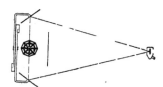

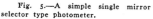

Fig. 5.—A simple single mirror selector type photometer.

Fig. 6.—The Matthews double mirror selector type photometer.

The simplest mirror type photometer is that shown in Figure 5. The mirror rotates on an arm, about the axis which passes through the center of the light source and the center of the photometer head. For any position of the mirror, light emitted by the source in that direction will be reflected to the photometer head. A screen prevents direct light from the source from striking the photometer head. The light enters the photometer head at a large angle with the normal, bringing in certain uncertainties, and causing some difficulty in shielding.

The Matthews double mirror selector, Figure 6, operates in the same manner as the single mirror type described above. It has the advantage, however, that the source is viewed simultaneously from opposite sides at equal angles from the vertical. This eliminates, to a certain extent, errors which may be encountered due to the fact that the candlepower of the source is not uniform about a vertical axis.

Another form of single mirror selector is that shown in Figure 7. The distance between the axis of rotation of the upper gear, and the axis of the horizontal portion of the arm supporting the lighting unit, is the same as the distance between axes of the upper and lower gears. By means of the train of gears, the unit is rotated about the mirror at a constant distance. The mirror, at the same time, is rotated just the correct amount to throw the reflected light along its horizontal axis of rotation, to

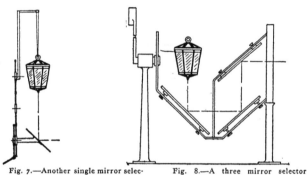

Fig. 7.—Another single mirror selector type photometer.

Fig. 8.—A three mirror selector type photometer.

a photometer head. In this device it is very difficult to rotate the unit under test.

A three mirror selector is shown in Figure 8. The three mirrors rotate as one unit about a horizontal axis passing through the center of the light source. At every position of the mirrors, light is reflected horizontally to a photometer head. This method has certain advantages, but also certain mechanical disadvantages.

The two mirror selector shown in Figure 9 is the type which has been designed and installed in the Physical Laboratory of the Westinghouse Lamp Company. The lighting unit is suspended from, or mounted on, the rotating head B. This head can be moved vertically to bring the light center to the proper position on the photometer axis lined through bearings H and J. Light from the unit, striking the large mirror C, is reflected to the small mirror D, and from there to a diffusing screen placed between

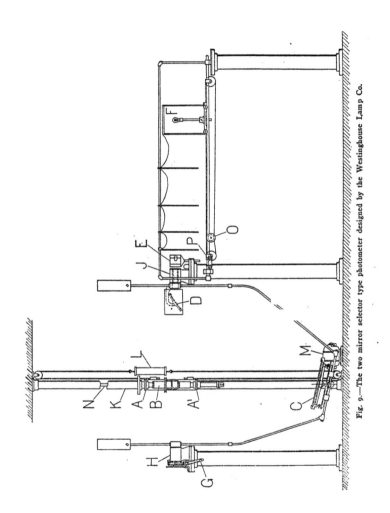

Fig. 9.—The two mirror selector type photometer designed by the Westinghouse Lamp Co.

it and the Lummer-Brodhun photometer head E. A working standard lamp in the housing at F, can be moved to a greater or less distance from the photometer head, to produce a brightness match. The mirrors C and D are rotated as a unit, about the axis which passes through the centers of bearings H and J, by turning the handle G.

Probably none of the photometers described is preeminently the best for all cases. However, a careful consideration of the problems of installation and required performance led to the adoption of this particular type as being the most desirable. It was required in this case that the photometer be installed on a building floor having approximately 10 feet head room. It was desirable to have the instrument take as small a floor space as might reasonably be arranged, although no definite limits were prescribed. It was necessary to build an instrument which could handle all kinds of luminaires from the smallest residential glass type to the largest street lighting units. In the latter case it was necessary to consider the weight of the unit as well as the size of the light source. Provision had to be made for mounting these units either in a pendent position or placed upright on a pedestal, as the case demanded. It was considered essential that a means be provided for rotating all units including the largest street lighting pendents and post tops. In addition to this it was desired that provision be made to transfer a complete unit from the distribution photometer to the ten-foot icosahedron,[1] or vice versa without disturbing the wiring, or the adjustment of the lamp in the unit. The distribution photometer room has been laid out to include a regular icosahedron 10 feet in height. The latter piece of apparatus is designed to be capable of measuring the total light output, and overall efficiency of any existing size of lighting unit by the same method as would be employed in using the Ulbricht sphere for this purpose.

A detailed description of the distribution photometer will show how the various conditions were fulfilled. This description follows:

[1] See paper presented at this convention by K. S. Weaver and B. E. Shackelford "The Regular Icosahedron as a Substitute for the Ulbricht Sphere" on page 290 of this

The rotating head which supports the lighting unit can be moved up or down on the vertical shaft K. The counterweight L makes this a fairly easy matter. Two handles are provided for clamping the head to the shaft. This one head is universal in that it can be used for either pendent or pedestal units. A pendent unit may be suspended from a suitable holder attached to the head, and the latter moved up along the vertical shaft to the proper position. In the same manner, a pedestal, or post top unit may be mounted on a suitable holder on top of the adjustable support, and the latter moved down until the proper position is reached.

The head is rotated at any desired speed by means of a direct current motor M, driving a vertical keyed shaft N, immediately behind the main shaft. The rotary motion is transmitted to the head by means of a round belt and pulleys.

Electrical connections are made by means of two wide slip rings and four brushes. One brush on each ring carries the lighting current, while the other is used for a voltmeter connection. It is considered that the drop in voltage between slip ring and lamp will be negligible. The shaft to which the rings are connected is hollow, and contains the wires leading to the lamp.

At points A and A¹ are flange connections of the same size and shape. The pieces which are attached above and below these points are built to accommodate certain lighting units. If a different type of unit is to be mounted in place, it is only necessary to disconnect a pair of flanges and connect the proper piece to support that particular type of unit. Electrical connections are automatically made by means of plugs and receptacles built into the flanged pieces. A sufficient number of pieces have been made up to take care of all ordinary lighting units. These can be added to at any time to accommodate new types. If the unit in question has a light center exceptionally far from the point of support, the upper or lower flange piece can be removed, depending on which one is not in use. This will give sufficient clearance so that the head can be raised or lowered to provide for all existing types of units, with an ample margin for future developments.

The icosahedron is equipped with upper and lower flange supports of the same type as those on the distribution photometer. Any unit, therefore, when once adjusted in place with its lamp, can be readily transferred from one photometer to the other by merely removing the flange bolts on one instrument, and bolting the flange in its place on the other. Electrical connections are thus automatically made, and no adjustments are disturbed.

It was found that the mirror C would have to be of such size as to make advisable a double arm support. The double bearing design as shown in Figure 9 was therefore decided on. This mirror is of plate glass, 20 inches in diameter. It is inclined at the proper angle from the vertical to reflect the light to the elliptical mirror D. The frame for the circular mirror is held to the main arms by a ball and socket and helical spring device. Adjustment of the mirror position is made by means of three screws near the edges of the mirror frame.

The elliptical mirror D, reflects the light to a translucent screen at the left end of the hollow bearing J. The mirror is mounted on a large ball and socket joint and is held in place by a set screw when the proper adjustment is obtained

The sheet metal hood which encloses the small mirror is painted a flat black. An aperture is provided for admission of light from the large mirror, but the hood shuts out most of the extraneous light. If it is found that a system of screening is needed for the large mirror, it is proposed to place a strip of black velvet, several feet wide, along the floor from about two feet in front of the lighting unit, straight back to the rear wall, five feet behind the unit. This will be extended straight up the wall, and out along the ceiling directly over the floor strip. Practically no extraneous light can strike the large mirror at the proper angle for reflection to the photometer head, except that which is reflected from the black velvet. This will of course be negligible.

The translucent screen at the left end of the hollow bearing is exactly ten feet from the light center of the source. This screen becomes a secondary source of light, illuminating one half of the Lummer-Brodhun photometer field. The other half of the photometer field is illuminated by means of the working standard lamp. The familiar arrangement of movable carriage and screens is shown. Neutral light filters may be inserted in holders provided approximately an inch from the photometer head, on either side of the latter.

The carriage is moved by turning the knurled handle O, which actuates the former through a string passing over pulleys. Photometer readings are taken at point P on a graduated metal tape

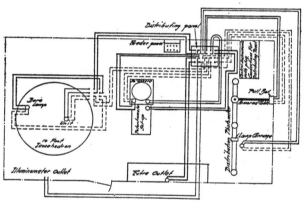

Fig. 10.—Layout of electrical control and supply system for the distribution photometer room. No. 14 wire, current leads,————, No. 8 wire, current leads,_____, No. 16 wire, voltage leads, - - - - -.

which also passes over suitable pulleys, and moves with the lamp carriage. This tape is graduated to read directly in candle-power.

The mirror arm is rotated by means of the handle G, which is connected to a worm and gear. The electrical control and meter table is placed immediately to the left of this point. The operator at this position can therefore rotate the mirror arm, and also read the angle between mirror arm and the vertical by means of a protractor on the left face of the worm gear.

The foregoing is probably a sufficiently detailed description of the distribution photometer proper. It may now be of interest

Fig. 11.—General view of Westinghouse two mirror selector type photometer.

Fig. 12.—Pendent type street lighting unit in position for test.

Fig. 13.—Post top type street lighting unit in position for test.

to describe, in a more general way, the layout of the entire photometer room, as there are several elements of flexible control
which are considered to be worthy of mention here.

The electrical control and instrument table, shown in Figure
10, controls circuits running to the distribution photometer, 10-
foot icosahedron and 30-inch sphere; and may be used to control
a circuit to the miscellaneous work table and to the potentiometer
table.

This control table is enclosed in a box, whose top and sides are
painted a flat white on the inside. Four concealed show case
lamps illuminate these white inside walls causing the meters to
be indirectly illuminated by a soft, thoroughly diffused light.
Space is provided for three large laboratory standard voltmeters
or ammeters, or more of a smaller type. A system of rheostats of
various capacities, with suitable switches, makes it possible to
handle a wide range of currents and voltages without resorting
to special connections.

Current is obtained from the feeder panel, and carried to the
distribution panel, control table, and room circuits by means of
wires terminating in plug connectors. The feeder panel in turn
can be supplied with any desired form of current from the main
laboratory switchboard.

Flexible wires, terminating in plugs, run from the control table
to the distributing panel. From the latter are run the various
circuits to carry current to the different pieces of apparatus.
From suitable points, voltage leads run back from the photometric
apparatus to the panel. At the icosahedron, a double pole double
throw knife switch enables the operator to throw current on
either the lighting unit or the bare lamp. A similar switch
connects the voltmeter wires to the proper point.

With this system, it is possible to feed current at the voltages
desired, to the various test and comparison lamps about the room.
It is possible to regulate the current and take voltmeter and am-
meter readings accurately. Tests can be made on lamps and
units of widely different sizes and characteristics. This can be
done with a minimum of labor required for mounting and adjusting test units and apparatus.

4

As these instruments are now designed, any sized unit which it is possible to handle on the distribution photometer will be within the capacity of the icosahedron, both in size of light source and in weight.

This means that practically any unit of ordinary proportions, not exceeding 30 inches in diameter nor 150 pounds in weight, can be readily accommodated. This will take care of all present sizes of lighting units of the type which would ordinarily be measured on a distribution photometer, and allow a considerable margin for possible future development.

Certain of the mechanical features of the distribution photometer described in this paper have been designed by Mr. H. I. Knecht of our Experimental Equipment Division.

DISCUSSION

W. F. LITTLE: A distribution photometer which has been in use at the Electrical Testing Laboratories for the past eight years is very similar to that described by the author. This photometer consists of adjustable rotator, two mirrors, hollow shaft through which the light is reflected, simple Lummer cube and photometer bar. It was originally designed as a recording photometer but experience showed a loss of time in interpreting the results. The rotator is moved up and down with a worm and gear, and the light source is brought to the photometer axis by projecting the filament onto an enlarged scale. All measurements are made from the light center; in this way the dimensions of the luminaire and the location of the glassware or accessory can be quickly and accurately determined in place.

The rotating arm which supports the French plate mirror is made of triangular construction in order to eliminate sag. In this way the double suspension is not necessary. In using the single suspension the mirror can be rotated through 360 degrees which is of considerable aid in measuring asymmetric reflectors or luminaires which cannot be rotated.

The reflection from the suspension arms is eliminated by the use of corrugated rubber matting with the corrugations running around the tubing.

The socket is revolved by beveled gears and continuous belt in such a way that the position of the rotator may be changed without interfering with or changing the speed of rotation.

The source of electrical supply is from the life test lines which is maintained constant to within $\pm \frac{1}{4}$ per cent. The comparison lamp is operated directly on the line with the test lamp operated from a variable auto transformer. In this way a standard may be quickly set and the two lamps varied together with the slight fluctuation in line voltage. With this method of wiring no electrical instruments are necessary. A variation of 2 per cent in the line voltage will show no perceptible photometric difference.

The scale range is from 4 to 120 candlepower and the range is further extended 10 to 1 by the introduction of a diaphragm over an opal glass in front of the comparison lamp carriage.

S. G. HIBBEN: I have always felt that it was essential for any manufacturer of lighting devices to have available some readily usable photometric facilities like these explained in the two papers. Any lighting service department is interested in knowing roughly the characteristics of the lighting devices they may be in position to recommend; and any manufacturer must know almost immediately, within an error of perhaps 5 per cent, the performance of units being in a process of design or manufacture.

It therefore seems necessary to carry on two general kinds of photometric work; one, perhaps, like the Electrical Testing Laboratories or others may do, which has an extreme accuracy,— the utmost possible accuracy; the other is the quick, commercial work. Devices like an icosahedron for integration or the distribution photometer, just explained, will be satisfactory for this second or latter class of work.

It is interesting to me to have worked with the authors on this development. One scarcely realizes the obstacles encountered. The icosahedron was of course developed in preference to a sphere on account of a limiting head room. One gets a larger integrating capacity with the icosahedron than with the sphere.

The ten-foot photometric axis on the distribution photometer, is an example of another space limitation, and it therefore seems necessary in publishing photometric data from either of these two devices, to make note of those conditions under which the test was made.

For example, I think on every distribution test, the photometric distance must be given and in addition (this is a vital point) the weight of the glass, especially if it be a diffusing globe or enclosing medium. It is fruitless to publish distribution curves or photometric data on a great many of our lighting devices and say, "This is a globe of a certain size and a certain shape, of a certain glass, which therefore has an output of 85 per cent." Any one who has seen the manufacture of those globes knows that a particular shape and glass may vary 15 or 20 per cent in photometric performance depending upon mechanical thickness of the walls of the glass.

Some other things in connection with these photometers must be watched: the glass of the mirrors for example must be very carefully chosen and supported, so that it will have no appreciable selective absorption, will not warp nor be hydroscopic, nor be subjected to wide temperature variations. The speed of rotation of the unit must be recorded. An automatic speed counter ought to be provided on the axis of the tested unit because the speed of rotation often has a bearing on the final candlepower results.

One point was not emphasized by the authors, namely; that the two devices just explained are used in conjunction, and a unit may be shifted from the distribution photometer immediately into the integrating icosahedron and back again for inter-test work.

A. H. TAYLOR: About ten years ago, we had occasion to design and build at the Bureau of Standards a distribution photometer, and after looking over the literature on the subject we decided on a double-mirror photometer similar in characteristics to that one given in Figure 6. It seems to me this is far superior in every way to a single-mirror photometer because it is very often difficult to rotate the source that you are measuring, and especially it is difficult to rotate it fast enough to get rid of the flicker that you may have, due to inequalities on the two sides of the luminaire.

Another case is an arc lamp. If any of you have worked on arc lamps on the distribution photometer, you know how mean they are to work with, because the arc in certain types is always wandering around the electrodes, and from one moment to another you may have a change of 50 per cent or more, which means that you have to go through your distribution curves a great many times, to be sure of anything that even approximates its distribution. So that any way by which you can get around some of the variables is of great benefit.

We find with a two-mirror selector that it cuts down this fluctuation a very great deal, where you do not rotate the arc, and I think you will find it is impracticable in most cases to rotate an arc lamp.

It is possible to use the two-mirror selector as a single-mirror selector if you so wish by merely covering up one of the mirrors with black velvet; there are certain angles that you can not reach with the two mirrors and you reach them in that same way, simply using one mirror and covering the other one with black velvet.

It is not difficult to get two mirrors that are almost identical in reflection, so that you do not introduce any errors due to the differences.

C. M. DOOLITTLE: The method of lighting the meters in the control table is interesting. The installation consists of three or four long, narrow lamps concealed in the box and lighting the interior, which is painted with white diffusing paint. That, it seems, is much better than the method of having miniature lamps immediately over the scale of the meter. There is no danger at all of seeing any brightly reflected image of the lamp filament in the mirror of the meter. At the same time a more cheerful appearance is produced. The average photometer room is usually painted black and is rather a dull and dingy place, and it is a little bit more cheerful for the operator who is watching the meters to have that somewhat illuminated space in front of him.

The front door of the meter table is painted white. That was really not intended to be that way. There is no object in having the door painted white, but that of course can be easily remedied.

I would like to emphasize again the point that Mr. Hibben brought out in regard to the ease of changing units from the distribution photometer to the icosahedron and vice versa. They are both in the same room and as you know, a large part of the time consumed in taking a distribution curve on a unit is required to set up the unit in place, and to set the lamp in the proper position in the unit. That is particularly true in cases where development work is being done on units, and it is necessary to mount the light in different positions, possibly in the position which is considered normal and then one or two inches above or below the normal position. Frequently, special jigs have to be made to hold the lamp in proper position. All that it is necessary to do to change over a unit completely from distribution photometer to icosahedron is simply to undo three bolts in a flange connection and carry the unit with the flange over to the icosahedron and bolt it in place there. The electrical connections are made automatically by means of a plug mounted in the flange.

Another point which Mr. Hibben mentioned on which I should like to lay further emphasis, is the value of these two instruments when used together as one unit for determining light distribution and unit efficiency. The distribution photometer is of course essential in that work, but especially when the unit is such as to produce a rather spotted distribution of light, it is extremely difficult to determine the efficiency, and the icosahedron or any large integrator will save a tremendous amount of time. The large integrator is of course very valuable for determining total light outputs and efficiencies of units, but on the other·hand it has to be supplemented by the distribution photometer to determine the form of the distribution. Together the two instruments form a most valuable complete set of apparatus for determining · light flux values on luminaires of all sorts.

C. C. COLBY, JR.: The authors would be ungrateful indeed if they failed to acknowledge Mr. Little's valuable help, in demonstrating for them on several occasions the photometer he has just described, which represents the type used as a basis for the design of the photometer described by the authors.

The title of our paper may perhaps be a trifle misleading, and I should like to emphasize the fact pointed out in the second paragraph of the paper; namely, that the actual principles involved in the design of our photometer are not new; the optical system, for example, is similar to that photometer mentioned by Mr. Little, and is also illustrated on page 66 of Bohle's Electric

Photometry and Illumination, London, 1912. The uniqueness of our photometer lies chiefly in the mechanical design.

Mr. Little has referred to the size of mirror C, Figure 9. With the present arrangement, the effective length of the optical path is 10 feet, and a light source with a maximum luminous dimension of 2 feet represents the largest size that can be photometered safely at this distance. The mirrors have been designed with a view to being able to take care of a unit with a maximum luminous dimension of 2.5 feet, in which case the bar with. the comparison lamp carriage would be moved to the right, and bearing J replaced by a larger hollow bearing, so as to increase the effective length of the optical path to about 12.5 feet.

Our scale and reading device are very similar to those of the photometer described by Mr. Little, the tape being graduated according to the inverse square law, and read by means of an index at the photometer head.

The distribution photometer is being used merely to determine the shape of the curve, the actual size being determined by calculation from the efficiency figure obtained by the use of the icosahedron. Accordingly, it is not necessary to hold any particular values of voltage or current for the test and comparison lamps, so long as the values chosen are held constant during the test, are less than the rated values for the particular lamps used, and give an approximate color match.

We are not prepared to dispute with Mr. Taylor about the relative merits of the photometers shown in Figures 6 and 9 for arc lamp photometry. We have not as yet had any arc lamps to photometer. While we feel, however, that our photometer will handle arc lamps in as satisfactory a manner as arc lamps can be handled, it was designed primarily for use with incandescent lamps, and we do not expect to do much arc lamp work with it.

We have chosen to try to rotate some of the larger units for the reason that we have been called upon to measure the distribution of quite a large number of units with panelled or rippled glass, and in either case it is difficult to secure average results unless the unit can be rotated. In case we cannot rotate units of this type and eliminate flicker sufficiently to secure trustworthy readings, we expect to take a number of curves in different vertical planes with the unit stationary and average the results.

THE REGULAR ICOSAHEDRON AS A SUBSTITUTE
FOR THE ULBRICHT SPHERE*

BY K. S. WEAVER AND B. E. SHACKELFORD**

SYNOPSIS: Increasing demand for efficiency tests of commercial lighting units has necessitated new and more expeditious methods of determining such values.

The spherical photometer would be a convenient instrument for such determinations. However, due to the large size of many of the units to be tested, an integrator of maximum size is desirable, and due to the difficulty in constructing a large true sphere, and to limited head room available, the icosahedron was thought to be more suitable for such a photometer.

The value of this solid as an integrator was determined by means of experiments on a 30-inch model, and as a result of these experiments a 10-foot icosahedron has been constructed for work with full sized units including the largest street lighting fixtures.

INTRODUCTION

Increasing interest has recently been shown in the performance and efficiency of nearly every type of modern lighting unit. This is particularly the case in view of the different effects on efficiencies of reflections and refractions in units of only slightly different designs. Accordingly there has resulted a considerable demand on illumination laboratories for tests of this character. The Illumination Laboratory of the Westinghouse Lamp Company has had so many requests of this nature that it has been found desirable to design new equipment for the work.

The method used in the past for determining the efficiencies of large luminaires has been to determine the light distribution and to calculate the efficiency by the use of the proper zonal factors. In many cases the distribution of light as well as the absolute efficiency was desired, but in the case of many of the units, of a more or less standard type, the general form of the distribution curve was known and the efficiency determination was the real object of the test. For such tests much time would have been saved if the efficiency could have been determined directly without

* A paper presented at the Annual Convention of the Illuminating Engineering Society, Swampscott, Mass., Sept. 25-28, 1922.

**Physical Laboratory, Westinghouse Lamp Co., Bloomfield, N. J.

The Illuminating Engineering Society is not responsible for the statements or opinions advanced by contributors.

resorting to the laborious method of calculation from the distribution curve data.

Moreover the proper determination of the efficiency of a unit by the distribution curve method is at times quite difficult and laborious. This difficulty may be due to a very high concentration in narrow zones or angles, to lack of uniformity in glassware particularly in large panelled post tops, or to certain other causes.

In such cases, errors of the order of 10 to 15 per cent are possible unless a great number of readings are taken in various vertical planes around the unit, and even after this laborious procedure, results are not entirely satisfactory. Hence it became desirable to investigate the possibilities of some sort of mammoth light flux integrator for this work. The use of the Ulbricht sphere naturally suggested itself. A study of the literature on the sphere included reports on several investigations of box integrators such as described by Grondahl.

PREVIOUS WORK ON PSEUDO SPHERE PHOTOMETERS

The box integrators[1] have the evident advantage over a true sphere of being easier to construct and more suitable for erection with limited head room. The units to be tested are so large that an integrator of the greatest possible size would be required for accurate work. Grondahl constructed an oblong box, painted it white on the inside and inserted a diffusing window in one face, the brightness of which was read by means of a Brodhun street photometer. He determined the illumination of the window due to a carbon lamp held in various positions in the box. A screen was used between the window and the lamp to cut off direct light from the lamp under test. The results were quite sensitive to the position of the lamp in the box.

He next tried an oblong box with truncated corners for the determination of the efficiencies of units of more or less concentrated distribution, and compared the results with values calculated from the distribution curves. Where the distribution was fairly uniform, the results were good, but in several cases where the light was concentrated in less than one hemisphere, errors of 6 to 7 per cent occurred.

---

[1] See "A Box Photometer" by L. O. Grondahl, TRANS. I. E. S., 1916, pp. 152. and "The Whitened Cube as a Precision Integrating Photometer," II. Buckley, *Jour. of the Inst. of Elec. Eng.*, London, Jan., 1921.

Buckley used a cube and worked out a very ingenious method for calculating the proper correction factor from the shape of the distribution curve. By the use of a concentrated beam of light he determined what he calls the contribution coefficient, or the relative amount of illumination on the observing window due to a given incident flux for various parts of the cube. In this way he mapped out the entire interior surface. From the contribution coefficients of the various parts of the interior of the cube and the shape of the distribution curve of the unit under test he calculated the amount of illumination of the observation window as a function of the distribution curve shape and hence determined a factor for each type of curve in order to reduce all readings to correct relative values. This method, although apparently quite involved, gave accurate results.

The box photometer while giving satisfactory results in the intercomparison of light sources of similar distribution did not promise sufficiently good results in the determination of the efficiencies of units with distribution as varied as that of those under test in this laboratory. The method of Buckley while giving accurate results seemed too much involved for conveniently handling a large amount of routine work. It was thought that the regular icosahedron might offer the advantages of the box photometer in ease of construction, and at the same time give results whose accuracy would be comparable with that obtained in the use of the true sphere. Accordingly in order to determine the advisability of erecting a full size integrator of this type, a 30-inch model was constructed for comparison with a 30-inch sphere. Models of large units were made on an appropriate scale for test in the 30-inch icosahedron.

GEOMETRY OF THE ICOSAHEDRON

Before describing the experimental work, a brief description of the geometrical properties of the regular icosahedron will be given. It is a solid bounded by twenty equal equilateral triangular faces and has twelve apexes and thirty edges. Three views of it are shown in Figure 1 together with the relative values of the principal dimensions. A development is shown in the upper half of Figure 2. This solid has the advantage of having all of its faces equal and of simple form which makes for ease in construction.

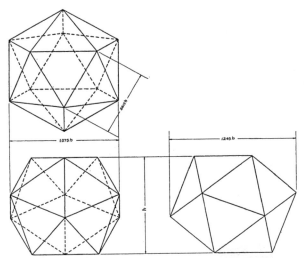

Fig. 1.—Geometrical form of regular icosahedron showing relative values of principal dimensions.

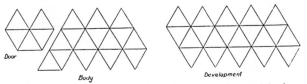

Door

Body

Development

Fig. 2.—Development of regular icosahedron and sections used in the construction of the 30-inch model.

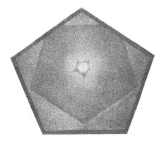

Fig. 3.—Showing light distribution in 30-inch icosahedron due to a beam projected on opal glass from outside.

Fig. 4.—Showing light distribution in the 30-inch icosahedron due to a narrow beam projected onto one side through the open door.

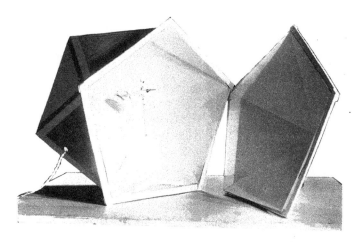

Fig. 5.—Set-up of model deep dome reflector in 30-inch icosahedron.

CONSTRUCTION OF THE 30-INCH MODEL

In the construction of the 30-inch model the twenty sides were cut out of heavy cardboard and bound together with binding paper into two sections as shown in the lower part of Figure 2.

The smaller section of five triangles was used for the door and the larger section for the body of the icosahedron. After the assembly had reached this point the panels were shellaced and given two coats of Kalco cold water paint. The resulting surface was quite satisfactory both in whiteness and diffusion. After the assembly was completed, wooden frames were built around the door and doorway giving a very stiff construction. The apex opposite the door was truncated and the observation window of white diffusing glass inserted here. The readings were taken by means of a Macbeth illuminometer sighted on the diffusing window.

THE LIGHT ACTION IN THE ICOSAHEDRON

The light action in the icosahedron is far from being as simple in theory as the light action in the sphere. While a complete mathematical analysis of the light action has not been attempted, a cursory examination of the action between two adjacent faces of the icosahedron shows that a given section of one face contributes a very different intensity of illumination at various points on the next face. The irregularities in the illumination on a face more remote are found to be less but still appreciable. The light action is brought out fairly well by Figure 3 and Figure 4. Figure 3 shows the distribution of light from a beam projected from the outside onto the diffusing window. The bright area around the window corresponds approximately to the region of maximum brightness determined by calculation. The practical effect of this condition has been to make it necessary at times in the tests on the model lighting units, to use screens larger than necessary to shield the glass window from direct light, otherwise when the area near the window is subjected to an illumination considerably greater than the average value on other parts of the icosahedron high readings would result. In the efficiency determinations the screens were made large enough to shield this area from direct light of the unit and this precaution seemed to eliminate the trouble.

Figure 4 shows the light distribution due to a concentrated beam projected into the icosahedron with the door removed. Slight differences in illumination on the various faces are shown here, the face on which the beam is incident is especially dark since it is illuminated only by reflections of a higher order than any of the other faces.

The observation window was placed at an apex rather than on a face because in the latter position the window would not receive any direct illumination, from the face in which it was inserted. A window of infinitesimal size inserted at an apex would receive light from only fifteen of the twenty sides, but practically, the finite size of the window results in its being illuminated in part by all of the faces of the icosahedron. In the large icosahedron which has now been constructed the observation window is no larger than the one used in the present work, hence the extra area to be shielded around the window is relatively much less. For this reason it is expected that the 10-foot icosahedron will give if anything, more satisfactory results than the 30-inch model.

EXPERIMENTAL WORK

For convenience in discussion, the experimental work may be divided into the following sections:

1. Test of effect of change in position of light source in the icosahedron on the readings, by means of a scale model of a Colonial Reflecto-Lux Post Top.

2. Determination of effect of light distribution by setting the model skirted cone enamelled steel reflector at different angles in the icosahedron.

3. Comparison of efficiencies of the model Post Top as determined in the 30-inch sphere and in the 30-inch icosahedron.

4. Comparison of efficiencies of model reflector as determined by the icosahedron and by calculation from distribution curves on unit and bare lamp.

1. For the first and third tests the model Post Top was made of such a size that it would bear the same relation to the 30-inch sphere as the full sized unit would bear to a 10-foot sphere. The

form of its distribution curve is shown in Figure 6 (f). A
Mazda B automobile headlight lamp was used as the light source.

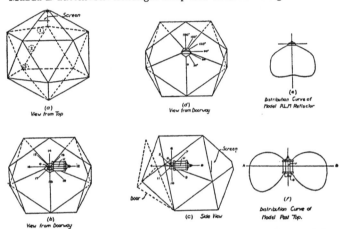

Fig. 6.—Chart showing position of model units during test in 30-inch icosahedron,
as well as the forms of the distribution curves of the model units.

The first series of readings were taken with the axis AB
(see Figure 6 (f)), horizontal and the position varied as shown
in Figure 3 (a). The unit was kept at the height of the window,
a little more than half way up from the lower face of the icosahedron. Table I gives the results of this test, each value given
representing the average of five readings.

TABLE I.

Position—see Figure 6 (a) Relative value of readings	I 8.44	2 8.80	3 8.45

The next test using the model Post Top was intended to show
the effect of change in direction of the light in the icosahedron.
For the various positions, the unit was rotated in a plane perpendicular to the window and to one side of it as shown in Figure
6 (c). Numbers 4 to 11 represent averages of five readings

taken with the center of rotation at "O." Numbers from 4a to 11a represent the same setting except that the light center "U" instead of "O" was kept fixed as the unit was rotated. Table II gives the results of these observations.

TABLE II.

Position	Relative value of readings	Position	Relative value of readings
4	8.30	4a	8.43
5	8.18	5a	8.39
6	8.70	6a	8.58
7	8.90	7a	8.81
8	8.80	8a	8.88
9	8.78	9a	8.80
10	8.74	10a	8.52
11	8.23	11a	8.74
Max. difference 0.72		Max. difference 0.49	

As is to be expected, the series 4a-11a show less variation than the preceeding series, the maximum difference of 0.49 representing about the maximum difference among individual readings on the same value. These results lead us to believe that the irregularities in the series 4-11 were due as much to change in the distance between light center and the walls, as to differences in the direction of the light. Positions 12-19 and 12a-19a were similar to the preceeding set except that the plane of rotation was parallel to the window and the center of rotation opposite the window Figure 6 (b). The readings 12-19 are with the center of rotation at "O" and 12a-19a with the center of rotation at "U," the light center of the unit. The results are given in Table III.

TABLE III.

Position	Relative value of readings	Position	Relative value of readings
12	8.42	12a	8.21
13	8.57	13a	8.42
14	8.37	14a	8.42
15	8.36	15a	8.34
16	8.34	16a	8.56
17	8.20	17a	8.11
18	8.15	18a	8.16
19	8.00	19a	8.39
Max. difference 0.57		Max. difference 0.45	

These results show irregularities slightly greater than the experimental error in readings but small considering the extreme variation in conditions.

2. For test number 2, a model skirted cone enamelled steel reflector was made representing a 14-inch reflector to the same scale as used for the Post Top. The angle of cut off was changed from the standard value to 65° in order to give a more concentrated distribution of light and hence exaggerate the irregularities inherent in the use of such a source. The distribution curve for the unit is given in Figure 6 (e). For the various positions, the unit was turned in a plane parallel to the window as shown in Figure 6 (d). The results of this test are given in Table IV.

TABLE IV.

Angle in degrees	0	30	60	90	120	150	180
Relative value of readings	7.02	7.07	6.94	6.98	7.02	7.02	7.16

The center of rotation of the unit was at the height of the window, a little more than half way up from the lower face of the icosahedron. These readings were taken by a different observer and the limit of error in individual readings is somewhat less than in the preceding work. The high value at 180° probably represents a real increase, and may be due to the fact that the unit is nearest to the upper face of the icosahedron. The set up for this test is shown in the photograph Figure 5. For the sake of comparison some results of investigators using true spheres may be interesting. Bloch[2] in determining the best location for lamps under test in the sphere obtained the following results, as shown in Table V.

TABLE V.

Source	SCP	Relative value	Source at center		h/r 0. 1	
			Readings	Relative value	Readings	Relative value
Carbon fil.	21.1	1	43	1	44.5	1
6 Amp. arc	225	12.1	473	11	550	12.3
15 Amp. arc	1000	47.3	1800	41.8	2000	46.3

This table shows the variation due to change in position of the source, and also indicates that the best position for inter-comparison of different sources is in the eccentric position. The

[2] Bloch; *Electrotech Zeit.*, 46, p. 1074, 1905

value h/r is the ratio of the displacement of the source from the center of the sphere to the radius of the sphere.

In their paper on the 88-inch sphere of the Bureau of Standards, Rosa and Taylor[3] describe a test to show the variation due to change in position of a 100-watt bare lamp from the center to a point near the wall. The extreme variation given is slightly over 3 per cent.

3. Tests 3 and 4 were intended to show what could be expected of the icosahedron in the determination of efficiencies of luminaires when all precautions were taken that are ordinarily observed in order to get the best results in the use of the true sphere.[4]

In the determination of the efficiencies of the model Post Top and the skirted cone reflector the unit was in position during both sets of readings in order to balance out as far as possible the effect of the light absorption of the unit. The bare lamp and unit were set up in positions of symetry with respect to the window. Three screens were used, one between the window and unit, one between the window and bare lamp, one between the unit and bare lamp.

The results of the test on the efficiences of the model Post Top as determined in the 30-inch icosahedron and in the 30-inch sphere are given in Table VI.

TABLE VI.

	Individual readings					Average
By 30-inch Icosahedron						
Bare lamp	5.35	5.05	5.25	5.30	5.23	5.24
Unit	2.51	2.49	2.52	2.48	2.53	2.50
Efficiency 47.7 per cent						
By 30-inch Sphere						
Bare lamp	3.40	3.40	3.37	3.22	3.40	3.36
Unit	1.52	1.56	1.57	1.53	1.55	1.55
Efficiency 46.2 per cent						

4. The efficiency of the model skirted cone reflector was determined by calculation from the distribution curves of the bare lamp and unit, in order to get a check independent of the sphere readings. The results are given in Table VII.

[3] "The Photometric Sphere, Its Construction and Use," Rosa and Taylor, I. E. S., 1916, Vol. 11, p. 453.

[4] "Notes on the Integrating Sphere and Arc Lamp Photometry," by N. K. Chaney and E. L. Clark, I. E. S., 1915, Vol. X, p. 1.

TABLE VII.

	Individual readings					Average
By Icosahedron						
Bare lamp	5.52	5.75	5.60	5.55	5.70	5.63
Unit	4.20	4.20	4.23	4.12	4.33	4.22
Efficiency 75 per cent						
By Distribution Curve Calculations						
Efficiency 74 per cent						

The test of the model Post Top in the sphere brought out the fact that the icosahedron is much more roomy and convenient to work in than a sphere of the same height. The arrangement of screens in the 30-inch sphere was found to be very troublesome due to the limited space hence it was thought desirable to test the model-skirted cone reflector by the distribution curve method in order to get a more accurate check on the finding of the icosahedron.

CONCLUSIONS

These efficiency determinations, especially the one on the model-skirted cone reflector are considered quite satisfactory for commercial work. The results of tests 1 and 2 representing exaggerated conditions, indicate that irregularities in distribution of light that would make efficiency determinations from distribution curves almost impossible will not seriously interfere with satisfactory determinations of efficiencies in the icosahedron.

That satisfactory results can be obtained in the icosahedron in spite of its theoratically irregular action is due partly to the multiple reflections which tend to wash out the irregularities, and partly due to the fact that under working conditions the illumination is not nearly so concentrated as that shown in Figures 3 and 4.

A high reflection coefficient for the walls is evidently of prime importance for the successful operation of the icosahedron, being even more vital to success than when a true sphere is used.

It is felt that as a substitute for the small spheres used in reflection factor[5] work where a concentrated beam is used as the

[5] "The Use of the Ulbricht Sphere in Measuring Reflection and Transmission Factors," Enoch Karrer, *Jour. Amer. Optical Soc.*, Jan., 1921, and "A Simple Portable Reflectometer of the Absolute Type," A. H. Taylor, *Jour. Amer. Optical Soc.*, Jan., 1920.

light source, the icosahedron would be far from satisfactory. As a matter of fact there would be little point in departing from the spherical shape for such instruments since a small sphere is about as easy to construct as a substitute solid. For large integrators, however, the icosahedron has two great advantages, greater simplicity in construction and greater roominess as compared with the true sphere.

As a result of this work, it was decided to construct an icosahedron 10 feet high for the testing of large lighting units. This has now been constructed. The panels are of heavy composition board; these are fastened together at the angles by means of strips of heavy sheet iron. The door includes five sides of the icosahedron as in the case of the model, thus allowing the largest units to be set up conveniently. In order to enable efficiency determinations to be made conveniently as a check on the distribution curve work, the icosahedron is provided with a coupling similar to the one on our new distribution photometer so that a unit may be removed from the distribution photometer and set up in the icosahedron by simply disconnecting this interchangeable coupling. The coupling carries the electrical contacts so that the unit is automatically placed in circuit when fixed in place.

It may be found advantageous to determine all efficiencies by means of the icosahedron even when a distribution curve is to be determined. In this case the distribution photometer work will involve finding the shape of the curve only, the relative values being reduced to the correct values according to the value of the efficiency. In this case, the maintenance of primary standards will not be necessary for the work, since the shape of the distribution curve and the efficiency will furnish complete data for the test. It will only be necessary for meters and lamps to remain constant for the duration of the test.

An extensive test of the large icosahedron is now being carried out and it is expected that fairly complete test data will shortly be available.

DISCUSSION

C. M. DOOLITTLE: Those of you especially who have been directly connected with the work of taking distribution curves will realize how difficult it is sometimes to get the proper values of efficiency by that method, particularly where the unit is at all irregular in its distribution of light. For instance paneled units and units of rippled glass, ribbed glass and irregular glass of all sorts, very frequently produce spots of light or light and dark streaks. While it is possible to determine the efficiency on such units by means of the distribution photometer, it entails a tremendous amount of work because it is necessary to take a large number of readings. The use of the icosahedron or any other integrating device means a great saving in time and labor.

Several days ago, someone brought up the question whether a unit with a very concentrated distribution of light could be accurately measured in a device of this sort. That can be answered in this way: the model was actually tested out with two units of very widely different distribution and found to be sufficiently accurate. For a unit of a very concentrated distribution, it might possibly be that a somewhat larger error could be expected, but in that case the icosahedron could be calibrated with a particular value for units of high concentration. It is not likely that units of that sort will be encountered in more than two or three cases perhaps out of one hundred, and for such cases it would be possible to take a very long series of distribution curve readings on a concentrated unit and then set that in the icosahedron and determine the calibration for units of that particular distribution. If there were any other sorts of distribution which were so very peculiar as to possibly throw out the calibration, the same method might be applied.

One very interesting application of the icosahedron is in connection with testing a number of units very quickly. A manufacturer of, say, diffusing globes of some sort, desiring to know the average distribution or efficiency values of his product, could easily send in five or ten globes or even more, for test. When one of these units is set up in place it is a very simple matter to take readings on a whole group of similar globes. You simply open the door of the icosahedron, slip another globe in place and take another set of readings, almost as quickly as sphere readings could be taken on large bare lamps.

A. H. TAYLOR: This paper is of considerable interest to me because I have done a great deal of work with the integrating sphere and I am interested to see how this compares in precision with the sphere. Just considering the thing purely from a casual viewpoint, it appears to me that the authors might get a still further increase in accuracy by rounding out the corners slightly with a little bit of plaster or something else to get away from all sharp angles.

On Page 298, the author has referred to some work which I reported on at a convention about six years ago, on an integrating sphere, in which I showed that by moving a lamp toward the door of the sphere we had a falling-off of sphere window intensity of a total of 3 per cent in the range that we moved it. This, by the way, was done by moving the lamp toward the wall which was screened from the sphere window.

In a publication of the Bureau of Standards which is now in press, I show that I account for this almost perfectly by reason of the fact that as you move the lamp closer to the wall in that way, more and more light falls on the areas that are screened from the window, and has to be reflected at least twice before any of it reaches the window. The theoretical curve I obtained from computations on that basis agrees almost perfectly with the observed curves, so I think that is the whole explanation.

I would like to ask the authors whether they can give us any estimate of the cost of the ten-foot icosahedron as compared with the ten-foot sphere?

F. E. CADY: It is very interesting to see the developments which have come in the way of integrating photometers since the time of the Matthews instrument, and the number of variations of the idea of the sphere some of which are referred to in the paper; nothing however is said in here regarding the method of using screens or the character of the window. I presume possibly the same type of window is used as that in the ordinary sphere, but I would like to ask the authors whether anybody has done anything in connection with the determination of the proper position of the screen. Of course, in a sphere where the mathematics are relatively simple, it is not difficult to compute the proper position of the screen, but if the mathematics of this icosahedron are not thoroughly understood, it seems to me there might be there a possible source of error, which should be taken into consideration.

F. A. BENFORD: In regard to the probable accuracy of this instrument, I think the lantern slide that we saw a minute ago explained what might possibly happen if you used, we will say, a lamp with a band refractor or some unit that gives very concentrated beams in some direction.

I think this instrument would have dubious values if you ran into extremes of photometric distribution.

C. H. SHARP: I suppose it may be said that the nearer any enclosure approaches to a sphere, the more perfect it becomes as an integrator, and the sphere being the ideal toward which they all approach, by taking twenty sides, a pretty close approach has been made to the sphere.

I think it would have been well if the authors had recalled the paper* presented by Dr. Carl Hering before this Society some thirteen years ago, in which he discussed mathematically different solids with respect to their efficiency as integrators, and he considered not only the regular icosahedron but a number of other solids with a smaller number of sides. If I remember correctly he had in mind using regular mirrors, not diffusing surfaces, considering that then the integration would be rather a summation of individual beam intensities, somewhat similar to what was made by the Matthews photometer, only lacking the complications of smoked glasses and so forth that the Matthews photometer had. It was a very interesting, very able article, and deserves mention in connection with this very practical paper on the regular icosahedron.

B. E. SHACKELFORD: I was very much interested in Mr. Taylor's comment relative to rounding the corners, because that is one of the things which we have felt, from the beginning, that we would possibly want to do but so far we have not made up our minds that it will be necessary.

Referring to Mr. Taylor's paper which is now in press, as to the reason for the change of position of the unit in the sphere, we have felt that the same reason held in our own case, and we did not bring out that point in the paper particularly because it is of the nature of a preliminary report and we hope later to

* Hering, Carl, "Measuring Spherical Candlepower by Averaging; the Equal Sub-division of the Sphere," TRANS. I. E. S., 1909, IV, p. 354.

take up some of these various effects. From a practical stand-point, that is not quite as important as it is from a technical view-point.

Referring to Mr. Cady's question as to the character and positions of the screens, we have done some work on that so far, the first result being that the screens are considerably larger than they would be in the case of a sphere. They very much more than cover the window, and a third screen is between the unit and the lamp. The unit is always in position, whether the lamp is being photometered in the unit or in the outside receptacle. As to the character of the window, we are using the same type of glass, and a similar set-up to that used on the sphere photometers. The full size apparatus has been up only a matter of a few weeks; naturally we have been able to get what you might call complete data on only a few points. We hope that when we accumulate further information we will be able to give a more complete report.

Referring to Mr. Benford's comments, I believe he mentioned it was his understanding that the window was in the side or in one of the sides. It is at one of the apexes, and it is very small of course, relative to the side. The edge of the sides is approximately 6.5 feet, so the window is more or less of a point.

We do get the rings and circles, but using the large screens and for the various units that we have tried, we have found that there appears to be less error and less cause of variation in the icosahedron determination than there is either in the distribution curve method, where it is very difficult at times to get zonal readings, or in the use of the available spheres. Perhaps Mr. Benford's own apparatus in cases of the kind gets around some of these troubles in a better fashion, but we feel that in the icosahedron we have certain practical advantages that make it a very useful piece of apparatus.

Dr. Sharp mentioned Dr. Hering's paper, and we should have made mention of that at the time this paper was written. We omitted that not because of any lack of feeling of obligation for this previous work, but because we were treating this from the start mainly as a practical proposition. A future more complete paper will have some discussion of the relation between the two.

ABSTRACTS

In this section of the TRANSACTIONS there will be used (1) ABSTRACTS of papers of general interest pertaining to the field of illumination appearing in technical journals, (2) ABSTRACTS of papers presented before the Illuminating Engineering Society, and (3) NOTES on research problems now in progress.

STANDARD SYSTEM OF LIGHTING FOR POST OFFICES

Establishment of a standard system of lighting in all post offices has been strongly recommended in a report just submitted to the Post Office Department by the Office of Industrial Hygiene and Sanitation of the United States Public Health Service. This report, the result of an elaborate survey conducted at the special request of Postmaster-General Work, maintains that millions of dollars will be saved through increased speed in the work of postal employees provided that a complete change is effected in the illumination of workrooms of post offices throughout the country.

Virtually all the work performed in post offices with the exception of that of laborers depends primarily upon the use of the eyes, making the question of illumination of paramount importance in decreasing or increasing its rapidity and accuracy. Conducting a thorough and technical study of lighting over a long period of time at two representative post offices, one modern and the other of the old type, the experts observed almost five thousand postal workers employed constantly in handling the mails both under artificial and natural light.

They found that illumination in many post offices was low in intensity and unsatisfactory in quality. In some instances it fell below the requirements provided by the state code of laws and below the artificial light furnished employees in private industries doing similar work. A study of the relationship between the volume of illumination and the strain on the eyes of workers revealed the fact that there were more eye defects among employees working under the average illumination of 2 to 3 foot-candles in the old post office than under the 3 to 4 foot-candles in the new post office. Sorting and separating mail in the basements of post office buildings was especially discountenanced except in extreme emergencies.

In the speed and accuracy tests it was found that for the letter separators there was an average increase in speed, or decrease in the times of separation, of at least 4.4 per cent when going from 3.6 to 8 foot-candles. Assuming that the same relative increase of speed would prevail in all the divisions of the post office where these tests were made under the same relative increase in illumination, an annual net saving, after deducting increased cost for lamps and energy, of about $109,000 was indicated in this post office alone.

The investigations recommended that there should be installed in the general workrooms and offices systems of totally inclosing units of the diffusing or light-directing type, giving a general intensity when first installed of 10 foot-candles everywhere on a horizontal working plane 45 inches above the floor. All local lighting should be done away with.

The lighting unit to be installed in the general workrooms of the post office should be of such quality of glass and of such shape and size that its surface brightness at any point would not exceed 2.5 cp. per square inch when used with an incandescent lamp or other source of light emitting 3,100 lumens. This unit should have an output of at least 80 per cent of that of the clear lamp and a spherical distribution of its candlepower such that at least 8 per cent of the light emitted by the clear lamp would be emitted by the unit through the zone from 0° to 30°, at least 28 per cent from 0° to 60°, at least 48 per cent from 0° to 90°, and at least 25 per cent from 90° to 180°.

The lighting unit for the offices of the post office should be of such quality of glass and of such shape and size that its surface brightness at any point would not exceed 2 cp. per square inch when used with an incandescent lamp or other source of light emitting 3,100 lumens. This unit should have an output of at least 80 per cent of that of the clear lamp and a spherical distribution of its candlepower such that at least 8 per cent of the light emitted by the clear lamp would be emitted by the unit through the zone from 0° to 30°, at least 23 per cent from 0° to 60°, at least 43 per cent from 0° to 90°, and at least 35 per cent from 90° to 180°.

Both the units for the general workrooms and the offices should be such in number and so spaced that the brightness of the units measured in lumens per square foot would not be more than one hundred times as great as the intensity of the illumination, measured in foot-candles, produced by them on a horizontal plane 45 inches above the floor.

Recommendations on the care of the lighting system and a disquisition on modern methods of illumination are included in the report.

LIGHTING IN FACTORIES AND WORKSHOPS

The third report of the Home Office departmental committee (England) marks a new stage in the treatment of the above subject. The two previous reports are first summarized. The primary recommendation contained in the first report stated that there should be a statutory provision in general terms requiring adequate and suitable lighting in every part of a factory or workshop and giving power to the Secretary of State to define such lighting. In the second report general requirements in regard to avoidance of glare, elimination of inconvenient shadows and absence of flicker were made. In the report recently issued the requirements in regard to glare are supplemented by an indication that when a brilliant source is covered by a small shade the brightness may considerably exceed 15 cp. to 20 cp. per square inch and should be treated as a source and its position with regard to the worker limited accordingly. The chief question considered in the third report is the degree of illumination needed for the actual carrying on of work. The committee presents a comprehensive schedule of industrial operation, divided into two classes described respectively as "fine work" and "very fine work," the former requiring 3 foot-candles and the latter about 5 foot-candles.—*Illuminating Engineer* (London), Vol. 15, No. 7.

MEASUREMENT OF LIGHT

BY J. W. WALSH

The fundamental photometric magnitude from the point of view of visual measurement is, according to the author, brightness and not illumination. The photometric unit is one of luminous intensity or luminous flux. Of the two possible systems of definitions based on these respective magnitudes, that in which the unit is maintained seems preferable because it follows the natural order of mental conception. The relation between the flux unit of brightness (the lambert) and the intensity unit (the candle per square centimeter) is pointed out.—*Philosophical Magazine*, December, 1922.

RECENT DEVELOPMENTS AND MODERN REQUIRE-MENTS IN STREET LIGHTING

BY H. T. HARRISON

In a paper presented before the British Illuminating Society the author refers to the great importance of good public lighting in promoting order and safety in the streets. The burden of rates is one reason why progress has been delayed, but improved street lighting need not always be at an increased expenditure. He also pointed out that the classification of streets on the basis of minimum horizontal illumination by joint committees on this subject in England and the classification by the street lighting committee of the N. E. L. A. were in close accord as to the amount of light.—*Electrical Times*, Dec. 28, 1922.

NEON-GLOW DISCHARGE LAMP ON ALTERNAT-ING-CURRENT CIRCUITS

BY R. A. BROCKBANK AND L. E. RYALL

After giving the various characteristics of this type of lamp by means of numerous curves, the author suggests further experiments that should be carried on to determine future applications.

Among these experiments would be a detailed study of the lamp for stroboscopic purposes at all frequencies, a study of the photometric properties of the lamp and experiments to determine the possibilities of its use for photographic purposes.—*Electrician*, Jan. 5, 1923.

TERMINAL LIGHTING DEVELOPMENT ON NEW HAVEN RAILROAD

BY G. F. JOHNSON

Railroad yard lighting is primarily space lighting, with the following problems injected, namely, the presence of deteriorating gases, movable cars causing shadows, lack of clearance between tracks to place poles, the prevention of glare that will interfere with the operation of the trains through or near the yards and the interference of the overhead propulsion current wires. Improved illumination in several of the yards on this road is the result of much study and experiment and is briefly discussed.—*Railway Electric Engineer*, January, 1923.

SOCIETY AFFAIRS

SECTION ACTIVITIES

NEW ENGLAND Meeting—February 9, 1923

On Friday, February 9th, Mr. J. M. Shute of the Edison Lamp Works addressed the New England Section at the Industrial Lighting Exhibit on "Motion Picture Signs and Colored Window Lighting."

The psychology and motives underlying the use of electric signs by motion picture theatres was very well explained by the speaker. Photographs of numerous signs of note were displayed showing the relation of the shape and type of the sign to the type of building.

In his remarks on the colored lighting of show windows, Mr. Shute told of the tests recently made showing the drawing power of different intensities and colored light on the passersby. Numerous tests showed that the higher the intensity the greater the number of people who stopped to look at the window. Changing the color of light accomplished the same effect.

In the discussion, Mr. R. R. Burnham called attention to the careful study the motion picture theatre owner had evidently made of his possible patronage and his methods of attracting patrons. This was, without question, a great factor in the success of this industry, and should serve as an example to the rest of us.

Mr. J. Daniels told of the progress of motion picture signs in Boston, and how old regulations had formerly prevented the installation of large signs. However, this condition has been changed and a number of large signs were in the process of erection.

NEW YORK Meeting—February 15, 1923

The New York Section held the February meeting in the auditorium of the Consolidated Gas Company Building, at which time were discussed two interesting papers.

"Some Observations in Europe,' by Dr. Clayton H. Sharp, Electrical Testing Laboratories, New York City, and "Comments on the International Standardization of Edison Lamp Bases and Sockets," by Mr. F. V. Magalhaes, New York Edison Co., New York City.

Dr. Sharp and Mr. Magalhaes were members of the American delegation to the International Conference of the International Electrotechnical Commission held last November at Geneva, Switzerland.

Dr. Sharp's talk covered the general activities of the Conference held in Geneva, the research conditions in England, and, mainly, the principle topic of the Geneva Conference, "The Standardization of the Edison Screw Base."

Mr. Magalhaes' talk also related to the subject of the probable standardization and international adoption of the Edison Lamp Socket and Base. Reference was made to Edison Standardization Bulletin, and a report of the A. S. M. E., also, the 1918 and 1920 theoretical reports of M. Zetter. This report did not take into consideration any of the financial and other inconveniences of the standardization, such as adapting lighting fixtures to accommodate the new base, abandonment of old machinery, and the purchase of new machinery. Mr. Magalhaes outlined the activities of the committee as follows: First day—Candlebra and miniature bases. Most complicated and least important. This subject was dropped for the time being. The discussion covered the possibility of deepening the thread. The opinion of the representatives from England seemed to be that the standardization of the Lamp would be most desirable. American representatives suggested that standardization of the Base be decided upon, and that the lamp standardization would automatically follow. Second day—The general opinion seemed to favor the standardization of the Edison Medium Screw Base. It was suggested that the term "universal" or "Edison" be used rather than American. The definite agreement was reached as follows: 1. Standardization on the Universal socket. 2. Questions of fact to be settled by experts of both nations, and report. 3. Another meeting to be held as soon as possible to finally accept socket.

There were in attendance some fifty members and guests. Prior to the meeting an informal dinner was held at the Gramercy Inn.

PHILADELPHIA Meeting—February 13, 1923

The Philadelphia Section met at the Engineers' Club on February 13 to hear a paper, "Reasons for Growing Interest in Show and Store Window Lighting," by Mr. J. R. Colville, Engineering Department of the National Lamp Works, Cleveland, Ohio.

Mr. Colville started his talk by a brief comparison between present day electric lamps and candle and oil illuminants, stating that the cost per candlepower was now very much less than it used to be. His second division was the development of the means of increasing the usefulness of the lamps by means of proper reflectors and reflecting surfaces. He called attention to the still common practice of using the later style lamps with old type reflectors for which they were not intended. For example, the use of a long neck nitrogen lamp in a glass shade intended for a carbon lamp, in which case the filament was wholly exposed to view. The third division of his talk referred to the collection of data of value to merchants and showing the benefits of good illumination. He stated that increasing the intensity of illumination added to the attractiveness of the store and show windows. As an illustration, two windows in the same store of unequal intensity were selected and the number of people in the street pausing before each were counted. The illumination of the two windows were then reversed and number of persons again counted. It was found that in both cases the greater number paused in front of the window which had the brighter intensity. In speaking of the use of color in the window he stated that caution was necessary; to simply use a color throughout the entire window was an attraction for a while but unless artistically done it was apt to be a failure. Good results can be obtained in connection with colored lamps if a flood lamp is used to bring out some particular feature of the window display. Flood lighting with high intensities can also be used as a means of eliminating the reflection which is frequently noticeable from the glass of the window.

Mr. Colville illustrated his talk by means of demonstration with a lighting booth. Statues, pictures and a minature show window with a display were shown and effects obtained by means of light coming from different directions and also with lights of different colors. By means of a chart he showed how the intensity of illumination could be increased by proper maintenance and by increasing the effectiveness of the reflectors and other reflecting surfaces.

Thirty-six members and guests attended a dinner at the Engineers' Club preceding the meeting at which there was an attendance of fifty.

CLEVELAND Meeting—February 15, 1923

The joint meeting of the Cleveland Section of the A. I. E. E. and the Cleveland Chapter of the I. E. S. was held on the evening of February 15, at Nela Park. Those attending were entertained at dinner as guests of the National Lamp Works.

The paper, "Modern Street Lighting Plans for a Large City," was presented by Mr. E. A. Anderson and Mr. O. F. Haas.

The paper clearly pointed out the need for more modern street lighting equipment. In this relation the benefits are the reduction of accidents and crimes along with 'its relation to other municipal enterprises in the interests of civic duty. The paper was profusely illustrated with lantern slides and charts.

The paper was quite generally discussed by Mr. C. C. Beckwith, Commissioner of Lighting and Heating, City of Cleveland, Professor H. B. Dates, Case School of Applied Science, Mr. W. E. Bush, Illuminating Engineer of the Thompson-Houston Company, London, England, and Mr. C. A. Atherton, National Lamp Works.

The Exhibition of Ornamental Street Lighting Equipment installed at Nela Park was inspected and an additional feature of the evening was the lighting of a 30-kw. incandescent lamp.

The meeting was attended by approximately a total of one hundred.

TORONTO Meeting—January 22, 1923

The Toronto Chapter meeting was held on January 22, in the Engineering Building of the University of Toronto. Professor G. R. Anderson gave an interesting paper, "Fundamental Principles in Illumination." A general discussion was held and fifteen members and guests were present.

COLUMBUS Meeting—February 8, 1923

At the organization meeting of the Columbus Chapter held on February 8, Mr. G. F. Evans and Mr. R. C. Moore were elected chairman and secretary of the Chapter.

COUNCIL NOTES

ITEMS OF INTEREST

At the meeting of the Council, February 8, 1923, the following were elected to membership:

Five Members

CHURCH, A. C., Pacific States Electric Co., 240 S. Los Angeles St., Los Angeles, Cal.

COGAN, D. E., General Electric Company, 84 State Street, Boston, Mass.

DRINKER, PHILIP, Harvard Medical School, Boston, Mass.

PATE, C. BERTREM, National X-Ray Reflector Co., 16 State St., Rochester, N. Y.

THOMSON, ELIHU, General Electric Company, Lynn, Mass.

Twenty-Four Associate Members

ADDIE, CHARLES E., Denver Gas & Electric Light Co., 900 15th Street, Denver, Colo.

ARNOLD, M. EDWIN, M. E. Arnold & Co., 1019 Cherry Street, Philadelphia, Pa.

BARR, TAYLOR M., General Electric Co., Schenectady, N. Y.

BEGGS, EUGENE W., Westinghouse Lamp Co., Bloomfield, N. J.

BROWN, HARRY, 113 St. Marks Place, New York, N. Y.

BUCKNAM, PAUL C., Pettingell-Andrews Co., Atlantic Avenue, Boston, Mass.

CLASEN, ARTHUR J., American Bank Note Company, Garrison Ave and Lafayette St., Bronx, N. Y.

DODDS, GEORGE, Homestead, Pa.

D'OLIVE, EUGENE R., Commonwealth Edison Company, 28 N. Market Street, Chicago, Ill.

ERICKSEN, THOMAS, Detroit Stove Works, 6900 Jefferson Ave., E., Detroit, Mich.

FOELLER, HENRY A., Foeller, Schober & Stephenson, Architects, Nicolet Bldg., Green Bay, Wis.

GOSLING, EDWARD P., Newport Electric Corp., 449 Thames St., Newport, R. I.

JOHNSON, L. B., General Electric Co., 719 Newhouse Bldg., Salt Lake City, Utah.

KIRLIN, IVAN M., National X-Ray Reflector Company, 400 Penobscot Bldg., Detroit, Mich.

KITTLE, ROBERT G., Benjamin Elec. Mfg. Company, 243 W. 17th Street, New York, N. Y.

MORROW, G. T., National X-Ray Reflector Company, 331 Fourth Avenue, Pittsburgh, Pa.

NELSON GEORGE E., National Lamp Works, 642 Beaubien Street, Detroit, Mich.

NYE, ARTHUR W., University of Southern California, 3551 University Ave., Los Angeles, Cal.

PIERSON, WM. V., Western Electric Co., 106 South Street, Baltimore, Md.

PRIGGE, JOHN, JR., 441 44th Street, Brooklyn, N. Y.

SANDOVAL, H. E., Pacific Gas & Elec. Co., 445 Sutter Street, San Francisco, Cal.

THOMPSON, R. B., Central Hudson Gas & Elec. Co., 129 Broadway, Newburg, N. Y.

FRIEDRICH, ERNEST G., Westinghouse Elec. & Mfg. Co., 3rd and Elm Sts., Cincinnati, Ohio.

WORSSAM, FRANK H., Pettingell-Andrews Co., 511 Atlantic Avenue, Boston, Mass.

One Student Member

MALONE, JAMES F., Polytechnic Institute of Brooklyn, Brooklyn, N. Y.

Six Sustaining Members

BANGOR RAILWAY & ELECTRIC CO., Bangor, Me.
W. E. Stooper, Official Representative.

GLOUCESTER ELECTRIC CO., 26 Vincent St., Gloucester, Mass.
Walter L. Brown, Jr., Official Representative.

LUNDIN ELECTRIC & MACHINE CO., 10 Thatcher St., Boston, Mass.
Emil O. Lundin, Official Representative.

NEWBURYPORT GAS & ELECTRIC CO., Newburyport, Mass.
J. Lee Potter, Official Representative.

UNION ELECTRIC SUPPLY CO., Providence, R. I.
Frank L. Falk, Official Representative.

UNITED ELECTRIC LIGHT CO., 73 State Street, Springfield, Mass.
J. Frank Murray, Official Representative.

One Transfer to Associate Membership

HAYNES, PIERRE E., 1089 Ellicott Square Bldg., Buffalo, N. Y.

The General Secretary reported the death of one associate member, JOHN DOYLE, Consolidated Gas Company, 128 East 15th Street, New York, N. Y.

CONFIRMATION OF APPOINTMENTS

As Members of the Committee on Nomenclature and Standards—E. C. McKinnie and William J. Serrill.

*As I. E. S. Representatives on A. E. S. C. Sectional Committee on Illuminating Engineering Nomenclature and Photometric Standards—*Howard Lyon and G. H. Stickney.

*As Individual Members on A. E. S. C. Sectional Committee on Illuminating Engineering Nomenclature and Photometric Standards—*G. A. Hoadley and M. Luckiesh.

NEWS ITEMS

Mr. Julius Daniels of The Edison Electric Illuminating Co., of Boston has recently been elected Secretary of the New England Section in place of Mr. C. A. Strong, resigned.

Mr. C. A. Strong formerly Secretary of the New England Section, has resigned from the Westinghouse Lamp Co.

Mr. Francis A. Gallagher of the Narragansett Electric Light & Power Co., Providence, R. I., was elected to the Board of Managers of the New England Section to fill the vacancy in the Board of Managers due to the changes.

Mr. Clarence L. Law, for many years Manager of the Bureau of Illuminating Engineering of the New York Edison Co., has recently been appointed to the post of Assistant to the General Commercial Manager, Mr. Arthur Williams.

On February 18, Professor W. J. Drisko gave a lecture, "Illuminants and Illumination," at the Institute Buildings in Cambridge. This lecture was one in the series of Sunday Science Lectures being conducted by the Massachusetts Institute of Technology.

On Thursday, Feb. 8, Mr. Henry Logan, Sales Engineer for the Holophane Glass Company, spoke in Detroit from the Detroit News Radio Station WWJ on the subject "The Romance of Illumination." Reports have been received that this talk was heard in New York City, Atlanta, Ga., and Milwaukee, Wis.

Mr. G. B. Nichols has recently opened an office at 300 Madison Ave., New York City, for general designing and consulting work, specializing on public buildings and institutions. Mr. Nichols was for-

6

merly chief engineer in the Department of Architecture, New York State, for ten years, and recently acted as engineer for the new central heating plant at Cornell University.

Mr. Samuel G. Hibben addressed the students of the University of West Virginia on the subject of illumination on February 15. In Montreal on February 22, he spoke before the engineering students of McGill University.

Mr. C. J. Russell, Commercial Vice-President of the Philadelphia Electric Co. and a member of the Society gave a talk before the Conference on January 18. The Electrical Conference is a monthly meeting which is largely attended by Electrical Contractors, Underwriters' Inspectors and Philadelphia Electric Company Inspectors.

Mr. G. Bertram Regar spoke before the Geographical Section of the Pennsylvania Electric Association at their meeting on February 20, at the Hotel Traylor in Allentown, on the subject, "Educating the Public on the Potentialities of Good Residence Lighting."

Mr. M. Luckiesh, National Lamp Works of Cleveland, addressed the New Business Co-operations' Committee of the O. E. L. A. at Columbus on the subject, "Modern Lighting," on February 21. At Rochester, on February 27, he spoke before the Rochester Section of the Optical Society on the subject, "Visual Illusions." The same lecture was repeated before the Brooklyn Institute of Arts and Sciences, on February 28.

PROFESSIONAL OPPORTUNITIES

ENGINEER: A recent engineering graduate to do development and testing work in illuminating engineering. A good opening for a young energetic man.   B-1

ILLUMINATION INDEX

PREPARED BY THE COMMITTEE ON PROGRESS.

An INDEX OF REFERENCES to books, papers, editorials, news and abstracts on illuminating engineering and allied subjects. This index is arranged alphabetically according to the names of the reference publications. The references are then given in order of the date of publication. Important references not appearing herein should be called to the attention of the Illuminating Engineering Society, 29 W. 39th St., New York, N. Y.

TRANSACTIONS
OF THE
ILLUMINATING ENGINEERING SOCIETY

| VOL. XVIII · | APRIL, 1923 | No. 4 |

Opportunity for International Co-operation on Technical Matters

IN THIS ERA of intensified nationalistic feeling, when many peoples are at swords' points, and keen political and commercial rivalry tends to separate even friendly nations, anything which contributes to mutual understanding and good-will is notable. So our international scientific and technical organizations have an importance greater than might appear from the published accomplishments in their special fields. Their revival since the war in spite of economic difficulties is one of the hopeful signs of the times, and in this revival members of the Illuminating Engineering Society have had an active part.

Difference of language is perhaps the most serious obstacle to mutual understanding between peoples; and as understanding is the first requisite for good-will and joint effort for advancement, any step which furthers it has wide significance. The formal meeting of the International Commission on Illumination held in 1921 made unexpected progress on technical questions, particularly in adopting international definitions in French for the most important terms used in photometry and illuminating engineering. It was significant, however, that while agreement could be reached on such a French text, there remained between the two great English speaking nations some differences of opinion as to its proper rendition into English.

The recent Geneva meeting of the International Electrotechnical Commission gave an opportunity for representatives of the two countries to meet and work out jointly an official translation of these definitions, and incidentally to arrange for closer co-operation in the further development of this work. In the past the several countries have proceeded more or less independently to build up a nomenclature suitable for their needs. When international meetings were held the problem was to reconcile the different practices thus established, and this was often difficult. It is now proposed that the active committees in each country be kept informed, through the central office of the International Commission on Illumination, of the work under way in other countries. For example, minutes of meetings of the I. E. S. Committee on Nomenclature and Standards will be regularly sent through the U. S. National Committee to the Commission offices in London for distribution to committees abroad which may be interested, and similar information regarding current work in other countries will be received through the same channels.

These arrangements for joint study of current problems should greatly facilitate the reaching of definite conclusions at the next meeting of the International Commission. It is therefore especially important that our own standard of nomenclature be put into the best possible form before that time, and all members of the Society are urged to submit suggestions and criticisms in accordance with the notice published in the February TRANSACTIONS. It is especially desirable to get the viewpoint of the practicing engineer and of others who actually have occasion to use the nomenclature. Having the advantage of such practical discussion and of working contact with our friends abroad, we should be able to make notable progress in removing some of the difficulties which have hitherto handicapped the interchange of ideas and information in our special field.

E. C. CRITTENDEN, *Chairman,*
Committee on Nomenclature and Standards.

REFLECTIONS

Artificial Light and Our Eyes

IN a recent lecture before the Royal Photograph Society of London, Dr. A. E. Bawtree made the statement "that mankind is being blinded by modern electric light." He laid emphasis on the damage done by the dangerous and invisible ultra-violet rays. It seems only just in considering a matter of this importance to cite some figures and to look at the matter from a common sense point of view.

The human eye was developed in a daylight environment and has been adapted to function satisfactorily under normal daylight conditions.

As we know, radiations from the sun as well as all other light sources consist of a wide range of wave-lengths. We have quite arbitrarily divided this range of wave-lengths into three parts; first, the visible range which we call light, this is a little less than one octave in extent; second, the range of wave-lengths longer than the visible which we call the heat rays or infra-red, and, third, the rays shorter than those visible to our eye which we call ultra-violet. We must not consider that these three divisions are separate. They are merely our names for certain parts of a continuous scale. We might make the same analogy on the piano by calling the notes lower than the middle octave infra-middle octave and the notes higher than the middle octave ultra-middle octave, but assigning such terms does not in any way make these three divisions separate and distinct things, and thus it is with the rays from the sun we have represented the continuous scale of ultra-violet rays, light rays and heat rays and our eye is normally healthy under this combination.

The progress of artificial light is not so much a history of the means of producing the light as it is a history of increasing temperatures. The pine knot gave a smoky orange flame of very much lower temperature than the sun. The early lamps and candles gave a hotter and, therefore, a brighter flame. Illuminating gas first as a flame and then with a mantle were still further advances in increasing the temperature of the source of light, and hence its brightness.

Then came the electric light. But considering it from the temperature point of view it is not a radical change, it is merely another step in the increase of the temperature of the source. In this case, however, instead of the source being a flame, it is a filament. Electricity is the method of heating the filament to a high temperature, and is, therefore, only a substitution of electrical energy as the heating agent in place of chemical energy which was the heating agent of all the previous flames.

The first incandescent lamps were no hotter and, therefore, no brighter than the brightest flames and the progress in temperature of light sources was continuous right through the change from chemical to electrical means. The electrically heated filament has, however, been steadily climbing in temperature from the carbon filament to the tungsten filament and finally to the high wattage gas filled incandescent lamps, which have reached at the present time their practical limit of temperature, as they are burned at a temperature near the melting point of the filament. Hence we may say the temperature of the ordinary artificial illuminant, such as used in the home has reached its maximum. This is, of course, not saying that efficiency may not further be increased.

Now the interesting point is that artificial light having climbed up the scale from a ruddy glow to the modern brilliant incandescent lamp, has not by any means approached either the brightness of the sun or the amount of ultra-violet rays coming to us from the sun. In taking the same quantity of light from the sun and from one of the modern incandescent lamps, we find, by measurement, that sunlight has over twice as much ultra-violet as the artificial light. Furthermore, outdoors on a bright day there is about 10,000 foot-candles of illumination. Indoors in a modern well lighted room there is about ten foot-candles. Therefore the amount of ultra-violet light that our eyes are exposed to under the natural outdoor conditions is 2,000 times greater than the ultra-violet light present in a modern well lighted room.

Our eyes normally do not suffer from the daylight exposure, hence how could they suffer from any of the usual sources of artificial light with only 1-2,000ths of the ultra-violet light.

Many physicians may cite cases where artificial light has hurt eye-sight more than the much brighter natural light, but this brings up another and a much more important point, and that is, the abuse of the eyes in using light. It is not our practice outdoors to stare at the sun when it is high, hence we have the feeling that it is quite harmless. But let us suppose that some happening, such as an eclipse, excites our curiosity so that many persons carelessly stare at the sun. A recent eclipse of the sun over central Europe caused over 200 cases of temporary blindness due to gazing at the sun without proper protection for the eye. This makes it quite evident that even sunlight is harmful when abused. The race has learned its lesson in the matter of the sun; that it is to be used, not stared at.

Artificial light may be abused in the same way by a person working with his eyes exposed to the direct glare of a filament for long periods. The symptoms and effect on the eyes are exactly the same as in the eclipse cases, although not nearly as severe. So it is seen that artificial light is really less harmful than sunlight in its quality but unfortunately it is abused by carelessness much more frequently than the sun, and hence has received quite undeservedly a bad reputation by certain physicians who have not analyzed all the factors involved.

There is, however, another type of artificial light which is true electric light. This differs from the incandescent electric lights, which are really only temperature sources of light, in that this is actual light production directly by electricity. These lights are called electric arcs. Arcs have two characteristics in which they differ from other artificial sources of light. First, they may be extremely bright, and, second, they can be made to emit even a larger proportion of ultra-violet in proportion to the visible light than sunlight does. Electric arcs, however, are not used as artificial light in the home or office, they are in a separate class and hence their case must be considered special.

The large amount of ultra-violet light which the special arcs contain is turned to very great advantage in photographic studios, and especially in motion picture studios, since it is the ultra-violet or actinic light that does chemical work, and thus greatly facilitates the taking of pictures where the exposure is perforce instantaneous. The problem of these abnormally bright and actinic lights in the studios has been carefully studied, and it has been finally determined that even in these extreme cases the harm done to the eye on occasional over-exposures is quite temporary and, though sometimes painful, leaves no after effects.

Furthermore, careful study of the use of artificial lights in motion picture studios has proved again that carelessness is the main factor which has caused what little trouble has been experienced. Many simple precautions can be taken which minimize any trouble that the actors have in working under these lights.

So we may conclude that the damage to eyesight by artificial light is a theory quite unsupported by the facts, but one which crops up perennially, due to the cases which are caused by abuse and carelessness, from which not even the sun is exempt. Much could be said in regard to eye troubles in modern life as being the result of a comparatively new and very different use of the eye by the human race that is, reading print, and especially fine print, which is far more different from the functions for which our eye was originally developed than the artificial light is from daylight.—By ELMER A. SPERRY, *The Sun*, March 16, 1923.

Artificial Light to Save the Colors of Tutankhamen

T O preserve the rich trappings of Pharaoh Tutankhamen from fading from their former glory scientists in London propose that the ancient objects be entirely illuminated with modern artificial light. Recent experiments made in England indicate that museum materials retain their colors longer when electrically lighted than when exposed to any form of nature's daylight.

Daylight contains damaging ultra-violet rays which are not so strong in most artificial lights. The best glass, they say, for use in cutting out these undesirable rays has a distinct yellow color which makes it scarcely practical for exhibiting purposes. Any kind of tinted glass merely delays fading but does not stop it.

Direct sunlight has been known to cause rapid fading, but these scientific experiments indicated that the diffused daylight for which modern museums are designed is six times as injurious as electric light.

Perceptible change of color in the wings of specimens of certain moths were were found in from ten to twenty-one days, in the fur of a tiger after one hundred and seventy-five days and in the coat of a brown horse and antelope after one thousand four hundred and eighty-five days.—*Science Service*.

Street Lighting Problems in 1823

THAT the street lighting problems of one hundred years ago were a matter of concern is suggested by this editorial apropos of a fire in Philadelphia, which appeared in *Freeman's Journal* of January 23, 1823.

"During the awful fire yesterday morning where were the men who are paid to watch our lives and property? They were not to be found. It was the earnest request of a great number of the citizens that they should light their lamps in the neighborhood of the fire—the call was repeated 'Watchman! Light your lamps!' But no one answered, and frequently, when the water deadened the light of the fire, the people knew not which way to turn to aid their friends in distress. The writer of this went to the Court House to get the request complied with, but it was not done. What is the reason our lamps are extinguished or rather not lighted when we have only a new moon, or only one in its quarter? In a time like the present we should have lights in our city until the day breaks, and the watchmen should by no means be permitted to leave their stands until that time."

Electric Signs in New York

THERE are 9,577 electric signs in New York, according to a census made recently by the New York Edison Company. These signs advertise every kind of business from undertakers, churches and bird stores to restaurants and flower shops. Over a million lamps are required to light them. Of these, exactly 947,623 are 10-watt lamps.

This is the first time such a count has ever been taken, and the findings will be valued by the electrical industry to which the number of lights on Broadway will always be of first interest.

Lamp-Socket Standardization

THE printed report on the International Standardization of Edison Screw Lamp Bases and Sockets, which was presented as an appendix to the report of the Lamp Committee of the Association of Edison Illuminating Companies last Fall at White Sulphur Springs, was presented and discussed at the International Electrotechnical Conference, held in Geneva, Switzerland, in the latter part of November.

The cabled and written reports on the results of the conference, now more fully supplemented by the verbal reports of the returned delegates, indicate that the printed report itself proved to be of the greatest value as a presentation of fundamental facts, as a basis of rational standardization. It made available for the first time a comprehensive and complete statement of the American practice and standardization which could be considered with the French and British statements already available. The report served so adequately as a basis of the discussion that the American delegates felt that all the efforts and expense made by the Association of Edison Illuminating Companies and the American Institute of Electrical Engineers, who prepared the report, were fully justified.

It is possible to report real progress in the efforts toward standardization of the medium screw base and the mogul (or Goliath). The candelabra and miniature bases of American and European manufacture are so widely divergent in diameter or number of threads, that action on the possible standardization of these bases was deferred to a later date.

The results of the discussion on the two days of conference are summed up below in the following conclusions or decisions which were adopted by vote:

(1) That effort shall be concentrated on the standardization of a universal type of socket.

(2) That through the good offices of representatives of the manufacturers present, the questions of fact which had been raised at the meeting shall be subjected to inquiry by a group of manufacturing experts of interested countries; that this group shall meet in Europe for the purpose of discussing the actual manufacturing dimensions

and tolerances which are requisite in such a universal socket.

(3) This group of manufacturing experts shall report back their findings to the National committees represented at this meeting, and to the I. E. C. central office for general distribution.

(4) That subsequently a meeting of this I. E. C. Advisory Committee shall be called, preferably for about the middle of 1923, at the discretion of the Central Office as to the time and place. The suggestion was made that this meeting be held preferably at The Hague in Holland in the month of May or early June, depending, however, upon the completion of the work of the manufacturing experts. The work accomplished by the manufacturing experts will be reviewed at the proposed meeting of the I. E. C. Advisory Committee, at which meeting as wide a representation as possible is suggested, including manufacturers, laboratories, standardization committees and representatives of lamp users.

Manufacture of Gas and Electric Fixtures, 1921

A RECENT communication of the Department of Commerce announces that according to reports made to the Bureau of the Census the value of products of establishments engaged primarily in the manufacture of gas and electric fixtures amounted to $42,890,000 in 1921 as compared with $42,268,000 in 1919 and $28,740,000 in 1914, an increase of 1 per cent from 1919 to 1921, and an increase of 49 per cent for the seven-year period 1914 to 1921.

This industry includes establishments manufacturing, as their products of chief value, gas fixtures, chandeliers, domes, burners, mantles, etc., electric fixtures, holders, electroliers, brackets, portables, etc. It does not include electric lighting fixtures to the value of $2,452,000 in 1921, $2,703,000 in 1919 and $3,384,000 in 1914 reported by establishments manufacturing electrical machinery, apparatus, and supplies as their product of chief value.

Of the 308 establishments reporting products valued at $5,000 and over in 1921, 97 were located in New York; 41 in Pennsylvania; 31 in Illinois; 25 in Ohio; 24 in California; 13 in Wisconsin; 11 in New Jersey; 8 each in Michigan and Minnesota; 7 each in Connecticut and Washington; 6 in Missouri; 5 each in Maryland, Massachusetts, and Tennessee; 3 each in Indiana, Iowa, and Oregon; 1 each in Colorado, Kansas, Kentucky, Nebraska, Oklahoma, and Rhode Island.

In December, the month of maximum employment, 10,680 wage earners were reported, and in August, the month of minimum employment, 8,787—the minimum representing 82 per cent of the maximum. The average number employed during the year was 9,419 as compared with 9,795 in 1919 to 10,913 in 1914.

The figures for 1921 are preliminary and subject to such change and correction as may be found necessary from a further examination of the original reports.

The statistics for 1921, 1919, and 1914 are summarized in the following statement:

	1921 [1]	1919 [1]	1914 [1]
Number of establishments..........	308	319	389
Persons engaged...................	11,735	12,379	13,649
Proprietors and firm members.....	203	221	320
Salaried employees...............	2,113	2,363	2,416
Wage earners (average number)...	9,419	9,795	10,913
Salaries and wages.................	$15,479,000	$14,292,000	$9,852,000
Salaries.....	4,419,000	4,490,000	3,348,000
Wages............................	11,060,000	9,802,000	6,504,000
Paid for contract work.............	41,000	65,000	67,000
Cost of materials..................	18,788,000	20,259,000	14,090,000
Value of products..................	42,890,000	42,268,000	28,740,000
Value added by manufacture [2]........	24,102,000	22,009,000	14,650,000

[1] Statistics for establishments with products valued at less than $5,000 are not included in the figures for 1921. There were 29 establishments of this class, reporting 23 wage earners and products valued at $77,000. For 1919, however, data for 22 establishments of this class, reporting 16 wage earners and products valued at $59,000, and for 1914 data for 71 such establishments, with 82 wage earners and products to the value of $200,000, are included in all items with the exception of "number of establishments."

[2] Value of products less cost of materials.

PAPERS

DETERMINATION OF REFLECTION FACTOR SURFACES*

BY W. F. LITTLE**

SYNOPSIS: Reflection Factor defined. Color and direction of light are two variables affecting the reflecting power of a surface. Difference between *specular* and *diffuse* reflection factors. Three methods described give the reflection factor by even illumination of the surface from all directions. The Reflection Gauge as an approximate method is mentioned. Discussion of effect of reflection factor upon illumination. The rate of increase of illumination for an increase in reflection factor of the walls of a sphere is shown in a curve.

"Reflection Factor" is the ratio of flux emitted by a surface to flux received by it; in other words, it is the light left after absorption.

No surface has a definite constant reflection factor under all conditions. There are two main variables affecting the reflecting power of the surface: Color of light and direction of light.

Reflection factor of a white surface will remain fairly constant irrespective of the color of the light, and in general the higher the reflecting power of a surface the nearer constant is the reflection factor with change in color of light. A white surface such as magnesium carbonate which has a reflection factor of 0.98 or 98 per cent, will have substantially the same reflecting power throughout, though it will fall off slightly toward the blue end of the spectrum. On the other hand, Mr. M. Luckiesh shows a considerable variation in reflection factor of different surfaces with different colors of light. A blue light on a red surface would be almost entirely absorbed, while a blue light on a blue surface or a red light on a red surface would be almost entirely reflected.

*A paper presented before the New York Section of the Illuminating Engineering Society, October 13, 1922.

**Engineer, Electrical Testing Laboratories, New York City.

The Illuminating Engineering Society is not responsible for the statements or opinions advanced by contributors.

There may be, also, a great difference between *specular* reflection factor and *diffuse* reflection factor. Specular reflection factor from a glossy paper may be in the neighborhood of 0.1, from a mat surface it is much lower, and from a mirrored surface it is in the neighborhood of 0.85; whereas the absolute reflection factor of all three may be identical.

The early tests of reflecting power of surfaces were made by assuming the surface to be a perfect diffuser. Light was then allowed to fall upon the surface at a given angle, the brightness measured at some other angle and the factor computed. These tests were approximate but for practical purposes were found fairly satisfactory. Next the exploration of the brightness of the sample from a number of angles, with the light incident from a number of directions was studied and by computation the reflection factor assigned. This method, if properly carried out, gives the most precise results and is the means of determining the standard.

In 1912 Dr. P. G. Nutting worked out his instrument,[1] known as the Reflectometer, consisting of a polorization photometer, a narrow ring with high reflecting surface on the inner side, a translucent diffusing window over one side of the ring, and the surface to be measured over the opposite side.. The translucent window is illuminated from the outside and located so that the photometer views simultaneously the center portion of the window for one field and the test surface for the other, the test surface being illuminated by the window. Theoretically, the window and test surface should be infinite in extent in order that the light will not vary between the surfaces; then if the test surface is 80 per cent as bright as the window, the reflection factor would be 0.80. Unfortunately, however, the window and surface were not infinite but finite and the angle of observation unfavorable for accurate results. It was with this instrument that Dr. Nutting assigned a value of 0.88 to magnesium carbonate.

For several years no attempt was made to investigate further the question of reflection factor until Mr. A. H. Taylor, then of the Bureau of Standards, found that by computation, surfaces not

[1] "A New Method and an Instrument for Determining the Reflecting Power of Opaque Bodies," P. G. Nutting, TRANS. I. E. S., 1912, VII, p. 412.

so white as magnesium carbonate were found to have a reflection factor above 0.90. He then worked out, mathematically, the reflection factor of the surface of the Bureau of Standards' sphere by measuring the sphere wall brightness for a given size lamp.

In January, 1920, Mr. F. A. Benford described his method of determining reflection factor by constructing a small sphere of magnesium carbonate, projecting a beam of light into it, measuring the wall brightness and then removing a portion of the sphere, which reduced the average reflection factor of the sphere. Then by again determining the brightness of the sphere wall, the reflection factor of the surface could be computed. Now by substituting for the open section of the sphere a test surface, its reflection factor can also be computed. Mr. Benford found the reflection factor of magnesium carbonate to be 0.97.

At the Annual Convention in 1920, two papers on this subject were presented—one by Taylor in which he used a small integrating sphere, measuring the wall brightness (1) when projecting a spot of light upon the test surface screened from the sphere window, and (2) when projecting the spot of light upon the sphere wall unscreened from the sphere window. The ratio of these readings thus taken is the reflection factor of the test surface. A second paper by Dr. C. H. Sharp and the writer reverses this process by illuminating the sphere wall with a small bright spot projected through a hole in the sphere and observing (1) the brightness of the test surface screened from the illuminated spot and (2) the sphere wall brightness unscreened from the illuminated spot. The ratio of these readings is the reflection factor of the test surface.

Mr. Taylor's method indicated a reflection factor for magnesium carbonate of between 0.98 and 0.99, whereas the values for the latter paper indicated a reflection factor of 0.98.

The three methods just described give the reflection factor by illuminating the sample evenly from all directions. Therefore the reflection factor derived is the coefficient of diffuse reflection.

The reflection factor of a number of diffusely reflecting surfaces was determined by the sphere method and used as standards to calibrate the Reflection Gauge. The Reflection Gauge is

a graduated half-tone, printed on paper having a reflection factor
of 0.85. Therefore a range can be measured which varies from
0.85 down to 0.08. This instrument, while only approximate,
furnishes a guide for illumination computation. It cannot be
used for the measurement of surfaces having a high specular reflection or for surfaces of considerable color saturation.

The effect of reflection factor upon illumination is very considerable. Assume that a light source in space gives off light
equal to "F" and produces an illumination at unity distance of
"I." If there is no surface to receive the light, there will be no
increase. Now place this light source in the center of a sphere,
the sphere walls will reflect the light back to other parts of the
sphere and increase "I," dependent upon the reflection factor of
the sphere. This equation developed is as follows:

$$I = F + Fr_1 + Fr_2 + Fr_3 + \ldots . Fr_n$$
$$= F + F\frac{r}{1-r}.$$

Now substitute for "r"—reflection factor of the sphere—a
value of 0.5, the above equation becomes $I = F + F \times 1$ or an
illumination of twice that which would be received upon a black
or totally absorbing surface. Now change the sphere surface to
0.90, a value which is entirely within reason, and we have

$$I = F + F \times \frac{0.9}{0.1} \text{ or } 10 \text{ times the illumination that would obtain}$$

in a sphere with black walls. Again assume the sphere to be made
of magnesium carbonate and substituting in the equation we
find an illumination 49 times that of the black sphere. This
woud mean an apparent efficiency of 4,900 per cent.

The same condition with a somewhat more complicated equation would apply in the illumination of a room. Take for instance
this hall. We will assume, as a matter of convenience, that its
dimensions are 50 feet x 80 feet with a 20-foot ceiling, the
reflection factor of the ceiling to be 0.80 and the wall to be 0.65
and the floor 0.15, and when occupied by persons wearing dark
clothes it would probably be somewhat less than 0.15. Without

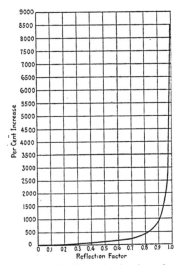

Fig. 1.—Per cent increase in illumination for change in reflection factor.

going through the entire computations, the average reflection factor of the ceiling, walls and floor would be approximately 0.55. Substituting in the equation, you will find an average illumination for the ceiling, walls and floor, of 2.2F, or an illumination of double what might be expected with black or totally absorbing walls, ceiling and floor.

In Figure 1 the rate of increase of illumination for an increase in reflection factor of the walls of a sphere is shown, and while this curve cannot be relied upon to give accurate results for rooms of odd dimensions, particularly with furniture in place, however it does serve to show the importance of reflection factor as related to illumination.

DISCUSSION

The papers by Messrs. Little, Hallett, Bryce and Jamme were discussed together. See page 357.

THE INFLUENCE OF PAINT ON INTERIOR ILLUMINATION*

BY R. L. HALLETT**

SYNOPSIS: This paper briefly discusses the illumination of building interiors, particularly from the standpont of secondary illumination or the illumination obtained by reflection of light from walls and ceilings. The effect of color and surface finish on reflection of light is discussed, and information is given concerning the reflection properties of paint made with various materials. Durability of paint films as affecting reflection of light is referred to, and the effect of the opacity of pigment as influencing reflection of light is brought out. The results of experiments showing the reflection factors of various kinds of paint and the durability of paints of different composition are given.

The usefulness of the interiors of buildings depends largely on their proper illumination. During the hours of daylight, illumination is often secured by allowing the daylight to enter the building interiors through suitable windows and skylights. During the hours of darkness and, in many cases, during the daylight hours also, it is necessary to depend on artificial illumination.

When we consider artificial illumination, we all realize that the production of artificial light always carries with it a certain cost, whether it be for fuel burned in a power plant or for the purchase of electrical current or gas.

The fundamental thought in connection with illumination, whether it be by daylight or artificial light, involves the utilization of the direct rays of light leaving the source of illumination, which may be the sun, the electric lamp or arc, or the gas flame, and striking that portion of the interior or the objects contained therein which are to be illuminated.

It is a well-known law of light that in case of a source of relatively small dimensions the illumination produced varies inversely as the square of the distance from the source.

*A paper presented before the New York Section of the Illuminating Engineering Society, October 13, 1922.

**Research Laboratories, National Lead Co., Brooklyn, N. Y.

The Illuminating Engineering Society is not responsible for the statements or opinions advanced by contributors.

In the case of direct sunlight, the decrease in the strength of illumination between the window through which the light enters and the portion of the interior which is to be illuminated is very slight, owing to the fact that this distance is but an infinitesimal part of the total distance from the source of light, or, in other words, the distance from the sun.

On the other hand, in considering illumination from an electric lamp or arc, or a gas flame, it is evident that the intensity of the illumination decreases very rapidly as the distance from the source increases. These fundamental considerations, which are very simple matters to you, as illuminating engineers, are brought out to show the importance of illumination and the important economic considerations which may be involved.

SECONDARY ILLUMINATION

By secondary illumination we mean light coming from the original source but reflected by the walls and ceilings of the interior in such a way as to be thrown back into the room, thus adding to the direct illumination and, in many cases, placing the light where it is most needed.

Generally speaking, the reflection of light from a surface is determined almost entirely by the color of the surface. This is evidently so because color is the physiological optical impression made on the human eye by the light rays which strike the colored object and which are, in turn, reflected to the eye.

Dark colored surfaces (the limiting color of which is absolute black) absorb a large amount of light which strikes them and do not reflect such absorbed light. Light colored surfaces, on the other hand (the limit of which is pure white), reflect a large amount of the light which strikes them, and such reflected light produces the impression of light colors on the human eye and tends to illumine the space into which the light is reflected.

In the painting of walls and ceilings, there are many questions involved, among which are light reflection, decorative effect, and the durability of the paint film. Under durability, we might include the ability of the paint film to indefinitely remain as a perfect unbroken structural coating, and also we might include the permanency of the decorative effect and the light reflecting properties.

In this discussion we are concerned principally with the question of light reflecting properties, but we must bear in mind that any paint problem, even if light reflection is of predominating interest, must also include the consideration of durability and decorative effect. From theoretical considerations, as fully substantiated by observation and innumerable tests and experiments we know that a white surface reflects more light than a darker one.

With the paint materials at our command, it is impossible to produce a paint which is either perfectly white or perfectly black. We can, however, approach these limits, and we may also produce various light and dark shades between white and black, varying in light reflecting value in accordance with the shade. It will, therefore, be seen that if we desire high illumination of an interior or the greatest economic utilization of a certain source of illumination, walls and ceilings should be painted white or in the lighter tints, such as light cream, blue, pink, ivory, etc., and if the darker colors are used, we must realize that they should be handled advisedly, and that by their use some of the illumination will be sacrificed.

While it is true that white is the color which has the greatest light reflection, we realize that a white surface is often trying to the eye, and, generally speaking, is not suitable for obtaining artistic results where the decorative effect must be largely considered.

Beyond pointing out the difference in light reflecting value of light and dark tints, I will not attempt to give definite figures for different shades and colors, but such figures from fairly reliable sources are available and may be used to good advantage in painting work in connection with illuminating problems.

SURFACE FINISH

There is one other paint characteristic which materially affects illumination, and that is the surface finish of the paint film. By surface finish we mean the degree of gloss or flat which results after the paint film has been applied and has dried. All degrees of gloss may be secured, varying from a very high gloss to an almost perfect flat. The degree of gloss of the surface finish is obtained by a proper mixture of the materials used in the making of the paint.

Extensive experiments have shown that for paint films of the same color, the total light reflection is substantially the same, no matter what degree of gloss or flat the surface may have. In the case of a gloss surface, the rays of light coming from the source strike the surface and are reflected from the plane of the surface in conformity with the laws of reflection, that is, the angles of reflection will be equal to the angles of incidence. For this reason, the more or less parallel or only slightly divergent rays of light striking a small area of a gloss surface are reflected together in concentrated form, often reflecting the image of the source of illumination, and almost invariably giving high light glare which causes eye-strain. Such concentrated reflection usually produces over-illumination in certain portions of the room, and under-illumination in some of the darker corners. The eye-strain caused by the high light glare usually results in an abnormal contraction of the pupil of the eye which gives the impression of lower illumination than actually exists.

On the other hand, rays of light striking a flat surface are broken up and diffused in all directions so that the glaring high lights are eliminated and the room does not have either over-illuminated or under-illuminated portions. In other words, the total light reflected by a gloss and a flat surface is substantially the same, but the secondary illumination, produced by the flat surface, is better because it is more uniformly distributed. We would point out, however, that a slight suggestion of gloss, sometimes called a slight eggshell finish, is more artistic than a perfectly flat finish, and, practically speaking, has about the same light reflecting properties.

In considering light reflecting qualities of paints, the white paints, and the light tints, consisting principally of white paints with a small amount of tinting material, are the most important, as they give the highest light reflection. Some recent carefully conducted experiments have been made to determine the relative light reflecting value of paints made with white lead, Titanox, zinc oxide and lithopone. These white pigments were selected as representing the pigments generally used in the manufacture of wall paints for interior use.

While it is true that some slight differences in the light reflection of various white pigments in the dry form may be detected, the object of these experiments was to determine the relative light reflection value of commercial paint mixtures. Paints of the highest class, in which these various pigments were used, were purchased in the open market. Only paints with the best reputation and selling for the highest price were included, and they were applied to panels in a regular workmanlike manner. The paints were secured and the panels were prepared at the Research Laboratories of the National Lead Company. The light reflection factors were determined by the Electrical Testing Laboratories of New York City.

The results of these tests were as follows:

Kind of paint	Light reflection factor
White lead	0.75 to 0.77
Titanox	0.79
Zinc oxide	0.77
Lithopone	0.75 to 0.78

In the case of the white lead paints and the lithopone paints, a number of samples were procured, but in the case of the Titanox paint and the zinc oxide paint, it was only possible to find one high-grade paint of each kind which contained only the particular pigment in question.

From these figures it will be seen that, practically speaking, there is no material difference in the light reflection factors of the best white paints as first applied, no matter which of the four pigments is used in making them.

PERMANENCY

The next important question is the permanency of the light reflection qualities.

There is more or less dust in the atmosphere in all buildings, and, to a certain extent, dust will adhere to the walls and ceilings of buildings interiors. As the dust is almost invariably of a darker color than the original paint film, the result is not only an unsightly condition, but a material decrease in light reflection value. The adherence of dust is probably not related to paint composition and probably has an equal effect on almost any paint which may be applied.

In the making of a practical paint it is necessary to use certain drying oil vehicles, and it is a well-known fact that all of these vehicles tend to turn yellow after the paint has been applied and has become dry, when the paint is not exposed to sunlight or strong daylight. This, of course, is the condition encountered in many building interiors.

While all of the vehicles yellow to some extent under these conditions, some of them yellow much more rapidly and to a much greater degree than do others. Linseed oil, perhaps the most important and most valuable drying oil known in the painting art, is, unfortunately, the greatest offender from the standpoint of yellowing when the paint film is not exposed to sunlight or strong daylight.

It, therefore, becomes necessary, in the manufacture of paints for interior use, to use the smallest possible amount of linseed oil in such paints and to design the vehicles specially to resist the yellowing.

All of the work which has been done on this subject points to the fact that the yellowing of paint films is almost entirely a question of the vehicle, and is not due to the pigment, because by the use of proper vehicles, paints which will satisfactorily hold their color when used for interior work are being made with all of the pigments, and such paints in commercial form are now on the market.

OPACITY

As reflection of light is a real light phenomenon, it is, of course, influenced by the opacity of the paint film as well as by its color. It is evident that a paint film which transmits the incident rays of light striking it will fail to reflect the light, and the result will be the same, whether incident light is transmitted or absorbed as far as light reflection is concerned.

For this reason, pigments having great opacity give the greatest reflection of light. The opacity or reflecting power of a pigment is largely due to its refractive index as compared to the refractive index of the vehicle in which it is used. Great opacity is obtained by having the greatest difference in refractive index between the pigment and vehicle. Such pigments as white lead, Titanox, zinc oxide and lithopone have a very high refractive index, and the difference in refractive index between them and the paint vehicles is quite great.

On the other hand, such paint materials, sometimes incorrectly called pigments, as barytes, China clay, asbestine, whiting, silica, magnesium carbonate and gypsum, have a comparatively low refractive index, differing little from that of the paint vehicles, and when used with the paint vehicles in the form of a paint film, they almost entirely lose their opacity, and paints made with them (even when they contain enough of these materials to produce mixtures of very heavy brushing consistency) become almost transparent. It will, therefore, be seen that, from the standpoint of reflection of light, the use of these extender pigments should be condemned, as they have comparatively no light reflecting properties.

DURABILITY

We could not complete a discussion of paints for any purpose without including the durability of the paint film itself. By durability we mean the ability of the paint film to maintain its original smooth unbroken condition and its firm adherence to the surface over which it is applied. In others words, a paint film should permanently remain a perfect unbroken part of the structure.

Practically speaking, there is little chemical change in dry paint films made with white lead, Titanox or zinc oxide. On the other hand, paints made with lithopone do sometimes show a change due to the oxidation of the pigment itself. You may be interested to know that lithopone is a composite pigment consisting of about 72 per cent of barium sulphate and 28 per cent of zinc sulphide. The zinc sulphide is susceptible to oxidation through action of the elements, becoming converted to zinc sulphate, which is soluble in water and, when exposed out of doors, is rapidly washed away by the rain. When applied to interior surfaces, zinc sulphide in lithopone paints slowly undergoes the same oxidation, forming water soluble zinc sulphate, but the action is much slower. As interior walls do not come in contact with rain, the water soluble zinc sulphate is not washed off, but, under some conditions, when the walls are washed to clean them, it will be found that a considerable portion of the paint film has become soluble and is removed. This was brought out sometime ago in a series of tests, where about fifty flat wall paints (being, at that time, practically all of the flat wall paints that were for sale) were applied to an interior wall in an office. At the end of a number of years, washing tests were made by rubbing each of the paints with a damp black cloth. Those paints having a pigment consisting of white lead or zinc oxide showed no water soluble material, but practically all of the paints containing lithopone did show water soluble material, and the degree to which the pigment could be removed with a damp cloth was in direct proportion to the amount of lithopone contained in the paint.

You all know that different substances expand and contract with changes of temperature, metals showing the greatest change in this respect. It may be a new thought that such materials as wood and plaster also expand and contract with changes of temperature, but such is the fact, although the changes are much less than with metals.

Furthermore, wood expands and contracts under different atmospheric moisture conditions, and shrinks, particularly across the grain, as it ages. Plaster also shrinks, particularly during the first few years of its life, so that there are real movements of expansion and contraction in wood and plaster surfaces.

If a paint film is to remain unbroken and is to adhere perfectly throughout its life, it must be able to follow the expansion and contraction of the surface over which it is applied. This requires a paint film which retains some elasticity so as to enable it to follow the movements of the surface without rupture and subsequent scaling of the film.

Most of the paint vehicles dry to a film which is relatively elastic. Such is not the case, however, with all of the paint films made up of both pigment and vehicle, as some of the pigments act on the vehicle in such a way as to produce hard and brittle

is obtained by having the greatest diff‹
between the pigment and vehicle. Suc
Titanox, zinc oxide and lithopone ha˙
index, and the difference in refractiv‹
the paint vehicles is quite great.

On the other hand, such paint ɟ
rectly called pigments, as barytes, C'
silica, magnesium carbonate and gɔ
low refractive index, differing little
cles, and when used with the pain
paint film, they almost entirely lo
made with them (even when they c
rials to produce mixtures of very
become almost transparent. It wil˙
the standpoint of reflection of li
pigments should be condemned,
light reflecting properties.

efractive index
s as white lead,
high refractive
tween them and

ɔmetimes incor-
ɔestine, whiting,
a comparatively
the paint vehi-
the form of a
ity, and paints
of these mate-
g consistenᴄ⋅˙

DURA

We could not complete a disc
without including the durability ɑ
bility we mean the ability of th‹
inal smooth unbroken conditior
surface over which it is appliec
should permanently remain a pe
ture.

Practically speaking, there
paint films made with white l
the other hand, paints made v
a change due to the oxidation
be interested to know that lith
sisting of about 72 per cent o
of zinc sulphide. The zinc s
through action of the element
phate, which is soluble in wate
is rapidly washed away by tʰ
surfaces, zinc sulphide in lith
same oxidation, forming wa

actic
tact
but,
then
film
som
pain
that
At th
rubbir
paints
showed
contain
degree
was in
the pai

slower. As interior walls do not come in contact the water soluble zinc sulphate is not washed off,
ie conditions, when the walls are washed to clean
e found that a considerable portion of the paint
soluble and is removed. This was brought out
a series of tests, where about fifty flat wall
that time, practically all of the flat wall paints
e) were applied to an interior wall in an office.
iumber of years, washing tests were made by
he paints with a damp black cloth. Those
ment consisting of white lead or zinc oxide
luble material, but practically all of the paints
did show water soluble material, and the
could be removed with a damp cloth
unt of lithopone contained in

and and contract
greatest change
such materials
changes of
are much

ent at-
cross
ring
nts

ly
d
e-
ie
d

is obtained by having the greatest difference in refractive index between the pigment and vehicle. Such pigments as white lead, Titanox, zinc oxide and lithopone have a very high refractive index, and the difference in refractive index between them and the paint vehicles is quite great.

On the other hand, such paint materials, sometimes incorrectly called pigments, as barytes, China clay, asbestine, whiting, silica, magnesium carbonate and gypsum, have a comparatively low refractive index, differing little from that of the paint vehicles, and when used with the paint vehicles in the form of a paint film, they almost entirely lose their opacity, and paints made with them (even when they contain enough of these materials to produce mixtures of very heavy brushing consistency) become almost transparent. It will, therefore, be seen that, from the standpoint of reflection of light, the use of these extender pigments should be condemned, as they have comparatively no light reflecting properties.

DURABILITY

We could not complete a discussion of paints for any purpose without including the durability of the paint film itself. By durability we mean the ability of the paint film to maintain its original smooth unbroken condition and its firm adherence to the surface over which it is applied. In others words, a paint film should permanently remain a perfect unbroken part of the structure.

Practically speaking, there is little chemical change in dry paint films made with white lead, Titanox or zinc oxide. On the other hand, paints made with lithopone do sometimes show a change due to the oxidation of the pigment itself. You may be interested to know that lithopone is a composite pigment consisting of about 72 per cent of barium sulphate and 28 per cent of zinc sulphide. The zinc sulphide is susceptible to oxidation through action of the elements, becoming converted to zinc sulphate, which is soluble in water and, when exposed out of doors, is rapidly washed away by the rain. When applied to interior surfaces, zinc sulphide in lithopone paints slowly undergoes the same oxidation, forming water soluble zinc sulphate, but the

action is much slower. As interior walls do not come in contact with rain, the water soluble zinc sulphate is not washed off,
but, under some conditions, when the walls are washed to clean
them, it will be found that a considerable portion of the paint
film has become soluble and is removed. This was brought out
sometime ago in a series of tests, where about fifty flat wall
paints (being, at that time, practically all of the flat wall paints
that were for sale) were applied to an interior wall in an office.
At the end of a number of years, washing tests were made by
rubbing each of the paints with a damp black cloth. Those
paints having a pigment consisting of white lead or zinc oxide
showed no water soluble material, but practically all of the paints
containing lithopone did show water soluble material, and the
degree to which the pigment could be removed with a damp cloth
was in direct proportion to the amount of lithopone contained in
the paint.

You all know that different substances expand and contract
with changes of temperature, metals showing the greatest change
in this respect. It may be a new thought that such materials
as wood and plaster also expand and contract with changes of
temperature, but such is the fact, although the changes are much
less than with metals.

Furthermore, wood expands and contracts under different atmospheric moisture conditions, and shrinks, particularly across
the grain, as it ages. Plaster also shrinks, particularly during
the first few years of its life, so that there are real movements
of expansion and contraction in wood and plaster surfaces.

If a paint film is to remain unbroken and is to adhere perfectly
throughout its life, it must be able to follow the expansion and
contraction of the surface over which it is applied. This requires a paint film which retains some elasticity so as to enable
it to follow the movements of the surface without rupture and
subsequent scaling of the film.

Most of the paint vehicles dry to a film which is relatively
elastic. Such is not the case, however, with all of the paint films
made up of both pigment and vehicle, as some of the pigments
act on the vehicle in such a way as to produce hard and brittle

films which tend to crack and scale after they have been applied for some time. It will, therefore, be seen that the production of a durable paint film requires the use of a pigment which will allow the paint film to remain sufficiently elastic. Many years of experience have shown that paints made with white lead are pre-eminently satisfactory from this standpoint, as they retain sufficient elasticity throughout their life, and practically always remain as an unbroken film adhering to the surface over which they are applied.

DISCUSSION

The papers by Messrs. Little, Hallett, Bryce and Jamme were discussed together. See page 357.

PAINT CHARACTERISTICS AND CONDITIONS TO BE CONSIDERED IN EFFECTING BETTER ILLUMINATION*

BY C. H. BRYCE**

SYNOPSIS: The author describes two general classes of paints—Oil Paints and Water Paints. In the first class is mentioned lead in oil, gloss interior, flat oil paint, gloss and flat enamels; in the second class, the calsomines. Proper preparation of the painted surface for light reflection is discussed. The use of paint under artificial light and daylight is considered as well as other factors relating to the artistic and decorative uses.

Some interesting data on the variation of reflection factors of various paints subject to different conditions are found in the accompanying tables in the paper.

In presenting this subject one should first classify the two general classes of paints used for this purpose—that is—Oil Paints and Water Paints. The former embracing such paints as lead in oil, gloss interior, or more commonly known as gloss mill paints, flat oil paints and gloss and flat enamels. The Water Paints, generally known as calsomines, comprising the latter.

Lead in oil will be described first as you are all more or less familiar with this product and no doubt have seen it used and heard it talked of by the painter and decorator. This material as received from the manufacturer is a paste product and is thinned with turpentine by the painter to produce a semi-flat finish for interior work. The fact that the painter controls the mixing of this paint leads to the possibility of materials being introduced that might be anything but beneficial from a light reflecting standpoint. No special effort has been made by the manufacturer to prepare this product so that it will not yellow materially with age. It is also very susceptible to changes due to sulphur gases and ammonia fumes. For these reasons this type of paint does not show as high a coefficient of reflection as do some of the others, particularly after a few months' age.

---

*A paper presented before the New York Section of the Illuminating Engineering Society, October 13, 1922.

**Chief Chemist, Benjamin Moore and Co., Brooklyn, N. Y.

The Illuminating Engineering Society is not responsible for the statements or opinions advanced by contributors.

The second type of oil paint, gloss interior, or gloss mill white, is received by the consumer prepared in liquid form ready for use. Its coefficient of reflection is high—it is not affected appreciably by sulphur gases nor ammonia fumes. The disadvantage of a paint of this character for small interior work lies in the fact that while its power of reflection is initially very high and is very consistently maintained, it suffers from specular reflection which is annoying and disagreeable to the occupants of small rooms. However, this disadvantage is materially subordinated where it is used for large work, such as lofts and factories where it is necessary to have a surface which does not collect dust and thereby become spoiled as a light reflecting surface.

The so-called flat interior oil paints are probably the most universally favored materials used for interior decoration. This is largely due to the fact that they maintain their color well, can be easily cleaned, give a soft velvety finish, are sanitary and give a high coefficient of reflection which is well maintained throughout the life of the paint. Their composition is quite variable, however, so that in making a selection for a job where illumination is considered, it would be advisable to secure from the manufacturer the records of reflection coefficients for his particular paints, or have them tested by a laboratory equipped to do this work. The preparation of samples for this work should be carried on carefully. They should be applied in conformity with the manufacturer's instruction for the particular paint to be tested, preferably on non-absorbent material. We have found that black light sheet-iron discs, $3\frac{1}{4}$ inches in diameter, are suitable as these fit the standard disintegrating sphere photometer aperture. A paint of this type should not be a dead flat paint, but should show a slight sheen when viewed at an angle of 30°. This sheen or slight gloss serves several important purposes—firstly, dust particles do not adhere to it—secondly, it is easily cleaned and consequently it has a more lasting high coefficient of reflection. It should dry sufficiently hard in several days to permit washing and yet be elastic enough to meet the expansion and contraction of the surface on which it is applied.

Now we come to a class of paints familiar to all of you and with which you have had varied experiences—that is—enamels—gloss and flat. You probably feel that I have left one out—I have purposely omitted eggshell enamel for the reason that no two people agree as to what the finish of an eggshell enamel should be. Some will select a slightly dull gloss, while others would favor the more nearly flat variety—so that it is just as well to eliminate it at present. The highest quality enamels found in the market to-day are those that are made with oil as the principal vehicle. The reason for this being the necessity of perfect ease of working under the brush, their long life or durability and the facility with which they can be rubbed down to any degree of finish and the fact that they can be painted over at some future time without deterimental effect on the new paint. While their coefficient of light reflection is very high, they are not the ideal finish for use to enhance light reflection on account of their inability to properly diffuse the light. The light reflected from their surface suffers from specular reflection and consequently produces glare which we all know is very objectionable. We have the same condition with gloss enamels made on a gum base. These are usually made with a very light colored, practically water white gum, known as Gum Damar, and are quick working—that is to say—they do not work as freely under the brush as do oil enamels. They maintain their color, however, for a much longer period and consequently show a high coefficient of reflection for a longer time. Owing to their poor brushing properties, they are not generally used on walls or ceilings, but are confined to small work such as furniture, etc. Their durability is not as great due to their tendency to crack and flake off easily.

We now come to the nearest approach to our ideal when we consider flat enamels. These are made on the same general formula as the oil gloss enamels, but contain a very much higher percentage of pigment; hold their color materially better and are capable of having their pigment modified to introduce a pigment which has in itself the highest coefficient of reflection of all pigments, that is, magnesium carbonate. This pigment has virtue in this respect, when used in a flat paint of this type, it might be claimed by some that a pigment such as this loses its power of reflection when mixed with oil (and this is true up to a certain point) but used as we have used it for the tests which we have made and the results which I will show you later, we have not used sufficient oil to entirely destroy the reflective power of this pigment and thereby cause light absorption, as would be the case if we tried to make a gloss enamel containing this pigment. An enamel of this composition has given us the highest initial coefficient of reflection and has maintained this high factor for a longer period than other paint product that we have known to be tested. We have had these tests running from 1916 to 1921 and are still continuing them with the hope that they will lead us further in producing a still more efficient paint of this character.

About three years ago we produced a paint for certain acoustical work along these lines. The pigment being straight magnesium carbonate—this is ground in a solution of rubber. When applied on a surface it dries flat, works well under the brush and is now used on acoustical treatments. Its coefficient of light reflection is the highest of any paint that we have had tested, running as high as 0.90. However, this paint in its present form is not suitable for general painting work as it will not stand rubbing or cleaning and collects dust readily.

We still have our calsomine group to consider. They are sold to the consumer in dry powdered form and made ready for application by mixing with either hot or cold water, depending on the character of the binder used in their manufacture. These paints give a flat finish, having a very high coefficient of reflection. They do not maintain this high power of reflection, however, as long as do the oil paints mentioned before, for the reason that they collect dust more readily due to their rough surface and, therefore, become darker in color. This, of course, is a disadvantage in the use of this class of paint where used to produce a light reflecting surface, but on the other hand it has an economical feature which must not be overlooked. It is easy of application, generally requiring but one coat to cover a surface. The cost of the material is low and the surface is ready for use in a couple of hours after application—and when a surface of this material has become discolored it can be recalsomined at much less expense than could a painted job be repainted.

I might say that in producing a painted surface for light reflection, it is very necessary that the surface be properly prepared.

In the first place the surface must be perfectly dry. It is a great mistake to take chances in this respect for blistering or peeling will inevitably result. The second step is to stop suction—this is done on wood by the addition of more oil to the flat paint for first coat and on plaster by the application of an impervious alkali resistent elastic size or surfacer. Several flat coats of paint should not be applied one on top of the other as this produces a finish coat that is too dry and flat and results ultimately in cracking and peeling. Some oil should be introduced in the intermediate coats.

Up to the present we have only considered white paints. While these are by far the most important and the best light reflecting surfaces, we still have to deal with certain other conditions, such as the predominating source of light—whether it be artifical or daylight, or both, and also the artistic and decorative side. It has been shown that the color of light in the room has a material effect on its occupants and consequently it is necessary to select colors to meet these requirements—but, in this selection it is also necessary to select those that show the highest light reflection for the particular tone desired. If a certain shade is not demanded and it is desired to tint the room and at the same time maintain a reasonably high coefficient of reflection, it is well to select a tint of a neutral green tone, made by mixing a blue, red and yellow (each should be of the highest brightness) in combination with any of the high light reflection whites mentioned before. In using a tint of this kind you secure reflection from ordinary lighting equipment that more nearly approaches daylight than with any other color. Any paint having an appreciably lower coefficient than 0.50 should not be considered from a light reflection standpoint.

Mr. M. Luckiesh, in a note[1] published in the TRANSACTIONS, gave some very interesting figures on the light reflection of a number of pigment colors, but unfortunately these colors from different manufacturers vary to such a large degree that we feel the necessity of going still further in these tests, considering more carefully brightness and clearness of tone, which undoubtedly influences the light reflection factor of the tints made from dif-

[1] "Reflection Factors of Powdered Pigments for Various Illuminants," M. Luckiesh, TRANS. I. E. S., 1922, XVII, p. 315.

ferent batches of color from the same and different manufacturers, but having the same chemical composition. However, we do know that by eliminating the use of black as a tinting color in tinted paints, their reflection factor is higher. This lowering of the reflection factor due to the use of black is unquestionably due to great absorption of light by the black.

In securing the data as shown in Table I, the original tests were made by the following method. (The figures however, have been corrected for comparison to read on a basis of 0.98 for magnesium carbonate): Each sample was placed in the center of an

TABLE I

REFLECTION FACTORS

Samples kept in light

Samples	Initial coefficient Jan. 12/16	Apr., 1916	May, 1916	June, 1916	Jan., 1917	Aug., 1921
1	0.85	0.80	0.80	0.80	0.81	0.78
2	0.87	0.81	0.81	0.81	0.81	0.79
3	0.86	0.79	0.80	0.81	0.81	0.76
4	0.87	0.87	0.85	0.85	0.85	0.74
5	0.82	0.78	0.79	0.79	0.75	0.59*
6	0.85	0.79	0.79	0.80	0.77	0.77
7	0.84	0.77	0.75	0.75	0.75	0.74
8	0.83	0.74	0.75	0.75	0.75	0.75

Samples kept in dark

Samples	Initial coefficient Jan. 12/16	Apr., 1916	May, 1916	June, 1916	Jan., 1917	Aug., 1921
1	0.85	0.80	0.80	0.80	0.80	0.70
2	0.87	0.81	0.87	0.87	0.81	0.71
3	0.86	0.79	0.79	0.79	0.77	0.715
4	0.87	0.81	0.80	0.75	0.73	0.64
5	0.82	0.80	0.80	0.80	0.80	0.72
6	0.85	0.80	0.77	0.77	0.74	0.74
7	0.84	0.72	0.74	0.74	0.74	0.69
8	0.83	0.72	0.74	0.74	0.74	0.70

*Sample destroyed by cleaning.

(1) A magnesium bearing flat enamel stippled; (2) The same enamel unstippled; (3) Flat lithopone oil paint stippled; (4) The same paint plain or unstippled; (5) Calsomine plain or unstippled; (6) Calsomine stippled; (7) Lead in oil thinned with turpentine plain; (8) The same stippled.

integrating sphere and illuminated by the light of a concentrated
filament lamp shining through an opening in the top of the sphere.
The light fell on the paint at an angle of 45 degrees. The interior
of the sphere received only such light as had first been reflected
by the paint. To effect a standardization a flat block of magnesium carbonate (commercial) was substituted for the sample.
The value of the coefficient of reflection of this block was taken
as 0.88 in accordance with the experiments of Messrs. P. G.
Nutting, L. A. Jones and F. A. Elliott.[2]

From Table I the results would indicate that the nature of the
surface of the paint where stippled or plain has an influence on
the reflection factor obtained. The plain painted surface showing
a slightly better degree of reflection. This may be due to the
slightly rougher surface of the strippled samples holding more
dust particles. I might explain to those that are not familiar
with the term stipple that what is meant is that the paint has
been patted while in the wet state with a long haired brush which
gives the appearance of stucco.

TABLE II

Sample	Initial Coeff.	Per cent	After one month Coeff.	Per cent	After four months Coeff.	Per cent	Color
1	0.92	100	0.865	94	0.85	92	White
2	0.88	100	0.80	91	0.80	91	White
3	0.87	100	0.76	87	0.765	88	White
4	0.855	100	0.80	93.4	0.80	93.5	White
5	0.845	100	0.815	96	0.805	95	White
6	0.83	100	0.74	89	0.74	89	White
7	0.825	100	0.74	90	0.74	90	White
8	0.82	100	0.77	94	0.75	91	White
9	0.81	100	0.745	92	0.75	92.5	White
10	0.76	100	0.66	87	0.67	88	Green
11	0.63	100	0.53	84	0.56	89	Green
12	0.61	100	0.535	88	0.53	87	Green
13	0.55	100	0.48	87	0.48	87	Green

Table II gives a prominent illuminating engineer's interpretation of the results of phometric reading obtained on over 100
different paints tested. The best nine samples were selected for

[2] "Tests of Some Possible Reflecting Power Standards," TRANS. I. E. S., 1914,
IX, p. 593.

comparison. You will note that the sample which gave the highest initial value when figured on a percentage efficiency basis does not show up so well, for, as you will note, it is still decreasing after 4 months' test. These samples were prepared somewhat differently from the average samples prepared for a test of this kind. Three coats of the paints were applied on concrete. This gives us a possible explanation of the very high factor shown, for as I mentioned before, three coats of flat paint applied one over the other, causes the finished coat to be very dry, thus giving more the effect of a dry pigment than the original paint would have when applied properly over a non-porous surface. It has been shown by Mr. H. A. Gardner[3] that the light reflection factors of dry pigments are considerably higher than paints containing them. So that it would be reasonable to suppose that these high figures could be explained in that way. I have shown you this chart so that when judging a paint from a photometric test that you will not be misled by a high initial coefficient factor, but rather be guided by this in combination with its maintenance of this high factor through a period of time.

In conclusion I would say that there are paints obtainable in the markets prepared as outlined before and if a little care is exercised in their selection, no difficulty need be encountered.

DISCUSSION

The papers by Messrs. Little, Hallett, Bryce, and Jamme were discussed together. See page 357.

[4] "Reflection Factors of Industrial Paints and Pigments," H. A. Gardner, TRANS. I. E. S., 1922, XVII, p. 318.

PAINT AS AN AID TO ILLUMINATION FROM THE MACHINERY PAINTING ANGLE*

BY L. E. JAMME**

SYNOPSIS: Efficiency of proper illumination fixtures at point of production enhanced by painting machine tools and machinery in a light color. Essentials to be considered in determining the characterstics of a proper finish. Selecting the color. The reason for light gray. Benefits accruing. How Illuminating Engineers can further their own efforts.

Efficient lighting has become so well recognized as a powerful influence in factory production that light colored walls and ceilings and proper illuminating fixtures are now commonly found in most modern industrial plants.

Picture these two efficiency factors, light colored walls and ceilings, and proper lighting fixtures, as the two sides of a triangle, and it becomes apparent at once that their effect will be enhanced through completing the triangle, by bringing more light to the actual point of production: the machines themselves.

Early in 1921, when a survey was made of a representative cross-section of the machine tool industry (which by the way, was extremely dormant just at that time), it was found that while no particular enthusiasm existed for a standard light colored finish, most manufacturers were "willing to be shown." In other words, if there developed on the part of users of machinery and machine tools, a sufficient trend toward some particular light colored finish which could be adopted as a standard the machine tool builders would give interested and possibly favorable consideration to the subject.

It might seem at first thought, that the simplest way in which to complete the triangle of brightness would be to paint the machines white. However, investigation quickly showed that white, aside from the ease with which it became soiled, had other disadvantages—notably the one of eye dazzle, produced by great areas of white below the center of vision—an effect similar to

*A paper presented before the New York Section of the Illuminating Engineering Society, October 13, 1922.

**Assistant Advertising Manager, Hilo Varnish Corporation, Brooklyn, N. Y.

The Illuminating Engineering Society is not responsible for the statements or opinions advanced by contributors.

snow blindness. So in choosing what seemed to be the logical light colored finish, there were several essential points to be considered: (a) It must be several steps away from white and yet possess a high coefficient of reflection; (b) It must be oil and gasoline proof and have a tough film that would stand constant wear, and wiping with gasoline or benzine; (c) It must have sufficient elasticity to withstand constant vibration, and the surface should be smooth and sufficiently glossy to permit easy wiping; (d) While glossy, it must not be bright enough to cause eye dazzle; (e) It must have sufficient hiding power to cover in one coat; (f) It must be acceptable from a sales standpoint.

No doubt you are all familiar with the Bulletin of Lighting Data compiled by Mr. A. L. Powell of the Edison Lamp Works, issued in May of this year. Reference to his table of reflection factors indicates that either buff or medium green should prove ideal. However, from a sales standpoint it was quickly found that neither was acceptable.

So it was determined after investigation and much experimenting that *one certain shade of light gray* embodied all of the essential points in a higher degree than white, cream, buff, ivory or light green,—with the exception that it had a reflection factor of only 33 per cent, when tested against the magnesia block. This was offset by the fact that it was a more acceptable color from a sales standpoint, than any of those just mentioned. At the same time it had sufficient hiding power to cover over black in one coat—a feature of importance in keeping the finishing cost at a minimum, at least where users of machinery who wished to refinish their equipment, were concerned.

As the development of the idea of a light gray finish progressed among users of machinery and machine tools, largely through the influence of Safety First Engineers and Welfare Divisions, many plants adopted this finish for repainting their equipment, and were quick to realize the benefits accruing in increased production, through lessened spoilage and rejections, and a better morale among operatives as a result of working in brightened surroundings.

And we might say, that experience is showing more and more every day, that there is a distinct trend toward the light gray finish, although the *particular* shade of light gray demanded by

Fig. 1.—The effect of a light finish on a huge Bliss Press is shown at the left. This machine is painted white, note the condition before painting at the right.

Fig. 2.—Light surfaces in the tool room of the National Cash Register Company, Dayton, Ohio. These machines are finished in a pea green shade. This concern was among the first to recognize the value of light colored finish on machines for precision work.

different users of machine tools, varies somewhat with the preference or whim of the individual. This variant however, could be overcome by the machine tool builders and machinery manufacturers adopting a *definite* shade of light gray as the standard finish in their industry.

And you gentlemen of the I. E. S., in your capacity as illuminating engineers, are in a position to render signal aid to production engineers in all branches of industry, through the advocacy of the light gray finish on machinery and machine tools.

DISCUSSION

F. G. BREYER: I believe it would be really interesting to the members of this, your valuable Society, to get some perspective on the historical side of this painting situation. You undoubtedly have all noticed lately when you go into a modern hotel or a modern building of any sort, the first thing that strikes you is that everything is painted. Ten years ago if you had gone into that same building and it was not all decorated up with wall paper you would have thought it was at least fifteen or twenty years out of date. Now it is out of date it is not painted.

There must be a number of reasons why that change has come about. With all due respect to my good friend Mr. Hallett, who earlier in the evening referred to the fact that the discoloration of paint was mostly the discoloration of the vehicle, I do not think anyone who has had any chemical experience at all in the laboratory but has at one time or another run into lead sulphide. We all know that in buildings inhabited by humans or animals hydrogen sulphide is present.

Paint is cheaper than wall paper, and the only reason paint had not developed up to ten years ago was the fact that the lead used would not stand up in the ordinary house. I have seen any number of instances, and you can all recall similar instances where lead paint used on the interior has not maintained its brightness. Those figures Mr. Bryce gave must have been exposed in some very ideal conditions, because it is not necessary to measure loss of reflection with a photometer, when you can go into a room and see dark streaks, especially in the dining room where the kitchen gases have come in.

If this development has occurred within the last ten years, and more particularly within the last five years, there must be some reason for it, and if you will go around and get an unbiased opinion from the paint manufacturers of the United States, the people who make paint and not pigment, you will find that 95 per cent are to-day using lithopone as a base for their interior pigment. (I do not believe I am overestimating it). Yet they can buy lead, lithopone or zinc all at nearly the same price.

Probably the largest paint manufacturer in the United States also produces lead, zinc and lithopone, and if you buy some of his flat wall paint or his factory enamel or factory white, you will find it is a lithopone paint. The reason for that must be that there is a real advantage in lithopone for his interior flat paints, and it is largely in the fact that the lithopone paint, first of all, will not change its color under adverse circumstances.

Fifty per cent of white lead may be all right, but if the other 50 per cent goes bad due to hydrogen sulphide you can not tell whether the paint that goes out will give satisfaction or not.

Well, let me give you the statistics, for example, and tell you really what the answer is. There is no question that white lead has been a wonderful pigment. It has undeniably good characteristics, and for exterior work has certain properties that at the present moment we do not seem to be able to do without. But it certainly has not been developed commercially yet for interior purposes. There is no reason why you should not use lithopone, a pigment that will not darken, first; because it gives better results and secondly, because pound for pound it goes further.

NORMAN MACBETH: Mr. Chairman, I want to ask Mr. Bryce if the reflection factors he gave were of four or five years ago, or have they been corrected to the later magnesium carborate standard?

C. H. BRYCE: They have been corrected.

NORMAN MACBETH: There is one other question brought in by Mr. Hallett about the reflection factor of paints, considering the absorption and transmission. Mr. Bryce stated that all the samples were on a flat sheet-iron base. It seems to me if that base had been plaster, those paint films having a reasonable degree of transmission would have shown higher factors than other paints which were more opaque. Plaster surfaces, I believe, would more nearly represent the average condition of use for interior finishes.

I would also like to ask Mr. Little if it would not be possible to select a sample color which under daylight would be the same color as a white surface under ordinary artificial light. It seems to me the paint people have specialized on an ideal white, penalizing paints which show yellowing, thereby setting up a false standard about the same as color printers do when using two or more colors of ink, on the assumption that the more it costs and the harder it is to do, the better it must be. If a certain paint does not yellow to a greater extent than a white would appear under the ordinary yellow artificial light I do not see how it would make any difference with paints used for interior wall and ceiling surfaces.

This also brings in another point that there should probably be a double set of reflection factors given for paints—factors for daylight and others for artificial light. Some of the blues here given which have a very low reflection factor under the ordinary artificial light used in the photometer might be considerably higher for daylight, and the same is true of yellows that would be higher with artificial light and not nearly so high for daylight.

R. L. HALLETT: I do not want to enter into any discussion of the relative merits of various paint pigments, as such a discussion would be very much out of place in a meeting of this kind, but referring to this question of discoloration, I would just like to say that in the many paint jobs I have seen where discoloration has taken place on all kinds of paint, sulphur discoloration has been present in so few instances as to be practically negligible. We hardly ever find hydrogen sulphide fumes in the atmosphere, and such discoloration is of small importance in the consideration of paint questions.

The discoloration which usually occurs, as I tried to point out in my earlier talk, is yellowing of the paint film due to the vehicle, and paint made with white lead and linseed oil will discolor the same as paint made with any other pigment and linseed oil, but I would like to say that for the last three or four years pure white lead paint for wall painting has been on the market in ready mixed form, made with specially prepared oils similar to those used for other wall paints. The original light reflecting value of this specially prepared pure white lead paint is about equal to the light reflecting value of other wall paints, as indicated in my paper, and the white lead paint holds its light reflecting value to about the same degree as do the others.

I was very much interested in the remarks that were made about a white plaster wall increasing the light reflection of a relatively transparent paint. Undoubtedly a white wall underneath, or an old white paint undercoating, would materially increase the reflection of the new paint, if the new paint were partially transparent and allowed some of the light to pass through. We must realize, however, that repainting is usually done over old paint, and we must consider the color of the old wall, which may or may not have a high light reflecting color. The condition of the old paint must also be considered, as if it is badly soiled it has lost much of its light reflecting qualities, and a wall is often repainted because it has become soiled and lost its light reflecting value.

WARD HARRISON: Most of the discussion thus far has centered around reflection factors higher than 70 per cent. For the walls in offices it is best to choose a color that runs between 40 per cent and 45 per cent, for if the walls are lighter than this, they are likely to be too bright for comfort. In our offices at Nela Park, we have set a deadline at 50 per cent for the reflection factor of the walls.

Regarding the depreciation values given, I was surprised to find that the drops in efficiency or reflecting power were small. These can be compared with the reflection factor charts for light buff, yellows and white. As you undoubtedly know, it is possible to go quite far into the yellow with only a small drop, and I imagine that some of the paints probably looked very yellow and would be condemned by the owner on that ground even though the reflection factor had dropped only 4 or 5 points.

As the speaker suggested, the painting of machinery is something which the Society can endorse most heartily in many instances, and I also believe that we should lend our influence toward the standardization of color and shade. If the manufacturer buys his machine tools from four or five companies and each article of machinery is an appreciably different shade of green, tan, buff, or gray, he will soon go back to ordering black finish. Is there any other serious objection to a light color?

S. L. CARMEL: I would like to ask Mr. Hallett one or two questions about the washing. He said that the lithopone paints when washed showed an injuring of the film. I would like to ask what kind of finish they had, whether flat, or what.

R. L. HALLETT: The paints which were tested were all flat paints, but I do not wish to intimate that such a change takes place rapidly with most lithopone paints. This test was carried on for several years before the change was noticed. I described the test to indicate that there is possibility of such change taking place in the paint when lithopone is used.

W. S. WILDING: In most discussions of light-reflecting characteristics of paint films I have noticed a tendency toward the theoretical and academic rather than the practical side. The talk and the figures seem to be based chiefly on *new* paint films, while those we have to deal most with in practice are *old* ones. How much light a given paint will reflect when first applied is not as important as how much it will reflect a year, or two years, or five years hence under actual practical conditions. What the industrial plant manager is trying to do with wall and ceiling paint is to maintain a given standard of lighting, whether with daylight or artificial light; and he wants this paint service at the lowest possible cost per square foot per year over a long term of years. The fact that a paint shows good initial results does not attract him if he knows that one year's time will tell a different story.

The few figures that have been presented on repeated tests on the same samples over a period of several years show surprisingly little deterioration in light-reflection, yet we all know that dust will accumulate on flat finish paint,—and dust is gray. One needs no instruments to tell the difference in lighting between two halves of a factory room, on one of which a flat finish paint is new and on the other a year old; the eye is sufficient.

To really determine what service a paint will render, long time tests should be made under actual conditions in the class of occupancy involved—rather than in practically dustless rooms. Such figures will not only be interesting, but will also be a safe basis for paint selection.

W. T. BLACKWELL: Is the gray for use on machinery obtained by the use of carbon black, or what?

W. F. LITTLE: Answering Mr. Macbeth, I know of no standard of artificial daylight. Sometime ago there was some agitation in regard to the use of Mr. D. McFarlan Moore's CO_2 tube as a standard but no official action was taken.

The color of the light used in determining the reflection factor is of great importance and should in all cases be coupled with the reflection factor.

While I have no data on reflection of gray paints of the same appearance there is little doubt that the reflection factor under the same light source will be the same regardless of the combination of pigments.

C. H. BRYCE: I believe some figures have been gotten at some previous time.

A. L. POWELL: Mr. Jamme, can you answer that question of Mr. Blackwell's?

L. E. JAMME: The question as I understand it, is whether gray produced from carbon black and white has the same reflection percentage as gray produced from white, green and red? I think that question is covered in Mr. Powell's Bulletin, issued last May.

W. T. BLACKWELL: I saw the Bulletin.

L. E. JAMME: We have proven through experiment that gray produced by mixing white and carbon black has a lower reflection factor than gray produced by mixing vermillion and green.

W. T. BLACKWELL: What is the difference?

L. E. JAMME: I could not tell you. I know there is a difference. That is a point which our chemist stressed to me recently. I give you the report by the Illuminating Engineering Bureau of the Westinghouse Lamp Co., of a set of readings made to determine the coefficient of reflection of Hilo "Lite-Gray" made from white and carbon black and Hilo "Lite-Gray" made from white, red and green. This showed the former to have a coefficient of reflection of 29, and the latter a slightly higher reflection factor of 30.

I believe it has been definitely established that where a particular shade of gray, produced by whatever combination, is identical with a particular shade of gray produced with some different different combination, the coefficient of reflection will be the same for both.

F. G. BREYER: It is perfectly ridiculous to talk about two grays differing, whether they are obtained by the mixture of black and white, or whether they are gotten by putting in a half dozen other paints. A gray is a reflection of your various wavelengths. It is non-selective. If you throw 100 per cent of red and 100 per cent of green and 100 per cent of blue on a surface and they are reflected 100 per cent, that is absolute white.

The definition of gray, a perfect gray, is a reflection of any percentage less than that, and tints off from total reflection of 100 per cent of each one of those down to a total absorption. Complete reflection is white. Non-selective reflection is gray.

The American Society of Testing Materials is now engaged in determining accurately by absolute methods the total percentage of light reflection and the percentage of saturation so that colors may be determined. We are doing it right along in our laboratory with a colorimeter.

BASSETT JONES (Communicated): The Section is to be congratulated on securing this interesting group of papers on a subject so important in interior illumination. There is no question that in most forms of interiors where more or less continuous eye work is done, such as offices or factories, the selection of the interior finishes is equally if not more important than the selection of any particular lighting fixture.

The importance of the selection of interior finishes is of even greater importance for daylighting. In artificial lighting, as the indirect component increases, the effect of the interior finish increases correspondingly.

As an illustration, the writer's drafting room ceiling is painted white kalsomine. Two years ago this ceiling was washed and rekalsomined, precautions being taken to prevent any change in the conditions of the lighting fixtures of the so-called semi-indirect type, by covering them with paper bags. The result was an increase of 30 per cent in the illumination on the drawing boards. In another case, changing the ceiling from buff to white, and the walls from light ecru to office wall green, increased the working illumination 50 per cent. In an effort to prove to a client the marked differences resulting from apparently slight differences in paint, two otherwise precisely similar rooms were treated, one in the light tones of lead and oil as selected by the architect and one as directed by the writer. The difference in the working illumination, in terms of the lower, was 32 per cent. These are merely typical cases.

The corresponding range of flux efficiency in modern commercial fixtures of the semi-indirect type is from about 75 per cent to 90 per cent—a maximum increase, in terms of the lowest, of about 20 per cent, and less than the value of a fresh coat of kalsomine on the ceiling—even less than the difference between a buff and a white ceiling, both fresh.

The problem before the paint manufacturer is, then, to produce inexpensive paints of durable character, showing high reflection factors when applied, both initially and at the end of a considerable lapse of time. This is a phase of paint-making that only recently has received any general attention. It has taken a long time for the manufacturers to realize the importance of the subject. The writer believes, although he may be wrong, that almost the first attempt to consistently study paints from the standpoint of illumination was begun by Benjamin Moore and Son in 1911. At any rate, samples were then set up for aging tests, on which the last readings were taken during the latter part of 1921.

In the early part of these tests, great difficulty was encountered in establishing a method of preparing samples that would give consistent results. The conclusion was reached that unless great care was exercised, and a ground of red lead on clean sheet-iron was employed, the results obtained from different samples were not comparative. For this reason much of the published data giving reflection factors of paints is valueless.

The writer has corresponded with the technical staff of several paint manufacturers and has found an entire lack in correspondence as to the method in making such tests. Even where an outside laboratory has been employed by more than one such manufacturer, the results obtained from different samples are rarely comparative due to radical differences in the preparation of such samples.

The manufacturer generally needs education in the reason for and the object of the test, and should be given a reasonable understanding of what the results mean and how they should be used, to the end that he will not unwittingly claim the impossible, and mislead his customers.

So far as the writer knows, no paint manufacturer has succeeded in setting up any constant and reproducible color standards —even for black and white. All color matching in paints is a guess—more or less accurate—depending on a number of uncontrolled variables. The determination of accurate reproducible color standards, while possible, is by no means simple, and necessarily must be based on purely physical measurements, using such device as that described by H. E. Ives in the *Journal of the Optical Society* for November, 1921. Such determinations are necessarily of a precise nature and must be carried out by skilled operators such as individual paint manufacturers are hardly prepared to employ. Such color standards must be maintained in a physical laboratory—not in a paint shop.

ILLUMINATING FLARES*

BY G. J. SCHLADT**

SYNOPSIS: The intensive character of the late war brought a well-deserved appreciation of the value of illuminating and signal flares. The perfection of the high candlepower flares suspended by parachutes meant that military operations no longer ceased with the fall of darkness. Service requirements for durability and dependability have been met by careful study and design.

The illuminating and visibility values of flares and signals are rated photometrically to insure their conformity to specifications. These flares and signals are considered as flares having a spherical candlepower distribution, thus giving maximum efficiency for visibility in all directions. Their light efficiency is high due both to temperature radiation and luminescence of metallic oxides formed during combustion.

INTRODUCTION

Illuminating Flares are now one of the subdivisions of military pyrotechnics. As history records pyrotechnics have long been used for military purposes and it may seem strange that in the days of wireless telegraphy, wireless telephone and high power search lights that we should have resource to such an ancient art as Fireworks.

Nevertheless, as a result of the late war, pyrotechnics signals and illuminating devices are now a regular part of the Ordnance equipment and are recognized by authorities of all countries as being the main dependance of military units when all other sources of communication and illumination fail.

On our entry into the war, we had as a guide the various devices that had proved their effectiveness on the war front. In order to meet the demands of the army for new and improved types there was established in 1917 a Pyrotechnic Development Branch, under the supervision of the Trench Warfare Section, Engineering Division of the Ordnance Department. In co-operation with the manufacturer this Branch successfully met the demands placed upon it and considerable credit is due to the men in charge of this work.

Because of its limited application in peace times, there was a very apparent lack of development in the art of Pyrotechny.

*A paper presented before the New York Section of the Illuminating Engineering Society, March 23, 1922.

**Ordnance Department, U. S. Army, Picatinny Arsenal, Dover, N. J.

However, during the war, considerable work was done in this country and abroad and work is still being carried on in order to place this art on a truly technical basis.

In this paper, a brief description will be given of several of the types of illuminating devices used in the service. It being understood that War Department data is of a confidential nature, the descriptions that follow are only general.

ILLUMINATING DEVICES

(a) *Ground Service*: Rifle Illuminating Shells—These shells are fired from a tube attached to the end of the service rifle. Types lately developed by our Ordnance Department are far superior to the types used during the war in regard to candlepower developed and more important in regard to being durable and dependable under the severe service conditions. Rifle illuminating shells proved their effectiveness in trench warfare in keeping the foreground illuminated at will. In this country alone, 1,000,000 rifle signals were made.

The field gun illuminating shell here described is a foreign type. See Figure 1. In principle, it operates the same as the rifle shell but differs in the construction details necessitated by the heavier shock of discharge, the rotative force of the shell and the requirement that it is able to function properly at various points as its trajectory.

An examination of the wire reinforced parachute gives an idea of the terrific strain placed on the parachute when it is called upon to put the brakes in a 5-pound illuminant traveling in some cases, 10 miles per minute.

(b) *Air Service*: The air service from the beginning of the war, extended their operations to all hours of the night and this required a complete line of signals similar to those used by the ground troops. In addition, they had the two special illuminating devices, the Wing Tip Flare and the Aeroplane Flare.

The wing tip flares are small flares held in special holder under the wings of the plane. These flares are ignited electrically from the cock-pit. Those used during the war burned for one minute with a candlepower of 10,000. It is necessary that these flares be dependable in action because they give the airmen a chance to select their landing place in case of necessity.

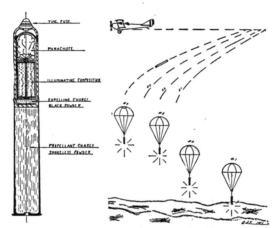

Fig. 1.—Comparative size of field   Fig. 2.—Airplane flares.
gun and rifle illuminating shells.

Aeroplane Flares were used in bombing raids for the illumination of targets and the uncovering of secret movements of troops. Aeroplance Flares are of different sizes and are usually slung on the underside of planes and are released in the same manner as drop bombs. They ranged in candlepower from 50,000 to 500,000.

In the early part of the war, these flares had a constant time mechanism, that is, they always functioned a certain distance below the plane. With the appearance of mobile search lights and large caliber anticraft guns, we find the airmen seeking the higher altitudes. This led to a demand for a time fuse, permitting functioning of the flare at any distance below the plane. This added a complication to a comparatively simple proposition because here we require a parachute to effectively decrease the rate of fall after it has attained in the maximum case a terrific momentum. Problems such as this are usually solved by compromising between the weight of the illuminant and the size of the parachute.

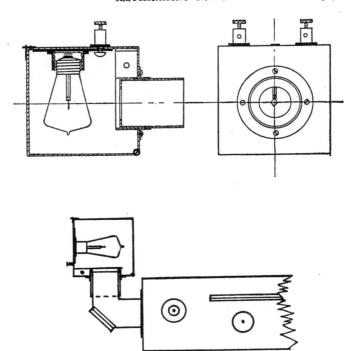

Fig. 3.—Calibrating device for Sharp-Millar photometer.

Aeroplane Flares properly designed have before functioning the same trajectory as drop bombs, that is, they are always to be found below the plane. With a number of high candlepower flares in the air at the same time, we have flood lighting on a large scale, with increasing illuminations on the ground due to the descending flares.

PHOTOMETRIC DETERMINATIONS

The demand for mixtures developing the highest efficiencies, both for our signals and illuminating flares, required a vast amount of work in a field conspicuous for its lack of data.

The efficiencies of mixtures are determined from photometric measurements. The mixtures, charged in cases, are burned in a special brick hood at the end of a photometric gallery 50 feet in length. At this distance measurements are made with standard size Sharp-Millar photometers, and without additional accessories, we can measure candlepowers ranging from 100 to 1,000,000.

Check readings are taken at intervals against standard mixtures and against a calibrating device here shown. This device fits over the sight tube of a standard size Sharp-Millar photometer and uses the same type of comparison lamp. At close to normal amperage, this device gives a reading at color match in the center of the scale. See Figure 3.

In determining the efficiency of a mixture, the weight of composition, the rate of burning and the candlepowers are noted. This gives us an efficiency rating expressed in candlepower seconds per ounce. This rating then gives us a basis for the comparison of mixtures and a basis for estimating the characteristics of the mixture in different size containers.

In the case of illuminating shells and aeroplance flares the illuminant is subjected to sudden shock. This requires the compression of the mixtures above a certain density, depending on their use, and this necessitates the determination of the efficiency characteristics over a wide range of pressures.

CONCLUSION

Illuminating flare mixtures are in general, a mechanical mixture of a compound that readily gives up to its oxygen and a material that readily combines with oxygen with the formation of a considerable amount of heat. Examples of oxygen carriers, as they are called, are barium nitrate and barium chlorate. Examples of metals capable of being easily oxidized, when finally divided are magnesium and aluminum.

Burning flares are considered as flames; light producing because of the reflection of solid particles raised to incandescence by the heat of reaction.

In examining the products of combustion of illuminating flares, we find the refractory oxides of barium, magnesium and aluminum. We have then two reasons for the high efficiency of these flares.

The first is the incandescence of oxides having high heats of formation and high melting points. The second reason has been given by Dr. E. L. Nichols as being due to the luminescence of the oxides of magnesium and aluminum. Perhaps we have here as Dr. Nichols says, a remunerative field for investigation. Dr. Nichols attributes the great atinic power resulting from the burning of magnesium to the extraordinary outburst of blue and violet radiation from the newly formed magnesium oxide. This should be true also of aluminum oxide, perhaps in a less degree. He gives the color temperature as 5,000°, but estimates the actual temperature as being only 2,000° C.

DISCUSSION

W. F. LITTLE: Mr. Schladt's paper has given us much interesting data and is a valuable contribution to the TRANSACTIONS.

During the period of the war many flares were measured at our Laboratories, the first of which produced in the neighborhood of 5,000 candlepower. As the method of manufacturing them advanced, so also did the candlepower, until flares having the same dimensions and duration of burning reached an average of 250,000 candlepower with exceptional ones reaching 400,000 to 500,000 candlepower.

Some flares were made reaching 1,000,000 candlepower but they were of larger dimensions. The average flare was 1¾ by 1½ inches, having a burning life of fifteen to thirty seconds. Some flares were submitted as large as 4½ inches in diameter and 30 inches in length, having a burning life of approximately eight minutes with candlepower averaging 300,000.

The method used in measuring the flares required the services of five persons, one to ignite the flare and record the time, two photometer operators, one for each of two photometers, and two recorders. In this way two series of readings could be taken on each flare with successive readings on each photometer at the rate of two readings in three seconds.

The flares were burned in a large stack which carried off the smoke. The candlepower measurements were made horizontally with the flares pointed toward the photometers. Photometers were placed at a distance of 50 feet from the flares.

A. B. RAY: My work during the war in the field of pyrotechnics was the development of colored smoke signals. Because of the success of flares for night signalling both the Army and Navy felt that pyrotechnic day signals would be of value. But tests showed that signal flares were not suitable for day use so we attempted to develop smoke signals which would have long range visibility and which could be used in the regular pyrotechnic devices such as rockets, rifle grenades, etc. Other special devices for carrying these signals were developed for aircraft use.

Colored smoke signals are a very recent development in the pyrotechnic field. They are produced by the rather unique method of volatilizing organic dyes. When, after all the other methods producing satisfactory smokes had failed, we suggested the use of dyes the suggestion was not received with enthusiasm. In fact the idea of burning up dyes to make smokes at a time when the United States could not get its stockings dyed was considered quite foolish. Nevertheless we went ahead with the work and found that a number of relatively cheap dyestuffs which could be procured in large quantities made very satisfactory smokes.

The dyes are mixed in suitable portion with a combustible and oxidizing agent and the mixture packed into a cardboard or metal container. The containers are perforated with a number of holes of a definite size and a special ignition system employed to ignite the smoke mixture at all these openings. This in itself was a very difficult thing to work out. The successful arrangement allows a large volume of deeply colored smoke to be evolved when the mixture is ignited. The heat of the reaction between combustible and oxidizing agent has to be carefully controlled otherwise the dyestuff will be burned up and a wisp of white smoke will be the result.

Several hundred dyes and colored compounds were tested during the investigation and a few were found which were satisfactory from price, quality and smoke producing standpoints. Excellent red, yellow, blue and green smokes were developed which are particularly suitable for signals because of their distinctness and long range visibility. A black smoke and smokes of intermediate shades were also developed.

The uses of these signal smokes are many. They are used by aircraft with very great success. The special smoke signal grenade which is operated exactly as an ordinary hand grenade is merely thrown overboard. For use in rockets, rifle grenades and Very pistol cartridges the smoke pot may or may not be attached to a parachute.

An interesting peace time use of these smoke signals has been made in informing the distant crowds at crew races regarding the order of the crews at points along the course. Signals fired at the mile mark in the order of red for Cornell, blue for Yale, etc., are self explanatory.

NON-MEMBER: Were the smokes very irritating?

A. B. RAY: No more so than ordinary smoke. A yellow smoke produced by volatilizing arsenic sulphide is toxic to a certain extent but none of the dye smokes are toxic or excessively irritating.

DAVIS TUCK: I would like to ask Mr. Schladt how they make powdered aluminum? I tried to set off some powdered aluminum and I could not get it to burn at all.

G. J. SCHLADT: The aluminum used for flares is in two forms. The first known as the straight powdered aluminum is made by spraying molten aluminum into a chilled atmosphere. The second known as flake aluminum is powdered aluminum put through rolling mills and combined with oil, so the flake aluminum has a large burning surface, and naturally oxidizes more quickly. Now as to the burning in the open air: Powdered magnesium will burn in the open air without any extra oxidizing agent, but the less active aluminum, in most cases, requires an oxidizing agent to start it—such oxidizing agents as potassium nitrate or barium nitrate. Barium nitrate is a very efficient oxidizing agent.

Mr. Ray speaks of smokes. We have continued working on smokes and at the present time there is quite a discussion as to the merits of inorganic and organic or dye smokes. I will not attempt to pronounce the technical names of the different dyes used for colored smokes, but I might say that we still find use for one inorganic compound, namely arsenic disulphide, for yellow smoke. We find we get fairly good and distinctive smokes; they are used for day signals, and as in the case of airplane flares, they are visible quite a distance. We use a black smoke whose basis is anthracene which is very effective as a black smudge against the skyline.

Mr. Stickney has just mentioned that he would like to hear some more about airplane flares. When the planes are flying at high altitudes, the plane preceding the bombers, drops a line of these flares—there are five or six in the air at the same time. You have in that case, flood-lighting at decreasing distances and intensity on the ground growing higher all the time. That is very effective for uncovering the secret movement of troops, or illuminating targets, ammunition dumps or cities.

R. W. Moot: Is the phenomena of luminescence appreciable in your flares?

G. J. Schladt: In the last part of my paper, I spoke of a paper by Dr. Nichols. That appears in the Transactions of the Illuminating Engineering Society, Volume 16, Number 7, p. 331. The title of his paper is "Luminescence as a Factor in Artificial Lighting." He attributes luminescence to be chiefly responsible for the extraordinary light performance of certain metals.

L. C. Porter: There is one question I would like to ask Mr. Schladt. Has he made any tests on the distances at which these flares are visible? I have done some work for the Navy Department on long range signals and we found we could get remarkable increases in distances by changing the color somewhat, which would increase the quality of the light. With a considerably lower candlepower, we could get greater distances at which the lights were visible, by using colors that would give more contrast to the background. I wonder, if in your flare work, you have any tests on the maximum ranges at which these flares would be visible at the different candlepowers and the different colors.

G. J. Schladt: Tests are under way at the present time. I presume you are speaking of colored signals.

L. C. Porter: Yes.

G. J. Schladt: With the colored signals, we try to get the maximum range and keep the purity of the colors. The colors we are using in signal work are red, green and white.

L. C. PORTER: What was the maximum distance you were able to see some of those flares at night?

G. J. SCHLADT: I suppose with high candlepower flares the maximum would be the geographical limit—the limit of the horizon, which would be from fifteen to eighteen miles. That would depend, of course, upon the height of the plane. For example, a high candlepower airplane flare at an elevation of 2,000 feet under favorable air conditions should be visible approximately thirty miles.

Have you made any long distance tests with your particular candlepower flares?

L. C. PORTER: We have signalled fifty-nine miles with approximately 300,000 candlepower.

G. J. SCHLADT: Speaking of the visibility of signals, I have a test in mind that was made in an informal way for daylight visibility. We have seen a 300,000 candlepower flare a distance of twelve miles and 40,000 candlepower at a distance of four miles; that is on a fairly clear day and we found that the following relation approximately applied, visibility (in miles) $= 0.02 \sqrt{cp}$. I see no reason why you should not see a 300,000 candlepower flare at night at the distance you speak of—fifty-nine miles, provided the flare has a high elevation.

NON-MEMBER: How high will the flares function?

G. J. SCHLADT: That depends upon the time mechanism. If an aviator is flying at 9,000 feet, he could get any distance below the plane he desired.

Referring to wing-tip flares we made two types, one with a red tinge and one a straight white, with the idea that the red-tinged flare would have the qualities for penetrating mists and so better assist locating of objects but the question there was whether a white flare of lower candlepower would not have been just as effective, because in mist, you have the back reflection or a glaring fog bank just the same as you have in automobile headlights. The white flare was approximately 15,000 candlepower and the red tinge flare about 9,000 candlepower.

G. H. STICKNEY: What is the use of the wing-tip flare?

G. J. SCHLADT: They are used in case of a forced landing at night. They are carried in holders under the lower planes. They are ignited by a powder squib ignited electrically by a push button. During the war they had a candlepower of 10,000 and a burning time of one minute.

The talk to-night was to have been on Illuminating Flares, but considerable work has also been done on the colored flares and colored signals.

I have some pictures here and if a lantern were available, we would have these pictures thrown on the screen, but these pictures which I will pass around will give you some idea of construction and the spectral distribution of these signals and flares. The white flare has a continuous distribution over the whole range of the spectrum.

These spectrograms were made with the idea of seeing if they could be used in drawing up specifications for the purity of the colors; the spectral distribution of the white flares may give some information to those who are interested in that line of work.

A. BERGMAN (Communicated): The star in an Illuminating Shell is of course the most important part. It is to get the benefit of the illumination thereof that the shell is fired and every possible means should be employed to make available the utmost number of cubic inches for the illuminating composition.

In determining the advantage of one type of shell over another it is, however, neither necessary nor correct to refer to candlepower tests of the stars, as this is an entirely different problem.

A number of suitable compositions might be developed or be in use and comparative candlepower tests of these should be made. Based on the data thus obtained the preferable composition should be adopted and used in the shell that has been found to possess the greatest mechanical efficiency.

In judging the value of the star, it is necessary to measure not only its candlepower but also to compute the candlepower seconds per cubic inch of composition it generates. This subject is fully treated in a paper entitled "Rockets and Illuminating Shells as Used in Trench Warfare." presented by me to the Illuminating Engineering Society of New York City, April 11, 1918.

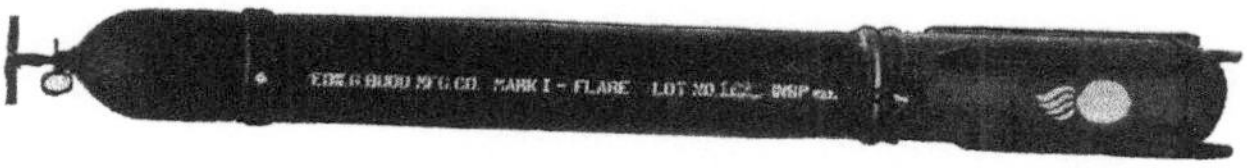

Fig. 4.—Airplane flare.

(1)

(2)

(3)

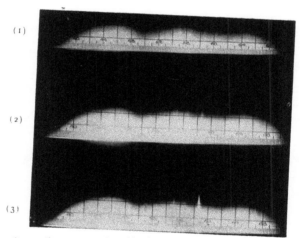

Fig. 5.—Spectrograms of a white flare. Length of exposure: (1) one second; (2)—five seconds; (3) thirty seconds.

It is an easy matter to make a star burn for a comparatively extended period of time, but this can only be done at the expense of the intensity of the light.

As there is little use in firing illuminating shell if they do not give satisfactory illumination and as, at least in the three-inch shell, there is not very much available space even in the best possible design, it is advisable to keep the time of burning down to a practical minimum so as to get the greatest possible candlepower from the star.

Based on this theory, I have always advocated fixing the burning time of the stars in three-inch shell at about 15 to 20 seconds, instead of making them burn for 30 to 40 seconds, (which, however, would meet with no difficulties if it would be preferred to have stars with longer burning time and less candlepower).

Furthermore, the longer a star is intended to burn, the bigger ought the parachute to be (so as to keep it suspended long enough) and a bigger parachute requires more space, which can only be obtained by decreasing the cubic contents of the star composition.

Still furthermore, the larger a parachute is the more resistance it will present to the air when it opens, and the more strain it will be exposed to;—the greater the strain is, the stronger must the parachute be and the stronger it is the more space it requires, which space must again be taken from the star.

The efficiency of the parachute can be judged by comparing its area with the weight it has to carry. It is obvious that the larger it is, the better it will keep the star suspended, but making the parachute too large should also be avoided as any increase beyond the necessary size requires space that could be used to better advantage by increasing the capacity of the star.

When designing the parachute it should be borne in mind that the faster the shell travels when it is made to burst, the greater is the strain put upon the parachute when this opens.

It is obvious that if a parachute can be made to function at all, a parachute arrangement can be made to function under any circumstances, but a parachute or set of parachutes constructed to open when the star is traveling at an excessive speed must be made a great deal more elaborate than when the parachute is expected to open when the star travels at a comparatively low rate of speed. Such elaborations require a certain amount of the valuable space that is available, and it is therefore contrary to good judgment to develop a shell with a parachute so designed and constructed that it can withstand the strain of being made to open at an earlier point on a shell's trajectory than is absolutely necessary.

It would of course be preferable to have a parachute that would function just as well at an extremely close range as at a long range, but this would call for a design that would require a comparatively very large percentage of all the available space inside a shell for the parachute arrangement, thus making the star (the most important part) correspondingly smaller.

Assuming we have a shell, the interior available space of which is so divided that enough space is left for the parachute to make it just large and strong enough to function satisfactorily at a range of say two miles, with all the balance of the space occupied by the star.

If we now make this shell burst at say one-half of a mile, the parachute will not stand up against the extra strain resulting from the increased speed by which the star is traveling when the parachute opens. (If it did stand up, it would indicate that the design was unnecessarily strong or elaborate for a minimum range of two miles). Hence, it would be necessary to modify the parachute design if it was found advisable to make one-half a mile the minimum range. This could be accomplished in two ways:

1st method. The parachute could be made smaller, which would result in the star dropping too fast, if the parachutes were originally just big enough to give satisfaction.

2nd method. The retarding devices or shock absorbers could be enlarged upon and the parachute made stronger, or more retarding parachute could be used. This would require more space,—that of necessity would have to be taken from the star.

Hence, it should in advance be determined what the minumum range should be at which the shell is intended to function; the velocity of the star when ejected at this range is known; the resulting strain on the parachute can be calculated or found by tests. Thereafter a suitable parachute design can be made, with a parachute that should be just big enough to prevent the star from dropping too fast and only strong enough (with proper margin of safety) to stand the expected strain. An Illuminating Shell, the parachute of which is designed following the above rule, should not be expected to function at a closer range than it is designed for.

The longer the minimum range is made, the more efficient can the design be made.

To determine which, of several types or designs, is the most efficient Illuminating Shell, the points to be considered should be divided into two main groups:

"A"—The design and general functioning of the shell.

"B"—The Illuminating composition.

"A" In judging the design and mechanical functioning of the shell the following points should be noted.

1st:—What muzzle velocity will the shell stand up against.

2nd:—How great percentage of failures if any are due to the mechanical parts not working properly, such as:

a.—Breaking up of the shell itself, due to weak body.

b.—Prematures: due to faulty design, or to such a design that so excessive care in loading is required that the human element plays too important a part.

c.—Prematures: due to weak interior parts collapsing from the set-back, causing pulverization of composition, friction heat and explosion.

3rd:—General functioning of the parachute.

a.—How great percentage of total and

b.—partial failure of the parachute.

c.—Is the rate of drop of the parachute excessively slow or rapid. If slower than necessary it indicates that unnecessarily large parachutes are used, resulting in decreasing the size of the star, and if too rapid it indicates the reverse.

d.—Efficiency of type of parachute:

Test: How many square inches of surface of parachute per cubic inch of space occupied.

e.—At what velocity of the shell will the parachute function.

4th:—Efficiency of principle employed in igniting star.
    Failure of ignition of star can be due to:
a.—Unreliable type of design.
b.—Unsuitable composition.
c.—Unsuitable method of loading.

It might sometimes be difficult to determine the causes of an ignition failure, but an experienced observer can generally tell. If due to the design or construction, nothing is seen after the burst of the shell. If due to the composition, the observer can generally see sparks or a short faint light, indicating that the first ignition charge did ignite but failed to ignite the illuminating composition itself.

5th:—What space is available for:
a.—The composition.
b.—The parachute.

6th:—The simplicity of construction should be judged.

7th:—How does the construction lend itself for mass production.

8th:—Cost of shell complet.e

"B"—In judging the value of the illuminating composition, the following points are to be noted:

1st:—Will the composition stand up against the shock of the setback, without exploding.

2nd:—Does it contain any ingredients that are liable to deteriorate or cause dangerous chemical reaction.

3rd:—What candlepower seconds per cubic inch of composition does it generate?

4th:—Are any of the ingredients difficult to obtain in quantity?

5th:—How will the composition sustain indefinite storage?

6th:—Is the finished star excessively difficult to ignite?

7th:—Does it have a tendency to break up and fall out of the container after being fired and when hanging suspended from the parachute?

After the type of shell construction has been decided upon and the illuminating composition to be used is selected, the following points are to be determined;

1st:—How long a time shall the star be made to burn, that is: Shall the star burn for say 15 seconds giving a very intense light, or shall it be made to burn for say 30 seconds giving only half the candlepower.

2nd:—With what rapidity the burning star may be allowed to descend after being ejected from the shell.

3rd:—What size and design of parachute will allow the star to descend with just the rate of speed decided upon.

4th:—How many cubic inches of space in the shell are required to take care of such a parachute.

5th:—The remaining available space will be utilized for the star, which now will be as large as the design of the shell permits.

In summing up the above these two points stand out as the main and most important. 1st:—How much space can be made available for the illuminating composition and, 2nd:—How many candlepower seconds per cubic inch does this composition generate?

ABSTRACTS

In this section of the TRANSACTIONS there will be used (1) ABSTRACTS of papers of general interest pertaining to the field of illumination appearing in technical journals, (2) ABSTRACTS of papers presented before the Illuminating Engineering Society, and (3) NOTES on research problems now in progress.

A SURVEY, COMPARISON AND CLASSIFICATION OF INDOOR ELECTRIC SIGNS*

BY P. SCHUYLER VAN BLOEM**

Neither electrical nor lighting engineers have given much attention to indoor signs. Perhaps it has been sub-consciously felt that the facts unearthed about the outdoor sign apply equally well to the indoor sign. To a certain extent this is true but the functional differences of the two classes of sign involve a radical separation of the principles of design.

The manufacturer of outdoor signs also makes indoor signs and generally has made his indoor sign but a miniature of the large exterior style. Accurate figures on the total size of the industry are not obtainable but there are less than ten concerns who specialize in indoor signs alone, while there seem to be about one hundred manufacturers who make some form or other of interior sign. Amongst these manufacturers are fixture makers, electricians, mechanics, glass houses, sign makers and painters, and others who have little capacity or ability to design really efficient indoor signs.

Lighting and power companies should be interested in the indoor sign field for the large number of these current-consuming devices make up, to a certain extent, the larger amount of current used by the few large signs. One large New York department store has a single outdoor electric sign, using 1,550 watts av-

*An Abstract of a paper presented before the New York Section of the Illuminating Engineering Society, April 20, 1922.
**President, Viking Sign Co., New York City.
The Illuminating Engineering Society is not responsible for the statements or opinions advanced by contributors.

erage, while on the interior of the same store there are 101 signs each using from 25 watts up. From actual computation the power used by the indoor signs amounts to 4,850 watts average. Thus, annually, the outdoor sign consumes approximately two million watt hours, while the indoor signs consume approximately eleven and one-half million watt hours. Ten orders from the books of a corporation specializing in the manufacture of indoor signs show that there were 1,800 signs represented, using 225,000 watts average, which, for six hours per day in a 300-day year, will have used approximately 405 million watt hours. Ten orders from the books of a corporation making outdoor signs show that the signs represented use approximately 109 million watt hours for the same period.

Many factors are bringing about the greater use of the indoor electric sign. The complexity of the modern building with its maze of corridors and passageways demands the rapid introduction of effective means of educating the public. Increasing competition among merchants is bringing about the greater use of indoor signs to induce sales, to promote good will and give customers every advantage of service and convenience. City ordinances, such as zoning laws and fire prevention laws, demand the use of a device to quickly attract the attention of people. The indoor electric sign has so far shown itself to be the most effective and persuasive medium for this purpose.

The most simple classification of this type of sign can be made according to the wording of the sign.

FUNCTIONAL CLASSIFICATION OF INDOOR ELECTRIC SIGNS

1. Safety sign
2. Directional sign
3. Explanatory sign
4. Announcement sign
5. Publicity sign
6. Merchandise display sign
7. Luminaire sign

Before comparing the several types of these signs it is necessary to set a reasonable standard to judge by. The following table briefly sets up this standard.

ESSENTIALS OF A SATISFACTORY INDOOR ELECTRIC SIGN

1. The sign must perform its function efficiently.
2. It must have position.
3. Its appearance should be pleasing.
4. The operating cost should be low.
5. The first cost should be low in proportion to the value.

From the above standard some of the commercial signs can now be judged. The following list gives the types in general use.

CLASSIFICATION OF INDOOR ELECTRIC SIGNS BY TYPE

1. Exposed lamp sign
2. Enclosed type
   - (a) Glass face, direct lighting
   - (b) Metal face
   - (c) Glass and metal face, indirect lighting
   - (d) Moving type
   - (e) Specialty

EXPOSED LAMP SIGNS

Signs of this type for indoor use range in size from one-inch letters to six or eight-inch letters. The large number of lamps required and the bold, heavy lettering often make this type of sign illegible and poor in appearance. Spotty illumination and short life of miniature lamps often make the first cost extremely high. Letters using small, special bulbs shaped to the character of the letter are not particularly graceful although they are sometimes very effective. Though operating and maintenance cost is comparatively high, this type of sign is often very good for publicity purposes.

TRANSPARENCY

This type of sign when carefully constructed as to letter proportion, color combination and even lighting becomes a very beautiful sign when new. In many cases, however, heat soon destroys the reflecting quality of the paint and very often carelessness in renewing lamps mars the interior painting and spoils the effect. The low cost of picturizing and the flexibility of letter arrangement makes this sign popular.

Special types of transparency such as the Decalcomania sign and photographic sign are expensive in first cost and expensive to

maintain. They are, however, admirably adapted for quantity
work and are often used in the advertising field.

Box shape, stenciled letter signs of this type when well made
and well lighted are very good. Letter design and spacing is unlimited, lamp breakage is low and when lighted by semi-indirect
methods using bold stroke letters, widely spaced, they become
very effective and economical.

The raised sheet glass letter sign is even better than a stenciled
letter type. Four to six-inch letters with wide strokes have good
long-distance legibility with little halation. Signs of this type are
less flexible than the stenciled sign but the better diffusion and
less halation offset this disadvantage. In smaller letters this type
of sign is less effective, unless a simple form of block or Egyptian
letter is used. The greatest disadvantage of this type of sign is
the lack of color and sparkle and the destructibility of the letters.

Prismatic glass raised letters are much better than the sheet
glass letters. Due to the peculiar form of the raised section
which is superimposed upon a thick back plate, a more brilliant
effect is produced with a given wattage than with any other sign
letter in use. With a back plate of either diffusing glass or
opalescent glass very good illumination occurs. Since the receiving surface is larger in square area than the projecting surface the largest amount of received light through refraction and
reflection is thrown out and this type of letter will pick up even
the slightest glow. Due to the method of manufacture sharp
contours and sharp lined character faces can be used, thus permitting the use of the more refined and artistic alphabet such as
the Greek and Roman. The sharp angular projections give the
letter an embossed appearance, producing a very pleasing effect.
Being made up of prisms the letter can be used equally well with
direct or indirect illumination, thus permitting a greater flexibility
in the design. This type of letter also gives more pure color and
is well adapted to the uses where color is important.

The "Polaralite" sign is one of the most popular on the market
due to its beauty and effective appearance. The illumination,
though at times slightly irregular, together with the novel effect

of no apparent visible means of illumination and its neat and clean appearance increases its popularity. As the position and length of the lamp filament relative to the matter to be illuminated is highly important the sign must be built around the lamp or combination of lamps to be used. Due to its indirect illumination the most suitable location for such a type of sign is dimly lighted corridors and rooms. Its effect in brightly lighted spaces is often lessened due to high reflection from its polished glass surfaces from other neighboring high wattage lamps. The operating and maintenance cost is low due to the use of standard tubular lamps with a rated life of 1,000 hours.

The "Viking T" type sign is very similar to the "Polaralite" except for the all metal surface. Indirect lighting is accomplished by admitting the light through the edge of the back plate of the prismatic letter. The feature of convertibility from standing type to a hanging type by turning the letter face increases its popularity. Colored letter effects may be obtained but no pictorial effects. The all metal construction makes this type practically indestructible and increases the popular demand.

CONCLUSION

In this paper there has been no attempt to take up the details of lamps, spacing and arrangement, methods of support, heating and ventilation or wiring. It seems best that such questions be left out from such a survey.

HOW BETTER LIGHTING INCREASED OUR PRODUC-
TION TWENTY-FIVE PER CENT*

BY JOHN MAGEE**

Three years ago the local light and power company came to us with a bargain to drive. And they drove it much to the mutual satisfaction of both parties. They said in so many words, "You have a plant here which we believe isn't operating at its full capacity because it is poorly lighted. We believe we can improve conditions. At the same time it just happens that we want to conduct some tests to decide which lighting arrangement is most adaptable to machine shop work. Let us come into your plant, install a meter and lighting equipment, supply the current, and your lighting costs will be nothing for 12 or 15 months—the period over which we want to make the tests. In return you can agree to supply us with production data."

I have heard it said that it is almost impossible to determine the effects of good lighting with anything like exactness at least. In fact there are factory executives to-day who pay but very little, if any, attention to improving their plant illumination. And probably their indifference is due in no small part to ignorance of just what savings can be accomplished with a slight rearrangement of fixtures which they already have and an intelligent study of the right candlepower bulbs to use.

Good lighting is so tied up with healthy working conditions that it is not difficult to foresee the time when it will be more closely controlled by laws and ordinances enforced by state and federal officers. But to my mind, factory executives have much more to gain, selfishly if you will, by taking the matter into their own hands now.

I started out to say that it was impossible to measure quantitatively the effects of good lighting. And I'll admit I believed this three years ago, but I don't to-day. The impossible has been

---

*Reprinted from *Factory*, February, 1923, p. 148.

**President, Detroit Piston Ring Co., Detroit, Mich.

The Illuminating Engineering Society is not responsible for the statements or opinions advanced by contributors.

done. It was done in our own plant. I am going to pass its good points along the line for the benefit of other managers who are looking for progressive ways to turn out more production with the same number of men and machines, thereby decreasing unit costs.

You know as a matter of fact, production and lighting are more closely tied up than you might imagine. It would not be going too far, perhaps, to say that lighting is to production what gasoline is to the automobile. The average automobile driver fills up his tank and never gives much thought to the engine so long as the car runs smoothly. Perhaps by buying some other kind of gasoline his motor would not carbonize so quickly, or he might show a better consumption per mile, or perhaps the motor wouldn't heat up so quickly. But no, gasoline is only a small part of the upkeep of the car and little attention is paid to it.

It's much the same with lighting. Compared with the amount of money it takes to meet payrolls every week, the light bill is a drop in the bucket. Supposing the bill does run twice as high some months as others, even after discounting for the varying lengths of daylight. Certainly the firm wouldn't have to go out of business—in fact I doubt if 99 per cent of concerns would even notice it.

All this is a little away from the subject, but what I am trying to make clear is that production is absolutely dependent upon good lighting. And the tests which we made will prove it.

And what happened? When the lighting company completed its tests we were so amazed with the results that it took no persuasion whatever for us to install the equipment permanently. Incidentally, we have recently acquired a foundry which enables us to get castings without depending upon outside concerns. And in this foundry we have put a lighting arrangement similar to the one in the shop. This fact in itself shows that we are completely sold on the large part good lighting plays in improved production and lower costs.

In Figure 1, I have endeavored to show just what occurred. By no means is it complete; that is, complete in the sense that it shows all phases of the improvements. But it does indicate at least the very tangible results secured in increased produc-

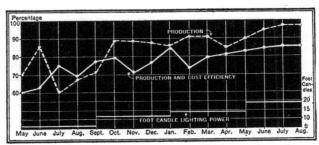

Fig. 1.—Production kept on the increase as lighting was improved.

tion, and also the increase in cost efficiency — a term which means little to the average man except to reflect in a general way a decrease in unit costs.

The department selected was one where there was considerable difficulty in lighting it properly on account of the close setting of belt-operated machines and the dust accumulation due to the grinding operations. The lighting arrangement which we were using at the time of the tests, consisted of individual drop lights at each machine. And with it we were turning out about 70 per cent production, basing the capacity at 12,000 piston rings a day. As nearly as we could estimate, the average intensity of illumination was about 1.2 foot-candles. Obviously, this was pretty low considering that a conservative estimate of plant capacity had been made.

One of the difficulties which arise in making any investigation lies in holding all the variable factors constant except the one which is to be studied. This is one of the reasons, I believe, why so little definite data are at hand showing the relation between good lighting and increased production. In fact you will notice from the curves that monthly production figures were hardly indicative of any regular rise or fall. On the contrary they zigzag most irregularly. This is something which it is practically impossible to get away from for there are many factors at work which affect the amount of goods turned out—changes in demand, poor raw material, delays in delivery, and so on.

KEEPING ALL FACTORS CONSTANT EXCEPT LIGHTING

Although no mean line is drawn for the curves below it takes no large amount of imagination to notice a distinct upward trend as the lighting intensity was increased.

In order to keep out of the results these outside factors which might offset production, the number of employees remained the same during the period of the test. The same was true of machinery and its accessories. In fact, the only factor which we wilfully varied was the lighting.

Furthermore, to keep out the influences of seasonal changes, we took a long period of time to make the tests—fifteen months, to be exact. To this we attribute the reliability of the tests obtained.

Another reason why but little has been written relative to production and better lighting is the fact that all experiences have been based upon the practise of one plant. And, says the executive across the street, "Smith has a different problem altogether from what I have. He turns out thousands of screws every day while we busy ourselves with producing half-ton casting. What can I profit by his experience?" Whether he can or not, I am not prepared to say, but I do believe that many times the idea of one man can be molded to fit the needs of another.

At any rate, I am going to describe briefly just what took place. As I stated before, we were working at about 70 per cent capacity. We took one department—25 feet by 100 feet, and provided it with thirty-five outlets spaced on 7.5-foot centers. These outlets were mounted 9.5 feet from the floor, and the whole system was so arranged that any size lamp up to 200 watts could be used.

Bulbs with diffusing caps rated at 100 watts were first tried over a period of four months. The average production during that time was a little over 13 per cent better than the old individual type of lamps. And the average intensity of light for the whole shop was about 6½ foot-candles.

And so it went; 150-watt lamps with standard dome reflectors pulled a production of 88.4 per cent or an increase of 17.9 per cent over the old installation; 200-watt lamps with dome glass top reflectors showed a 96.3 per cent production which was actually a 25.8 per cent increase. The intensity averages for these two last tests were 9 foot-candles and 14 foot-candles respectively.

We analyzed the results as we went along, but before we were half through with the job it was perfectly obvious that there was going to be a neat little saving here. So we laid aside the figures for an instant and started estimating in another direction.

Like all companies, we had a certain amount of spoilage, accident hazards, and so forth, which, no doubt, could be laid to inadequate lighting. A few minutes spent in adding and subtracting revealed this fact: that the costs of all the spoilage, accident hazards, and so on, which represented wastes, could be offset by a 1 per cent increase in production. And the 200-watt installation gave us a 25 per cent increase!

While the increased cost of the new lighting system over the old one was 48 per cent, this only represented about a 2 per cent increase when measured in terms of the monthly payroll.

In other words, the maintaining of a good lighting system is but a small fraction of the general operating expenses. And because it has such an important effect on production it is folly to think of curtailing its expense.

I doubt if I could say off-hand just how much more business we have been able to get because of the lower unit costs due to a large volume of production.

Should we resort to the ancient, though convenient method of counting on our fingers, there would be some difficulty in finding enough fingers on two hands, even if we counted by thousands of rings a day.

Because of the marked decrease in overhead expense we could get our costs down to a point where we could sell at a figure which brought an increased volume. And this volume is a factor which still further reduces overhead. Improved lighting was an eye-opener to the many savings better methods of management can make. And we are becoming increasingly aware of these better methods every day.

SOCIETY AFFAIRS

SECTION ACTIVITIES

The March meeting of the Philadelphia Section was held at the Engineers' Club on the evening of the 13th. The paper, "Church Lighting," was to have been given by Mr. A. L. Powell, Illuminating Engineer of the Edison Lamp Works of Harrison, N. J.; but on account of sickness Mr. A. B. Oday of the same company presented the subject.

Mr. Oday started his talk by calling attention that in the churches in Medieval times no artificial light was required as services were held during the day. In later times various forms of illumination were gradually introduced as the demand for artificial lighting increased. He then described the various forms of lighting which are now available, stating that in the Ritualistic churches direct lighting is generally the most appropriate, either using single units or a number of units grouped on a fixture. Also that by means of reflectors on the rear of the chancel arch higher intensity can be obtained to bring out this, the most important part of the church. In this style of architecture it is frequently possible to conceal the lamps behind the roof beams.

In the Evangelical churches any style of lighting may be used appropriately but indirect lighting may be used to an advantage.

The speaker stated that not infrequently the poor lighting effects which were noticeable in some churches were due to lack of co-operation between the architect and the illuminating engineer and hoped that better results might be obtained in the future. He also spoke on the possibility of using colored lighting and mentioned the result obtained by this method in St. Mark's Church in New York. The use of a spot light for illuminating the speaker could also be used to an advantage at times.

The lighting of stained glass windows was an object which frequently necessitated considerable skill to produce good results on account of the different densities of the colored glasses. This might be done either by flood lights, where the space is available, or locally by means of smaller units.

Mr. Oday also spoke on the advertising illumination of churches. This may be either in the form of signs or simply illuminating the bill boards which contain the various notices of the services. He stated that in the interior wiring provision should be made for the use of a stereoptican lamp.

Maintenance of the lighting in churches is an important item and one which should be seriously considered by the architect and illuminating engineer. If the fixtures are not accessible, or the lamps can be replaced only by considerable trouble and expense, the burnt out lamps will be permitted to remain for some time with the resultant poor effect and illumination.

The talk was illustrated by a number of lantern slides showing good and bad lighting of both interior and exterior of various churches.

The paper was followed by a short description by Mrs. Mary H. Greenewalt on "The Lighting Score for the Episcopal Service and Mechanism for Interpreting It," which method consisted in using colored lights which are varied in accordance with the different parts of the service.

Forty attended the dinner preceding the meeting, at which there were fifty-five members and guests present.

NORTHERN NEW JERSEY Meeting—March 7, 1923

The Northern New Jersey Chapter of the Illuminating Engineering Society was organized on the evening of March 7, 1923, at a meeting held at 8:00 P. M. in the auditorium of the Public Service Terminal, Newark, N. J.

The meeting was called to order by Mr. D. McFarlan Moore, who acted as temporary chairman. In some preliminary remarks Mr. Moore emphasized the benefits which the members of the Illuminating Engineering Society derive from the activities of the organization.

Mr. Ward Harrison, President of the I. E. S., was then introduced by Mr. Moore as the first speaker of the evening. President Harrison spoke of the great progress which the Society has made during the past year, and stated that membership had increased this year more rapidly than at any time in its history. Mr. Harrison further stated that since only about two per cent of the people in the country or any city are responsible for about ninety-eight per cent of the progress made, he felt confident that with the showing made at this first meeting, this section will in the near future be able to show great progress towards better lighting conditions.

Mr. P. S. Young of the Public Service Electric Co. was the next speaker and spoke in part as follows:

"I am glad to have heard the President of the Society state that it has now reached its third and greatest stage of development because I agree with him that the Society will do its best work when the knowledge of proper illumination is disseminated broadly through all classes. * * * * I believe it means that there is going to be some very useful work done in New Jersey. I believe that useful work means not only a better commercial development of lighting but it means better health for the people and greater satisfaction for everyone in the lighting industry. I think we are all going to derive great benefit from it and will feel the satisfaction of doing something worth while. I think that as individuals are banded together for work that they know is beneficial the standard of their own work is improved and they get a great deal more pleasure out of their work."

The next speaker was Colonel L. T. Bryant, Commissioner of Labor of the State of New Jersey. Colonel Bryant stated that after having been head of the Department of Labor in New Jersey for so many years he realized the very great necessity of improved lighting conditions in our industries, especially in the factories, and that the Department of Labor has a very direct interest in having the factories properly lighted. It is better for the eyes, it affects the morale of the laboring man, it has a close bearing on the number of accidents, it has a bearing on the health and efficiency of the employees, and undoubtedly increases the output of the factory. Colonel Bryant pledged the fullest co-operation of the Department of Labor in any movement that will impress upon the minds of factory owners the necessity for better lighting conditions.

Mr. Preston S. Millar of the Electrical Testing Laboratories and Mr. F. Rosseter, President of the Master Electricians Association of Newark, were the next speakers. Mr. Millar gave a short historical sketch of the Society, and discussed its aims and objects, and reviewed some of its past accomplishments. Mr. Rosseter stated that the organization which he represented was very glad to see this Chapter being formed and would do everything they could to make it a success. He thought that by getting together in a Chapter of this kind the wiring contractors could learn more about the illuminating engineer's viewpoint and the illuminating engineer would learn more about the wiring contractors' end of the business.

The following officers were elected: Chairman, Mr. W. T. Blackwell, Public Service Elec. Co., Newark, N. J.; Secretary, Mr. H. C. Calahan, Tri-City Electric Company, Newark, N. J.; Treasurer, Mr. George Davis, Davis Electric Company, Newark, N. J. Messrs D. McFarlan Moore, F. Rosseter, Paul Jaehnig, Ellsworth Francisco and R. R. Young were elected on the Board of Managers.

Messrs. D. McFarlan Moore and F. Rosseter were appointed on a committee to prepare a constitution and by-laws which will be presented to the Board of Managers at a later date.

Among the other speakers was Mr. Herman Plaut, Vice-President of the National Council of Fixture Manufacturers. Mr. Plaut emphasized the benefit which the manufacturers and electrical contractors, and other people interested in illumination, could derive from membership in the Society and attendance at its meetings. Other speakers were Mr. Geo. H. Stickney, Past-President of the Society; Mr. W. T. Blackwell, General Lighting Representative, Public Service Co.; Mr. Samuel G. Hibben, General Secretary of the Society; and Mr. Rowland H. Leveridge, Department of Labor, State of New Jersey.

On the motion of Chairman W. T. Blackwell, Colonel L. T. Bryant was elected Honorary Member of the Northern New Jersey Chapter of the Society.

Mr. A. L. Powell, Chairman, New York Section, was unable to attend but forwarded a letter which was read to the meeting and in which Mr. Powell wished the new Chapter success and hoped that the New York Section and Northern New Jersey Chapter would work in close harmony and to their mutual advantage.

NEW ENGLAND Meeting—March 30, 1923

On March 30, a joint meeting of the New England Section of the I. E. S. and the Boston Section of the A. I. E. E, was held at the Engineers' Club, Boston. The speaker of the evening was Mr. M. Luckiesh, director of the Laboratory of Applied Science, Nela Park, Cleveland. His subject was the "History of Artificial Lighting and its Influence on Civilization."

Mr. Luckiesh's talk was illustrated by numerous slides showing luminaires of past ages starting with the firebrand of prehistoric man and culminating in the present day incandescent lamp. Slides were shown of modern types of lighting installations made possible by the present day lighting sources.

The feature of the evening was the display of the 30,000-watt incandescent lamp developed at the Research Laboratories of the General Electric Company for Maude Adams. This lamp excited much interest amongst the audience, and received a great deal of publicity in the local newspapers.

An interesting discussion on the future of lighting took place at the close of the meeting in which members of the faculty of M. I. T. took an active part.

Many architects were present at this meeting and commented on the excellence of the talk. About one hundred and twenty members of the two Societies were present including a number of the members of the faculty of the Massachusetts Institute of Technology and engineers from the large engineering firms in New England.

NEW YORK Meeting—April 3, 1923

The meeting scheduled for March was held on the evening of April 3 at the American Art Gallery, 57th and Madison Ave., New York City. Prior to this meeting the members of the Section inspected the Grand Central Art Galleries at Grand Central Terminal. The systems of illumination of both of these art galleries were informally explained during the tour of inspection.

A paper, "New Developments in Art Gallery Illumination," was presented by Mr. Lawrence X. Champeau, of Kirby-Champeau Co., Inc., of New York City. An interesting discussion followed and eighty-five members and guests attended the meeting.

COLUMBUS Meeting—February 20, 1923

The Columbus Chapter met on the evening of February 20 at which Mr. Ward Harrison, President of the I. E. S. gave the principal address.

The Officers of the Chapter are as follows: Chairman, Mr. G. F. Evans; Secretary, Mr. R. C. Moore; the Board of Governors consist of the Chairman, Secretary, and Messrs. F. C. Caldwell, Karl Piez, and Dr. G. E. Sulzer.

COUNCIL NOTES

ITEMS OF INTEREST

At the meeting of the Council, March 8, 1923, the following were elected to membership.

One Member

PENCE, DAVID C., Illinois Electric Co., 313-315 San Pedro St., Los Angeles, Calif.

Twenty-Four Associate Members

BENSI, R. J., South American General Electric Co., 560 Av. De Mayo, Buenos Aires, Argentine, South America.

BLIVEN, JOSEPH EDWARD, Empire Theatre Co., 394 Bank Street, New London, Conn.

BROWN, DAVID SEYMOUR, New York Telephone Co., 104 Broad Street, New York, N. Y.

CAMP, H. M., The Tennessee Electric Power Co., 620 Market Street, Chattanooga, Tenn.

GRAAFF, ANTONIUS DE, Philips' Glowlampworks, Ltd., Eindhoven, Holland.

EVERSON, R. W., Westinghouse Lamp Co., 1321 Candler Bldg., Atlanta, Ga.

HOELLER, OTTO A., Central Ill. Public Service Co., Beardstown, Ill.

HOLMAN, E. R., Solar Lighting Co., Inc., 219 East Fourth St., Los Angeles, Calif.

KERSHAW, K. O., Panama Lamp & Com. Co., 757 S. Los Angeles St., Los Angeles, Calif.

KNIGHT, CHARLES, Eastern Malleable Iron Co., Bridge Street, Naugatuck, Conn.

KRAPF, EDGAR W., F. Huffer Co., 1208 Liberty Avenue, Brooklyn, N. Y.

LYLE, A. ERNEST, Canadian Westinghouse Co., Aberdeen Ave., Hamilton, Ontario, Canada.

MOSSGROVE, J. R., Erner & Hopkins Electric Co., 146 N. Third St., Columbus, Ohio.

NALAND, C. W., The Union Electric Co., Abilene, Kansas.

O'LEARY, JOSEPH J., Westinghouse Electric & Mfg. Co., 111 W. Washington St., Chicago, Ill.

OLMSTEAD, HENRY C., Western Electric Co., 500 S. Clinton St., Chicago, Ill.

SMITH, JAMES Y., Westinghouse Elec. & Mfg. Co., 165 Broadway, New York, N. Y.

SMITH, RAYMOND A., Crescent Electric Co., 150 Jefferson St., East., Detroit, Mich.

SYLVESTER, ELMER LE MUR, Edison Lamp Works of the G. E. Co., Harrison, N. J.

VAN VOLKENBURG, RAY, Student, University of Michigan, 523 Packard St., Ann Arbor, Mich.

WHITE, WILFRED FRENCH, Westinghouse Elec. & Mfg. Co., Notre Dame St., South Bend, Indiana.

WILLIAMS, R. O., Harter Mfg. Co., 1850 Fulton St., Chicago, Ill,

WORK, W. M., Westinghouse Electric & Mfg. Co., George Cutter Works, South Bend, Indiana.

YOUNGLOVE, G. WILSON, Student, University of Michigan, 1308 E. Ann St., Ann Arbor, Mich.

Three Sustaining Members

AMHERST GAS COMPANY, 11 Pleasant St., Amherst, Mass.
J. J. O'Connell, Official Representative

CORNING GLASS WORKS, Corning, N. Y.
E. C. Sullivan, Official Representative.

IRVING & CASSON—A. H. DAVENPORT CO., 573 Boylston St., Boston, Mass.
David Crownfield, Official Representative.

The General Secretary reported the death, on March 21, 1922, of one associate member, JOHN WILLIAMSON, Peoples Gas Light & Coke Co., 122 South Michigan Blvd., Chicago, Ill.

The General Secretary reported the transfer of HENRY LOGAN, 1284 Nicholson Ave., Lakewood, Ohio, to full membership.

CONFIRMATION OF APPOINTMENTS

As member of the sub-committee on Revision of the Code of School Lighting—E. W. Commery.

NEWS ITEMS

Dr. Richard Ulbricht, known by all illuminating engineers as the inventor of the Ulbricht Integrating Sphere, died on January 13, 1923 at the age of 74 years. Dr. Ulbricht was eminent as an electrical engineer, having in 1894 built the first large polyphase station in Germany. For 27 years he gave instruction in the Technical High School in Dresden. From 1910 to 1919 he was President of the State Railroads of Saxony. His first publication of the integrating sphere appeared in 1900. The first publication was followed by others in which the theory of the sphere was very thoroughly investigated mathematically. The illuminating engineering profession owes a great deal to Dr. Ulbricht for the sphere photometer which is now so universally used and generally recognized as the ideal form of integrator and which may be said to be the greatest single contribution to the art of photometry which has been made since Bunsen brought out the grease spot principle.

The Senior Electrical Engineers of Yale University are taking their annual inspection trip under the direction of Professor Warner and the Senior Electrical Engineers of the University of Pennsylvania under the direction of Professor Fawcett, visited the Edison Lamp Works, Harrison, New Jersey, on March 21st and 28th respectively. They inspected typical lamp factories and were given some talks and demonstrations by members of the Lighting Service and Commercial Engineering Departments.

Dr. Elihu Thomson, of the General Electric Company, Lynn, Mass., was tendered a dinner in the Boston City Club on March 29 by more than seventy-five friends and associates in the General Electric Company, and from Harvard University and the Massachusetts Institute of Technology, on the occasion of his seventieth birthday anniversary. He was presented with a silver loving cup, the presentation being made by Mr. Walter C. Fish, formerly manager of the Lynn Works, who served as toastmaster. There were addresses by Professor A. E. Kennelly, of Harvard University; Professor Dugald C. Jackson, of the Massachusetts Institute of Technology; Professor Comfort Avery Adams, of Harvard University; Mr. George E. Emmons, of Schenectady, Vice-President of the General Electric Company; Mr. J. R. Lovejoy, of New York City, also a Vice-President of the company; Mr. Herman Lemp, of Erie, Pa., and Mr. F. P. Cox, of Lynn, Mass.

The Play Work Shop of the Polytechnic Institute of Brooklyn in conjunction with the Department of Electrical Engineering, is presenting a course on stage lighting under the direction of Mr. A. L. Powell. Mr. Powell is giving the introductory lectures and demonstrations and several of the leading stage lighting men will supplement these talks. Inspection trips are planned to representative theatres in order that actual working conditions may be observed. A very comprehensive outline has been prepared and the subject is being discussed in greater detail. This is probably the first time that an Engineering School has given a special course of this nature treating one particular field of lighting. Although lectures are given outside of regular hours for classes there are approximately one hundred and fifty men in attendance and considerable enthusiasm and interest is displayed.

Mr. M. Luckiesh, National Lamp Works of Cleveland, spoke at the annual banquet of the Lynn Section of the A. I. E. E. on March 31 on the subject, "The Fairyland of Clouds." On April 12 he spoke before the Kiwanis Club of Newark, N. J. on the subject, "Artificial Light and Civilization." The same lecture will be repeated before the Royal Canadian Institute on April 21, at Toronto, Canada.

Mr. W. H. Rademacher gave a series of talks on illumination during the last week in March and the first two weeks in April before Rotary and Kiwanis Clubs, Chambers of Commerce and similar organizations in several towns in Michigan.

Dr. Edward P. Hyde, who organized the Nela Research Laboratories in 1908, and who for the past few years has occupied the position of Director of Research of the National Lamp Works of the General Electric Co., Cleveland, Ohio, has tendered his resignation to take effect June 30 of this year. Dr. Hyde, who has been active in scientific and technical affairs for a number of years, has decided to take a prolonged rest abroad. He will temporarily discontinue many of his activities in the scientific and engineering societies, but will retain his office of President of the International Commission on Illumination until its next plenary meeting which is scheduled to be held in this country in 1924. Dr. Hyde is a Past-President of the Illuminating Engineering Society and has taken a great deal of interest in its affairs.

Mr. Samuel G. Hibben was married to Miss Ruth H. Rittenhouse of Montclair, N. J., on Saturday evening, April 14, 1923. After an automobile trip through the South, they will reside at 8 Bradford Place, Montclair, N. J.

Mr. C. M. Masson of Los Angeles, California has been spending the winter in New York City in the study of special illumination problems, at the Illumination Bureau of the Westinghouse Lamp Co. and at the Electrical Testing Laboratories.

Mr. A. S. Turner has been for the past six weeks on a trip through Georgia, Alabama, Tennessee and Texas, looking after special lighting problems as well as giving lectures in the principal cities.

Mr. O. L. Johnson was recently appointed Commercial Engineer by the King Manufacturing Co., 53 West Jackson Blvd., Chicago, Ill.

Announcement of a new book, "Ultra-violet Radiation," by Mr. M. Luckiesh, Director of Applied Science, Nela Research Laboratories of Cleveland has been recently made by D. Van Nostrand Co. of New York City.

PROFESSIONAL OPPORTUNITIES

ILLUMINATING ENGINEER: To take charge of the Illuminating Engineering Department and to act as consultant in residence work with a large Eastern Public Utility. Give age, education, experience and salary expected.   E-1

ILLUMINATION INDEX

PREPARED BY THE COMMITTEE ON PROGRESS.

An INDEX OF REFERENCES to books, papers, editorials, news and abstracts on illuminating engineering and allied subjects. This index is arranged alphabetically according to the names of the reference publications. The references are then given in order of the date of publication. Important references not appearing herein should be called to the attention of the Illuminating Engineering Society, 29 W. 39th St., New York, N. Y.

	AUTHOR	DATE	PAGE
Annalen der Physik		1923	
Über die Beständigkeit der Phosphor-			
eszenzentren, I—	Hans Kuppenheim	Jan. 31	81
II—			113
Central Station			
The Street Lighting of Greater Cleve-			
land—	A. E. Suker	Feb.	228
Lighting for Public Eating Places—	J. L. Stair	Feb.	231
Chemical Abstracts			
Application of Artificial Daylight to			
Laboratory Purposes—	S. H. Groom	Feb. 10	366
The Action of Light on Mesonitro-			
anthracene—	J. Battegay, Ph.		
	Brandt, and J.		
	Moritz	Feb. 10	366
Action of the Arc Light on Aqueous			
Oxalic Acid Solutions in the Pres-			
ence of Ferric, Chromic and Mer-			
curic Chloride; of Uranyl Acetate			
and of Iodic Acid—	H. Kunz-Krause		
	and P. Manicke	Feb. 10	367
Comptes Rendus			
Sur la perte de lumiere a Paris et			
dans les environs—	Louis Besson	Jan. 15	180
Nouvelles experiences sur le phenomene			
de Broca et Sulzer (ondulation de			
fatigue)—	Emile Haas	Jan. 15	188
Electrical Merchandising			
Lighting the Show Window—Color Op-			
portunities for the Merchant—	M. Luckiesh	Feb.	3120

TRANSACTIONS
OF THE
ILLUMINATING ENGINEERING SOCIETY

| VOL. XVIII | MAY, 1923 | NO. 5 |

Individual Responsibility in Making I. E. S. Activities Beneficial

BECAUSE OF the fact that from the very start an unusually high standard has been maintained for the material that appears in the TRANSACTIONS, we can proudly boast that the official publication of our Society forms the most important and valuable collection to lighting literature in the world.

The bulk of the burden of providing the necessary material to maintain this high standard, falls upon a few committees, one of which is the Papers Committee. The Committees, therefore, should have the fullest co-operation of the entire membership. They should receive from time to time frank suggestions; ideas for the improvement of the TRANSACTIONS and for material that is to be included in them.

One of the big events in our Society affairs—one that we all look forward to—is the national Convention. The program for the 1923 annual meeting is rapidly taking form and in order to make this Convention program, "of, by and for the membership," we want your early suggestions.

Do you know of or have you been making a recent valuable investigation? If so, notify the Papers Committee.

Have you heard of some remarkable development in lighting devices or applications? If you have, send a brief note concerning it to the Committee.

Can you suggest a subject that combines a practical idea with an entertainment feature? We need it to round out the program.

Can you suggest anything that you believe should be investigated—a contemplated research, or ideas that you think would be the basis for a real contribution to lighting literature?

The Committee has on hand a list of possible subjects for papers that have come in from various sources. The nature of the subjects so far received, indicate that material will be available to make the coming Convention—from the standpoint of the Papers program—a wonderfully interesting and valuable one.

We have a selection of papers of a purely scientific nature—those that will record in minute detail some of the most recent technical phases of our lighting science.

A number of excellent subjects have been received that might be designated as practical subjects—discussing the correct applications of good lighting principles.

There are some ideas for semi-scientific discussions, as well as a number that might logically be called "general interest topics." A few of the latter can easily take on the nature of a demonstration lecture and thus be made to present valuable information in a very entertaining way.

It is the purpose of the General Papers Committee to provide for the fall convention a perfectly balanced program, one that will give the proper consideration for both scientific and so-called practical subjects. The program will not be crowded. Ample time will be reserved for discussions—a most essential part of the program. The tentative program for the fall convention is rapidly being completed.

Suggestions from every member of the Society as to how the program can be improved are urgently requested. Ideas for papers to be included should be forwarded to the Committee chairman without further delay if they are to be considered. Most every year topics of value must be omitted because they have been sent in too late.

J. L. STAIR, *Chairman,*
Committee on Papers.

REFLECTIONS

Important Advances in Street-Lighting Equipment

DREARY standardization of street-lighting equipment has been threatened of late, and therefore it is refreshing to find that a step forward is actually being taken, one in a direction already familiar, the automatic operation of sub-stations combined with out-door apparatus. The progress referred to has been achieved in Kansas City and is described by A. E. Bettis in the *Electrical World*, February 10, 1923.

Time was not many years ago when the suggestion of placing even switches and transformers out of doors was regarded as a dangerous heresy, stamping the engineer who proposed it as a harebrained innovator who should promptly be suppressed lest he break in on the beautiful symmetry of catalog equipment. But the world moves, and as time has gone on more and more apparatus has gone out of doors, sometimes indeed more than under the local conditions might always be desirable. The radical departure in Kansas City lies in placing the constant-current transformers and their immediate equipment on the line poles at convenient outlying points for the supply of the street-lighting circuits. To feed and control these a single 4,000-volt, three-phase, four-wire circuit is run out of each of seven general sub-stations. The several phases are distributed as 2,300-volt feeders in three symmetrical territories around each sub-station. In the automatic street-lighting units the line feeder is connected as usual, but is controlled by a time clock which operates the master controller, energizes the circuit and puts the equipment into operation. The same time clock takes the street-lighting circuits off by opening the control circuit and thus checking the activities of that particular feeder. The ordinary constant-current transformer used in this way is rated at 20 kva., smaller units being employed in some of the outlying districts. Each of these 20-kva. installations is good for about forty 600-cp. lamps or a proportionately greater number of 400-cp. or smaller sizes. As there are more than six thousand lamps in all, of which more than half are of 600-cp., the thorough subdivision of the transformer system is self-evident.

This automatic service does two very important things. It eliminates a serious amount of equipment in the sub-stations, necessarily well filled with apparatus for other purposes and, what is even more important, it gets rid of the long and troublesome dead runs of wire on the street-lighting circuits. Any one who is familiar with the complication of laying out street-lighting circuits which have to be operated from one or a few sub-stations will realize the value of pole equipment for handling comparatively short circuits automatically. In this way one can get the maximum number of lamps per actual mile of circuit and avoid contingencies which might put out of action a large portion of the circuit at once.

The other outstanding feature of the Kansas City installation is the underground cable work. The city was not provided with the conduits adequate for the new system even if it had been desired to use them, and it was desired to get rid of overhead wires in so far as this was humanly possible in a growing installation. The difficulty was met by a special lead-covered cable laid over with a triple jute covering impregnated with insulating compound. The chief purpose of the jute covering is to check danger from electrolysis. The cables are paper-insulated, carrying a single No. 8 conductor, and the experience of a year with this cable, of which 750,000 feet are now under ground, has been entirely satisfactory. It is laid directly in earth in a narrow trench 12 inches or 14 inches deep, and at street intersections an additional protection is installed in the form of a 2-inch fiber contact. Special lamp posts are used only where there are no trolley poles handy. Where these are available a special bracket carries the lamp. This arrangement of cable for the underground supply of street lamps has proved to be reasonably cheap. The cost of trenching, placing and filling is running less than 10 cents a foot in clear ground, while the typical total installation cost is reckoned at about 23 cents per foot exclusive of street crossings. Very little trouble has been had with the underground system, and experience indicates that no serious difficulty is to be anticipated.

Of course, the critical period for any such installation is after some years of service, and future results will be watched with interest. Enough has been done, however, to show not only that a thoroughly practical system of automatic pole transformers for street-lighting service can be used but that supply by underground cables laid simply in the earth is here, as it it has been proved in various places abroad, an entirely satisfactory means of getting the wires out of the way.

Back to Candles

ELECTRICITY is blinding mankind. If modern man wishes to preserve his eyesight he must take a long step backward, stop using electric lights and return to the use of candles. This is the statement of an English scientist.

Fifty per cent of men, he says, wear spectacles to-day, 20 per cent of women and many children. The principal reason for this condition, according to the scientist, is that the eyes of man for many generations have been accustomed only to the soft light of the open fire or the candle's gleam and are not fitted to withstand the brilliancy of electric light.

There is something to be said for this theory. Long exposure of the eyes to glaring light is trying and undoubtedly fosters visual troubles. It is true also that because of better lighting facilities people use their eyes for more hours out of the 24 and weary them proportionately. On the other hand the effort to use the eyes in a poor light, particularly for close work, is swiftly productive of bad vision.

The solution hardly lies in a return to candles. It lies rather in a wise softening of the electric glare and in a more sane observance of hours of rest from both light and eye-strain. More than all it lies in a careful observance of proper placing for all lighting fixtures with relation to the work in hand.—Editorial, *Ithaca Journal-News*, April 4, 1923.

New York City Has Novel Electric Sign

AN electric sign of novel and artistic type has been erected along the canyon of white lights on Broadway, New York, to announce a popular motion-picture entitled "The Covered Wagon." The wagon and entire upper portion of the sign are lighted by ten 500-watt flood-lamps, and on the front of the wagon is a lantern which is an exact facsimile of the type used in 1849. A lighting effect which simulates water running under the wagon is obtained by six stereopticons automatically operated by motors which turn a color wheel with water painted on it. Each lamp is a 2,000-watt nitrogen-filled unit specially made for stereopticon purposes. In the center of the water is a small sign in raised bronze letters reading "Paramount Pictures." These letters are set in a box flush with the water and are illuminated by indirect lighting which causes the letters to stand out as if they were set in the water.

Relative Effectiveness of Direct and Indirect Illumination

NO SUBJECT connected with illumination has been the cause of more controversy than the relative merits of direct and indirect lighting. The quarrel has not been merely a commercial one such as might be reasonably expected, but a technical one also, and in that aspect there is less reason for variations of opinion. A number of so-called practical studies have been made dealing with the properties of light falling directly on the page and falling upon it after reflection from the walls and ceiling, and utterly diverse data have been obtained. The probable cause of these striking discrepancies is the fact that any investigation of this sort tends to emphasize psychological rather than physical values, values therefore determinable only in vague impressions of comfort and discomfort, of ease and difficulty, of fatigue and facility.

In any tests made upon the eye other factors than the ordinary ones of shade perception and acuity of necessity enter. The particular kind of work attempted, the extent to which the several illuminations are pushed above the limit for moderate acuity, the contrast of the things observed and the length of time for which the observations may be continued, all are things which enter the final judgment as to the sufficiency of the illumination received. There is added to these causes of uncertainty a still greater one in the element of unconscious suggestion in the manner of the test or the instructions given for carrying it out. It would not be difficult to make a given group dissatisfied with almost any system of illumination without an expressed word of disapprobation. As a matter of fact there can be no sweeping decision between direct and indirect lighting. Some of the worst and some of the best examples of illumination alike belong to each.—Editorial, *Electrical World*, April 28, 1923.

PAPERS

ARTISTIC COLOR LIGHTING IN MOTION PICTURE THEATRES*

BY S. L. ROTHAFEL**

SYNOPSIS: Mr. Rothafel's lighting effects at the Capitol Theatre are the result of study of the psychology of the audience. He has found that the ideal lighting arrangement is soothing to the mind and nerves and at the same time stimulating to the imagination. As illustration of his methods he describes the way he builds up lighting effects for various units, or parts, of a program.

The theatre of to-morrow, which will be very different from that of to-day, Mr. Rothafel predicts will depend almost entirely upon lighting for scenic effects, and this lighting will be controlled from one unit.

I am going to make a confession to you before I start, and that is, I don't know anything at all about lighting, and don't know anything about color. It just happens—that's all. I am going to try to tell you how it happens. If I were to attempt a discourse on the technical part of what we are doing, (I will use the personal pronoun) I am doing, I am afraid I would get into deep water.

The lighting, the equipment, and the various things that we have done, came to me just as naturally as the music has come, and before I launch into the lighting, and the way we use it, I think it only fair to talk a little bit about the why and the wherefore of it.

What we are doing in lighting is a practical application of the psychological. We have studied the audience, or the people with whom we are doing business and whom we desire to entertain. We are merely applying that which we have found through hard work and study.

We are a peculiar nation. We are living very, very rapidly. We are a nation with a perpetual sandwich in our hand. In other words, we are a cafeteria nation. We live rapidly. We never quite get set. We can't do as they do across the water. We

*A paper presented before the New York Section of the Illuminating Engineering Society, January 11, 1923.

**Managing Director of the Capitol Theatre, New York City.

The Illuminating Engineering Society is not responsible for the statements or opinions advanced by contributors.

don't have two and a half or three hours for our dinner, and hour luncheons, and we don't stop in the afternoon to have tea. We are going all the time.

And what we have done, as culminated in the Capitol Theatre, is this: we have taken you and them out of this tremendous turmoil, we take you in our arms as it were, by the service which places you in a seat as quickly as we possibly can. Then we take your tired and frazzled nerves, and we try to calm them and bring them down to their normal level. We entertain you, and there isn't any factor more conductive to getting that result than the lighting.

We have in the Capitol a rather unique system of lighting. As I told you before, I know nothing of lighting, as you know it, and as the artist knows it, but what we have done is this:

We have organized an equipment—probably the finest lighting equipment in the world. We have so organized, amplified, and intensified it that we are able to use this phrase,—"We paint with light." Whatever we do is not mysterious. It isn't intricate. There is nothing that we do that no one else can not do. We use the same gelatin that you all use; only four colors. There is red,—the different grades of it. There is blue. There is amber and there is green. We very rarely ever use what is known as white light, that is, a naked light.

Now comes the rehearsal. My art director has conferred with me early in the week and we have arranged a certain number of units. For instance, this week we are doing a little scarf dance. We haven't any scenery, except one little bit of fabric, a drop which is painted in relief, as it were, with a suggestion of an arch, the shadow and a pool, and the rest is simply neutral. That's all the scenery we have in the entire bill this week. We don't believe in scenery. We create and paint our scenes with light.

Now it is our business to create a mood and to interpret the scarf-dancer. I am going to show you how we do it.

We start from the very back wall, which is a unit,—the back unit,—which you call psychic, those of you who are acquainted with that. We start there, taking our pattern from nature. We have a unit which is called Number eleven. I call for Number eleven in blue. I use a certain number of blue. The boys through a course of training know exactly what I want, and I get it on my dimmer plate. It comes up slowly to where I want it, and then I say, "That's good." Then I work from Number twelve, which is a unit above it. I pour in a little green, and then from my side lamps and my strips, or whatever unit I want to use, I may pour in a little amplification, or a little reinforcement, in order to get that dim slightly hazy effect that I want, which is obtained by throwing light upon light, or by crossing rays and a certain amount of diffusion.

Now this is all done in the rear of this fabric ground—and from an ordinary nondescript piece of tapestry begins to loom up the shadow of the outline of the composition.

Then in front I begin to use diffusing light. On top of that, for instance, I may begin to sneak in a little bit of red and imagine that I am seeking the fringe of the setting sun behind me, just about to go down. That gives me red and yellow. I let my imagination play and begin to color the front of that drop to any proportion that I want. Then, to create a more realistic effect, I take a stereopticon and place it off and throw in a ray directly across just to give that little pool a little rippling,—enough to suggest that it is water—not obvious, not blatant, not cheap, but just enough so that it begins to simmer. Then in front of the whole thing we simply throw a couple of long pieces of chiffon that go into the wings or into the fly. There we have our background. Now that's for the small proscenium.

I have spoken to you about subtle effects, about appealing to the imagination. That is the true charm of all entertainment, appealing to your imagination—to make your imagination play. As you sit and watch motion pictures in your subconscious self you are supplying the color. You are supplying the words, the music, the sounds. You feel the emotions, according to your respective capacities.

The future of this kind of work will be in that one thing, to suggest so subtly—just to give you an outline and let you supply the balance. That is what we doing at the Capitol.

In the theatre of the future, the lighting is going to play a very important part. You are not going to see a single light. You will be transported from one mood to another simply by the change and combination of color and lighting. There will be no decorative work or plastered ornamentation in this theatre. The building is going to be shaped somewhat like an egg, just a plain, neutral ceiling side. There isn't going to be any balcony —just a big group of chairs, and all around there will be light. There will be a wainscoting and certain units throughout the house are going to project rays to portions of that surface and they are going to be reflected again to you so that there will be an ideal form of lighting.

The stage equipment will be an absolute revolution in stage lighting. The advancement of the incandescent lamp, the advencement of projecting with the aid of reflectors, the advancement of the automatic machinery to control this, will absolutely revolutionize lighting as we know it to-day.

The old-fashioned border lights and the old-fashioned footlights will be gone. They will be superseded by tubular units, in which there will be a revolving cylinder, or two or three revolving cylinders, controlled by automatic machinery, and these cylinders will be composed of a very fine, highly sensitized prism —very luminous. This prism will pass over the object and will then be diffused down so that we will be able to control with one unit of lamps almost any shade or any combination of colors that we could possibly wish for.

Gelatin is a thing of the past. We are going to create and make our own colors, and we need them. Toward that end we are now experimenting with certain fluids mixed with liquid, which we pour into a receptacle and which we can color instantly to any grade of red, blue or pink. We can bring it down to absolutely pitch black and then take it away again to the most delicate hues of blue, and then plain—almost water-like. We get a little milky effect, a frost, that is perfectly delightful.

Now what I think we need in the near future more than anything else is better equipment to get better control. That is the thing that is essential in lighting. The rest is creative—the making of something that you can pour out to those to whom you wish to pour it, by light, and that is what we are trying to do at the Capitol—impressionistic lighting. We are not trying to do anything different. We don't intend to blaze any new paths. We are simply trying to express ourselves in light, in color, to enhance, as it were, or to bring forth to you the mood that we are working on.

I spoke about the novelty effects. One novelty effect that we did last year that comes to me very readily was probably the greatest novelty that I did last year, an effect called, "The Song of India." It was an interpretation of Rimsky-Korsakoff's "Chanson Indoue." I interpreted it in the following way. I have been to India, and I pictured, as I saw it there, one of the little quays right outside of Calcutta, with a little knoll, and a little, drooping palm. I saw the rocky coast and I saw this young Indian prince standing there gazing out on to the waters. The waves were breaking over the rocks, and I saw these little sea-nymphs come out of the water and call mysteriously, calling him until he went, he knew not where, but he went. He went into the water and disappeared. Now here is how we accomplished that.

We used a plain white drop as our sky. We colored it a little blue, just so that it made a horizon. Then we gave it a little tinge of lighter blue. Then we crossed over a very fine ray of frost, just enough to create a sort of a fog, rather a mist. Then we put a roll in front and we covered this roll with a little bit of cheese-cloth, just so as to give me a surface to project on. Then I got a sea scene from a moving picture film, made a loop and put it in the machine in a way so that the same scene would keep coming around and around and you wouldn't see any flash of different scenes as you so often do in moving pictures. We didn't have time to take six or seven minutes of the surf breaking over, so we used this little loop in the machine and continued it around, projecting it right on top of that scene, matching off, if you please, the very edges of the scene itself. Then we cut the sky and matched that off, so that we projected it with a very high-powered lamp and a long focused lens in order to get all the intensity possible. Then we projected it on this little bit of cheese-cloth, this sort of hedge, and there we had surf breaking over the rocks all the time you saw it, and from behind this little hedge, girls simply danced and they danced as though they were in the water. They went up and down and the surf kept breaking all the time, and the effect was quite an illusion. Probably some of you saw it. And you see it wasn't so difficult at all. Simply an idea which we were able to put across, because we had worked hard and studied our projection and knew the possibilities, and we weren't afraid to try.

I want to tell you that if I could paint to you what I see in the future of this work you would think I was a raving maniac. I stop painting pictures of the future, because I fear they will put me away some day and I don't want to be put away yet. And I see it just as clearly as I saw the Capitol fifteen years ago.

You are going to see pictures projected into space. You are going to see lighting effects that will take the breath right out of your body. You are going to be awed. You are going to "ah" yourself into a state of non-resistance. You are going to see orchestras of one hundred and fifty or two hundred and fifty men, and a new instrumentation of an entirely different form. You are going to see productions where there isn't going to be any stage at all, just a big sweep. You are going to see the scene which is to be portrayed brought out into the theatre itself and you are going to sit right in the middle of it, and you are going to be part of that performance, just as we are trying to make you a part of our performance to-day. We don't tell this to a great many people, but I am telling you. You come into the Capitol Theatre and are just as much a part of our entertainment as the actor performing. That's the psychology of it. I make you play with us, and when I can do that—that's all I want.

I made a prophecy in Boston that the lighting of the theatre of to-morrow would be controlled from one unit, the same as the piano. I don't mean by that that they are going to strike notes—the thing that they did in Paris, where every note meant a light—that is a lot of bunk, but I do believe in mood, I believe profoundly in that. How many times have you witnessed the finale of a big selection—they bring up a light and the combination of eye and ear is much greater. If you could see that same climax without the lights coming up and giving it the general boost, if you could see it alone and flat, why you would be amazed at the difference.

DISCUSSION

The papers by Messrs. Rothafel and Hartmann were discussed together. See page 426.

LIGHTING EFFECTS ON THE STAGE*

BY LOUIS HARTMANN**

SYNOPSIS: The author speaks of the value of lighting in the theatre and of the lack of uniform outfits in different theatres necessitating a complete electrical equipment for every play. He describes his method of lighting interior and exterior scenes of dramatic productions, calling attention to the difficulties encountered, such as the impossibility of obtaining pure color effects and the danger of over-illumination.

He uses reflex glass as a medium for softening light, on both reflectors and lenses. In some cases foot lights are dispensed with and overhead lighting alone is used.

The most desirable stage lighting is simple in construction and not radical in effect. The possibilities in the lighting of the drama are unlimited when combined with intelligence and imagination.

We all know that there are numerous possibilities in the field of lighting; but I am going to say a few words about the part of it that I know best, and love best,— the lighting of the drama.

My real experience began with David Belasco about twenty-two years ago; although I had been employed in theatres years previous to this time, it was not until I came to work for David Belasco that I awoke to the realization of what light means to the stage, how valuable it is, and how much it assists the drama. Volumes could be written on this subject, but they would only serve as the expression of an idea; as text books they would contain very little of value.

When we speak of light in the commercial field, it is generally treated as a slide rule proposition; it resolves itself into a thing of mathematics—of course there are some exceptions to this, but I mean in general. There are several well known systems which have proved successful, their application being a product of well worked out formulae. In theatre lighting we have no formula; it is replaced by a truism that can be expressed in one sentence: "Love your work." This seems easy enough as there is a certain fascination about it; but with a great many new comers the strain of long rehearsals and lack of sleep dampens the enthusiasm they felt at first.

*A paper presented before the New York Section of the Illuminating Engineering Society, January 11, 1923.

**Lighting Engineer, The David Belasco Productions, New York City.

The Illuminating Engineering Society is not responsible for the statements or opinions advanced by contributors.

Stage lighting cannot be treated as a subject by itself; it is but a component part of a structure and of itself has no value. But if the play and its accessories are well conceived, proper lighting is a matter of great importance as it practically creates the atmosphere for the scenes. To obtain the best results, the lights require intelligent handling and good electrical equipment. It is very essential to have smoothly working dimmers, and a sufficient number of them, so each lamp or unit of lamps may be controlled separately. This is the only way by which the proper balance of light can be maintained; without this balance, the lighting looks just what it is, a number of separate units throwing blotches of light, creating bad shadows and sharp contrasts. If the proper equipment is not available, it is better to light a scene with the foots and borders than to try to use paraphernalia which cannot be fully controlled. The lights when badly handled only tend to draw the attention of the audience and detract from the effect the player is trying to create. Everything on the stage can be done in so many different ways; and it is far better to do it in a simple manner than to spoil it by an elaborate attempt which cannot be carried out.

I have found it necessary to have a full electrical equipment for every play, as there is no uniformity of outfit in theatres. Even when all the equipment is carried, the effect is not always the same when a play is moved to another theatre. The change is generally caused by the foot lights. They are very necessary and essential when properly installed, but in most instances they are badly planned; some are not hooded at all, while others have one row under the hood (generally the colors), and the white row is either half way under the hood or entirely outside of it.

For dramatic productions no light from the foots should strike the proscenium—much less illuminate the proscenium; and when properly installed, the audience should not be aware of their existence. The light from the foots should be so directed that it does not strike the ceiling of the setting. For this purpose I have found reflectors very efficient. There are several types which serve this purpose, and by painting parts of the reflector black I have been able to place the light about where I wanted it. In most instances I have found the small round bulb 25-watt lamp to give sufficient light for the foots when used in a scoopette reflector.

For a ceiling strip, I use the same reflector and lamp, fixed in a special strip. This may be tilted at any angle, takes up very little room, and does not throw the light on the ceiling. The ceiling gets all the illumination it requires from natural reflection. All scenes to be effective require light and shade. By using merely foots and ceiling strip, the scenes are flat, there being no contrast. Baby lenses have made it possible to get this contrast in interiors. This form of lighting was worked out in the Belasco Theatre and used for the first time in Mr. Warfield's play "The Music Master." One of the scenes where this was especially effective was where Von Barwig holds up the lighted lamp to see Helen's face. I'll never forget the many times Mr. Warfield lifted up that lamp during rehearsals while we struggled to bring up the babies at the right time.

For these baby lenses I found a concentrated filament lamp put on the market for use in small stereopticons for the home. It was rated at 50 candlepower. I built a housing for it, fitted it with a 5-inch x 9-inch focus lens, and to control it I used a small dimmer having fourteen steps. We called it a baby lens, and this name has stuck to it. There are thousands of them in use to-day. Of course their candlepower has been increased. We have them in hoods of different shapes using from a 1½-inch lens to a 5-inch lens and ranging from 50 watts to 1,000 watts. We set them behind the drapery and up and down the sides of the scene on special frames and brackets, each one controlled by a separate dimmer operated by men who are trained for weeks in handling them. On the front of the balcony we have reflectors in special housings, also on separate dimmers. In this way we get the light from all angles, and when it is balanced by proper dimmer regulation, almost any effect may be obtained with this outfit.

I have found reflex glass a very effective medium for softening the light. I use it on both reflectors and lenses. It is superior to frosted gelatin for this purpose. Mr. Belasco managed to get some very fine results with the old arc lenses. We used to soften the light with a slide having graduated thicknesses of mica, each piece cut V-shape with the edges tapered. Still the lamps were cumbersome and could not be placed to the best advantage.

When the tungsten lamp was put on the market Mr. Belasco told me to keep after the lamp manufacturers to get something that would prove more effective for our purpose. The American Lamp Company turned out two lamps for us which were very valuable at the time. One was a 6-volt concentrated filament lamp with two filaments. The other was a 60-volt lamp in a G. 40 bulb rated at 200 candlepower. I placed four of these in a bunch light; they looked like the sun—in those days. The light had an excellent quality and the diffusion was good, the filament being long; in fact, the light was superior to the modern 1,000-watt lamp so far as diffusion is concerned. In those days the Ward Leonard Company made hundreds of special rheostats for us. All this entailed a greater cost than if we had used stock materials, especially in the lamps, as we had to carry a large supply or wait three to four weeks for them, which was out of the question. In the theatre you need things quickly.

In some of Mr. Belasco's plays we discarded the foot lights entirely. "Peter Grimm" was lighted from over-head by strips and babies. Nine men were rehearsed for two weeks and then it took six weeks on the road before the lighting was finally perfected. "Marie Odile" also was lighted without foots. The scene of this play was laid in a convent. The effect of the light was as though it came through a large Gothic opening over the door. Foots would have ruined the atmosphere of this scene.

Whatever Mr. Belasco has tried in this line has been for the purpose of obtaining atmosphere for the play. Nothing is done in haphazard fashion; everything is the outcome of a preconceived plan. The original idea is worked on and experimented with until the best result has been obtained. The results are not always satisfactory to him, but if they do not come up to his expectations, it is the fault of a condition that cannot be remedied. You cannot always have perfection in all things on the stage, as there are numerous obstacles which present themselves. In planning a production the first details are worked out from the models, which are made to a scale of one-half inch to the foot. The artist tries to make his scenes as effective as possible, and where there are several of these scenes in a production, the working room on the stage is very cramped. This is one of the reasons why the lights can not always be placed where they will give the best effect. The only thing to do in a case of this kind is to take the available space and experiment until you have obtained the best results you can get under the prevailing conditions. I have seen Mr. Belasco cut out an entire scene and re-arrange the play when the conditions for lighting proved too unsatisfactory. Scenery is nothing but canvas and paint and appears as such when badly lighted. The reason we spend so much time with the scenery and the lights is that we realize their imperfections, regarding them as necessary evils. When they become so obtrusive as to detract from the play, we eliminate them.

The tendency to-day is to use too much light. The high wattage lamp has brought about this condition. I can remember when gas was the means of illumination in theatres, and the effect of gas light on the scenery as a whole was better and softer. It was impossible to over-illuminate a scene in those days. The contrasts were worked out in the painting. To-day we depend upon the light to accomplish this. In illuminating a stage in a large theatre it is not good to bring up the foot-lights so high that the expression on the player's face may be seen from the last row, a feat you could not accomplish by day-light. It is a mistaken idea to think that an actor's face must always be in a bright light. It all depends upon what he is doing. To work in semi-light is an aid to the actor at times.

Another difficulty experienced to-day lies in the color values. It is almost impossible to obtain pure color, either in the pigments used on the scenery or in the gelatines used on the lamps. The colors used on the scenery are dull and lifeless. When a color should be vibrant, it is instead flat and dirty looking. This is caused to some extent by the fire-proofing solution, which contains ammonia, an alkali that is ruinous to certain colors. Cobalt blue, an excellent medium for a sky, looks like a dirty white wash after it has been fire-proofed. In some instances the fabric is fire-proofed before the paint is applied but the result is about the same. The anilines used in the gelatine mediums also are poor, and it is impossible to obtain a blue without a purple or green tinge. This is often the reason why the color values are unsatisfactory. We have to take the best we can get and make the most of it. I have tried to procure glass to take the place of gelatines, but could not find sufficient uniformity, especially in the blue. The light blue was effective, but the dark blue varied in shade even when all the pieces were cut from the same sheet. Colored glass is blown in large sheets and varies in thickness, which causes this difference.

For the lighting of exteriors I have found reflectors of twelve inches and over of great value when fitted with proper spill-rings; but they are harder to handle in some cases than lenses, the light leaving the reflector with such a wide spread that it is difficult at times to kill reflections. In some cases I have made the spill-rings very long, at other times I have put long flippers on the side which could be closed in. The conditions vary as to space so that one must continually experiment. Generally they can be overcome in one way or another. To light a scene where the lights remain stationary during an entire act is simple, when compared with an act that has several changes of light.

One of the most difficult changes is to reduce gradually the illumination of an interior scene in a manner to simulate the setting of the sun and the approach of twilight. To keep these graduations perfect requires a great deal of time and patience. The lights inside and out must be on certain steps of the dimmers at a certain time; and to get the same tempo they must always be on a given point when a particular line of the dialogue is spoken. This change would be easy if the man operating it could see it, but the set is boxed in and he is working on the side where he does not see. By putting the switch-board under the stage with a hood in the apron, the man can look through—but this has proved unsatisfactory for a number of reasons. It was tried in the Century and abandoned. The Metropolitan Opera House still has this system, but the same men are operating it to-day who operated it when it was first installed.

Imagination is the theory of the theatre. People do not come to the theatre to see reality. On the stage you must exaggerate to be convincing. The public comes to the theatre to be fooled, but there is a vast difference between fooling an audience and insulting its intelligence. Use your imagination, but be logical. In other words you must make the audience see things as you want them to see them, and if your imagination is great enough you can convey any mood and make them feel it. Radical departures seldom succeed in the theatre, as the audience does not want to do any guessing. If the things you are doing appear false, you lose the attention of your public. A clever stage director guards against this. If the dialogue drags he brings in something to divert unconsciously their attention; a shift of light may change the entire mood, creating a mental change,—in short a play on the senses. You have watched a good magician. He has everything timed, nothing is left to chance. He draws your attention to something, but all the time he is talking. After you leave the theatre you don't remember what he said, but you remember his tricks. Still, in most instances, it was the talking that made the trick possible. In drama the process is the same—except that it is reversed. You remember the play but forget the trick. Big mechanical devices are generally easy to construct and the effects they create are valuable. It is the little fine touches that are really difficult, and in the end make the greatest impression, although you are not aware of it.

I could take up your time by describing to you in detail the different forms of apparatus now in use in the theatre, but you are familiar with lenses and parabolic mirrors. The housings for them depend upon the size of the lamp you wish to use, to give them proper ventilation without having leaks of light, and to make them as compact as possible. The shapes may vary to suit different conditions. But it is the intelligence you display in handling them that really counts—something that can come to you only through experience.

Good apparatus should be simple, the simpler the better. The easier it is to handle the less time is consumed in getting results,—a great factor during rehearsals when time means so much. In big productions, scene and light rehearsals take an entire week, sometimes longer, the men working night and day. Even where the expense is of no consideration, the players become nervous through waiting, so anything that will save time is of the utmost importance.

In placing the lighting equipment, it is well to have enough apparatus to be able to put it in every conceivable place where a light could possibly be used,—having it wired and connected to the switch-board. If it is not used it can be taken down, but if used the time saved is valuable.

As I have said, there are untold possibilities in the field of lighting; and not only in the theatre but everywhere that light is used. To realize this you have but to look at the number of stores where they make use of colored light in the windows, projected either by lenses or reflectors. This is an idea borrowed from the theatre. This is one of the reasons why I believe that what is done in the theatre can have a powerful influence, by stimulating the imagination, thus creating a sense of the idealistic and applying it to the commonplace.

DISCUSSION

WILLIAM HALL: I thank you, Ladies and Gentlemen, but I am absolutely unprepared. I am simply stepping forward to take some one else's place who has disappointed you.

I am very much pleased with two of the statements heard here this evening, one of them that even though we of the stage appear old-fashioned, they have to refer back to our old-fashioned methods to gain effects. Let me take you back to 1887 when I first started as a gas boy, at the old Niblo's Garden.

Can you imagine climbing out on the grooves (they shifted the scenes on and off stage in these) twenty feet from the stage level, with a pair of calcium light cylinders a little bigger than a fire extinguisher strapped to your back, a calcium light reflector fifteen inches in diameter fastened on your chest, holding a set of colors between your knees, and forty men changing the colors at the same time for the dance group pictures which occurred generally every sixteen bars of the music.

I have always claimed that the electric switchboard and dimmer equipment is only the evolution of the old gas table or switchboard, not a new creation. With the gas table of other days we accomplished the same results you obtain to-day—in a crude way, we will admit, but with wonderful results. I will state without fear of contradiction, that with the combination of the old gas lighting equipment and the calcium light apparatus, we have given productions that even in this era of advancement have never been equalled, that is so far as stage spectacular effect is concerned.

Understand this, we had producers in those days who were absolutely striving for accomplishment. Their hearts and souls were in the work, irrespective of the financial outcome. We had our Irving, our Davenport, our Henderson and Kiralfy, people who produced shows for the productive effect, not everything for the financial return. They went at things in a different way than they do to-day. Commercialism did not enter into it.

A few of my very pleasant experiences. As a gas boy in Niblo's Garden, to make dark changes the lights were turned down to a blue flame by a by-pass valve system; frequently the draft would blow out the flame during the change, to light a border light it would then be necessary to take a long pole, say thirty feet long, with an alcohol torch fastened on end and touch the gas tips. Perhaps the gas would have been turned full on, then you would see a flash of flame the entire length of the border, some fifty feet. (Laughter). But we did not have any fires. But my what a noise! Just like a flash light powder going off on a large scale. The audience get frightened? No, they got used to it I guess. (Laughter).

Recalling my traveling with Fanny Davenport through the country. That was in 1895. Electricity was just being introduced in the theatres then—we had with us what was called the Meyerhoffer lamp. The Universal Stage Lighting Company is the successor of that company and far advanced away beyond the Meyerhoffer period I am pleased to say. The Meyerhoffer device was of the hand feed type, to feed taking up the amount of carbon consumed it was necessary to place your hand generally up to a height above your head in an opening inside of the hood and turn the feed screw.

On this trip in the West, San Diego, California, was the place I think, looking for electricians as operators. I could not get anybody but a trolley car motorman (laughter); of course, he was an electrician. (Laughter). Understand the situation; Fanny Davenport was playing in one of Mr. Hartman's productions, the place had to be kept very quiet. As the operator (trolley motorman) would reach up to feed up on the carbon, his natural inclination was to stamp on the floor as if clanging his bell. (Laughter). Can you imagine the effect of this interruption on Fanny Davenport? I came very near losing my position over the incident.

Along those lines and over such obstacles we advanced with electricity. Once for a choke coil for five hundred and fifty volts I used up an old bed spring taking the spiral springs and fastening them together, quite an accomplishment for those days, and naturally I was proud of it, placed the coil in the cellar of the theatre with a switch on a post alongside of it. The arc from the slow break of the switch had me hesitating though about handling it; so rigged up a set of pull strings extending to a pocket on the stage floor and opened and closed the switch from that safe point. Got kind of careless though toward the end of the week and while proudly exhibiting the device to the power house electrician thoughtlessly placed my hand on those five hundred and fifty volts and was knocked clear across the cellar. (Laughter). We had practical lessons in those days.

An interesting statement this evening was that they had natural phenomena in the theatre to-day and how naturally these results and effects were accomplished. Of course in this era the motion picture has cut out a lot of our effects. Mr. Rothafel described how he obtained a water ripple effect to illustrate a song with the motion picture machine. The first way we made water ripple effects was with a large metal barrel with holes or slots cut horizontally; into the inside of which we would place a couple of calcium light burners and revolve the barrel around the lights. By this same method of operation we had cloud effects painted on glass revolving in front of lens lamps.

Mr. Rothafel has told you of some very natural effects they obtain to-day. Well, I guess we had some natural ones in the old days too. With the Savage Grand Opera Company for Carmen the rocky pass scene in the third act the back drop had a waterfall painted on it for the full length ending in a sort of ravine. Behind the drop we had a projective water fall effect. Set in front was a runway for Don Jose's entrance on horseback. Well, as the horse made his entrance he would invariably stop to take a drink out of the waterfall. (Laughter).

In traveling through the country with a show I carried apparatus for a combination of gas and electricity. The devices would have a gas burner for the calcium or if electricity was available, remove the gas burner and place the arc carbon burner in the same hood. So you see, we were quite resourceful in those days. Admitting an advance in the lighting of the present day theatre, I would say that if you have any idea of going into the field of stage illumination, I would recommend you gather all the data possible on how we accomplished results in the old days. For spectacular lighting, I have not up to the present day and I have been associated with some of the largest productions ever staged in my capacity as department head with the New York Calcium Light Co., seen a production to equal any of the productions as given at the New York Theatre like "The Man in the Moon," "The King's Carnival" or "Million Dollars." The Hippodrome had some wonderful spectacles, but even they do not compare with spectacles of the past.

Mr. Rothafel has described the wonderful new theatre he has in mind, with electric lamps set around like footlights surrounded with glass prisms eliminating the coloring of lamps or sheet gelatines for color effect and how he would have a mechanical system to turn these colored prisms of glass in front of the lamps for desired color changes. As Mr. Rothafel extended you an invitation to visit him, I will also invite you to come any time to my office and I will show these same colored glasses as used in the Eighties set around arc and gas burners on an individual revolving base with a string attached so we could turn any color we desired in front of the burner; once again illustrating my contention that you have to go some to show anything new in stage lighting to anyone of the gentlemen who are here to-night and associated with the stage as long as I have been. I still contend you are adapting a different method of doing something simpler than the way we did it. We of the stage do not wish to retard you in any way, we welcome you to show us something new, come over to our Company, talk over what you have in mind, we will show you how we did it and help you all we can. That is all I have to say. Thank you.

W. D'A. Ryan: I have been very greatly interested in the two excellent addresses delivered to-night. While I am not in a position from experience to discuss theatre lighting, I have had some experience in the use of color on a large scale.

Color is just as important in giving pleasure as music, when properly handled. The first experience that I had in color was possibly twenty-five years ago at Bass Point, Nahant. I tried out there, on artificial clouds created by specially-prepared shells for making smoke, the effect of color on the public, and was greatly pleased to see what wonderful drawing power that it had.

At the time of the Hudson-Fulton Centennial. you probably recall some of the color effects used on the river, and the illumination of Niagara Falls in 1907, but the climax in color on a large scale was at San Francisco in 1915 at the Panama-Pacific International Exposition.

We had one color effect alone—spectacular—aside from the general color used, covering possibly nine million square feet of surface, a battery of searchlights of a combined beam candle-power of a little over two billion, six hundred million,—visible eighty or ninety miles away, as far as Sacramento. It is admitted by the Exposition officials that that color effect, the coloring of the Exposition, the light at night, and particularly the effect of this scintillator, did more than any other one thing to make that Exposition a financial success in the face of the war.

There were forty-eight searchlights, manned in such a way that they could control for various combinations of color, and over three hundred specially prepared shells, and the effects were led up to a final climax in which five hundred shells were thrown into the air in one second, one half being four and a half inch. We had a locomotive, largest Pacific type, to generate steam for certain effects, and the final was a climax of all the effects, and it never failed to impress the people.

Following the Exposition, the first case that I know of the use of color in a theatre, for the house proper, that is, distinguishing it from the effects used on the stage, was in Growman's Theatre in Los Angeles. They followed some of the effects similar to those used in the Exposition, where we brought out in very fine silhouette many of the decorative features of the building.

I have noted a number of instances since, but may say that in general they are too harsh. They go too much to strong color rather than tints and shades, and too much contrast and not particularly happy combinations. That is just a general statement. There are some theatres very beautifully treated, but I think the tendency in the window lighting also is to get too harsh effects in the combinations, and it is like anything developing new —it will start in crude and finally become refined.

Now, in conclusion, I would like to say that there is not anything more important to the public to-day than the art of colored lighting, and I believe that every effort should be put forth to study that art and develop it to a high degree of perfection within the coming years.

M. J. LEVY: I was very much interested in what Mr. Rothafel said. I have known him for probably ten years. I remember when he first came here, when he went to the Strand Theatre and there tried to impress the owners with the idea of having colored lighting in the auditorium. As the theatre was pretty well finished by that time, we could not help him out. I had designed the electric lighting installation of this house as I have of three hundred other theatres throughout the country.

When Mr. Rothafel built the Rialto Theatre in New York he came to me and said, "Now I want something here that I can operate, the same as an artist plays on the piano," and he had drummed the architect full of his notions—full of three-color auditorium lighting. The architect did not quite grasp him and sent him to me, and we carried out the lighting of that theatre as best we could, with what the builder and the architect would give us. We had a great deal of help from them, but we had a great deal of trouble with the different artisans, plasterers and particularly with the painter. We could not get the painter to put on the color we wanted. In fact, I do not believe I have been able to get a painter to put on color so that we could really show the color values of our lamps. As a rule, they put on a lot of browns and a lot of gold, particularly in domes and around cornices, and we have a great deal of trouble, but Mr. Rothafel at that time worked very hard to get that theatre in such shape that he could bring out his color lighting scheme. That was the first theatre of that kind. That is a matter of about eight or nine years ago.

Then Mr. Rothafel brought out the Rivoli Theatre, where that style of lighting was again used, and now in the Capitol Theatre.

A great many theatres have since been equipped with that kind of lighting, but I have found that in most places it is not properly used. A man like Mr. Rothafel with his artistic sense would use it properly. He has an idea of values. He does not smear it on, but I have found in a great many other theatres that the lighting effects are left to a stage electrician. Now there are a great many highly intelligent stage electricians, but the average man who gets into a house of that kind has not the necessary intelligence to follow the music, or follow the stage productions with the lights. It takes a man of a higher calibre, as is proven by the fact that a man like Mr. Rothafel directs that feature in his own theatre.

There is a statement of Mr. Hartman that is very much like the statements that I have made time and time again, on the subject of stage lighting. One great trouble with the stage has been that when the nitrogen lamp was introduced it was in the thousand watt unit. It is a pity that the one hundred watt, or even the fifty watt, had not been perfected long before the thousand watt, because the thousand watt unit, with all its glare, was used by the stage electricians because of its peculiar white effect, and as a consequence, stages are now crowded with thousand watt units, which give you simply glare. Stage illusion is entirely out of the question.

I know of one instance where, after we had given a theatre a very good equipment, the stage electrician came along and he said, "Do not want anything like that. I am going to put thousand watt units all over the place." He did, and on the opening night his back drop was unquestionably a back drop. It was not a scene. You could see every paint mark on that back drop. Where you expected to find a castle on a hill, with a road running up to the castle, you saw absolutely nothing but a flat piece of painting, and I think that the high intensive units should be used with much more skill and care than at present. It would be better if they were not used at all. Smaller units that diffuse the light would make the stage illusion complete. A great many stage managers and stage electricians forget that what you see on the stage is, as Mr. Hartman says, intended to fool the public. It is intended to create an atmosphere in which the public sees something that is not actual. And so the lighting unit should be kept down. There should not be quite so much of the glare. There should not be quite so much of the intent to create daylight on the stage, because it cannot be done, without overstepping the mark. You cannot get artificial daylight, as daylight appears to you in the day time. It is a rather difficult proposition, and the best you can do is to create an illusion that will make you think it is daylight.

DAYLIGHT ILLUMINATION ON HORIZONTAL, VERTICAL AND SLOPING SURFACES*

REPORT OF THE COMMITTEE ON SKY BRIGHTNESS,

H. H. KIMBALL, *Chairman*

SYNOPSIS: The report summarizes sky-brightness and daylight-illumination measurements made during the year ending April 6, 1922. For ten months the measurements were made in a suburb of Washington that is comparatively free from city smoke. During the other two months, one in summer and one in winter, the measurements were made in the smoky atmosphere of the city of Chicago.

The measurements were made as nearly as possible with the sun at altitudes above the horizon of 0°, 20°, 40°, 60°, and 70°. From the sky-brightness measurements the resulting illumination on vertical surfaces differently oriented with respect to the sun, and on surfaces sloping at different angles and in different directions, has been computed. These computed values have been utilized, in connection with daylight-illumination measurements, to construct charts showing for latitude 42° north, illumination intensities for each hour of each day of the year as follows:

(1) On a vertical and on a horizontal surface, from a cloudy sky.

(2) On a horizontal surface and on vertical surfaces facing the eight principal points of the compass, from a clear sky.

(3) On a horizontal surface and on vertical surfaces facing the eight principal points of the compass, from the sun and clear sky combined.

The illumination on sloping surfaces from skylight and from solar and skylight combined has been summarized in tables.

The application of these data to the lighting of working space in a building through saw-tooth roof construction is show. It is pointed out that with a clear sky the larger proportion of the illumination should result from the reflection of light from the outside roof surface through the window opening, rather than by the direct transmission of skylight through the window.

With a cloudy sky the illumination on a horizontal surface is considerably more than twice that on a vertical surface, due to the fact that the region of maximum brightness is in or near the zenith.

With high sun, as at midday in summer, the illumination from a cloudy sky averages higher than the illumination from a clear sky, except on a vertical surface facing the sun. This is not the case with low sun.

The maximum illumination from a clear sky on vertical surfaces is a little in excess of 1,400 foot-candles, and occurs on surfaces facing the sun from early June to early September, between the hours of 8.30 A. M., and 3:30 P. M.

The minimum illumination from skylight is on a vertical surface facing away from the sun. At Chicago in the smoky Loop District the illumination from a cloudless sky on such a surface averages about two-thirds the illumination at Washington on a similar surface from a clear sky.

*A report presented at the annual convention of the Illuminating Engineering Society, Swampscott, Mass., September 28, 1922; later revised and extended.

The Illuminating Engineering Society is not responsible for the statements or opinions advanced by contributors.

The total (solar + sky) illumination generally increases on surfaces sloping toward the south until the angle of slope reaches 20°, except with low sun during the summer months. The maximum is about 11,000 foot-candles at noon in midsummer.

At Washington the illumination from a clear sky on both horizontal and vertical surfaces varies between 150 and 60 per cent of the average values; from a cloudy sky, between 200 and 30 per cent; from a sky partly covered with white clouds, on a horizontal surface three to four times, and on a vertical surface two to three times that from a clear sky; with rain falling, about half that from a cloudy sky.

SKY-BRIGHTNESS AND DAYLIGHT-ILLUMINATION MEASUREMENTS

In a report of this committee[1] presented at the Rochester meeting in September, 1921, the program of sky-brightness measurements was outlined, and some preliminary results were given. The program of a full year of sky-brightness measurements was completed on April 6, 1922. During four weeks ending with August 13, 1921, and a second four weeks ending with February 2, 1922, the measurements were made in the city of Chicago. During the remainder of the year they were made in a suburb of the city of Washington that is practically free from city smoke.

As was explained in the previous report, at Washington the photometer was mounted on a stand inside a small shelter that was painted white on the outside and flat black on the inside. The upper edge of the sides of the house is on a level with the center of the elbow tube of the photometer when the latter is horizontal. This exposure permits measurements of the illumination from skylight on both horizontal and vertical surfaces. With a clear sky, however, illumination measurements on vertical surfaces have been confined to surfaces facing in azimuth 0° and 45° from the sun, as at greater azimuths the blackened inside walls of the shelter reflect too much sunlight to the photometer.

It may be well to recall to mind some details of the sky-brightness measurements. Figure 1 is a stereographic projection of the half of the sky on either side of the sun's vertical. The sun's position is indicated by the letter X. The horizontal straight line represents the horizon, and above it are lines of equal altitude 10° apart. Extending from the zenith to the horizon are azimuth circles also 10° apart.

[1] TRANSACTIONS, Illum. Eng. Soc., Oct., 1921. Vol. XVI. pp. 255-275. *Mo. Weather Rev.*, Sept., 1921, 49, pp. 481-488.

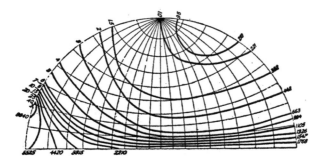

Fig. 1.—Sky brightness in millilamberts. Sun's position indicated by X. Washington, D. C., cloudless sky, ground covered with snow.

A complete series of sky-brightness measurements consists of three photometric readings on each of the points at 2°, 15°, 30°, 45°, 60°, 75°, and 90° above the horizon, and on azimuth circles 0°, 45°, 90°, 135°, and 180° from the sun, covering half the sky only. Unless the sky is cloudy the point that falls nearest the sun on azimuth circle 0° is usually too bright to measure with the screens at our disposal. There are therefore 102 photometric readings in each series. In addition, with a clear sky the intensity of the total illumination from the sun and sky is measured on a horizontal surface, and on a surface normal to the direct solar rays, and the illumination from skylight alone is measured on these two surfaces and also on vertical surfaces facing 0° and 45° in azimuth from the sun. If the sky is cloudy the illumination is measured on a horizontal surface, and on vertical surfaces facing 0°, 45°, 90°, 135°, and 180° in azimuth from the sun. There are therefore eighteen photometric readings (six sets of three readings each) in each series of illumination measurements. It is usually a matter of chance whether the readings are on azimuth circles to the right or to the left of the sun. Unless there is inequality in the cloud or haze distribution, or in the character of the earth's surface on the two sides of the sun's vertical, the sky brightness on the two sides should be symmetrical.

TABLE I.—NUMBER OF SERIES OF SKY-BRIGHTNESS MEASUREMENTS
WASHINGTON, D. C.

Solar altitude	0°	20°	40°	60°	70°	Total
Clear sky.....................	9	62	122	23	11	227
Thin clouds.......................	2	34	36	16	......	88
Partly cloudy sky..............	1	21	57	24	17	120
Cloudy sky......	6	30	26	21	4	87
Rain or snow.....................	......	3	6	8	......	17
Total.......................	18	150	247	92	32	539

CHICAGO, ILL., FEDERAL BUILDING

Solar altitude	0°	20°	26°	40°	60°	Total
Clear sky........................	6	13	3	8	3	33
Thin clouds.........	......	8	5	2	6	21
Partly cloudy sky................	5	9	......	7	9	30
Cloudy sky....................	4	6	3	1	1	15
Total.................	15	36	11	18	19	99

UNIVERSITY OF CHICAGO

Solar altitude	0°	20°	29°	40°	60°	Total
Clear sky.......................	2	11	3	9	10	35
Thin clouds......	.....	9	5	2	2	18
Partly cloudy sky.................	6	4	......	5	6	21
Cloudy sky....................	2	7	2	1	1	13
Total.......................	10	31	10	17	19	87

The attempt is made to obtain complete series of sky-brightness
and illumination measurements when the sun is 0°, 20°, 40°, 60°,
and 70° above the horizon. On account of the length of the day
the readings at 0° and 20° solar altitude are generally omitted in
midsummer, and in winter the sun does not reach an altitude
much in excess of 40°. In fact, at the Federal Building, Chicago,
in January, the average altitude of the sun at the time of making
the noon readings was 26°, and at the University of Chicago, at
the end of January and early in February it was 29°. When rain
was falling, only sky-brightness measurements up to an altitude
of 60° could be made by pointing the photometer out of a window.

Table I gives the number of series of sky-brightness measurements obtained at Washington and Chicago with the sun at the
altitudes indicated and with the different types of sky. In all

Solar altitude	Azimuth from sun	Point in sky where brightness was measured							Zenith brightness
		Altitude							mL.
		2°	15°	30°	45°	60°	75°	90°	
CLEAR SKY, WINTER									
0°	0	12.66	9.61	4.33	2.40	1.44	1.15	1.00	27.1
	45	4.40	4.76	2.83	1.98	1.36	1.15		
	90	2.43	2.76	2.08	1.47	1.20	1.10		
	135	2.99	3.43	2.41	1.58	1.17	1.03		
	180	3.83	4.20	2.74	1.77	1.25	1.08		
20°	0	23.52	......	9.78	4.27	2.26	1.42	1.00	281
	45	8.91	5.54	3.76	2.57	1.75	1.27		
	90	3.95	2.60	1.69	1.30	1.10	1.06		
	135	3.92	2.34	1.42	1.01	0.87	0.88		
	180	4.38	2.55	1.42	0.99	0.81	0.79		
40°	0	8.29	6.81	9.44	......	2.72	1.53	1.00	544
	45	4.56	3.45	2.77	2.21	1.76	1.36		
	90	2.62	1.66	1.12	1.00	0.89	0.95		
	135	2.51	1.33	0.82	0.68	0.66	0.75		
	180	2.76	1.44	0.81	0.60	0.58	0.69		
CLEAR SKY, SUMMER									
20°	0	21.10	......	10.85	4.06	2.34	1.40	1.00	400
	45	7.72	5.90	3.91	2.74	1.94	1.31		
	90	3.42	2.81	1.78	1.35	1.07	1.14		
	135	2.48	1.86	1.23	0.83	0.72	0.82		
	180	2.85	2.13	1.19	0.85	0.74	0.76		
40°	0	7.35	6.19	8.43	......	2.75	1.58	1.00	803
	45	4.15	3.13	2.69	2.45	1.88	1.42		
	90	2.13	1.41	1.13	0.97	0.99	1.04		
	135	1.74	1.13	0.74	0.60	0.62	0.74		
	180	1.83	1.11	0.68	0.54	0.53	0.68		
60°	0	2.08	1.74	1.84	2.51	......	1.65	1.00	1,650
	45	1.74	1.53	1.30	1.53	1.67	1.41		
	90	1.16	0.88	0.72	0.74	0.88	0.99		
	135	0.95	0.61	0.47	0.50	0.58	0.75		
	180	1.00	0.64	0.47	0.50	0.54	0.72		
70°	0	1.45	1.29	1.13	1.60	3.30	......	1.00	2,300
	45	1.25	0.90	0.89	1.07	1.33	1.53		
	90	0.77	0.65	0.56	0.54	0.62	0.89		
	135	0.62	0.42	0.37	0.38	0.44	0.66		
	180	0.58	0.39	0.31	0.31	0.41	0.61		
CLOUDY SKY, WINTER									
20°	0	0.39	0.59	0.80	0.93	1.03	1.04	1.00	989
	45	0.40	0.55	0.67	0.84	0.97	0.97		
	90	0.31	0.56	0.68	0.81	0.90	0.92	Max.=2,211	
	135	0.31	0.49	0.66	0.80	0.86	0.94	Min.=245	
	180	0.32	0.54	0.67	0.83	0.94	0.98		

there are about 55,000 photometric readings of sky brightness and 9,000 photometric readings of illumination intensity at Washington and 19,000 and 2,200, respectively, at Chicago.

In Tables II and III are summarized the sky-brightness measurements made at Washington and Chicago, respectively, during

TABLE III.—Average of Sky-Brightness Measurements Expressed in Terms of Zenith Brightness. Chicago, Ill.

| Place | Solar alti- tude | Azi- muth from sun | \multicolumn{7}{c}{Point in sky where brightness was measured. Altitude} | | | | | | | Zenith bright- ness |
			2°	15°	30°	45°	60°	75°	90°	mL.
\multicolumn{11}{c}{CLEAR SKY, WINTER.}
University	0°	0	9.80	8.04	3.67	2.03	1.38	1.08	1.00	19.4
		45	3.04	4.06	2.14	1.50	1.13	0.87		
		90	1.03	2.36	1.79	1.46	1.29	1.07		
		135	1.31	2.46	1.84	1.18	0.78			
		180	0.80	2.84	3.34	1.64	1.22	0.96		
Federal Building	0°	0	1.80	4.41	4.00	1.21	1.48	1.13	1.00	19.7
		45	0.85	2.11	2.04	1.54	1.36	1.10		
		90	0.80	1.29	1.27	1.24	1.03	0.99		
		135	0.88	1.88	1.00	1.43	1.05	1.08		
		180	1.11	2.12	1.85	1.66	1.25	1.09		
University	20°	0	16.38	. . .	9.18	4.25	2.21	1.22	1.00	345
		45	6.36	5.00	3.59	2.58	1.93	1.33		
		90	2.51	2.22	1.59	1.30	1.40	1.05		
		135	2.06	1.50	1.25	0.98	0.95	0.85		
		180	2.03	1.75	1.14	0.89	0.72	0.77		
Federal Building	20°	0	13.56	. . .	9.86	4.37	2.34	1.55	1.00	340
		45	5.30	4.87	3.37	2.59	1.96	1.16		
		90	1.98	1.93	1.70	1.35	1.31	1.00		
		135	1.24	1.27	1.01	0.75	0.66	0 72		
		180	1.32	1.45	1.08	0.79	0.72	0.82		
University	29°	0	13.65	18.80	. . .	3.72	2.61	1.37	1.00	433
		45	5.32	4.11	3.50	2.59	1.92	1.32		
		90	2.59	1.96	1.49	1.07	1.02	1.03		
		135	1.89	1.65	1.01	0.82	0.71	0.78		
		180	2.13	1.78	1.05	0.82	0.77	0.71		
Federal Building	26°	0	7.81	15.73	12.14	4.96	2.44	1.46	1.00	511
		45	3.84	4.12	4.02	2.85	2.04	0.94		
		90	0.95	1.20	1.39	0.93	0.79	1.00		
		135	0.73	0.72	0.73	0.73	0.73	0.89		
		180	0.65	0.80	0.70	0.58	0.63	0.75		
\multicolumn{11}{c}{CLOUDY SKY, WINTER.}
University	20°	0	0.29	0.52	0.86	1.06	1.07	1.08	1.00	614
		45	0.30	0.47	0.73	0.89	0.96	0.99		
		90	0.29	0.41	0.63	0.86	0.91	0.97	Max.	1,074
		135	0.30	0.42	0.58	0.71	0.86	0.88		
		180	0.25	0.37	0.56	0.73	0.94	0.91	Min.	84
Federal Building	20°	0	0.38	0.56	0.83	1.02	1.23	1.30	1.00	500
		45	0.33	0.61	0.56	0.78	0.93	1.10		
		90	0.42	0.64	0.74	0.92	1.06	1.10	Max.-	1,034
		135	0.34	0.43	0.60	0.98	1.13	1.08		
		180	0.33	0.56	0.80	0.84	0.70	0.93	Min.	132

TABLE III. (CONTINUED)—AVERAGE OF SKY-BRIGHTNESS MEASURE-
MENTS EXPRESSED IN TERMS OF ZENITH BRIGHTNESS:
CHICAGO, ILL.

Place	Solar altitude	Azimuth from sun	2°	15°	30°	45°	60°	75°	90°	Zenith brightness mL.
CLEAR SKY, SUMMER										
University......	8°	0	12.29	...	5.64	2.37	1.18	1.19	1.00	231
		45	5.46	5.09	2.65	1.83	1.17	0.90		
		90	3.43	2.37	1.35	1.07	0.95	0.76		
		135	2.62	2.18	1.36	0.96	0.72	0.84		
		180	2.73	2.69	1.64	0.91	0.79	0.86		
Do	20°	0	14.35	...	11.44	5.07	2.71	1.62	1.00	528
		45	5.91	5.04	3.88	2.62	1.86	1.32		
		90	2.77	2.26	1.61	1.25	1.09	1.04		
		135	1.69	1.60	1.04	0.73	0.68	0.72		
		180	1.58	1.78	1.04	0.71	0.64	0.70		
Federal Building ..	20°	0	15.08	13.73	13.58	6.19	2.79	2.20	1.00	419
		45	4.76	4.81	4.54	3.27	1.97	1.37		
		90	1.65	1.56	1.32	1.06	0.91	0.90		
		135	1.50	1.51	0.91	0.73	0.71	0.80		
		180	1.20	1.13	0.76	0.54	0.55	0.59		
University	40°	0	6.29	5.67	7.28	...	2.66	1.63	1.00	810
		45	3.72	3.07	2.70	2.57	2.16	1.49		
		90	1.68	1.41	1.15	1.09	0.96	0.98		
		135	1.73	1.05	0.77	0.73	0.72	0.83		
		180	1.26	0.97	0.61	0.48	0.50	0.62		
Federal Building ..	40°	0	4.50	5.56	6.35	...	3.23	1.76	1.00	900
		45	1.51	3.09	3.38	2.69	2.17	1.37		
		90	1.16	1.19	0.99	...	0.95	...		
		135	0.67	0.57	0.49	0.43	0.49	0.70		
		180	0.97	0.78	0.51	0.41	0.47	0.68		
University......	60°	0	1.75	1.88	2.06	2.44	...	1.67	1.00	1920
		45	1.32	1.23	1.25	1.38	1.52	1.30		
		90	0.84	0.72	0.64	0.82	0.88	0.88		
		135	0.80	0.69	0.52	0.52	0.63	0.82		
		180	0.65	0.55	0.45	0.50	0.50	0.66		
Federal Building ..	60°	0	1.63	1.88	2.19	3.24	...	1.83	1.00	1360
		45	1.36	1.33	1.11	1.54	1.36	1.40		
		90	0.94	0.67	0.61	0.58	0.72	1.09		
		135	0.96	0.69	0.56	...	0.63	...		
		180	1.00	0.63	0.39	0.38	0.43	0.60		

summer and winter months with a cloudless sky, and during
winter months with the sun 20° above the horizon and the sky
covered with clouds. It will be noted from Table I that at Chicago most of the sky-brightness measurements with a cloudy sky
were obtained when the sun was at an altitude of 20°.

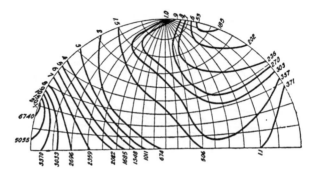

Fig. 2.—Sky brightness in millilamberts. Sun's position indicated by X. Federal Building, Chicago, Ill., cloudless sky with dense smoke.

Referring again to Figure 1, the irregularly curved lines are lines of equal brightness that have been drawn to represent the brightness of the sky at Washington on the morning of February 17, 1922, with the sun at an altitude of 20°, and the ground covered with newly fallen crusted snow. The figures on the left above the sun represent the brightness of the sky with reference to the zenith brightness; the figures on the right and at the bottom of the figure, the brightness of the sky in millilamberts. The sky 90° from the sun and in his vertical was a deep blue and unusually dark. Near the horizon it was unusually bright on account of the reflection of light to the atmosphere from the snow surface, and especially beyond 90° in azimuth from the sun.

Figure 2 shows the brightness of the sky as measured from the top of the dome on the Federal Building, Chicago, Ill., on the morning of January 16, 1922, with no clouds in the sky, but heavy smoke in the lower atmosphere. The sun was at altitude 20°, and the ground was covered with snow, as was the case at Washington on February 17, but the snow was not clean. Compared with Figure 1, Figure 2 gives a brighter zenith, a point of minimum that is less bright, and a horizon beyond azimuth 90° from the sun only about one-fourth as bright.

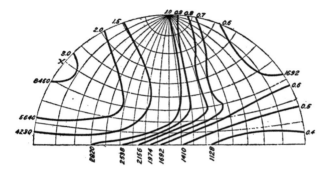

Fig. 3.—Sky brightness in millilamberts. Sun's position indicated by X. Washington, D. C., sky covered with dense haze.

Figure 3 represents the sky brightness at Washington on the morning of July 5, 1921, with the sky covered with dense haze, but without clouds, and the sun 40° above the horizon. The sky is much brighter than a clear blue sky, except near the horizon opposite the sun.

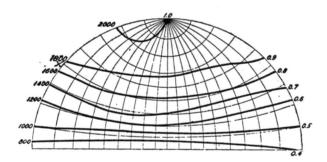

Fig. 4.—Sky brightness in millilamberts. Sun's position indicated by X. Washington, D. C., sky covered with dense clouds.

Figure 4 represents the mean of all the sky-brightness measurements at Washington with the sun 40° above the horizon, and the clouds so dense that neither blue sky nor the sun could be seen. The brightest point is near the zenith, and there is little variation in brightness with azimuth. The zenith and in general the sky opposite the sun is brighter when covered with clouds than when clear, but near the horizon and in the vicinity of the sun the clear sky is much the brighter. Thin clouds, and clouds that partly cover the sky, increase its brightness much the same as does haze.

ILLUMINATION FROM SKYLIGHT ON HORIZONTAL AND VERTICAL SURFACES

During the past year the energies of the committee, aside from the observational work, have been directed principally to computing from the clear-sky-brightness measurements, as summarized in Tables II and III, the resulting illumination on vertical surfaces facing in azimuth 70°, 90°, 135°, and 180° from the sun. The process is a simple one. As explained in the 1921 report,[1]

TABLE IV.—ILLUMINATION FROM SKYLIGHT, WASHINGTON, D. C.

Solar altitude	On horizontal surface	On vertical surface						Mean	Zenith brightness
		Azimuth between normal to surface and sun's azimuth							
		0°	45°	70°	90°	135°	180°		
		Foot-candles							mL.
		CLOUDY SKY							
0°. . . .	15.2	5.6	5.8	. . . .	6.4	6.7	7.1	6.3	15.8
20.2°. . . .	726	295	280	. . . .	273	273	272	279	989
41.0°. . . .	1,505	614	608	. . . .	615	622	606	613	2,000
61.4°. . . .	2,150	881	941	. . . .	977	932	939	932	3,600
71.4°. . . .	2,950	1,142	1,103	. . . .	1,118	1,122	1,203	1,138	4,840
		CLEAR SKY, SUMMER							
20°	840	1,252	1,028	803	526	316	293	. . . .	400
40°	1,340	1,454	1,325	932	686	417	358	. . . .	803
60°	1,600	1,420	1,255	923	751	559	486	. . . .	1,650
70°	1,600	1,291	1,074	903	754	542	475	. . . .	2,300
		CLEAR SKY, WINTER							
0°	67.8	64.6	63.7	. .	39.6	39.2	31.5	. . . .	27.1
20°	683	1,042	873	562	393	295	257	. . .	281
40°	977	1,121	936	690	505	325	295	. . .	544

[1] TRANS., Illum. Eng. Soc., Vol. XVI, p. 267; Mo. Weather Rev., Sept., 1921, 49, p. 485

the sky is divided into zones of equal angular width about a point on the horizon 90° in azimuth from the illuminated surface, and the horizontal component of the illumination from each zone is determined. The sum of the illumination from all the zones gives the total skylight illumination on the vertical surface.

Table IV summarizes the results of these computations from Washington measurements, and also the illumination measurements, for both clear and cloudy skies. The cloudy-sky measurements have been confined to skies with so dense a cloud layer that the position of the sun could not be seen. No seasonal variation in the illumination intensity is apparent. With clear skies the computations have been made for both midsummer (June to August) and midwinter (December to February) conditions.

These data have been plotted on Figure 5 (summer conditions) and Figure 6 (winter conditions) with the solar altitude as abscissas and illumination intensities as ordinates. By interpolating between the curves it is possible to determine the illumination intensity for both summer and winter conditions on a vertical surface facing at any desired azimuth from the sun, and with the sun at any desired altitude. For spring and fall months a straight-line interpolation has been made between winter and summer values.

Measurements with the sun on the horizon were made during the winter months only, and these measurements have been used for summer as well. With the sun at altitudes 20° and 40° it will be noted that the zenith brightness in winter is approximately 70 per cent of the corresponding brightness in summer. The percentage of winter to summer illumination is somewhat greater than this, since the brightness of the sky near the horizon in terms of the zenith brightness is greater in winter than in summer.

In the *Monthly Weather Review* for November, 1919, **47**, pp. 770-771, are given the altitude and azimuth of the sun for the 21st day of each month and for even-hour angles of the sun from the meridian, for latitudes 30°, 36°, 42°, and 48° north. Using the azimuths and altitudes for latitude 42° N., in connection with

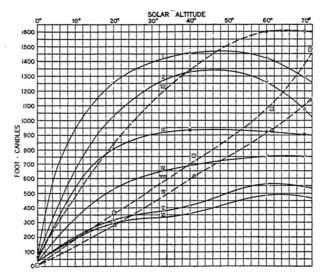

Fig. 5.—Curves of summer skylight illumination intensity on different surfaces.
Curve I.   Clear sky.   Vertical surface facing 0° in azimuth from sun.
Curve II.   Clear sky.   Vertical surface facing 45° in azimuth from sun.
Curve III. Clear sky.   Vertical surface facing 70° in azimuth from sun.
Curve IV.   Clear sky.   Vertical surface facing 90° in azimuth from sun.
Curve V.   Clear sky.   Vertical surface facing 135° in azimuth from sun.
Curve VI.   Clear sky.   Vertical surface facing 180° in azimuth from sun.
Curve VII. Clear sky.   Horizontal surface.
Curve VIII. Cloudy sky. Horizontal surface. (Note: Double the intensity
scale).
Curve IX.   Cloudy sky. Vertical surface.

the illumination intensity curves of Figures 5 and 6, Figures
7 to 18 have been drawn. Latitude 42° N. was selected because many important industrial districts are near this latitude,
and also because measurements made at Washington, latitude
38° 56′ N., and Chicago, Ill., latitude 41° 53′ N., represent
fairly well the sky brightness at this latitude east of the Mississippi River. Farther west the clear sky is generally a deeper
blue and not so bright. Probably clear skies average brighter
in low than in high latitudes, especially in winter. This conclusion is supported by Little's measurements made near Key

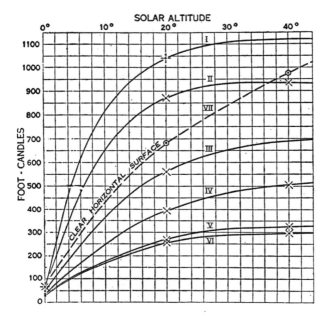

Fig. 6.—Curves of winter skylight illumination intensity on different surfaces.
Curve I. Clear sky. Vertical surface facing 0° in azimuth from sun.
Curve II. Clear sky. Vertical surface facing 45° in azimuth from sun.
Curve III. Clear sky. Vertical surface facing 70° in azimuth from sun.
Curve IV. Clear sky. Vertical surface facing 90° in azimuth from sun.
Curve V. Clear sky. Vertical surface facing 135° in azimuth from sun.
Curve VI. Clear sky. Vertical surface facing 180° in azimuth from sun.
Curve VII. Clear sky. Horizontal surface.

West, Fla., in February, 1918, which give for the zenith sky
brightness with the sun at altitudes averaging 22.8°, 42°, and
53°, 390, 780, and 1,150 millilamberts, respectively; while meas-
usements made by him from a ship off Long Island, N. Y., in
October, 1917, give for the sky brightness with solar altitudes
averaging 20.8°, and 41.2°, 296 and 495 millilamberts, respectively. The latter are somewhat lower readings than those obtained at Washington in winter with similar solar altitudes, while
the Key West measurements are considerably higher.

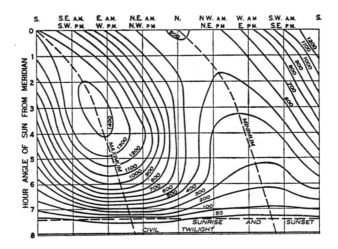

Fig. 7.—Variations in skylight illumination on vertical surfaces differently oriented at latitude 42° north on July 21. Cloudless sky. Foot-candles.

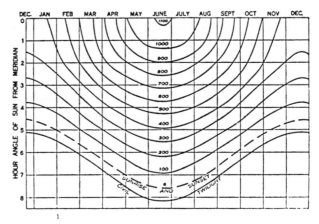

Fig. 8.—Illumination from a cloudy sky on a vertical surface at latitude 42° north. Foot-candles.

Figure 7 shows the variations with the hour of the day in sky-light illumination on vertical surfaces, such as the walls of buildings, differently oriented, on July 21, at latitude 42° N. with a clear sky. The maximum illumination is, of course, on a vertical surface facing the sun. It faces about east-northeast at sunrise, east at about 7:30 A. M., south at noon, west at about 4:30 P. M., and about west-northwest at sunset. The minimum illumination is on a vertical surface facing 180° in azimuth from the sun, or about west-southwest at sunrise, west at about 7:30 A. M., north at noon, east at about 4:30 P. M., and about east-southeast at sunset. Taking into consideration the hours between 7 A. M. and 5 P. M., which cover the usual working day, in the morning vertical surfaces facing northwest are most unfavorably oriented for illumination from a clear sky, and in the afternoon vertical surfaces facing northeast. On the other hand, surfaces facing northwest are favorably oriented for skylight illumination in the afternoon, and those facing northeast, in the morning.

From Table IV and Figures 8, 9 and 10 we derive the following:

(1) With a cloudy sky the illumination on a vertical surface is practically independent of the orientation of that surface.

(2) With a cloudy sky the illumination on a horizontal surface is considerably more than twice that on a vertical surface, due to the fact that the point of maximum brightness of a cloudy sky is in or near the zenith.

(3) With high sun, as at midday in summer, the illumination from a cloudy sky exceeds that from a clear sky except on vertical surfaces facing the sun. This is not true with low sun, however.

The eight figures, 11 to 18, inclusive, give the illumination from clear skies on vertical surfaces oriented as indicated. Were it not for the fact that clear skies in July, August, September, October, and November, are on the average whiter and therefore brighter than clear skies in May, April, March, February, and January, respectively, the lines of equal illumination intensity would be nearly symmetrical on each side of a vertical line representing June 21.

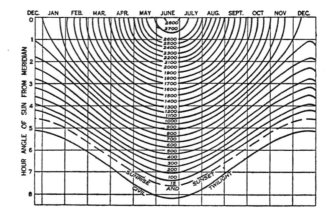

Fig. 9.—Illumination from a cloudy sky on a horizontal surface at latitude 42° north. Foot-candles.

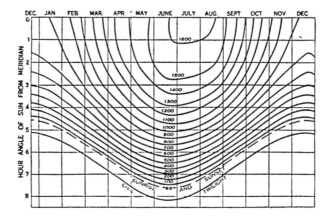

Fig. 10.—Illumination from a cloudless sky on a horizontal surface at latitude 42° north. Foot-candles.

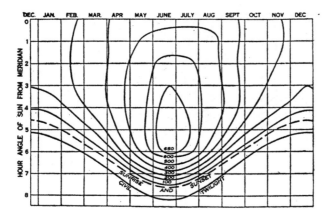

Fig. 11.—Illumination from a cloudless sky on a vertical surface facing north at latitude 42° north. Foot-candles.

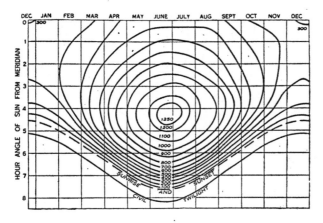

Fig. 12.—Illumination from a cloudless sky on a vertical surface facing northeast, A. M., or northwest, P. M., at latitude 42° north. Foot-candles.

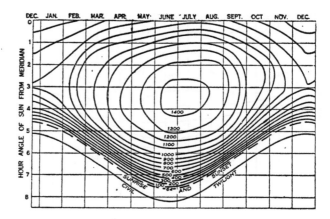

Fig. 13.—Illumination from a cloudless sky on a vertical surface facing east, A. M., or west, P. M., at latitude 42° north. Foot-candles

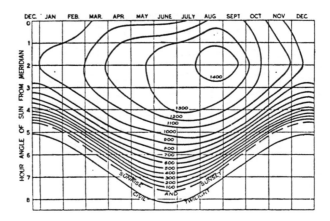

Fig. 14.—Illumination from a cloudless sky on a vertical surface facing south-east, A. M. or southwest, P. M., at latitude 42° north. Foot-candles.

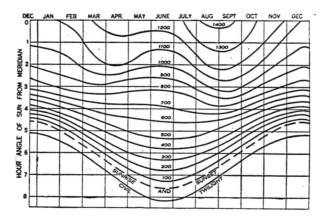

Fig. 15.—Illumination from a cloudless sky on a vertical surface facing south at latitude 42° north. Foot-candles.

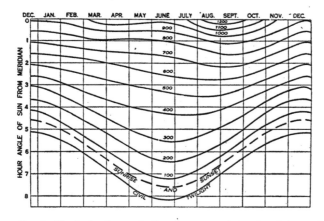

Fig. 16.—Illumination from a cloudless sky on a vertical surface facing south-west, A. M., or southeast, P. M., at latitude 42° north. Foot-candles.

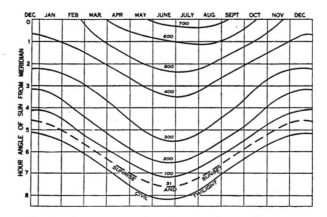

Fig. 17.—Illumination from a cloudless sky on a vertical surface facing west, A. M., or east, P. M., at latitude 42° north. Foot-candles.

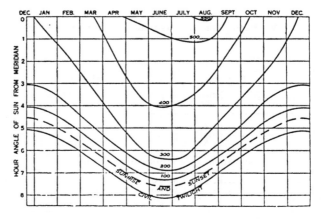

Fig. 18.—Illumination from a cloudless sky on a vertical surface facing north-west, A. M., or northeast, P. M., at latitude 42° north. Foot-candles.

These figures, like Figure 7, show the low intensity of clear-sky illumination during the morning hours on a vertical surface facing northwest and during the afternoon hours on a vertical surface facing northeast. The maximum illumination, slightly in excess of 1,400 foot-candles, occurs late in August and in early September at midday, on a vertical surface facing south; from the end of July to the latter part of September, on a vertical surface facing southeast at 10 A. M., and southwest at 2 P. M., and from early in June to the middle of August, on a vertical surface facing east at 8:30 A. M., and west at 3:30 P. M. Illumination on a vertical surface facing north is good throughout the day in summer but poor in winter.

In the report for 1921[4] it was shown that at the Federal Building, Chicago, which is in the smoky Loop District, the illumination on a vertical surface facing 180° from the sun is only about two-thirds as intense as at Washington, while at the University of Chicago the sky-brightness and the illumination measurements differ but little from the corresponding Washington measurements. Table V gives a summary of comparisons between Washington and Chicago illumination measurements on a horizontal surface, and on a vertical facing 180° from the sun, under winter conditions.

TABLE V.—RATIO, CHICAGO/WASHINGTON ILLUMINATION FROM WINTER SKIES

ON HORIZONTAL SURFACE					
Federal Building			University of Chicago		
Solar altitude	Clear sky	Cloudy sky	Solar altitude	Clear sky	Cloudy sky
0°	0.49	0.90	0°	0.48	0.69
20°	1.05	0.77	20°	1.03	0.56
26°	0.84	0.82	29°	1.14	0.76
ON VERTICAL SURFACE FACING 180° FROM THE SUN					
0°	0.42	0.85	0°	0.46	0.55
20°	0.69	0.54	20°	0.84	0.56
26°	0.62	0.88	29°	0.97	0.98

[4] *Mo. Weather Rev.*, Sept., 1921, 49, p. 482 and p. 486.

The darkening effect of the smoke is rather more pronounced in winter than in summer on a vertical surface facing away from the sun. The effect is slight at both seasons of the year on surfaces facing the sun when no clouds are present, except when the sun is near the horizon. The effect is closely related to the velocity of the wind. With light wind, and especially when the sky is covered with clouds, the smoke sometimes forms a cover or blanket of great thickness which cuts off practically all the daylight. A dark day results, and artificial lighting is necessary outdoors as well as in. No such days are included in the sky-brightness measurements for Chicago here considered, although on January 4, with the sun 20° above the horizon, the zenith brightness was only 150 millilamberts, and 2° above the horizon it averaged only 37 millilamberts, while a measurement of the illumination on a horizontal surface gave only 34 foot-candles. A comparison with the corresponding data of Table IV shows that the zenith brightness was 15 per cent and the illumination on a horizontal surface 5 per cent that for Washington with average cloudy conditions and the sun 20° above the horizon. The measurements show that the smoke cloud varied greatly in intensity during the period of observation.

With a cloudless sky, and solar altitude 20°, in winter the intensity of direct solar illumination at normal incidence at Chicago averages about half the intensity at Washington; in summer, in the Loop District, with solar altitudes 20° and 40°, about three-fourths as intense.

At Washington, with a clear sky, the illumination measurements on both a horizontal and on a vertical surface vary between 150 per cent and 60 per cent of the values given in Table IV. With a cloudy sky the variation is between 200 and 30 per cent. When rain is falling, the illumination is about half as great as the average for cloudy skies; with a sky partly covered with clouds, the illumination on a horizontal surface may be from three to four times as intense, and on a vertical surface two to three times as intense, as the corresponding illumination from a clear sky given in Table IV.

From seasonal averages of sky brightness for Davos Platz, Switzerland, given by Dorno,[5] it appears that when expressed in terms of the zenith brightness the sky at Davos Platz opposite the sun is brighter than at Washington. The zenith brightness in winter averages more than 50 per cent brighter, and in summer a few per cent less bright at Davos Platz than at Washington. On the whole, Davos Platz skies when free from clouds are brighter in winter and less bright in summer than at Washington. Probably the increased brightness in winter is due in part to reflection of light from the snow-covered surface.

TOTAL SOLAR AND SKY ILLUMINATION

In the *Monthly Weather Review* for November, 1919, **47**, p. 785, Table XIV, are given the illumination equivalents of solar energy expressed in heat units, with the sun at different altitudes. These equivalents were derived from simultaneous readings made at Mount Weather, Va., in 1913-14, with a pyrheliometer and a photometer. The photometer had its uncompensated test plate exposed horizontally, and the error, due to the oblique angle at which the sun's rays were received, was unknown.

TABLE VI.—ILLUMINATION EQUIVALENT OF 1 GRAM-CALORY PER MINUTE PER SQUARE CENTIMETER OF SOLAR ENERGY WITH THE SUN AT DIFFERENT ALTITUDES

Air mass. . . .	1.06	1.10	1.50	2.00	2.50	3.00	3.50	4.00	4.50	5.00	5.50
Solar altitude	$70°0$	$65°0$	$42°7$	$30°0$	$23°5$	$19°3$	$16°4$	$14°3$	$12°6$	$11°3$	$10°2$
Foot-candles .	7,040	7,020	6,880	6,740	6,650	6,580	6,520	6,460	6,410	6,370	6,320

In the measurements made at Washington in 1921-22 a *compensated* test plate was used, and the certificate furnished by the Electrical Testing Laboratories, New York, shows no appreciable error due to an obliquity in the angle of incidence of the sun's rays. Illumination intensities were measured with the test plate horizontal and also normal to the incident solar rays, but the latter measurements were given twice the weight of the former. Comparison of these measurements with simultaneous pyrheliometric measurements give the illumination equivalents of Table VI. These are considerably higher than the equivalents determined at Mount Weather, and particularly with low sun, as one would expect.

[5]Dorno, C. Himmelshelligkeit, Himmelspolarisation und Sonnenintensität in Davos 1911 bis 1918. Veröffentlichungen des Preusischen Meteorlogischen Instituts., Nr. 303. Abhandlungen Bd. VI, Tabellen 4A und 6.

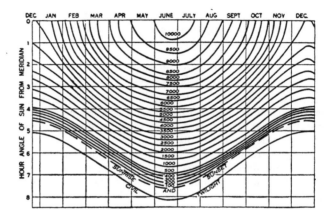

Fig. 19.—Total daylight illumination on a horizontal surface with a cloud-less sky at latitude 42° north. Foot-candles.

By means of the equivalents of Table VI and the solar radiation intensity at normal incidence for latitude 42° N., given in the number of the *Review* above quoted (p. 773, Table Va), the solar illumination intensities of Table VII have been obtained.

TABLE VII.—SOLAR ILLUMINATION INTENSITY AT NORMAL INCIDENCE AT LATITUDE 42° NORTH, WITH A CLOUDLESS SKY (EAST OF THE MISSISSIPPI RIVER)

Day	Hour angle of sun from meridian							
	0	1	2	3	4	5	6	7
Dec. 21..........	7600	7300	6640	5190	2460	......	......	......
Jan. 21..........	8120	7890	7290	6040	3760	......	......	......
Feb. 21..........	9140	9040	8440	7450	6140	2460	......	......
Mar. 21..........	9270	9110	8710	7910	6700	4650	720	......
Apr. 21..........	9230	9060	8800	8300	7350	5860	3600	......
May 21..........	9070	8990	8630	8140	7480	6260	4700	1200
June 21..........	9080	9000	8740	8220	7430	6420	4880	2160
July 21..........	9070	8990	8670	8140	7550	6330	4830	1200
Aug. 21..........	8810	8710	8390	7830	6880	5460	2990	......
Sept. 21..........	8910	8760	8510	7710	6500	4590	720	......
Oct. 21..........	8510	8420	7960	6910	5220	2100	......	......
Nov. 21..........	8120	7890	7290	5960	3390	......	......	......

Representing the illumination intensities of Table VII by I_n, the illumination on a horizontal surface, I_h, and on a vertical surface, I_v, may be obtained by the equations

$$I_h = I_n \sin a \qquad (1)$$
$$I_v = I_n \cos a \cos a \qquad (2)$$

where a is the altitude of the sun, and a is the difference between the sun's azimuth and the azimuth of a line normal to the vertical surface. The surface will be illuminated by the sun only when the value of a is less than 90°.

Adding the values of I_h to the skylight-illumination values for corresponding days and hours given on Figure 10, we obtain the total daylight illumination on a horizontal surface for a cloudless sky of average brightness at latitude 42° N., which is charted on Figure 19.

Similarly, by adding the values of I_v for vertical surfaces facing the eight principal points of the compass to the skylight illumination for corresponding days and hours given on Figures 11 to 16, inclusive, we obtain the total daylight illumination on vertical surfaces facing south; southeast A. M., or southwest P. M.; southwest A. M., or southeast P. M.; east A. M., or west P. M.; northeast A. M., or northwest P. M.; and north; as given on Figures 20 to 25, inclusive.

It is to be noted that with north solar declination all vertical surfaces receive direct solar radiation during only a part of the day. During the remainder of the day the total daylight illumination is the same as the skylight illumination on Figures 11 to 18, inclusive.

The data of Figures 1 to 25, inclusive, assume that the surface under consideration has an unobstructed exposure to the sky. Where a part of the sky is cut off by adjacent buildings or other obstructions, the shading effect of such obstructions may be determined by the method given in the previous report.[6]

This shading effect, and also the reflection of daylight from surrounding objects, will receive more detailed consideration in a later report.

[6] TRANS., Illum. Eng. Soc., Vol. XVI, p. 270; *Mo. Weather Rev.*, Sept., 1921, 49, p. 486.

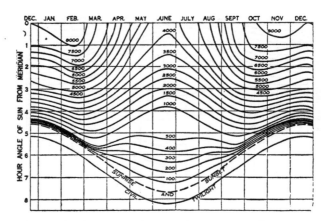

Fig. 20.—Total daylight illumination on a vertical surface facing south with a cloudless sky at latitude 42° north. Foot-candles.

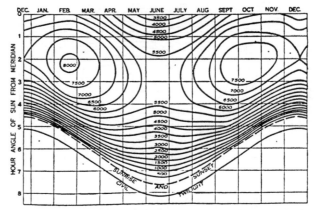

Fig. 21.—Total daylight illumination on a vertical surface facing southeast, A. M., or southwest, P. M., with a cloudless sky at latitude 42° north. Foot-candles.

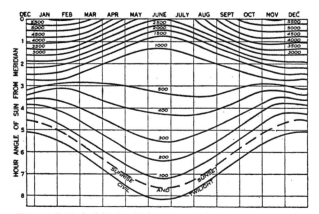

Fig. 22.—Total daylight illumination on a vertical surface facing southwest, A. M., or southeast P. M., with a cloudless sky at latitude 42° north. Foot-candles.

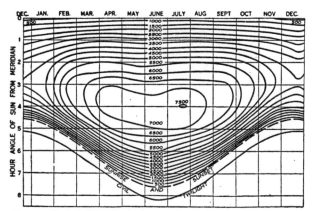

Fig. 23.—Total daylight illumination on a vertical surface facing east, A. M., or west P. M., with a cloudless sky at latitude 42° north. Foot-candles.

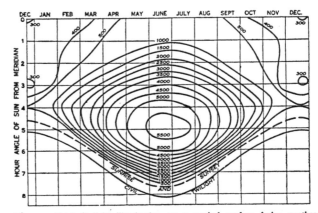

Fig. 24.—Total daylight illumination on a vertical surface facing northeast, A. M., or northwest P. M., with a cloudless sky at latitude 42° north. Foot-candles.

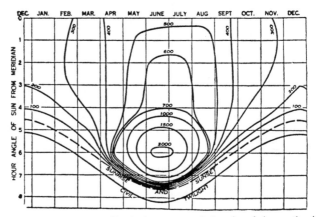

Fig. 25.—Total daylight illumination on a vertical surface facing north with a cloudless sky at latitude 42° north. Foot-candles.

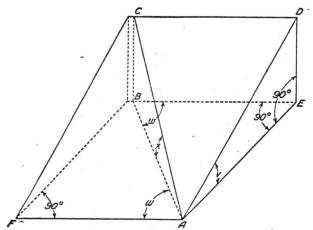

Fig. 26.—Determination of the angle of incidence of solar rays with a sloping surface.

DAYLIGHT ILLUMINATION ON SLOPING SURFACES

The intensity of solar radiation on surfaces sloping in different directions should be of importance to agriculturalists and engineers.[7] Illumination intensities on such surfaces are of especial interest in connection with the lighting of industrial plants by means of the so-called saw-tooth-roof construction.

The computation of direct solar illumination on sloping surfaces presents no special difficulties. Let v, Figure 26, represent the angle between the sloping surface and a horizontal surface; w, the difference between the azimuth bearing of AF, the intersection of these two surfaces, and AB, representing the sun's azimuth; and x, the angle between the intersections of a plane in the sun's vertical with the sloping surface and with a horizontal surface. Then

$$\tan x = \sin w \tan v, \text{ and } a' = a + x \qquad (3)^8$$

[7]*Mo. Weather Rev.*, Nov., 1919, 47, p. 781.

[8]In Daylight *vs.* Sunlight in Sawtooth-Roof Construction., *Transactions American Society of Mechanical Engineers*, 40, pp. 603-625, W. S. Brown derives this equation as follows:

$$\frac{AE}{ED} = \frac{\cos v}{\sin v} \quad \therefore \quad AE = ED \cot v.$$

Similarly, $AB = ED \cot x$; and $\sin w = \dfrac{AE}{AB} = \dfrac{\cot v}{\cot x}$, from which $\tan x = \sin w \tan v$.

where a is the altitude of the sun and a' is the angle between the incident solar rays and the sloping surface.

To obtain the intensity of solar illumination on a sloping surface we have only to substitute a' for a in equation (1).

We may also obtain a' by first determining the latitude and longitude of a point at which a horizontal surface is parallel to the sloping surface, by the method given in the *Monthly Weather Review*, November, 1919, **47**, p. 781.[9] Making allowance for the difference in time represented by the difference in longitude of the sloping surface and its parallel horizontal surface, we may obtain directly from an altitude table the altitude of the sun at the latitude of the horizontal surface, and therefore the angle a' which the incident solar rays make with the sloping surface at any hour of any day of the year.

The computation of the *skylight* illumination on sloping surfaces requires the replotting of the sky-brightness measurements for each surface considered.

Figure 27 shows the data for a clear sky with the sun at altitude 40°, projected on a surface for which 90° $-w = 45°$ and $v = 10°$ (surface 80° out of the vertical and facing 45° in azimuth from the sun). In this case the zenith of the sky falls 10° from the zenith of the sloping surface. The line of the horizon from azimuth $+ 45°$ to $- 135°$ with reference to the sun, and the lines on which the sky-brightness measurements are to be plotted, have been determined by means of the methods given under Solution of Problems in Stereographic Projections" (pp. 52-58), in General Theory of Polyconic Projections, by Oscar S. Adams, United States Coast and Geodetic Survey, Special Publication No. 57, Serial No. 110.

[9] NOTE.—In lines 15 and 16 from the bottom of the second column of the page referred to, the words "longitude" and "latitude" should be interchanged. The difference in latitude between the sloping surface and its parallel horizontal surface is given by the equation

$$\tan \Delta \phi = \frac{\cos a'}{\cot v},$$

and the difference in longitude by the equation

$$\sin \Delta \lambda = \sin a' \sin v$$

where a' is the azimuth in which the sloping surface faces, and v its angle of slope. When a' 0° or 180°, $\sin \Delta \lambda$ 0, and $\tan$ $\tan v$. That is, the sloping surface and the parallel horizontal surface have the same longitude, and the difference in latitude equals the angle of slope, v.

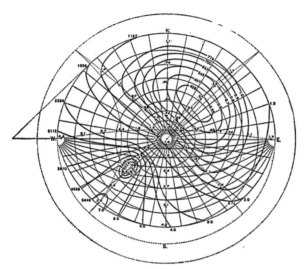

Fig. 27.—Stereographic projection of sky-brightness measurements on a sloping surface.

It is to be noted that the location of the line of the horizon consists in passing a circle through two given points, when the center of the circle falls on a given line. Or, it may also be determined by passing a circle through three given points, as is the location of the lines on which the sky-brightness measurements are to be plotted.

On Figure 27 the brightness of the entire sky is shown with the exception of a spherical lune which falls below the plane of projection on the side WNE., and for which the maximum width is 10° at N. The sky-brightness values that have been obtained by measurement on the half of the sky on one side of the sun's vertical have been plotted on both sides of this vertical.

In Tables VIII and IX is given the total (solar + sky) illumination on surfaces sloping in southerly directions, as indicated. In Tables X and XI is given the skylight illumination on surfaces sloping in northerly directions as indicated. In Table XII is given the ratio of the total illumination to the sky illumination on surfaces facing opposite each other in azimuth.

TABLE VIII.—TOTAL ILLUMINATION ON SURFACES SLOPING SOUTH

Date	Hour angle of the sun from meridan							
	0	1	2	3	4	5	6	7
	Foot-candles							
	Surface sloping 10° from horizontal							
Dec. 21	5,220	4,810	3,800	2,280	656	......	......	......
Jan. 21	6,000	5,590	4,560	2,910	1,140	......	......	......
Feb. 21	7,890	7,520	6,280	4,480	2,240	486	......	......
Mar. 21	9,250	8,800	7,660	5,880	3,660	1,560	82	......
Apr. 21	10,230	9,780	8,700	7,040	4,910	2,640	720	......
May 21	10,690	10,280	9,110	7,530	5,570	3,340	1,420	154
June 21	10,980	10,530	9,460	7,830	5,820	3,730	1,720	280
July 21	10,820	10,390	9,270	7,610	5,700	3,460	1,500	195
Aug. 21	10,050	9,660	8,580	6,920	4,790	2,620	710	......
Sept. 21	9,150	8,730	7,730	5,910	3,720	1,680	98	......
Oct. 21	7,550	7,220	6,090	4,280	2,250	440	......	......
Nov. 21	6,060	5,640	4,590	2,890	1,070	......	......	......
	Surface sloping 20° from horizontal							
Dec. 21	6,300	5,820	4,690	2,890	900	......	......	......
Jan. 21	7,070	6,600	5,460	3,600	1,490	......	......	......
Feb. 21	8,870	8,500	7,140	5,120	2,910	600	......	......
Mar. 21	10,050	9,550	8,250	6,310	3,940	1,630	47	......
Apr. 21	10,770	10,310	9,050	7,250	4,950	2,540	630	......
May 21	11,060	10,490	9,230	7,510	5,400	3,100	1,180	141
June 21	11,260	10,640	9,460	7,750	5,610	3,360	1,380	232
July 21	11,180	10,590	9,320	7,580	5,570	3,220	1,190	156
Aug. 21	10,590	10,090	8,890	7,120	4,830	2,540	620	......
Sept. 21	9,930	9,450	8,310	6,360	3,980	1,680	61	......
Oct. 21	8,510	8,130	6,890	4,920	2,630	550	......	......
Nov. 21	7,130	6,660	5,500	3,580	1,380	......	......	......
	Surface sloping 30° from horizontal							
Dec. 21	7,220	6,680	5,410	3,440	1,140	......	......	......
Jan. 21	7,790	7,360	6,190	4,150	1,820	......	......	......
Feb. 21	9,690	9,220	7,720	5,680	3,250	720	......	......
Mar. 21	10,640	9,990	8,590	6,550	4,110	1,700	50	......
Apr. 21	11,100	10,340	9,040	7,120	4,870	2,390	470	......
May 21	10,960	10,320	8,990	7,260	5,100	2,740	875	134
June 21	10,970	10,320	9,150	7,420	5,200	2,960	1,040	230
July 21	11,060	10,400	9,120	7,300	5,220	2,870	880	156
Aug. 21	10,910	10,210	8,900	7,020	4,650	2,650	500	......
Sept. 21	10,520	9,860	8,660	6,650	4,180	1,800	62	......
Oct. 21	9,250	8,790	7,420	5,380	2,940	650	......	......
Nov. 21	3,050	7,410	6,230	4,160	1,680	......	......	......

TABLE IX.—TOTAL ILLUMINATION ON SURFACES SLOPING TOWARD SOUTHEAST OR SOUTHWEST.

Date	Hour angle of sun from meridian														
	A. M. for surface sloping SE.; P. M., SW.							P. M. for surface sloping SE.; A. M., SW.							
	7	6	5	4	3	2	1	0	1	2	3	4	5	6	7

Foot-candles

Surface-sloping 10° from horizontal

Date	7	6	5	4	3	2	1	0	1	2	3	4	5	6	7
Dec. 21				86	2,530	3,920	4,60	4,80	4,120	3,086	1,620	92			
Jan. 21			756	1,420	3,260	4,680	5,490	5,580	4,90	3,740	2,130	63			
Feb. 21		171	2,120	2,980	4,950	6,500	7,420	7,30	8,180	6,760	3,500	1,610	80	48	
Mar. 21		1,240	3,339	4,340	6,399	7,960	8,800	8,890	8,140	7,930	4,840	2,790	907	305	186
Apr. 21		2,090	4,160	5,90	7,600	8,990	9,730	9,970	9,90	8,320	6,90	3,890	1,850	850	235
May 21		2,380	4,520	6,620	8,470	9,400	10,690	10,480	9,90	8,900	6,90	4,580	2,530	1,180	160
June 21	595	2,380	4,240	6,520	8,266	9,886	10,520	10,770	10,050	8,570	6,000	4,90	2,920	902	
July 21	713	2,210	3,330	5,540	7,520	9,640	10,690	10,660	9,920	7,850	4,920	4,90	2,640	330	
Aug. 21	412	1,130		4,386	6,439	8,966	9,710	9,800	9,150	6,890	3,380	3,870	1,890	63	
Sept. 21		190	676	2,710	4,740	8,010	8,730	8,820	8,130	5,040	2,110	2,850	839		
Oct. 21				1,320	3,239	6,310	7,110	7,186	6,530	3,756		1,520	184		
Nov. 21						4,700	5,540	5,540	4,970			94			

Surface sloping 20° from horizontal

Date	7	6	5	4	3	2	1	0	1	2	3	4	5	6	7
Dec. 21				1,260	3,360	4,880	5,60	5,559	4,90	3,310	1,590	22			
Jan. 21			1,139	2,040	4,90	5,740	6,30	6,310	5,30	3,920	2,060	67			
Feb. 21		83	2,790	3,990	6,020	7,560	8,340	8,120	7,150	5,380	3,250	72	157	38	
Mar. 21		1,40	3,956	5,290	7,370	8,820	9,470	9,160	9,60	6,580	4,400	90	485	418	140
Apr. 21		2,40	4,670	6,590	8,400	9,720	10,399	10,479	9,610	7,500	5,30	3,100	1,540	540	190
May 21		2,90	4,790	6,910	8,660	10,000	10,800	10,179	9,90	7,770	6,90	3,90	1,920	459	120
June 21		2,90	4,790	7,180	9,000	10,340	11,040	10,720	9,610	8,139	5,910	3,820	1,700	459	
July 21		2,59	4,270	7,160	8,90	10,220	10,920	10,610	9,90	8,020	5,359	3,110	1,110	57	
Aug. 21		1,90		5,220	7,370	8,910	9,439	9,980	9,90	7,459	4,539	2,300	500		
Sept. 21		26	1,010	3,510	5,720	7,310	7,970	9,60	8,90	6,700	3,160	1,200	163		
Oct. 21				1,880	4,160	5,730	6,480	7,90	6,870	4,999	2,059	81			
Nov. 21						5,730	6,480	6,360	5,90	3,959					

Surface sloping 30° from horizontal

Date	7	6	5	4	3	2	1	0	1	2	3	4	5	6	7
Dec. 21				1,90	4,180	5,740	6,280	6,086	4,960	3,386	1,530	175			
Jan. 21			1,490	2,60	5,090	6,586	7,40	6,800	5,530	3,970	1,940	325	74	46	
Feb. 21		34	3,90	4,850	6,940	8,410	8,950	8,480	7,299	5,220	2,900	859	450	230	120
Mar. 21		1,090	4,600	5,910	8,090	9,500	9,930	9,466	8,140	6,240	3,880	1,600	90	90	180
Apr. 21		2,90	5,150	7,159	9,040	10,340	10,90	10,010	8,700	6,850	4,560	2,240	90	90	110
May 21	467	2,90	5,386	7,386	9,120	10,360	10,966	10,140	8,900	7,040	4,860	2,610	90	81	
June 21	81			7,490	9,286	10,570	10,970	10,270	9,070	7,240	5,020	2,770	90	410	
July 21	470	2,90	5,240	7,59	9,180	10,559	10,970	10,240	9,050	7,186	4,859	2,700	90	90	
Aug. 21		1,90	4,350	6,790	8,770	10,530	10,530	9,830	8,630	6,786	4,540	2,180	90	90	
Sept. 21	470	350	3,340	6,060	8,110	9,650	9,650	9,370	8,040	6,286	3,959	1,660	455	58	

TABLE X.—SKYLIGHT ILLUMINATION ON SURFACES SLOPING NORTH

Date	Hour angle of sun from meridian							
	0	1	2	3	4	5	6	7
	Foot-candles							
	Surface sloping 10° from vertical							
Dec. 21	310	310	285	221	115	......	......	......
Jan. 21	320	330	310	259	160	......	......	......
Feb. 21	340	352	348	340	293	112	......	......
Mar. 21	421	428	406	429	436	327	48	......
Apr. 21	520	540	562	564	550	485	346	......
May 21	554	600	660	685	641	650	590	228
June 21	580	640	758	770	740	790	719	365
July 21	580	635	733	730	712	700	630	288
Aug. 21	570	605	610	635	629	560	371	......
Sept. 21	468	480	462	494	487	354	72	......
Oct. 21	368	386	372	374	331	147	......	......
Nov. 21	333	343	328	270	167	......	......	......
	Surface sloping 20° from vertical							
Dec. 21	355	350	320	236	112	......	......	......
Jan. 21	362	365	355	276	172	......	......	......
Feb. 21	390	400	410	395	294	120	......	......
Mar. 21	489	498	471	516	486	416	46	......
Apr. 21	619	635	638	640	642	589	412	......
May 21	660	720	782	805	759	764	625	241
June 21	698	777	888	901	956	879	750	365
July 21	691	760	874	860	834	817	720	312
Aug. 21	671	710	715	730	730	660	418	......
Sept. 21	542	560	538	558	555	411	60	......
Oct. 21	430	446	460	420	338	122	......	......
Nov. 21	378	394	366	278	182	......	......	......
	Surface sloping 30° from vertical							
Dec. 21	377	380	345	255	120	.....	......	......
Jan. 21	398	407	384	302	172	.....	......	......
Feb. 21	431	450	441	424	341	147	......	......
Mar. 21	569	575	524	572	582	451	47	......
Apr. 21	754	760	742	753	745	651	409	......
May 21	801	873	913	882	898	871	643	200
June 21	830	920	1,035	1,028	1,035	1,004	770	337
July 21	837	917	998	990	980	962	655	217
Aug. 21	828	870	840	837	850	735	442	......
Sept. 21	631	645	596	630	626	465	58	......
Oct. 21	473	497	502	472	378	189	......	......
Nov. 21	413	423	417	323	200	......	......	......

TABLE XI.—SKYLIGHT ILLUMINATION ON SURFACES SLOPING NORTHEAST OR NORTHWEST

Date	Hour angle of sun from meridian														
	7	6	5	4	3	2	1	0	1	2	3	4	5	6	7
	A. M., sloping NE.; P. M., sloping NW.							P. M., sloping NE.; A. M., sloping NW.							
	Foot-candles														
	Surface sloping 10° from vertical														
Dec. 21	..	..	..	122	292	380	380	336	317	298	186	108	..	..	..
Jan. 21	..	..	..	220	355	418	418	351	330	307	245	142	..	..	..
Feb. 21	..	..	152	457	503	478	445	375	360	338	308	236	97	..	..
Mar. 21	..	80	630	780	660	614	544	465	425	373	370	325	245	42	..
Apr. 21	..	650	985	990	920	790	700	569	527	465	410	375	320	220	..
May 21	450	1,150	1,300	1,240	1,100	1,010	815	624	575	540	485	425	430	330	150
June 21	600	1,350	1,500	1,325	1,383	1,070	900	661	600	590	533	478	475	400	180
July 21	450	1,220	1,500	1,300	1,130	1,200	870	660	600	585	510	440	450	380	120
Aug. 21	..	900	1,110	1,080	1,080	1,000	830	640	588	520	465	430	355	275	..
Sept. 21	..	108	700	860	760	670	655	529	470	420	400	360	260	47	..
Oct. 21	..	..	185	496	597	562	529	402	358	335	310	230	97	..	..
Nov. 21	..	..	..	268	376	444	488	372	345	308	242	148	..	..	..
	Surface sloping 20° from vertical														
Dec. 21	..	..	..	170	337	405	440	391	370	308	225	105	..	..	..
Jan. 21	..	..	..	230	402	480	525	418	385	340	257	145	..	..	..
Feb. 21	..	..	202	502	580	580	565	440	420	380	345	250	98	..	..
Mar. 21	..	78	640	755	785	705	660	547	495	425	370	360	245	38	..
Apr. 21	..	720	980	1,050	1,010	935	810	676	625	550	485	440	355	225	..
May 21	450	1,050	1,356	1,280	1,190	1,125	935	747	690	655	560	490	500	360	130
June 21	600	1,350	1,500	1,420	1,300	1,210	1,070	800	730	685	630	530	570	450	165
July 21	450	1,220	1,500	1,320	1,300	1,195	1,020	792	715	675	605	508	530	400	115
Aug. 21	..	900	1,080	1,140	1,220	1,080	920	760	685	610	535	480	390	240	..
Sept. 21	..	105	680	935	1,010	803	750	624	560	492	450	375	262	44	..
Oct. 21	..	..	185	500	684	680	638	488	460	425	365	260	91	..	..
Nov. 21	..	..	..	247	424	518	556	440	394	340	275	125	..	..	..
	Surface sloping 30° from vertical														
Dec. 21	..	..	..	220	320	430	480	420	387	340	255	135	..	..	..
Jan. 21	..	..	..	285	455	550	580	448	422	375	294	164	..	..	..
Feb. 21	..	..	202	530	660	645	592	491	450	428	400	277	107	..	..
Mar. 21	..	69	625	840	975	980	790	640	570	486	425	385	290	42	..
Apr. 21	..	755	910	950	1,200	1,050	920	824	785	650	525	475	385	240	..
May 21	450	1,050	1,320	1,350	1,380	1,300	1,100	888	830	800	665	530	535	390	120
June 21	600	1,350	1,500	1,450	1,360	1,350	1,180	941	855	840	738	595	600	445	220
July 21	450	1,220	1,500	1,400	1,400	1,320	1,160	949	870	835	695	575	535	420	140
Aug. 21	..	900	1,085	1,270	1,320	1,200	1,070	928	850	720	620	510	425	270	..
Sept. 21	..	100	740	960	810	857	910	730	627	535	485	420	310	48	..
Oct. 21	..	..	215	580	784	750	847	524	505	470	415	275	107	..	..
Nov. 21	..	..	..	326	510	620	590	484	440	380	285	167	..	..	..

TABLE XII.—RATIO OF TOTAL ILLUMINATION, T, TO
SKY ILLUMINATION, S

CLOUDLESS SKY, LATITUDE 42° N.								
Date	Hour angle of sun from meridian							
	0	1	2	3	4	5	6	7
T, vertical surface facing south; S, vertical surface facing north								
Dec. 21..........	29	27	22	22	16	......	......	......
Feb. 21..........	28	26	22	17	12	8	......	......
Apr. 21..........	13	12	10	7	4	1.4	0.3	......
June 21..........	9	7	5	3	1.1	0.3	0.2	0.1
Aug. 21.	12	11	9	6	3	1	0.5	
Oct. 21..........	25	23	20	15	11	5	......	......
T, surface sloping south 10° from horizontal; S, sloping north 10° from vertical								
Dec. 21..........	18	16	13	10	6	......	......	......
Feb. 21..........	23	21	18	13	8	4	......	......
Apr. 21..........	20	18	16	12	9	5	2	......
June 21..........	19	16	12	10	8	5	2	0.8
Aug. 21.	18	16	14	11	8	5	2	
Oct. 21..........	20	19	16	11	7	3	......	......
T, surface sloping south 30° from horizontal; S, sloping north 30° from vertical								
Dec. 21..........	19	18	16	14	10	......	......	......
Feb. 21..........	22	20	18	13	10	5	......	......
Apr. 21..........	15	14	12	10	6	4	1.2	......
June 21..........	13	11	9	7	5	3	1.4	0.7
Aug. 21..........	13	12	11	8	6	4	1.1	
Oct. 21..........	20	18	15	11	8	3	......	......

	HOUR ANGLE OF SUN FROM MERIDIAN														
Date	7	6	5	4	3	2	1	0	1	2	3	4	5	6	7
	T, surface facing SE., A. M., or SW., P. M. S, surface facing NW., A. M., or NE., P. M.								T, surface facing SW., A. M., or SE., P. M. S, surface facing NE., A. M., or NW., P. M.						
	Surfaces vertical														
Dec. 21..........	...	...	...	28	29	28	24	20	12	7	2	0.7	...	...	...
Feb. 21..........	...	...	30	30	28	28	24	19	10	4	0.9	0.6	0.5	...	...
Apr. 21..........	...	14	17	22	21	17	13	9	4	0.9	0.6	0.4	0.3	0.3	...
June 21..........	6	8	11	16	14	11	9	6	1.6	0.7	0.5	0.3	0.3	0.3	0.3
Aug. 21..........	...	11	15	18	18	15	12	8	3	0.9	0.6	0.4	0.3	0.3	...
Oct. 21..........	...	...	24	25	25	24	20	16	9	4	0.9	0.6	0.5	...	...

TABLE XII.—(*Continued*)—RATIO OF TOTAL ILLUMINATION, T, TO SKY ILLUMINATION, S

Date	HOUR ANGLE OF SUN FROM MERIDIAN														
---	7	6	5	4	3	2	1	0	1	2	3	4	5	6	7
T, surface sloping 10° from horizontal; S, 10° from vertical															
Dec 21	...	...	...	8	14	13	15	14	11	8	6	2	...	...	...
Feb. 21	...	...	8	13	16	19	21	20	15	11	7	4	1.2	...	...
Apr. 21	...	6	10	15	18	19	18	17	13	10	6	4	2	0.5	...
June 21	3	6	10	14	16	17	17	16	11	8	5	4	2	0.9	0.4
Aug. 21	...	4	9	13	16	17	18	15	11	8	6	4	2	0.4	...
Oct: 21	...	...	7	12	15	19	20	18	12	9	6	3	1.0	...	...
T, surface sloping 30° from horizontal; S, 30° from vertical															
Dec. 21	...	...	...	12	16	17	16	14	10	8	5	0.8	...	...	
Feb. 21	...	...	14	18	18	20	20	17	12	8	4	1.6	0.9	...	...
Apr. 21	...	8	12	15	17	16	14	12	9	7	4	2.4	0.6	0.3	...
June 21	4	7	9	13	13	13	13	11	8	5	4	1.9	0.7	0.3	0.3
Aug. 21	...	7	10	13	14	14	12	11	8	6	3	1.7	0.5	0'3	...
Oct. 21	...	...	12	16	16	17	17	16	8	6	4	1.4	0.7	...	...

Table VIII shows that in general on surfaces sloping south and with south solar declination the total illumination increases with v. With north solar declination the illumination reaches a maximum in the middle of the day when v equals about 20°, and decreases as v increases with the sun near the horizon.

Table IX shows that in the morning on surfaces facing southeast, and in the afternoon on surfaces facing southwest, there is an increase in the total illumination with increase in v, except near midday in midsummer with v greater than about 20°. Also in the morning, on surfaces facing southwest, and in the afternoon, on surfaces facing southeast, the illumination generally decreases with increase in v, except near midday with south solar declination.

Tables X and XI show an increase with v in skylight illumination on vertical surfaces sloping northward, as one would expect. It must be remembered, however, that with saw-tooth construction a very considerable part of the skylight is cut off by shading.

This is unimportant when considering the total illumination on surfaces sloping in a southerly direction, but becomes important in connection with the skylight illumination on surfaces facing towards the north, since it is the brightest part of the sky that is cut off.

Let it be assumed that the ridges of the saw-teeth of the roof are horizontal, and of infinite length, and let $\theta =$ the maximum angular width of the spherical lune of the sky cut off. Then Table XIII gives the percentages of decrease in the skylight illu-

TABLE XIII.—SHADING EFFECT IN SAW-TOOTH ROOF CONSTRUCTION

90°—w.	Solar altitude				
	20°	40°	60°	70°	
	PERCENTAGE OF SKYLIGHT CUT OFF				
°	Surface 10 degrees out of vertical				°
180	32	24	20	17	10
135	26	25	18	17	10
90	25	19	17	15	10
	Surface 20 degrees out of vertical				
180	43	37	31	25	
	Surface 30 degrees out of vertical				
180	39	33	24	21	20
135	38	28	23	21	20
90	34	26	24	18	20
180	53	44	33	31	30
135	50	41	33	28	30
90	46	37	33	30	30
	Surface 60 degrees out of vertical				
0	19	11	6	6	10
45	14	9	6	5	10
90	9	7	5	4	10
0	53	39	23	21	30
45	44	32	22	21	30
90	28	22	18	16	30
	Surface 80 degrees out of vertical				
0	11	6	3	3	10
45	9	5	3	3	10
90	4	3	2	2	10
0	35	29	18	15	30
45	32	21	13	12	30
90	16	12	10	9	30
0	64	54	36	30	50
45	55	42	28	27	50
90	30	23	20	19	50

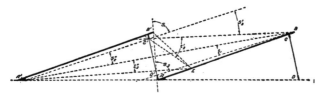

Fig. 28.—Cross section of a saw-tooth roof.

mination,[10] due to shading by the adjacent saw tooth.

Table XII shows that during most of the working hours of the day (except in midsummer) the total daylight illumination on surfaces sloping southward exceeds by more than tenfold the skylight illumination on surfaces sloping northward.

Computations from the sky-brightness data given in TRANS-ACTIONS, Illuminating Engineering Society, Vol. XVI, p. 260, Figures 6 and 7, show that skylight illumination on vertical or sloping surfaces facing away from the sun is about twice as intense when the sky is covered with thin clouds or haze or partly covered with white clouds, which is its usual condition, as when clear. The total (solar + sky) illumination on surfaces facing the sun is usually diminished by the presence of haze or clouds of the above character. In consequence, when the angle w lies between about 45° and 135°, the ratios of Table XII will be diminished, on the average, by at least one-half, and will vary in value from their maxima with clear-sky conditions, given in Table XII, to about 2 for a sky completely covered with dense clouds. This will be made clear from a comparison of the sky-brightness data of Figures 4 (opposite p. 259) and 13 (p. 263) with Figures 6, 7 and 8 (p. 260) and 11 (p. 262), TRANSACTIONS, Illuminating Engineering Society, Vol. XVI, October, 1921, and Figure 3 (p. 442) of this paper; and by reference to the illumination intensities of Table IV (p. 443) of this paper.

These ratios are of use in computing the daylight that can be made available for illuminating working space in a building through saw-tooth-roof construction. In general, there are two sources from which the light may be obtained as follows:

[10]The percentages of Table XIII have been computed from Figure 27 and other similar figures. See also TRANS., Illum. Eng. Soc., Vol. XVI, p. 270 and *Mo. Weather Review*, Sept , 1921, 49, p. 486.

(1)   Light from the northern sky incident at the working space, or reflected thereto from the ceiling of the saw-tooth roof (sky angles S_1 to S_3, and S'_2 to S''_2, respectively, Figure 28).

(2)   Solar and skylight reflected from the outside surface of the saw-tooth directly to the working space, or through a secondary reflection from the ceiling of the saw-tooth roof (roof angles t_1 to o, and t'_2 to t''_2, Figure 28).

It is to be understood that the angles here shown are cross-sections of spherical wedges.

Assuming the ratio of the intensity of the total light reaching the southerly-sloping roof surface of a saw-tooth window, AB (Figure 28) to the light received from the sky on the northerly-sloping window surface C'D' (Figure 28) to be 4, Brown[11] computed the relative values of (1) and (2) to be 14.6 and 6.1, respectively.

Let us consider a saw-tooth construction that gives a window surface facing north and sloping 20° from the vertical, and a roof surface sloping south 20° from the horizontal. Let the latitude be 42° north, the sky clear, the date March 21, and the hour 10 A. M., or 2 P. M., apparent time. The solar altitude will be 40° and its azimuth 41°. Disregarding shading, Tables VIII and X give 8,250 and 471 foot-candles for the illumination intensity on the roof surface and the window surface, respectively, the ratio of the former to the latter being 18.

The ridge of the adjacent saw tooth would cut off from the window a spherical wedge near the horizon for which the average value of θ would be about 10°. The window is facing 139° from the sun, and from Table XIII it is estimated that the shading by the roof diminishes the skylight illumination at the window surface 20 per cent. The roof between B and E will be in sunlight, and between E and A it will be illuminated by skylight only. We may therefore disregard the small quantity of light this latter can reflect to the under side of A'B'. The skylight from a spherical lune near the southern horizon for which θ averages about 30° will be cut off from BE by the adjacent saw tooth. From Table XIII we estimate that the illumination from skylight will be decreased by about 27 per cent. Therefore, the available sky illumination on the north-sloping window surface is 471×0.80

ᵃLoc. cit., p. 620.

$= 377$, and on the south-sloping roof surface it is $1,100 \times 0.73$ $= 800$ foot-candles. The total illumination on the south-sloping roof surface is 7,950 foot-candles, and its ratio to the illumination on the window surface is $7,950/377 = 21$. Substituting this value for 4, we obtain for the relative values of (1) and (2) 14.6 and 32, respectively. Or, if we suppose the sky to be covered with thin clouds, or partly covered with white clouds the values become 11.6 and 16.

Apparently, therefore, for clear-sky conditions, or even for the most usual sky conditions, when thin clouds or haze, or scattered white clouds are present, most of the daylight received through a saw-tooth-roof window will be from the reflection of skylight and sunlight from the roof of an adjacent saw tooth. In cloudy weather, however, nearly all the light will be received from the northern sky.

No attempt has been made to express the illumination intensity at the working space in absolute units. In order to do so, it is necessary to know the average of the solid sky angles S_1 and S_3, and of S'_2 and S''_2 (Figure 28); the brightness of the sky included in each of these angles, from which the sky illumination may be computed; the solar illumination intensity on the roof surface BE; the solid roof angle t_1 and the average of the solid roof angles t'_2 to t''_2; the coefficients of reflection of the surface of the ceiling A'B', and the roof, BE; the solid angle subtended by the ceiling at the working space, and the angle at which light is incident at the roof or the ceiling, and is received at the working space either directly or by reflection from the outside roof or the inside ceiling.

Of the above factors the brightness of the sky and the intensity of the solar illumination are given with reasonable accuracy in this paper for latitude 42° N. The remaining factors depend upon the design of the saw-tooth roof and its window openings, and must be determined for each individual case.

During the winter months, in a smoky city like Chicago, disregarding the probable decrease in the reflecting power of the ceiling of A'B', and the roof surface BE, the absolute values of (1) and (2) can not exceed two-thirds and one-half, respectively of their values in a comparatively smokeless region.

ABSTRACTS

In this section of the TRANSACTIONS there will be used (1) ABSTRACTS of papers of general interest pertaining to the field of illumination appearing in technical journals, (2) ABSTRACTS of papers presented before the Illuminating Engineering Society, and (3) NOTES on research problems now in progress.

NOTE ON THE INTEGRATING SPHERE REFLECTOMETER

BY JOHN W. T. WALSH[*]

A convenient form of absolute reflectometer consists of an Ulbricht sphere with a small aperture which is covered by the surface under test so that this receives an almost perfectly diffused illumination when a beam of light is projected on to the sphere wall. If the test surface be screened from the illuminated patch on the sphere, the flux received by the test surface is b lumens per unit area if the brightness of the general surface of the sphere is b lamberts. It follows that the brightness of the test surface is ρ b lamberts *if this surface is perfectly diffusing,* ρ being its reflection factor. Thus ρ may be found by comparing the brightness of the test surface with that of the sphere wall.

It is clear that the same result holds when the test surface is a perfect mirror, for in this case the brightness of the surface is that of an image of a portion of the sphere wall seen in the mirror. The conclusion has been drawn from this result that the method is universally applicable, whether the test surface be a perfect diffuser or not,[1] but a closer consideration of the problem will show that this cannot be the case.

Let A, Fig. 1, be an element of surface which is not perfectly diffusing and let IA represent a ray of light incident at the angle *i* in the plane of the paper. Let AM (also in the plane of the paper) be the direction of specular reflection of IA and let AE be any other direction (not necessarily in the plane of the paper).

[*] M. A., M. Sc., F. Inst. P., The National Physical Laboratory, England.

[1] C. H. Sharp and W. F. Little, Illuminating Engineering Society., N. Y., TRANS., 15, 1920, p. 804. E. Karrer, *Opt. Soc. Am., J.*, 5, 1921.

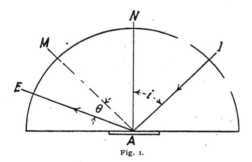

Fig. 1.

Then the brightness of A when viewed along AE is proportional
to the illumination of A due to the light IA and to some function
of the angles NAI, MAE and NAE.

It is legitimate to consider the special case when this function is independent of one of these angles,[2] and for convenience
in calculation it will be assumed that this function depends on
the angles NAI and MAE only and may therefore be written
$f(i, \theta)$ where $i = $ NAI and $\theta = $ MAE. This assumption implies
that the brightness of the surface is the same in all directions
equally inclined to the direction of specular reflection of the
incident light.

Now referring to Fig. 2, let A be an element of surface, and
let AY be the normal. In this case let AE, the direction from
which the surface is viewed, be in the plane of the paper, and
let IA, the direction of the incident light, make an angle ϕ with
the plane of the paper. Also let the projection of IA on the
plane of the paper make an angle ψ with AY. Then the angle
EAI is given by $\cos$ EAI $= \cos \phi \cos (\psi + \epsilon)$.

Also, it is clear that if AI_1 be the direction of specular reflection
of IA,

$$\cos \theta = \cos \text{EAI}_1 = \cos \phi \cos (\psi - \epsilon) \dots \dots (1)$$

The angle of incidence of the light is given by

$$\cos i = \cos \text{YAI} = \cos \phi \cos \psi \dots \dots \dots \dots (2)$$

so that if the surface A receive perfectly diffused flux, (i. e., flux
of which the density is the same in all directions) the illumina-

[2] If the function is independent of all three the surface is a perfect diffuser.

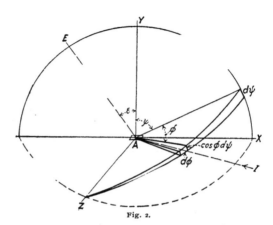

Fig. 2.

tion E_i due to the light incident within the cone whose solid angle is $\cos \phi \, d\psi \, d\phi$ and whose axis is IA (see Fig. 2) may be written

$$E_i = F \cos i \cos \phi \, d\psi \, d\phi$$
$$= F \cos^2 \phi \cos \psi \, d\psi \, d\phi \dots \dots \dots (3)$$

The brightness of A in the direction AE, due to this illumination is, then $B_i = Ff(i, \theta) \cos^2 \phi \cos \psi \, d\psi \, d\phi$

From a consideration of Fig. 2 it will be seen that the total brightness of A in the direction AE is given by

$$B = 2F \int_{\phi = 0}^{\phi = \frac{\pi}{2}} \int_{\psi = -\frac{\pi}{2}}^{\psi = +\frac{\pi}{2}} f(i, \theta) \cos^2 \phi \cos \psi \, d\psi \, d\phi \quad \dots \dots (4)$$

If the surface be a perfect diffuser, $f(i, \theta)$ is a constant ($= \rho$) and the above expression reduces to $\pi \rho F$ as is to be expected since the total flux incident on the surface per unit area, $i.\,e.$ the total illumination, is πF.

If, however, the surface be not a perfect diffuser, $f(i, \theta)$ is not independent of i, and θ, i.c. of ϵ, and *the brightness of the surface will depend on the angle of view*, and will not be independent of it as has been generally assumed hitherto.

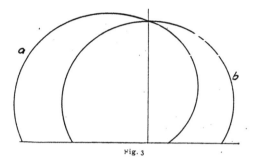

Fig. 3

The degree of importance of this effect in particular cases may be gauged by assuming a special form for $f(i, \theta)$. This function should clearly have a maximum value when $\theta = 0$ and it should not become negative for any value of θ.

It also seems desirable to assume a form for $f(i, \theta)$ which leads to a value for the reflection factor of the surface (ratio of emitted flux to flux received) that is independent of the angle of incidence of the light. The function $m(1 + k \sec i \cos \theta)$ fulfills these conditions, i being the angle of incidence and θ, as before, the angle between the direction of view and the direction of specular reflection.

The polar curve of such a surface, when $i = 45°$, and $k = 0.7$, (a) in the plane of the incident light and (b) in a perpendicular plane, is shown in Fig. 3, from which it will be seen that this is a very reasonable distribution to assume for such a surface as that of slightly glazed paper.[3]

It is easily shown that with this form of $f(i, \theta)$ the reflection factor is independent of i for the flux incident at the surface per unit area is clearly $F \cos i$ while the total flux emitted is (Fig. 2)

$$2\int_{0}^{\frac{\pi}{2}} \int_{-\frac{\pi}{2}}^{+\frac{\pi}{2}} (F \cos i / \pi)\, m\ (1 + k \sec i \cos \theta)\, \cos \phi \cos \psi \cos \phi\, d\psi\, d\phi$$

$$=(2Fm/\pi) \iint \left\{ \cos i \cos^2 \phi \cos \psi + k \cos^3\phi \cos \psi \cos(\psi + i) \right\}\, d\psi\, d\phi$$

[3] See A. P. Trotter, Ill. Eng. (London) 12, 1919, p. 260.

$$=(2Fm/\pi)\left\{\frac{\pi}{2}\cos i + k\,\frac{\pi}{3}\cos i\right\}$$

so that the reflection factor is m $(1 + \frac{2}{3}k)$.

Now if a surface of this kind is viewed in a constant direction AE making an angle ϵ with the surface, it is clear from Fig. 2 that the brightness in this direction, due to the flux received by the surface in the direction IA is, if the incident flux be perfectly diffused (*i. e.* of the same density F in all directions), represented by

$$F\cos i\, m(1 + k\sec i\cos\theta)\,\cos\phi\cos\psi\,d\psi\,d\phi$$
$$\text{where } \cos i = \cos\phi\cos\psi$$
$$\text{and } \cos\theta = \cos\phi\cos(\psi - \epsilon).$$

It follows that the total brightness in the direction AE is

$$2\int_0^{\frac{\pi}{2}}\int_{-\frac{\pi}{2}}^{+\frac{\pi}{2}}(Fm/\pi)\left\{\cos^2\phi\cos\psi + k\cos^2\phi\cos(\psi - \epsilon)\right\}d\psi\,d\phi$$

$$= (2Fm/\pi)\left\{\frac{\pi}{2} + k\,\frac{\pi}{2}\cos\epsilon\right\}$$

$$= Fm\,(1 + k\cos\epsilon).$$

The extreme values for the brightness of the surface are thus in the ratio of $1 : (1 + k)$ at glancing aspect and normal aspect respectively. The true value of the diffuse reflection factor of this surface is, actually, the ratio of the total reflected to the total incident flux. The former is clearly

$$Fm\int_0^{\frac{\pi}{2}}(1 + k\cos\epsilon)\,2\pi\sin\epsilon\cos\epsilon\,d\epsilon.$$

$$= \pi Fm\,\left(1 + \frac{2}{3}\,k\right)$$

while the incident flux is πF so that the value of ρ is $m(1 + \frac{2}{3}k)$ as before. This is the value obtained by a measurement of brightness at the angle $\cos^{-1}(\frac{2}{3})$. A measurement of the normal brightness gives a value of ρ which is in error by no less than 20 per cent for the case when k = 1.[4]

[4] The consistency in the value of ρ obtained by Dr. Sharp and Mr. Little (*loc. cit.*) in the case of polished opal glass is, presumably, due to the fact that they made all their measurements at the same value of ϵ.

In view of the very large errors given by the method in the case postulated, where the departure from perfect diffusion is by no means extreme, it seems extraordinary that a correct result should be obtained for a true mirror. Further consideration *a priori*, however, shows that in this case, although at first sight, the surface behaves according to rule, in reality it does not do so except when certain conditions, not usually taken into account are fulfilled. It is clear, for instance, that unless the *reflection factor of the mirror be the same for all angles of incidence of the light*, different values of ρ will be obtained for different angles of aspect and these will not, in general, be equal to the ratio of the flux reflected to the flux incident on the surface.

It is interesting to note that the source of error just investigated provides a complete explanation of the low values of reflection factor obtained in the case of magnesium carbonate by the use of the Nutting reflectometer. It is not unreasonable to assume that for this substance $f(i, \theta) = m(1 + 0.2 \sec i \cos \theta)$ since this gives a reduction of brightness of only 5 per cent when $\epsilon = 45°$ and the light is incident normally.[5] In this case the expression for B becomes

$$mF (1 + 0.2 \cos \epsilon)$$

while $\rho = (1 + \frac{2}{15})m = 1.13m.$

If $\epsilon = 75°$ $B = 1.05 \ mF$

so that the measured reflection factor is too small by about 8 per cent and this is of the same order as the discrepancy actually found.[6]

CONCLUSION

In measuring the diffuse reflection factor of any except a perfectly diffusing surface it is necessary to ensure that the measurement made gives a correct value for the whole luminous flux reflected from the surface.

It is not justifiable to assume that the brightness of a surface which is not perfectly diffusing is the same at all angles of view when the illumination is diffused.

[5] See F. Thaler, *Ann. d. Phys.*, 11, 1903, p. 996.

[6] A. H. Taylor, Bureau of Standards, Bull. 16, 1920, p. 425.

SOCIETY AFFAIRS

SECTION ACTIVITIES

NEW YORK Meeting—April 19, 1923

The April meeting of the New York Section was held on Thursday afternoon, April 19, 1923, at five-thirty, in Room 2 of the Engineering Societies Building, 29 West 39th Street. The subject "The Application of Artificial Daylight to the Hospital Operating Room" was presented by Mr. Norman Macbeth of the Artificial Daylighting Company, New York City. A demonstration and display of daylighting luminaires was presented in conjunction with this talk. The Kny-Scheerer Corporation also displayed their operating room luminaire. This device showed the method of an auxiliary gas light which might be utilized in the event of any interruption in the electric service.

At five o'clock prior to the meeting a buffet luncheon was served to the members and guests. The time of the meeting was selected as a matter of experiment and judging from the size of the attendance of one hundred members and guests, the idea met with approval and will no doubt be tried at other section meetings.

CHICAGO Meeting—April 26, 1923

A noon-day meeting of the Chicago Section was held on April 26 and began with a luncheon at the Chicago Engineers' Club at twelve o'clock. Following the luncheon, the meeting was continued in the rooms of the Western Society of Engineers, where Mr. C. A. Atherton, National Lamp Works of Cleveland, delivered his paper, "Some New Ideas in Electric Sign Advertising," which was illustrated by means of lantern slides. The paper was very ably presented and was followed by an interesting discussion. The meeting adjourned at 3:15 P. M. There was an attendance of sixty members and guests.

PHILADELPHIA Meeting—April 10, 1923

At the meeting of the Philadelphia Section held at the Engineers' Club on April 10, Mr. G. Bertram Regar of the Philadelphia Electric Company presented a paper, "Selling the Idea of Better Lighting," which had been read before the convention of the Pennsylvania Electric Association by him last September. While primarily a central station subject the paper contained considerable material that was of interest to the lighting industry.

Mr. Regar compared the lighting load with the appliances and power load and stated that the former had not reached as high a standard as the latter. Electrical appliances have become a necessity in the home but the lighting installation is frequently subject to criticism. Poor lighting is not so much a matter of stinginess as it is poor planning and carelessness.

Lighting has three basic principles: First—*Utilization*. Successful business methods demand that the lighting be good, and good lighting is efficient, because the predominating factors have been treated in a scientific way. Second—*The Physiological Effect*. With the recognized advantages, both from the commercial and the industrial standpoints of a high intensity of illumination, if the eyes of our people are to be conserved, then careful consideration must be given the subject of proper direction, diffusion and color of lighting. Third—*Psychological Effect*. With the advancement of civilization, we have naturally been educated to the comforts and refinements of our present-day lives, and the artistic effect of lighting, therefore, plays a prominent part, as to whether it is pleasing or distasteful.

Mr. Regar then considered the aspects of residence and store lighting stating the characteristics which it should have and bringing out the places where improvements can frequently be made. Under the heading of industrial lighting he pointed out the necessity for insisting that this be sufficient not only to avoid accidents but to increase plant output.

In an appeal for better lighting the speaker stated that an intensive campaign for better lighting should be inaugurated; setting forth the arguments that good lighting brings about increased production, aids accident prevention, conservation of vision, improved aesthetic effects, happiness, health and many other advantages. It should be done nationally by articles in trade publications, and a concentrated effort by national advertisers. Automobile Associations should be asked to co-operate with a view of getting better highway lighting. State authorities should be asked to adopt school and factory lighting codes. City officials should be made to realize that good street lighting is a means of protection.

The paper was illustrated by lantern slides showing interiors of stores with examples of good and bad lighting. An interesting discussion was held and the meeting was attended by thirty members and guests.

MICHIGAN Meeting—May 7, 1923

The organization meeting of the new Michigan Chapter was held on the evening of May 7th in Detroit. Twenty-two members were present and a very gratifying interest was shown.

The following officers were elected: Chairman, James M. Ketch; Secretary-Treasurer, A. L. Lent; Messrs. Clarence Carson, H. H. Higbie, Ivan N. Kirlin, Harold Shaw and George Wagschal were elected to the Board of Managers.

After the regular business the probable activities of the new Chapter were discussed and they comprised such projects as State School, Industrial, and Headlighting Codes, Cooperation with Electrical Contractors, Architects and Consumers in the use of better lighting, and research standardization of special lighting problems in the automobile industry.

COUNCIL NOTES

ITEMS OF INTEREST

At the meeting of the Council, April 12, 1923, the following were elected to membership:

Four Members

GUNNISON, FOSTER, Cox, Nostrand & Gunnison, Inc., 337 Adams Street, Brooklyn, N. Y.

KILBY, KARL E., Coleman Lamp Co., 222 N. St. Francis Ave., Wichita, Kan.

KNIGHT, A. R., University of Illinois, Urbana, Ill.

WHEELER, HARRY E., 45 E. 55th St., New York, N. Y.

Twenty-Three Associate Members

ALEXANDER, L. M., University of Cincinnati, Cincinnati, Ohio.

BLACKMORE, CHARLES T., Stone & Webster, Inc., 147 Milk St., Boston, Mass.

CAMPBELL, CLARENCE J., Westinghouse Lamp Co., 165 Broadway, New York, N. Y.

CLARKE, FRANK A., Westinghouse Elec. & Mfg. Co., South Bend, Ind.

COBBY, E. V., Pacific Telephone & Telegraph Co., 807 Sheldon Bldg., San Francisco, Cal.

DOANE, FRANCIS H., International Correspondence Schools, Scranton, Pa.

DOLBIER, F. VAN BUREN, Philadelphia Electric Co., 132 S. 11th St., Philadelphia, Pa.

DOYLE, WALTER H., Central Hudson Gas & Electric Co., 129 Broadway, Newburgh, N. Y.

ENGELFRIED, HENRY O., Consolidated Elec. Lamp Co., 88 Holton St., Danvers, Mass.

HECKER, LOUIS M., Commercial Light Co., 127 N. Dearborn St., Chicago, Ill.

HERRMANN, W. S., Westinghouse Elec. & Mfg. Co., 1535 Sixth St., Detroit, Mich.

JAQUET, GEORGE E., Garden City Press, Gardenvale, Quebec, Canada.

KINSEY, FREDERICK S., Westinghouse Lamp Co., 165 Broadway, New York, N. Y.

LEWIS, FRANK L., Ivanhoe-Regent Works, G. E. Co., 108 S. 59th St., Philadelphia, Pa.

MARTIN, ALLEN J., 272 Engineering Bldg., University of Michigan, Ann Arbor, Mich.

MUDGETT, GUERNSEY F., Westinghouse Elec. & Mfg. Co., South Bend, Ind.

THOMPSON, H. R., Westinghouse Elec. & Mfg. Co., 515 Hanna Bldg., Cleveland, Ohio.

TRAWICK, SAMUEL W., JR., New Orleans Public Service, Inc., 201 Baronne St., New Orleans, La.

WEILER, EDWARD W., Edison Lamp Works, Harrison, N. J.

WEINER, PAUL S., Van Dyk & Reeves, Inc., 167-41st St., Brooklyn, N. Y.

WINETSKY, MICHAEL, Public Service Electric Co., 71 Murray St., Elizabeth, N. J.

WRIGHT, HOWARD L., National X-Ray Reflector Co., 750 Prospect Ave., Cleveland, Ohio.

ZIMMERMAN, J. HAROLD, Westinghouse Lamp Co., 165 Broadway, New York N. Y.

Two Sustaining Members

FRANKLIN ELECTRIC LIGHT CO., Avenue A, Turners Falls, Mass.
C. E. Bankwitz, Official Representative.

GAS & ELECTRIC IMPROVEMENT CO., 77 Franklin St., Boston, Mass.
Philip B. Jameson, Official Representative.

One Re-Election to Associate Membership

HALLOCK-GREENEWALT, MARY, 1424 Master St., Philadelphia, Pa.

Two Transfers to Full Membership

HANLAN, JAMES P., Public Service Gas Co., 80 Park Place, Newark, N. J.

LYON, HOWARD, Welsbach Co., Gloucester, N. J.

The General Secretary reported the death, on March 9, 1923 of one associate member, FRANK L. SAMPLE, National X-Ray Reflector Co., Boston, Mass.

CONFIRMATION OF APPOINTMENTS

As Chairman of the General Convention Committee—W. D'A. Ryan.

As members of the Committee on Membership—H. M. Camp and L. B. Johnson.

As an Advisory Committee to co-operate with the N. E. L. A. in arranging a course of instruction for men in lighting departments of various central stations—H. H. Higbie, Chairman; E. M. Alger, Earl Anderson, C. C. Munroe and T. W. Rolph.

NEWS ITEMS

BULLETIN OF THE COMMITTEE ON LIGHTING LEGISLATION

Massachusetts Industrial Lighting Code:

The Massachusetts Department of Labor and Industries has voted to adopt a lighting code for factories, workshops, manufacturing and mechanical establishments for the purpose of protecting employees from accidents and eye strain due to inadequate or faulty lighting. At present the code is not mandatory but the Labor Department has issued notice that the code will become mandatory on Jan. 1, 1924.

The Code is divided into two parts. Part I consists of six rules as follows: general requirements, intensity required, protection from glare—shading of lamps, distribution of light, entrance and exit lighting, classification of intensity grades. Part II contains notes and recommendations.

Both Part I and Part II follow very closely the lines laid down in the New York State code of lighting factories and mercantile establishments as amended May 1, 1922, except that in Part I the Massachusetts code contains an additional rule on entrance and exit lighting, based partly on the exit and emergency lighting rule of the I. E. S. (American Standard) Code, as follows:

Entrance and Exit Lighting: Lighting shall be provided in all stairways and exits of factories and in all the passageways appurtenant thereto, independent of the regular lighting of the working space. The lighting circuits for the stairways and exits should extend inside any working room where twenty or more persons are regularly employed, so as to light the immediate entrance to the stairway or exit. Such lights shall be served from a source not subject to failure because of the failure of the circuit fuses for the room lighting, and preferably from an independent connection extending back to the main service entrance for the building. In case unusual danger may exist on account of type of building, nature of the work, crowded conditions, or lack of suitable exit space, the Commissioner of Labor and Industries may require such lighting to be further extended within the working space; and that independent service shall be ensured by connection to a separate source of supply without or within the building.

The intensity rule of the Massachusetts code does not go as far as that of the New York State code in that the former does not contain a table of minimum intensities for detailed industrial operations and processes. On this point the Massachusetts code states that the "assignment of industrial operations (intensity grades) shall be determined by the Commissioner of Labor and Industries subject to review by the Department upon application of anyone affected thereby."

The glare rule follows the same form as that used in the earlier state codes and is open to the criticism of being too general. It is to be regretted that this rule was not made more specific to conform with the glare rule in the I. E. S. code (American Standard).

In the preparation of the code the Department of Labor and Industries had the assistance of a state committee of which Dr. Louis Bell was chairman. Dr. Bell also represented the I. E. S. Committee on Lighting Legislation.

SECTIONAL COMMITTEE ON I. E. N. ORGANIZED

Thirteen men, consisting of six representatives of producers, three representatives of consumers, and four representatives of general interests, constitute the personnel of the Sectional Committee on Illuminating Engineering Nomenclature and Photometric Standards, one of the projects officially before the American Engineering Standards Committee.

The Illuminating Engineering Society has been named sponsor for this project. The men who constitute this committee, and the organizations which they represent, follow: American Gas Association, W. J. Serrill; American Institute of Electrical Engineers, A. E. Kennelly; Bureau of Standards, A. S. McAllister; Illuminating Engineering Society, Howard Lyon and G. H. Stickney; National Committee of International Commission on Illumination, Louis Bell; National Committee of International Electrotechnical Commission, C. O. Mailloux; National Council of Lighting Fixture Manufacturers, E. C. McKinnie; National Electric Light Association, C. H. Sharp; Optical Society of America, E. C. Crittenden; American Physical Society, E. P. Hyde; Individuals, G. A. Hoadley and M. Luckiesh.

STANDARDIZATION OF TRAFFIC SIGNAL COLORS

Forty-two men, representing the manufacturers and users of traffic signals, federal and state governmental departments, associations interested in the prevention of traffic accidents, and representatives of the general public, are now at work on the drafting of a national code on the proper colors for traffic signals, which it is expected will not only cut down the annual loss of life through traffic accidents, but will eliminate many of the existing irritations to motorists and to the operators of steam and electric railways.

This work is being carried on under the auspices of the American Engineering Standards Committee whose approval of a code or standard insures its ultimate acceptance and observance throughout the country. The American Engineering Standards Committee is composed of seven departments of the U. S. Government, the principal technical, industrial and engineering societies and individual business concerns interested in standardization.

The sectional committee drafting this code is made up of seven representatives of the manufacturers of traffic signals, nine representatives of the purchasers of such equipment, three representatives of the users of traffic signals, twelve representatives of governmental bodies, five technical specialists, and six insurance representatives.

Mr. Charles J. Bennett, State Highway Commissioner of Connecticut, who represents the American Association of State Highway Officials, has been selected chairman of the sectional committee. Dr. M. G. Lloyd of the U. S. Bureau of Standards, who is the representative of both the Bureau and the American Society of Safety Engineers, is vice-chairman, and Mr. Walter S. Paine, Research Engineer of the Aetna Life Insurance

Co., who is the representative of the National Safety Council is secretary of the sectional committee.

PLAN BETTER LIGHTING IN SCHOOLS TO SAVE THE EYES OF YOUTH

The lighting, building, education, health, and social agencies of the country have joined hands in an effort to develop a nationally accepted code for school lighting which will correct the conditions partially responsible for the defective vision of 10 to 20 per cent of the school children.

The formulation of this code is being carried on under the auspices of the American Engineering Standards Committee, a federation of national organizations, government departments, and other agencies interested in standardization, whose official approval of a standard or code insures its ultimate acceptance by the principal interests concerned.

The conditions that make such a code necessary have been summarized as follows by a committee of the Illuminating Engineering Society:

"Examinations of thousands of school children, extending over many years, have shown that from 10 to 20 per cent of the children suffer from defective vision, the result largely of continued use of the eyes in close work under unhygienic conditions. It is well established that defective vision is progressive and is therefore found to a larger extent among the older children.

"Many of the factors contributing to defective vision of children are closely connected with school life, and to this extent the causes are preventable and may be removed. Modern educational methods impose severe requirements upon the immature eyes of children and create the need for the very best working conditions.

"It is therefore essential that the lighting of school buildings, both natural and artificial, should be of the best design. The status of the art of illumination is so well established that it is entirely feasible and practicable to prevent eye-strain by the proper design of school buildings and the installation of suitable lighting equipment.

"Economically, it is found that, in general, children with defective vision are retarded in their progress in school life, and also enter upon their life work seriously handicapped. It is right, therefore, that a state should concern itself to protect and conserve the vision of children from an economical, as well as a humanitarian standpoint."

The American Engineering Standards Committee has appointed the American Institute of Architects and the Illuminating Engineering Society joint sponsors for the code on school lighting. The sponsors will organize a representative sectional committee to formulate the code and will provide for the publication of the code after it has been approved by the A. E. S. C.

The organizations that are already co-operating in this work are: American Gas Association, American Institute of Architects, American

Institute of Electrical Engineers, American Public Health Association, American Society of Safety Engineers, American Medical Association, Illuminating Engineering Society, National Committee for the Prevention of Blindness, National Bureau of Casualty and Surety Underwriters, National Education Association, U. S. Bureau of Education, Department of Interior, U. S. Public Health Service, Treasury Department, U. S. Women's Bureau, Department of Labor, U. S. Bureau of Standards.

GRAND ILLUMINATION AT WASHINGTON PLANNED FOR MYSTIC SHRINERS

Brilliant illumination and showers of light will flood the capital city during the entertainment of more than 300,000 Shriners who will come from all parts of the United States to Washington in June. A Garden of Allah will be erected in front of the White House, and a blaze of light will usher in the street pageant of the Masons. Committees of electricians are now working out the details for the gala electrical effects.

Many special electrical settings will be installed upon the wide elm-lined avenues of the nation's official headquarters, and the parade of costumed Shriners, uniformed patrols, bands, and dignitaries will pass through a flood of light.

Pennsylvania avenue, down whose historic lanes Presidents have ridden to attend their own inaugural ceremonies for many generations, and whose wide street spreads toward the Capitol, lit by batteries of searchlights, will be brilliantly illuminated for this convention.

CONVENTION PAPER INCREASES WINDOW LIGHTING

It is very gratifying to note how the disinterested work of the I. E. S. is helping to increase the lighting load of the central stations. The following reference to the Sturrock-Shute paper presented at the last convention in Swampscott appeared in the *Electrical World* of April 21, 1923.

> "The connected load in window lighting of a large department store served by the Fitchburg (Mass.) Gas and Electric Light Company was increased nearly 200 per cent as a result of presenting the customer with a paper on show-window lighting which was read at the last convention of the Illuminating Engineering Society. After offering the paper to the customer for perusal, the central-station sales department and the department store's window decorator conducted a series of tests which proved to the user's satisfaction that it would pay to use a higher intensity in his windows than had formerly been employed."

1923 CONVENTION AT LAKE GEORGE

At recent meetings of the Council the time and place of the Seventeenth Annual Convention of the Illuminating Engineering Society have been decided, and members of the 1923 Convention Committee have been appointed. The convention headquarters will be at the Fort William Henry Hotel at Lake George, N. Y., and September 24 to 28 are the days chosen. The activities of the committee are in charge of the following: Mr. W. D'A. Ryan, Chairman; Mr. H. W. Peck, Vice-Chairman; and Mr. H. E. Mahan, Secretary.

Plans are now being formulated and full publicity will be given to the members of the Society, and present indications point to a very successful convention. It is hoped that a large number of the members will reserve these dates and make every effort to attend the convention.

THE MEMBERSHIP DRIVE

Under the able leadership of Mr. G. Bertram Regar, Vice-President, the membership drive for the past eight months has been the most successful in the history of the Society. Reference to the Society Affairs in the TRANSACTIONS will show that nearly two hundred new members have been elected.

The splendid work already accomplished can be greatly aided by a bit of *personal effort* on the part of *each individual member.* There must be men among your associates who should be members in the I. E. S. Take up this challenge and send to the chairman of the committee names of prospective applicants; or, better still, send the signed application with check for entrance fee and current dues. Applicants for the grade of *associate member* for the balance of the fiscal year should remit $3.75 for dues and $2.50 for entrance fee; for the grade of *member*, $7.50 for dues and $2.50 for entrance fee.

Application blanks and membership literature can be secured from members of the committee or from the General Office.

CHICAGO ELECTRICIANS' APPRENTICES COURSE

A series of talks under the supervision of Mr. J. L. Stair, National X-Ray Reflector Company, was given during the latter part of March before apprentice classes consisting of six hundred construction electricians and stage electricians at the Washbourne Continuation School, Chicago.

The apprentices are required by their Unions to attend school one day out of every two weeks for which they receive regular pay from their employer. The course continues for four years, and after the course is over, the apprentices are required to take examinations to become journeymen and receive cards from the Local Union.

In the series of talks, historical facts concerning the development of artificial lighting were presented, together with some of the fundamental principles in lighting practice, and examples of modern installations. The subject was treated in a purely educational way with demonstrations and lantern slides.

PERSONALS

Mr. J. L. Wolf, formerly secretary of the Lighting Fixture Dealers' Society of America, and secretary of the Cleveland Electrical League, is now in the retail fixture business as the Wolf Lighting Company.

Mr. George H. Stickney, Past-President of the I. E. S., has recently been honored by an election to membership in the Alpha Chapter of the Sigma Xi fraternity at Cornell University. Mr. Stickney has been identified with work in illuminating engineering since his graduation from Cornell in 1896.

SUGGESTIONS TO AUTHORS OF PAPERS

For

PRESENTATION AT THE ANNUAL CONVENTION

Final Acceptance Date

The manuscripts of papers to be preprinted for the Annual Convention must be in New York by *August 1* if authors wish to see proof before publication. Papers will be accepted from this date until *August 15*, but no proofs of them will be submitted to the authors for approval.

Synopsis

A short synopsis (200 words or less) should accompany each paper. This is required for publication with the paper and also for publication in advance notices of the convention.

Illustrations

All illustrations for half tone cuts should be submitted as original photographs upon glossy paper.

Diagrams should be drawn upon white paper, or blue cross-section paper which will be supplied by the Committee upon request. Blue lines upon the cross-section paper do not reproduce photographically and, therefore, all lines which are to be shown in the finished graph should be inked in. Blue prints and photostat prints do not make good cuts.

Captions and descriptive matter should be typewritten and attached to the photographs or diagrams.

Lettering upon the face of the diagram should be as simple as possible, and large enough to reproduce when the diagram is photographed and reduced to one-quarter size.

Copyright

Papers presented before the Society are considered to be the property of the Society and are copyrighted upon publication in the Society's TRANSACTIONS. Permission may be given by the Committee to reprint any paper in part or in full subsequent to presentation provided proper credit is given to the author and to the Society.

ILLUMINATING ENGINEERING SOCIETY
Committee on Editing and Publication,
29 West 39th Street, New York, N. Y.

ILLUMINATION INDEX

PREPARED BY THE COMMITTEE ON PROGRESS.

An INDEX OF REFERENCES to books, papers, editorials, news and abstracts on illuminating engineering and allied subjects. This index is arranged alphabetically according to the names of the reference publications. The references are then given in order of the date of publication. Important references not appearing herein should be called to the attention of the Illuminating Engineering Society, 29 W. 39th St., New York, N. Y.

IN MEMORIAM
Louis Bell
1864-1923

Resolution Adopted by the Council of the
Illuminating Engineering Society.
June 28th, 1923

The Officers and Council of the
Illuminating Engineering Society
hereby record their sense of loss in the
passing of their friend and co-labourer
Dr Louis Bell. His death removes
an active, enthusiastic and effective
contributor to the science and art of
illumination; a pioneer in this as
in other fields of scientific endeavor;
a leader in establishing illuminating engineering as a distinct specialty; an authority of recognized distinction in this field; and the third
President of this Society.

The many friends of Dr Bell
will long cherish the memory of the
cultured gentleman whose rare blend
of comprehensive knowledge, never
failing humor and loyal comrade-
ship endeared him to all who were
privileged to know him.

For the Council

Ward Harrison
President

Samuel G. Hibben
General Secretary.

LOUIS BELL
1864-1923

TRANSACTIONS
OF THE
ILLUMINATING ENGINEERING SOCIETY

Vol. XVIII	July, 1923	No. 6

In Memoriam
LOUIS BELL

BY A. E. KENNELLY

THE late Dr. Louis Bell was so well known to members of the Illuminating Engineering Society, as a Past-President, leader and author, that a brief account of his personality, career and accomplishments is sure to be welcome to many.

Louis Bell was the grandson of Governor Samuel Bell of New Hampshire, who represented that state for two terms in the U. S. Senate at Washington, D. C. His father was General Louis Bell, who fought in Grant's army during the Civil War. General Bell, when 28 year old, commanded the attack on Fort Fisher, and was killed in action during the last few minutes of the final successful rush on the trenches, just before the close of the war. The news of his death came as a terrible shock to his young wife, in the family home at Chester, N. H., and she survived him only six months, leaving little Louis, then one year old, (born Dec. 5th, 1864), and his sister four years older. The two children were brought up at Chester, by their grandmother, Governor Samuel Bell's widow. In remote village life, these two children were thrown much on their own resources, and mingled but little with the outer world. Louis was gifted mentally, and was given his bent in study, with full encouragement by his fond grandparent. The boy was an omnivorous reader, even at the age of eight, and was consumed with a thirst for knowledge of all kinds, but particularly for scientific knowledge. I used to wonder, in later years, how Dr. Bell became so well versed in the bible and bible history, until I learned that he was frequently allowed to choose, on Sunday mornings, between going down the hill to church, and staying home to learn a chapter of scripture by heart. Little Louis, with his swift memory, always chose the latter alternative.

When Louis was twelve years old, it was decided that he should be taken from his native village and sent to school at the Phillips Exeter Academy, Exeter, N. H. Thereafter, he only returned to Chester for brief and occasional visits; but he was very deeply attached to his childhood's home and he always desired to be carried back there after his death. He was rather a lonely lad at school, brilliant but living apart. His letters to his sister show how close was their attachment, and how much he depended on her guidance. When she died untimely at twenty-one, he was distracted with grief. He would never trust himself to speak of her in after life, even to his most intimate friends.

Louis entered Dartmouth, the college of his family, in 1880, and took his A. B. degree there is 1884. In his studies, he distinguished himself in chemistry and physics, and took post-graduate college work in physics. The Professor of Astronomy there allowed Louis to live at the College observatory, and to use the telescopes for observations, during two summers. Bell always loved a telescope and admired those who knew how to use them.

Bell was always a most loyal and devoted alumnus of Dartmouth, giving, throughout his life, unstintedly of his time and efforts to her aims. From Dartmouth, he went to Johns Hopkins University, where he studied under Rowland, and where he took his Ph.D. degree in 1888.

The subject of his thesis was "The Absolute Wave Length of Light." At the Manchester British Association Meeting in September 1888, Dr. Bell presented a joint paper by Prof. Henry A. Rowland and himself on "Explanation of the Action of a Magnet on Chemical Action."

After graduating at Johns Hopkins, Dr. Bell went to join, for a year, the faculty of Purdue University, as Professor of applied electricity, or what is now generally known as electrical engineering. This was a new chair at that time in Purdue. He organized the electrical engineering instruction there on a sound basis, and in 1890, he went to New York, as editor of the "Electrical World," which has since continued to be a prominent electrical engineering weekly journal. At that time, it was owned by W. J. Johnston. It was during this period that I first became acquainted with Dr. Bell as the editor of the "Electrical World." Ever after that interview, I was his admirer and friend. Bell was always a very clear and forceful writer. It was fortunate for the "Electrical World" that its editor, in Dr. Bell, was not only a prominent electrical engineer, but also a good scholar in English. He left the chief editorship in 1892, to enter the service of the General Electric Co., but he maintained literary connection with the "Electrical World", editorially and otherwise, up to his death. Anyone acquainted with the personal characteristics of his literary style, could readily discern articles by him. He had the gift of writing in such a manner as to compel attention and understanding. There was frequently a background of quaint humor in his articles, that tempted the reader to follow, even when the subject was dry and difficult.

It was about 1892, that the alternating-current motor began to enter commercial development. Dr. Bell, as chief engineer of the then newly organized power-transmission department of the General Electric Co., designed and installed some of the first polyphase power-transmission plants in the country. It was at Redlands in California, that one of these early induction-motor plants was installed, and Bell attended to it in person. One of his induction motors was direct connected to an air compressor of considerable power. Up to that time, alternating-current motors had not been notorious for powerful starting torque under load, and the superintendent of the Redlands compressor looked with some disdain on the new induction motor, which he considered much two small for the work. Bell, who had made thorough factory tests of the motor, knew what it could do when started under load, and impressed upon the superintendent the great importance of not closing the motor switch at the receiving end of the polyphase line, unless the outlets of the compressor were open. In his disdain of the new induction motor, however, the compressor man forgot these injunctions, and one day he closed the starting switch of the motor with the compressor outlets tightly shut. Bell was at the other end of the line at the time, near the generators, and saw the load come on at his switchboard. He could follow what happened at the motor by watching the instruments. The motor got under way for a short interval, evidently heavily loaded. Then there was a groan at the generator, and the automatic circuit-breakers opened at the motor end of the line. At the compressor plant, the motor had carried the compressor to such a point that, with no relief available, the crank shaft had broken, and the coupling bent. It took some time to repair the damages, but after that, the compressor men always treated that induction motor with great respect, and described it as "heap big medicine."

In 1892, Dr. Bell published (with Oscar T. Crosby) "The Electric Railway," which was a pioneer book on the subject. In 1896, he himself wrote "Power Transmission for Electric Railroads," and in 1897, "Electric Power Transmission," both of which books were soon widely known.

In 1895, he entered the profession of consulting engineering in Boston. He continued in this work for the rest of his lfe. At first, his work dealt mainly with electric power transmission, but afterwards he took up electric lighting more particularly, and later on he specialized in electric illuminating engineering. He published, in 1902, the text book "The Art of Illumination," which became a classic in that field. He lectured frequently on illumination both at Harvard University, the Massachusetts Institute of Technology and elsewhere. He was President of the Illuminating Engineering Society in 1908, and was an active member of several of its important committees. He continued to work for the Society up to the last. He was also, for about ten years, a Vice-President of the Illuminating Engineering Society of Great Britain.

Although he was always engaged in engineering work, he was keenly interested in pure science, especially in optics and optical astronomy. His book "The Telescope," published in 1922, is delightful to read, both from the practical and historical, as well as the purely optical view points. He was a member of the American Astronomical Society, and of the Board of Visitors of the Harvard Observatory.

He was a member of the American Academy of Arts and Sciences, was on its Rumford Medal Committee, and contributed various papers to its volumes, particularly on optics. A list is appended of his papers in the Proceedings of the Academy, and in those of the Illuminating Engineering Society, only. No attempt will be made here to give a list of his contributions to technical literature generally, because their name is legion. The fertility of his inventiveness is also shown by the fact that he took out more than forty patents, mainly in the applications of optics and illumination.

During the Spanish American War, Dr. Bell was Technical Officer of the Volunteer Electrical Corps, dealing with electrical harbor and Atlantic coast defenses from the Chesapeake Bay to the Canadian Border.

In the Great War, he was a member of the Advisory Committee on the Council of National Defense. He was specially occupied with telescopic sights for naval guns. He also developed a very practicable system of invisible signalling by ultra-violet light.

Dr. Bell married Sarah G. Hemenway of Somerville, Mass. in 1893, who with his son Louis and two grandchildren, survives

him. Bell was a home-making and home-loving man, shining
most brightly when acting as host. He was a witty and gracious
conversationalist, loving, for its own sake, a humor which never
degenerated into satire or sarcasm.

In his leisure hours, which were rather few and far between,
Bell enjoyed rifle and revolver shooting. He was always a good
marksman, and indeed maintained, in that way, a family tradition
of several generations. The Bells were always good shots. He
was a member, and had been the President, of the Massachusetts
Rifle Association. He also held at one time (New York, 1892) the
American amateur revolver-shooting championship. In 1903, he
was one of a team of the American Revolver Association which
won an international shooting match with a similar French team.
He would often spend his Saturday afternoons on the rifle range
at Walnut Hill, Mass., where he was very popular and had many
friends. He looked upon the rifle rather as an instrument of
scientific precision in target hitting, than as a deadly weapon. He
was, on the other hand, however, an eager student of military
history.

Some two years ago, his usually robust health broke down with
an attack of pneumonia and after that it failed rapidly; but his
courage was invincible. At the last, he had great hopes, although
in reality quite illusory hopes, of recovery. He passed away,
happily, during sleep, on June 14, 1923. His final resting place
is among the tombs of his kindred, at the quiet village cemetery
of Chester, in the farming country of New Hampshire, where
many of the old present homesteads were celebrated in colonial
times.

Five days after his death, the posthumous honorary degree of
Doctor of Science was awarded to him by his alma mater, Dartmouth College. Dr. E. M. Hopkins, the President, delivered on
that occasion the following allocution, which Dr. Bell's numerous
friends will heartily endorse:

"Louis Bell, teacher, writer, investigator; pioneer in the development of electrical transmission; authority in the fields of illumination and optics; fruitful in the work of making the accumulations of the laboratory of service to mankind; whose scientific achievements have .been combined with wide interest in literature and art, and who, as a writer and lecturer, has brought literary finish and quiet humor to the exposition of sound common sense; loyal alumnus of the college, and ever interested in her welfare".

PUBLICATIONS BY DR. LOUIS BELL IN THE PROCEEDINGS OF THE AMERICAN ACADEMY OF ARTS AND SCIENCES

"The Physiological Basis of illumination." Vol. 43, No. 4, 1907.

"Note on some Meteorological Uses of the Polariscope." Vol. 43, No. 15, 1907.

"On the Opacity of Certain Glasses for the Ultra-violet." Vol. 46, No. 24, 1910.

"On the Ultra-violet Component of Artificial Light." Vol. 48, No. 1, 1912.

"Types of Abnormal Color Vision." Vol. 50, No. 1, 1914.

"The Pathological Effects of Radiant Energy upon the Eye." (with F. H. Verhoeff and C. B. Walker), Vol. 51, No. 13, 1915.

"Ghosts and Oculars." Vol. 56, No. 2, 1920.

"Notes on the Early Evolution of the Reflector." Vol. 57, No. 4, 1921.

PUBLICATIONS BY DR. LOUIS BELL IN THE TRANSACTIONS OF THE ILLUMINATING ENGINEERING SOCIETY

"The Illumination of the Building of the Edison Electric Illuminating Company of Boston" (with L. B. Marks and W. D'A. Ryan). Vol. II, No. 7, 1907.

"Coefficients of Diffuse Reflections." Vol. II, No. 7, 1907.

"Response to the Address of Welcome to Convention." Vol. IV, No. 7, 1909.

"The Principles of Shades and Reflectors." Vol. IV, No. 8, 1909.

"Street Photometry."   Vol. V, No. 5, 1910.

"Photometry at Low Intensities."   Vol. VI, No. 7, 1911.

"Report of Committee on Progress" as Chairman of Committee. Vol. VI, No. 7, 1911.

"The Pathological Effects of Radiation on the Eye" (with F. H. Verhoeff). Vol. XVI, No. 9, 1921.

"Report of Sub-Committee on Glare" as Chairman.   Vol. XVII, No. 10, 1922.

REFLECTIONS

Big Lights for Night Flights

GREAT electric beacons capable of generating a beam of a half billion candlepower will guide pilots of transcontinental mail planes in their night flights.

One of these huge lights, to be the western terminal of a chain of beacons, is on its way today to Cheyenne, Wyo,. from the Brooklyn plant of the Sperry Gyroscope Company.

The non-stop flights are to be so timed that east of Chicago and west of Cheyenne the flying in either direction will be by daylight. Between those two cities will be what aviators call a "dark belt."

As the pilot passes over Chicago at sunset he will peer ahead, and some hundred miles distant will see the 500,000,000 candlepower beacon whirling around at a low angle at three revolutions a minute. He will direct his flight to the source of this light.

When he has passed it he will glance backward, guiding himself by the light he has left behind until he picks up the rays of the next series of beacons. Five of these lights will be spaced at 200-mile intervals, covering the dark zone, approximately 1,000 miles. They will be at Chicago, Iowa City, Iowa; Omaha, North Platte, Neb., and at Cheyenne.

Two lights are already in place. The apparatus shipped yesterday is the third, and two more will be dispatched soon. Less powerful beams set at stations about twenty-five miles apart along the air lane will illuminate fields where emergency landings may be made. *The Evening Mail,* June 13, 1923.

Electric Lamps Will Aid Sun in Growing of Laboratory Plants

A STATEMENT issued yesterday by Dr. William Crocker, research director of the new Thompson Institute for Plant Research which Col. William B. Thompson is establishing in Yonkers at a first cost of more than $500,000, gave further details of the plan by which powerful electric lamps are to supplement sunlight in growing plants. Eventually the institution is to cost $2,500,000.

"This new institution, with its gardens, greenhouses and laboratories," said Dr. Crocker, "is to be to plants and flowers what the Rockefeller Institute is to humanity. In other words it is to study and try to cure diseases of plants and flowers and other vegetation."

Dr. Crocker has had wide experience as head of plant physiology of the University of Chicago. On the board of trustees are the business group, consisting of Col. Thompson, C. C. Dula and Theodore Shulz, and the scientific group, including Prof. John Coulter of the University of Chicago, Prof. L. R. Jones of the University of Wisconsin, Dr. Raymond Foss Bacon, consulting chemist of New York City, and Dr. Crocker.

Col. Thompson said recently: "In another century this country must feed, clothe and shelter several hundred millions of people instead of 100,000,000 as now."

The laboratories are to be located on a nine acre plot at 1086 North Broadway, Yonkers.

In these chambers the temperature, humidity, carbon dioxide concentration of the air and the quality, intensity and duration of light can be adjusted at will and automatically maintained. In other words, the experimenter can manufacture climate and atmosphere to suit the needs of the experiment. *New York Herald*, June 24, 1923.

Non-Fading Light Needed to Preserve Colors

THERE are numberless examples of paintings which have faded beyond all semblance of their pristine selves. Very serious loss of color has taken place in many museum specimens which it was imperative to preserve as nearly as possible in their original state. In an endeavor to combat this deterioration Sir Sydney Harmer of the British Museum has within the last few years conducted a long series of experiments to determine the character of color changes produced by the exposure of pigments to sunlight or other light, the cause of these color changes in so far as it may be ascertainable, and particularly the possible remedies. The experiments were made by exposing pigmented and other objects under plain and tinted glasses with a control series under black glass, the exposure lasting for several years. The screening glasses covered a range from practically light window glass to a strongly absorbing glass similar to what used to be known as

Euphos ordinary yellow green, cutting off all the ultra-violet and most of the violet and blue. This last-named glass had decidedly too strong a color to be a satisfactory permanent housing. Years before this same ground had been, for the same good purpose, pretty thoroughly explored by Dr. Russell and Sir William Abney.

The first fact which was clearly established was that, on the whole, fading from the action of light does not ordinarily take place in the absence of oxygen and moisture. This immunity probably does not extend to some of the very fugitive dyes like erythrosine, but the relation is generally true that where there is no oxygen and no water there is no fading. Moreover, as might be anticipated from our general knowledge of photochemical reactions, the strongest fading effects were produced by direct sunlight, which was at least a score of times more injurious than the strong devised daylight. The control experiments, which freely let through a large part of the heat associated with the illumination, showed plainly that the effect of heat as such was substantially negligible.

Fading due to artificial electric light was much less than even with devised daylight—about one-sixth as much according to the experiments—the electric light used for the comparison being a powerful gas-filled incandescent lamp. Such a lamp is by no means free from ultra-violet radiation; hence there is the probability that a toning screen which would hold down the radiation on objects to something substantially within the limits of the visible spectrum would be of material assistance in still further preventing fading. Indeed, artificial daylight, when not abnormally rich in blue, gives promise of displaying colored objects as in a museum to better advantage in many respects than natural daylight. However, one must remember that while a great deal of photochemical activity has been charged up to the ultra-violet light, pretty nearly every part of the spectrum has under favorable circumstances a chemical effect upon something.

From a practical standpoint it would seem to be worth while to work out a source of non-fading light and then take up wherever possible the additional problem of preserving valuable objects with a still greater probability of success by sealing them from the air. There is probably no such thing as absolute prevention of all fading, but if the danger could be put far off down the centuries by means which now can be easily applied the end would seem well worth the effort. *Electrical World*, May 12, 1923.

200,000,000 Tungsten Lamps Sold in 1922

THE total sales of tungsten filament incandescent lamps (excluding miniature lamps) in the United States during 1922 amounted to slightly over 200 million lamps. This is a 25 per cent increase over the 1921 sales of 160 millions, and is within one per cent of the 202 millions sold in 1920.

There were less than three million carbon lamps sold in 1922, compared with six million in 1921 and nine million in 1920. It is apparent that the carbon lamp will soon disappear from the market; it is now a negligible item in the total lamp sales.

Lamp manufacturers have developed a coloring material for lamps which has been very satisfactory. It is sprayed on the lamp, is weatherproof and does not fade. It is much better than the artificial coloring material previously used, which was applied by dipping the lamp in the coloring solution. It faded rapidly and was not weatherproof.

The sprayed color is even more uniform in color than natural colored glass bulbs which vary in color due to variation in density and thickness of individual bulbs. Sprayed colored lamps can be supplied more quickly than natural colored glass lamps, as the sprayed color can be quickly applied to clear lamps in stock, whereas the natural colored glass lamps usually have to be made to order and often the bulbs have to be specially blown. Only four colors are supplied, which are a standard shade of red, blue, green and yellow.

The sprayed colored lamp is no more expensive than dipped colored lamps and is cheaper than natural colored glass. The sprayed color can also be satisfactorily used on Madza C lamps up to and including the 150-watt size.

Colored lamps are less efficient than clear or frosted lamps and should therefore be used only for decorative purposes. *N. E. L. A. Lamp Committee Report.*

PAPERS

NEW DEVELOPMENTS IN ART GALLERY ILLUMINATION*

BY G. T. KIRBY AND L. X. CHAMPEAU**

SYNOPSIS: In this paper the authors point out the importance of the proper lighting of art galleries, that is, to light each object as nearly as possible in the same manner as when the object was executed, without interfering with the architecture of the room or devoting special space for individual objects.

They explain how with deflectors and prismatic glass they control the daylight bringing it to the objects at the proper angle, and maintaining it at that angle and intensity throughout the day and the seasonal changes.

The authors further explain how with special lenses and deflectors they succeeded in producing artificial light of nearly a uniform intensity the full height of the picture zone, that compared favorably with the daylight. This was accomplished without sacrificing the architecture of the room and without glare.

If we have made a success of the lighting of The American Art Galleries, as we believe we have, it is due to the fact that we know what Art Galleries are; what the needs of the lighting of Art Galleries are and with careful study and many experiments, we have brought about a good sound result. Of course, you often start with a theory and I am mighty proud of the fact that years ago—many more years than most of you here can boast of—I was an undergraduate in the school of Mines and had the rare good fortune of studying under two great men among other great men. One was Prof. Hallock (then Professor of Physics) who is now dead. The other was Professor Pupin, then Professor of Electrical Engineering, as he is today, one of our greatest, if not our greatest, scientist.

*A paper presented before the New York Section of the Illuminating Engineering Society, April 3, 1923.

**Kirby, Champeau Co., Inc. ,New York City.

The Illuminating Engineering Society is not responsible for the statements or opinions advanced by contributors.

515

Hallock was always "bugs" on lighting. It was his fad and fancy in every way. Some of you may remember that not long after the beautiful library up on Morningside Heights was built, the lighting there was very poor. So Hallock placed right in the center of that dome a tremendous globe and he threw on to that globe from four or eight corners of the rotunda, lights which were reflected down upon the writing desks and gave a general soft illumination to the entire library.

Well, I thought that was a great stunt. So it happened that a few years afterward when I got some lighting ideas in my own head, I went up to Hallock and he and I had a heart-to-heart talk as to whether I was an awful fool or somewhat of a wise man and whether I should go ahead on it.

Shortly after, Hallock died and Cushman helped me out in a lot of these experiments. I had the craziest ideas that a young man ever had. I thought we could put up in this town a forty-story building. However, big art galleries need to be near the sidewalks because ofttimes when we have an exhibition such as we have on now, there are literally thousands of persons coming in every day to see them, and you can't carry a crowd as big as that in elevators; so you must have, as you see we have, wide stairways leading up to the rooms in order to give ready access to and from them.

Real estate is so expensive in New York that I thought we would combine the art galleries on the second floor and occupy the space above. So my bughouse plan was to put a forty-story building on the plot but have about ten stories over the second story vacant, the idea being that the daylight would come in from all four sides and come down through the skylights and light up the rooms and objects therein. It would have been all right if the building had been out on a plain, but I promptly, in my theory, forgot that adjoining buildings would soon go up around us and that those side lights would be shut off.

It is very important in showing artistic property to be able to show it in its true color value. The artist goes out into the field or into a studio to paint; he does not do it by artificial light. His color values, his color balance is as the daylight shows them. Of course, his picture, or tapestry or fabric, whatever it may be that he is working on, may be equally beautiful under a yellow gas

light or an old-fashioned carbon filament or whatever the illuminant may be, but, even though it is beautiful, it is seldom if ever under
such artificial light shown in the same way as the artist painted
it and as the artist intended it to be seen.

Another thing which one should always remember is that the
light coming from directly over a picture or even a tapestry or
any art object, ofttimes throws shadows from the pigment or
from the material in a way that the artist never intended, distort-
ing his drawing and giving it a very different effect from what the
picture was intended to give. All you have to do to prove that
very important point is to notice the thousands of advertising
signs as we now see them when we walk up and down our streets
and avenues—first we see a young lady putting on a stocking;
then we go about five feet more and she is taking it off, or something like that. You know what they are better than I. But that
is the whole thing; it is the angle from which you look at it, the
angle from which the light strikes those lines.

So, in illuminating a beautiful artistic object, you have to be
very careful that the rays of light, whether they are daylight or
artificial, strike the object in the way that the artist intended
they should strike it and in the way that the artist painted.

When you walk through these galleries—as you are going to
do very shortly—and have the pleasure not only of seeing our
lighting devices, but of seeing the greatest collection of highly
artistic property which for many and many a year has been gathered
together (The Solomon Collection) just note how clearly not alone
the color values come out, but how clearly you see the picture, as
if daylight instead of electric light were the illuminant.

You, of course, are only secondarily interested in daylighting,
but you should be very much interested in it because one great
thing in art galleries is to have the lighting so that when you
change from your daylight to your artificial light or from your
artificial to daylight, it will hardly be observed. Many times in
these Galleries, when the crowds come up in the daytime and it
suddenly becomes dark—as it has several times this spring— we
switch on the artificial light and the people in the Galleries at the
time do not know whether there has been any change or not.
That is a very good test of your lighting efficiency.

Of course, in daylight, (if I may speak of that for a moment)
as we all know, the artist is striving for what is called "North

light." It really does not make much difference where the light comes from, whether it is north, south, east or west. What he is really striving for is a reflected light, a light that comes from a million and one different points. As old Hallock used to say: "What he is striving for is the million reflectors, little dust motes in the atmosphere that reflect the sun," because, after all, the sun is the source of our daylight and it would be very easy to give an absolutely correct daylight at all times were it not that, as the old colored preacher proclaimed, "the sun do move."

It was only the day before yesterday that I went into one of our best known galleries here in the city, and let us say there (indicating) was the north wall, this (indicating) the south, behind me is east, of course, and that (indicating) is west. It was about one o'clock and the sun at this time of the year is somewhat in the south, as you know. That north wall which was covered with prints was simply bathed in a glorious yellow sunlight, and that wall (indicating) was in shadow. Well now, it was beautiful as an object of sunlight, but to show anything on the walls was simply absurd because that wall (indicating) should have been of the same illumination as that (indicating) and the east and the west.

What we have done here, as Mr. Champeau will show you, much more plainly than I can because he is the practical man, is really to construct louvers which, in effect at all hours of the day, redirect the varying sunlight and give it a constant direction. It would be easy if the light were always coming down from above, then you could diffuse it without any difficulty. As it is, it comes up here low in the east and we have horizontal deflectors which let the same amount of light through as it reflects to the opposite wall; and we have vertical deflectors which do the same for the east and the west walls which the horizontals do for the north and south.

It might very well be described as a lot of cubes. You have seen eggs packed in such pasteboard boxes; that is exactly what we have over our skylights, the sides are parallel when the sun is east and west, but as the sun passes from east to west the vertically pivoted deflectors are kept at right angles to a vertical plane passing through the sun and the sides of the cubes are then "lop sided." The material of the sides of those little compartments instead of being opaque are of diffusing substance, so that

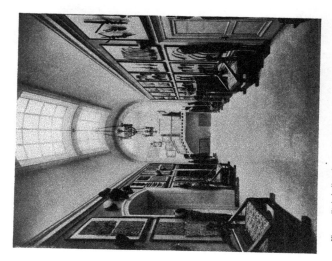

Fig. 2—Stair hall and corridor. The main source of light at night is behind the ceiling sash while the ornamental chandeliers with their shaded bulbs give a pleasing touch of light, and appear to be the only source of light.

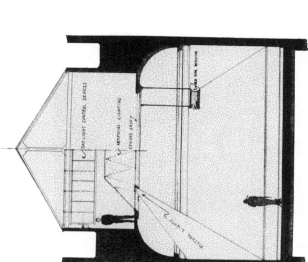

Fig. 1—Diagram of a typical installation of devices for day and night lighting of galleries.

Fig. 3—Showing end and side of typical gallery by daylight; note the uniform lighting of walls and lack of intense light usually found in center of galleries.

Fig. 4—Same gallery as above viewed at night.

some of the light goes through to the wall facing the sun and an equal amount is reflected to light the wall facing away from the sun.

In other words, the light is scattered, and that is what you want to do. You want to break it up so that no direct sunlight will reach the gallery.

This apparatus serves the purpose very well but the downward component of light is still too great, lighting the center of your room more than the walls, and you would have with that lighting, an effect very much the same as you have in this room. You are sitting here in a light that you can read by, that you can see each other by, but your walls are comparatively dark. In the main Gallery which Mr. Champeau and Mr. Day will show you, that light as it comes down is taken by prismatic glass and thrown to the four walls. So that, instead of coming directly down, it comes down to that glass and then it is bent and sent to the walls. Therefore, in that room you will find a practically even illumination, not alone of the walls, but of the floor and it is done without hanging that hideous and unnecessary canopy from the top or by putting in any of the many devices which are eye-sores and which you see in so many museums.

Now we come to the night lighting which is equally if not more important, because it has to do not alone with galleries where you must have your top lights for the color values of which I have spoken, but also for homes, for places where pictures and tapestries and other beautiful things are shown. What we have done in the Galleries is just as adaptable for homes as it is for the Galleries. In point of fact, we have just installed in the home of one of our young men of good fortune and good philanthropy as well, reflectors to light his million dollar tapestries.

The great trouble with most of our reflectors which help in the illumination of our walls in galleries and other places is that we have to drop them down from the ceiling. When we do that, we are doing something which is bad in many ways. Primarily, it looks bad. You don't like to see something hanging down in that way. Next: It cuts right around the wall so that if you were looking at a large picture or a tapestry or hanging, you have that line of your reflector everlastingly in your eyes. Next, and just as important: In the daytime, when the light comes from your skylight, you have the shadow of that reflector thrown on the wall thereby putting the shadow on the picture or whatever is there, and you can't cut it out. Even here where there are no shades to draw, where these louvers or deflectors take the place entirely of shades, and where we can cut out all the light or a large part of it any time we want to, there is always that shadow where there is a hung reflector. So you will find that our reflectors are right up in the ceiling where the skylights are. In point of fact, they make an ornamental border for the skylight, it depends of course upon the proportion of your room to where those reflectors are placed.

If those reflectors were the only thing you had you would find your room dark in the center. You must, therefore, in addition to those reflectors, place your artificial lights over your skylights to give the general illumination of the room. You can project through your lower skylight, enough light from reflectors above to light your walls, but to do that is expensive and it does not give you the ease of control that you have, nor does it give you the big point that we are striving for in picture or other art object illumination and that is this: To have that object equally illuminated both as to the bottom and the top—the whole picture, let us say, to have an equal illumination—so that in the photometric test, you would find the same intensity of light at the top, bottom, in the middle and the sides, because that is the way that object was made, painted, woven or whatever the case might be.

As we know, and as the diagram which will shortly be shown to you graphically shows, when light starts out from a point, it spreads out and most reflectors do necessarily spread the light to give a wide enough path of that light to take in the entire picture and, as it spreads, the intensity diminishes and the consequence is that you have the top of your picture which is near the reflector over-illuminated and the bottom of the picture which is away from the reflector under-illuminated. What we do to overcome that is simple; there is nothing complicated about it. After months of investigation on the part of Mr. Champeau and myself of galleries and other places here and abroad and our experimental gallery (which we built up in the woods and tried out for three years with all kinds of devices) we merely put those things together so that they would work.

First of all, there is the lamp. That lamp is set in front of a reflector so that we get not only the rays from the lamp itself (from the incandescent filament) but also coming back through the glass from the reflector behind it.   Then they come to the prismatic glass which is in front of both the filament and, of course, the reflector behind the filament, and when it strikes that prismatic glass, the rays are brought together nearly parallel so that they come to the object in nearly parallel lines.   Therefore, you have practically the same intensity and that means that in lighting your picture or your tapestry or your bit of furniture, or metal or your Chinese vase, you are giving that object the same intensity of light at its bottom, its side, its middle and every other part and without casting those deep shadows which you would get if you had a single source of light, because these reflectors are put in series, forming a continuous band of light (directed at the proper angle), about the room.   Instead of having a light like this (indicating) which would throw a shadow to the left or to the right, you have them coming from both sides which makes the object stand out as the artist intended it to.

Well, that is the whole story.   It is just as simple as anything possibly could be and you will see it, you experts, with your own eyes.   It is just as effective with flat objects as it is with round ones.

The great sculptor of the Lincoln statue in the Lincoln Memorial in Washington, is pleading that we should do something like this with his figure down there, to the end that it might stand out as he designed it in his studio, and as he saw a small object stand out here.   Our great dealers in art objects feel that now they have something which is simplicity itself.

I remember well, not so very many years ago, I was sitting in a little studio of an art dealer on Bond Street in London.   He was pulling out picture after picture to show me.   There was a dark Rembrandt that he had and he wanted all the light he could get on it.   First he let one shade go, then another, to bring in all the light from the skylight.   Pretty soon, he showed me a picture of a snowstorm.   He pulled one shade, then another, to get a subdued light, because naturally the intensity for the dark Rembrandt was entirely too much for the snow scene.

Well, with our lighting, all he would have to do would be either to touch a button and have a little electric control which would turn his louvers so that the light would be right or he could simply do as we do here, turn a little crank, and in a second's time, he would have the illumination that he wished.

It is nothing wonderful. There has been no great discovery. It is just a simple application of simple things which you men know a great deal more about than Mr. Lawrence X. Champeau and I have ever thought of knowing. But it has been done by coordinating the experience of one who has been brought up as I have, practically in an art gallery, with one who has been brought up with a great practical knowledge of light and lighting as one of our foremost photographers, and I deem it a rare good fortune that in being able to carry out these plans first of all, I, who had somewhat the genesis of the idea, should have had Mr. Champeau cross my path at a time when he could take up the development of these ideas and give them the practical application which they have been given, and latterly in having again come in contact with Mr. Harry Day, who as an architect knows proportions and distances and design, to take Champeau's ideas and my ideas and help us to give not only lighting which is at least the best yet, but also proportion of room and proper design so that we have an artistic and perfect home.

On behalf of the American Art Association, I welcome you here. I hope you will have a good time seeing not alone the lighting, but the exhibition. On behalf of Mr. Champeau, Mr. Day and myself, I trust that in this new system you will find something of interest and perhaps much of improvement.

DISCUSSION

L. H. Graves (Communicated): Inasmuch as the Grand Central Art Gallery lighting was also inspected on the evening of Mr. Kirby's paper on the lighting of the American Art Gallery and time did not permit of a description and discussion of the lighting of the former, it seems appropriate to offer a brief description of this work. In the Central Galleries there is no elaborate attempt at the control of daylight by the use of a system of louvres as described by Mr. Kirby, though because of the location of the galleries and the arrangement of skylights exceptionally good daylight is provided. The more interesting features are the several schemes of artificial lighting employed in the various individual galleries. There are schemes of direct skylight lighting with units

Fig. 5—Fountain room, Grand Central Art Galleries.

Fig. 6—Type of X-Ray reflector used for direct lighting.

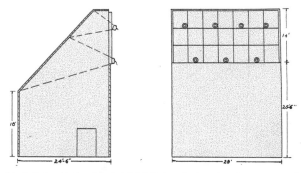

Fig. 7—Gallery F, Grand Central Art Galleries. Equipment used; seven 500-watt flood lighting units.

Fig. 8—Gallery F, Grand Central Art Galleries.

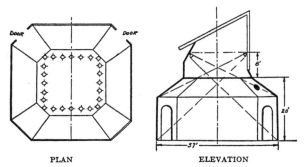

PLAN   ELEVATION

Fig. 9.—Fountain Room, Grand Central Art Galleries. Equipment used; twenty-four No. 33 X-Ray units with 200-watt Madza Daylight lamps.

designed to control and distribute the light as required, also schemes of indirect lighting in the smaller narrow galleries where no skylights are provided and also the rather unique application of flood lighting in one gallery where, with a vertical skylight, projector equipment is mounted outdoors in such a way as to flood the sloping ceiling opposite. This gave a system of diffused light in the gallery very much as when daylight is admitted.

In the fountain room (See Figures 5 and 9) and in the two galleries with the flat ceiling skylights, where the reflectors are mounted above for direct lighting, projector type X-Ray reflectors were used with 200-watt daylight Madza C Lamps. (See Figure 6) The units were located around the edge of the skylight and by use of the swivel joint mounting, were adjusted to direct the light on to the opposite walls. The incidental spill light, together with some diffusion caused by directing the beams through the skylights, gives ample general illumination in the room. The effect with the overlapping of light on the skylight and the use of the daylight lamps is very similar to that attained with daylight. It is indeed very pleasing and most successful for this type of interior. From twenty to twenty-four units were used over the skylight in each of these rooms.

In the smaller connecting galleries it was important to get a diffused artificial light high in intensity during the day time as well as at night, for, because of the construction of the building, it was impossible to work in skylights. Total indirect lighting fixtures . of simple design were used. Finished in a light tone to match the ceiling, the fixtures themselves are hardly noticeable.

In Figure 7 is shown section drawings of Gallery F in which the skylight was in one of the vertical walls. The location of the seven 500-watt flood lighting units is shown outside the room. As mentioned above the light is directed on to the sloping ceiling opposite and diffused over the entire interior. The effect is very pleasing and the results from a lighting point of view are entirely satisfactory. With the use of blue daylight lamps in the other galleries the effect was that of diffused north light, while in this gallery with the use of the 500-watt flood lighting lamps one might imagine the room flooded with warm diffused sunlight.

The equipment and scheme employed in the artificial lighting on this entire installation was so simple, yet the results so very satisfactory and so efficiently obtained, that I believe it offers an example which may be readily copied in many galleries old and new throughout the country that are now poorly lighted either because of the antiquated equipment or improper application.

AESTHETIC AND UTILITARIAN VALUE
OF LUMINAIRES*

EMILE G. PERROT**

SYNOPSIS: In order to properly appreciate the aesthetic value of an object, it is necessary to have a clear understanding of the nature of beauty,—for mere sense gratification—namely of sight or hearing does not constitute the essential attribute of beauty.

According to ancient Greek philosophers, "beauty is a characteristic of any object composed of various elements that produce a *unity* of effects upon the sensation of the beholder."

Now there are certain laws that seem to be followed by all works of painting, or sculpture, or architecture, that the consensus of opinion of mankind has judged beautiful. These laws seem to be in accordance with the working of man's mind, when he is striving to create something pleasing to his senses, and they may be grouped under three major heads, of Unity, Grace, and Proportion.

In designing luminaires, therefore, these laws should govern, and further, as the lighting fixtures should form an integral part of the whole architectural scheme, the illuminating engineer should have a knowledge of the architectural styles and periods, so that a method of lighting will be in harmony with the style of the room or building.

While there are certain problems such as the lighting of offices and factories, in which the object of the light provided is purely utilitarian, and where the illuminating engineer has much good work to do in providing efficient and serviceable conditions of illumination, there are higher fields for his activity in connection with the lighting of buildings of distinction, where aesthetic and architectural considerations must prevail.

In my talk to you this evening on the subject announced, it is my purpose to treat it from the Architect's view point. Being concerned primarily with architecture, I am naturally interested in seeing that the element of beauty is maintained in all that pertains to making a building beautiful.

As all the fine arts have for their distinctive object the expression of the beautiful, it might be well for us to consider briefly what constitutes the beautiful, without investigating the philosophy of beauty.

An object admitted to be beautiful is one that is calculated to awaken a noble emotion. Mere sense, gratification, namely of sight or hearing, does not constitute the essential attribute of beauty. For instance, a painter may spread upon his canvas an array of meaningless colors, and however delicate the shades, if

*A paper presented before the Philadelphia Section of the Illuminating Engineering Society, March 14, 1922.

**Architect, Philadelphia, Pa.

The Illuminating Engineering Society is not responsible for the statements or opinions advanced by its contributors.

it is mere color and nothing more, we cannot call the result beautiful, neither can we call the single chord of music sounded upon an
organ, though pleasing to the ear, beautiful.

The reason we cannot apply the epithet beautiful to such
colors or sounds, may be best understood by the following comparisons—If we will consider how essentially different is the effect
produced upon us by a painting of the seashore or a forest, and
by the mere sense, gratification produced by any shade or combination of colors; or again, if we try to realize the difference
between the effect of a Beethoven symphony and the pleasure
given by a chord of music or succession of chords, we shall instantly realize that they belong to different categories.

The difference is not merely one of intensity of pleasure, but
that of the nature of the pleasure. The former is, properly speaking, beautiful; the latter is pleasing, but not beautiful in the strict
sense of the term.

Now there are certain laws that seem to be followed by all
works of painting or sculpture or architecture that the consensus
of opinion of mankind has judged beautiful. Not only are these
laws deducible in painting and sculpture and architecture, but
the working of the same laws or others closely analogous to them
can be found in good literature and good music. These laws
seem to be in accordance with the working of man's mind, when he
is striving to create something pleasing to his senses, or that has
the quality of beauty, and they may be grouped under three
heads—Unity, Grace and Proportion.

The first of these is so universal and so important that com-
pliance with it has often been recognized as the sole necessity of
beauty. Pythagoras and Aristotle voiced it in Greece over two
thousand years ago, and almost every philosopher since has recorded it and restated it when dealing with the subject of beauty.
Beauty, according to these authorities, is a characteristic of any
object composed of various elements that produced a *unity* of
effects upon the sensations of the beholder. This definition covers
only a small part of the whole field of what men call beautiful;
it neglects the entire emotional and associative value of beauty.
It considers beauty merely as a matter of the senses rather than
of the heart.

However, allowing all the onesidedness of this definition, it is found that Unity is fundamentally essential to beauty, and may be defined as the manifest connection of all the parts in a whole, being the quality of an object by which it appears as definitely and organically one single thing.

A simple illustration will make the meaning of Unity clearer. For example, draw twelve lines at random, as shown in the illustration, Figure 1,—there is no evident connection among them—there is no unity; but if they are drawn as in Figure 2, unity appears—they constitute a whole by virtue of their arrangement. If now, instead of straight lines, we give the parts shapes that are pleasing, we add grace, as in Figure 3.

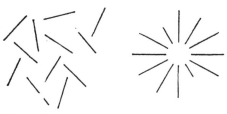

1. Random Lines-No Unity   2. Unity by Virtue of Their Arrangement.

3. Grace Added to Unity by the Shapes Given to the Parts.

Figs. 1, 2 and 3

By grace we mean that quality of form that produces in the human mind, through change of direction of the contour or outline of an object, the sense of perfect ease and consequent satisfaction.

Curved forms or outlines possess the characteristic of being more graceful than those composed of straight lines, the "S" curve is called the "curve of beauty" because its everchanging curvature is particularly fascinating.

The Greeks used it to the full, and along with it discovered the value of gradually changing the curvature in every line they used. There is scarce a Greek vase or a Greek moulding or ornament which has any circular curves at all, their outline starting in nearly a straight line, becomes more and more curved throughout its length, and finally ends in an abrupt curve. The inside curve of the thumb on your hand being a near approach to this outline.

This is in striking contrast with Roman mouldings and ornaments, which are always composed of circular curves.

Of equal importance to grace is the quality of proportion. Broadly speaking, this is the quality possessed by any object whose several parts are so related to each other as to give a pleasing impression. It is primarily a quality of the relationship of all the units in an object, rather than a quality of the units themselves. We will later see wherein this quality is manifest in our lighting fixtures.

While there are certain problems such as the lighting of offices and factories, in which the object of the light provided is purely utilitarian, and where the illuminating engineer has much good work to do in providing efficient and serviceable conditions of illuminations, there are higher fields for his activity in connection with the lighting of buildings of distinction, where aesthetic and architectural considerations must prevail.

If he really wants to be of great service to the architectural profession and to the community, he should understand how to adapt a method of lighting so as to be in harmony with the style of a room, its furniture, and the general scheme of decoration.

He should know something of the history of architecture, and the meaning of the traditions by which the Architect is guided.

He should appreciate the fact that the light provided must not only serve to illuminate the table, but to reveal panels, mouldings, ornaments and color.

If the illuminating engineer will study these things, his knowledge of the technical possibilities of the various illuminants, of shade and reflectors, etc., should be of great value. In the same way, it may be conceded that some Architects do not appreciate sufficiently the aims of illuminating engineering, and it would be well for them too, to receive more information about illumination.

There are many instances in which the combined efforts of the Architect and the Illuminating Engineer would lead to great results and for the more important problems of lighting work co-operation of the two professions is necessary.

In my talk to you this evening I am going to emphasize the view point that considers the lighting fixtures an integral part of the whole architectural scheme, rather than attempt to go thoroughly into the details of the fixtures themselves.

So before considering actual installations, it will be well to have some understanding of the great architectural styles and the sub-division of these styles, known as periods.

It is customary, and indeed fitting, to commence the study of historic art with its beginnings in Egypt, although there has been left little of its influence in modern art. Here and there we find an example, such as the Egyptian Hall in Wanamaker's Store, which is a good example. The lotus plant is the principle natural motive used as a basis of the ornamentation of columns, etc. Walls and columns are decorated with a wealth of sculpture, symbolic or hieroglyphic in character fashioned within incised outlines or cut in low relief.

GREEK ART

The architecture, sculpture and painting of the people of Greek lands, who, gathering together the influences of Asia and Africa, developed a national art which marks the beginning of European civilization and culture. It was a style rich in beauty, yet keenly practical and full of originality, vigor, truth, intelligence, color, moderation and self-restraint.

The importance of Greek architecture in subsequent evolution cannot be over-estimated. Greek architecture is fundamentally the basis of all modern architecture, in that from it sprang the architecture of Rome, and from that later, the architecture of the Renaissance, which supplanted the Gothic idea.

In Greek architecture, furthermore, it is possible, for the first time, to perceive the origin of a multitude of architectural forms with which we are daily surrounded today—mouldings, ornamented motifs and the immortal "Greek orders" themselves—forms which

have come down to the present day, while those of ancient Egypt
and Assyria did not live beyond the confines of their lands, or after
the downfall of their empires.

The Greek evolved the "Classic Ideal" in architecture, an
ideal of such purity and nobility and perfection that it has constituted the standard through the ages, and is today the fundamental of architectural design.

Greek architecture, elementally, is a column-and-lintel architecture, highly developed as time went on from the severest Doric
orders to the most ornate Corinthian orders.

The names of the three orders are—Doric, Ionic and Corinthian. Each of them presents a different series of proportions,
mouldings, features, and ornaments, though the main forms of the
building are the same in all. The column and its entablature being
the most prominent features in every such building, have come to
be regarded as the index or characteristic from an inspection of
which, the order can be recognized, just as a botanist recognizes
plants by their flowers.

THE ARCHITECTURE OF ROME

The architects of Rome took Greek architecture and elaborated
it, introducing in addition, and highly developing, the use of the
arch.

The old Greek Doric order did not appeal to the sophisticated
Romans, to whom it doubtless appeared too severe and too
primitive. Their corresponding form was the Roman Doric column
also called Tuscan. They made but little use of the Ionic, but
appropriated and highly embellished the Corinthian.

Most characteristic of the Roman development of architecture
was the combined use of column and arch, later a favorite theme
for the architects of the Italian Renaissance.

Roman carving and ornamentation was rarely so refined or
pure as similar work of the Greeks, but was usually more decorative.
The Romans were lovers of inscriptions, and, in their architecture,
began to pay more attention to secular buildings, both public and
private, than had previously been accorded them. Public works,
such as aqueducts and bridges, became architectural monuments,
as well as theatres, baths, and triumphal arches, while the private
residences, or villas, became luxurious and elaborate to a degree,
and were filled with paintings, statuary, bronzes and other works
of art, including Greek antiquities.

Architecture was fast coming into a closer relationship with the people, ceasing to occupy its earlier position of exclusive consecration to the gods.

There were Roman temples, to be sure, but there were an even greater number of Roman secular buildings which have played as important a part in the subsequent development of architecture as the earlier monuments of Greece.

BYZANTINE AND ROMANESQUE ARCHITECTURE

Before the final downfall and dismemberment of the Roman Empire in the year 455 of the Christian Era, with all the elaborate civilization it had developed, there grew up two types of church architecture which struggled on through the Dark Ages, sustained by the warmth of religious enthusiasm, and, in their way, keeping the lamp of architecture burning until times more propitious for its further development.

These two styles are known as Byzantine and Romanesque— the first of which, reaching a high development in itself, led to nothing else, and the second of which, by reason of its vital structural merits, grew directly into the great Gothic style, which was to completely fill the architectural stage until the coming of the Renaissance in Italy in the year 1400.

GOTHIC ART

In the eleventh century we. see the beginning of that wonderful development of Gothic art which originated in France and was an outgrowth and refinement of Romanesque methods.

Gothic architecture is remarkable in that it is dually a structural architecture and a decorative architecture, with both of these essential aspects existent in equal proportions. The most important single thing to remember in considering Gothic architecture is that it may be closely likened to an organic growth. Its development was as natural and as consistent as the growth of a tree, rising up, putting forth branches, and these, in turn, putting forth leaves.

Let us endeavor to summarize the evolution of the Gothic church or cathedral, from its beginning in the vaulting achievements of the late Romanesque builders.

The typical plan took the form of a great cross, with three short arms and one long arm. The entrance was at the end of the long arm, and gave access directly into the great central nave, flanked by side-aisles. The arms of the cross formed the transept, and a great tower rose at its intersection with the nave, or there were twin towers rising above the entrance front. The remaining arm of the cross was the apse, or sanctuary. There were other types of plan, but the cross was the most usual.

Architecturally, the plan was carried out with an intricate diversity of which only Gothic architecture could be capable. The walls of the nave, above the lower side-aisles, were carried on columns and pointed arches; the side-aisles, also arched and vaulted were supported, outside, by buttresses to take the lateral thrust. Above these, on the exterior, rose flying buttresses to take the thrust of the nave arches, and everywhere there was opportunity for pinnacles, turrets, grotesques, gargoyles, niches with images of saints, and all the profusion of Gothic detail. Within, the building was lofty and mysterious, richly and dimly lighted by tall, pointed windows fitted with stained glass—perhaps a magnificent rose window at the rear end of the nave. Everywhere, too, carved niches and holy images, intricate carving, dull color in polychrome or textiles.

Gothic architecture is often nicknamed "perpendicular architecture," which is reasonably descriptive, inasmuch as the horizontal entablature, with its frieze and cornice, forms no part of the Gothic idea, wherein all members mount ever upward, climbing one upon the other in one magnificent expression of altitude. Columns, arches, vaults, windows, pinnacles, buttresses, towers— all point upward—even the details of tracery and the niches for images point upward.

It is this sense of upward motion, reaching often to the heights of the sublime, which has made Gothic architecture essentially the architecture of the church, rendering, as it does, a remarkable expression of spiritual nobility in architectural terms.

The Gothic architecture of the important Northern countries can be sub-divided into three distinct styles—for instance, in England we have the style known as "Early English" or 13th Century Gothic. The next development was the "Decorated Gothic" (Fourteenth Century) and in the Fifteenth Century we have the style known as "Perpendicular," which is noted for a particular development of vaulting known as "fan vaulting."

The architectural style called the "Tudor" is to be applied to buildings under the reign of Henry VII, Henry VIII, Edward VI and Mary.

It was a transitional period, the flat-pointed arch was a conspicuous characteristic, and the Gothic forms of design were being gradually superseded by innovations of Italian Renaissance.

Following the Tudor style came the Elizabethan and Jacobean styles.

In France we see the same influences at work in the Gothic period, at the beginning of Gothic we have the fully developed Gothic rib vaulting in all its simplicity, corresponding to the Early English styles, the best example being Notre Dame at Paris. Then the fully developed 13th Century Gothic, which is the highest expression of Gothic art in the world, having for its examples such great cathedrals as Rheims, Amiens and Chartres, known as the Rayonnant (wheel-like). While the 14th and 15th Centuries, or late Gothic, deteriorated into an art of redundant detail, yet full of vigor and full of expression, called the Flamboyant (flame like), corresponding to the perpendicular style in England. So we can go through the same categories of style in other countries of Germany, Belgium and Spain.

RENAISSANCE ART

With the revival of the study of classic arts and letters in the Fifteenth Century, began Renaissance Art. Developing first in Italy, particularly in the City of Florence, it spread throughout the peninsula and eventually over all Europe and to America.

The Influences of classic study showed itself in the outlines and masses of the grandest churches, civic buildings, and in the humble details of the simplest home. Borrowing from Greece and Rome, the idea of the lintel, column arch, vault and dome, it applied these features sometimes structurally, but too often decoratively, and in its later developments with little taste and true architectural meaning.

In Italy the Renaissance architects used column and arch, or pilaster and arch, extensively, built splendid domes, and showed great fancy for surface decorations. Construction played a subsidiary part to design, as is evidenced by such frank expedients as the introduction of the rods between the supports of arches and vaults, to take the thrust, which could not be met in the design.

There was a general use of the Roman entablature, elaborate pediment and general profusion of ornamental detail. The so-called "Arabesque" decorations of the Renaissance pilaster is one of the most characteristic single details of the style. Another Renaissance design of frequent occurence is the statuary niche, with the upper portion in the form of a shell.

From Italy the style was carried to France, and the transitional style evolved was known as the style of "France I" or Francois Primier." The office building located at the S. W. Corner of 15th and Market Street is a very good modern adaptation of this style. To other countries in Europe the style was carried—to the Netherlands, Germany, Spain and England, developing in each country a peculiar style of its own, according as the national character found its expression. For instance, in Spain we find it strangely and richly blended with Moorish influence.

In England the Renaissance found its highest expression in the reign of Queen Elizabeth, so that the term "Elizabethan," when applied to a country house, is synonymous with the term "English Renaissance." The more important buildings of this period in England were distinctly formal and dignified, with little of the spontaneous diversity of the Italian Renaissance. Italian forms, however, constituted the basis of design, and deeply influenced all subsequent English architecture. The English Renaissance development of greatest interest today, by reason of its importance in the evolution of the modern country-house, was the Elizabethan manor. The country free from internal wars, the government powerful and protective, the element of defence became increasingly less in evidence. The houses became more livable, more comfortable and "modern" in character. Increased facilities for the manufacture of glass brought about the design of beautiful leaded windows. The interiors were rich in carved woodwork, and floor coverings came into use. The Elizabethan, or English Renaissance country-house, was an important step in the development of the country-house of today.

The decadence of Renaissance art is to be seen in the Rococo or Baroque development of the last three quarters of the Seventeenth Century. Ornament was developed to an extravagant and tastelessly disproportionate degree—architectural forms were distorted and preverted in a thousand fantastic and impossible vagaries. Structural principles were ignored, and decoration was the main feature, not the embellishment of Baroque buildings.

It has been the habit of most architectural critics to sweepingly condemn all Baroque architecture, but such condemnation is neither intelligent nor merited. Granted that the style may be proved fundamentally illogical on many scores, it evolved many forms of permanent beauty and value, and was, if nothing else, an essentially decorative style, later developed along more rational lines in some phases of the French style of Louis XV. Despite the usual dismissal, then, of the Baroque or Rococo style as a mere architectural curiosity, entirely decadent, and even artistically immoral, it will be found more valuable to place it as a distinct expression of a peculiar idea, and an undeniably interesting page in the sequence of the architectural styles of the past.

In the latter part of the Eighteenth Century, a reaction from the extravagance of Baroque was inevitable. And the reaction took the form of the Classic Revival in France and in England. Some feeling of this revival manifested itself in this country even as late as the first two decades of the Nineteenth Century. And while there is at the present period no feeling in design so sweeping or general as to be called a "Classic Revival," Classic derivations are everywhere apparent, and almost invariably in such monumental buildings as are desired to express qualities of dignity and permanency, such as capitols, post offices, libraries, museums, banks, and the larger railroad stations.

THE CLASSIC REVIVAL IN FRANCE

Observation of the Classic Revival is best begun in France with a momentary survey of the progress of architecture immediately preceding the period of XIV.

The reign of Louis XIV came to a close in 1714 and the architecture of the period, as well as that of the preceding period, was pompous, elaborate, grandiose. Buildings of the time show a conflict between Renaissance order in design and Baroque extravagance, and all carried out in what was called (most appropriately) "The Grand Manner."

The succeeding period, that of Louis XV, is often called the "Rococo" period, because the "rock-and-shell" style reached its height at this time. "Louis Quinze work is practically synonymous with Rococo, the fanciful rock-and-shell curves that, like some fungous growth, invaded all branches of decorative art with amazing recklessness and rapidity." The characteristics of the period were distinct in their nature, though elaborate and various in form. Curved lines and intricate foliation appeared in all designs, and lack of symmetry was considered a desirable achievement. Importation of many works of art from China at this time added Oriental fantasies to the already fantastic Baroque-Rococo style, which grew increasingly extravagant throughout the reign of Louis XV. Much decorative work of the period is by no means without merit, but the style was too frivolous to effect any permanently great architectural expression.

Despite the intense interest in the Rococo style, its very extravagance finally became so wearisome and distracting that the reaction of the Classic Revival set in with the reign of Louis XVI (1774-1793), and Classic forms became increasingly popular until the close of the "Empire" period, in 1814.

THE CLASSIC REVIVAL IN ENGLAND

The Classic Revival in England commenced with the Georgian period (1714-1820) immediately following the Dutch influences of the reign of Queen Anne, and lasted, though waning somewhat in its Classicism toward the end, until the beginning of the Victorian era in 1837.

The Eighteenth Century Classic Revival in England reached its height during the reigns of the Georges, notably in the works of the Brothers Adam (1760-1820), in the reign of George III. The Adams, however, were not the first architects to design in the "Classic Taste."

Following Sir Christopher Wren and Inigo Jones, the great English Renaissance architects of the Jacobean period, William Kent, who died in 1748, produced works which were more in the nature of a "Classic Revival" than of the style of the Renaissance.

GEORGIAN COLONIAL IN AMERICA

With such widespread enthusiasm for Classic ideas in architecture in England, it is not at all surprising that this should have crossed the ocean to the American colonies, creating and moulding the style which should accurately be called "Georgian Colonial."

The "American Classic Revival" was a distinctly different development, coming, as it did, largely from France, and at a considerably later date than the Georgian Classic influences.

The Georgian Colonial types of American architecture took different forms in the North and the South, especially in the treatment of dwellings. In the North there is noticeable a great Georgian Classicism of detail, rather than of general form. New England doorways, windows and interior woodwork followed Classic formulae, rendering Palladian windows and Greek orders with an honest carpenter's technique. Independence Hall is one of the best examples of this type.

THE CLASSIC REVIVAL IN AMERICA

The American Classic Revival, as distinct from the Georgian Classic inspiration, came about largely through the development of friendly relations with France and the distaste for things English during the War of 1812. So closely allied, indeed, is this American Classic Revival to the contemporary style of France, that it has often been called "American Empire." The popularity of the Ultra-Classic left a number of interesting monuments in this country, of which, perhaps, the purest example is to be found in the old Stock Exchange Building, Dock and Walnut Streets, the Custom House, and grandest of all, the main building of Girard College.

Having thus reviewed the leading Historic styles, we will consider the aesthetic value of lighting fixtures to buildings of some of the great styles, noting the manner in which the fixture manufacturer has conformed the style of the fixtures to the architecture of the building.

The utilitarian aspect of fixtures, of course, is based entirely on the study of efficiency in illumination. Much effort has been put forth to obtain satisfactory results with the least expenditure of energy, and we are at once confronted with the claims of this or that special device or type of fixture as being superior to any other make.

The recent tendency of incandescent lamp manufacture has been toward high concentration of filament in quest of economy in current consumption. We may consider that the development has occurred in two stages. The first stage of lamps as represented by the carbon filament in vacuum bulbs are now considered extremely wasteful in the current they consume, yet they were in the early days a scientific marvel and object of interest to the extent that they were hung about on stringers or mounted bare on chandeliers just as so many open flame gas jets. This might be called the age of bare lamps, and the glow from these carbon lamps was so mellow as to cause no discomfort to the eyes of the observers. Then came the second stage of lamps, which consisted of the tungsten metal filament in a vacuum bulb, and the intrinsic brightness as well as the economy in current consumption of the lamp unit was greatly increased. With this improvement various types of shades appeared for either protecting the eyes from the direct rays of the lamp or concentrating it for particular uses, so that now some type of reflector is essential for proper light diffusion. With the advent of the nitrogen gas filled bulb with its high candle power and economical current consumption, the problem of diffusion was solved by the invention of reflectors, so shaped that the concentrated rays of the lamp are split up and scattered, thus producing a soft light.

Without attempting to discuss the merits of any particular type of lighting unit, a consideration of the more popular and meritorious ones will be in order. The illustrations I am about to show you have been obtained in most instances through the courtesy of the various fixture manufacturers, and without their co-operation I should have been unable to make the presentation of the subject as thorough as it is.

The simplest method of obtaining high efficiency in direct lighting is with the single unit of sufficient intensity and proper diffusing reflector, to give as near uniform conditions of lighting as is possible with artificial illumination.

I show on the screen views of interiors of buildings in which the various types, all known to you, are used. The installations are supposed to represent what is good practice with each type, so that it is possible to judge of the architectural value of the fixtures as well as the utilitarian.

The Americolite Company's single unit is a very artistic luminaire, as are also those of the Brascolite, in their various treatments as to styles to harmonize with the architecture of the building, as the illustration of the lobby of the Hotel Marion, Little Rock, Arkansas, shows. See figure 4.

Fig. 4—Luminaires in Lobby, Hotel Marion, Little Rock, Ark.

Fig. 5—Roman Catholic Church of Gesu, Philadelphia, Pa.

Fig. 6—Detail of Luminaire in Church of Gesu.

The Celestalite unit also can be obtained in simple designs or more ornamental ones, to conform to the architecture of the building, as shown by the illustrations.

In considering lighting installations in churches, two distinct methods of lighting are at hand—indirect or semi-indirect. The installations of indirect lighting in the Eberhart Memorial Church, Mishawaka, Indiana, is a notable example of an indirect system in which the architectural style has been carried out in the fixture, the style being Gothic, National X-Ray Reflectors being used.

The Cathedral of St. Helena, Helena, Montana, is a very good example of indirect lighting, supplemented with direct light from concealed X-Ray reflectors inserted in the crown of the vaulting.

The use of standards with X-Ray reflectors for lighting show rooms, stores, hotel lobbies, etc., is well exemplified in the lighting scheme of the Commodore Hotel, New York.

The second method of lighting churches, that is, by the semi-indirect method, is well exemplified in the installation recently completed at the Roman Catholic Church of Gesu, Philadelphia. The Church is one of the largest in the country, having a nave vault 100 feet high and 78 feet clear span, and a length of 147 feet. The original lighting consisted of side wall gas brackets and electric standard on the pew divisions, a very unsatisfactory scheme. This has all been superseded by ten large luminaires, hung from the vaulting, the bottom of which are 30 feet from the floor. The chandeliers are arranged in two rows, 46 feet apart, with the fixtures spaced 22 feet from each other on each side. Each fixture has 1400 watts, arranged with one large central unit of 500 watts and twelve small ones of 75 watts each. The average foot-candle illumination is about $1\frac{1}{2}$ foot-candles, and is very evenly distributed, the maximum being $1\frac{3}{4}$ foot-candles. The building is more than amply illuminated at this intensity, there being no doubt that with a lower foot-candle intensity the illumination would be satisfactory. Each fixture is controlled with two separate circuits so that it is possible to distribute the lighting at will, according to the needs. See figures 5 and 6.

ABSTRACTS

In this section of the TRANSACTIONS there will be used (1) ABSTRACTS of papers of general interest pertaining to the field of illumination appearing in technical journals, (2) ABSTRACTS of papers presented before the Illuminating Engineering Society, and (3) NOTES on research problems now in progress.

THE PRESENT STATUS OF VISUAL SCIENCE*

BY L. T. TROLAND**

The Monograph, "The Present Status of Visual Science," recently published as a Bulletin of the National Research Council, is officially the product of the Monograph Sub-Committee of the Committee on Physiological Optics of the Council, although the present writer must acknowledge responsibility for its contents and form.

A word may first be said concerning the Committee on Physiological Optics. This committee was the outcome of an original conference held between Dr. Augustus Trowbridge, then chairman of the Division of Physical Sciences of the Research Council, and Professor F. K. Richtmyer in Chicago early in 1921. Suggestions there developed recognizing the need of a survey and coördination of work in the field of physiological optics, took concrete form in a meeting of a group of men interested in the subject in New York shortly afterwards. This meeting resulted in the formation of a formally authorized committee consisting of Mr. Adelbert Ames, Professor W. T. Bovie, Dr. P. W. Cobb, Mr. L. A. Jones, Dr. W. B. Lancaster, Dr. P. G. Nutting, Mr. I. G. Priest, Professor J. P. C. Southall, Dr. L. T. Troland and Professor F. K. Richtmyer, chairman. The committee held numerous meetings during 1921 and 1922 and discussed problems relating to the progress of visual science. Valuable conferences were held between the committee and other groups of scientists interested in optics, physiology or allied subjects, one of the meetings being at Woods Hole and another at the Harvard Medical School. One of the most important practical results of the work of the committee consisted in the forma-

*Research Council Monograph.
**Dept. of Psychology, Harvard University, Cambridge, Mass.
The Illuminating Engineering Society is not responsible for the statements or opinions advanced by contributors.

tion of a section of the Optical Society of America relating to physiological optics and guaranteeing the inclusion in the annual program of this Society of a considerable number of papers dealing with this subject. Another very important enterprise which is at present under way under the editorship of Professor J. P. C. Southall is the translation into English of Helmholtz's great Handbook of Physiological Optics; an enterprise which is officially sponsored by the Optical Society, but the initiation of which may be attributed to the Research Council Committee. A third undertaking was the preparation of the Monograph to which the present note refers.

The purpose of this Monograph was to outline briefly the general field of research denoted or connoted by the term "physiological optics," to evaluate the present state of knowledge in this field, and, if possible, to suggest the most profitable lines for future research. In spite of the relatively small scope of the subject in comparison to the whole body of science this project was so ambitious a one that it could scarcely be carried through with any hope of perfection of execution. However, an attempt was made to sketch the situation in a way which may prove helpful to those interested in the subject.

It is a striking feature of visual science that it involves in an almost equally balanced degree the three separate sciences of physics, physiology and psychology, so that workers upon visual problems are nearly always hindered by lack of acquaintance with one or more of these general subjects. The Monograph in its published form attempts to combine the various factors derived from physical, physiological and psychological studies into a coherent system and to show how researchs undertaken from seemingly discrepant points of view may fit together harmoniously. No attempt is made to summarize established facts in detail, although in each department of the subject the discussion is made sufficiently concrete to render the argument substantial. The principal purpose of the brief summaries of extant results is to reveal the vacant spaces which exist between them and thus to indicate needed lines of research.

The Monograph is divided into five chapters. The first chapter deals with the position of visual optics among the sciences, including a brief historical sketch to establish the necessary perspective, this sketch being followed by a study of certain general characteristics of present visual knowledge, including its peculiar weaknesses. The second chapter deals with the fundamental conceptions and methods of visual science. An attempt is here made to specify the ultimate factors in the problem of vision, including an analysis of the general physiological system which is involved, its relation to the stimulus and to consciousness. The principal methods, psychological, physiological and physical, which are available to the visual investigator are next considered, and the final section of the chapter deals with the utility and requirements of theories in visual research.

The last three chapters of the Monograph are concerned with concrete problems in the science. Chapter III considers problems in the analysis of visual experience, or the purely psychological aspects of the subject. In this chapter a discussion is offered of the system of colors, regarded as psychological qualities, together with color nomenclature. The general properties of the visual field and of visual space are also outlined, important lines of research being suggested in connection with both of these general topics. The fourth chapter deals with the physiological and other physical factors in vision reviewed entirely apart from their psychological or conscious concomitants. This chapter contains six sections dealing in order with visual objects and stimuli, the dioptric and allied processes of the eye, the retinal stimulation, the afferent nerve excitation and conduction, the central processes in vision, and oculomotor mechanisms. Under each of these topics there is a sketch of existing knowledge combined with suggestions for further research.

The fifth and last chapter deals with the salient problems of visual psychophysiology, which concern relations between the psychological and the physiological factors discussed in the two preceding chapters respectively. The psychophysiological problems are divided as follows: brilliance vision, chromatic vision, form vision, motion vision, visual relations essentially involving time, visual relations essentially involving pattern or position, and the explanation of visual psychophysical correlations. In this chapter, as in previous ones, facts are outlined mainly for the purpose of suggesting the gaps which exist between them. The writer makes no pretense to complete comprehensiveness or accuracy in the presentation of facts or problems, although he has attempted to cover the field in as adequate a way as is possible within the scope of 120 pages. The references to literature number 268, but literature subsequent to the year 1920 is not considered. The discussion is limited primarily to human vision.

It is hoped that the above comment may serve to bring the Monograph to the attention of members of the Illuminating Engineering Society and others who are interested in the progress of knowledge concerning vision.

GLARE AND ITS RELATION TO EYE SIGHT CONSERVATION*

BY F. C. CALDWELL**

I am very pleased to come into this movement for eye-sight conservation, and I hope to bring to the work some advantage because of my connection with illuminating engineering.

When our Director asked me to speak, it seemed that it might be best to pick out one phase of the relation of illumination to eyesight conservation and to dwell on that in the hope that those of you who have not given much thought to this relation may have firmly fixed a few definite ideas.

That good illumination is one of the most important factors in the conservation of eyesight is a truism as old as the hills. I doubt not that among the earliest recollection of all of us are admonitions against reading in insufficient light and instructions concerning the direction from which the illumination should come. Indeed such admonitions and instructions indicate two of the three most important defects found in lighting—that is inadequate illumination and improper distribution. We were also told in our earliest years not to look at the sun, and without reflections upon the sun as an illuminant, there was indicated the third and perhaps the most serious of the defects of bad lighting—glare.

It is because glare is not only the most common and serious defect in lighting at the present day, but also because it is the least

*Address delivered before the Eye Sight Conservation Council, Feb. 6, 1923.
**A.B., M.E., Professor of Electrical Engineering, Ohio State University; member of Council, Illuminating Engineering Society; member Board of Directors, Eye Sight Conservation Council.

The Illuminating Engineering Society is not responsible for the statements or opinions advanced by contributors.

understood, that it has been chosen as the subject of the paper which I am to have the pleasure of giving this evening.

Before examining in detail the problem of glare it will be well to remind ourselves of some of the facts concerning the development of illumination. Lighting is one of the oldest of the human arts, but until recently one of the slowest to develop. So far as their characteristics as light-sources were concerned, there was very little difference between the candles or whale-oil lamps of our grandmothers and the selected fagots of the cave dwellers of fifty-thousand years ago. Even the gas and kerosene oil flames of the end of the last century, even carbon incandescent lamps were only a little larger and a little brighter. But the curve of development which for so many milleniums had run along almost horizontal, was commencing to trend upward and since the beginning of the present century with the development of the tungsten filament and the gas-filled lamp, it has risen even more sharply. One can safely say that in the past half century there has been more development in lighting than in all the ages before since man first picked a brand from the fire to light his way.

These facts of history are brought out to emphasize the magnitude of the task which lies before us in becoming accustomed in so short a time to these enormous changes in the condition of lighting. The three most striking differences in light sources that have come to pass are—greatly increased light-flux or candlepower, greatly increased brightness and greatly reduced cost, thus making possible very widely extended use of artificial light, and at the same time vastly magnifying the problems involved in its application. This extension of artifical illumination has brought with it a burden for the eye, which all too quickly it must learn to bear.

This is a point which should be particularly emphasized, the human eye has gone along through countless years, with little change in what was demanded from it. It worked through the daytime, for the most part looking at rather large objects, and when it became dark, it generally ceased to function. Now, within just a few years this has all changed. The use of the eye for fine work, for reading and for fine manufacturing operations, has been very greatly increased and then in addition, we are demanding that it do much of its work under artificial light which, furthermore, in many cases is far from good.

The change of lighting facilities has been so sudden, that men, trying to use the new illuminants in the same way that they have used the old thru countless generations, are likely to produce bad results. Of old, the problem of lighting was little more than the problem of getting the work near enough to the puny flame so that it could be seen, and few and far between were those who did work which needed even so much light. It is not strange, therefore, that many make sorry work of it when they set out unaided and without study to provide themselves with the new light.

The importance of glare has been recognized by regulations against it in the industrial and school lighting codes, drawn up by the Illuminating Engineering Society and followed more or less closely in the laws of several states.

Glare is not an easy phenomenon to define. It will perhaps be best to study first its different forms and characteristics and then to sum up our observations in a definition. The magic number "three" runs thru this subject in a rather remarkable way. We have already observed three changes in light sources and three defects in lighting. Now we may note three elements in the production of glare. These are excessive brightness, excessive volume of light flux and excessive contrast.

Though excessive brightness is probably the commonest and most generally recognized source of glare, the other two are also deserving of careful study. The seriousness of glare from too great brightness has enormously increased during recent years on account of the introduction of the tungsten filament lamps, especially the gas-filled type with spiral filament. Even the earlier lamps were too bright to make their use without screening devices desirable, but the consequences of operating the present types in this way are quite intolerable except where the lamps are hung very high. Even where globes or deep reflectors, which hide the lamps, are used, consideration must be given to the quality of the glass. If it is of too low a density, the brightness may still be too great for safety and comfort to the eyes. On the other hand many kinds of glass which are dense enough to cut down the brightness adequately, absorb too much of the light. The best makes of glass available when used in large enough globes, pass about eighty percent of the light while still sufficiently reducing the brightness. With the right kind of glass the brightness decreases as the size of the globe increases. Thus a 100-watt lamp in a 14-inch globe would give so low a brightness as to be satisfactory for any location, while the same lamp in a 6-inch globe would give a brightness about equal to that of the clear sky and would be at about the limit even when placed well up in the room where it would not constantly come into the field of vision of the occupants. With some types of diffusing glass, the brightness is not uniform, but is greatly increased at a spot in line with the filament. In such cases the above reasoning does not hold and the maximum brightness must be reckoned with.

The second source of glare, that is excessive volume of light flux, is likely to occur wherever powerful light sources, even of comparatively low brightness are in the field of vision and close to the observer. The influence of amount of flux on glare is also shown in the fact that a large light of a given brightness will be more glaring than a small one of the same brightness. On one occasion the writer experienced a quite uncomfortable case of this type of glare when at a birthday party, the hostess seated the guests at small tables with a birthday cake carrying a number of candles in the center of each table. While the brightness was low and the total light flux was not large, the fact that its source was so near, resulted in a relatively large entrance of flux into the eyes. The familiar cases of glare from a window, opening to the sky and of reflection from snow and water also involve the element of large light flux.

Glare of the third type—that due to excessive contrast, is well illustrated by the difference between the glare from an automobile headlight in the daytime amid bright surroundings, giving little contrast and the same light at night, when surrounded by black darkness, with a maximum of contrast. A window which, placed in a white wall, might cause no discomfort, might become trying if the wall were painted a dark color. Again a luminaire so designed as to light up the ceiling around it may not appear glaring, while another of the same brightness which leaves the ceiling dark may be trying to the eyes.

In this discussion the three sources of glare have been considered separately, and it is indeed possible to produce glaring conditions where only one of the above causes is present in a notable degree. On the other hand it is very common to find two of these causes present or even all three in a single instance of glare.

Having given a source of glare, due to either of the above causes, there are three conditions not associated either with the light source itself nor the condition of the eyes which determine its seriousness to an observer. These are first the angle which the source of glare makes with the line of vision of the observer. Thus if light is nearly overhead no glare will be experienced, no matter how bright or how large it may be; then as the angle decreases the unpleasant sensation increases till, when the eye is directed toward the source of glare the effect will be at its maximum. It is also true that for a given angle the glare is greatest if the light is below the eye and least when it is above. For the latter case the effect of the glare is not generally serious except for large and bright light sources, when the angle between the light and the line of vision is greater than 30 degrees. The line of vision is generally taken as horizontal, for while it may be directed downward for work, it generally reaches the horizontal frequently for rest or for distant vision.

The second condition determining the seriousness of glare is the distance away of the light-source. It is quite obvious that at a sufficient distance, any source of glare would cease to be trying. The effect of the third condition is equally obvious, namely, the period during which the eye is exposed to the glare. Thus any case where the exposure is continuous is much more troublesome than one where it is intermittent, with short periods of exposure.

Evidently the seriousness of any case of glare will be largely determined by the condition of the eye as to pupillary area and other circumstances affecting its sensitiveness. The glare upon coming into a brightly lighted room from the dark is a familiar experience.

When we consider the effects produced by glare we find another of our series of trinities. These effects are temporary impairment of vision or blinding, discomfort, and eye fatigue, often leading to injury. Impairment of vision comes both from the reduction of the pupillary area and from the direct effect of the light entering the eye. It is often more serious than is ordinarily realized. The discomfort, which protects the eye by compelling it to turn away from the sun or other excessively bright lights is an important natural safety measure. It is however generally the light of lesser flux or brightness which is not so readily recognized as a source of glare, that is most likely to cause fatigue and injury to the eyes.

To complete this discussion of glare, mention should be made of a special though common and important case, namely, glare due to reflection from polished surfaces. As such reflections generally come from below, their effect is magnified by the greater sensitiveness of the eye to light from this direction. Common cases of this are reflection from bright materials upon which work is being done, glass or polished wood table tops and glossy paper. In the latter case an effect is produced which is known as veiling glare. The bright light reflected from the surface of the paper to the eye largely masks that coming from the type and greatly increases the eye strain involved in reading. Much effort has been made in recent years to encourage the use of paper with a mat surface, especially in school-books, so as to reduce this troublesome phenomenon.

It will now be in order to attempt a definition of glare. Perhaps as good a one as any is that "glare is any condition of light entering the eye which produces blinding, discomfort or injury." Glare has also been tersely defined as "light out of place."

From what has been said it will be readily understood that glare, unlike intensity of illumination, is not easy to measure, nor can very satisfactory specifications be drawn up as to permissible limits. A fairly successful approximation to such specifications has, however, been made in recent editions of the industrial lighting codes mentioned above. Here cases of potential glare have been classified according to harshness into ten grades. These grades depend upon combined candlepower and brightness and upon the type of luminaire or light source. Then tables are given showing the grade of potential glare permissible in different kinds of lighting and in different positions relative to the eyes of the workers.

To sum up the discussion of glare then, we have the following group of trinities:—

Three defects of lighting—inadequacy, bad distribution and glare.

Three elements in the production of glare—excess of brightness, excess of light flux or excessive contrast.

Three conditions determining the seriousness of glare—angle above the horizontal, distance away and period of exposure.

Three effects of glare—blinding, discomfort and fatigue or injury.

SOCIETY AFFAIRS

SECTION ACTIVITIES

NEW ENGLAND Meeting—May 11, 1923

At the May meeting of the New England Section which was held on the evening of May 11 at the Industrial Lighting Exhibit in the Rogers Building, Mr. Guy Lowell presented a paper, "Illumination from the Architect's Point of View."

For some time it has been apparent that some method should be evolved whereby the viewpoint of the architect on illumination might be interchanged with that of the illuminating engineer. There has often been an unfortunate tendency to consider the illumination and lighting outlets for buildings as incidental details, and to solve the problem by rule of thumb or mere routine, often to the neglect of the wonderful utilitarian and aesthetic values of high grade lighting. On the other hand, there is tendency of the lighting engineer to lose sight of the aesthetic side of illumination in his desire to provide an efficient installation from the engineering point of view.

From the standpoint of lighting, Mr. Lowell divided buildings into two classes, those built for commercial or industrial purposes and those built for recreation. In the commercial or industrial building, the architect is glad to accept the advice of the engineer regarding the lighting. In buildings erected for recreation, such as museums, residences, etc., Mr. Lowell hoped that some method would be evolved whereby the engineer might more closely absorb the ideas of the architect and install lighting in accordance with the general artistic treatment.

Many problems and their solution were related, several of them being quite daring and spectacular. In the Boston Museum of Fine Arts it was found desirable to install a window equipped with artificial daylight to obtain the proper effect in a display located some distance from any windows. The interest and upkeep on the artificial window was found to be much less than if an actual window were constructed.

At the close of the talk an interesting discussion was held in which representatives of the fixtures houses took part. There were in attendance about fifty members and guests.

PHILADELPHIA Meeting—May 17, 1923

The Philadelphia Section met at the Engineers' Club on the evening of May 17 to hear a paper, "Modern and Spectacular Lighting" presented by Mr. W. D'Arcy Ryan, Director of the Illuminating Engineering Laboratory of the General Electric Co., Schenectady, N. Y.

In view of the general and popular interest in the paper the section observed "Ladies' Night." Preceding the meeting dinner was served to fifty members and guests at the Arcadia Cafe.

Mr. Ryan described the lighting effects at the Panama-Pacific Exposition and also the illumination of the Brazilian Centennial Exposition held during the past winter at Rio de Janeiro. The paper was especially interesting on account of the approaching Sesqui-Centennial Exposition to be held in Philadelphia in 1926, and was illustrated by a large number of lantern slides.

Col. John Price Jackson, Chairman of the Sesqui-Centennial Committee also spoke on behalf of the proposed exposition and showed slides of the Centennial Exposition which was held in 1876 at Philadelphia. There were in attendance one hundred and seventy-five members and guests at the meeting.

NEW YORK Meeting—May 11, 1923

A joint meeting of the New York Section and the Northern New Jersey Chapter was held in Paterson, N. J., on the evening of May 11. Preceding the meeting dinner was served to sixty-five members and guests at Faust Restaurant which was followed by an inspection of the lighting installation of the Cooper Hewitt Mercury Vapor lighting system in use at the National Silk Throwing Co. of Paterson.

The meeting was held in the Chamber of Commerce Rooms and two interesting papers on silk mill lighting were presented. Mr. W. J. Winninghoff read a paper, "Silk Mill Illumination with Mercury Vapor Lamps" by Mr. C. F. Strebig, Cooper Hewitt Electric Co., Hoboken, N. J., and Mr. H. W. Desaix, Watson-Flagg Engineering Co., Paterson, N. J., presented a paper, "Silk Mill Illumination with Incandescent Lamps."

There was considerable discussion of the papers and the attendance eighty-five members and guests.

NORTHERN NEW JERSEY Meeting—June 11, 1923

A joint meeting of the Newark Master Electricians' Association and the Northern New Jersey Chapter was held on the evening of June 11 at the Eagles' Home in Newark, N. J.

The meeting was addressed by Mr. S. G. Hibben, Manager of the Illumination Bureau of the Westinghouse Lamp Co., on the topic, "Practical Electric Lighting." In presenting the subject, Mr. Hibben discussed the following points: Light, and how it is measured; Characteristic distribution of light from various types of luminaries; Planning lighting installations; and the Value of good lighting.

There were eighty-six members and guests in attendance at this joint meeting and a general discussion was held by the members and guests present. A rising vote of thanks was tendered Mr. Hibben for the presentation of the talk.

TORONTO Activities

At the Toronto Chapter meeting on March 26, Mr. G. G. Cousins of the Hydro Electric Power Commission gave an interesting talk on "Characteristics of Glassware."

Mr. F. T. Groome presented a paper, "Difficult Problems in Illumination" before the chapter on the evening of April 23.

The annual meeting was held on May 28 at which the following officers for the coming year were elected: Chairman, Mr. W. H. Woods; Secretary, Mr. J. T. Scott; Executive Committee, Messrs. R. D. Albertina, R. M. Love, W. Orr, and G. R. Anderson, ex-officio.

COUNCIL NOTES

ITEMS OF INTEREST

At the meeting of the Council, May 10, 1923, the following were elected to membership:

Six Members

BAYLISS, ROGER V., American Gas Accumulator Co., Elizabeth, N. J.

CROSBY, JOSEPH G., Whalen Crosby Electric Co., 140 N. 11th St., Philadelphia, Pa.

ELLIOTT, E. LEAVENWORTH, Cooper Hewitt Electric Co., Hoboken, N. J.

HOBBS, LEONARD A., St. Louis Brass Mfg. Co. & Brascolite Co., 1331 W. 7th St., Los Angeles, Cal.

TAGGART, RALPH C., Dept. of Architecture, Capitol, Albany, N. Y.

VAN GILLUWE, FRANK, Western Electric Co., 301 E. 8th St., Los Angeles, Cal.

Sixteen Associate Members

BALKAM, HERBERT H., Consumers Power Co., 252 W. Main St., Jackson, Mich.

COLLAR, OLCOTT N., Sargent & Lundy, 1412 Edison Bldg., Chicago, Ill.

DESKINS, HIRAM T., Kilgore Electric Co., Box 371, Williamson, W. Va.

ETESON, FRANKLIN C., Blackstone Valley Gas & Elec. Co., 231 Main St., Pawtucket, R. I.

FLOWERS, DEAN W., St. Paul Gas Light Co., 51 E. 6th St., St. Paul, Minn.

GLAMEYER, WILLIAM, JR., United Electric Light & Power Co., 514 W. 147th St., New York, N. Y.

GROSSBERG, ARTHUR S., Albert Kahn, 1000 Marquette Bldg., Detroit, Mich.

HALE, H. S., Westinghouse Elec. & Mfg. Co., 717 S. 12th St., St. Louis, Mo.

LOCKER, FRANK H., Detroit Edison Co., 2000 Second St., Detroit, Mich.

MATCHETT, FRED D., Victor Electric Supply Co., 131 Jefferson Ave., E. Detroit, Mich.

QUIVEY, WYLLIS E., Benjamin Electric Mfg. Co., 847 W. Jackson Blvd., Chicago, Ill.

ROBINS, ORRIN A., Electric League of Columbus, 9 E. Long St., Columbus, O.

STREBIG, CHARLES F., Cooper Hewitt Electric Co., 95 River St., Hoboken, N. J.

TIMM, EDWARD W., Western Electric Co., 458 Milwaukee St., Milwaukee, Wis.

TURNBULL, T. S., Tallman Brass & Metal, Ltd., Wilson St., Hamilton, Ont., Canada.

WAGSCHAL, GEORGE, George D. Mason & Co., 508 Griswold St., Detroit, Mich.

One Transfer to Full Membership

MAHAN, HOWARD E., General Electric Co., Schenectady, N. Y.

The General Secretary reported the death, on April 22, 1923, of one associate member, C. N. Jelliffe, American Light and Traction Co., New York City.

CONFIRMATION OF APPOINTMENTS

As members of the Committee of Tellers—E. L. Bradbury, Chairman; D. W. Atwater, R. H. Maurer, H. H. Millar, and A. S. Turner.

As members of the General Convention Committee—Henry W. Peck, Vice-Chairman; H. E. Mahan, Secretary; Alexander Anderson, H. Calvert, Julius Daniels, E. Y. Davidson, Jr., Frank H. Gale, S. G. Hibben, Preston S. Millar, G. Bertram Regar, W. M. Skiff, J. L. Stair, G. H. Stickney and E. D. Tillson.

As member of the Committee on Research—Ernest F. Nichols.

The Council accepted with regret the resignation of Dr. Edward P. Hyde as Chairman of the Committee on Research.

The Council accepted the report of the Committee on Time and Place, designating the Convention headquarters at the Fort William Henry Hotel, Lake George, N. Y., September 24 to 28, 1923.

The Council approved the amendment of By-law (a), Section 1, Article IV, of the Constitution to read as follows: The entrance fee for Members and for Associate Members shall be $2.50. Remittance for entrance fee and current dues shall accompany the application.

At the meeting of the Council, June 28, 1923, the following were elected to membership:

Three Members

Curtis, Kenneth, National X-Ray Reflector Co., 235 W. Jackson Blvd., Chicago, Ill.

Gould, Herman P., Eastman Optical Shop, Inc., 12 Maiden Lane, New York, N. Y.

Odenath, Harry E., Sears Roebuck & Co., 4640 Roosevelt Blvd., Philadelphia, Pa.

Twenty-two Associates

Besinsky, Vaclav, Czechoslavak League of Electrotechnics, 1 Palackého Ulice, Prague VII, Czecho-Slovak Republic.

Brown, Willard C., National Lamp Works of G. E. Co., Nela Park, Cleveland, Ohio.

Domoney, Earl R., Consumers Power Co., 134 S. Washington St., Saginaw, Mich.

Donahue, Rev. Joseph N., Columbia University, Portland, Ore.

Dunn, J. M., Radio Appliance Co., 123 Pleasant St., Morgantown, West Va.

Fleming, E. F., Central Electric Co., 316 W. Wells St., Chicago, Ill.

Gray, Samuel McK., Electrical Testing Laboratories, 80th St. & East End Ave., New York, N. Y.

Haas, O. F., National Lamp Work of G. E. Co., Nela Park, Cleveland, Ohio.

Hannum, J. E., Eye Sight Conservation Council of America, 1206 Times Bldg., New York, N. Y.

Hartman, Harris V., New York Edison Co., 130 East 15th St., New York, N.Y.

Hill, Marvin, Benton, Kansas.

Hinton, James W., Westinghouse Lamp Co., 1005 Market St., Philadelphia, Pa.

Humez, J. F., Macbeth-Evans Glass Co., 5-134 General Motors Bldg., Detroit, Mich.

Humphrey, Arthur F., Biddle-Gaumer Co., 3846-56 Lancaster Ave., Philadelphia, Pa.

Johnston, Richard J., George Cutter Works of Westinghouse Elec. & Mfg. Co., South Bend, Ind.

Kase, Daniel B., Rumsey Electric Co., 1007 Arch St., Philadelphia, Pa.

Laughton, Abbot A., Athol Gas & Electric Co., 426 Main St., Athol, Mass.

Mausk, Raymond E., National Lamp Works, Nela Park, Cleveland, Ohio.

Nichol, H. G., Jr., Macbeth-Evans Glass Co., Chamber of Commerce Bldg., Pittsburgh, Pa.

Norris, George T., Philadelphia Electric Co., 1000 Chestnut St., Philadelphia, Pa.

Wilson, Elmer D., 209 Clinton Avenue, Newark, N. J.

Wise, John E., University of Wisconsin, Madison, Wis.

One Associate Member Reinstated

Kato, K., Tokio Electric Co., Kawasaki-Machi, Kanagawa-Ken, Japan.

One Transfer to Associate Membership

Trimming, Percy H., Dominion Flour Mills, Ltd., Montreal, Canada.

The General Secretary reported the deaths of two members and two associate members: Dr. Louis Bell, 120 Boylston Street, Boston, Mass.; Professor A. G. Webster, Clark University, Worcester, Mass.; and Mr. Uhl M. Smith, Bureau of Standards, Washington, D. C.; Mr. C. A. Strong, 79 Milk Street, Boston, Mass.

CONFIRMATION OF APPOINTMENTS

As members of the General Convention Committee—N. R. Birge, B. S. Beach, W. T. Blackwell, S. H. Blake, A. D. Cameron, S. E. Doane, W. L. Robb, C. P. Steinmetz, C. D. Wagoner, D. B. Taylor, H. F. Wallace, F. H. Winkley and L. A. S. Wood.

As member on the Advisory Committee, Engineering Division, National Research Council—Dugald C. Jackson.

As Chairman of the Committee on Research—Ernest F. Nichols.

The General Secretary presented a report of the letter-ballot on the amendment of By-law (a), Section 1, Article IV, of the Constitution showing a concurring vote of the majority of the entire Council.

The Council approved by letter ballot the granting of a petition for the organization of a chapter covering southern California, with headquarters in Los Angeles, to be known as the Los Angeles Chapter.

Committee Reports

The General Secretary presented the report of the Committee of Tellers which met on May 29, 1923. The Council instructed the General Secretary to give the names of the newly elected officers to the technical press.

NEWS ITEMS

ADOPTION OF THE INDUSTRIAL LIGHTING CODE IN PENNSYLVANIA

According to the May *Bulletin of Information* issued by the Industrial Board, Department of Labor and Industry, Commonwealth of Pennsylvania, the Industrial Lighting Code which has been in the process of revision for a year was adopted by the Industrial Board at its May 10, 1923 meeting.

The revised code is based on the National Lighting Code adopted by the American Engineering Standards Committee as American Standard. The public hearings developed the necessity for changing some of the intensity requirements provided in the National Code, and for providing minimum and recommendatory intensities for a number of industries not covered by the National Code. Rules on Emergency Lighting and Protection against Explosion are also included in the revised draft.

In addition to the general rules, the code contains two tables. Table I is to be used in determining the illumination necessary for industrial establishments. Table II gives the minimum and recommendatory intensities of illumination for various industries and occupations. The following is an abstract of the code as adopted:

"RULE 378. DISTRIBUTION OF LIGHT.

(a) (A-1) Lamps shall be installed in regard to height, spacing, reflectors or other accessories to secure a good distribution of light on the work, avoiding objectionable shadows and excessively sharp contrasts.

"RULE 379. EMERGENCY LIGHTING.

(a) (A-1) All ways or egress of means of escape in establishments wherein persons are employed after darkness shall be provided with a reliable emergency electric lighting circuit, of a type to be approved by the Commissioner of Labor and Industry.

Such emergency lighting shall have a minimum intensity of 0.50 footcandle on the space.

"RULE 380. INTENSITY OF ILLUMINATION.

"RULE 381. (a) Gas, vapor and dust-proof lighting fixtures shall be provided at all places where explosive gas, vapor, or dust accumulate."

The Industrial Board was assisted by the following persons in the drafting of this code:

Earl A. Anderson, Illuminating Engineering Society, Cleveland; Maurice L. Crass, Grasselli Chemical Co., Cleveland; E. Y. Davidson, Jr., Macbeth-Evans Glass Co., Pittsburgh; H. B. Harmer, Philadelphia Electric Co., Philadelphia; B. E. Hatch, Westinghouse Electric Co., Philadelphia; W. J. Hart, Jones and Laughlin Steel Corp., Pittsburgh; Howard Heslip, Duquesne Light Co., Pittsburgh; Ward Harrison, Illuminating Engineering Society, Cleveland; J. J. Minnick, Westinghouse Electric Co., Philadelphia; W. E. Megraw, H. H. Robertson Co., Pittsburgh; A. A. McLean, Travelers Insurance Co., Pittsburgh; J. W. Pollock, Baldwin Locomotive Works; Miss G. M. Pugh, Consumers' League; Wm. J. Serrill, Illuminating Engineering Society, Philadelphia; Charles Thomas, American Bridge Co., Ambridge, Pa; Walter C. Titus, Jones and Laughlin Steel Co., Pittsburgh.

G. BERTRAM REGAR WINS DOHERTY PRIZE

In 1910 Mr. Henry L. Doherty provided for the annual presentation of a gold medal to the author of the best paper presented before a section of the National Electric Light Association. The prize this year was won by Mr. G. Bertram Regar of the Philadelphia Electric Company, his paper being entitled "More Business through Better Lighting." It is understood that this is the first time that the award of the Doherty prize has been made to the author of a paper dealing with lighting. It will therefore be doubly gratifying to members of this Society to note that this important prize goes to so prominent a member of this Society and that the subject of lighting has thus received such favorable attention at the hands of central station men.

Mr. Regar's distinguished services to the Society include his successful administration of the Committee on Membership during the past two years, during the latter of which he has served also as Chairman of the Lighting Sales Bureau of the National Electric Light Association. Once more attention is thus drawn to the fortunate coincidence of public interest with the interests of the light and power companies in the development of improved illumination. Altruism and commercial interest go hand in hand and many authors like Mr. Regar are promoting the public interest while promoting their private interests through activities directed at the improvement of lighting conditions.

NEW OFFICERS

At the June meeting of the Council the report of the Committee of Tellers was read and accepted. According to the returns reported by the committee the following have been elected to the offices indicated:

General Officers

President, Mr. Clarence L. Law, New York City.
General Secretary, Mr. Samuel G. Hibben, New York City.
Treasurer, Mr. Louis B. Marks, New York City.
Vice-President, Mr. D. McFarlan Moore, Harrison, N. J.
Directors, Messrs. James P. Hanlan, Newark, N. J., Howard Lyon, Gloucester, N. J., and H. F. Wallace, Boston, Mass.

Chicago Section Officers

Chairman, Mr. F. A. Rogers; *Secretary*, Mr. E. J. Teberg; *Board of Managers:* Messrs. A. L. Arenberg, W. S. Hamm, N. B. Hickox, W. E. Quivey and E. D. Tillson.

New England Section Officers

Chairman, Mr. Walter V. Batson; *Secretary*, Mr. Julius Daniels; *Board of Managers:* Messrs. Cyrus Barnes, A. W. Devine, W. S. Fitch, R. W. Hosmer, and J. A. Toohey.

New York Section Officers

Chairman, Mr. L. J. Lewinson; *Secretary*, Mr. J. E. Buckley; *Board of Managers:* Messrs. S. K. Barrett, H. W. Desaix, E. E. Dorting, J. R. Fenniman, and E. H. Hobbie.

Philadelphia Section Officers

Chairman, Mr. H. Calvert; *Secretary*, Mr. J. J. Reilly; *Board of Managers:* Messrs. H. B. Anderson, G. A. Hoadley, M. C. Huse, Howard Lyon, and E. L. Sholl.

NATIONAL ILLUMINATING COMMITTEE OF GREAT BRITAIN*

In the Journal of the Institution of Electrical Engineers, May, 1923, (London), there appears a report of the Chairman of the year 1922, and in view of its interest to members of the I. E. S., is reprinted below.

In February last (1922) the provisional Definitions of Photometric Terms and Units proposed by the British National Committee were published together with a prefatory note and have been officially adopted by the three constituent Societies. They also form the basis of a set of Photometric Definitions shortly to be issued by the British Engineering Standards Association as part of a comprehensive set of Electrical Engineering terms.

The Definitions in question, whilst agreeing with the decisions of the International Commission on Illumination held in Paris in 1921, go considerably further and are in some respects at variance with a set of Definitions approved in July, 1922, by the American Engineering Standards Committee.[1] The occasion of a visit by Dr. Clayton H. Sharp to this country in December last was seized upon to discuss these Definitions with one so largely instrumental in the drafting of the American Definitions. Dr. Sharp kindly consented to attend a Meeting of the Nomenclature Sub-Committee, and as a result of this interchange of views, the Sub-Committee are now considering how the proposed Definitions can be amended so as to minimize the points of difference between this country and the United States.

A preliminary list of Symbols has also been prepared by the Nomenclature Sub-Committee and, after submission to the British Committee, these have been communicated to a number of interested Societies, publication being deferred until their criticisms, if any, have been considered.

Dr. C. O. Mailloux (U.S.A.) and Mr. K. Edgcumbe (Gt. Britain) were asked by the central office of the National Illumination Commission to prepare an English translation of the French official text of Terms and Definitions adopted in Paris in 1921. A meeting was held in this country and the translation agreed upon. The text forms an Appendix to this Report.

At the 1921 Paris Meeting of the Commission an International Committee on Automobile Headlights was appointed and Mr. K. Edgcumbe was subsequently nominated by the British National Committee as their Representative thereon. A fairly complete set of recommendations having been drawn up in the United States by a Committee under the Chairmanship of Dr. Clayton H. Sharp, the subject was discussed with that gentleman on the occasion of his visit to this country, and at a subsequent interview with Mr. Perrin of the Ministry of Transport the question was raised of how the British National Committee could best serve the interests of this country in connection with Automobile Headlights. It appeared that the most useful course would be to appoint a Sub-Committee to consider the recommendations which had already been published in other countries, with a view, if possible, of arriving at common agreement through the medium of the International Headlights Committee.

*See *Institution Notes*, No. 30, page 11, January 1922.
[1] Illuminating Engineering Nomenclature and Photometric Standards, American Standard, approved July 11, 1922 by A. E. S. C.

In view of the fact that a large and increasing part of the work of the British National Committee relates to standardization, it was decided, with the approval of the three constituent Societies to ask the British Engineering Standards Association to form a sectional Committee on Illumination to which such matters could be referred. It is proposed that this Committee should deal solely with standardization or similar questions referred to it by the British National Committee, all international matters being dealt with by the National Committee as heretofore.

K. EDGCUMBE.

January, 1923.   *Chairman.*

PHOTOMETRIC DEFINITIONS

Official Translation of the French Text

Luminous flux.—Is the rate of passage of radiant energy evaluated by reference to the luminous sensation produced by it.

Although luminous flux should be regarded, strictly, as the rate of passage of radiant energy as just defined, it can, nevertheless, be accepted as an entity for the purposes of practical photometry, since the velocity may be regarded as being constant under those conditions.

The unit of luminous flux is the lumen.—It is equal to the flux emitted in unit solid angle by a uniform point source of one international candle.

Illumination.—The illumination at a point of a surface is the density of the luminous flux at that point, or the quotient of the flux by the area of the surface when the latter is uniformly illuminated.

The practical unit of illumination is the lux.—It is the illumination of a surface one square metre in area, receiving a uniformly distributed flux of one lumen, or the illumination produced at the surface of a sphere having a radius of one metre by a uniform point source of one international candle situated at its centre.

In view of certain recognized usages, illumination may also be expressed in terms of the following units:—

Taking the centimetre as the unit of length, the unit of illumination is the lumen per square centimetre; it is known as the "phot." Taking the foot as the unit of length, the unit of illumination is the lumen per square foot; it is known as the "foot-candle."

$$1 \text{ foot-candle} = 10.764 \text{ lux}$$
$$= 1.0764 \text{ milli-phot}$$

Luminous intensity (candlepower).—The luminous intensity (candlepower) of a point source in any direction is the luminous flux per unit solid angle emitted by that source in that direction. (The flux emanating from a source whose dimensions are negligible in comparison with the distance from which it is observed may be considered as coming from a point.)

The unit of luminous intensity (candlepower) is the International Candle, such as resulted from agreements effected between the three National Standardizing Laboratories of France, Great Britain and the United States in 1909.[*]

This unit has been maintained since then by means of incandescent electric lamps in these laboratories which continue to be entrusted with its maintenance.

[*] These Laboratories are: the Laboratorie Central d'Electricité in Paris; the National Physical Laboratory in Teddington, and the Bureau of Standards in Washington.

SUPPORT THE MEMBERSHIP DRIVE

The membership drive under the direction of Vice-President G. Bertram Regar of Philadelphia is proving fruitful of splendid results. A record of newly elected members of nearly two hundred and fifty is reported to date which is very gratifying to the officers and members of the I. E. S.

The drive will be earnestly continued for the next three months and the committee hopes to receive at least one hundred new applications during this period. To achieve this goal means the hearty cooperation and loyal support of *each individual member.*

The Constitution provides that membership dues shall date from the quarter of the fiscal year nearest the date of notice of admission to the applicant. For the last three months of the fiscal year the dues for the grade of *associate member* amount to $1.88 and for the grade of *member* $3.75; the entrance fee is $2.50.

Application blanks and membership literature can be secured from members of the committee or from the General Office.

Have *you* sent in a signed application? Have *you* sent to the chairman, Mr. G. Bertram Regar, 1000 Chestnut Street, Philadelphia, names of *prospective applicants?* Show your loyalty and interest in the membership drive by securing a new member.

I. E. S. CONVENTION AT LAKE GEORGE

The 1923 Convention of the Illuminating Engineering Society is to be held September 24 to 28 inclusive at Lake George, N. Y., famous the world over as the most picturesque resort in America. The lake is 32 miles in length, its width varies from three quarters of a mile to four miles, dotted with many islands and surrounded by majestic mountains. The headquarters of the Convention will be the Fort William Henry Hotel, possessing every modern convenience and attraction. Golf, tennis, boating and bathing are splendidly provided for.

Lake George is seventy miles from Albany, accessible by railroad and unexcelled automobile highways. A wealth of historical interest including old forts and battlegrounds prevails in the immediate vicinity.

A unique program of entertainment is being provided and unusual spectacular lighting features are being planned. It is hoped to combine business and pleasure at this Convention in a manner to enable all visiting delegates and particularly the ladies to enjoy the wealth of scenic beauty so abundant in this wonderful country of mountains and lakes.

A well balanced program of Commercial and Technical papers is being prepared by the Committee on papers under the direction of Mr. J. L. Stair of Chicago.

COMMITTEE ACTIVITIES

The General Convention Committee held its first meeting at the Society Headquarters on June 7, 1923. The general program of entertainment and other features of the convention was outlined by Chairman Ryan and a tentative schedule of events was adopted.

Convention Headquarters at Fort William Henry Hotel, Lake George, N. Y.

The Boat Landing.

View of Boat Landing from hotel.

Lake George, N. Y.

At the meeting of the Sectional Committee of the A. E. S. C. on June 4, 1923 the following officers were elected: Mr. L. B. Marks, Chairman; Mr. Sullivan W. Jones, Vice-Chairman; and Mr. W. F. Little, Secretary. This sectional committee is considering the revision of the Code of School Lighting.

A joint session of the Committee on Lighting Legislation and the Sectional Committee of the A.E.S.C. convened on June 4, 1923 at which time the discussion of the revised Code of School Lighting was held. The revised Code will be presented at the coming convention in September for general discussion by members of the I. E. S.

The Committee on Nomenclature and Standards met at the Society Headquarters on June 22, 1923 to consider the report to be presented at the convention and a summary of the changes in the American Standard which have so far been agreed to in previous meetings of the committee.

OBITUARY

Prof. Arthur Gordon Webster of Clark University, an eminent physicist, died from self-inflicted bullet wound on May 15, 1923. This tragic occurrence, which means a great loss to science caused amazement among Professor Webster's friends and associates, for he had shown no signs of depression and must, it is believed, have taken his life as the result of a sudden impulse. He had attained great success in many scientific lines and was recognized as one of the world's greatest authorities on electricity and sound. He was born in Brookline, Mass., was graduated from Harvard in 1885 and later studied in Paris, Berlin and Stockholm. After spending a year as instructor at Harvard, he entered the service of Clark University, where he was successively docent in physics, assistant professor of physics, professor of physics and director of the physical laboratory, holding the last-named offices at the time of his death. Professor Webster was the author of several important works on electricity and dynamics and during the war was made a member of the Naval Advisory Board of Scientists, of which Thomas A. Edison was chairman. All who knew him will mourn the loss of a steadfast friend and a delightful personality, and the electrical industry will sadly miss the fruits of his brain and the enthusiasm he inspired for scientific research.

Mr. Uhl M. Smith, of the Bureau of Standards, Washington, D. C., was instantly killed on April 21, 1923, when a Martin air service bombing plane nose-dived into the Great Miami river at Dayton, Ohio.

Mr. Smith was at Dayton in connection with some work on Colors for Traffic Signals in which he had taken a leading part. At the Bureau he was in charge of work on automobile lighting in general, and was a member of the I. E. S. Committee on Motor Vehicle Lighting.

BIOGRAPHY OF CLARENCE L. LAW

PRESIDENT-ELECT I. E. S.

1923-1924

Mr. Clarence L. Law, newly-elected President of the Illuminating Engineering Society, was born in New York City, April 15, 1885. His education was received in New York City elementary and high schools, supplemented by private tutors.

Mr. Law became connected with the electrical industry November 1, 1906, when he entered the employ of The New York Edison Company as a special inspector. A year later he was made Special Agent; in July, 1910, he was advanced to Manager of the Bureau of Illuminating Engineering, and was recently appointed Assistant to the General Commercial Manager, which position he now holds.

During Mr. Law's period of service with The New York Edison Company, he specialized in the field of illumination from the time when it became necessary to demonstrate the most economical and efficient use of light by means of improved and perfected lamps. Mr. Law's duties, during his association with The New York Edison Company, covered work pertinent to illuminating. engineering, large building oversight, surveys, lamp development, special outdoor decorative and spectacular lighting.

Mr. Law is affiliated with a number of prominent organizations, including several contributing to the development of the electrical industry. He is at present Chairman of the Board of Trustees of the Association of Employees of The New York Edison Company; Director, New York State Commission for the Prevention of Blindness; Member, American Association for the Advancement of Science; the Architectural League of New York; the New York Electrical League; the American Institute of Electrical Engineers; a former member of the Executive Committee of the Commercial Section of the National Electric Light Association. Mr. Law has also served on a number of Committees of the N. E. L. A., and is Past-Chairman of the Metropolitan New York Section.

Mr. Law joined the Illuminating Engineering Society in 1912, and has served in the capacity of Secretary of the New York Section; Acting Chairman, New York Section; Vice-President, for two years; General Secretary, for five years; and Director for one year, and is also a member of several committees. He has presented papers before the Society on different subjects pertaining to illuminating engineering.

In local movements, in New York City, Mr. Law is active in welfare work, and is connected with civic organizations such as The Merchants' Association, the Fifth Avenue Association and the Broadway Association.

CLARENCE L. LAW
PRESIDENT-ELECT I. E. S.
1923-1924

PERSONAL MENTION

Mr. F. M. Feiker, formerly vice-president of the McGraw-Hill Company, Inc., and more recently on leave of absence as special agent of the Department of Commerce at Washington will after his return from Washington, be associated with the staff of the Society for Electrical Development, New York City. As a result of the appointment of Mr. Feiker the various branches of the electrical industry served by the society will secure the benefit of his broad experience and background, for he will be available to act as a special counselor to engineers, manufacturers, central stations, jobbers, contractor-dealers and publishers. His special training and wide knowledge in the engineering publishing and public relations field of many industries, qualify him eminently for such consulting work. Mr. Feiker will retain a consulting relation to the McGraw-Hill Company, and he will continue in a similar capacity his relation to the problems of personnel and organization of the Department of Commerce at Washington.

Dr. A. S. McAllister, engineer physicist, Bureau of Standards, who during the past two years has been liaison officer of the U. S. Bureau of Standards and the Federal Specifications Board assigned to the headquarters of the American Engineering Standards Committee at New York City, has been recalled to Washington for special work, by Secretary Hoover of the Department of Commerce. Dr. D. R. Harper 3d, physicist of the Bureau of Standards, has been assigned to the American Engineering Standards Committee succeeding Dr. McAllister.

Mr. W. E. Clement, commercial agent of the New Orleans Public Service Co., was recently elected president of the Electrical League of New Orleans.

Mr. Walter H. Johnson, senior vice-president of the Philadelphia Electric Co., was elected president of the National Electric Light Association at the recent convention held in New York City.

Mr. Dudley Farrand, assistant to the president of the Public Service Corporation of New Jersey, was recently elected vice-president in charge of industrial relations.

Mr. Edwin F. Guth is president of the Edwin F. Guth Company, a new organization of the united interests of the St. Louis Brass Mfg. Co. and the Brascolite Co., recently formed in St. Louis, Mo.

Prof. Harris J. Ryan of Stanford University, California, was elected president of the American Institute of Electrical Engineers at its annual business meeting, held in New York last May.

Mr. J. R. Fenniman has recently been elected assistant treasurer of the Consolidated Gas Co. of New York City.

Mr. Franklin S. Terry, chairman of the advisory board of the National Lamp Works, Nela Park, Cleveland was elected vice-president of the General Electric Company at a meeting of the board of directors held in New York City on June 22. 1923.

Mr. B. G. Tremaine, vice-chairman of the advisory board of the National Lamp Works, Nela Park, Cleveland was elected director of the General Electric Company on June 22, 1923.

Mr. Samuel G. Hibben addressed the Canadian Electrical Association at their convention in Montreal on June 22, 1923 upon the subject of the industrial lighting codes and the reasons for their adoption by the various states.

Mr. D. W. Atwater presented an industrial lighting lecture, explaining the Pennsylvania legislation, at Ridgway, Pa., before the industrial plant engineers of that territory.

GENERAL OFFICE NOTES

The Code of Lighting Factories Mills and Other Work Places has been reprinted as Bulletin No. 331, U. S. Department of Labor, Bureau of Labor Statistics. Copies of this bulletin can be obtained from Mr. Ethelbert Stewart, U. S. Commissioner of Labor Statistics, Washington, D. C.

A request to publish the Code of Lighting, Factories Mills and Other Work Places has been granted M. Remy Delauney, Editor in Chief of the Bulletin of Labor Inspection, Paris, France. It is very gratifying to learn that a translation of the I. E. S. Code will appear in an early issue of the bulletin.

TRANSACTIONS for October, November and December, 1922, and January, 1923, are out of print. Please advise the General Secretary of any of these issues for sale and price will be quoted.

WALSH PAPER DISCUSSIONS

Through an inadvertence a paper entitled "Note on the Integrating Sphere Reflectometer" by John W. T. Walsh, was printed in the Abstracts Section of the May number without an explanatory note.

This paper is being presented to the membership under the auspices of the Sub-committee on Reflection Factor Measurements, and this committee invites written discussion of the paper which will be presented to Mr. Walsh when received, and the paper and discussions with Mr. Walsh's rejoinder will be brought before the convention at Lake George in September.

Discussions should be sent to Dr. Clayton H. Sharp, Chairman, 80th Street and East End Avenue, New York City.

ILLUMINATION INDEX

PREPARED BY THE COMMITTEE ON PROGRESS.

An INDEX OF REFERENCES to books, papers, editorials, news and abstracts on illuminating engineering and allied subjects. This index is arranged alphabetically according to the names of the reference publications. The references are then given in order of the date of publication. Important references not appearing herein should be called to the attention of the Illuminating Engineering Society, 29 W. 39th St., New York, N. Y.

TRANSACTIONS
OF THE
ILLUMINATING ENGINEERING SOCIETY

| Vol. XVIII | September, 1923 | No. 7 |

The Coming Convention at Lake George, N. Y.

LAKE GEORGE, "Queen of American Lakes," has been chosen as the place for the 1923 Convention of the Illuminating Engineering Society. This particularly happy selection will afford delegates to this year's meeting opportunities to combine business and pleasure amid incomparable scenery and ideal climate with all the adjuncts of diversified recreation. The Convention dates are September 24 to 28 inclusive. Special plans for the entertainment of the ladies are being made including boat rides on the lake, bridge, dancing, etc. An elaborate program of spectacular lighting and fireworks is also scheduled.

The Fort William Henry Hotel, Convention Headquarters, is one of the show places of this section. Situated at the south end of the lake it is near the site of old Fort William Henry, rich in historical lore connected with the early settler days and on the site of many bloody battles between the white man and the Indians. Lake George is at an altitude of 331 feet and is 70 miles from Albany. A beautiful motor trip from Albany or Schenectady is afforded through some of the most imposing scenery to be found south of the Adirondacks proper. Lake George in fact is the southern gateway to the Adirondack Mountains.

It is bordered on both sides for miles of its forty mile course with high wooded hills and towering cliffs. Dotted with many islands and innumerable bays, a vast panorama of sky, land and water greets the visitor from the wide veranda of the hotel looking up the lake to the north.

Lake George is reached via the Delaware & Hudson Railroad from Albany which road also operates the commodius steamers that ply up and down the lake. At the head of the lake are the ruins of famous Fort Ticonderoga. It is indeed a land abounding in profusion of natural beauty, a land of great scenic attractions, history and romance and alluring charm.

The Convention comes at a time when the full glory of the autumn foliage should be in evidence adding new colors to a spot which has aptly been called a Summer Paradise.

W. D'A. RYAN, *Chairman*

1923 General Convention Committee.

REFLECTIONS

The Year's Progress in Illumination

THIS report by the Committee on Progress, presented on page 583 of this issue, covers developments, not necessarily improvements, in the whole field of illuminating engineering as reported in the scientific and technical press. It constitutes what amounts to an annual history of the subject.

Motor Vehicle Regulations

THE Committee on Motor Vehicle Lighting in the report for the present year summarize their activities. Specifications for rear lamps have been drawn up and adopted, receiving the endorsement of the Society of Automotive Engineers. The Headlighting Specifications have become a tentative American Standard. The Conference of Motor Vehicle Administrators, consisting of the New England, Middle Atlantic States and Ohio, are approving headlighting devices under the Standard Specifications. The California Headlight law is the standard specifications with but slight modification. International relationship has been established through the Committee Chairman with the International Commission on Illumination, and work is in progress with England, France, and Switzerland. No further change can be made in the Specifications except through the Sectional Committee of the A. E. S. C.

Pageant Street Lighting

THE advent of the new spray-colored or diffusing bulb Mazda lamps has made possible some remarkable decorative effects in the spectacular lighting of streets for pageants and festivals.

One of the latest and most striking of such engineering accomplishments is described by the author, being the decorations of Washington, D. C., during the National Shriners' Convention of June 1923. This convention paper of extreme interest is contributed by Mr. Samuel G. Hibben of the Westinghouse Lamp Co. of New York City.

Preliminary Studies in the Response of Plants to Artificial Light

SEVERAL thousand vegetable seedlings raised in flats and a larger number of flowering plants started from cuttings, raised in pots, were placed under ten 500-watt, 110-volt Mazda C clear lamps from March 1, 1923 to April 4, 1923. The lamps were turned on at 9:00 p. m. and automatically shut off at 2:00 a. m. during the five weeks of the test. In addition to the artificial light the plants were exposed to sunlight. Check plants under sunlight only were duplicated in size and variety. As a result the plants under artificial light grew more rapidly and the flowering plants bloomed approximately 8 days earlier than the check plants. The chemical tests show approximately the same amount of chlorophyll in both groups of plants. A progressive series of illustrations as well as curves plotted on charts show the remarkable growth and production of forced plants.

Light, one of the most important external factors in the growth of the plant, may in a few years be supplied economically to the commercial grower which will mean that the crops, both vegetables and flowers will be raised in a shorter period, that there will be earlier productions in the spring and also in the bringing of the crops in on scheduled dates.

The author of this paper Prof. R. H. Harvey of University of Minnesota, has carried on an investigation which will prove very fruitful.

Carbohydrate Production and Growth in Plants Under Artificial Light

THE optimum conditions of light intensity for the growth of a great variety of plants have been determined using continuous artificial light. The intensity requirements of many plants is such that growth in artificial light alone is practicable in northern regions where winter sunlight is low and unreliable. Plants such as Easter Lilies can be speeded up in time of blooming to bring them into the market at a certain date. The use of artificial light makes it possible to force blooming of two varieties at the same time so that hybrids may be more easily produced. There is a correlation between the intensity of continuous illumination and the quantity and nature of the carbohydrates produced in photosynthesis. The production of male or female flowers by some dioecious plants is dependent upon light intensity.

This convention paper is a result of an investigation conducted at Columbia University under the direction of Prof. Hugh Findlay and should be extremely interesting to layman as well as the illuminating engineer.

Some Experiments on the Speed of Vision

THE question of the time required for light to make an adequate visual impression has recently become of fundamental interest to industrial lighting.

This "time of impression" will depend upon many conditions of vision, such as: (a) the size and distance of the object; (b) its brightness compared with that of its backgrounds; (c) the general level of brightness; (d) whether vision of form is necessary, or the simple "picking-up" of the object, and; (e) the reduction in effectiveness of the image of the object due to images of other objects seen immediately before and after it.

The working out of a complete law of the "speed of vision" will be a tremendous task, when representative variations of each of these factors have been considered. The present paper is a contribution to this work, and shows in what way the speed of impression of the eye changes with change in the brightness-level, under various conditions as to the size of the test-object, as to the necessity for seeing form as against the simple recognition of the presence of the object, and as to the presence or absence of confusion due to the images of other objects seen immediately before and after the test-object.

Dr. Percy W. Cobb has contributed an interesting paper, which is of value to industry at large.

Depreciation of Lighting Equipments Due to Dust and Dirt

IN this convention paper Mr. E. A. Anderson set forth a report of tests under service conditions to determine the relative depreciation or loss in efficiency of lighting equipments due to the accumulation of dust and dirt. Comparison tests were made under forced rates of dirt accumulation in an effort to determine the feasibility of obtaining quick comparisons between the depreciation rates of different equipments. Consideration of the possibilities of a simple comparison standard for predicting depreciation rates in a particular installation is included.

Further Studies of the Effect of Composition of Light on Important Ocular Functions

IN a paper presented to this Society in 1921 results were given showing the eye's acuity, speed of discrimination, and power to sustain acuity for spectrum lights of a high degree of purity, made equal photometrically at the test surface. The efficiency of the eye with regard to these three important functions was found to be greater in the mid region of the spectrum than towards either end. Spectrum lights equalized as to brightness, however, show a considerable difference in saturation. It became important, therefore, to determine whether the difference in the results obtained should be ascribed to differences in saturation or whether, for example, some hues are more favorable than others as a background for the discrimination of black test letters or printed characters. In the present investigation the spectrum lights were equalized both as to saturation and luminosity and the tests repeated for the same observers. An advantage, not so great but still considerable, was found for lights in the mid-region of the spectrum.

The investigation was supplemented by the fatigue test used in our earlier investigations. That is, the power to sustain a acuity was obtained before and after three hours of reading with the page illuminated by red, yellow, green, and blue filtered lights (dipped lamps), equalized in luminosity and saturation at the point of work. A comparison of the results before and after reading shows that the eye held its power to sustain clear seeing best under the yellow light. The test surface and reading page were illuminated by the same light and were carefully matched for each color in hue, saturation and brightness. A spectrophotometric analysis was made of the light reflected from the reading page. The yellow light was found by test also to show the least tendency to produce discomfort.

This convention paper is a continuation of a series of studies by the well known authorities, Dr. C. E. Ferree, and Dr. Gertrude Rand of Bryn Mawr College.

Artificial Illumination in the Iron and Steel Industry

THE convention paper, by Mr. W. H. Rademacher, describes an interesting phase of illuminating engineering.

During the last decade the application of artificial light in the Iron and Steel Industry has undergone marked changes with a distinct trend toward betterment. The modern incandescent lamp has rapidly displaced other forms of illuminants and in conjunction with modern reflecting equipment is today recognized standard. Altho much of the work involved in this industry is of a rough nature and does not necessitate lighting intensities of a relatively high magnitude, the requirements are nevertheless far from important. Chief among the credits to the account of modern lighting are safety insurance and the twenty-four hour day, attended by the successful coping with keen competition and the affection of economies in production.

The selection and application of equipment for the various areas embraced in plant structure are exceptionally important problems, dictating as they do the success or failure of the resultant illuminating effect.

In this paper the requirements of the various sections and operations are treated in detail, recommendations being offered as to the best practice. Photographs illustrating the application of the modern principles discussed accompany the text.

Railway Car Lighting

THE subject of Railway Car Lighting is treated in an excellent manner by Mr. G. E. Hulse. Limitations encountered in the problem of supplying illuminations to cars. Amount of energy available limited, due to car being on the move. Position of lighting fixtures determined by car construction, preventing flexibility in placing units.

Maintenance of reflecting and transmitting surfaces more difficult than in most other situations.

Means of Lighting—

Gas—incandescent mantle.

Electricity—axle driven generator with storage battery.

Standardization of Car Illumination—

The postal car lighting tests of 1912 determined and standardized.

The amount of illumination necessary for postal clerks to properly handle mail.

The types of reflectors best suited for such use.

Based on these test results, the Railway Mail Service issued specifications for lighting of postal cars, giving definite values for their illumination and other details, such as mounting height of lamps, and angle of cut-off.

These specifications can be applied without the necessity of further investigation, in the case of the change in interior design of postal cars, the type of reflector available, or the type of light source available.

Coach Lighting Tests of 1913—

Determined the amount of illumination obtained with the possible arrangements of fixtures, and the available types of reflectors, bowls and lamps.

The results of these tests are still in use as the basis for designing lighting installations in practically all classes of cars.

Arrangements of Fixtures for Various Types of Cars, and the Resulting Illumination—

Coaches, Dining cars, Sleeping cars, Postal cars, Business cars, Baggage cars, Parlor Smoking cars.

Types of Glassware Used and the Efficiency of Installation with this Glassware.

Illumination values obtained. The illumination obtained runs lower than illumination values used in office or factory installations, but seems to be ample for the conditions under which it is used.

Daylighting from Windows

THIS convention paper, by Messrs. H. H. Higbee and G. W. Younglove appeals to the architect as well as the illuminating engineer.

The purpose of this paper is to present a considerable amount of experimental data covering various points of practical importance and great interest with respect to daylighting of interiors, upon which actual and detailed quantitative information appears to be meagre. The investigations which yielded these data are still in progress, so it is deemed unwise to draw conclusions yet. The data here presented cover actual utilization co-efficient for typical daylight illuminations, together with curves and co-efficients representing distribution of the daylight under a variety of conditions;

also, data on the effect of width of mullions or columns between windows upon distribution of illumination in the room, on the relative effect and efficiency of light from various portions of the window area, on the effect of various methods of controlling the light from windows by means of shades and blinds, and on the effect of dirt accumulations on windows.

Salient Features in Power Station Lighting

IN a paper to be presented at the 17th Annual Convention, Mr. R. A. Hopkins discusses some interesting problems pertaining the central stations.

Unusual problems are met in the lighting of power stations on account of individual arrangements of equipment, severe service conditions and exacting requirements. The successful lighting system must be reliable, economical, easy to maintain and adequately suited to the specific local requirements which requirements are found to differ throughout the station. The most reliable and economical source of energy is usually the station auxiliary bus. The distribution wiring should be of the particular quality best suited to meet power station conditions and should be designed to give the best possible voltage regulation consistent with economy. An emergency lighting system should be provided and of several possible arrangements the one giving greatest dependability should be selected. All equipment such as cabinets, switches, receptacles, lamps, globes, shades and reflectors should be carefully selected to give maximum operating convenience, long life and high efficiency. A thorough survey of a large number of existing first-class power stations gives data for the solution of a number of typical station illumination problems so selected that the designing engineer may extend the data and conclusions given to meet the requirements of any ordinary station.

Working with the Architect on Difficult Lighting Problems

IN this paper, the authors, Messrs. Augustus D. Curtis and J. L. Stair, have contributed some interesting problems that appeal to the engineer as well as the architect.

The cost for the lighting of a building is not represented by the cost of the luminaires, but is measured by the satisfactory character of the illumination effects produced in the building.

The necessity for early consultation between the architect and the lighting man in the planning of the lighting features of a building is emphasized as well as the responsibility of the lighting man in developing in himself an appreciation of architectural values so as to most intelligently work with his architectural colleague.

Some specific examples of difficult lighting problems are given to illustrate some of the advantages to be derived by considering the lighting as a component part of the structure.

The Relation of Illumination to Production

THIS convention paper by Messrs. D. P. Hess and Ward Harrison is a report of extensive tests on the time required for the inspection of parts of roller bearings under various levels of illumination from 5 to 20 foot-candles. Over 7,000,000 separate pieces of material were inspected during the test period. The types of lighting employed as well as the illumination levels were found to have an important bearing on the output of the department. Cost data on the lighting and the value of increased production are included in the paper.

Some Principles Governing the Proper Utilization of the Light of Day in Roof Fenestration

THE light of day is usually considered a wholesome and vital requirement in our buildings. But it often needs modification, especially with regard to direct sunlight.

Sources of daylight with which the architect has to work are briefly described as regards their intensity, direction and seasonal and diurnal variation.

The usual requirements for natural illumination are outlined and some of the general principles governing the utilization of the latter in roof fenestration set forth—the entering daylight being analyzed as consisting of:

(a) Sunlight directly admitted to the working space.

(b) Light from the sky only, directly admitted to the working space.

(c) Combined light from sun and sky (or from portions of the sky alone) diffusely reflected from adjacent interior or exterior surfaces.

Methods of evaluating each are described, and examples worked out for certain usual types of roof fenestration,—these being divided into the two following general classes:

Class 1. The *one way type* in which the directly entering light comes largely from a single half or side of the "sky dome."

Class 2. The *two way* or *opposed type* in which the directly entering light comes from both halves or sides of the "sky dome." Horizontal roof openings may be considered as an extreme of Class 2.

Some of the advantages and disadvantages of each class are summarized.

The interesting treatment of the use of daylight is contributed by Mr. W. S. Brown.

Proposed Revised Code of Lighting School Buildings

THE Preliminary Draft of the Proposed Revised Code of Lighting School Buildings will be presented before the Annual Convention at Lake George. It is hoped that a complete discussion of this Code will be contributed by the members present. The Committee on Lighting Legislation, L. B. Marks, *Chairman*, and the Sub-Committee on School Lighting, M. Luckiesh, *Chairman* have assisted in the work of preparing this draft.

The present revision of the Society's Code of Lighting School Buildings is being carried out under the rules of procedure of the American Engineering Standards Committee under the joint sponsorship of the Illuminating Engineering Society and the American Institute of Architects.

Since the code was originally issued in 1918, changes in lighting practice have made necessary a revision of the rules and standards previously adopted. Moreover there has been an insistent demand by school architects, school superintendents and others identified with the lighting of school buildings for more definite specifications in regard to both natural and artificial illumination.

The present revision aims to bring the code up to date and to modify and amplify the rules and text in accordance with experience gained since the original code was issued.

In the proposed revision the code is divided into three parts: (1) *Rules;* (2) *Why the fulfillment of the rules is important;* (3) *How to comply with the rules.*

There are eight rules as in the original draft.

The standards of illumination required have been raised considerably and specifications of definite requirements under the rule relating to glare have been added following the precedent set in the CODE OF LIGHTING FACTORIES, MILLS AND OTHER WORK PLACES (AMERICAN STANDARD). A limiting ratio of maximum to minimum illumination has been set in the rule relating to the distribution of artificial light. Reflection-factors are specified in the rule relating to color and finish of interior. The rule relating to exit and emergency lighting has been made more specific and is based upon the specifications adopted by the Building Exits Code Committee of the A. E. S. C. The subject of blackboards is treated in a separate rule.

Unit Costs of Industrial Lighting

IN a paper to be presented at the coming Convention, Mr. Davis H. Tuck of the Holophane Glass Co. of New York states that the cost of an industrial lighting system may be divided into two parts. (a) Installation Cost, (b) Operating Cost.

By unit cost is meant the cost per unit of light. The unit adopted is the footcandle per square foot and the unit cost is in cents per foot candle per square foot.

Unit costs of installation and operation of various actual systems of lighting in industrial plants are shown. The value of unit costs is in comparing the economy of installation and operating costs of various types of lighting and in arriving at a quick estimate of the cost of any industrial lighting installation when the area to be illuminated, the foot candle intensity to be obtained and the type of equipment to be used is known.

By a study of the factors entering into the unit costs of installation and operation, it has been possible to materially decrease both the installation and operating costs without sacrificing the quality of the light.

Colored Lighting

COLORED lighting has assumed a very important position in illuminating engineering in recent years. However, many people are uninformed as to the methods of obtaining color in lighting, and the media which are available. In this paper the spectral limits of the various colors are given also the relative luminosity of the various portions of the spectrum from a 150 or 200-watt gas-filled lamp. The characteristics, advantages and disadvantages of various colored media are discussed. These media include colored glass lamps colored glass accessories, gelatine filters, and colored lacquers and spray coatings for lamp bulbs. The transmission factors of many samples of such media have been measured, and are given in this paper. The data show that many of these colored media are much less efficient than they could be in order to produce satisfactory colors. The need for standardization be of such media is very evident from the data given. Two examples of recent large installations of colored lighting are described, with connected-load data. A bibliography of colored lighting is also included.

This interesting convention paper is contributed by Messrs. M. Luckiesh and A. H. Taylor of the National Lamp Works of Cleveland, Ohio.

The Determination of Daylight Intensity at a Window Opening

THIS convention paper by H. H. Kimball, briefly reviews the Sphotometric measurements of sky brightness, and the determinations of the intensity of daylight on horizontal, vertical, and sloping surfaces, which are given in full in the reports of the Committee for 1921 and 1922. Most of the measurements were made in a comparatively smoke-free suburb of Washington, but some were made in a smoky section of the City of Chicago.

A method of determining the extent of the shading of window openings by neighboring buildings, first given in the report for 1921, is reviewed and extended. The advantage of laying out cities so that the streets run NE–SE instead of E–W and N–S, is pointed out.

Determinations of the reflecting power of surfaces of different kinds, and photometric measurements of their brightness under both cloudy and clear-sky conditions, are also summarized.

The data thus brought together, namely, the brightness of the sky, the intensity of daylight, the shading effect of near-by buildings and other objects, and the reflecting power, or the brightness, of different surfaces, is utilized to compute the intensity of daylight at a window opening under given conditions.

Recent Developments in Nomenclature and Standards

THIS report of the Committee on Nomenclature and Standards sets forth the progress made during the year in the revision of the "Illuminating Engineering Nomenclature and Photometric Standards" previously prepared by the Committee and approved as American Standard by the American Engineering Standards Committee. This revision has been carried out with a view to making the recommendations of the Committee practicable and applicable in the everyday work of the illuminating engineer. In accordance with this purpose the report presents for consideration several questions on which it is desired to have discussion and the advice of those interested before a final decision is reached.

Among these questions are the definition of the terms "light" and "lighting," the matter of "brightness" and the units to be used in measuring it, and the use of the term "luminaire." The Committee will be glad also to receive any suggestions supplementing the discussion presented at the Convention.

Testing Colored Material for Fastness to Light

IN this convention paper the author, Mr. H. S. Thayer, describes a series of experiments on various colored materials exposed to light sources of different kinds. A number of illustrations are used, consisting of comparisons spectra of the sun, the mercury arc, and the violet carbon arc, and also the commercial form of the violet carbon arc.

Several tables of interesting data, and one set of curves showing variations in the intensity of sunlight with time of lay and season of year, are included.

Solutions of a Street Lighting Problem

APPROXIMATELY three months ago a letter with a questionnaire and a blue print of the plan and photograph of the street, were sent to a selected list of representative street lighting specialists throughout the country. Eleven solutions of the problem were received which are incorporated in this symposium and should serve as a basis for discussion. These answers were coordinated by the Papers Committee in order that a uniform presentation might be made. The essential data, however, is given exactly as submitted.

The Visibility of Radiant Energy

A NEW determination of visibility of radiant energy has been made by the Bureau of Standards in co-operation with the Nela Research Laboratories. The step-by-step method was used, an equality-of-brightness method with little or no hue difference in the two parts of the photometric field. The apparatus and method are briefly described. Energy values were based upon radiometric and spectrophotometric measurements made at the Bureau, checked by an independent color temperature measurement at the Nela Research Laboratories.

Comparisons are made between the results of the present investigation and those previously obtained by the step-by-step and flicker methods. A revision of the I. E. S. adopted visibility values is proposed which results in better agreement with the average experimental data and still gives the same wave-length center of gravity for light of a color temperature of 2077°K as is given by Ives's physical photometer solution.

This convention paper by Messrs. K. S. Gibson and E. P. T. Tyndall will be equally interesting to the illuminating engineer as the physicist and is a valuable contribution to the present literature.

Electric Lights Advance Plant Blooming and Seed Time

VIOLETS in July, poinsettas in August, midwinter irises, dahlias in May, radishes that do not seed, cosmos fifteen feet tall and other similar miracles are being produced by the United States Bureau of Plant Industry as a result of recent experiments and discoveries of the laws of plant growth, according to L. E. Theiss in *World's Work*.

Plants bloom and fruit solely in response to the length of daylight occurring at their normal seasons of maturity, and by using electric lights to increase the number of light-hours until they correspond with the number of hours of sunlight at the season of a given plant's time of blooming, the plant will bring forth blossoms, regardless of the time of year. Conversely, if the long days of midsummer are made shorter by putting the plant into a dark place after a given number of hours of daylight, the plant will bloom out of season if its normal time of blossoming is one of natural short days.

Working with this knowledge, the experts of the Bureau have caused plants to blossom at all seasons of the year; they have advanced and retarded the fruition of vegetables, and by synchronizing the blooming-times of flowers and plants that normally bloom months apart, they have opened a new field for crossbreeding and development of hitherto unknown species.

The value of these discoveries lies in the future, according to Mr. Theiss, who cites experiments with tobacco as showing their commercial importance.

Maryland Mammoth tobacco seedlings grown in hot-houses in Maryland, where the winter days are short, flower when they are no larger than other tobacco plants. Artificial lengthening of the daylight hours produced plants that reached the desired mammoth proportions before seeding, and further experiments showed that if the plants were grown in Florida in winter, the light periods and climate were favorable both for growth and seed production. In other words, the electric lights showed the way for cheap and abundant Florida-grown seed.

"New crops, new things to eat, that are now non-existent and even undreamed of, will as assuredly come as daylight follows dawn," writes Mr. Thiess. "Things we can hardly even imagine, will come from these discoveries of why plants flower and produce fruit."

PAPERS

THE YEAR'S PROGRESS IN ILLUMINATION*
1922-1923

1923 REPORT OF THE COMMITTEE ON PROGRESS

"A man may proceed on his path in three ways: he may grope his way for himself in the dark; he may be led by the hand of another, without himself seeing anything; or, lastly he may get a light and so direct his steps."

—Francis Bacon's *"De Dignitate et Augumentis Scientarium"*

The world of chemistry has been stirred this year by the announcement of the discovery of a new element called "hafnium," which by reason of its position in the periodic table next to the rare earths may prove to be a factor in the great field of light emission. The confimation of Einstein's theory of relativity by data obtained at the last eclipse registers another triumph for astronomy and food for thought for the physicists. But, while a large amount of experimental research work has been going on in those fields in which the illuminating engineer is most interested, and considerable additions have been made to the store of special knowledge in these fields, no new light source has flashed across the horizon, nor have there been any radical changes in the

*A Report to be presented before the Annual Convention of the Illuminating Engineering Society, Lake George, N. Y., September 24-28, 1923. The Papers and Discussions included in our Transactions are not, in general, referred to in this Report, it being taken for granted that members keep themselves advised of the contents of the Transactions.

The Illuminating Engineering Society is not responsible for the statements or opinions advanced by contributors.

efficiencies of the light sources in use at present or in the general principles and methods of illuminating engineering.

The phenomena of the arc in gases and metallic vapors have attracted the attention of many investigators and apparently afford great opportunities for the study of those fundamental problems which lie at the very basis of molecular and atomic structure and hence, of the explanation of matter itself. A number of references to work in this field will be found under the heading "Arc and Vapor Tube Lamps".

. The physiological action of light both from the standpoint of vision and from that of therapy and biology has also received some attention. Hardly a year passes, and this year is no exception, which does not produce a new theory of vision, while the study of the eye and the mechanism of seeing continues unabated.

Though tables of recommended foot-candle values have been available for sometime for almost all cases of interior illumination, such as school lighting, store lighting, factory lighting, etc., little of a classified character has been heretofore published on values for street lighting. This deficit has now been filled for streets of cities up to 100,000 population. The entire table is too extensive to reprint in the report, but may be found in the original publication. The great strides made in public appreciation of streetlighting are apparent[1] when one reads the following taken from a New England paper of 1816, said to represent the most serious public thought of that date; "(1) A theological objection.—Artificial illumination is an attempt to interfere with the Divine plan of the world which had pre-ordained that it should be dark during nighttime. (2) A medical objection.—Emanations of illuminating gas are injurious. Lighted streets will incline people to remain late out of doors, thus leading to increase of ailments by colds. (3) A moral objection.—The fear of darkness will vanish, and drunkenness and depravity increase. (4) Police objection.—Horses will be frightened and thieves emboldened. (5) Objection from the people.—If streets are illuminated every night, such constant illumination will rob festive occasions of their charm."

[1] Gas Journal, Aug. 2, 1922, p. 270.

The experiments and investigations on the effect on production in various industrial processes of increased illumination are beginning to show results in a quite general tendency toward higher illumination values in recent installations for the lighting of shops and factories. Additional data have been obtained in this field and are noted in the report. A very encouraging sign is the awakening of the Government Post Office Department to the importance of lighting as indicated by the extensive study of lighting conditions in post offices, mentioned in the section on interior illumination. The duration and apparent thoroughness of the tests give considerable weight to the conclusions and recommendations which may be found applicable to interiors in other lines of business where conditions are correlative.

Further encouragement is to be found in a closer cooperation between the luminaire industry and the architects' guild.[2] Two representatives of the National Council of Lighting Fixture Manufacturers have been appointed on the Structural Committee of the American Institute of Architects. They will work on lighting in connection with plans for dwellings. In England, a paper on "Illuminating Engineering and the Architect" was read[3] before the Royal Institute of British Architects last fall in which the benefits to be derived from fuller cooperation between these two branches of engineering are strongly emphasized.

In the past the "cost of living" statistics published in Canada[4] have been misleading in that part referring to lighting, because heating data have been included. Hereafter, these items will be separated and the result will show a decrease in the average cost of lighting, per family, since 1914. General statistics showing the extent of electric lighting both for residential and industrial purposes in the United States and the world at large have been made available.[5]

The passing of the old cable ship, "Faraday", recalls the fact[6] that it was one of the first ships to use electric lighting. Arc lamps

[2] Lighting Fixtures and Lighting, Dec. 1922, p. 18.
[3] Electrical Review, Dec. 1, 1922, p. 840.
[4] Electrical World, June 30, 1923, p. 1540.
[5] Electrical Merchandising, Feb. 1923, p. 3082.
[6] Electrician, Mar. 9, 1923, p. 250.

were employed for this purpose when the French Atlantic cable was being laid in 1879.

At several places in the report, material has been mentioned which belongs in previous years but which had not been available to the Committee. In order to make the report as complete a picture of illuminating engineering as possible, it has been decided to include such information with the letters "N. P. R." after the reference to indicate "not previously reported". As usual the Committee has received information from the engineers in charge of lighting in various of the larger cities and the thanks of the Committee are extended to them and to the numerous journals whose pages have been so generously consulted.

Respectfully submitted,

FRANCIS E. CADY, *Chairman*
GEO. S. CRAMPTON
WM. E. SAUNDERS

INDEX

GAS

The origin of the name "gas" has been traced[7] to Jan Babtist Van Helmont of Brussels, among the last of the alchemists and one of the earliest contributors to modern chemistry. He was one of the first to understand that there exist other aeriform bodies differing from ordinary air and stated: "This spirit, up to the present unknown, not susceptible of being enclosed in vessels, not being capable of being reduced to a visible body, I call by the name of 'gas'". It was supposed that Helmont took the name from the Dutch word "geest", but it has been pointed out[8] that a study of the writings of Helmont, which were in Latin, indicates that he took the word from the Greek word "chaos".

Installation of a high pressure gas system in the Ancient Royal Burgh of Culross has aroused some interest as it is claimed that in a coal tar factory there, gas was first made some 140 years ago. The Earl of Dundonald in the course of experiments with the tar noticed the inflammable nature of the vapor during the process of distillation. Later he met Murdock and the subject was discussed. It is believed this was the origin of the scheme for the manufacture of gas which Murdock subsequently developed.

It is said that Fredonia, New York,[9] has the credit of being the first place to use natural gas for lighting purposes. The first well was drilled in 1826, the year Lafayette made a tour of the United

[7]Gas Jour., Dec. 27, 1922, p. 810.
[8]Licht und Lampe, Aug. 24, 1922, p. 406.
[9]Amer. Gas Jour., Sept. 23, 1922, p. 20.

States and the tavern where he was entertained was illuminated by gas piped from this well.

As a result of the fact that superheating of the air and gas in a burner increases the flame temperature considerably, and that the smaller the mantle the greater its light (proportional to the gas consumed), and the greater its strength, there is a growing tendency in England to convert gas lamps from the "universal" size mantle to the superheated cluster type of small mantles.

According to recent statistics,[10] gas is now used in 8,800,000 incandescent burners. In Japan there are reported[11] 1,217,094 installations of gas lighting.

In a paper before the German Illuminating Engineering Society it was stated that changes in the constituents of the gas supplied, occasioned by lack of coal during the war period and after, had resulted in a hotter and narrower flame for which prewar mantles were not fitted. Mantles must now be of smaller diameter to insure the fabric being in the hottest part of the flame.

Calorific Standard

A second progress report[12] has been made by the Joint Committee on Efficiency and Economy of Gas, of the Railroad Commission of California. The Commission carried out extensive investigations and came to the general conclusion that "in fixing a standard of gas quality to be applicable to all California manufactured-gas plants, the fundamental object to be achieved is to supply gas which will make possible the best service to the consumer at the least cost". The report concludes that the present 570 B. t.u. standard is not the best and that a 550 B. t. u. standard would be more satisfactory. On August 20, 1922[13] the Public Service Commission of New York adopted an order prescribing for gas companies operating in the City of New York B.t.u. standard in place of the obsolete candlepower standard. A monthly average must be maintained of not less than 537 B.t.u. and there must be no daily average of any three consecutive days of less than 525 B.t.u.

[10]Gas Age-Rec., Dec. 16, 1922, p. 828.
[11]Gas Jour., June 20, 1923, p. 744.
[12]Gas Age-Rec., Dec. 16, 1922, p. 819.
[13]Am. Gas Jour., Jan. 27, 1923, p. 70.

Burners

The Bureau of Standards[14] is conducting tests on the development of a standardized gas burner for dwellings. Improvements in gas burners[15] used for street lighting have been made in England. It was found that the "intermediate" mantle was the most economical one to use considering the amount of light required and the breakage due to vibrations from the heavy motor-truck traffic. Further, with the "intermediate" mantle, it is much easier to adjust the flow so as to get a higher efficiency than is possible with the "universal" mantle. A burner using two mantles has the centers of the mantles and the bunsen tube in one straight line and this arrangement was found to be well adapted for use in conjunction with automatic controllers. The same type is used for burners equipped with three to six mantles, except that where more than two are employed, the outlets are arranged in zig-zag fashion. This facilitates replacement in case of breakage. Tests have shown an average candlepower of 344 between 0 and 20° to the horizontal in a direction at right angles to the long axis of the burner and an average of 191 candlepower at right angles to the short axis. The burner was consuming 18 B.t.u. per candle hour or 13.76 cu. ft. of 450 B.t.u. gas. These results should be compared with those from a six-light circular cluster of the old type which gave an average of 250 candlepower between 0 and 20°. The new type tends to concentrate the light in two directions,—namely, up and down the street. To assist in the development of a maintenance scheme, the burner parts have been worked out so that there shall be interchangeable component parts such as gas-nipples and adjustors, air regulators, bunsen tubes, etc. capable of being assembled in various sizes to make up any particular burner desired.

Experiments to determine how much light from one mantle is cut off by the opacity of an adjacent mantle showed[16] for the particular mantles tested that one allowed 19 per cent of the light of the other to pass through. This would obviously depend on the texture of the mantle. An investigation has been made[17] on the

[14]Am. Gas Jour., Sept. 23, 1922, p. 20.
[15]Gas Jour., Mar. 7, 1923, p. 614.
[16]Gas Jour., Mar. 7, 1923, p. 614.
[17]Licht und Lampe, Aug. 24, 1922, p. 403.

effect of the mass of the primary air on the luminosity and gas consumption of inverted burners, keeping the other conditions constant and only changing the amount of the primary air. The experiments indicated that the luminous intensity depends on the gas consumption and corresponding temperature of the gas flame.

Mantles

A new element called "hafnium,"[18] has been discovered. It appears to have an atomic weight of about 180 and is chemically allied to the rate earths of the thorium group, to titanium and zirconium. This relationship suggests the possibility of its practical application in the production of gas mantles. A type of silk mantle[19] has been reported which is said to be stronger and more durable than any so far produced. Another new mantle not yet perfected is entirely textureless and yet is said to give the full luminous intensity obtainable from gas by the ordinary means. It consists of finely divided particles of oxide of thorium bound with a soluble salt of that metal and encased in collodion.

It has been found that the ash of incandescent gas mantles emits radioactively in the α and β rays.[20] The β rays are photographically active and may be absorbed by tinfoil 0.01 mm thick to the extent of 57 per cent. The α rays are strongly ionizing and are absorbed by the same tinfoil to the extent of 84 per cent.

The bump test to determine the resistance to shock of tungsten filament incandescent lamps is a familiar procedure but the testing of the tensile strength of a gas mantle is a much more complicated and delicate task. An apparatus designed[21] for this test comprises a vertical cylindrical float surrounded by water in a circular container. The float carries a cup of melted wax into which the lower edge of the mantle to be tested is placed and the wax allowed to solidify around it. The mantle is thus attached to the float. By running off the water slowly, the mantle takes an increased proportion of the weight of the float which drops when the mantle breaks and automatically shuts off the water. The height of the remaining water is then a gauge of the breaking weight. Tests

---

[18]Sci. News-Letter, Mar. 10. 1923, p. 6.
[19]Gas Jour., Mar. 21, 1923, p. 751.
[20]Chem. Zentralbl., Feb. 22, 1922, p. 440. N. P. R.
[21]Proc. Phys. Soc. of London, Dec. 15, 1922, p. 46.

have shown that some mantles can support five hundred times their own weight.

The development of a mantle[22] for use with acetylene gas has been slow in spite of many experiments. Some success has been attained with high pressure burners but to make a mantle last 500 to 600 hours, chemically well-purified acetylene was necessary. A mantle has been reported which is designed to insure complete combustion of either purified or crude acetylene and not to carbonize under any operating pressure. The burner consists of four parts; a tip screwed into a mixing tube on which is screwed a cap with gauge carrying the mica chimney and mantle, the whole having a height of 4½ inches. In the jet is screwed a four-hole disk from below bearing a rod; at the upper end is a screw thread instead of the usual cone entering the orifice plate. This gives a vortex motion to the issuing gas and assists in mixing it with the air injected through openings in the tube below the level of the orifice. Laboratory tests have indicated an average of 111.6 candles per cu. ft. per hour for burners rated at ⅓ and ¼ cu. ft. or almost six times as much light as from an open flame acetylene burner.

Auxiliaries

A French semi-automatic gas lighter[23] utilizes a pilot light with independent feed line and a container in which the gas accumulates until it reaches a certain pressure. The time required is the time during which it is desired to have the light out. In another model, the gas is cut off at the source and turned into the main again when illumination is desired. In either model the pressure of the gas operates a cut-in and cut-out float. In the first type, provision for extinguishing the light is made by accumulating a small amount of the flowing gas until the pressure is right. The time required for this is that during which the light is in operation. The regulating device is contained in a small cylinder attached to the lamp post.

[22]Gas Age-Record, Aug. 5, 1922, p. 178.
[23]Sci. Amer., Feb. 1923, p. 116.

Kerosene

A new competitor in the field of portable oil lamps is made in the form of an old fashioned candle and holder. Its fuel is kerosene and it is claimed to furnish 100 hours light from a teaspoonful of the combustible.[24].

INCANDESCENT ELECTRIC LAMPS

While experimental work on incandescent electric lamps is still being carried on, there have been no striking developments during the past year. The following table shows[25] recently compiled data on the annual and per capita consumption of incandescent lamps in the United States and several foreign countries.

	Population	Number of Lamps	Lamps per capita
United States	112 million	205 million	1.83
Switzerland	4	6.5	1.62
Germany	57	50	0.88
France	41.5	30	0.72
Austria	6	4	0.67
England	44	20	0.45
Italy	40	15	0.38
Hungary	7.25	2.7	0.37

For the fiscal year ending June 30, 1923 the government ordered 1,355,000 incandescent filament lamps, a number smaller than for several years preceding. 85 per cent of this number were large tungsten lamps as distinct from miniature tungsten and carbon. In 1922 the corresponding per cent was 79. 15.4 per cent of the large tungsten lamps were of the gas-filled type. 1706 representative lamps were life tested at the Bureau of Standards. Mill-type tungsten lamps have replaced carbon in Navy work.

Manufacture

The demand for carbon lamps[26] has dropped to such an extent that the number sold in 1922 was only about 1.5 per cent of the total number of lamps purchased in that year in the United States and only 1/3 the number used in 1920. In the case of tungsten lamps, the distribution according to size and type has apparently reached a point where differences from year to

[24]Pop. Mech., June 1923, p. 934.
[25]Jour. of A.I.E.E., June 1923, p. 659.
[26]Report of Lamp Committee, N.E.L.A., June 4, 1923.

year are only fractions of a per cent. For the past three years, the vacuum type has composed 79.5 per cent and the gas-filled, 20.5 per cent of the total number distributed, with variations of only 2.2 per cent. The most popular sizes have been the 40-watt, 25-watt, and 50-watt in the vacuum type and the 75-watt and 100-watt in the gas-filled type and this relation has held true for the past three years. However, the 50-watt vacuum lamp has been still gaining. The average efficiency has again increased by about the same percentage as last year and is now 11.5 lumens per watt. The distribution according to voltage shows 88.2 per cent in the 115-volt class with a falling off again in the 230-volt list. The standardization of lamp voltages continues to progress, over 91 per cent of those reported being 115, 110 or 120, in this order of preference.

The growth in the use of miniature tungsten filament lamps has been large in the last few years[27] and in 1922 it reached almost 29 per cent of the total used output (large and miniature). Of this number about 18 per cent were for flashlight service, about 10 per cent for Christmas tree decoration, about 70 per cent for automobile lighting and the rest mainly for signaling purposes.

In the days of the carbon lamp,[28] renewing or refilling burned out lamps was quite a common procedure, but the tungsten filament mounting and treatment is such that the problem was practically dropped except in Germany. A new method has been developed in Italy and is already in commercial use. In this method, the special feature is that all the operations necessary for winding the new filament are carried on outside the lamp to be renewed. The hole in the bulb is made no larger than would be necessary for exhausting an ordinary new lamp. A special mounting of the filament is required and is in the form of a small spiral crown. The arms holding the filament have sufficient resiliency to allow the whole to be folded up with an "umbrella-like movement" when the carrier system is introduced into the bulb.

The importance of the question of international standardization of lamp bases is such that attention should be called to the discussion of this subject under "Reflections" in the April number of the Transactions.

[27] Elec. World, Dec. 23, 1922, p. 1416.
[28] Electrician, May 25, 1923, p. 569.

The use of the iron wire auxiliary to control violent fluctuations
in the power supply was a most potent factor in the practicability
of the Nernst filament and has been in use to regulate the voltage
supply for railway headlights for some time.   A foreign experimenter has worked out data[29] from which may be determined the
wire necessary to take care of a number of cases.

According to reports from Russia[30] vacuum lamps are now
being made at the Moscow lamp works at the rate of 10,000 per
month and gas-filled lamps at the rate of 250 per day.

A company has been formed to manufacture in Hungary
special types of lamps such as those for Xmas Tree decoration.
About 250 million glass bulbs for incandescent lamps are now
being produced yearly in Germany.[31] of which 20 per cent are for
export.   All bulbs are blown individually by hand, operation of
the automatic machine having been found to be more expensive
than manual labor.

Types

In the "mill" type lamp which has an especially rugged construction, a 10 watt size[32] in a straight-sided bulb of either clear
or blue glass with a concentrated filament has been developed.
It is suggested for use in sign lighting.   A lamp is announced from
Canada[33] which in a number of sizes is claimed to have a tested
life of 1500 hours.   No data are presented as to whether this
longer life is due to operation at lower efficiency and hence at a
lower temperature, or to actual improvements in materials or in
methods of construction which would give this life at higher
efficiencies corresponding to a 1000 hour life in the lamps as ordinarily manufactured.

A novel application of the electric lamp to advertising sign
purposes has appeared in Germany.[34]   The lamp has a tubular
bulb from 1 to 2 feet long depending on the length of the advertisement.   The text of the latter is fitted in blue glass letters in a
glass frame in the bulb.   Over the lines of the letters, a fine tung-

[29]Rev. Gen. d'Elec., Mar. 24, 1923, p. 477.
[30]Elec. Rev., June 29, 1923, p. 1013.
[31]Zeit. f. Ver. Deut. Ing., May 26, 1923.
[32]Elec. Record, Mar. 1923, p. 162.
[33]Elec. Review, Dec. 29, 1922, p. 992.
[34]Helios, Dec., 24, 1922, p. 4161.

sten filament is laid and kept in position by a small holder. The color of the letters makes them visible by day and the filament illuminates them by night, also furnishing general illumination if desired.

Properties

The relationship in incandescent filament lamps between the efficiency and the cost of operation, including the cost of the lamp, was pointed out in the late 90's, but no practical application on an extended scale has come to the attention of this Committee until this year. As a result of a study of the subject, taking into account different power rates and lamp prices,[35] the Hydro-Electric Power Commission of Ontario, Canada, has adopted efficiencies for incandescent lamps for general use on hydro-systems which will result in an average life of 1500 hours. It was found that the rigorous solution of the problem of the most economical efficiency was too complicated for general application throughout the province, and the value decided upon was taken as a compromise.

The experimental determination of the electric and photometric charactertistics of tungsten lamps has been extended to the gas-filled type ranging from 50 to 1000 watts. Formulae[36] of the type, $cp = AV^k$, where A and k are constants and $cp.$ is the maximum horizontal candlepower, show variations in k from 1.73 to 3.45, the values generally being higher for the higher $cp.$ lamps. The 50- and 100-$cp.$ lamps were made up with argon gas, the rest with nitrogen. The differences in the values of k were attributed to differences in the purity of the tungsten. The spherical reduction factor (scp/mhc) was found to be 0.978 for the lower $cp.$ lamps and 1.04 for the 600- and 1000-$cp.$ sizes. The ratio of the resistance hot to that cold ranged from 15 for the 50-$cp.$ to 16 for the 1000-$cp.$ lamps. Data were obtained on lamps with opal bulbs.

The physical properties at high temperatures of many materials can be studied if they can be mounted in the form of filaments and heated electrically. In such cases, knowledge of the losses due to conduction of heat at the junctions of the leading-in wires and at the supports is desired. The theory of such losses[37] and its appli-

[35] Elec. World, Mar. 17, 1923, p. 645.
[36] Rev. Gen. d'Elec., Aug. 19, 1922, p. 245.
[37] Jour. of Frank. Inst., Nov. 1922, p. 597.

cation to tungsten filaments in a vacuum has been worked out both for long and short filaments. Expressions have been derived for the temperature, resistivity, emission intensity, brightness, and thermionic distributions near a cooling junction for tungsten for the conditions just referred to. The relation of the distribution curves for short filaments to those for long filaments has been obtained and it was found that equations and conclusions developed for cylindrical or wire filaments apply equally well to filaments with rectangular and other cross sections when the proper substitution is made for the radius of the cylindrical filament.

A further study[38] has been made of the rate of evaporation of tungsten in the form of an incandescent filament in a vacuum and in a gas. It was found that at a temperature of 2950°K for the vacuum experiments the rate of evaporation per square centimeter is independent of the diameter from 0.05 to 0.25 mm., but is over 40 per cent greater for fine grained than for very coarse grained wires. In nitrogen, the rate varied from 2 per cent to 5 per cent of that for the same wire in a vacuum and in argon, from 1.3 per cent to 3 per cent greater than in a vacuum.

Specifications to be used in the purchase of incandescent lamps have been published[39] by the Union des Syndicats de l'Electricite in France. For vacuum lamps, the *cp* rating on the bulb is for the maximum intensity in a plane perpendicular to the axis. Lamps are to be tested on three points: quality of materials and conditions of manufacture, candlepower and watt consumption, and life.

ARC AND VAPOR TUBE LAMPS

Ever since the introduction of the mercury arc and the Moore tube as lighting sources there has been marked interest in this vapor-tube type of light production. The very nature of the source and the character of its construction seem to afford unlimited opportunities for experiment and development. Reference to previous issues of this report will show repeated allusions to research in which lamps of this type are the direct object of the experimental work or a most important auxiliary. Involving as they do luminescence and phosphorescence, the continuous and

[38]Phys. Rev., Mar. 1923, p. 343.
[39]Rev. Gen. d'Elec.. Aug. 26, 1922, p. 277.

the line spectra with the possibility in the way of a monochromatic source of the maximum of luminous efficiency, it is not surprising that they require an increasing amount of space in this report.

Another contribution to the theory of the electric arc has been made,[40] based on a study of the thermionic emission from the cathode, the current carried by positive ions and the cathode fall, ionization in the region between the electrodes and the anode fall. The view is supported that the arc seems to be dependent on an adequate supply of electrons at the cathode, their escape from it being made possible by a sufficient ionization of gas near it to form a space charge.

Types

Carbon arc lamps designed especially for moving picture work[41] have been brought out for an arc voltage of 160. They are said to burn for 20 hours with one pair of carbons. The result is obtained by an unusually long arc (5 to 6 cm.) burning in its own combustion gases in a comparatively small glass globe. By using one horizontal-light[42] carbon and two vertical auxiliary carbons, which conduct current of displaced phase, another arc for projection use has been announced. Effectively a single phase current results and it is claimed that a higher intrinsic brilliancy and less flicker is obtained than with the ordinary two carbon arc. Still another carbon arc is a reminder of the old Jablochkoff candle. It is a small lamp fitted with an Edison base, has two carbons which are parallel and held in a simple but substantial clip sleeve and the arc is maintained by a magnet. It is started by a spark and operates at a current of from 8 to 10 amperes. An increase in the life and efficiency of luminous arc lamps[43] has been accomplished by making the electrodes in compressed square or oval cross sections.

Experiments on the size of the cathode spark of the carbon arc[44] have shown that if the anode is placed far enough from the cathode so that its radiation no longer influences the size of the

[40]Physical Review, March 1923, p. 266.
[41]Helios, June 10, 1923, p. 1299.
[42]Elek. Zeit. 44 April 12, 1923, p. 335.
[43]Electrical World, January 6, 1923, p. 27.
[44]Zeit. f. Physik, Oct., 7, 1922, p. 71.

cathode spot, the latter is then proportional to the current strength. The constant current density of the spot was found to be 470 amps. per sq. cm. The high frequency alternating-current arc behaves differently from either the low frequency or the spark discharge in that it has a pseudo-stationary range at the middle part of the arc stream.[45] This conclusion has been derived as the result of stroboscopic experiments on the light emission in the arc stream with various electrodes at a frequency of 6000 complete periods per second and a current of about 10 amperes. Solid and cored carbon electrodes were employed and tests were also made with rods of Fe, Cu, or Ni as one electrode. For short arcs the variation of intensity of the different spectral lines was found to be very closely in phase with the arc current. In a long arc if the part near the electrodes was excluded, the middle region gave a spectrum which would not suffer any variation with the current oscillation and was as steady as a D. C. arc.

Other experiments on the arc between metallic electrodes with alternating current indicate that the current passes only from the hot electrode to the cold whatever the material of which the electrodes are made.[46] By the use of a water cooled electrode and two other electrodes, a pulsating current, but in a single direction, was obtained from a 220 volt A. C. circuit.

Properties

Among the physical properties of the carbon-carbon arc[47] is a repulsion effect upon the poles. Experiments have shown that this effect increases with the current ranging from 0 to less than 10 dynes for currents up to 20 amperes. It does not vary much with the arc length but is strongly influenced by the purity of the carbon, metallic salts causing an increase in the pressure on the cathode and a lessening on the anode. Using a spectro-photometer and the optical pyrometer method, another determination has been made of the temperature of the crater of the carbon arc and other radiation from the flaming arc[48]. The carbons were burned in an automatic projection lamp wherein the electrodes were placed at an angle to each other. The temperature was found

[45]Il Nuovo Cimento, Jan., 1922, p. 59. N.P.R.
[46]Journal de Physique et le Radium, Nov., 1922, p. 389.
[47]Phil. Mag., Oct. 1922, p. 765.
[48]Zeit. f. Tech. Physik, No. 2, 1923, p. 66.

to be dependent on the current strength and the material of the electrodes. Two pure graphite carbons gave the highest blackbody temperature. At a normal specified operation of 0.746 amps. per sq. cm. the temperature was 3775° which with the Beck effect ran up to 3900°. The negative electrodes with pure carbon showed the same temperature as the positive crater. Heating the anode to 1100° caused an effective drop of 70° in all colors. In the normal arc a zone of highest intensity occurs just in front of the electrodes. More work on the ultra-violet spectrum of the carbon arc[49] has resulted in a number of lines not previously observed which correspond with prominent lines in the hot-spark spectra studied by Millikan. In addition, values of λ have been checked for a number of the prominent lines. Reference should be made to an elaborate discussion[50] of the Goertz-Beck high intensity arc lamp for searchlights which gives data on the spectral distribution, the surface brightness, the rate of burning of the carbons, the influence of the core diameter on the surface brilliancy, etc. as well as information on the mechanical construction of the searchlight unit.

Mercury Vapor

Some interest was aroused in technical circles by a report in a French newspaper[51] of a new light source which was said to be in the form of a vacuum tube 6 m. long, 7 mm. in diameter, coated on the inside with a phosphorescent material and exhausted to 0.02 mm. It was said to give a light comparable to that of ordinary sources and a power consumption of only 15 watts. This information was subsequently repudiated[51a] by the French professor who was credited with the invention.

In every form of vapor arc lamp, the elimination of foreign gases from the various parts of the lamp is a necessary step in the manufacture. The effect of such gases in the mercury arc lamp has aroused enough interest to stimulate a research on the subject[52]. Using a spherical glass container 40 cm. in diameter, an investigation was carried out with a discharge which was unstratified and

[49]Proc. of Royal Soc., Jan. 1, 1923, p. 484.
[50]Zeit. f. tech. Physik, No. 4, 1923, p. 138.
[51]Le Matin, March 30, 1923.
[51a]Elec., Apr. 27, 1923, p. 465.
[52]Zeit. f. Physik, May 26, 1923, p. 254.

of medium width and voltage requirement. With the introduction of a foreign gas, the voltage drop increased in proportion as the heat conductivity of the gas was greater than that of the mercury vapor, and in proportion to the increased energy loss of the electrons by collisions of the molecules of the foreign gas. The order of gases with reference to the voltage loss was Hg, Ar, CO_2, NH_3, N, H, O, SO_2. The greater the energy loss by collision just referred to, the greater the voltage required to start the arc.

The mercury arc in quartz can be operated at intensities high enough to give a continuous spectrum in addition to the mercury lines.[53] This effect has been adapted in a compact illuminator especially for monochromatic purposes operating on 110 volts and with either A. C. or D. C. current. The dimensions of the effective light spot are 1.25 x 0.25 inches. A mica filter is provided for observation of the extreme ultra-violet and an adjustable slit and light-tight holder for other filters. With a resistance in series, the lamp operates as a low pressure arc giving only the strongest spectral lines. At full intensity, it changes to a high pressure arc. Additional study from the standpoint of applicability[54] has been made of the quartz mercury arc having an inert gas such as neon, argon, helium, etc. introduced into the tube as described in the 1920 report.[55] A number of models have been constructed suitable for operation on either A. C. or D. C. One D. C. lamp is designed for 115 volt, 4 amperes with a difference of potential at the electrodes of from 80 to 85 volts. The initial current is 11 amperes which falls in 30 seconds to 6 amperes and in 5 minutes to 4 amperes.

On the basis of new experiments on the mechanism of the mercury vapor arc,[56] it has been shown that the anode fall is independent of the current strength and the temperature of the anode as long as the vapor pressure remains constant, and that it is probably influenced by the material of the anode and its form. It has been evaluated both by probe measurements and by measurements of the energy emitted at the anode. The anode fall has been found to approach,[57] with increasing vapor pressure

[53] Jour. of Opt. Soc. of Amer., Dec. 1922, p. 1066.
[54] Revue d'Optique, June 1922, p. 304.
[55] Trans. I.E.S., *15*, 1920, p. 439.
[56] Zeit. f. Physik Feb. 26 1923, p. 378.
[57] Ibid, May 26, 1923, p. 287.

and current strength, a limiting value of 3.72 volts which agrees
with the difference between the ionization potential of 10.39
volts and the exciting potential of 6.67 volts. In hydrogen, the
anode fall amounts to 17.6 volts and is about 1.2 volts larger than
the ionization potential. In argon and neon, the anode fall changes
with increasing current strength at the lowest exciting voltage.
Data have been obtained[58] on the dissociation temperature and
vapor pressure conditions in mercury vapor. A 100-ampere rectifier flask was employed as the arc vessel. In the lower part of
the cooling chamber the cross section of the velocity of a mercury
vapor jet under full operation was found to be 4.5 x 10^3 cm. per sec.
The calculation of the temperature in the axis of the positive
column gave results which for full load lay between 1000° for a
10-ampere and about 10000° for a 500-ampere rectifier. A study
has also been made[59] of the energy dissipation at the cathode spot
in the mercury of the cathode.

Data are available on the infra-red radiation from the quartz
mercury arc[60] in the range of wave lengths from 0.70 to 4.0μ. Differ-
entiating between the radiation from the quartz tube alone and
from the mercury, it was found that the output from the former
increased more rapidly than from the luminous mercury and that
radiation decreased in intensity throughout the tube in passing
from the anode to the cathode. Over a range of from 40 to 120
volts, the mercury radiation formed 30 to 55 per cent of the total
and this proportion was increased when the light was filtered
through a plate of quartz. The absorption of the fused quartz for
mercury radiation was about 38 per cent.

A quiet jet of green vapor plays between and forms the arc
stream when mercury is the positive, and carbon the negative
electrode of an arc.[61] The jet is always perpendicular to the crater
which is a circular depression (3 to 5 mm.) surrounded by an oxide-
of-mercury ring. Experiments on the velocity of the stream showed
10 to 30 meters per second. The rate of mercury consumption
with known current and crater diameter was also determined. The
current density was of the order of 300 to 1000 amps. per sq. cm.
The crater area was found to vary directly as the current and the

[58]Ibid, Nov. 17, 1922, p. 260.
[59]Zeit. f. Physik, Oct. 7, 1922, p. 74.
[60]Ibid, No. 6, 1922, p. 353.
[61]Nuovo Cimento, Jan. 1922, p. 31 N.P.R.

variation of voltage with arc length at constant current was approximately linear. In this arc the cyanogen spectral lines are absent. If a meniscus of mercury is used for the anode[62] and a very thin sheet of metal or carbon used for the cathode and the whole properly regulated, the cathode will be punctured and a small flame surmounted by a high tuft of whitish fumes will be seen to play above the hole.

Comparison of the ultra-violet radiation from a quartz mercury arc and the sun at the zenith at Davos, Germany,[63] indicated that at 100 cm. distance, the arc had 3.7 and at 50 cm., 12.2 times the ultra-violet intensity of the sunlight.

Mercury and other Vapor

A systematic study of twenty-four illuminating gases[64] when excited by an electrical discharge at low pressure has shown that the light emitted per unit energy consumed varied from 0 in the case of arsenic vapor to 17 candles per watt for sodium vapor. The light was measured from a limited portion of the vacuum tube and the corresponding electrical energy was determined by measuring the potential drop and the current in this part of the tube. The maximum observed efficiencies of neon, mercury, and sodium were 1.8, 10, and 17 mean spherical candles-per-watt respectively. These three proved to be the only elements which gave as high a luminous efficiency as that given by the ordinary illuminants. It was shown that it is very improbable that any gas composed of polyatomic molecules will have high efficiencies.

The fall of potential between electrodes where the cold anode lies within the region lighted by the cathode has been determined for arcs of mercury, zinc, cadmium, and magnesium. For mercury, the value, 9.20 ± 0.15 volts was obtained for currents from 5 to 20 amperes; i.e., for this region the fall of potential is independent of the current strength. This was found true for anodes of iron, nickel, and platinum. The zinc arc at small current strengths gave 8.80 and for large currents (10 to 30 amperes) 10.53 volts; the cadmium arc gave 9.9 volts which was constant for currents from 10 to 20 amperes; the magnesium arc showed 12.3 volts for 10 amperes and 12.45 volts for 30 amperes.

[62]Ibid, Oct. 1922, p. 165.
[63]Meteorol. Zs. *39*, 1922, p. 303.
[64]Physical Review, Feb. 1923, p. 210.

Further information is available on the characteristics of the neon glow lamp.[65] These seem to vary with the type of construction, character of the electrodes, distances between them, etc. The voltampere characteristics and spectra have been worked out for two types designed for direct current, one with a star shaped, the other with a conical spiral electrode. Similar experiments have been made on a lamp with the conical spiral electrodes designed for alternating current.

For use as a source of monochromatic radiation for high in-intensities[66] a quartz sodium-potassium vapor arc lamp has been found satisfactory. It is made by joining two quartz spherical bulbs about 3 cm. in diameter with a short length of quartz tubing 5 mm. bore, each bulb having a long neck attached which carries the electrode wire, an iron rod 4 mm. in diameter. The liquid alloy of sodium-potassium (two parts by weight of sodium and one of potassium) is run into one bulb and both are then exhausted. On passing an electrical discharge through the tubes with the alloy as cathode, the oxide on the surface disintegrates and the surface becomes clean. The alloy can then be made to flow into the other bulb. Using direct current, the lamp works on a minimum applied voltage of 30, although after the arc is struck, which is accomplished by tilting, the fall of potential is only 10 volts with a current of 1.5 amperes. The current should be kept below 2.5 amperes to prevent browning. The lamp does not require continuous pumping while it is working as the alloy absorbs all gasses, particularly nitrogen and hydrogen, while the current is passing. The potassium lines are found to be very faint compared to the sodium lines. A sodium vapor tube having a nickel disk anode 2.5 cm. in diameter and 8 cm. from a tungsten filament 0.038 cm. in diameter, in a 18 cm. pyrex bulb was found to have properties shown in the following table:[67]

volts	10	14	19	30	100	200
Discharge:						
amperes	0.081	0.223	0.525	0.810	0.475	0.510
Candlepower of glow	5.1	27.2	115	201.5	163.5	269.5
Watts/cp. of glow	0.16	0.12	0.09	0.12	0.29	0.38

[65]Electrician, Dec. 1, 1922, p. 626; Jan. 5, 1923, p. 4.
[66]Phil. Mag., Nov., 1922, p. 944.
[67]Physical Review, Feb. 1923, p. 209.

In gas-filled electric discharge tubes in which an electric discharge is started by supplying current to the electrodes,[68] the starting potential is reduced by combining magnesium or beryllium or their alloys with the metal, e.g. iron or aluminum, forming the electrodes. Thus magnesium or beryllium may be mechanically applied to the electrodes, and material disintegrated from the electrodes by the discharge is prevented from forming a desposit upon the walls of the discharge tube by suitably shaping the electrodes, e.g., by partly surrounding the auxiliary metal or alloy by the electrodes.

LAMPS FOR PROJECTION PURPOSES

As in the case of other light sources, developments in this class have been more in the direction of minor improvements and increased information as the result of experimental work.

Hand Lamps

The focusing type of hand flashlights has been amplified[69] by providing one end with a bull's eye or outer convex general service lens giving a diffused light over a wide radius and the other end with the spotlight obtained by the focusing reflector and lens. The two lights may be operated independently or both at the same time. Hand lamps are now used for so many purposes[70] that it is difficult to formulate tests which will be sufficiently general to cover all requirements. However, some interesting information was obtained from experiments on lights made by eleven manufacturers. Each lamp was tested with its own battery and lighted for 35 seconds four times an hour, first for eight hours a day, then for twenty-four hours a day, to determine life. The main conclusions were: except in the case of intermittent use, long two or three cell flashlights should be used as the greater life more than compensates for the increased cost; large diameter lenses and reflectors should be avoided except possibly for illumination at a considerable distance; concial reflectors usually give a better distribution than parabolic.

[68] Jour. of Soc. Chem. Ind., Oct. 31, 1922.
[69] Elec. Merch., June 1923, p. 3445.
[70] Elec. Jour., Apr. 1922, p. 150, N.P.R.

A small hand lamp[71] has been brought out in France devised especially for the reading of graduations on scales such as those on optical instruments. It employs a flashlight bulb operating on a dry battery, the light being diffused by a ground glass and in turn reflected from a mica sheet placed at an angle of 45° to the glass plate.

A hand lamp of the general miners' class[72] had been approved by Lloyd's for ships carrying oil having a flashing point of less than 150°; for vessels holding a passenger certificate and using oil fuel; and for employment in petrol tank steamers. The outer case is made of a special aluminum alloy and has a locking device to prevent unauthorized opening of the lamp. The accumulater case is of steel and an alkaline electrolyte is employed which is said to be absolutely unspillable. The continuous lighting capacity is about ten hours.

Automobiles

Additional lights are finding their place in the equipment of the latest models in automobiles. Among these may be mentioned the "courtesy"[73] light which silhouettes the arm of the driver when it is extended to signal cars in the rear; the side-door light which illuminates the step when the door is opened; the "backing" light, convenient when going from the garage to the street or when turning on a poorly lighted street.

Recognition of one of the fundamental principles of illuminating engineering[74] is seen in a headlight controller which dims the light gradually instead of making the abrupt change which frequently causes difficulty in seeing on the part of oncoming drivers. The device is operated from the steering wheel by touching a lever on the steering post. Directive action in automobile headlights[75] has in the great majority of cases been accomplished by means of glass lenses with modified surfaces backed by smooth reflectors of the parabolic type but results can be obtained by properly modifying the surface of the reflector itself and this idea is incorporated in a headlight which spreads the beam laterally by

[71] Rev. Gen. de l'Elec., Apr. 21, 1923, 130D.
[72] Elec., Sept. 1, 1922, p. 248.
[73] Elec. Merch., Apr. 1923, p. 3252.
[74] Pop. Mech., Dec. 1922, p. 879.
[75] Jour. of A.I.E.E., July, 1923, p. 754.

means of vertical flutes in the mirror surface, the latter having a hyperbolic rather than a parabolic contour which gives a slightly increased beam depth.

Another departure from the usual type is to be found in a head light which produces two distinct fields of illumination,[76] one of high intensity for distant observation and the other of moderate intensity for side and fore-field illumination. This is accomplished by the combination of a 6 inch paraboloid reflector with an oblique reflecting projector, all the useful light being projected to a vertical slit 0.5 inch in width and 4 inches long and remaining below a horizontal plane 42 inches above the ground. Additional side and fore-field illumination is provided by an independent small side lamp integral with the housing which can also be used as a parking lamp.

A modification of the usual auto tail light is a unit which carries two lamps,[77] a green one which is lighted continuously except when the clutch is thrown out or the brake applied, when it is extinguished and a red lamp lighted. When the car is at a standstill, the red light serves for parking purposes. Two arrows show red to indicate whether the car is going to turn to the right or to the left.

In cooperation with various state authorities,[78] the Bureau of Standards has been carrying on experiments to determine the proper illumination of auto license plates and their visibility under different conditions of illumination. This work has not yet been completed.

An anti-dazzle light bill has been prepared for introduction[79] in the British Parliament. One of the provisions of this bill is that where the range of lamps exceeds 150 feet, a beam must be used which falls below the vision of drivers of other vehicles and of approaching pedestrians.

A special headlight has been developed for mine locomotives.[80] The main body of the lamp has been shaped so as to ward off blows caused by falling coal, is mounted on a sub-base with a three-point suspension and made dust-proof and moisture-proof. Arrangements for hand-focusing are provided.

[76] Jour. of Soc. of Automot. Eng., Jan. 1923, p. 3.
[77] Pop. Mech., Jan. 1923, p. 73.
[78] Jour. of A.I.E.E., Nov. 1922, p. 818.
[79] Elec. Rev., Jan. 12, 1923, p. 66.
[80] Elec. Rec., Jan. 1923, p. 28.

Signalling

The number of incandescent lamps used in railway signal work[81] is increasing. The types most commonly used are the 3.5-volt, 0.3-ampere and the 13.5-volt, 0.25-ampere. Through track relays, the lamps light when a train is approaching and are extinguished after it has passed. In this way, batteries are conserved. The use of "sun-valves" has also been introduced. These valves automatically turn the current on and off with the approach and cessation of darkness. Similar applications are being made in the small range-lights and buoy-lights of the lighthouse service.

Automobile traffic has so enormously increased the duties of traffic officers that various devices have been developed from time to time to make their work more effective.[82] One of the latest is a signal lamp the size of a watch which lies in the palm of the hand and is supported by a strap at the wrist. It receives current from a small pocket storage battery weighing only twelve ounces, the connecting wires passing through the sleeve of the coat. The lamp is automatically lighted or extinguished by the raising or lowering of the arm. The life of the battery is claimed to be five years. An innovation in street signalling devices[83] has a bell which automatically rings when the signal operates to show a change in the traffic direction, thus giving an audible as well as a visible notice. The apparatus is about 7.5 feet high and displays a red or a green light at night and the usual stop and go signs by day. The controls are such that the automatic feature may be switched off at will to meet any emergency conditions.

A combined signal and aerodrome portable landing lighthouse[84] has been developed in France. It sheds light through 180° within a radius of nearly a half mile and the light is cut off at so low an altitude that, it is stated, an airman coming to ground would not see the source until his head was on the level with the lamp. The main lens is composed of twenty-one dioptric elements and is enclosed in a metal lantern carried by three adjustable bearings on a four-wheeled truck. The light source is a D. C. arc taking 130 amperes at 60 volts, the carbons being arranged for either automatic or hand feed. In order to make it visible to airmen flying at

[81] Elec. Rec., Jan. 1923, p. 38.
[82] Elec. Merch., July 1923, p. 3493.
[83] Illus. World, Jan. 1923, p. 743.
[84] Elec. Times, Sept. 21, 1922, p. 246.

high altitudes, the top of the lantern carries a fourth order lens (250 mm. focal distance) built up of six dioptric elements and subtending 360°. At the center of this lens is fitted, in an adjustable carrier to permit exact focusing, a 500-watt incandescent lamp wired to flash synchronously with the main beam. The top of this upper lens is closed by a ventilating glass dome. The light of the main beam is said to be visible to the naked eye at a distance of 60 miles. The light from the upper beam is visible under the same conditions at all flying heights as soon as the upper limit of the main beam has been passed.

International regulations covering ships' navigation lights[85] were prescribed in 1910 when it was provided that the starboard green light and the port red light should be visible at a distance of at least two sea miles. No rules were given as to the candlepower of the lights and quality of the colored glass to insure this visibility. This has been remedied as the result of a test conducted in England from which it was deduced that the candlepower at the source should be not less than 12, and 60-watt or preferably 100-watt lamps were recommended. It was found that in practice 85 per cent to 90 per cent of the light was absorbed by approved colored screens.

Miners Lamps

Doubts as to the effectiveness of contact breakers[86] in miners' lamps when the glass is broken led the Bureau of Mines to carry out a number of tests. The first consisted of ten 0.65-ampere, 2-volt lamps with the glass bulb removed. The filament was surrounded by an explosive mixture and energized by a 2-volt battery. The second test used the same number and type of lamps but smashed the bulbs after the filaments were lighted. The third test was the same as the first but used a different surrounding explosive gas. The fourth test was made with only one lamp to show how many times the filament could be lighted when exposed to an inflammable gas mixture. The results proved conclusively the necessity of an adequate current interrupting safety device. In fifty-five tests in which the naked filament was exposed either

[85]Elec., July 28, 1922, p. 88.
[86]Elec. Rec., Sept. 1922, p. 170.

at the beginning or after smashing the bulb, fifty-one ignitions occurred, all obtained by normal voltage impressed on the filament.

In British coal mines, the electric safety lamp[87] is replacing the old Davy lamp at the rate of about 22,000 per annum. At the end of 1921 the numbers were approximately 605,500 oil-flame (Davy) and 268,500 electric, the latter number having increased during the past year to nearly 300,000. The experimental work on "pillarless" electric safety lamps[88] which has been conducted by Miners' Lamps Committee during the past two years has reached the stage where the Committee has felt justified in recommending that lamps of this type be accepted for tests. The Secretary of Mines has approved the recommendation. Lamps with a working voltage over 2 must be fitted with a contrivance for interrupting the current automatically in the event of breakage of the well-glass. The object of the "pillarless" lamps is to avoid both the loss of light due to the four or five brass rods or pillars surrounding the glass and the detrimental contrasts and shadows. Six lamps of this type made by different manufacturers were approved shortly after,[89] four of which are required to maintain a light of not less than one candlepower all around a horizontal plane for not less than nine hours and to give a light of not less than 15 candlepower over an arc of 45° in a horizontal plane. The total weight of such lamps must not exceed the specified figures (5.5 to 6.75 pounds). In Australia, coal miners' representatives from the north, south, and west districts of New South Wales[90] have asked for assistance in getting the use of electric lights in the mines. The question is under advisement by the authorities.

Projection

A carbon arc for moving picture projection[91] has both carbons horizontal, the crater of the positive carbon facing away from the projector aperture and toward a convex mirror through the center of which passes the metal-coated negative carbon. The rays of light from the positive crater are caught by the reflector and sent back through the projector aperture. No condensing lens is used

[87] Elec. Rev., Apr. 27, 1923, p. 662.
[88] Elec. Times, Aug. 3, 1922, p. 92.
[89] Elec., Dec. 29, 1922, p. 751.
[90] Elec. Rev., Oct. 20, 1922, p. 570.
[91] Mov. Pic. World, June 9, 1923, p. 525.

with this system which is claimed to give an even screen illumination and to be especially efficient for long focal-length projector lenses, since the mirror is located at a good distance from the aperture.

The heat evolved in a searchlight[91a] is considerable and the temperature of the various parts of the apparatus is an important factor in their design. A study of the temperature distribution over the mirror of a 36-inch, 150-ampere, high intensity instrument showed values ranging from 30°C near the outer edge to 85°C near the center. By studying the isothermals, when plotted, shadows were noted and corrections made which, it is expected, will materially increase the life of the mirror.

STREET LIGHTING

The following statement is quoted from the Report of the Lighting Sales Bureau of the N.E.L.A. for 1922-23[92]: "The year just closed has seen more street lighting improvements made than have been reported in any previous annual period and there is every prospect that the coming year will see still further increases". The trend in the development of ornamental street lighting equipment has been toward higher mounting and larger units. One reason given is the demand for increased illumination due to modern traffic conditions and because the efficiency of the gas-filled tungsten lamp increases rapidly with its size. Installations have been made during the past year involving ornamental standards with mounting heights varying from 16 to 30 feet equipped with lamps in sizes up to 25,000 lumens. Some attention is being given to alleys, always inclined to be dark and frequently the scene of crime.

The Report of the Street Lighting Committee of the American Society for Municipal Improvements[93] contains a table of "street lighting practice" in which the recommendations for main business thoroughfares in cities of 100,000 population or larger are 10,000 to 50,000 lumens per post with mounting heights 14 to 25 feet and spacing 80 to 150 feet. For similar streets and cities from 20,000 to 100,000 population, 10,000 to 25,000 lumens per post are pro-

[91a]Gen. Elec. Rev., Aug. 1922, p. 498.
[92]Report of Lighting Sales Bureau, N.E.L.A., June 1923.
[93]Jour. of A.I.E.E., Feb. 1923, p. 178.

posed. This indicates the keen interest of municipal authorities in the desirability of improved street lighting. Before the war the Bureau of Standards undertook a comprehensive study of street-lighting service of all types,—gas, electric and other special classes.[94] It was to include the problems of design of street-lighting systems from the illumination standpoint; the distribution of gas and electricity for street-lighting purposes; methods of operating and maintenance; and the technical and engineering features of contracts. This study, which was interrupted by the war, has been actively resumed.

The following is quoted as part of a discussion at a meeting of the local branch of the Psychological Corporation[95] at Washington: "Street lighting must be based on common sense, but no good sense ignores scientific facts. Modern street-lighting makes a pretty effect, but it is inefficient in these days of congested fast traffic and high taxes. It would pay any city to have its lighting arrangements planned by a specialist—an illuminating engineer in consultation with artists and the local lighting companies". The last sentence contains a doctrine which has been preached by this society for a good many years.

The installation of additional gas street lamps was made in eleven cities in December of last year[96] and in thirty-five cities in January and May of this year.[97]

A legal measure which promises to facilitate improved highway lighting has been passed in New York State.[98] It allows the counties of the state to appropriate money for the illumination of highways. The importance of such lighting is becoming more seriously recognized in all parts of the country.

Information has been received from engineers in a number of the larger cities and incorporated in the following descriptions of specific installations.

Western Coast

Seattle has just completed a new installation of sixty-four single light units of the 6,000-lumen, 20-ampere incandescent

---

[94]Elec. World, Apr. 7, 1923, p. 806.
[95]Science-News-Letter, March 10, 1923.
[96]Am. Gas Jour., Dec. 9, 1922, p. 667.
[97]Gas Age-Rec., Jan. 20, 1923, p. 98; Am. Gas Jour., July 14, 1923, p. 43.
[98]Jour. of A.I.E.E., Dec. 1922, p. 1033.

lamp type mounted on high poles designed by the city.   These poles have replaced center span units.   In San Francisco, changing over from gas to electric lights has been slowly progressing.   There is at present on foot a movement to put some ornamental lamps in "Chinatown", these being of a character that will convey by their very outline the oriental idea.   The Chinese are not at all adverse to spending a considerable sum of money in order to secure the installation of this lighting system which will be in the nature of "white way" lighting.

Altadena, a suburb of Pasadena, Calif., is made up mainly of country estates.[99]   The problem of street-lighting was actually one of high-way lighting and was solved by the use of refractor units suspended by a central span arrangement with inconspicuous wires.   The light height is 28 feet above the road and the spacing 540 feet.   Two hundred seventy units have been installed in a lighting district covering about four square miles.

Attention is being given to improving the lighting in the outlying districts and those business streets not already included in the present ornamental lighting area of Los Angeles.[100]   Pendant units will be employed having 20-ampere series incandescent filament lamps in closed-bowl refractors consisting of two pieces of pressed crystal glass nested one within the other.

Middle West

Denver celebrated the completion of its ornamental street llighting[101] by a carnival last December.   A new city ordinance[102] has provided for the lighting of alleys and the police claim this is responsible for the gradual decrease in the number of burglaries and hold-ups in the downtown district, as well as materially protecting the policemen on their beats.

Mandan, N. D., a city of only about 4400 inhabitants, has installed[103] an ornamental system of 447 standards or one to each ten persons.   One-hundred-candlepower lamps in urn-shaped globes 10.5 feet from the ground are used in the residential sections and 12-foot standards with 250-cp. lamps in the business district.

[99] Jour. of Elec. and W.I., Jan. 15, 1923, p. 55.
[100] Public Works, Apr. 1923, p. 135.
[101] Jour. of Elec. and W.I., Jan. 15, 1923, p. 68.
[102] Elec. World, June 9, 1923, p. 1376.
[103] Elec. World, Oct. 14, 1922, p. 829.

About 9.5 miles of streets are lighted. A residential boulevard in St. Paul[104] is lighted by a system consisting of 400 ornamental refractor lanterns on standards 120 feet apart and carrying 400-cp. tungsten incandescent lamps. In order to show the advantages and possibilities of improved lighting, a one mile "sample" installation of twenty 400-cp. lamps has been placed[105] on part of a twelve mile rural "white way" between Kenosha and Racine, Wis. If satisfactory, the entire district will be so lighted. An ornamental lighting system in the city parks of Green Bay, Wis.,[106] has been made possible by the response to a suggestion by the City Park Commissioner that fifty people donate $100 each to the city to be used for lamp posts in the parks. At least fifty 10-foot standards with octagonal lanterns will be installed.

The outstanding development in street illumination in the city of Milwaukee in the year 1922 was the entire elimination of electric arc lamps and of gas and gasoline lamps. The city is now lighted entirely by series incandescent electric lamps of various sizes. There are now in service approximately

$$
\begin{aligned}
3550 &- 100 \text{ cp. lamps} \\
1970 &- 250 \text{ cp. lamps} \\
2750 &- 400 \text{ cp. lamps} \\
1360 &- 600 \text{ cp. lamps} \\
530 &- 1000 \text{ cp. lamps}
\end{aligned}
$$

Attention should be called to the lighting of Kansas City, Mo.,[107] which was referred to in the May issue of the Transactions. Appropriation has been made for the rejuvenation of the street lighting in St. Louis, Mo.[108] Approximately 23,000 gas street lamps will be replaced by about 50,000 electric lights. Some experimentation is to be carried out to determine by actual use what type of lamp and reflector gives the most satisfactory lighting. One unit to be tried out has a special refractor lens placed on the sidewalk side of the lamp to limit the area of illumination, light the sidewalk without throwing light into the houses and illuminate the area of the street well over to the other side. Maplewood, a

[104]Central Station, Jan. 1923, p. 208.
[105]Elec. Rec., Jan. 1923, p. 9.
[106]Ibid, April, 1923, p. 252.
[107]Elec. World, Feb. 10, 1923, p. 321.
[108]Ibid, June 23, 1923, p. 1491.

suburb of St. Louis,[109] has inaugurated a "white way" consisting of one hundred and fifty 250-cp. series incandescent filament lamps mounted on ornamental iron standards spaced 50 to 80 feet apart.

Oak Cliff, a suburb of Dallas, Texas,[110] is putting in a "white way" with 31 standards each carrying a 10,000-lumen lamp. El Paso's business district is being equipped with an ornamental lighting system, the standards being furnished by business men.

Central States

Part of a plan for relighting Lansing, Mich.,[111] has been completed. It is claimed that this is the first case in the United States where the street lighting system was designed on the basis of architectural uniformity. On the main business streets, two-light standards will be used with 10,000-lumen lamps placed 20 feet from the ground. Secondary business streets will have single light standards 15 feet high with the same luminous flux output while boulevards will have 13 foot standards and 4,000-lumen lamps. For the minor residential districts, the 2500-lumen size will be employed with a mounting height of 11 feet 6 inches.

A system at Lima, Ohio,[112] which was referred to in the 1921 Report has been completed and details are now available. It is said to be noteworthy as a tendency on the part of municipalities to treat ornamental lighting as a part of a scheme to beautify the city as a whole. Another interesting feature is the graduation of candlepower, mounting height and spacing in accordance with the requirements of the individual streets. In the public square, the posts are 15 feet high with 1,000-cp. lamps and special refractors which throw the light toward the center of the square. On the main business street, they are 13 feet 3 inches high and carry 400-cp. lamps. On the minor business thoroughfares and in the residential sections, the posts are 1 foot lower and carry 250-cp. lamps. Downtown the spacing is 50 feet but throughout the balance of the system, 100 feet is the standard distance. Twelve hundred forty-five units are involved, extending over 16.2 miles of streets covering the downtown section and four of the principal residential thor-

[109]Elec. World, May 19, 1923, p. 1169.
[110]Elec. Rec., Apr. 1923, p. 251.
[111]Elec. World, Mar. 3, 1923, p. 511.
[112]Public Works, Sept. 1922, p. 150.

oughfares. Newark, Ohio,[113] clamis to be one of the first cities to use the new 25,000-lumen tungsten filament lamps. They are mounted 30 feet above the sidewalks in special bowl refractor pendants arranged to give a maximum illumination at 15° below the horizontal.

The advantage of good street lighting in advertising a city which is in the path of through auto traffic has been recognized by Freemont, Ohio.[114] In redesigning the lighting system, the following ideas were incorporated: a closer spacing of lighting units on all residential streets to eliminate dark shadows and create an atmosphere of safety; a bright "white way" in the central business district which will attract trade and dispel an impression of deadness in this district after dark; adequate lighting for the principal thoroughfare from city limit to city limit. The old lighting consisted of 328 lamps mostly 4-ampere magnetite arc. The new series has 602 incandescent filament lamps ranging in size from 100 to 600 cp. The latter size is used in the main street on 13.5 foot pressed steel standards spaced 80 feet apart on each side of the street and on the main through highway at a mounting height of 20 feet and at distances to give one lamp for 150 feet of street. Rippled globes with refractors are used for the glassware. For the less traversed streets 400-cp., 15-ampere lamps are employed in rippled globe, dome refractor units on mast arms at heights 20 to 25 feet above the ground. On residential streets 250-cp. lamps are suspended from mast arms while 100-cp. lamps with radial wave reflectors are used in alleys and outlying districts.

In Cleveland, Ohio, increases in lighting equipment have been as follows: eleven 4-ampere magnetite arcs, one hundred sixty-four 600-cp., 20-ampere incandescent filament lamps and twenty-eight 1,000-cp. "white way" lamps. A new "white way" on East 105th Street has been added to that already established on Euclid Avenue. It consists of sixty-eight 1500-cp. units, fifty-nine 1,000-cp. units and nine of the 600-cp. size.

Eastern Cities

Buffalo is carrying out the plan developed sometime ago.[115] Two hundred inverted magnetite arcs have been installed on

[113] Report of Lighting Sales Bureau, N.E.I.A., June, 1923, p. 27.
[114] Elec. World, July 14, 1923, p. 73.

Niagara Street and 800 more will be placed this year in the business district. Syracuse, New York, has designed a thoroughly up to date street lighting system.[116] Four types of lighting are called for, involving business streets, main and secondary thoroughfares, and residential districts. Seven hundred five-lamp ornamental poles and two thousand 4-ampere luminous arcs on overhead lines will be abandoned. The new lights will be supplied from underground circuits. The new plan provides for cast iron posts with two 6.6-ampere inverted luminous arc lamps with 100-foot spacing and 18.5 feet mounting height. On side streets one-light posts alternate with a mounting height of 14.5 feet. For main thoroughfares, cast iron posts will be used carrying 400-cp. 7.5 ampere series incandescent lamps in large lanterns, the posts being 125 to 200 feet apart and with a light source 12.5 feet from the ground, and similar equipment and spacing for residence sections.

There has been no increase in the number of incandescent lamps in Philadelphia since June 1922. However, at scattered locations throughout the city arc lamps have been erected to the number of 406. Arrangements are being made for a comprehensive scheme of lighting by high candlepower incandescent lamps, experiments for which have been going on for sometime, to be installed for ten miles on Broad Street, In the business district of Boston,[117] 76 magnetite arc lamps mounted 14.5 feet above the sidewalk on posts of the boulevard type have been installed. The lamps are placed opposite each other with 75 to 85 foot spacings.

In New York City at the close of the year 1922, there were 81,731 lamps in service on the streets and in the parks of which 7789 were gas, 12 were naphtha, and 73,930 were electric. This is a decrease of 1223 gas lamps and an increase of 3614 electric lamps from the number in service at the beginning of this year. The total number of lamps installed was 4563 and of this number 1345 were 25-watt or 60-cp. units inclosed in a vermilion colored globe used in connection with the improved fire-alarm signal system and 18 were 25-watt lamps used in connection with the new traffic signal stations. In the latter six lamps are installed in each station, two of which burn continuously during the hours public street

[115]Elec. World, Feb. 24, 1923, p. 473.
[116]Ibid, May 26, 1923, p. 1232.
[117]Elec. World Aug. 5, 1922, p. 294.

lamps are required to burn and four burn intermittently during the same hours. In the spring of this year, the city officially opened its mammouth boardwalk along the beach at Coney Island, lighted by 158 pairs of 200-watt incandescent filament lamps of which one lamp on each of 79 posts burns all night and the others burn from official lighting time to 1:00 a. m.

A new and modern system has been installed in Charlotte, N. C.[118] Gas-filled incandescent lamps in refractor globes constitute the light sources.

Canada

Comparing street lighting on the basis of candlepower per 100 of population,[119] the following figures show Canadian practice:

Pembroke	1758 cp. per 100 of population
Hamilton	1000 " " " " "
Toronto	920 " " " " "
Chatham	900 " " " " "
Sarnia	770 " " " " "
London	700 " " " " "

Pembroke is lighted by 230 lamps, 50 of 1000-cp., 103 of 600-cp. and 17 of 100-cp.

Great Britain

Practice in England is indicated by the following table[120] suggested in a report of a joint committee and referred to in a discussion before the English Illuminating Engineering Society.

Character	Minimum Illumination in Foot-Candles
Important Streets	.06 to 0.1
Good Class District	.04 to 0.06
Average London District	.025 to 0.04
Residential London District	.01 to 0.025
Poorer Class District	.01 and below

In another discussion on street lighting before that Society,[121] it was brought out that in London the height of centrally suspended lamps averages 25 to 27 feet with a spacing of 100 feet. Special fittings and columns were designed for the lighting of St. James' Park New Road, London.[122] The system involves ten lamps,

[118]Elec. Rec., Dec. 1922, p. 390.
[119]Elec. News, June 1, 1923, p. 67.
[120]Ill. Eng., Jan. 1923, p. 6.
[121]Ibid, p. 1.
[122]Ibid, Sept. 1922, p. 257.

each containing a cluster of 16 super-heated burners providing an intensity of approximately 1000-cp. and consuming about 2.25 cu. ft. of gas each. The columns are approximately 180 feet apart and the height from the ground to the mantle is 22 feet. Lighting and extinguishing is handled from a tap at the base of the column controlling a pilot light. Reference should be made to a paper presented before the Public Works, Roads and Transport Congress in England[123] by the distributing department of a London gas company. In regard to research the paper says: "The endeavor .... to achieve a satisfactory system of public lighting under these conditions resolves itself into: (1) the experimental determination and specification of a burner suitable for use under the extreme conditions met with in its subsequent use in the streets; (2) the assembling of component parts for the bulk manufacture of such standardized burners; (3) the verification from time to time of the duty rendered by burners supplied in bulk, and (4) further experimental work tending to enhance the efficiency of the appliance". Data and curves are included showing results of tests made by the company on street lighting equipment.

Birmingham has 22,000 public gas lamps.[124] These include a number of high pressure units, some centrally hung, some on 30 foot columns on island junctions in main thoroughfares. On certain islands, three low pressure lamps are installed, the highest and middle ones being fitted with ruby glass and used as "safety first" traffic regulators. In testing burners for city lighting, it is required that they must not pass 0.05 cu. ft. of gas less or more than a predetermined amount. Mantles have to pass certain gauges.

Having decided to continue gas lighting, Leamington, England,[125] has adopted new square lanterns with inverted fittings and automatic lighting and extinguishing. In Glasgow,[126] lighted mainly by flaming arcs, the mounting height is 20 to 25 feet and the spacing 120 to 150 feet. In a section of one of the narrow streets about 40 feet broad lighted by one 350-watt gas-filled tungsten lamp in a directive fitting 27.5 feet above the ground, values of horizontal illumination 1 meter above the road were found to be as follows: under lamp on center line, 0.95 fc.; 30 feet

[123]Gas-Age-Rec., July 14, 1923, p. 37.
[124]Ill Eng., Jan. 1923, p. 24; Gas Jour., Jan. 10, 1923, p. 85.
[125]Gas Jour., Aug. 30, 1922, p. 481.
[126]Ill. Eng., Jan. 1923, p. 1.

along the street, 0.41; 60 feet further, 0.44; 120 feet further, 0,067. In another stretch of street with similar fittings 150 feet apart, there was a maximum of 0.52 fc. and midway between the lamps 0.107 fc. The total number of lamps in Edinburgh[127] at the end of the fiscal year, May 1922, was 1690. Modified lighting as compared to pre-war times has been accomplished by reducing the candlepower instead of limiting the number of lamps.

OTHER EXTERIOR ILLUMINATION

Developments in exterior illumination have been mainly in the direction of sign or display lighting and the problems connected with transportation in one form or another. The increased use of airplanes for commercial purposes has necessitated special consideration of the lighting requirements of this type of navigation.

The Committee on Practical Illumination Questions of the German I. E. S. has recommended the following for illumination values[128] in general exterior lighting such as that of open parkways, factory yards, railroad yards and docks:

Location	Mean Illumination	Minimum Illumination
Railroad trackyards	0.2–0.5	0.1–0.3
Same near factory sidings	0.5–1.5	0.2–0.5
Streets and Parks		
Light traffic	0.5–1.5	0.05–0.3
Heavy traffic	1.5–5	0.3–1
In front of railroad stations	5 –10	1 –2

These values are for horizontal illumination at a height of 1 meter above the ground.

Pageants

The Tercentennial Exposition in Gothenburg, Sweden, offered another opportunity to employ the expressive possibilities of lighting.[129] As a result the artistic lines and the architectural beauty of the buildings are said to be more evident at night than in daytime. At the portal of the Sports Hall advantage has been taken of a small lake to produce a beautiful effect, as it mirrors in rainbow streams the vari-colored lights outlining the building

---

[127]Elec., Sept. 8, 1922, p. 266.
[128]L. u. L., Apr. 26, 1923, p. 207.
[129]Pop. Mech., Aug. 1923, p. 246.

and the adjoining promenade. In another building lights placed in geometrical design behind towering columns give it the appearance of a glowing jewel guarded by pillars of steel and surmounted by a crown of fire.

One of the attractions of the Rocky Mountain Electric Exposition[130] held at Salt Lake City was a huge "Regional Arch" in colored lights and jewels. A curtain containing some 15,000 "jewels" and suspended about 25 feet above the ground carried at each end a rosette with three shields containing the official emblems of the six Rocky Mountain states. Each rosette carried a lightning burst and together with the curtain were illuminated by two hidden batteries of ten 18 inch arc searchlights provided with colored filters.

Again this year the Capitol City[131] was elaborately lighted, the occasion being the Shriners' Convention. At night Pennsylvania Avenue from the Capitol to the Treasury Building presented a veritable ceiling of lights—red, green, yellow and white,—divided at intervals by curtains of light extending below the main canopy. Prominent buildings were floodlighted and Lafayette Square was brilliantly illuminated as the Court of Honor.

Reference should be made to the spectacular lighting of the Centennial Exposition at Rio de Janiero, which was described in the October 1922 issue of the Transactions.

Buildings

Display lighting by neon tubes[132] has been employed in Paris for some years. What is said to be the first example of outlined lighting by this illuminant in England is that of the tower of a building in London. The glass tubes are 30 mm. in diameter and bent to the shape of the arches and other main lines of the structure. When lighted they appear as broad bands of orange color but when extinguished they are hardly perceptible. The tubes are connected in a series of 13 distinct groups, each with its own transformer.

The problem of providing light for a one-day-a-week outdoor market is taken care of in an English town by a temporary wiring

[130]Southw. Elec., Aug. 1922, p. 23.
[131]Elec. World, June 23, 1923, p. 1490.
[132]Elec. Rev., May 11, 1923, p. 746.

lay-out,[133] the erection and dismantling of which requires the labor of one man for one day. Each of forty to fifty stalls are supplied with one 50-watt lamp, the lights being on until 9 p. m.

A novel instance of protective lighting[134] is found in a Detroit building in which the exterior spaces between the supporting columns are strongly illuminated by 200-watt lamps in reflectors set in the building 24 feet above the sidewalk. In this way the dark shadows of the columns which might conceal loiterers and cause apprehension in the minds of passing pedestrians are obliterated.

Signs

Not a little general street illumination in cities[135] comes from illuminated or self-luminous signs. It is reported that while less than 10 per cent of city showings are illuminated, it is expected that this number will be increased to 20 per cent before the end of 1923. Some statistical data have been obtained [136] on the number and size of signs in eight cities ranging from 10,000 to 350,000 in population. The data show an average of one sign and 202 sockets for every 735 persons. Practice in the number of signs and of sockets was found to vary widely in the different cities. The situation in New York where another survey of this kind was made, was referred to in the April 1923 number of the Transactions.

Again the claim to be the largest electric sign in the world has appeared.[137] The letters of this sign are 56.5 feet high, the vertical parts being 12 feet wide and the horizontal, 10 feet wide. The outside dimensions of the structure are 153 feet by 75 feet and the weight is 100 tons. It is said to be legible to the naked eye at 8 miles and visible at a distance of 30 miles.

The slow progress of sign lighting in England[138] is attributed to antiquated restrictive legislation. Piccadilly Circus, undoubtedly the most conspicuous place in London, is not as yet comparable to Broadway in its display lighting. One factor which has a great effect on the lighting situation is that the Crown, which owns

[133] Elec Times, Dec. 21, 1922. p. 586.
[134] Pop. Mech., Dec. 1922, p. 848.
[135] Signs of the Times, Jan. 1923, p. 20.
[136] Jour. of A.I.E.E., Dec. 1922, p. 1033.
[137] Signs of the Times, Mar. 1923, p. 64.
[138] Signs of the Times, Mar. 1923, p. 58.

a large amount of property in the shopping district, will not allow any illuminated signs on buildings erected on these sites, and has refused to alter existing laws.

Aerial

What is reported to be the first aerial lighthouse in this country[139] has been put in operation at College Point, Long Island, under the supervision of the United States Lighthouse Service. The source is a 14-inch navy type searchlight throwing its beams upward at an angle between 45° and 60°. The cross country service to be inaugurated by the Post Office Department[140] will necessitate a string of beacons stretching across the country. Lighted emergency fields will form a continuously lighted highway from Chicago to San Francisco, the part of the route covered at night. The most powerful lights will be located at each of the regular flying fields,—Chicago, Iowa City, Omaha, North Platte, and Cheyenne. Each will be of 600,000,000 cp. and will swing slowly around on a tower mounting, being visible 50 miles away. The intermediate lights will also be on towers and have a visibility range of 30 miles. As a final safeguard in cases of compulsory low flying, flashing traffic lights directed upward will be located every three miles. The huge field at Chicago will be outlined with lights spaced 200 feet apart. In one corner a large well-lighted arrow pivoted like a weather cock will give the pilot wind directions. On top of the hangars a floodlight will throw a pattern of light on the field. This is placed high enough to prevent glare in the eyes of the pilot.

An interesting mathematical study[141] has been made of the probability of detecting an airplane by a searchlight beam with a range of 10 km. and a beam solid angle equal to 2°. A probability of 600 to 1 is deduced that the airplane will not be illuminated. At 1 km. the probability is reduced to 3600 to 1. If the solid angle of the beam is reduced to 1°, the probabilities for the two ranges are 9200 to 1 and 22,900 to 1 respectively. Taking into account the effects of duration of illumination, of beams of different solid angles and different intensities and also the influence of the move-

[139]Popular Mech., July 1923, p. 58.
[140]C. R., Sept. 18, 1922, p. 466.
[141]C. R., Sept., 18, 1922, p. 466.

ment of the beam in sweeping the sky, it was found, as would be expected, that it is to the interest of the aviator to fly as high as possible. In a typical case the probability of his escaping detection is increased in the ratio of 100 to 14 when the flying height is increased from 1,000 m to 5,000 m.

Bridges

The lighting of the new memorial bridge at Springfield, Massachusetts,[142] combines architectural fitness and beauty with good engineering practice. Four central towers 80 feet high capped with sectional duffused globes 6 inches in diameter, each containing four 500-watt incandescent filament lamps, provide general illumination. The sides of the bridge are equipped with 50–ampere inverted magnetite arcs on posts 18 feet high. Special attention has been given to the planning of the lighting system[143] of the new bridge over the Mohawk River at Schenectady. Concrete obelisks will form the support for bronze lamp holders and brackets. These obelisks will rise to a height of 21 feet above the parapet of the bridge and will be spaced at 100 feet intervals. The light units, 1500–cp. 6.6–ampere incandescent filament lamps in eight-panel globes, will be 17 feet above the roadway. Another example of bridge lighting,[144] although in this case only temporary, was that during the celebration of the opening of the new Broadway Bridge at Little Rock, Ark. Four batteries, each consisting of two 1,000-cp. and three 5 00-cp. units, were employed to illuminate the structure and placed about 100 feet away on both sides of the bridge and the river. To bring out the center span, additional floodlights were installed about 250 feet away.

Transportation

The importance of good lighting in railway yards, shops and roundhouses was emphasized in a discussion[145] at the January meeting of the Society of Terminal Engineers. Yard lighting is primarily space lighting with the following problems injected; namely, the presence of deteriorating gases, movable cars causing

[142]Elec. World, Feb. 17, 1923, p. 378; Mar. 10, 1923, p. 578.
[143]Elec. Record, Feb. 1923, p. 116.
[144]Elec. World, April 7, 1923, p. 825.
[145]Railway Elec. Eng., Jan. 1923, pp. 2, 17.

shadows, lack of clearance between tracks to place poles, the prevention of glare that will interfere with the operation of trains through or near the yards and in some cases the presence of overhead propulsion-current wires. One solution has been found in the use of 1,000-watt lamps in silvered glass reflectors of the floodlighting type. Searchlights of the 400-watt size are also used. It has been felt that while flood lamps are not a "cure all", their use has resulted in a large increase in efficiency, decrease in accidents and a minimum of expense.

To enable passengers to enjoy scenic effects at night, it is planned by a transcontinental system [146]to equip the ends of its observation cars with batteries of powerful searchlights. They will be arranged to cover 160° with sufficient height and depth to illuminate cannons, mountains, rivers and lakes along the right of way. The use of colored screens is also proposed.

The Merchant Shipping Advisory Committee in England[147] has recommended the illumination of lifeboats upon launching. Equipment for this purpose provides two 4-volt lamps mounted in the ends of battens pivoted upon the gunwale of each boat. A battery furnishes current enough to maintain the illumination for 24 hours. The lamps are switched on through the agency of a float which acts as soon as the boat touches the water. The arms holding the lamps are adjustable to permit a larger or smaller area of illumination.

INTERIOR ILLUMINATION

What is believed to be the earliest recorded example of the "indirect" system of lighting is credited to Queen Victoria.[148] About 1890 at her suggestion, the Durbar Room at Osborne was illuminated entirely by "deflected" light, not a lamp or fitting or any source of light being noticeable. One of the most novel uses for artificial daylight is that of providing illumination[149] for the examination of fabrics and articles found in the tomb of King Tutankhamen at Luxor, the object being to avoid the necessity of bringing the materials out into the open air and sunlight in order to distinguish the shades and colorings.

[146]Pop. Mech., Oct. 1922, p. 514.
[147]Elec. Rev., May 25, 1923, p. 831.
[148]Elec. Rev., Sept. 8, 1922, p. 346.
[149]Pop. Mech., Aug. 1923, p. 201.

It has been computed[150] that there are four times as many artificial lighting hours in winter as in summer. One and a half hours cover the use of light in the average residence in June while six and a half hours is the average time for December. Occasionally during the year, tables of recommended foot-candle values have been published for various industries[151] ranging from canning factories and packing houses to woodworking plants as well as for all kinds of small stores. In general, they show higher values than heretofore. Additional data on the effect of increased illumination on output have been obtained[152] including the results of the post office tests referred to later. In nine different types of industrial work an average increase from 2.3 fc. to 11.2 fc. gave an average production increase of 15.5 per cent. A test in a Lancashire coal mine[153] on behalf of the Institute of Industrial Psychology, England, showed over 14.5 per cent increase in output as the result of using six times the ordinary illumination as given by miners' standard lamps. This was in spite of the very considerable increased weight of the higher candlepower lamps.

It is reported[154] that many central stations are adding a lighting expert to their personnel. These engineers are endeavoring to cooperate with architects, consulting engineers and building owners to insure as far as possible the installation of good illumination for lighting systems in large buildings of both the commercial and public type, recommended foot-candle values for which have also been worked out.

Public Buildings

Since the largest part of the work done in the postal service involves vision under artificial illumination, it is gratifying to have lighting conditions in post offices made the subject of a government investigation.[155] A discussion of this work was given in the Transactions for March 1923.

The lighting of the new building of the London County Council[156] gives an illustration of present English practice for large

[150]Gas Age-Rec., Jan. 20, 1923, p. 84.
[151]Elec. Rec., Feb. 1923, p. 76 and Apr. 1923, p. 247; Indus. Eng., Nov. 1922, p. 511.
[152]Elec. World, June 30, 1923, p. 1530.
[153]Elec. Rec., May, 1923, p. 315.
[154]Report of Lighting Sales Bureau, N. E. L. A., June 1923, p. 18.
[155]Elec. World, Feb. 24, 1923, p. 470; Mar. 24, 1923, p. 673.
[156]Elec. Times, July 20, 1922, p. 53.

public buildings. In the main council chamber are four bowl fittings 5 feet in diameter and 7 feet tall of bronze supported by bronze chains about 40 feet above the floor levels. In each bowl are ten 300-watt lamps. On the marble staircase six bronze standards 7.5 feet high, carrying 22-inch opal glass globes containing 300-watt lamps, provide a decorative as well as impressive illumination. Bowl ceiling luminairies are used in the gallery surrounding the staircase. The corridors are lighted by small bronze ceiling fittings to the number of about 500. A large bronze lantern lights the terrace entrance hall while cast bronze pendants with 20-inch bowls furnish light for the street entrance hall. In the lobbies to the entrance hall are special bronze ceiling fittings with shallow satin-finished glass dishes, the top reflector with the lamp holder being sunk into the roof.

The Educational Committee of the London County Council has decided [157] to replace, in those elementary school buildings largely used at night, the old upright mantle burners by super-heated inverted burners adapted to existing fittings. A type of luminaire will be employed which carries a cluster of small mantles without glassware and equipped with a deep type of enamel shade so as to screen thoroughly the eyes of teachers and pupils from glare.

Canadian practice in hotel lighting[158] may be seen in what is claimed to be the largest hotel in the British Empire. Approximately 5000 lighting units are employed. One-hundred-watt gas-filled incandescent lamps are used in the bedrooms, 25-watt vacuum incandescent lamps in the bathrooms and corridors, and 200-watt lamps in the shops. The glass was especially designed for the hotel.

A new large hotel in this country[159] has over 11,000 lamps. Among the novel features of this installation may be mentioned the use of wrought iron in some of the public rooms. One luminaire in the men's cafe has a full rigged ship of the galleon type in the center of an elaborate wrought iron frame work, which also carries a number of candelabra fittings. Another large wrought-iron luminaire is in the palm or sun room. In the library is a luminaire in the form of a terrestrial globe, while localized lighting is provided by two brackets over the fireplace and by floor and table lamps. In the writing room portables are combined with ceiling

[157]Gas Jour., Oct. 18, 1922, p. 176.
[158]Elec. News, Aug. 1, 1923, p. 43.
[159]Ltg. Fix. & Ltg., July 1923, p. 15.

pieces and localized two-light units over each desk. A six-light luminaire of colonial design is used in private sitting rooms along with table lamps.

Lighting equipment and methods have improved so much in the last twenty years that redecoration of a public building is not complete without a rejuvenation of the lighting system.[160] A case in point is that of a church whose trustees desired to retain the beauty of the original decorations and hence had the new luminaires planned with this point in view. Ten dark bronze pendants were made, each fitted with twenty-eight 25-watt candle-flame tinted lamps. Brackets for the vestibule and balcony were designed to harmonize with the elaborate wooden paneling which is a feature of this church. In changing over from a gas to an electric system an English church[161] used the original ring fittings but suspended from each ring four 100-watt lamps in glass shades. The illumination values were increased from an average of 0.5 fc. with practically no light in the upper sections of the auditorium to more than 1.5 fc. with a range from 0.8 to 3. Two 100-watt pip-frosted lamps on brackets were used to light the chancel and the illumination at the middle of the choir stalls was increased from 0.7 to 7.5 fc. The reredos was lighted with concealed carbon lamps and the lectern, provided with a specially designed luminaire, showed an illumination of 1 fc. A novel item was the use of two green 40-watt lamps by which the organist gives the time for the choir to whom he is invisible. They are fixed on either side of the chancel arch but out of sight from the nave. Signal lights at the organ were installed so that the organist may be notified of the entrance of the choir in the case of processional hymns.

A prize fighting ring requires a lighting system similar to that of some exhibitions and pageants; i.e., relatively concentrated lighting over a small area.[162] In one case good results were obtained by installing a 1000-watt lamp 14 feet above the center of the ring which was 20 feet square. Eight glass reflector units with 100-watt bowl frosted lamps were mounted on a 14-foot square conduit frame 12 feet high. These lamps were carefully tilted to keep the glare from the eyes of the occupants of the press bench located

[160]Lighting Fixtures & Lighting, Feb. 1923, p. 20.
[161]Elec. Rev., May 4, 1923, p. 700.
[162]Elec. Rec., Mar. 1923, p. 182.

around the ring, and were provided with a piece of glass at the bottom to protect the eyes of the spectators. The illumination was 45 fc.

A new building, erected by the various packing interests of a western city for industrial expositions and live stock shows, has its arena lighted by RLM dome reflectors with 500-watt lamps.[163] They are arranged to be lowered 12 feet from the ceiling when extra high illumination is desired as in judging cattle. Normally they are suspended at the ceiling which is 39 feet high and produce an illumination of 8 fc. In the lowered position 16 fc. are available. At the outer edge of the arena, tests showed 9 fc. The corridors below the seats are lighted by various angle type reflectors equipped with 200-watt lamps mounted on columns and side walls.

Galleries

The fading of colored objects in museums is a serious loss in many cases.[164] To ascertain corrective measures, if they exist, a study was made involving exposures on 2515 days during the eight years from 1914 to 1921 on fugitive colors (chiefly synthetic dyes). Maximum fading was caused by direct sunlight and diffused sunlight was found to be more destructive than artificial light. Tinted glasses varied in their protective effect according to their relative power to absorb the violet and blue rays, but most of such glasses were objectionable because they altered the appearance of the objects viewed through them. The best glasses only delayed fading; they did not prevent it. Pigments made with oil fade less rapidly than when made up as water color. It was concluded that artificial light is desirable for museums in order that the fading of objects may be delayed. This subject was referred to in the Transtions for March and July, 1923.

A recent art gallery lighting installation[165] is said to be different from and yet embody the good points of two methods already in use elsewhere, one where the lighting is accomplished from the sides and the other where the illumination simulates actual daylight conditions coming uniformly from overhead. In the case referred to, three rooms have horizontal skylights while that in the fourth is vertical. Above the glass are placed 150- and 200-watt

[163]Elec. Rec., June 1923, p. 383.
[164]Jour. Royal Soc. of Arts, 71, 1923, p. 144.
[165]Elec. Rec., July 1923, p. 3.

blue bulb lamps in mirrored glass projectors. They direct the beam of light on the opposite rather than on the adjacent wall. The skylight glass conceals the light sources but does not materially modify the distribution. This general scheme is similar to that of the Cleveland Museum of Art described before the Society some years ago. The walls of the gallery are neutral in tone, non-glossy and of low reflecting power. There is enough stray light to make the skylight luminous and produce general illumination in the central portions of the rooms. Outside of the vertical skylight are placed seven weather-proof type floodlighting projectors with 500-watt clear lamps. The light beams are directed against the opposite wall, painted a light cream color, from which they are diffused and illuminate the rest of the room. The foot-candle values at the 12.8- and 4-foot levels measured 7.2, 5.2, and 3.1 fc. on one wall and 4.6, 3.2, and 2.2 fc. on the other wall. In the galleries having the horizontal skylights at the same height levels, the foot-candle values were found to be 3.5, 2.8, and 2.3 fc. Some long narrow galleries are lighted by single rows of totally indirect units.

Theaters

The three-color indirect system[166] continues to be the dominant feature in the lighting of moving picture theaters. In order to eliminate eye-strain and make the pictures as clear and sharply visible from the side seats as those directly in front, a movie screen has been constructed[167] with a white surface embossed with a multitude of small squares producing an appearance similar to that of a waffle-iron. The accuracy of the design and construction of the checkered surface determines the character of the diffusion and the uniformity of the illumination. In connection with the design of the lighting system for a moving picture theater,[168] data were obtained on the brightness values of parts of the picture itself. Sunlight on the white clothing of one of the actors was represented in the picture by a brightness of 3.0 millilamberts; a reproduction of a letter written on white paper, 6 mL.; title background, 0.06 mL.; with the projection machine running but no picture, the screen brightness was 7.0 mL. The screen brightness as the result of general house illumination was only 0.01 mL.

[166]Mov. Pic. World, Oct. 7, 1922, p. 512.
[167]Pop. Mech., Jan. 1923, p. 95.
[168]Jour. of A. I. E. E., June 1923, p. 573.

In a new theater which is used for both musical and moving picture purposes,[169] an effort has been made to embody the principles discussed in a paper before this society in 1920.[170] These involved permissible general illumination in the auditorium and graded illumination from the lobby inward while the pictures are shown, together with means for providing satisfactory illumination when the hall is used for concerts. In the auditorium the main object of interest from the lighting standpoint is a large crystal chandelier. In this are mounted 16 concealed 75-watt units throwing light onto the ceiling from which it is diffused and provides an average horizontal illumination of 0.2 fc. with a range from 0.09 to 0.52. The brightness of the luminaire glassware is on the average 0.33 mL. The chandelier itself is softly illuminated to a point which prevents it from being silhouetted as a dark mass against the ceiling, by a few unconcealed lamps operated at a very low voltage. For low general illumination when a concert is being given, thirty-two 150-watt lamps, also concealed in the chandelier, throw light on the ceiling while ninety-six of the same size lamps may be similarly used to provide a high intensity of illumination. For brilliant effects or special gala occasions, direct and special lighting with scintillating of the crystals is produced by thirty-six 15-watt and three hundred eighty-four 25-watt lamps in candles and sockets following the contour of the chandelier bowl. Further to equalize the indirect lighting from the ceiling, 100-watt and 150-watt lamps are concealed above the tops of the doors leading from the balcony to the corridors. The mezzanine is lighted chiefly by lights in the cove. All light sources within the region occupied by the seats are completely concealed from the spectators except the few small units previously mentioned. The graduation of the illumination from the entrance to the interior of the auditorium is shown by illumination readings as follows:

Main lobby just inside entrance.......... 15   fc.
Center of lobby....................... 2   fc.
Near orchestra foyer.................... 1   fc.
Main vestibule just inside door........... 0.23 fc.
Orchestra foyer just inside door........... 2.1   fc.
Orchestra foyer near aisle entrance........ 0.09 fc.
Central portion of main floor............. 0.04 fc.

[169] Jour. of A. I. E. E., June 1923, p. 569. .
[170] Trans. I. E. S., 15, 1920, p. 645.

The high level lighting of the auditorium gave an average illumination of o.95 fc.

Present concert hall lighting practice in Germany may be seen in the re-equipment of the Philharmonic in Berlin.[171] The design was turned over to a committee of the German Illuminating Engineering Society, the former lighting having proved unsatisfactory. Six mirror reflectors with 100-watt lamps mounted 17 feet high resulted in a maximum illumination of 65 lux (6 fc.), a minimum of 20.3 lux (1.9 fc.), and a mean of 43 lux (4 fc.). The light output has been increased almost three times with less than half the former wattage.

Complimentary colors are the foundation on which has been built the lighting of one of the most elaborate moving picture theaters as yet constructed.[172] When an object is lighted on one side with a maximum of intensity of one color, the shadow formed behind the object, instead of appearing black, is lighted with a minimum intensity of the complimentary color. Color has also been depended upon by the architect to give an impression of depth where structural limitations prevent actual depth. Primary colors have been used for decoration to a surprising extent. In the main auditorium all lighting sources are concealed, there being no luminaires except those beneath the balcony. Every place where a light source of which 10,400 in varied shapes and wattages are employed may be hidden, holds its battery of spotlights or trough of lamps. Designed especially for the purpose are 780 baby spotlights and lamps operated by a motor driven dimmer. Four colors —red, green, blue, and amber, in the order named—focused in one direction, increase from zero intensity to a maximum and then fade. From the opposite direction minimum intensities, of red and of blue, light the shadows, each color extending over one cycle of the maximum colors, a complete cycle requiring 16 minutes. The color variations are almost limitless. These lights are also used to produce atmospheric effects in connection with stage numbers or musical interpretations.

In an elaborate new theater devoted to vaudeville,[173] in which the modern French note dominates the decorative scheme, lumi-

[171] L. u. L., Mar. 15, 1923, p. 141.
[172] Jour. of Elec. & W. I., Feb. 15, 1923, p. 131.
[173] Lighting Fixtures & Lighting, May 1923, p. 13.

naires have been obtained in harmony with the rest of the surroundings. In the grand hall are five large crystal chandeliers containing 60 lights mounted in tiers and set around the body of the luminaire. Pendants in other parts of the interior are duplicates on a smaller scale. A distinctive feature of the bracket lights in the main hall is an illuminated oval made of crystal with a light back of the oval, while above it are four candle arms draped with crystals. These brackets are placed around the hall while in four corners are torchieres made with carved wood standards and a three-tier effect of candles with crystal-draped arms. Supplementing the luminaires and brackets are many floor lamps. Ten-light luminaires are placed at intervals around the mezzanine and twelve-light glass arm chandeliers are installed in the ladies' room. Under the soffit of the balcony the luminaires are of an umbrella shape about 3 feet in diameter and consisting of a finely detailed metal form filled in solid with crystals, the frame itself being three rows of crystal garlands, pentalogs, etc. In the dome and procenium promenade three-color cove lighting is used. In the dressing rooms, specially designed luminaires adjustable up and down add to the convenience of the actors. In the grand hall is a combination lighting standard and clock 10 feet high in the Italian renaissance style.

Some novel and bizarre effects have been worked into the lighting of an English moving picture theater.[174] The decoration is in the "New Art" style, the colors being daring in harmony and quaint in application. Groups of peculiarly designed lanterns and erratic colorings hang from ceiling bays, and wall brackets and suspended balloon lights and an aerial fountain add to the uniqueness. The ceiling lights are an integral part of the design of the ceiling instead of hanging therefrom and additional light is obtained from the bull's-eye windows in the frieze and from bases around the balcony front. On the exterior a cupola, which dominates the building, is floodlighted by 24 lamps in addition to 6 flambeaux.

During the past few years stage lighting has kept pace with the improvements in the lighting of the rest of the theater. A system developed on the continent[175] is said to simulate convincingly natural atmospheric phenomena on a single background by a combination of electrical and optical effects in conjunction with an

[174]Elec. Rev., Oct. 20, 1922, p. 561.
[175]Ibid, Mar. 16, 1923, p. 406.

artificial horizon which occupies the whole background of the theater. Floodlights, projectors and spotlights are employed to illuminate the acting area. The system referred to employes a 1500-watt lamp provided with a set of two deflectors, the upper mirror reflecting rays which reach it from beneath, the lower one utilizing the lateral rays. The projectors may be suspended above the stage and colored disks employed to vary the effect. These spotlights have a telescopic lens to concentrate and focus sharply on the object of interest. By adding a second lens and objective, sets of spotlights can be used for projection purposes and will throw definite pictures on a plane surface. Indirect lighting is used for the footlights to avoid glare. The "artificial horizon" may have the shape of a cupola or a surface of a cylinder and is a brilliant white. It can be erected as a permanent structure or as a movable screen. Thousand-watt tubular lamps in special housings containing colored glass slides are built up in tiers, to make possible all sorts of color effects from moonlight to the reddish glow of dawn. Cloud pictures as well as other atmospheric effects may be projected on the artificial horizon.

Transportation

Approximately 15,000 lamps are required to take care of the 4,000 rooms of the Leviathan, "the queen of the American Merchant Marine".[176] Festoons and decorative lighting are also provided for. In the public places, such as the social hall, winter garden, swimming pool and dining room, the lamps are concealed in cornices. In addition to these cornice lights, the social hall has a large glass skylight with lamps above it. Lighting equipment is provided for the stage in this room. On this vessel, particular attention has been given to the illumination in the boiler and engine rooms.

In the report of the Committee on Locomotive and Car Lighting of the A. R. A.[177] at this year's convention, it was recommended that a smaller bulb be used for the 15-watt locomotive cab light. There is an increased use for train lighting of the 25-watt, 30-volt, gas-filled lamp and it has been suggested that both the 25-watt and the 100-watt lamps be added to the list of recommended sizes. To

[176]N. E. L. A. Bulletin, Apr. 1923, p. 209; Ltg. Fix. & Ltg., Apr. 1923, p. 13.
[177]Railway Elec. Eng., July 1923, p. 211.

Offices

In remodeling the lighting of a large office and showroom,[179] it has been found possible to replace 1068 gas mantles by only 252, the gas consumption being reduced from 1068 to 1008 cu. ft. per hr. and still obtain twice the light previously available. Ten large wrought-iron luminaires equipped with 16 burners carrying three mantles each have been converted into the semi-indirect type by putting white glass panels in the lower cast iron bracket and in the lower outside ring. A 9-mantle burner has been employed and the floor illumination increased from 3.5 fc. per sq. ft. to 6 fc. In the clerical space eighteen 10-light luminaires were used giving an illumination of 4.5 fc. at desk height. These fittings have been re-modelled to the indirect type and equipped with a 9-mantle burner. The result has been 8.5 fc. at desk level.

Stores

A survey[179a] of lighting conditions in the retail stores of a large middle western city showed only 25 per cent with an illumination of 5 fc. or better and only 2 per cent with illumination of 10 fc. Since then this condition has been changed so that in cigar stores the illumination has been raised from 6.9 to 14 fc.; in grocery stores from 3.8 to 8.1 fc.; in drygoods stores from 5.8 to 7.5 fc.; and in tailor shops from 3.4 to 8.1 fc., and in other stores up to 10 fc. Tests of the illumination provided by a new installation in a department store [180] showed on the first floor an average of 12 fc. directly beneath the luminaire and 6 fc. directly between units. Similar readings on the second floor gave 10 and 4 fc. The luminaires were pendant enclosing globes of three-layer cased glass and were located one in each bay and down the centers of the aisles at 24-foot intervals. The average ceiling illumination was 22 fc. on

[178]Ibid, Feb. 1923, p. 45.
[179]Gas Age-Rec., Jan. 13, 1923, p. 37.
[179a] Elec. Rec., Jan. 1923, p. 11.
[180]Ibid, Apr. 1923, p. 248.

the first floor and 18 fc. on the second.   Mounting heights were
12.5 and 11.5 feet, respectively, 750-watt lamps being employed
on the first floor and 500-watt lamps on the second.

The use of pedestals for indirect lighting luminaires[181] has become more or less common in the case of public halls, restaurants,
etc., but an English store has adopted this method of avoiding
obstructions on the ceiling.   The standards are of wood 7.5 feet
high fitted with three mirror reflectors and especially designed by
the architect to harmonize in finish and character with the other
woodwork.   The main units are supplemented by other mirror
reflectors on the tops of shelves.   In a room on the top floor lighted
by a skylight, reflectors placed above the glass provide a night
illumination of the same intensity as that of daylight.

A different window lighting effect for every night in the year[182]
is obtained in an eastern store by the use of 16 circuits in the
window, 12 overhead and 4 at the floor, and 9 flashers.   Spotlights
with ball and socket joints and fitted with 200-watt lamps give
light at any angle and at any desired spot.   Each unit has four
color effects available at any angle in a hemisphere.   One of the
luminaire display circuits is arranged so that when the window and
sign lights are put out, one or more luminaires may be made to
flare up against an attractive background.

Factories

That there is still a great deal of improvement possible in all
classes of lighting in England,[183] both in the industrial and residential spheres, is indicated in an address on the subject of salesmanship and lighting before the Electrical Development Association.   The speaker asserted that there are probably only 30 per
cent of the factories in that country which "have proper lighting
equipment."   He spoke of one case in which this condition had been
corrected.   A factory manager had requested a demonstration of
artificial lighting which would approximate daylight as given
through a north skylight.   The daylight measurement showed 20
fc. and by using 200-watt lamps in industrial reflectors on 10-foot
centers, an illumination of 15 fc. was obtained and accepted as

[181] Elec., Jan. 26, 1923, p. 95.
[182] Electragist, Aug. 1923, p. 29.
[183] Elec. Times, Oct. 19, 1922, p. 357.

satisfactory. While great strides have been made in the lighting of offices and public buildings, generally speaking the intensity of illumination is far too low and the standard of equipment used very poor. There are still in existence tens of thousands of drop pendants.

Comparison of production[184] in a textile-spindle shop under daylight and under directed artificial light indicated 25 per cent increase in the latter case. Straightening of the spindles after grinding is a hand operation which it was found difficult to perform satisfactorily under conditions of varying daylight. An ordinary shaded lamp was tried but was not satisfactory because of reflections on the spindle. An angle reflector with a 150-watt lamp solved the difficulty when a silvered reflector-cap diffuser was placed over the side and end of the lamp so as to shade in the direction of the work. The result was no glare or disturbing reflections. As all production of the plant centered on the spindle straightener, an increase in his work resulted in an increased plant output.

Numerous methods have been worked out in the past for increasing the brightness of the field of view of the microscope.[185] Another step in this direction is the use in the coaxial illumination systems, of an "azimuth" illumination screen, where by "azimuth" is meant a direction perpendicular to the axis of the microscope.

General

Colorimetric chemical analysis by the use of "indicators" has been difficult under artificial light[186] but some of the troubles have been remedied by the use of artificial daylight. Tests have indicated an accuracy with the latter in some cases equal to, but in many cases even greater than by daylight.

Some data on comparative values of oil and electric light have been obtained[187] by the Agricultural Department of the University of Wisconsin. A working farm foreman did his usual routine work by kerosene oil lamp one night and by electric light the next. The results are shown in the following table:

[184]Elec. World, Dec. 30, 1922, p. 1450.
[185]Phys. Zs., Feb. 15, 1923, p. 91.
[186]Analyst, Oct. 1922, p. 424.
[187]Elec. Rev., May 25, 1923, p. 807.

Operation	Minutes for Operation	
	Oil	Electric Light
Stabling cows	4.	3
Cleaning mangers	9.5	7
Weighing and feeding grain	31.75	12
Feeding silage and hay	39.25	35

It should be noted that there are a good many factors which should be taken into account when drawing conclusions from these data.

LUMINAIRES

The adoption of the term "luminaire"[188] to replace the word "fixture" in the lighting industry seems to be very slow in spite of the efforts of this Society. At the winter convention of the National Council of Lighting Fixture Manufacturers,[189] it was pointed out "that the word 'fixture' was a handicap to the expansion of the industry" but instead of replacing it by the word "luminaire" it was suggested that the term "lighting equipment" be its successor.

In the exhibition of luminaires at the convention just referred to, the shower type was quite generally represented but the candelabra idea still predominated. A decided advance was noted in wrought-iron pieces. More attention is being given to special designs for individual types of public buildings.[190] Thus for church lighting, a finish in the Gothic is given to luminaires made especially for churches in the Gothic architecture. A growing use of torchiers is indicated by the new designs in this class of lighting unit.[191]

Gas

A fitting which converts an open-flame gas jet into a mantle-type burner[192] has a goose-neck which is screwed on in place of the open flame outlet and is of such dimensions that it will fit inside the standard shades or globes. The shade holder may be used for direct or indirect lighting depending on whether it is placed below or on top of the globe. In the latter case, it is suspended by a threaded knob on top of the goose-neck.

[188] Lighting Fixtures and Lighting, March 1923, p. 26.
[189] Ibid, Feb., 1923, p. 15.
[190] Ibid, Dec., 1922, pp. 5, 6.
[191] Elec. Rev., Aug. 1922, pp. 97, 102.
[192] Pop. Mech., Sept. 1922, p. 330.

An inverted super-heated gas luminaire of the cluster burner type for general commercial use has been worked out on the basis of being dust– and insect-proof.[193] The air is drawn through three tubes from the base of the corona band which carries the glass globe into a mixing chamber on the center down rod leading to the injector tube. The burner is fitted with three mantles controlled by one tap and may be used interchangeably with any of the three standard sizes of mantles. The unit is said to give a light of from 500 to 700 cp. In order to prevent dust collection above gas lumi-naires, a special burner[194] deflects the heat from the gas flame so that it is projected at an angle outward from the bowl. A small mica baffle suspended below the burner catches the mantle if it breaks or drops from the holder.

Street Lighting

The increased size of incandescent lamps available for street lighting has hastened the retirement of the cluster standard in favor of the single light unit.[195] For the latter a much more at-tractive architectural design of the post is possible and it is now largely used for "main street" lighting. A distinct change is also noticeable in the character of the glassware which has expanded from a globular form to the urn-shape, while the tops are covered either with reflector canopies or the glass is so designed as to re-direct light thrown upward and formerly wasted. There has been an increased use of alabaster rippled globes.[196] A new glassware has a surface made up with a number of rectangular protuberances which tend to diffuse the light without materially altering its direction, and is made in clear crystal or slightly opalescent material. The trend toward panelled lanterns continues and several types are now available. A new type of bowl refractor for pendant units has a distribution which tends to build up the light laterally while at the same time adequately illuminate immediately below the unit. Experiments are being made on a design for an assymetrical refractor which will build up the light longitudinally allowing only a small portion to be directed across the street. New designs of ornamental post-top units have also been brought out,

[193]Gas. Jour., Oct. 25, 1922, p. 226.
[194]Pop. Mech., Jan. 1923, p. 113.
[195]Elec. Rec., Apr. 1923, p. 217.
[196]Report of Lighting Sales Bureau, N. E. L. A., June 1923.

the light distribution being controlled by upper and lower parabolic reflectors arranged to give a maximum distribution 20° below the horizontal.   These are said to be particularly adaptable for 50 foot to 100 foot spacing.   A new highway unit, totally enclosed, has the lamp at the focus of a parabolic reflector which redirects the light vertically downward upon two refracting prism surfaces and thence to the roadway in two directions.

Industrial

Through the cooperative efforts of a number of reflector manufacturers, glassware makers, and illuminating engineers, an addition has been made to the type of equipment for industrial lighting which has become more or less standardized.   It consists[197] of an enameled steel reflector with a glass diffusing globe which completely surrounds the lamp.   Sections of the upper portion of the steel reflector are cut away permitting about 7 per cent of the light to reach the ceiling and thereby avoid harsh contrasts.   It is available in two sizes, the reflectors of which are 18 inches and 20 inches in diameter respectively.   The smaller size takes 100– to 200-watt lamps, the larger, 300– to 500-watt.   A standard socket adapter makes the larger reflector adaptable for smaller lamps by extending the light center length.   The brightness is given[198] as 2 to 5 per sq. in. and the efficiency as 65 per cent.   Both the bowl and the shallow dome type of industrial reflectors[199] have been adapted for use with the so-called "mill" type incandescent lamp on pendant or portable cords or for permanent local lights.

A special heavy duty reflector[200] has been designed for industries where it is desired to have strong illumination concentrated on machines with enough "spilled" light to illuminate the aisles and passageways.   The unit is nearly 12 inches in diameter, 6 inches high and carries an adjustable holder so that it may be used with lamps of various sizes from 75 to 200 watts.   With a 200-watt clear lamp, 1200 cp. are obtained directly beneath the reflector, 850 to 900 cp. at 30° to the vertical and 400 cp. at 45°,   It will be noted that there is a rapid falling off after 15° and the cut-off is at 60° to eliminate glare.

[197] Elec. Rec., Feb. 1923, p. 108; Central Station, April 1923, p. 300.
[198] Elec. World, Jan. 6, 1923, p. 27.
[199] The Electragist, Aug. 1923, p. 50.
[200] Elec. Rec., Mar. 1923, p. 183.

When the standardized steel dome reflector was first brought out[201] tests of seven standard types covering the mechanical, thermal and illuminating properties showed quite a variation. This condition has been remedied to such an extent that it is felt that the choice of these reflectors may be made on the basis of their thermal and mechanical characteristics since the illuminating properties are very closely the same for different makes.

An effective method of making the letters stand out[202] has been employed in a sign made up for use in interiors such as banks, offices, etc. The surface of the letters is flooded by light through an aperture just below them while light from the same lamp reflected by a mirror is diffusely transmitted through them. Due to the double illumination, the contrast between the letters and their background is quite sharp and the legibility clearly heightened.

Luminaires

A dual distribution of light is obtained from a semi-indirect luminaire which has a bowl of an ordinary semi-indirect type 13 inches in diameter of heavy density tinted opalescent glass. At the bottom, a circular section has been removed[203] and replaced with a piece in the nature of a lens made of crystal glass. The latter tends to concentrate the downward light into a relatively narrow beam such as might be desirable for lighting a dining room table. A decorative husk covers the lamp socket and the whole is suspended by a chain.

Prismatic glass is used[204] in a totally enclosed semi-direct decorative bowl luminaire made in two parts. The exterior surface is smooth and plain, control of the light rays being effected by an interior prismatic construction designed to minimize glare and give wide distribution. The top diffuser, besides protecting the lamp and the prismatic surfaces from the accumulation of dust, diffuses the upward rays which light the ceiling. The lower diffusing zone has a permanent enameled inner surface with an ornamental exterior design.

A luminaire[205] designed especially for the kitchen has a ceiling fitting and holder of white porcelain enamel or steel. Set screws to

[201]Elec. World, Aug. 12, 1922, p. 330.
[202]Pop. Mech., Jan. 1923, p. 72.
[203]Elec. Rec., Nov. 1922, p. 303.
[204]Elec. Rev., Dec. 15, 1922, p. 923.
[205]The Electragist, Aug. 1923, p. 48.

hold the glassware are avoided by a spring holder with no dirt accumulating projections. A white diffusing glass 9 inches in diameter with a 5-inch filter permits the use of lamps up to the 150-watt size.

Hospitals

An apparatus designed especially for use in operating rooms of hospitals, clinics, etc., has been brought out in Germany.[206] A gas-filled projector type of incandescent lamp has its filament adjusted in a parabolic mirror so that the outgoing rays are parallel and fall on two plane mirrors from which they are reflected to the spot under examination, one mirror reflecting rays from the upper half, the other from the lower half of the reflector. The latter is surrounded by a cylindrical cover to prevent stray light. Arrangements are provided for color filters to give daylight quality to the illumination. Another hospital light designed to throw a powerful but cool and shadowless illumination on the subject[207] employs a 100-watt lamp and covers an 18 inch ring on the table. It consists of a dioptric lens fitted at the center of a metal saucer around the rim of which are a number of silver reflectors. The lamp holder at the top of the fitting has a quick release catch for changing the lamp and is adjustable to any standard type of lamp. The bottom of the luminaire is closed by a screen of tough glass. A lamp arranged especially for therapeutic work[208] has an adjustable hood which may be raised to a height of 7 feet, is mounted on a lazy tongs and is counterbalanced. The 1500-watt incandescent lamp can be moved for focusing in four positions. At the lowest position an even diffusion of light results while at the highest a markedly hot area is apparent.

Portables

A new adapter[209] which may be easily attached to a floor or table pedestal or used to convert an oil lamp to an electric or to make a lamp out of a vase has a top part in the form of an inverted mirror bowl reflector. Below this is an inverted cone type of diffusing glass which throws out sideways and downwards, light

---

[206] L. u. L., Feb. 1, 1923, p. 58.
[207] Elec. Rev., Oct. 13, 1922, p. 536.
[208] Elec. Rec., Nov. 1922, p. 302.
[209] Elec. Rec., Oct. 1922, p. 220.

from two or more lamps in sockets just below it. The adapter is intended for use with art glass or cloth shades and is equipped for lamps from 75- to 200-watt in size in the reflector and 15- or 25-watt frosted lamps in the outer sockets.

An exceedingly novel table lamp[210] has a shade and standard made of cows' horns, the latter having been rubbed down with emory until transparent enough to show their natural color by transmitted light. The finished product is said to be of a very pleasing and mellow glow.

A small lamp which can be used as a clamp lamp,[211] a stand lamp or a suspension type for the wall is made to swivel and angle and can be placed and will stay in any position at will. It has a ball and socket joint and a counter-weighted base which holds the lamp firmly either in the vertical position or swung through an arc of 90°. The clamp works on a flat screw. Another portable lamp for the living room[212] has an arm supported by a thick braid which may be fastened to the wall by a thumb tack or hook and adjusted by sliding it up or down on the braid. The arm can be fixed so that the lamp lies flat against the wall as a wall lamp or extended over a chair, desk or dresser. The lamp shade is made up in a combination of braid and mica tinted in colors to make the whole look artistic.

An electric lamp for use around automobiles or by campers[213] is made up like an old-fashioned trainman's lantern. It is provided with three standard unit dry-cell batteries concealed in what would have been the oil reservoir. It can be set or hung anywhere and has a practically unobstructed light beam except in the direction of the base.

Accessories

For use in shades an impregnated silk is available[214] which is said to be stiffer, to have improved qualities of diffusion, and to be more readily cleaned than material ordinarily used, as well as being practically non-inflammable. So much effort has been expended in designing luminaires so that the lamps will be kept free from dust and dirt that it is not surprising that some energy has

[210]Pop. Mech., Oct. 1922, p. 604.
[211]Elec. Merch., Jan. 1923, p. 3063.
[212]Elec. Rec., July 1923, p. 42.
[213]Elec. Merch., July 1923, p. 3510.
[214]Illum. Eng., Nov. 1922, p. 307.

been deflected to produce a protector for the handsome silk or other cloth lamp shade which adorns a table lamp.[215] This protector is made of a transparent durable material which is said to have the transparency of glass and is impervious to the finest particles of dust. These covers range in size from 18 to 30 inches in diameter, and can be obtained in white, old rose, blue, or gold colors. A hanger for an ornamental dining room luminaire of the general dome type[216] is made of wood fiber finished in antique silver.

In order to convert a direct lighting luminaire[217] to an indirect without discarding the old chain or reflector, a device has been worked out which has the form of a bent rod carrying an ordinary plug at one end which screws into the existing socket; at the other end is an inverted socket into which the lamp is screwed, and which carries the reflector or globe holder. The whole is concealed by a silk shade.

The electric-light switch plate[218] as mounted on the walls of a residence is ordinarily finished in brush brass or bronze and is not decorative in character. Attention has been given to this small feature of the room equipment and plates are now being hand-painted in four color combinations and radium treated to make them easily found in the dark. Another opportunity for decorative improvement has been found in the pull chains for floor and table lamps.[219] These chains formerly ending in a button or knob may now be fitted with a colorful object such as an owl or a humming bird, a bit of rare rock or amber or the chain may be a string of beads of various colors. A still more elaborate touch is the switch operated in the old-fashioned bell-rope style. This switch pull is a long-woven silk ribbon hanging against the wall and reaching to the ceiling. It is made in mulberry, rose, old gold, and other colors.

A screwless shade holder[220] for luminaries has on the inner surface of the holder three spring fingers. These fingers are forced inward against the outside lip of the bowl by means of a simple locking ring. To facilitate the installations of window lighting reflectors a square steel tubing already wired is obtainable.[221] The

[215]Elec. Merch., June, 1923, p. 3453.
[216]Elec. Merch., Jan. 1923, p. 3063.
[217]Ibid, Aug. 1922, p. 109.
[218]Ibid, Nov. 1922, p. 114.
[219]Ibid, Apr. 1923, p. 3245.
[220]Lighting Fixtures and Lighting, Mar. 1923, p. 42.
[221]Pop. Mech., Dec. 1922, p. 898.

contractor installing the equipment merely sends the dimensions and location of the outlets to the manufacturer and receives the tubing prepared so that all that is required is the connection of the wires at the end to the lighting main and the placing of a few hanger screws.

An automatic electric time switch[222] which turns street lamps on and off at predetermined times has been amplified so that it takes care of seasonal variations on the lighting schedule. The device follows the sun and automatically controls the clock mechanism which operates the switch.

A very simple device for locking incandescent lamps[223] against removal by theft has come out in Germany. A hole is bored through the shell of the lamp socket and the threaded part of the lamp base. A split ring has an inside projection which fits into these two holes. As the lamp is screwed in, this projection follows the threads until the hole in the base is reached when it snaps in. The lamp must be broken to remove it. Numerous other devices to prevent stealing have been brought out in that country. One of these[224] has an elaborate arrangement of adjusting rings, each carrying on its circumference ten letters which must be oriented properly to permit the lamp to be removed after it is once set.

A change has been made in the condenser which is part of the projection apparatus employed with incandescent filament lamps in moving picture work.[225] Sufficient glass was taken from the center of the condenser to make the two halves slightly offset the images of the filament, thereby interposing on the curtain the bright bands of one field on the dark bands of the other. An evenly illuminated surface is claimed together with a greatly in-. creased picture illumination due to greater possible concentration of light on the film.

Cleaning and Ventilation

The effect of dirt in general on decreasing the efficiency of lighting installations[226] is a matter of common knowledge but a recent investigation undertook to classify the kinds of dirt which

[222]Elec. News, Aug. 15, 1922, p. 41.

[223]E. T. Z., Apr. 19, 1923, p. 367.

[224]Elek. Anz., Sept. 16, 1922, p. 1167, Helios, Sept. 17, 1922, p. 3119.

[225]Gen. Elec. Rev., Jan. 1923, p. 51.

[226]Indus. Eng., Jan. 1923, p. 46.

accumulate in representative industries, and to determine the effect of cleaning. Typical lamps were obtained from automobile, paint and varnish, chemical and storage battery industries, machine shops, factories, transportation offices and warehouses. Sixty per cent of the companies visited did not clean their lamps at all, 33 per cent cleaned at infrequent intervals, and only 7 per cent cleaned frequently at regular intervals. Four separate classes of dirt were encountered—dust and dry dirt, smoke, etc.; oily dirt and grease; paints tar, varnish, and pitches; and acid fumes—given in the order in which they most frequently occur. The average per cent increase in illumination after wiping was 77.7, 84.7, 37.2 respectively; and the average per cent increase after washing was 78.4, 47.0, and 67.2, no measurements being made on the lamps coated by acid fumes.

The effect of dirt is just as important in reducing the illumination due to natural lighting passing through windows, skylights, etc., as in the case of artificial illuminants. When clean the various ribbed, rippled, pebbled, and clear glasses employed in industrial plants transmit from 70 to 90 per cent of the incident light. The accumulation of soot dust, rust etc., reduces this transmission very quickly. Tests on seven sample glasses just as they were taken from shops and after they were cleaned showed that the daylight illumination in these plants would be increased from 4 to 15 times if the entire glass were cleaned.[227]

Additional evidence has been obtained in a northern city[228] of the desirability of not ventilating street globes containing 400- to 600-cp. lamps. Ventilated globes allowed the entrance of snow and caused lamp failures. The non-ventilated units were found to radiate enough heat from their exposed surface to avoid harmful temperature rises.

PHOTOMETRY

The photometric work done by the German Bureau of Standards (P. T. R.) in 1921 involved testing 47 hefner lamps,[229] 40 carbon and 335 metal filament lamps, 3 glimm and 1 mercury vapor lamp, 3 kinds of carbons for direct current arcs and 2 metallic reflectors. Forty-two materials were examined for light lost by

[227] Jour. of Frank. Inst., Oct. 1922, p. 546.
[228] Elec. World, May 26, 1923, p. 1222.
[229] Elek. u. Masch., Dec. 3, 1922, p. 571.

reflection and absorption. The carbon lamps and 235 of the metal filament lamps were used as standards. Work has also been done[230] on the proposed light unit based on black body radiation at 1700° absolute and wave-length, 0.656. The result indicated the desirability of using a higher temperature and shorter wave-lengths.

At the National Physical Laboratory, England, the work on ships' navigation lights has been continued.[231] An experimental building has been constructed for the investigation, by means of scale models, of problems in the natural and artificial lighting of buildings.

Units

To realize the black-body standard of light[232] a special furnace has been constructed in which the carbon heater tube operates in a vacuum and has cooled copper conductors fitted into its slotted ends to make as positive a contact as possible. Contact pressure is obtained by an elastic ring of heat-resisting steel embedded in the entrance of the tube.

The hefner lamp is the official standard for maintaining the unit of light in Germany.[233] The effect of barometric pressure on the light intensity of the hefner has been determined a number of times, notably in 1911 when the work was done in a pneumatic chamber. A new investigation has been made at four different altitudes— Vienna, (165 m.), Böckstein (1125 m.), Maserboden (1965 m.), and Sonnblick (3100 m.). Three lamps certified by the P. T. R. were used in the test and the results showed an average change of 0.0004 h.c. per mm. change in the barometric pressure, exactly confirming the data previously obtained in the pneumatic chamber.

Instruments

A new lumen-meter or lux-meter has been brought out in Germany.[234] An effort has been made to combine the convenience of the most portable types already in existence with the accuracy of the more complicated instruments. Errors of measurement are said to be of the order of 5 per cent to 10 per cent. The comparison lamp is held in a small box and operated at reduced voltage

[230]Zs. f. Instr., Mar. 1922, N. P. R.
[231]Elec. Rev., July 6, 1923, p. 37.
[232]Zs. f. Beleuch., 28, 1922, p. 76, u.s.w.
[233]Elek. u. Masch., Oct. 29, 1922, p. 511.
[234]L. u. L. Apr. 26, 1923, p. 207.

giving a reddish light but long life and constancy. It illuminates the whitened interior of a box which at the end opposite the lamp is closed by a piece of blue glass in contact with a milk glass plate. The resultant color matches that of a tungsten lamp at normal voltage. The box is divided into two parts separated by a partition containing a series of small circular holes, and a disk with a corresponding series of the same sized holes. Moving the disk by a knurled wheel operating a worm changes the extent to which the holes in the disk coincide with those in the partition and thus changes the amount of light which reaches the milk glass screen, in steps which can be calculated. This screen is seen in one half of the circular observation field, the other half being a white (barium sulphate) screen which receives the illumination to be measured. The field is observed at an angle of 35° or 40°. To increase the range from 10 to 100 lux, the barium sulphate receiving screen is replaced through the rotation of a disk, by one which reflects only 10 per cent of the incident light, while higher values may be measured by using other screens of still greater absorption power. In this way illuminations of 1 to 500 lux may be measured. Lower illuminations are taken care of by interposing a dark-blue glass plate in the comparison lamp beam. The total range of the instrument is given as from 0.01 to 500 lux. The whole is contained in a portable case which also carries the dry battery for operating the comparison lamp.

Improvements have been made in the shadowmeter[235] (Schattenmesser) described in the Report for 1920[236] and used for illumination measurements. A thin curved "shadow caster" makes the angular diaphragm and a "shadow seeker," consisting of a needle placed perpendicular to the center of the disk, determines the direction in which the individual light sources throw shadows on the working plane. The cube instead of the sphere[237] for total luminous flux measurements has been the subject of study in England for some years. An extension of this idea is a so-called lumen comparator which consists of a rectangular box painted white on the interior and divided into two cubical portions by a central diaphragm which also divides the observation window. The lamps to

[235]Zw. f. Beleuch., 27, 1921, p. 109, M. P. R.
[236]Trans. I. E. S., 15, 1920; p. 473.
[237]Ill. Eng., Oct. 1922, p. 283.

be compared are placed in the two divisions and comparisons made between the relative brightness of the two windows.

Another photometric device[238] is a modification of the integrating photometer described in the Report for 1919 (Trans. I. E. S. 14 1919, p. 370) and employs a square or rectangle as the geometric outline to support the diffusing plates instead of the circle.

Photometric equipment[239] for use with the Ulbricht sphere designed along somewhat similar lines to those of a foreign apparatus referred to in the 1918 Progress Report has been made in this country. A suitable comparison lamp is enclosed in a rectangular box whose interior surfaces are painted with the same paint as that in the sphere. In the end of the box toward the photometer head is an opal glass window, in front of which moves vertically a metal slide with a V-shaped opening and back of which are two broad horizontal bars forming an adjustable space between them. Thus the area of the window may be altered continuously and in turn the illumination on the photometer screen from which it is some 30 cm. distant. The slide with the V-shaped opening carries three photometric scales. The apparatus has been modified by removing the opal glass from the window in the box and placing the lamp out of the line of view of the photometer which then receives its illumination from the back wall of the box.

A spectrophotometer,[240] constructed primarily for the measurement of spectral reflection from surfaces, consists of a hollow sphere containing four incandescent electric lamps illuminating indirectly a sample and a standard white. The light from the two fields is reflected through a rotating sectored diaphragm and into a spectrometer. By this means the ratio of the light reflected by the sample to that reflected by the standard is measured in the different wave lengths. Another spectro-photometer measures the spectral transmission of liquids. It consists of a constant deviation photometer in combination with an "exponential" or "variation of thickness" photometer. A unique feature of the latter is the use of two totally reflecting, partially immersed rhombs, so that one of the two beams of light employed is diverted through a variable thickness of liquid depending on the thickness of the rhombs. The instrument is

[238]Elek. Zeit., July 12, 1923, p. 665.
[239]Jour. of Frank. Inst., Oct. 1922. p. 543.
[240]Jour. of Opt. Soc. of Am., Jan. 1923. p. 98.

direct reading in transmissivity, log of transmissivity, and wavelength. It does its own computing automatically and a spectral curve may be determined and plotted in a few minutes. It is expected that this instrument will be of special value in technical examination of oils, dye solutions, sugar solutions, etc.

Heterochromatic Photometry

It is well known that where the sources compared are considerably different in color[241] corresponding to black-body distributions at wide intervals of temperature, the Crova method involves errors of appreciable magnitude. A modification of this method permits comparison where the color differences corresponding to temperature differences are as much as 2,000°. A standard is employed whose temperature can be determined and hence the relative intensity at different temperatures computed for the Crova wavelength 0.582μ. The ratio of the emission at this wave length of the unknown to that of the standard, if raised to the 1.015 power, will give the correct ratio of the luminous intensity to within less than 2 per cent. This formula can be used for a standard whose temperature lies between 2000° and 3000°. A more elaborate formula based on the De Lepinay method using measurements at two wavelengths is suggested for standards whose temperature limit is beyond 3000°.

Another solution of the difficulties of heterochromatic photometry[242] is offered in a color-match photometer, a three-color mixture instrument in which the primaries are of equal luminous value and of just sufficient purity to reproduce by their mixture all of the common illuminants. The optical arrangement is such that the illumination of the comparison field remains constant as the color is varied to match the standard or test illumination. A 400-watt projection type incandescent lamp is employed in which the parts of the filament are in one plane. An image of the filament is focused on a ground glass window in a hexagonal box with whitened interior walls. The light diffused by the window helps to illuminate one side of the box, the remaining illumination coming from the other walls. Opposite the window the box is closed by an extended white surface illuminated by the source to be measured. This surface and the side of the box are observed by means of a photometric

[241] Compte Rendu, Oct. 23, 1922, p. 688; Rev. d'Opt., Feb. 1923, p. 42.
[242] Jour. Opt. Soc. of Am., Jan. 1923, p. 75; Mar. 1923. p. 243.

prism and an optical train. The composition and intensity of the light falling on the diffusion chamber are controlled by colored glasses and a variable aperature slide. This instrument is said to make possible the determination of the most difficult part of monochromatic analysis measurements, namely the illumination values of the color and its constituent white. Methods and diaphragms have been developed for transforming trichromatic data to other sets of trichromatic primaries and to the monochromatic system, and the graphical methods are applicable to measurements of illuminants.

Another comparative study has been made of the flicker and equality-of-brightness methods of color photometry.[243] Four spectral lights were photometered against a 32 cp.—4.85 wpc. carbon lamp by the two methods with illumination of 12.5, 25, and 50 mc. The rise in sensation in just noticeably different steps of brightness was determined for each of these lights and intensities with the same observers and state of adaptation of the eye. The effect of variation in intensity was determined by using seven values including the three previously mentioned. One intensity was found for each pair of lights at which agreement occurred with the most sensitive speed of rotation of the flicker disk. Values of this intensity were found to be widely separated for the four pairs of lights. Each pair of lights was rated by the equality-of-brightness method with a length of exposure equivalent to that of the individual exposures used in the flicker method. In these cases, agreement of the two methods was obtained within the limits of judgment between 0.4 and 1.5 per cent for the flicker and 1.3 to 2.7 per cent for the brightness method. It was concluded among other things that "no differential summation effect is produced by the succession of exposures used in the method of flicker."

For the measurement of the light absorption of solutions,[244] a compensation principle employing two photoelectric cells enables the null method to be used. With this arrangement and a mercury vapor lamp as a source, the absorption of solutions of potassium chromate in caustic potash, copper sulphate in ammonia, azobenzene in alcohol, for the mercury lines between 0.254 and 0.579μ was determined.

[243]Psych. Bul., Feb. 1923, p. 87.
[244]Zs. f. Physik. Chem., 100, 1922, p. 208.

Computations

Using mathematical processes[245] based on the theory of the integrating sphere, a series of formulae have been developed for the determination of the effect on illumination of diffuse reflection from walls and ceilings of enclosures of various shapes. Corrections have been worked out for the effect of corners and other irregularities. Another monographic chart[246] has been worked out for the solution of problems of illumination design, based on a method of computation referred to in the 1921 Report of this Committee. The chart is said to simplify the operation of computation and eliminate the use of all tables of coefficients and angles without sacrificing any of the accuracy of the method. The latter is equally applicable to direct or indirect lighting, either by using the distribution curve of the luminaire, or foot-candle values in one plane, and enables the determination either of the proper height of the luminaire or the necessary luminous intensity of the unit.

PHYSICS

In that branch of physics of particular interest to the illuminating engineer, the subject of luminescence seems to have been given an unusual amount of attention during the past year. The subject has been studied from the standpoint not only of its production but also of the many factors which influence its character. The nature of fluorescent light is such that it is not unreasonable to suppose that the next big advance in the development of an efficient source will come from this great class of light producers.

Light Sources

Further experiments on the emissivity of platinum and tungsten[247] using the micropyrometer method, have shown that the emissivity is independent of the temperature for platinum between room temperature and $1710°C$ and for tungsten between $2000°$ and $3200°C$. The values at $\lambda = 0.647\mu$ were 0.348 for platinum and 0.49 for tungsten; at $\lambda = 0.536\mu$, they were 0.363 for platinum and 0.49 for tungsten. The melting point for tungsten was also determined and found to be $3370°C \pm 50°$.

Several years ago it was discovered[248] that a fine wire through which is passed the discharge of a high voltage condenser vaporizes

[245] E. T. Z., Oct. 12, 1922, p. 1262.
[246] Elec. World, Apr. 7, 1923, p. 801.
[247] Zs. f. Physik., June 16, 1923, p. 63.
[248] Proc. of Nat. Acad. of Sci., July 1922, p. 231.

as if exploded, with a flash of light, the radiation of which corresponds to that of a black body at a temperature of about 20,000°K. A spectral study of this light, using wires of Fe, Cu, Ag, Au, Mg, Zn, Cd, Al, Sb, Pb, Ni, T, gave for all a continuous spectrum of the same general intensity.

So much comment has been made from time to time on the ultra-violet radiation in the ordinary artificial light sources[249] that information on the amount in sunlight is of value and interest. Measurement on this subject, corrected for absorption in the solar atmosphere and using the positive crater of an arc for a comparison source, gave the following results:

Wave-length	Ratio Sun/Arc	Temp. Abs. of Sun
0.3940μ	40	6016°
0.3620	50	5931
0.3143	79	5832
0.3022	112	5959
0.2922	134	5970

The temperature of the arc crater was taken at 3750° absolute. The sun temperature was computed assuming the energy distribution of the arc to be that of a black body at the same temperature.

The color temperature of daylight[250] has been determined by the rotary dispersion method. The comparison source had a color temperature of 2848°K. For the northwest sky, the color temperature varied from 6870°K for a uniformly overcast condition to 24150°K for a clear blue state. A horizontal surface illuminated by the whole sky showed a color temperature of 6500°K for a bright overcast condition, 7225°K for a broken overcast condition with pale blue on one day to over 25,000°K on another day with pale blue above and the horizon hazy. On a day with direct sunlight and a bright light haze in the atmosphere, the color temperature was 5300°K. Measurements of the brightness of the moon[251] and the color temperature of moonlight gave for total light 0.24 c. per cm.² and a temperature of 4125° absolute. Taking the color temperature of sunlight as 5600° absolute, the difference indicates the selectivity of the observed area of the moon, the reflecting power for total light being 0.07.

[249]C. R., July 17, 1922, p. 156.
[250]Jour. of Opt. Soc. of Am., Jan. 1923, p. 78.
[251]Nature, Apr. 21, 1923, p. 532.

A continuation of the earlier work on the effect of temperature on the current in discharge tubes[252] has been obtained with gases other than air. With hydrogen and carbonic anhydride, the increase in current with increase of temperature was found to be much more marked than with air. The temperatures employed ranged from $-180°$ to $275°C$.

A report has been made to the Optical Society of America on stellar and planetary radiometry[253] which reviews knowledge obtained since 1920 on the radiation intensity and temperatures of the planets. A bibliography is included.

Luminescence

On the subject of luminescence, reference should be made[254] to a Bulletin of the National Research Council which discusses present knowledge and contains an elaborate bibliography.

If tungsten is used as an anode[255] in a concentrated solution of H_2SO_4 with an applied voltage of 60, a faint but continuous luminescence appears about the tungsten. The effect seems to be associated with the conversion of WO_2 and W_2O_5 found at low voltages, into WO_3. The luminescence is of the same nature as that observed when pyrophoric WO_2 is oxidized in the air. In solutions where the oxide film is at once dissolved as in 30 per cent NaOH, no luminescence was found. A preliminary study of 70 substances[256] to see what part of the ultra-violet spectrum excites fluorescence revealed the fact that the oxides (20) and simple chlorides (8) were not excited. A few substances (7), including Zn and Cd compounds, fluoresced faintly while the uranyl compounds fluoresced strongly. For the last group, the effective exciting spectrum extended from 0.55μ to 0.35μ only, while for the others it went continuously to 0.2μ except in a few special cases.

There are impurities in the Missouri calcites[257] which include magnesium, zinc, and the rare earths. Phosphorescence has been excited in these calcites by the molybdenum arc whose characteristic wave-length is 0.379μ. Exposure in a mineral cabinet for about twenty years has caused an abatement but not an oblitera-

[252]Elletrotecnica, Aug. 15, 1922 p. 507
[253]Jour. of Opt. Soc. of Am., Dec. 1922, p. 1016.
[254]Bul. Nat. Res. Coun., No. 5, 1923, p. 5.
[255]Zs. f. Physik, Mar. 5, 1923, p. 14.
[256]Phys. Rev., Dec. 1922, p. 552.
[257]Am. Jour. of Sci., Apr. 1923, p. 3141.

tion of the phosphorescent property. The yellow calcite shows a thermophosphorescence between 60° and 180°C which is destroyed by heating to about 150°C for less than 8 hours. Samples so heated until the phosphorescence had disappeared showed a strong luminescence when heated to 300–400°C. It has been found that in titanium oxide[258] the free oxide present determines the character of the luminescence produced by heat. Marked changes in color can be produced between red heat and 1200°C by a slight adjustment of the oxyhydrogen flame used in heating. Above 1200°C all excessive radiation disappears and the spectrum coincides with that of a black body both in distribution and brightness. Luminescence has been produced in chemical reactions[259] where a non-reacting compound is present. Pure oxydisilin gave a faint green luminescence when oxidized in an acid solution of $KMnO_4$. It was then treated with a solution of rhodomine B and the resulting red leaflets oxidized with acid $KMnO_4$, when an intense luminescence appeared. Isoquinoline red gave similar results.

Of a large number of compounds of the type RMgX (Grignard Reagents)[260] about thirty have been found to exhibit chemiluminescence on oxidation with O_2. It has been found that concentrations of the solution affect the intensity; that magnesium is a necessary constituent; slight changes in temperature have little effect; most aliphatic compounds are non-luminous while many aromatic are; changing the reacting halogen affects both the color and the intensity of the emission; the oxidation of p–$ClC_6 H_4MgBr$ is said to be probably the brightest case of chemiluminescence on record; a large number of these compounds and their oxidation products are fluorescent in ultra-violet light. Additional work on p–bromophenylmagnesium bromide[261] showed the spectrum of chemiluminescence to lie within the limits 0.350μ and 0.520μ. It was found to be distinct from the associated fluorescent spectra. Photographs of two kinds of fluorescence were obtained for the oxidation product of this compound.

The luminescence emitted by the vapors of benzene[262] when subjected to a Tesla discharge has a fragmentary carbon spectrum

[258] Phys. Rev., June 1923, p. 713.
[259] Naturwiss., 11, 1923, p. 194.
[260] Phys. Rev., Feb. 1923, p. 203.
[261] Jour. Am., Chem. Soc., Feb. 1923, p. 278.
[262] Jour. of Chem. Soc., Mar. 1923, p. 642.

at ordinary pressures and the boiling point, while aniline emits a short continuous spectrum. On reducing the pressure, benzene emits a spectrum of band groups. A close relation is apparent between the Tesla-luminescence, the fluorescence, and the absorption spectra of benzene. On passing ozone through a tube electrically heated to 200°, luminescence was obtained[263] which diminished as the purity of the gas was increased. The light consisted of a band of light from about 0.460μ to 0.560μ. The width of the band was independent of the concentration, pressure, temperature, and thickness of the layer observed. The light emitted was approximately proportional to the concentrations of CO and O_3.

Some observations on the coloration of glass by β-rays[264] indicated that the fluorescence is due to a change in the molecular aggregration of the substances. Three pieces of glass tubing were treated with radium emanation until they were colored a deep brown and ceased to fluoresce. Further treatment did not alter the color. They were then placed in an oven maintained at 110°C. They immediately began to fluoresce, and continued to do so for 13 minutes. Further heating at higher temperatures caused renewed fluorescence but for shorter and shorter periods. The results indicated that complete decoloration would occur at 500–600°C. The annealing temperature of these glasses was about 550°C. As both coloration and decoloration were accompanied by fluorescence, it seems probable that the latter was due to the change in the state of the molecular aggregations. The general fluorescent properties of cellulose and its derivatives[265] have been shown to vary with different specimens. Materials supposed to be and taken for the same by the manufacturers and the expert are by no means the same from the standpoint of their fluorescent activities. By using a brilliant reflected light[266] derived from an arc or a 400-watt incandescent lamp and a special arrangement of the cardiod dark field condenser of an ultra-microscope, the fluorescence of phycocyanin of the blue-green algae as well as a red fluorescence of the diatoms has been demonstrated. The visible fluorescence of the diatoms is confined to certain vacuoles which, by transmitted light, appear a pale greenish yellow.

[263] Ann. d. Physik, June 29, 1922, p. 527.
[264] Phil. Mag., Apr. 1923, p. 735.
[265] Jour. of Soc. Dyers Colorist, Aug. 1921, p. 201, N. P. R.
[266] Science, Aug. 3, 1923, p. 91.

Phosphorescence continues to be a subject of study and investigation.[267] Further work has been done on the decay of phosphorescence using the photoelectric light-summational method with special reference to the time measurements and to the effect of temperature. While studying the fluorescence of various substances excited by ultra-violet light,[268] it was found that the peel of the mandarin orange contains a material which fluoresces with a bright yellowish green color. The intensity is such that its spectrum can be photographed in half an hour with a direct vision spectroscope. The exciting radiation was found to be of wavelength shorter than 0.405μ. When lumps of sugar are broken flashes of light appear, and this phenomenon is known as triboluminescence.[269] It was originally thought that this light had a continuous spectrum but experiments have shown that it is discontinuous and has most of the bands belonging to nitrogen but to no other spectrum. Lumps of sugar placed in a glass jar with an air pressure of from 4 down to 0.1 cm. of mercury, when shaken against the walls to break them, showed a much more intense triboluminescence than at atmospheric pressure.

Photoelectricity

It has been found possible to increase enormously the sensitiveness of the potassium photoelectric cell by increasing the voltage between the potassium anode and the platinum cathode.[270] For red light $\lambda = 0.630\mu$, an increase in voltage from 20 to 210 multiplied the sensitiveness by 1045 and for violet light ($\lambda = 0.462\mu$) by 1595. In order to employ such a high voltage, it was necessary to raise the potential gradually for hours or even days at a time. In forming the cell, it was found that each time the voltage was increased, there was a strong "darkness current" at first which diminished to zero for lower voltages. The relation between the photoelectric current and the illumination was found to be $I^z = ML$, where I is the photoelectric current, L the flow of radiant energy and Z and M are constants. Z was found to vary with wave-length and the voltage applied to the cell. Using this cell, another determination for the exponent, C_2 in the Wien radiation

[267] Ann. d. Physik, Jan. 31, 1923, pp. 81, 113; Zs. f. tech. Physik, No. 2, 1923, p. 53.
[268] Jap. Jour. of Physics, 1, No. 5, 1922.
[269] C. R., June 19, 1922, p. 1633.
[270] Zs. f. Physik, Nov. 17, 1922, p. 215.

law gave a value 14385. In this work, it was found also that carbon·
radiates throughout the wave-lengths investigated, which included
the ultra-violet as far as 0.316μ, as a gray body.

In connection with some experiments on the photoelectric
theory of vision,[271] it was observed that an aqueous solution of
potassium ferrocyanide exhibited a pronounced photoelectric
effect when exposed to a carbon arc. Further investigation of the
effect of concentration of the solution indicated that the negative
radical is probably responsible for the photoelectric action and that
the increase in activity of the stronger solutions is to be attributed to
the effect of a possible association of the molecules of the solute. In
studying the photoelectric effect in crystals,[272] it has been found
that the use of single crystals considerably simplifies the phenomena. Curves showing the relation between current and field
intensity of the light seem to be independent of the wave-length
and of the direction of the luminous radiation. Experiments on a
large number of crystals indicate that the saturation field intensity
is related to the refractive index for red light. Other experiments
have shown that for low voltages the potential drop near the electrodes is greater than in the middle of the crystal. The effects of
crystal structure, chemical constitution and atomic weight on the
spectrophotoelectric sensitivity of the chlorides, bromides, and
iodides of thallium and silver and the iodide of lead have been
studied. The photoelectric action of these salts was found to be
confined to a narrow region of the spectrum.[273] The effect of the
atomic weight is to shift the maximum of the reaction toward
longer wave-lengths.

In a photoelectric cell having a tungsten filament as an anode
and the usual sensitive alkaline layer for a cathode, it has been
found that a much larger current will flow, when the cell is illuminated, if the tungsten filament is hot than if it is cold,[274]. Furthermore, the resistance of the cell is lower. When the filament is
gradually heated, the sensitivity rises to a maximum value, after
which further rise in the filament temperature results in a decrease
in the sensitivity. A proposed explanation is that, when cool, the
filament is coated with a thin layer of alkaline atoms which, as the

[271]Phil. Mag., May 1923, p. 805.
[272]Physik. Zs., Oct. 15, 1922, p. 417.
[273]B. of S. Sci. Paper No. 456, Nov. 8, 1922, p. 489.
[274]Phys. Rev., Feb. 1923, p. 210.

filament heats, tend to neutralize the space charge in the tube and thereby influence the photoelectric current.

The relatively low light sensitivity of selenium strips,[275] whose dark conductivity is large, can be increased about forty times by the use of very thin layers of selenium (0.5μ thickness). However, if by very careful purification of the material, the dark conductivity of the material is very considerably diminished (in some cases less than 10^{-4} of the ordinary), then the sensitivity to light will be found to be already high, even greater than in the case of the very thin layers, and practically independent of the layer thickness, so that the thinning produces no increase.

Properties of Materials

More data on the reflection coefficient of magnesium[276] has been obtained by the P. T. R. If the incandescent light is perpendicular, the reflection coefficient depends on the angle of emission. Relative values of red and green light are shown in the following table:

Angle from the vertical	Relative Reflection Coefficients	
	Red	Green
0°	100	100
15	99	99
30	96	98
45	91	95
60	85	89
75	74	79
90	58	62

The diffuse reflection value for total light as obtained previously was 0.955.

By exposing rapid photographic plates[277] for an hour and a half in the ocean at a depth of 3300 feet, evidence of light was obtained, but another plate exposed at a depth of a mile showed no impression. Care was taken to avoid effects produced by luminous animals. Trials made with ultra-violet light for the sterilization of water have shown that it is operative over only a comparatively limited range of even clear water.

[275]Zs. f. Feinmech., *30*, 1922, p. 169.
[276]Zs. f. Instr., May 1922, p. 131, N. P. R.
[277]Phot. Jour. of Am., Feb. 1923, p. 79.

Further work has been done on the effect of temperature on the transmission of glasses for ultra-violet radiation.[278] A series of both crown and flint glasses in two thicknesses, 3 mm. and 6 mm., was examined using a quartz spectrograph. The range of temperature was 300° to −180° C, and the change seemed relatively small in most cases. On cooling cyanin and pinacyanol[279] from room temperature to that of liquid air, the absorption for wave-lengths in the region of 0.543μ to 0.644μ became less at the lower temperatures for the former except at 0.595μ and at the shorter wavelengths for the latter.

It has been found that ultra-red light restores diamonds,[280] which have been altered in their properties by the action of ultra-violet light, to their original conditions more quickly than when they are kept in the dark. A considerable amount of work has been done on the effect of light on chemicals.[281]

The relations between brightness, opacity, and approximation to pure whiteness of pigments and paints have been studied with a colorimeter and cryptometer.[282] It was found that the addition of ultra-marine blue to paints containing very bright pigments which are low in blue brightness decreases their brightness somewhat but makes them far more nearly nonselective and greatly increases their hiding power. This question is of considerable importance in the case of paints used in coating Ulbricht spheres.

Much effort has been spent on trying to discover the principles underlying the action of light in discoloring pigments.[283] In the case of lithopone, experiments have indicated that the discoloration is due to the phosphorescence of the ignited zinc sulphide in the pigment caused by the presence of small quantities of foreign metals which give with the zinc sulphide colored sulphates on exposure to light. The subsequent disappearance of the color is due to the oxidation of these sulphides to the corresponding oxides. The presence of Pb, Mn, and Cu causes the pigment to become gray rapidly. Lithopone absolutely free from foreign metals showed no discoloration at all. The drying time of linseed oil and

[278]Zrch. sc. phys. et nat., Sept. 1922, p. 355.

[279]Zs. f. Physik. Chem., 100, 1922, p. 266.

[280]Physik. Zs., Oct. 15, 1922, p. 304.

[281]Chem. Abs , Feb. 10, 1923, pp. 366, 367.

[282]Paint Mfgs. Assoc. of U. S., Circ. A 173, 1923, p. 153; Jour. of Fr. Inst., July 1923, p. 69.

[283]Jour. Soc. of Chem. Indus., Feb. 16, 1923, 150A.

varnish[284] has been found to be influenced by colored light. Films of raw linseed oil exposed to light diffused through amber, blue, red, and ground glass were dry in 14 days or less, while under green and plain glass plates they were still tacky. Those kept in the dark were as wet as at the beginning. Sunlight dried the films in two or three days. An initial exposure of a half hour to ultra-violet light was found to accelerate greatly the drying of raw linseed oil. Further experimentation is going on to determine the practical application of this work in the varnishing of automobiles.

A new field of work for the physical chemist which is being more and more explored and which in its bearing on diffusion is of importance to the illuminating engineer[285] is the molecular scattering of light in various media such as gases and vapors. Additional experiments have been performed on the scattering of light by dense vapors and gases not obeying Boyle's law.[286] For this case the results indicated that the scattering power per unit value is proportional to the square of the density of the substance and to its compressibility. A thermodynamical investigation of the scattering in liquid mixtures led to the conclusion that the light scattering arises in two distinct ways: first, to the spontaneous fluctuations in the composition of the mixture; and second, due to local fluctuations in the density. The effect of any increase in temperature on the light scattering by liquids such as benzene, methane, and naphthalene[287] has been shown by experiment to be an increase in the intensity of the scattering which passes through a high value at the critical temperature falling off again rapidly for temperatures above the critical.

Color

Attention should be called to an elaborate report on colorimetry[288] made by the committee on that subject to the Optical Society of America. This includes a bibliography. A new color measuring instrument[289] depends on the measurement of the reflected or transmitted light in three colors—red, green and blue. It was developed especially to identify colors in the Oswald system

[284] Paint Mfgs. Assoc. of U. S., Circ. No. 172, 1923, p. 148.
[285] Jour. of Chem. Soc., Dec. 1922, p. 2655.
[286] Phil. Mag., Jan. 1923, pp. 113, 213.
[287] Jour. of Phys. Chem., June 1923, p. 558.
[288] Jour. of Opt. Soc. of Am., Aug. 1922, p. 527.
[289] Zs. f. tech. Physik, No. 4, 1923, p. 175.

and measures either transparent or opaque bodies. At the bottom of an inclosed and internally blackened box are two closed and adjacent surfaces illuminated by the same light source. The illumination of one can be continuously varied. They are observed by a photometric train, and the object to be measured is placed over one of them. If the object is colored, the comparison is made in succession through each of three colored glasses in the ocular of the photometer. A color filter,[290] which is said to transmit light only between o.410μ and o.570μ, can be made of a saturated solution of three parts of toluidinblau and one part of filterblaugrun. A sheet 10 mm. thick of this solution reduces the light to 1/50 while a 30 mm. sheet reduces it to 1/100. For sealing quartz windows to glass bulb lamps[291] such as mercury vapor or incandescent filament, it has been found possible to use silver chloride and make a vacuum-tight joint. Silver chloride melts at 455°C, does not give off gas in any quantity and does not decompose readily with time.

PHYSIOLOGY

The present status of visual science[292] carried up to January 1921 is discussed in a Bulletin of the National Research Council.

Theories

A theory of color vision which agrees with the Young-Helmholtz theory[293] as to the relative processes associated with vision has been derived from the Hering theory by assuming that the cones of the retina contain three light-sensitive substances similar to visual purple—the pigment of the rods—but less stable than that substance. The antagonistic action of complimentary colors is assumed to be produced by a nervous mechanism. Blue rather than yellow is assumed as a "katabolic" color. Another theory of color vision[294] is based on the following fundamental hypothesis: first, vision is produced by the emission of photoelectrons from a light substance occurring in both rods and cones which shows the selective photoelectric effect; second, differences in luminosity depend on the number of electrons emitted while differences in color depend on their velocities; third, chromatic vision is possible only

[290]Physik. Berichte. Oct. 15, 1922, p. 1012.
[291]Science, Aug. 4, 1922, p. 147.
[292]Bul. of Nat. Res. Coun., Dec. 1922, Vol. 3, Pt. 2.
[293]Psych. Rev., Jan. 1923, p. 56.
[294]Jour. of Opt. Soc. of Am., Oct. 1922, p. 813.

in the cones and rhodopsin being found in the rods alone is concerned only with achromatic vision. It acts as a sensitizer to dim
light and in the presence of the sensitizer, the maximum of the
curve of photoelectric sensitivity is shifted to wave-length o. 53μ.
This theory is said to explain the main facts of color vision, after-
images, and color blindness reasonably well. Additional data have
been obtained on the luminous sensitivity[295] of color material in
reference to the explanation of color vision.

Eye

An investigation made on the interpupillary distances of some
400 persons[296] gave a mean value of 63 mm. for men over 18 years
of age and 61 mm. for 50 women. The data showed no definite
change with age of the average interocular distances. The average
apparent diameter of the iris for all individuals tested was 12mm.
The pupillary reaction to wave-lengths in an equal energy spectrum
at high and low intensities for the light and dark adapted eye has
been studied in man, pigeon, and alligator.[297] The curve for man
is similar to the ordinary visibility curve with a maximum at o. 554μ
for the high intensity and o. 534μ for the low intensity, light adapted eye. For low intensity of the dark adapted eye, the value was
o. 514μ. For the pigeon, the corresponding wave-lengths were
o. 564μ, o. 544μ, and o. 524μ; and for the alligator, o. 544μ, o. 514–
o. 504μ, and o. 514μ. The pupillary motor values are comparable
with those of rod and cone vision for radiation of high and low intensity. The reaction time of the human pupil was found to be
between 5 and 6 seconds.

Measurements of retinal sensitivity[298] in a test of 101 subjects
showed individual differences far in excess of the accidental variations for the same individual and the same conditions. Correlation
between retinal sensitivity and a rating based upon the ophthalmic
portion of the Government Air Service medical examination for
flying status indicates that no general relation exists between
retinal sensitivity and disqualification. There is not much in the
literature on vision [299] concerning actual experimental work in
which accomodation and convergence of the eye are differentiated.

[295]Zs. f. Physik. Chemie, 100, 1922, p. 537.
[296]Trans. of Opt. Soc., 23, 1922, p. 44.
[297]Am. Jour. of Physiol., Mar. 1, 1923, p. 97.
[298]Jour. of Exper. Psych., Aug., 1922, p. 227.
[299]Ibid, June, 1923, p. 222.

Tests to determine at low intensity of illumination the accuracy of convergence as compared to that of accommodation showed the average relative error for accommodation alone to be 1/23 in contrast to 1/58 for accommodation and convergence. The illumination was 0.0117 fc. As the illumination was decreased, there was a gradual increase in error for both accomodation alone and with convergence, down to an intensity of 0.001 fc. At a point near the lower threshold for vision, accommodation alone breaks down while convergence shows little change in the gradual rate of increase of error. The results indicate further that fatigue has a greater effect on the intrinsic accommodation muscles of the eye than upon the extrinsic converging muscles.

In an investigation on the speed of retinal response[300] as a function of field brightness and light distribution, seven subjects were employed. Rather large variations in sensitivity were found and somewhat different variations with the changes in illumination and distribution of light in the visual field. The following equation expresses the results where t represents the threshold time of stimulus, B the brightness of the experimental field, k and B_0 being constants depending on the individual and on other experimental conditions,

$$1/t = k \log B/B_0$$

Vision

The importance of binocular vision in the work of aviators and for drivers of automobiles[301] has led to an attempt to redetermine the threshold of binocular perception of distance among persons of normal vision. Three uniform threads, 0.5 mm. in diameter, having lead bobs attached to their ends, were suspended with the bobs immersed in oil. Each of the outer threads was attached to an adjustable block provided with a scale indicating the distances from the center thread. A uniform time for observation was obtained by the use of a shutter. The threshold values were found to correspond to retinal displacements of less than the diameter of one cone for all observers. A standard signal arm as used on the railroad[302] measures 10 in. x 4 ft. 9 in. and subtends an angle of 5 minutes at a distance of 1085 yards. It can be read accurately by a person of normal vision up to 1.5 miles. A test of the vision of

[300] Jour. of Exper. Psych., Apr. 1923, p. 138.
[301] Am. Jour. of Physiol., May 1, 1923, p. 561.
[302] Br. Jour. of Ophth., 6, 1922, p. 319.

locomotive drivers, signal men, and station masters showed that a man with vision as given by the ophthalmic formula, R. 6/18, L. 6/18, both 6/12, can with both eyes open read these signals accurately at 1500 yards and that, provided vision with both eyes open is not below 6/12, color perception is not dangerously lowered.

Caused by a desire to know why a greater intensity of natural lighting than of artificial lighting seems to be required for the same kind of work, [303], additional comparative data have been obtained on the intensities desirable for reading using both the direct and indirect lighting systems. The room used for the experiments was 30 ft. x 18.5 ft. and 13 ft. high with a ceiling and upper half walls white, lower half and floor drab. The work was done at night. For direct lighting, opaque white enamel reflectors were used, placed 7.5 feet above the floor. For the indirect lighting, opaque bowl and silver reflectors were used and hung 2 feet below the ceiling. Current through the lamps was adjusted so that the illumination was approximately the same for either system. Five different values of the current were used to produce variations in illumination from one which was just too low for comfortable reading to one which was just too high. Twenty-six observers took the test and recorded their impressions. The results showed that an average of 2.8 fc. was considered "just right" under the direct illumination while 4 fc. were required under the indirect. The result was attributed to the effect of the greater equality of distributed illumination in the field of view under the indirect system, causing pupillary contraction and requiring a higher intensity for comfortable reading. This condition was considered to be similar to the conditions occurring in the use of natural lighting.

Another determination of the visibility curve[304] by the "step by step" equality-of-brightness method has been made between the wave-lengths 0.430μ and 0.740μ, using 52 observers of which number, 14 made measurements between 0.49μ and 0.68μ only. Luminosity values were obtained with a Brace spectrophotometer by moving the collimator slit. The results were somewhat higher in the extreme blue and red[305] than those previously obtained by

[303] Elec. World, Dec. 9, 1922, p. 1268.
[304] Jour. of Opt. Soc. of Am., Jan. 1923, p. 68.
[305] Jour. of Wash. Acad. of Sci., Mar. 4, 1923, p. 88.

this method, but they still indicate a real difference between the data obtained by this method and those obtained by the flicker method.

Visual Acuity

It has been shown that a reduction of the visual acuity[306] is always followed by an apparent diminution in size and at the same time the removing ·and bringing nearer together of the objects observed. On the other hand, a sudden increase in visual acuity produces an apparent increase in size. A proposed explanation is based on the assumption that maximum impulses to accommodate are liberated in such a manner that the distinctness of the image is not blurred when the conditions of the formation of the image are altered. An elaborate analysis has been made of the physiological limits to the accuracy of visual observations and measurements, using data and information already available and a form of field suggested[307] for instruments used in measuring color and intensity permitting of a higher degree of accuracy than has been heretofore possible.

Color

An experimental method of evaluating the spectral colors[308] in terms of the three primary colors as used in the color triangle has been worked out. The values obtained gave wave-lengths 0.675 up to 0.5625μ in a straight line and wave-lengths 0.530 to 0.430μ on another line, the two forming sides of the triangle. The color of the vacuum tungsten lamp as given by this triangle was found to be 19.6 red, 30.1 green and 50.3 blue. Daylight from a north sky was 8 red, 20 green and 72 blue.

The situation and extent of the region of the spectrum which is yellow[309] has been determined by 12 men and 12 women, all accustomed to judge colors. The extent and position of the yellow zone was found to vary from individual to individual. For one person, it lay at 0.577 – 0.583μ; for another at 0.583 – 0.585μ; and for a third at 0.5875 – 0.590μ. For the twenty-four persons, no radiation was conceived as pure yellow beyond the limits of 0.596μ at one end and 0.574μ at the other. The extent of the spectrum

[306] Jour. of Opt. Soc. of Am., Aug. 1922, p. 597.
[307] Phil. Mag., July 1923, p. 49.
[308] Ibid, Jan. 1923. p. 169.
[309] Jour. of Physiol., Mar. 21, 1923, p. 181.

varied between 0.001μ and 0.012μ and was on the average less for the women than for the men. The results indicated that no monochromatic light exists which gives the impression of being yellow to normal eyes when adapted to a weak (1 m. lux) neutral illumination and the radiations exposed are not chosen stronger than is necessary to render possible the distinguishing of colors. It has been quite common practice[310] to take 0.4μ as the dividing line between the visible radiation and the ultra-violet. Observations (by one observer) to determine this point have confirmed the practicability of using this wave-length, although shorter wavelengths were seen, since there was a very marked drop in sensitivity for light below 0.4μ.

As a result of experiments on the selectiveness of the eye's chromatic response to wave-length[311] and its change with change in intensity, it has been concluded that at the threshold the eye is most sensitive to green, yellow-green, blue, blue-green, yellow, red, and orange in the order named, with a dark surrounding field; with a light surrounding field, the order changed to blue-green, green, blue, yellow-green, red, yellow, and orange. At the point of maximum saturation with a dark surrounding field, the eye was most sensitive to yellow-green, then to green, orange, blue, blue-green, red, and yellow; and with light surrounding field, to the yellow-green, green, orange, blue-green, blue, red, and yellow.

An investigation to determine whether the selection and combination of colors according to the Munsell system[312] is in accord with the actual preference of various classes of persons was carried out with 18 business men, 15 graduate men students, 25 women (not students), and 25 commercial artists. The relative agreement of the colors in five sets of colored papers was first determined and then the colors and their combinations were analysed according to the Munsell notation. The results showed that complimentary color combinations were most highly preferred with some exceptions; that the commercial artists agreed no more closely among themselves than did the persons in the other groups; and that the preferences of the commercial artists did not agree so closely with the preferences of non-artists as varying groups of the latter agreed with one another.

[310] Am. Jour. of Physiol. Opt., Apr. 1923, p. 145.
[311] Jour. of Exper. Psych., Oct. 1922, p. 347.
[312] Psych. Bul., Feb. 1923, p. 85.

What factors influence an individual in deciding that a light is gray and what constitutes a normal gray? In order to answer these questions, experiments have been performed[313] on the effect of field size and shape, brightness, exposure, duration of dark adaptation, and order of presentation of stimuli. The stimuli employed were obtained by filtering the light of an incandescent lamp through color-matching glass. The results showed that the stimulus of gray as determined by the method used is independent of the first four conditions. The first few stimuli presented as short exposures after long dark adaptation will momentarily be called yellow, even though they would be called gray or blue on long exposures. If the stimuli are presented in increasing or decreasing order of color temperature and in rapid succession, the result depend on the order of presentation.

It has been known for years that, at low intensities of illumination, the visible spectrum, as viewed by a dark adapted eye, appears colorless. A new set of data have been obtained[314] on the relative energy necessary in the different parts of the spectrum in order to produce a colorless sensation in the eye. It was found that the curve representing the visibility of the spectrum at very low intensities has exactly the same shape as that for high intensities involving color vision but is shifted a distance of 0.048μ toward the red. A proposed possible explanation is that the same substance, i.e. visual purple, whose absorption maximum in water solution is at 0.503μ, is dissolved in the rods, where its absorption maximum is at 0.511μ, and in the cones, where its maximum is at 0.554μ. Tests for color blindness of 547 students by the Raleigh method[315] showed no abnormality of color vision among the women and only 14 among the men, who were very or partially abnormal.

An investigation of the effect of fatigue in one eye on the perception of color by the other[316] resulted in the discovery that each color of the spectrum produced a reflex effect which affected three points—red, violet, and greenish yellow. Each reflex curve has three deflections below normal. In this case, the left eye was fatigued and measurements of critical frequencies of flicker were made with the right eye which was kept in daylight adaptation.

[313] Jour. of Opt. Soc. of Am., Jan. 1923, p. 73.
[314] Jour. of Gen. Physiol., Sept. 20, 1922, p. 1.
[315] Proc. Royal Soc. (A), Dec. 1, 1922, p. 353.
[316] Jour. of Opt. Soc. of Am., Jan. 1923, p. 61.

Experiments were also made with the right eye fatigued. Both sets of data indicated that every ray of color stimulated all three sensations by reflex action, thereby causing a sensation of whiteness underlying and inseparable from color.

Action of Light

A great deal of attention has been given to the action of radiation,[317] both visual and ultra-violet, on plants and animals, not only for the purpose of increasing general scientific knowledge but also from a therapeutic and prophylactic standpoint. The action of ultra-violet light on egg white[318] has been studied for colloidal solutions of globulin, albumin and fibrinogen. The experiments showed an increased coagulation temperature and a reaction toward alcohol. Viscosity increased in proportion to the duration of irradiation.

The movement of animals when exposed to light (phototropism) has been shown to be due in the case of the lowest forms (Amoeba)[319] to inhibition of movement toward the more highly illuminated side. In other unicellular forms, there are primitive receptors which serve to produce in unoriented species rapid changes in luminous intensity on the sensitive tissue side. In the flat worms, the receptors (eyes) are of such a nature that illumination from different directions results in different locations of the stimuli, each of which produces special series of orienting reactions. The action of the salt water king crab (Limulus),[320] when exposed to light under laboratory conditions, is plus phototropic. The phototropism may be modified or obliterated by fright, hunger, stereotropism, photokinesis and other unknown stimuli. The rate of locomotion varies directly as the luminous intensity, being 178 cm. per minute for an illumination of 8,000 m.c. and 157 cm. for 900 m.c. These reactions are said to be satisfactorily explained by the tropism theory. Snail embryos have been found to present a wide range of individual resistance to the action of ultra-violet light.[321] Resistance increases with advancing age. A study has been made of the effect of intermittent light on the tachina fly[322] which is usually highly

[317]Physiol. Abs., Apr. 1923, p. 72.
[318]Pflüger's Archiv, Nov. 11, 1922, p. 540.
[319]Jour. of Opt. Soc. of Am., Jan. 1923, p. 61.
[320]Jour. of Gen. Physiol., Mar. 20, 1923, p. 417.
[321]Jour. of Exper. Zool., Jan. 5, 1923, p. 1.
[322]Am. Jour. of Physiol., Apr. 1, 1923, p. 364.

phototropic. When exposed to continuous light from two sources, the rays of which cross at right angles, the flies were found to move toward a point between the sources, the location of the point depending upon the relation between the illuminations received. With an illumination of 35 m.c., the stimulating efficiency of intermittent light in the orientation of this fly varies with the flash frequency. At flash frequencies from 50 down to 10 per second, the stimulating efficiency was higher than that for continuous light, the maximum being at 15 per second. At a flash frequency of 2 per second, it was lower than that of continuous light and at frequencies of 5, 60, and 160 per second, approximately equal to that of the steady light.

Investigations of the absorption of ultra-violet radiation by living tissues,[323] some physiological solutions and some spectral glasses, showed that in the case of the glasses, flint is more effective than crown. In the case of the tissues of the eye, the lens is the most effective screen and next to it, the cornea. The data on eye absorption have been verified[324] by another investigator who found that the combined tissues of the eye absorbed ultra-violet radiations up to about 0.3134μ and that formalin changes the absorption. The eyes used in the latter experiments included four human eyes, and others taken from cows, pigs and sheep.

ILLUMINATING ENGINEERING

Daylight Saving

In spite of the protest of the rural population[325] Great Britain again went on daylight saving time, April 22, 1923 to continue until September 16. In France[326] a compromise was effected by changing the basis of time comparisons from Paris to Strasbourg, which had the effect of advancing the clocks a half hour the year round. Until ten years ago Greenwich time was used in France, but at that date a shift was made to Paris time which advanced the clocks at least ten minutes.

Daylight saving time was adopted in Niagara Falls, Ontario[327] over vigourous opposition, but was dropped in Baltimore, Maryland.[328]

[323] Australian Med. Jour., Sept. 2, 1922.
[324] Jour. of Opt. Soc. of Am., Aug. 1922, p. 605.
[325] Gas Jour., Mar. 28, 1923, p. 818.
[326] Clev. Plain Dealer Sunday Apr. 29, 1923.
[327] Mov. Pic. World, May 5, 1923, p. 39.
[328] Ibid, May 12, 1923, p. 116.

Light Sources

The relatively low efficiency of artificial light sources compared with the relatively high efficiency of luminous organisms such as the firefly has emphasized the importance of these organisms as sources of light production. Those interested in this subject should look up an elaborate review[329] of recent literature which contains a bibliography.

The problem of how the firefly and other luminous organisms produce their light is progressing toward a solution.[330] It has already been shown that the light-producing reaction is one involving the oxidation of a substance called luciferin in the presence of a thermolabile material called luciferase, which apparently acts as a catalyst, and is necessary for light emission. The oxidation product, oxyluciferin has now been shown to be reducible to luciferin by the passage of an electric current through the solution. It may also be reduced at cathodes of oxidation-reduction cells of the NaCl-Pt-Pt-Na$_2$S type, and also at metal surfaces such as Al, Mn, Zn and Cd which liberate nascent hydrogen from water. A method of producing continuous luminescence is suggested, using a slow stream of hydrogen and oxygen. A large palladiumized surface would continually reduce oxyluciferin which would just as continuously reoxidize in the presence of luciferas and oxygen. Thus a continuously operated luminous lamp could be made. It is suggested that the steady luminescence of bacteria may be due to the continuous oxidation of luciferin and reduction of oxyluciferin in different parts of the bacterial cell.

Further work on the minimum concentration of luciferin,[331] which is visible, has shown that one portion in from 4 to 40 billion parts of sea water gives on oxidation light visible to the unaided, but dark adapted eye. A similar experiment with luciferin showed that one part in from 800 million to 8 billion parts of sea water will oxidise a stock solution of luciferin with visible luminescense.

A report to the Smithsonian Institution[332] of explorations in the wilds of Costa Rica describes a type of beetle which is said to emit almost without interruption a light so brilliant that one or two im-

[329]Erge. der Physiologie *21*, 1923, p. 166.
[330]Jour. of Gen. Physiol., Jan. 20, 1923, p. 275.
[331]Science, Apr. 27, 1923, p. 501.
[332]Sci. News Letter, July 28, 1923, p. 10.

prisoned in an inverted tumbler will illuminate a moderate sized room sufficiently to make print readable. The beetles are called "carbuncles" and the light differs in color in different individuals, being mostly yellow but sometimes green and occasionally ruby red.

A practical application of the idea that insects are more attracted to violet and blue light than to red or white light[333] has been reported from a western city. Electric globes colored a light red have been used in lunch rooms, dancing pavillions, bathing beaches, country club houses, and street cars with a resulting freedom from insects, the number of which previously had been very annoying.

A study of the causes of fouling ships' bottoms[334] has indicated that color is an important factor. Barnacles, especially, seem averse to light colors. In general, the fouling was found to be greater on dark than on light painted bottoms, the difference between white and black paint being especially marked.

Photography

The high luminous intensity when a wire filament is exploded by the application of a high potential[335] has inspired the development of a new light source for instantaneous photography. Experiments with tin and copper wires not being fruitful, mercury was successfully substituted. The mercury filaments were made by drawing mercury into a glass capillary tube each end of which contained a hair wire electrode sealed in with sealing wax. With lengths of 10 mm and diameters ranging from 0.172 to 0.324 mm. and an applied voltage of 80, the duration of the flash ranged from less than 10^{-5} seconds with a very feeble light to 8.0×10^{-4} seconds with a very bright flash. The duration was found to depend on the diameter and length of the filament and on the thickness and wall of the capillary.

In connection with photomicrographic work the ratio of actinic or chemical surface brightnesses[336] of various light sources has been worked out as well as an apparatus and method for determining the absolute brightness in terms of the Hefner. The absolute properties of mat reflective surfaces used in this work have also been studied.

[333] Scien. American, Mar. 1923, p. 179.
[334] Pop. Mech., May, 1923, p. 760.
[335] Jap. Jour. of Phys. I, No. 9, 1923, p. 97.
[336] Zeit. f. Instr., Dec., 1922, p. 349.

Legislation

Attention should be called to references in the notes and abstract portions of several issues of the Transactions on legislation involving lighting in various states. In Wisconsin[337] the new state electric code which went into effect the first of the year provides that, "gas-filled incandescent lamps of more than 100-watt size cannot be used in the common type of socket having paper linings." Special sockets of porcelain are required. The load allowed to branch sockets is increased to 1000 watts provided that fuses not heavier than 15 amperes are in the circuit. The Rhode Island Legislature[338] has passed an act apparently especially directed toward the City of Providence, providing that, "In all theatres . . . ordinarily lighted by electricity there shall be in addition . . . an independent means of light which shall be (either) gas from a commercial gas distribution system (or) a storage battery system controlled either automatically or from . . . the lobby or front of the theater, or such other system or means of lighting as shall be approved by the City Council of Providence. In case such lighting system shall be gas, at least two lights on each side of the main body of the theater and on each side of all balconys, as well as a sufficient number of additional lights to illuminate the stairways, hallways, and lobbies shall be kept burning at all times when such theater is open to the public."

In the Annual Report of the Chief Inspector of Factories and Workshops of Great Britain,[339] an account is given of improved lighting conditions on docks and wharfs where cluster lamps have been found to be satisfactory. In South London, the report states, "It is unusual to find an entirely satisfactory system of lighting." In 78% of the underground workrooms visited, the natural lighting was insufficient to illuminate the whole room. However, there is some evidence of increased attention being paid to correct methods of illumination. Several cases of lighting by the indirect system are noted including a lithograph factory, plating works, and a tobacco works.

[337]Elec. Rec., Jan., 1923, p. 5.
[338]Gas. Jour., Dec. 6, 1922, p. 636.
[339]Elec., Aug. 4, 1922, p. 122.

Plants

The problem of photosynthesis or the action of sunlight on the green leaves of plants[340] to produce sugar, starch and wood out of carbon and carbon-dioxide was discussed at a joint meeting of chemists and botanists at the Boston meeting of the A. A. A. S. A proposed explanation is that the first step in the process is the breaking down of the carbonhydrogen molecule into a very large number of enormously reactive substances. These substances either rearrange and react with each other or with some other substance in the cell, possibly with CO_2. Experiments on the growth of plants under artificial light[341] from seed to seed have been reported as successful. Wheat, oats, barley, rye, buckwheat, potatoes, and several vegetables growing from seed, under artificial illumination only, produced seed of good quality which germinated well. The illumination was obtained from tungsten filament incandescent lamps burning 24 hours a day. One set of lamps was said to be enough to produce an ordinary crop, such as cereals. Spring wheat produced ripe seed in about 90 days. All the plants tested, except cabbage, bloomed and no variety seemed to require any particular period of illumination to cause blooming. Four ranges of light intensity were used, and a number of plants bloomed under all of them although the illumination was continuous. The temperature was maintained automatically for cereals at 14° C, by blowing in outside air.

The increased velocity of germination of seeds[342] when exposed to moonlight was indicated by some experiments in which the effect was thought to be due to the action of the moonlight on diastase. A test of the latter point, using crushed mustard seed showed a 15 per cent increase in sugar yield over seed not exposed. At certain periods moonlight is plane polarized and a similar set of experiments on the seed, using polarized daylight gave similar results. Care was taken in this work to have the temperature conditions the same for exposed and unexposed packages of seed. The experiments are being continued.

The stimulation of plant growth[343] by artificial light has been tried in Germany. Over a plot of lettuces 5 lamps were arranged

[340]Sci. News Letter, Jan. 13, 1923, p. 8.
[341]Science, Sept., 29, 1922, p. 366.
[342]Nature, Jan. 13, 1923, p. 49.
[343]Elec. Rev., Mar. 30, 1923, p. 511.

so that the light was diffused as uniformly as possible. The lamps were turned on for six hours each day beginning at dusk. Cabbage-lettuces illuminated from the middle of November had after 12 days about two and a half times as many fresh leaves as those not illuminated. After seven weeks of prolonged illumination a comparison made between the plants on the illuminated plot and those on the plot not illuminated showed a 50 per cent superiority in weight of the former over the latter in the green state and 68 per cent in the dried state. The effect was equally good with beans and vetches. Illuminated strawberry plants yielded as early as the middle of March as compared to four weeks later for non-illuminated fruits.

Societies

In the plan for the future as laid out by the British Illuminating Engineering Society[344] it is proposed to organize extensive propaganda with a view to interesting municipal bodies, educational authorities, owners of workshops and factories, etc., in illumination; arrange lecture demonstrations in various parts of the country for the benefit both of those technically concerned with lighting (architects, contractors, etc.,) and improved classes of consumers; form a fund for researches and especially aid experiments to be conducted by the various committees of the Society; expand the size of the Journal by including special contributions on selected subjects; initiate propaganda in the Colonies and Dominions. This program indicates that much of the work of our Society has met with the approval of our neighbors who are incorporating it in their plans.

A joint committee has been set up by the British Commercial Gas Association and the Society of British Gas Industries[345] to consider matters affecting gas lighting. Among other points for consideration are the standardization of the essential parts of the burner; the elimination of shoddy burners, mantles, and fittings; the maintainance and development of gas lighting.

The French National Committee on Illumination[346] met last November. Among other subjects discussed was the question of making a bibliography of all the work connected with lighting. A

[344]Ill. Eng., July 1922, p. 218.
[345]Gas. Jour., Nov. 1, 1922, p. 286.
[346]Rev. Gen. d'Elec., Dec. 9, 1922, p. 873.

meeting of the Sub-Committee on Automobile Headlights[347] was held on March 21, this year, and took up the question of the French codes on this subject. A report of the English National Committee was referred to in the July 1923 issue of the TRANSACTIONS of this Society.

At the Tenth Annual Convention of the German Illuminating Engineering Society[348] the program included among others the following subjects: "Our present conception of the nature of light"; "Illuminating engineering lectures and papers"; "The present and future of gas lighting"; "Impressions of a course of study on illuminating engineering in the United States of America".

The Advisory Board of the Engineering Foundation[349] has received and is thoroughly investigating the feasibility, cost, and probability of support of a suggested experimental investigation of the relation of quality and quantity of illumination to efficiency. It is felt that the determination of this relationship by disinterested parties will be very valuable in industry.

General

The largest telescope in the world, the 100 inch reflector at Mount Wilson Observatory[350] collects 160,000 times the light received by the eye.

The photo-electric cell is one of the numerous illustrations[351] of the possibility of research work having direct practical application. Means of applying the cell, either of the selemium, alkali metal, or thalofide type, to the control of industrial processes have been proposed and specific cases worked out. The general method is to produce such a change in the intensity of illumination of the cell as to produce a corresponding change in the current sufficient to operate a relay which in turn controls the proper valves, switches or other means necessary to effect the desired reaction. Among the optical properties suggested for use are absorption power for white light, selective absorption power, index of refraction, and rotation of polarized light.

[347] Ibid, May 12, 1923, p. 769.
[348] L. u. L. Sept., 21, 1922, —. 443 and Oct., 5, p. 463. Zeit. f. Bel. Sept. 15, 1922, p. 121.,
[349] Report of Engineering Foundation, Apr., 1923, p. 23.
[350] Sci. News Letter, Dec., 9, 1922, p. 8.
[351] Jour. Soc. Chem. Indus., Feb., 16, 1923, p. 127A.

The detection between ordinary artificial pearls and natural pearls[352] can be made by mechanical means, but the Japanese cultivated pearl has defied such detection except by destruction, until recently.   Under the light from a mercury vapor lamp the Japanese pearl shows a distinct translucent opalescence, while the natural pearl, though opalescent is opaque.   This distinction is evident even to a layman.

The meteorological service of the city of Paris[353] has conducted experiments on the effect of smoke in causing loss of light in the city and suburbs.   The total actinometer or lucimeter of Bellani was employed to make the measurements.   Observations were made on eight stations and the quantity of light received in 24 hours was determined at each one and compared with that which should be received.   It was found that where the wind was such that the smoke from the city was blowing over the suburb the loss of light was as much as 25 per cent.

The Institute of Technology at Carlsruhe[354] inaugurated its school of light in June 1922.   This department has four rooms darkened and provided with photometric equipment.   The first has three photometer benches, 3 m., 4 m., and 4.8 m. long respectively as well as an Ulbricht sphere.   The other rooms have arrangements to extend the length of the photometer benches, and one has a Rousseau photometer as well as other equipment to measure illumination and distribution.   A fifth room is equipped with a number sockets to try out different lighting units.   A number of rooms are provided for special work such as that of physiology of the eye, glare, influence of diffusion, etc.   The equipment includes two sets of storage batteries which provide the electric current. The buildings and observation tower will permit the photometric and physical study of light from the sky.

JOURNALS AND BOOKS

The American Builders Magazine has started a department[355] in which, in addition to other information, will be given information for wiring luminaires, and arrangements and suggestions for decorative lighting.

[352]Sci. Amer., Apr., 1923, p. 227 and Chem. News., Aug., 11, 1922.
[353]Comp. Rendu., Jan. 15, 1923, p. 180.
[354]Rev. Gen. d'Elec., Jan. 20 1923, p. 24D.
[355]Elec. World, Feb., 24, 1923, p. 472.

The Germany Illuminating Engineering Society[356] has designated the technical journal "Licht und Lampe" as the official organ for the publication of its proceedings, in the place of the Zeitschrift fur Beleuchtungswesen which owing to economic conditions has been compelled to suspend issue. Dr. Lux, the former editor of the Zeitschrift fur Beleuchtungswesen will continue to edit the Society's work as presented in Licht und Lampe.

Books

"Colour Vision," Prof. W. Peddie, (London: E. Arnold & Co., 1922). Pp. xii + 208. See "Nature," June 16, 1923, p. 799.

"Colour and Methods of Colour Reproduction," Dr. L. C. Martin, William Gamble, (London, Glasgow and Bombay: Blackie & Sons, Ltd., 1923). Pp. xiii + 187. See "Nature," June 16, 1923, p. 799.

"L'Eclairage," E. Darmois, (Paris: Gauthier-Villars & Cie, 1923). Pp. xv + 276.

"The Elementary Principles of Lighting and Photometry," J. W. T. Walsh, (London: Methuen & Co., Ltd., 1923). Pp. xvi + 220. See "Nature," June 16, 1923, p. 804.

"L'Arc Electrique," Maurice Leblanc fils, (Paris, Les Presses Universitaires de France, 1922). Pp. 131. See "Nature," June 16, 1923, p. 805.

"L'Eclairage et le Demarrage electriques des Automobiles," R. Bardin, (Desforges: Paris, 1922). Pp. 64.

"Die Fabrikation und Berechnung der Modernen Metalldrahtglühlampen," F. Knepper, (Leipzig: Hachmeister & Thal, 1922). See "Helios," Sept. 3, 1922, p. 428.

"Heliotherapy," Dr. A. Rollier, (London: Oxford Medical Publications, Henry Frowde & Hodder & Stoughton, 1923). Pp. 288.

"Die Heterochrome Photometrie in Theorie und Praxis," R. von Voss, (Julius Springer: Berlin, 1919).

"Incandescent Lighting," S. I. Levy, (London: Sir Isaac Putnam & Sons, Ltd., 1922). Pp. 124. See "Gas Journal," Dec. 20, 1922, p. 734.

"Lamp Shades—How to Make Them," Olive Earle, (Dodd, Mead & Co.).

[356] L. u. L., Nov., 1922. p. 483.

"Motion Picture Projection," J. Slater, (London: The Bioscope Pub. Co., Ltd.) Pp. 165.

"The Romance of the Gas Industry," O. E. Norman, (Chicago: A. C. McClurg & Co.) Pp. 200; illus., 51.

"Die Stereoskopie im Dienste der Isochromen und Heterochromen Photometrie," C. Pulfrich, (Berlin: Julius Springer, 1922).

"Visual Illusions, Their Causes, Characteristics and Applications," M. Luckiesh, (New York: D. Van Nostrand Co., 1922). Pp. 246 + xv. Ill. 100.

"Ultraviolet Radiation," M. Luckiesh, (New York: D. Van Nostrand Co., 1922). Pp. 247 + xiv. Ill. 12.

"The Book of the Sky," M. Luckiesh, (New York: E. P. Dutton & Co., 1922). Pp. 236 + xi. Ill. 16.

"Light and Color in Advertising and Merchandising," M. Luckiesh, (New York: D. Van Nostrand Co., 1923). Pp. 263 + xx. Ill. 37.

LIGHTING THE SILK INDUSTRY WITH INCANDESCENT LAMPS*

BY H. W. DESAIX**

SYNOPSIS: This paper is based on observations, investigations and experiments conducted by the author to secure for the Silk Industry a practical system of illumination to replace the present inadequate and impractical methods.

Each process in the manufacture of silk fabrics presents a problem in itself and each has been so treated.

Owing to the paper being prepared for an audience of laymen technical terms and expressions were purposely omitted where possible.

During the early months of this year a survey was made of various silk manufacturing plants for the purpose of obtaining data for the better lighting of plants in this industry. This survey was later supplemented by experimental installations from which tests and further observations were made, eventually leading up to some very gratifying results.

As the idea seems to suggest itself, this presentation is in the nature of a report of this investigation. Time, however, will not permit of detailed data being given for each step in the process of manufacturing silk fabrics, but what appears to be the most important will be dealt with.

An inspection revealed most of the plants using pendant drop cords, hung about two feet over the working plane of looms and a trifle higher over accessory machines. In some instances bare 50- and 60-watt Mazda lamps were in use; but in the majority tin cone or flat shaped painted reflectors were in use. This latter system seems to be the conventional means of lighting silk weaving plants.

*A paper presented before the joint meeting of the New York Section and the Northern New Jersey Chapter of the Illuminating Engineering Society, Patterson, N. J., May 11, 1923.

**Watson-Flagg Engineering Company, Paterson, N. J.

The Illuminating Engineering Society is not responsible for the statements or opinions advanced by contributors.

WEAVING

In some weaving plants, drop cords were in use being even with the bottom of the arch of the loom using bowl type, enameled steel reflectors, with 50- and 60-watt type "B" Mazda lamps, and in some instances mill type lamps. This latter system is unquestionably an improvement over those mentioned above. It has its faults, however, being expensive in comparison to the result obtained; lamps do not as a rule burn their natural life, due to constant handling, striking of reflectors against loom arch: and because of theft. There is also a tendency on the part of operators to install larger lamps in the sockets, with the well-known result; glare. Another fault of the local lighting system for weaving, and in fact other processes, is eye strain of operators due to the constant change of looking into comparative darkness as compared with the light on the warp.

Illuminating engineers, heretofore, have recommended a general lighting system to supplement the local to overcome the last objection and to provide proper lighting for aisles. This has always seemed to the mill executives a duplication of system and extra avoidable expense; very often it resulted in an unpopular lighting engineer.

On the other hand, general lighting systems in weave sheds have been attempted but have proven unsatisfactory because objectionable shadows were cast by loom arches, and an insufficient amount of light in between the heddles at which point ends must at times be picked up and passed thru.

Conferences with mill executives, weaving foremen and operators themselves, led to the opinion that if a system of illumination could be worked out that would incorporate the advantages of the concentrated result of the improved local system, and the further advantages of the usual general system, at a reasonable cost of installation and upkeep, consistent with the desired result, it would fill a long felt want and at once become popular. With this idea in mind, experiments and tests were started.

For the purpose, a room was used containing 24 plain broad silk looms, arranged in 4 rows across, a wide aisle in center, with 6 looms to a row. The room was typical of most that are used for the purpose, the ceiling was of open wood beam construction, with walls and ceiling painted white, except for a distance 4 feet up from the floor where the walls were painted light grey. It was located on the ground floor necessitating the use of artificial light all day.

The lighting system in use was the customary drop cord with tin cone shaped painted reflectors, and 50-watt mill type Madza lamps. Test showed this system as delivering an average of 11 foot-candles on the warp, but with the undesirable effects outlined above.

In order to be satisfied that previous opinions were well founded a general system was next installed employing RLM dome steel enameled reflectors, with 200-watt·"C" bowl enameled Mazda lamps, spaced on centers over each loom station or weavers alley; this spacing meant 13'3" between units along the rows and 7'0" across and 10'0" between rows on each side of center aisle. Mounting height 12'0" from floor or approximately 9'0" above working plane. Uniform average intensity of 10 foot-candles was secured.

From general appearances this system seemed to be quite satisfactory. It certainly eliminated all the objections of the local system; but while sufficient light was delivered for general purposes there was not enough to see to pass ends through the heddles in the harness. Shadows were encountered from the loom arches and by the operators when leaning over the work. This system also failed to provide an upward component to enable the transmission equipment, such as shafting, hangers and pulleys to be adjusted.

An analysis was then made of the above systems, the obstacles to be overcome and the nature of the work done in a weaving plant; from this it was concluded that the system desired must contain the following factors:

A varied intensity of illumination. With a higher intensity on the looms as compared with drives and aisles.

Adequate intensity for all purposes.

Freedom from objectionable shadows.

Freedom from glare, both direct and by contrast.

Must provide sufficient light on walls with a well distributed upward component, in order to have interior or room give a bright cheerful appearance.

The cost of installation and upkeep must be reasonable in comparison to result and must pay a return on the investment.

Uusually the required intensity is known, but in this case the value in use did not appear sufficient, making it necessary to work the reverse of regular practice; i.e.—effect to cause.

After a thorough study of the various types of luminaire the market afforded one was selected. This unit consisted of a rugged prismatic glass reflector 11⅞″ in diameter and 5⅞″ in height, open top and bottom. The distribution curve showed it to be of the intensive type, with sharp cut off at 60° from the vertical, falling away sharply after 15°. Supported by a three prong holder, designed on the ice tong principle which grips the lower flange of the reflector. An adjustable stem is provided with marked calibrations for 75- 100- 150- and 200-watt lamps; it made an ideal unit for the purpose of the test. These units were hung on the same spacings and mounting heights as were the RLM reflectors used in the previous experiment.

With 200-watt type "C" clear Mazda lamps, readings were taken of 30 foot-candles on the warp tapering to 15 foot-candles in the aisles and on the drives, which satisfied the first requirement of varied intensity or "spilled" light. See Figure 1.

The amount of light was entirely sufficient; in fact, at first was considered too high, but later experiments disproved this. Entering ends thru harness could be done with ease and plenty of light. Arch and operators shadows were not at all objectionable because of high intensity.

With practically the entire ceiling moderately illuminated glare by contrast was not encountered as the tunnel-like appearance was illuminated; and transmission equipment could be oiled and adjusted easily. This excellent upward component is very desirable for the sake of its cheerful appearance, especially where women are employed, as in this industry. Because the light cut off occurs at 60° direct glare was practically nil.

Quite naturally, certain objections were raised, such as the possibility of glass breakage, but this was not considered serious, because the units were hung at a height from the floor which removed them from the sphere of activity and being of rugged construction would require a hard blow to cause breakage.

Another objection was the collection of dirt or dust, but it was pointed out that if dust did collect on top to any extent the condition could not be any different than that of an opaque reflector.

Fig. 1—Unobjectionable varied intensity, no confusing shadows, and well lighted interior in a weave shed. Illustrates "spilled" lighting principle.

Fig. 2—Winding and Quilling room lighted at 12 foot-candles, using prismatic glass reflectors spaced 12 x 12 feet.

FIG. 3—Night photograph of dye house. Vapor proof units spaced 10 x 10 feet with 150-watt Mazda C lamps, resulted in intensity of 12 foot-candles with equal distribution on vertical as well as horizontal surfaces.

The fact that dirt might collect would have a tendency to induce shorter periods between cleanings. Being designed with reflecting prisms, dirt or dust does not impair reflecting efficiency unless mixed with oil as the oil excludes the air film which normally prevents optical contact. In fact, dust and dirt is something which should not exist to any great extent in a silk weaving plant.

The results obtained were well worth the effort expended, as it was the unanimous opinion that a system of illumination had been worked out that fullfilled all the necessary requirements of silk weaving.

Later this system was installed in the entire plant which made it possible to determine costs. It was found that the cost of installation, using a modern conduit wiring system with convenient switching arrangement equalled one-half a cent per foot-candle per square foot.

WINDING AND QUILLING

When winding, the broken ends are detected by the stopping of the swifts, and in quilling are detected by the raising of the quides, making it very easy to detect broken ends or spent spools from any position in the aisle. For these operations RLM dome steel reflectors with 200-watt Madza "C" lamps spaced on 12 ft. centers hung 12 ft. high will give a general illumination of about 8 ft. candles. An intensity sufficiently high for ordinary purposes.

It has been found advisable in some instances that a higher intensity is required for which the same equipment is recommended as for weaving, spaced at a distance not to exceed mounting height which will deliver a general intensity of 12 foot-candles and also provide a more cheerful room appearance. See Figure 2.

WARPING

In warping rooms, where horizontal warpers are employed, a recommended system is to treat each warper as an individual unit using the "spilled" lighting principle. This requires the same type of luminaire as suggested earlier for weaving. These should be spaced, one over the creel, one over the reed and one over the beam centered in all cases with the width of the machine and use 200-watt, Mazda "C" lamps. One of the factors to consider in warping is that work is done both in a horizontal and vertical plane. The system suggested will furnish a sufficient intensity of illumination to both surfaces and spill suffiicent light on aisles, and driving gear. Intensities should be about 30 foot-candles on the work and taper off gradually.

If Swiss warpers are installed in the same room with horizontal warpers the same system may be employed for uniform appearance, but where they are set up in a separate part of the mill it has been found more satisfactory to use two units with extensive type reflectors and the same size lamps, due to the more compact form of this type of warper. The creel is smaller, the reed wider, and the mill itself not over three yards in circumference, practically eliminating the vertical working plane. Intensity for Swiss warping should be 12 to 15 foot-candles.

THROWING

The throwing branch of the industry is one requiring very careful consideration when planning the lighting system. Most throwing plants run twenty-four hours per day, which means that the night shift works entirely under artificial illumination.

High intensities are not so necessary because of the nature of the process, as it is a comparatively simple matter to detect broken ends or empty spools. Excellent results have been obtained by using a general system of illumination employing RLM steel dome reflectors with 200-watt bowl enameled lamps spaced 16 to 18 foot-centers and hung 12 to 14 ft. from the floor. This will provide about 4 or 5 foot candles. Higher intensities, which are advisable, are secured by placing two units in each alley, between machines hung 12 ft. from floor, using RLM steel dome reflectors and 200-watt bowl enameled Mazda lamps, giving an intensity of approximately 10 foot-candles.

If glass reflectors are used a pleasing upward component, adding to the appearance of the room, making it more cheerful is secured.

It should not be forgotten, in any event, to provide properly diffused illumination for the entire room including all aisles in order to relieve eye strain.

DYEING

Another branch of the silk industry that presents quite a problem for the illuminating engineer is in the dye house.

On the basis of a recent investigation a system was installed in a modern dye house.

Because of the humid atmosphere existing in buildings where dyeing is done, vapor-proof enclosing globes should be used to protect lamp base and socket parts from corrosion. Ordinary vapor-proof globes give the same light distribution as bare lamps and do not protect against glare from excessively bright lamp filaments.

For this reason a prismatic glass vapor-proof reflector globe, designed to screw into a cast iron or lead alloy holder containing lamp socket, is often used with 150-watt type "C" clear Mazda lamp.

The distribution produced by this unit has the maximum candle power at 55° from the vertical which provided a uniform intensity on both vertical and horizontal surfaces, a necessary factor. A sharp light cut off at 60° eliminates objectionable glare.

Units spaced in bays 10 x 10 feet and mounted 10 feet from the floor, result in a uniform initial intensity of 12 foot-candles. See Figure 3.

An important point in dyeing is the proper matching of colors. Customers usually send a sample of color wanted or specify shade from Standard Color Card.

For this purpose a small room should be provided so constructed that north skylight will be available. At times when this is not to be obtained or in dark hours, the best sustitute is a unit producing artificial daylight, consisting of a heavy gauge copper hood with a one piece pot glass filter with the color in the glass. Applied color will fade. Regular incandescent lamps are used which are usually supplied with the unit to insure proper size and type.

It is suggested that cheap unreliable substitutes be avoided as the cost of a proper type of color matching unit is made many times by the saving in having goods dyed with proper shades.

CONCLUSION

Other phases of the silk industry deserve consideration, such as ribbon weaving and Jacquard broad silk weaving, but the machinery in these branches differ widely in various plants, especially with ribbon weaving and it might be suggested to treat each problem as it presents itself; although it might be said that it is almost impossible to use anything else but a well designed local system for either of these branches, except in a very few cases.

In conclusion, it might be added that practically no silk plants can operate without the use of artificial light and when compared with the amount of money represented by the raw and finished material handled in the course of only one year, the cost of the best possible lighting system is very negligible.

ABSTRACTS

In this section of the TRANSACTIONS there will be used (1) ABSTRACTS of papers of general interest pertaining to the field of illumination appearing in technical journals, (2) ABSTRACTS of papers presented before the Illuminating Engineering Society, and (3) NOTES on research problems now in progress.

DESIRABLE DISTRIBUTION OF LIGHT AND BRIGHT-NESS FOR THE DINING-ROOM

BY M. LUCKIESH* AND MARGARET S. FULLERTON**

An investigation was conducted to ascertain the most desirable distribution of illumination in a typical dining-room, 15'-4" wide and 21'-6" long. The walls were of two tones of warm gray, the upper walls being some lighter than the lower. The ceiling was a very light warm gray, commonly termed cream white. The rug was taupe with a suggestion of rose shade. A white cloth was provided for the table and experiments were conducted with and without this cloth. A bowl with brightly colored flowers added a final touch to the table. Reflection-factors of these various areas are presented in Table I.

TABLE I

REFLECTION-FACTORS OF IMPORTANT AREAS

	Per cent
Table (without white cloth)	15
Table (with white cloth)	66
Ceiling	61
Upper Walls	53
Lower Walls	45
Rug	11

The fixture used, was so equipped that four steps of illumination were obtained. Step 0 was constant throughout the experiment and consisted of one 75-watt lamp providing "direct" light,

*Director, Laboratory of Applied Science, National Lamp Works of G. E. Co., Cleveland, Ohio.
**Laboratory of Applied Science, National Lamp Works of G. E. Co., Cleveland, Ohio.
The Illuminating Engineering Society is not responsible for the statements or opinions advanced by contributors.

confined to the top of the table, and three 25-watt lamps providing a "semi-indirect" component. Step 1 consisted of Step 0 illumination plus three 25-watt lamps providing a component of "indirect" lighting. Step 2 consisted of Step 0 plus three 50-watt lamps providing "indirect" lighting. Step 3 added to the illumination of Step 0 three 25-watt lamps and three 50-watt lamps providing "indirect lighting. Foot-candle readings were taken on the table and on the ceiling at four different points.

TABLE II
ILLUMINATION INTENSITIES ON IMPORTANT AREAS
Foot-Candles

Step	On table	At various points*on ceiling							
		Without cloth on table				With cloth on table			
		1	2	3	4	1	2	3	4
0	11	1.0	.3	0.1	0.1	2.0	0.5	0.2	0.1
1	12	6.0	1.9	0.4	0.4	7.0	2.2	0.5	0.5
2	13	11.0	2.9	0.5	0.5	12.0	3.0	0.7	0.6
3	14	16.0	4.2	0.8	0.8	17.0	4.5	1.0	0.9

*Point 1 was directly over the table and about a foot from the center of the room. Point 2 was about 4 feet from the center of the room. Point 3 was about 6 feet from the center of the room. Point 4 was about 7½ feet from the center of the room.

The brightnesses of the various points of interest are presented in Table III.

TABLE III
BRIGHTNESSES OF VARIOUS SURFACES IN THE ROOM IN MILLI-LAMBERTS

Step	Without cloth on table				With cloth on table			
	Table	Lower Walls	Upper Walls	Ceiling (Pt. 1)	Table	Lower Walls	Upper Walls	Ceiling (Pt. 1)
0	1.72	——	——	0.66	7.75	0.1	0.11	1.3
1	1.94	0.19	0.28	3.98	8.5	0.21	0.32	4.5
2	2.04	0.29	0.39	7.2	9.15	0.32	0.48	7.85
3	2.26	0.39	0.52	10.76	9.9	0.43	0.57	11.19

The observer was asked to indicate his preference of these four lightings under each of two conditions, namely with and without a white cloth on the table: and from two viewpoints under these two conditions, namely, at the table and away from the table. When the observer first made his choice there was no white cloth on the table. His preference was determined under this condition with the observer away from the table, taking a general or mental photograph of the room as a whole. Next his preference was determined when he was seated at the table without a white cloth. The white cloth was then put on the table and the observer asked to state his preference under this condition both away from the table and at the table. Forty observers, male and female, were used but no sex difference was noted. The results are shown in table IV.

TABLE IV

A Summary of References

Step	Without white cloth		With white cloth	
	Away from table	At table	Away from table	At table
0	4	3	2	2
1	10	18	12	15
2	19	14	23	14
3	7	5	3	9

According to Table IV when the observer was viewing the room as a whole and there was no white cloth on the dining-table, Step 2 was the most desired although Step 3 was a close second. In Table III, it is seen that under the illumination of Step 2 the table and ceiling are the brightest areas and the walls are less bright. These brightness contrasts are just great enough to give pleasing prominence to the center of the room, namely the table. Its desirability above the other steps is apparently due to the fact that in the other steps, the brightness contrasts are too great. However, from a viewpoint at the table, under this same condition, Step 1 is most preferred.

When the white cloth is added to the dining-table, the same two steps, namely Step 2 when the observer is away from the table and Step 1 when observer is seated at the table, were found to be the most preferred. The percentage of observers choosing these two steps, however, is higher than under the condition without the cloth. That is, the extremes of contrast are even less preferred with a cloth on the table. In Table III it is seen that the white cloth has added enormously to the brightness of the table but not, in proportion, to the other places in the room. This makes the brightness contrast between the table and the other surfaces in the room so great under the extremes of illumination that they rank very low in preference. In Step 1 and Step 2, the brightness contrasts are not as great.

The distribution of brightness determines the pleasantness of surroundings. In this experiment, it is found that the desirable brightness distribution is obtained by the lighting conditions of Step 1 and Step 2, the middle steps of the range shown. The combination of illumination intensities and reflection-factors is responsible for brightness distribution. The experiment shows that in a given dining-room there are definite ratios of upward to downward light which are much more preferred than extremes of this ratio. In another room whose reflection-factors differed greatly, the same range of ratios obviously would not produce the most preferred distribution of brightness. However, in any room there is a ratio of upward to downward light which produces approximately the same brightness distribution as found desirable in this experimental room. In other words, there is always a ratio of upward to downward light which produces a pleasing effect but it varies with the decorative scheme of the room.

WESTINGHOUSE LAMP COMPANY'S TRAINING COURSE*

BY A. R. DENNINGTON**

Realizing that the future growth of the industry depends upon the men connected therewith, the Management of the Westinghouse Lamp Company has arranged a systematic Training Course for new men entering the Organization.

An Educational Committee consisting of Mr. T. G. Whaling, General Manager, Chairman and Messrs. E. L. Callahan, Sales Manager, H. D. Madden, Supt. of Equipment Dept., R. E. Myers, Chief Engineer and F. M. Wicks, Manager of Works, as members, was appointed by the Advisory Board.

The Educational Committee, after careful consideration of the needs of the Company, recommended the organization of a Training Course covering a period of three months for general instruction in lamp manufacture and supplemented by three months of training along commercial lines, for salesmen, or along manufacturing lines for men taking up work relating to lamp production and development.

Schedules covering each week of the training period were made up and are being put into effect, so that each student is transferred to a new phase of the work at regular intervals. The student's observations on each assignment during the first three months are directed by a series of questions and note books are provided in which he is encouraged to write the salient facts he has learned.

The Course is planned to give men taking up commercial activities, actual training with agents, thus extending to sales work the practical experience method which has been successfully used in the past in training men for manufacturing and engineering work.

*Owing to the increased interest in the training of Illuminating Engineers and Lighting Salesmen by the Central Stations, this brief note is very timely.—*Committee on Editing and Publication.*

**In charge of Educational Department, Westinghouse Lamp Co., New York, N. Y.

The Illuminating Engineering Society is not responsible for the statements or opinions advanced by contributors.

During the week July 30th to August 4th, when the manufacturing divisions were closed, a special schedule consisting principally of lectures, was arranged and speakers from outside, as well as within the organization, gave talks on various subjects relating to illumination and the commercial phases of the incandescent lamp business. As this forms a somewhat unusual plan for giving men who have recently joined the Organization special instruction, which should be helpful to them and to the Company, the schedule for the week is given:

TRAINING COURSE SCHEDULE: JULY 30TH TO
AUG. 4TH, INCLUSIVE

Monday, July 30th.

8:45 A. M.—"Fundamentals of Illumination", Mr. C. M. Doolittle, Illumination Bureau.

10:00 A. M.—"Work of our Syndicate Department," Mr. Fred Kinsey, Mgr.

1:00 P. M.—"Merchandising Department," Mr. P. C. Pfenning, Mgr.

3:00 P. M.—"Work of Illumination Bureau," Mr. S. G. Hibben, Mgr.

Tuesday, July 31st.

8:45 A. M.—"Lighting Specifications and the Foot Candle Meter," Mr. S. G. Hibben.

10:00 A. M.—"Electrical Testing Laboratories Inspection Service," Mr. Preston S. Millar, Gen. Mgr.

1:00 P. M.—"Calculation of Illumination for Office."

Wednesday, August 1st.

8:45 A. M.—"Store Window Lighting," Mr. D. W. Atwater, Illumination Bureau.

10:00 A. M.—"Show Window Trimming," Mr. R. MacWilliams," Adv. Dept.

1 P. M.—"Calculation of Illumination for Factory."

Thursday, August 2nd.

8:45 A. M.—"Illumination from Central Station Standpoint," Mr. W. T. Blackwell, Illuminating Engineer, Public Service Electric Co.

10:00 A. M.—"George Cutter Equipment," Mr. C. H. Stahl, Illuminating Engineer, George Cutter Works.

1:00 P. M.—"Merchandising of Westinghouse Elec. & Mfg. Co.," Mr. M. C. Turpin, Mgr. Sales Promotion Section, W. E. & M. Co.

3:00 P. M.—"Interior Store Arrangement," Mr. R. B. Ely, Mdse. Div.

Friday, August 3rd.

8:45 A. M.—"Commercial Engineering Problems," Mr. A. E. Snyder, Asst. to Commercial Engineer.

10:00 A. M.—"Lighting Units and Glasswares," Mr. S. G. Hibben.

1:00 P. M.—"Salesmanship," Mr. F. W. Loomis, Illuminating Engineer, Duquesne Light Company.

2:00 P. M.—"Sales Experience,"—Mr. R. G. Reynolds, Alexander Hamilton Institute.

3:00 P. M.—"Calculation of Illumination for Drafting Room."

Saturday, August 4th.

8:45 A. M.—"Calculation of Illumination for Street Lighting."

All schedules are arranged with the idea of giving the student a broad view of the business and getting him to appreciate the correlation of the work of the various sections and the great advantage of cooperation, to individuals as well as to departments.

Two lectures are scheduled for each week of the entire Training Course. One of these lectures is to cover a manufacturing subject and one is to relate to some commercial matter. These will be given by a department head or by others who are qualified to talk upon the subject chosen.

The students are under the direction of foremen or those who are duly qualified in the section in which the students are assigned. Reports on the students' activity, interest, application and other general characteristics are made by the foreman at the end of the assignment in each section. These reports are summarized in the Educational Department, on a record blank, for each student. The file of student records is open for inspection by anyone, including the students. In this way each student may see what record he has made and is encouraged to correct any undesirable traits he may possess.

Though the Training Course has been arranged to meet the needs of men graduating from college, it is equally adaptable for men who have satisfactory qualifications, especially for sales work, even though they may not have had the advantages of college training.

The visitor's first glimpse of Lake George, when approaching by motor or railroad train. The lake is 40 miles in length and contains hundreds of beautiful islands.

Fort William Henry Hotel, Headquarters of 17th Annual Convention, situated on shore of Lake George facing up the lake between high rows of imposing mountains.

View of Lake George to the north as seen from the verandas of the hotel.

The Formal Garden and Pergola with adjoining buildings, in which are housed a
gift shop, tea-room, and sun parlors.

SOCIETY AFFAIRS

CONVENTION NOTES

The Seventeenth Annual Convention of the Society will be held at the Fort William Henry Hotel, Lake George, N. Y., September 24, 25, 26, 27 and 28, 1923.

GENERAL CONVENTION COMMITTEE

W. D'A. Ryan, *Chairman*

H. W. Peck, *Vice-Chairman*

H. E. Mahan, *Secretary*

Alexander Anderson	G. Bertram Regar
B. S. Beach	W. L. Robb
N. R. Birge	W. M. Skiff
W. T. Blackwell	J. L. Stair
S. H. Blake	C. P. Steinmetz
H. Calvert	G. H. Stickney
A. D. Cameron	D. B. Taylor
Julius Daniels	R. B. Thompson
E. Y. Davidson, Jr.	E. D. Tillson
S. E. Doane	C. D. Wagoner
F. H. Gale	H. F. Wallace
Samuel G. Hibben	F. H. Winkley
Preston S. Millar	L. A. S. Wood

SUB-COMMITTEES

Entertainment

A. D. Cameron, *Chairman*

N. R. Birge

S. H. Blake

C. A. B. Halvorson

Publicity

A. F. Dickerson, *Chairman*

B. S. Beach

R. C. Rodgers

C. D. Wagoner

Finance

F. H. Winkley, *Chairman*

Alexander Anderson

Attendance and Transportation

J. F. Anderson

W. T. Blackwell

H. Calvert

L. C. Conant

G. G. Cousins

Julius Daniels

C. G. Eichelberger

G. F. Evans

F. A. Gallagher, Jr.

K. W. Mackall

Preston S. Millar

C. C. Munroe

G. B. Regar

W. M. Skiff

Frank C. Taylor

E. D. Tillson

L. E. Voyer

PROGRAM

MONDAY, SEPTEMBER 24, 1923

10:00 A. M. to 12:30 P. M. . . . . :.Registration
2:00 P. M. to 5:00 P. M. . . . . General Session
Evening. . . . . Entertainment and Informal Dance

TUESDAY, SEPTEMBER 25, 1923

9:00 A. M. to 12:00 P. M. . . . . Papers Session
2:00 P. M. to 5:00 P. M. . . . . Papers Session
Evening. . . . . A Night of Light and Color

WEDNESDAY, SEPTEMBER 26, 1923

9:00 A. M. to 11:45 A. M. . . . . :. . . . . . Papers Session
12:30 P. M. . . . . Motor Trip and Lunch in Woods
Evening. . . . . :. . Papers Session

THURSDAY, SEPTEMBER 27, 1923

9:00 A. M. to 12:00 M. . . . . Papers Session
2:00 P. M. to 5:00 P. M. . . . . Papers Session
Evening. . . . . Banquet

FRIDAY, SEPTEMBER 28, 1923

9:00 A. M. . . . . Section Development Conference

No special entertainment program has been planned for Friday, September 28th, but opportunity will be afforded the delegates and guests to make inspection trips to Fort Ticonderoga, Saratoga Springs, Albany, Schenectady (including the General Electric Co. Works and the Illuminating Engineering Laboratory) and other points of interest.

1923 CONVENTION PAPERS PROGRAM

MONDAY AFTERNOON, SEPTEMBER 24, 2:00 P. M.

Address of Welcome, Hon. George R. Lunn
Response to Address of Welcome
President's Address, WARD HARRISON
Survey of the year's work—Report of General Secretary, SAMUEL G. HIBBEN
The Year's Progress in Illumination—Report of Committee on Progress, F. E. CADY, *Chairman*
Twelve Solutions of a Street Lighting Problem

TUESDAY MORNING, SEPTEMBER 25, 9:00 A. M.

Pageant Street Lighting, SAMUEL G. HIBBEN
The Relation of Illumination to Production, D. P. HESS and WARD HARRISON
Salient Features in Power Station Lighting, R. A. HOPKINS

TUESDAY AFTERNOON, SEPTEMBER 25, 12:30 P. M.

Boat Trip with Session of Committee Reports

Motor Vehicle Lighting Regulations—Committee on Motor Vehicle Lighting, CLAYTON H. SHARP, *Chairman*

Recent Developments in Nomenclature and Standards—Committee on Nomenclature and Standards, E. C. CRITTENDEN, *Chairman*

Progress of the Tentative Code of Luminaire Design—Committee to Co-operate with Fixture Manufacturers, M. LUCKIESH, *Chairman*

How to Make the I. E. S. a Truly National Body—Committee on New Sections and Chapters, D. MCFARLAN MOORE, *Chairman*

TUESDAY EVENING, SEPTEMBER 25, 7:45 P. M.

A Night of Light and Color

"Light"—The Designer, M. LUCKIESH

Outdoor Spectacular Lighting Display

WEDNESDAY MORNING, SEPTEMBER 26, 9:00 A. M.

Research Problems—Report of Committee on Research, E. F. NICHOLS, *Chairman*

Visibility of Radiant Energy, K. S. GIBSON and E. P. T. TYNDALL

Some Experiments on the Speed of Vision, PERCY W. COBB

The Colorimetry and Photometry of Daylight and Incandescent Illuminants, I. G. PRIEST

Further Studies of the Effect of Composition of Light on Important Ocular Functions, C. E. FERREE and G. RAND

WEDNESDAY EVENING, SEPTEMBER 26, 7:30 P. M.

Artificial Illumination in the Iron and Steel Industry, W. H. RADAMACHER

Working with the Architect on Difficult Lighting Problems, Messrs. A. D. CURTIS and J. L. STAIR

Railway Car Lighting, G. E. HULSE

Depreciation of Lighting Equipment due to Dust and Dirt, E. A. ANDERSON

THURSDAY MORNING, SEPTEMBER 27, 9:00 A. M.

Determination of Daylight Intensity at a Window Opening, Report of Committee on Sky Brightness, H. H. KIMBALL, *Chairman*

Daylighting from Windows, H. H. HIGBIE and G. W. YOUNGLOVE

Some Principles Governing Proper Utilization of the Light of Day in Roof Fenestration, W. S. BROWN

Lighting for School Buildings—Preliminary Draft of Revised Code, Report of Committee on Lighting Legislation, L. B. MARKS, *Chairman,* and Sub-Committee on School Lighting, M. LUCKIESH, *Chairman*

THURSDAY AFTERNOON, SEPTEMBER 27, 2:00 P. M.

Colored Light, by M. LUCKIESH and A. H. TAYLOR

Production and Growth in Plants under Artificial Illumination, R. B. HARVEY

The Response of Plants to Artificial Lighting, HUGH FINDLAY

Unit Costs of Industrial Lighting, DAVIS H. TUCK

Testing Colored Material for Fastness to Light, H. S. THAYER

FRIDAY MORNING, SEPTEMBER 28, 9:00 A. M.

There will be held a Section Development Conference for Officers and Committee Chairmen of Sections and Chapters and all others interested.

ENTERTAINMENT

The entertainment features of the convention will include automobile trips to points of historic interest, a boat ride on Lake George, and a motor trip to Tripp Lake Lodge with luncheon in the woods.

For those who play golf, tournaments have been arranged at a time which will n ot conflict with the regular business sessions. Cards, teas, motor-boat trips, and a musical for the ladies are also included in the program. The banquet will be held Thursday night, to which all delegates to the convention are invited.

The Entertainment Committee has planned several novel features which will be in the nature of surprises.

The advanced registration indicates that this will probably be the largest convention in the history of the Society. Inasmuch as the committee is making special arrangements for the ladies, it is hoped that delegates will bring their wives where possible.

CONVENTION BRIEFS

Tuesday evening has been set aside as "A Night of Light and Color."

Detailed plans for special lighting have been completed and promise to surpass in spectacular effect and all around interest anything yet attempted by the Society. A battery of great searchlights forming a scintillator of moving colored beams, will be the big event of the Convention. This scintillator will be located on the wharf in front of the Fort William Henry Hotel, and will send a multitude of powerful rays into the sky in a fan-like formation. By means of colored slides and a well drilled corps of operators, the beams will be constantly changed in colors, one blending into another to give a startling and beautiful rainbow effect.

An elaborate program of fireworks is another feature, a high point of which will be the explosion of the largest bomb used in such displays. A jeweled emblem of the Society, illuminated in various colors by floodlights is another thing of interest on which the Committee is working. This will be placed on the ground near the hotel.

There should not be a dull moment for the ladies as automobile rides, dancing, golf contests, teas, motor-boat rides, musicales, bridge tournaments and other features have been prepared for their entertainment.

The Philadelphia Section wishes to formally advise all other Sections that it intends to take the attendance prize at the national convention at Lake George, N. Y. The Cleveland Chapter won the prize, a silver-band gavel, last year and have sent word that it will be retained by them this year. The New York and New England Sections promise large delegations, and also the Chicago Section will be well represented. Canada will send a number of delegates from the Toronto Chapter.

During the evenings of late September at Lake George light-weight overcoats and wraps will be needed.

SECTION ACTIVITIES

CLEVELAND CHAPTER

The following men have been elected as officers of the Cleveland Chapter for the coming year:

Chairman, Prof. H. B. Dates; Secretary, Robert A. Fulton; Chairman Papers Committee, E. W. Commery; Board of Managers, H. L. Wright, C. L. Dows, J. W. Beam, E. W. Commery, P. C. Pfenning.

NEWS ITEMS

At the annual convention of the Nation Council of Lighting Fixture Manufacturers, held in Buffalo, June 26 to 28, Mr. Herman Plaut, of New York, was elected president for the ensuing year.

Capt. D. W. Blakeslee, Electrical Engineer, Jones & Laughlin Steel Corporation, Pittsburgh, Pa., married Miss Margaret K. Steel of Pittsburgh on June first.

Mr. Frederick J. McGuire, of the Department of Water Supply, Gas, and Electricity, New York City, has again planned, in cooperation with the Department of Education, to continue this year classes of instruction in "Practical Artificial Lighting" in the evening trade schools in Brooklyn and in New York.

GENERAL OFFICE NOTES

A new Membership List was published during the summer, which is a revision of the 1919 list. Reply postal cards were sent to every member and only a small percentage were returned to the General Office. Care was taken to make the new list as accurate as possible. A number of errors have been reported, and it is requested that members advise the General Secretary immediately of any change of address, or other items that should be corrected.

An errata page will be included in a future issue of the TRANSACTIONS.

TRANSACTIONS for October, November and December, 1922, and January, February, March, April and May, 1923, are out of print. Please advise the General Secretary of any of these issues for sale and price will be quoted.

COUNCIL NOTES

ITEMS OF INTEREST

At the meeting of the Executive Committee on July 27, 1923 the following were elected to membership:

Two Members

HATCH, B. E., Westinghouse Elec. & Mfg. Co., 1442 Widener Bldg., Philadelphia, Pa.

SAVAGE, ARTHUR H., A. H. Savage Co., 914 Pioneer Bldg., St. Paul, Minn.

Eighteen Associate Members

ALEXANDER, GEORGE L., International General Electric Co., Schenectady, N. Y.

BELL, HOWARD H., General Electric Co., Schenectady, N. Y.

BOTSFORD, C. J., Westinghouse Lamp Co., 165 Broadway, New York, N. Y.

FUCHS, THEODORE, JR., Edison Lamp Works, Harrison, N. J.

GAWAN, LOUIS B., Utah Power & Light Co., 132 Main St., Salt Lake City, Utah.

GOTTSCHE, A. L., Westinghouse Elec. & Mfg. Co., George Cutter Works, South Bend, Ind.

HILL, GEORGE A., National X-Ray Reflector Co., 235 W. Jackson Blvd., Chicago, Ill.

HOWARD, HAROLD W., General Electric Co., Schenectady, N. Y.

KINNEY, RAYMOND C., Western Electric Co., Hawthorn Plant, Dept. 6723, Chicago, Ill.

LABELLE, PHILIP R., Shawinigan Water & Power Co., 621 Power Bldg., Montreal, Quebec, Canada.

NORRIS, B. H., General Electric Co., Schenectady, N. Y.

PROFFATT, CHARLES P., Universal Elec. Co., 701 Asbury Ave, Ocean City, N. J.

SIMONS, S. A., Save Electric Corp., 254-36th St., Brooklyn, N. Y.

SWEENEY, GEORGE J., United Elec. Light & Power Co., 514 W. 147th St., New York, N. Y.

WERNERT, A. L., Texas Central Power Co., 1st Nat'l Bank Bldg., San Antonio, Tex.

WILSON, EDWIN S., Philadelphia Electric Co., 3100 Kensington Ave., Philadelphia, Pa.

WINKLER, CLEM R., Arkansas Central Power Co., 115 W. 4th St., Little Rock, Ark.

ZIEME, HARRY, Penn Public Service Corp., 222 Levergood St., Johnstown, Pa.

Two Sustaining Members

Frink, I. P., Inc., 24th St. & 10th Ave., New York, N. Y.
William H. Spencer, Official Representative.

Western Electric Co., 195 Broadway, New York, N. Y.
E. J. Dailey, Jr., Official Representative.

The General Secretary reported the deaths of two Associate Members: L. H. Plaisted, Columbus Railway Power & Light Co., Columbus, Ohio; C. A. Strong, 79 Milk Street, Boston, Mass.

At the meeting of the Executive Committee on August 30, 1923, the following were elected to membership:

Two Members

Bergman, Axel G., Ordnance Engineering Corp., 170 Broadway, New York.

Wible, Harvey M., Westinghouse Elec. & Mfg. Co., George Cutter Works, South Bend, Ind.

Eleven Associate Members

Allen, Francis P., Elec. Light & Power Co., of Abington, 64 Charles Street, N Abington, Mass.

Bremmer, R. C., Pacific States Elec. Co., 61 North 5th St., Portland, Ore.

Dudley, M. E., M. E. Dudley Co., 1044 Garfield St., Lincoln, Neb.

Glass, David Hasler, Jr., Buick Motor Co., 604 Garland St., Flint, Mich.

Gowdy, Robert Clyde, University of Cincinnati, Cincinnati, Ohio.

Johnson, Wilber M., National Lamp Works of G. E. Co., Nela Park, Cleveland, Ohio.

Kent, Charles N., 12 East 46th St., New York, N. Y.

Smith, Harold C., Catton Neill & Co., Fort St. & Beretania, Honolulu, Hawaii.

Thompson, Willard W., Hixon Electric Co., 308 Dover St., Boston, Mass.

Tracy, Charles R., Penn. & Ohio Power & Light Co., P. O. Box 58, Youngstown, Ohio.

Wright, George Ellery, University of Illinois, Urbana, Ill.

One Transfer to Full Membership

Alden, Walter A., Westinghouse Elec. & Mfg. Co., 420 S. San Pedro St., Los Angeles, Calif.

One Transfer to Associate Membership

Nodell, W. L., 1405 Eighth Ave., Brooklyn, N. Y.

ILLUMINATION INDEX

PREPARED BY THE COMMITTEE ON PROGRESS

An INDEX OF REFERENCES to books, papers, editorials, news and abstracts on illuminating engineering and allied subjects. This index is arranged alphabetically according to the names of the reference publications. The references are then given in order of the date of publication. Important references not appearing herein should be called to the attention of the Illuminating Engineering Society, 29 W. 39th St., New York, N. Y.

*Not previously reviewed.

*Not previously reviewed.

TRANSACTIONS
OF THE
ILLUMINATING ENGINEERING SOCIETY

| Vol. XVIII | October, 1923 | No. 8 |

Service

THE Accomplishments of the Illuminating Engineering Society, since its inception seventeen years ago, have been noteworthy in its direction of advancing the knowledge of the science and art of Illuminating Engineering. That its endeavors in this field have been recognized is evidenced by its growth and continuous success. It should not, however, rest upon these achievements alone, but these activities should be accompanied by an earnest effort to be also of material value in creating agencies through which the information and data acquired may be sent broadcast to the thousands of interested individuals in the lighting industry spread out over the country.

The Constitution of the Society provides a means by which this may be accomplished. In fact, it offers two ways: First—the society publication, TRANSACTIONS, an excellent means of spreading the knowledge, which is a reference and permanent record of our activities; Second—and of extreme importance, through the medium of meetings of the Sections and Chapters. Not enough stress can be laid upon the latter method. As these meetings increase in importance, attendance and popularity, the service to the industry increases in like proportion. Yet there is not a sufficient number of these Sections and Chapters, as they should be established in all of our larger cities, and it is for this purpose solely that great emphasis is now being placed upon this activity. The importance of this issue must not be underestimated for the proper functioning of the Society as a whole depends principally upon its Sections and Chapters and through their success only can our organization establish a very definite standing in the lighting industry and increase our opportunities to be of actual service to all branches.

It is primarily our function to disseminate useful information that will aid in bringing about improved lighting conditions. This can more readily be accomplished by a closer cooperation of the various Sections, Chapters and Local Representatives throughout the country. The progress made by us has a material bearing upon the atmosphere of a community, and it is our great responsibility, therefore, to see that all channels through which our information may be disseminated are taken advantage of. And, by close communion of these agencies we can more readily accomplish our object of rendering a real service to the industry. Progression, not retrogression should be our constant aim, which might be reduced to a formula:—

$$(S+C)^n = (s+a+r) \backsimeq \text{TRUE SERVICE}$$

transcribed meaning, *Sections, plus Chapters, to the nth power, should yield Suggestions, plus Action, plus Results, which again reduced, yields TRUE SERVICE.*

Foreword on Proposed Activities of Committees

BY CLARENCE L. LAW

AT the suggestion of the Committee on Editing and Publication a communication was sent to each Committee Chairman, asking for a brief statement on the proposed activities of their Committees for the coming administrative year.

The response to this request was very gratifying and the reports that follow will give some idea of the proposed work which the Committees will undertake.

The functions of the Board of Examiners and the Finance Committee are outlined in the Constitution and therefore, no statement from these Chairmen has been made.

It is hoped that all Committee Chairmen will feel free to attend any or all of the Council Meetings, held each month during the year, so that they may have the opportunity of reporting personally on the progress of their Committee as well as to form some idea as to how the Council functions.

Council Meetings are held the second Thursday of each month, at the headquarters of the Society, unless otherwise arranged. Committee Chairmen, when attending these Council Meetings, should be prepared to report briefly, as the time at the disposal of the members is limited.

Committee on Papers

The Committee on Papers is planning to render as much service as possible to Sections and Chapters. As a step in this direction we are now at work compiling a list of available sources of speakers and papers.

Several of the manufacturers, laboratories and other organizations have men who frequently travel about the country. Many of these men are recognized leaders in the development of the art of illumination. It is believed that in general they would be glad to address sections or chapters as opportunities arise.

The Committee on Papers will send each Section, Chapter and Local Representative a list of men who are willing to volunteer for this activity, the subjects on which they are best qualified to talk, and will further endeavor to develop some scheme whereby the official representative is kept informed of travelling itineraries, etc.

To expedite publication of the TRANSACTIONS it is planned to more closely coordinate the work of the general committee with the sectional committees. When papers of interest to the Society as a whole are presented before sections they should appear in the TRANSACTIONS. It has been somewhat difficult in the past for the office to obtain copies for consideration. This is a matter calling for closer cooperation on the part of Section Chairmen.

It is hoped to make the program at the annual convention as far as possible of direct practical value to the rank and file of the membership. It is believed that a few papers of the highest grade are preferable to a long program treating many subjects. To this end the Committee will welcome suggestions as to topics that are believed to be timely and of the greatest value to the greatest number. With these suggestions in hand your Committee will endeavor to obtain a paper from the best authority on that particular phase of the art.

It seems desirable to have more discussions of papers than has sometimes prevailed. It is in discussion that the various sides of a given question are brought forth. For this to take place it is necessary that the papers be available for study by the members in advance of the convention. The attempt will therefore be made to carry out this and a special effort to secure discussion from those qualified.

The Committee bespeaks your interest, cooperation and frank expression of opinion on its work.

ALVIN L. POWELL, *Chairman*

Committee on Editing and Publication

The TRANSACTIONS of the Society are perhaps the biggest single effort which is undertaken by the Society. The direct expense of printing and publication is about one third of the Society's expenses and if the General Office expenses for labor, printing and postage in this connection is allocated to the proper cause, the expense of the Society for the Transactions will probably be found to be about half of its income.

Realizing this and that the TRANSACTIONS are the medium for reaching all of the membership (it is probable that less than one half of the membership is affiliated closely with the sectional or chapter activities) it is the Committee's hope that the TRANSACTIONS may be made, even more than in the past, a medium for the dissemination of news of the industry in addition to its function as the archives of the Society's deliberations.

The attempt will be made this year in cooperation with the Committee on Papers to issue the TRANSACTIONS on a definite schedule. The membership in general can help this program in two ways.

First, by sending to the General Office when occasion permits, notes on items of interest to lighting men, clippings from other papers for our Reflections section and news of the various unusual installations which are being completed from time to time, or notes about men of the industry;

Second, by correcting and returning promptly manuscripts for discussion. A copy of the TRANSACTIONS is in a sense like a picture puzzle, almost useless unless all the parts are available and in their proper positions.

The Committee bespeaks your active cooperation in these two particulars and promises that as this is effected, it will be reflected in a corresponding improvement in the numbers of the TRANSACTIONS which come to you.

NORMAN D. MACDONALD, *Chairman*

Committee on Lighting Legislation.

During the year 1923-24 the Committee will be occupied largely with a campaign to bring the revised Code of Lighting School Buildings to the attention of legislative bodies, school boards, architects, and others interested in enactments, rules, and regulations in regard to school buildings.

It is planned to issue a simplified version of this Code, much abbreviated, for the use of those who find the present Code too technical.

The campaign of education to be conducted in connection with the introduction of the School Lighting Code and the adoption of this Code by the states will be much the same as the campaign that was conducted in the case of the Factory Lighting Code. The work on both of these Codes will be carried on along similar lines under the following headings:

(a) Preparation of a guide to inspectors and cooperation with states, municipalities and others in interpreting the rules, regulations and provisions of the Codes.

(b) Lectures by representatives of the Committee, to school authorities and factory representatives.

(c) Service on international, national, state and municipal committees.

(d) Representation at public hearings on the Codes.

(e) Circularization to legislative bodies in those states that have not put the Codes into effect.

The Committee plans to continue the campaign to further the general recognition of the importance of light as a factor in accident prevention, and in particular to bring about the inclusion of the item of lighting as a prominent clause in all industrial accident insurance policies.

LOUIS B. MARKS, Chairman.

Committee on Membership

The splendid work of the Committee on Membership aided by the members in general resulted in the election of 289 new members during the past year,—an increase of 58.5 per cent over the members elected during the preceding year.

The aim of the Society, namely, "The dissemination of knowledge relating to the theory and practice of Illuminating Engineering" can best be accomplished by an ever-increasing membership in the Society.

A knowledge of the principles of Illuminating Engineering brings about the realization of the great benefit it engenders to mankind and the creative force it is to more and better business.

The Committee on Membership has signified its willingness to serve again in that capacity, and the Society may rest assured that they will do everything in their power to surpass the splendid record just made. It seems only fair to expect that the members in general will do their part and the procuring of one new member by each member of the Society would mean the doubling of its membership and would evidence on their part a serious interest in the Society's welfare.

The Chairman sincerely hopes for this cooperation, and will be very glad to give any information or aid to any member having a prospective member in view. He further desires to express his appreciation to all those who aided in the past year's activities.

G. BERTRAM REGAR, *Chairman.*

Committee on Motor Vehicle Lighting.

The Committee on Motor Vehicle Lighting feeling that the matter of Specifications for Laboratory Tests of Headlighting Appliances has reached a stage where for the time being no further revisions are called for, proposes during the coming year to devote itself chiefly to a consideration of methods whereby the character of headlighting which is contemplated in the testing specifications may be generally attained on the road. The Committee recognizes that the use of approved equipment on headlights does not in any way insure proper road illumination and the absence of glare. It believes that while in proportion to the number of cars, lighting conditions on the road have been greatly improved, yet they are still far from being satisfactory, and wishes to bend its efforts chiefly toward assisting to the best of its ability in improving these road conditions. A number of propositions are before it for consideration with respect to education of the public and to a more thorough and intelligent enforcement of the laws. Just what success the Committee will have in these endeavors is of course problematical.

While turning its chief endeavors in the above direction, the Committee does not intend to fail to give attention to any points in which its previous work can be improved, and will welcome suggestions toward this end as well as toward the perhaps broader problem of improved road conditions.

CLAYTON H. SHARP, *Chairman.*

Committee on Nomenclature and Standards

The principal duty of the Committee on Nomenclature and Standards during the year will be to complete the revision of the booklet on Illuminating Engineering Nomenclature and Photometric Standards so that the revised edition can be adopted as American standard practice. It will then be presented for consideration at the 1924 meeting of the International Commission on Illumination in the hope that further progress may be made in securing international agreement on names and definitions.

The Society has had success in combining the efforts of men whose activities lie along widely divergent lines, such as professional practice, technical and scientific research, manufacture, sales, and engineering of the most practical sort. Successful cooperation of such groups is to some extent dependent upon a common technical language which facilitates the interchange of ideas. The work of this committee is directed primarily toward the development of such a common language so that we may all understand each other.

Proposed changes in the former standards have been set forth in the Committee's Reports for 1922 and 1923, and the Committee would like to have comments or suggestions from all classes of members before final action is taken on the changes.

Some Sections and Chapters have found it worth while to have talks on the elementary principles of illumination, including methods of design, calculation, and measurement. Discussion of these subjects is likely to bring out points in which the standard nomenclature appears to be lacking in clearness or precision, and if so the Committee will be glad to hear about them. If the definitions are not clear or the terms not usable, now is the time to improve them. Positive and constructive proposals are of course most helpful, but criticism of weak points will be welcome, and the Committee will do its best to remedy deficiencies which may be pointed out.

E. C. CRITTENDEN, *Chairman.*

Committee on Progress

It is the intention of the Committee to continue the practice of previous years in preparing periodically for inclusion in the TRANSACTIONS of the Society an Illumination Index, covering allarticles on the subject to be found in periodicals available to the Committee.

The usual procedure will be followed in preparing the annual report of the Committee, which at the last Convention was known as the "Year's Progress in Illumination", the material for which is obtained from articles published in the technical and scientific press and from engineers in charge of street lighting of various large cities.

It is the intention to include in next year's report among the references under the appropriate headings the various subjects which have appeared in the Transactions or have been presented before the Society. It is hoped in this way to make the report a still more complete record of activities in illuminating engineering.

FRANCIS E. CADY, *Chairman*

Committee on Research

The recent Convention Report of this Committee looks forward as well as backward, and there is nothing which I·could say by way of program of work for the coming year that would be different from the work outlined in that Report.

ERNEST F. NICHOLS, *Chairman*

Committee to Cooperate with Fixture Manufacturers

In 1922 a tentative Code of Luminaire Design was prepared by this Committee and published. It was submitted to many fixture manufacturers, designers, illuminating engineers, etc. Very little criticism has been received but from these a few minor changes are in contemplation.

It is the aim to add two more sections to this Code, one dealing with artistic fixtures, which embody proper lighting principles for various purposes, and the other, a section on the influence of installation on fixture design.

The Chairman and Secretary of this Committee have work under way in regard to the first of the proposed sections. When the work is done on both sections, it is the plan to incorporate it into the Code of Luminaire Design and to publish this complete for final distribution.

During past years members of this Committee have been individually active in spreading the gospel of proper fixtures and proper lighting principles among fixture designers, fixture manufacturers, fixture dealers, and glassware manufacturers. The Committee members collectively and individually have attended meetings of these various organizations including the Illuminating Glassware Guild. They have given lectures and have aided in various demonstrations.

It is the plan to continue such activity. The Committee has been an effective means of contact with fixture manufacturers and the Illuminating Glassware Guild as well as others. The work that it is trying to do is very important but there are many difficulties in the way. Fortunately, these difficulties are becoming less formidable each year.

M. LUCKIESH, *Chairman*

Committee on Sky Brightness

The program for the year 1923-24 contemplates the inauguration of qualitative measurements of daylight, perhaps with the Rotatory Dispersion Colorimetric Photometer designed by Mr. I. G. Priest of the Committee.

The Committee also hopes to avail itself of daylight intensity records that the U. S. Public Health Service expects to obtain from photo-electric cells it has purchased for installation in Washington. These should give more detailed information respecting the extreme variations of daylight intensity with weather conditions than can be obtained from eye readings of a photometer.

There is also a probability that measurements of sky brightness may be made during balloon and airplane flights, and correlated with determinations of the dust content of the atmosphere.

Data are available for extending the charts published in the May 1923 number of the TRANSACTIONS, giving the intensity of daylight on horizontal and on vertical surfaces in the United States east of the Mississippi River at Latitude 42° North, under both clear and cloudy sky conditions, to other latitudes and regions in the United States. This will not be undertaken, however, unless the demand seems to justify the considerable labor involved.

The Committee would be glad to receive suggestions as to ways in which it may serve the interests of members of the Society.

H. H. KIMBALL, *Chairman*

REFLECTIONS

Stimulating Plants and Animals

EXPERIMENTS at Columbia University show that plant growth proceeds successfully under electric light. Flowers and vegetables subjected to several hours of artificial illumination outgrew those of the same planting which received only daylight.

Many professional poultry raisers use electric light night and morning to increase the egg production of their hens. The New Jersey experiment station at New Brunswick reported last month that hens stimulated by artificial light which kept them awake after sundown and caused their morning to begin before sunrise in winter laid at least ten eggs more a season that their sisters whose quarters lacked modern improvements. Most poultrymen who use artificial light in their henhouses think that the margin is larger and that keeping their hens awake pays good profits.

Scientists may see in this proof of the stimulating power of light, but the practical poultryman has a simpler explanation. Chickens have hearty appetites and are everlastingly scratching around for food. They go to sleep early in winter simply because they cannot hunt food in the dark. In the morning they are more hungry than sleepy, and the first crack of dawn finds them ready for breakfast. Hence, say the practical poultrymen who disregard science, hens that put in a long day at work in winter are better nourished and consequently lay more eggs that those working from sun to sun on short days.

In primitive days man slept from sun to sun because darkness kept him from doing anything else. Now eight hours is the standard sleeping night as well as the standard working day. But such an eminent scientist as Edison thinks that sleep is a good deal of a habit and greatly overdone.

Perhaps a common explanation of all these diverse phenomena may be found in the philosopher's saying that labor is the price of life. The tree labors in growth, the fowl labors in search of food, the scientist in search of facts. Each advance of knowledge narrows the gulf that separates the animal from the vegetable. The bean vine follows the sun round the bean pole as inevitably as the moth follows the candle round a room. Similarly light seems to stimulate the physical and chemical processes of the tree or plant. Plants, like man, work better in light than in darkness. Their roots reach out for more nourishment, their cells function more rapidly. Sleep, as the physiologists long ago noted, is a slowing down process. Artificial light keeps these Columbia plants awake and working.

But labor is not the whole of life. These experiments on light and growth will not be completed until attempts have been made to perpetuate the species through specimens that have been subjected to artificial light. Florists may get splendid specimens by this method; but the plants may be weakened in vitality in spite of their size and showiness.

With regard to human beings it has been suggested that physical recuperation is the least of sleep's benefits. Some psychologists hold that undue loss of sleep tends to bring on mental and nervous ailments, because the subconscious mind, which works while we sleep, has lacked time enough to sort, check up, erase or accept the impressions and rationalizations stored up by day. It is entirely possible that in plants, beasts and human beings alike too little sleep may seriously weaken the individual in ways not at once apparent. Whether it does so in the case of plants no doubt will soon be determined.—*Editorial, New York Herald, Oct. 29, 1923.*

PAPERS

ADDRESS OF THE PRESIDENT*

By Ward Harrison

For the subject of this address I would refer you to a much
quoted passage from the Constitution of the Society, a clause
which is reprinted on the face of every issue of our Transactions;
namely, "That the object of this Society shall be the advancement
of the theory and practice of illuminating engineering and the
dissemination of knowledge relating thereto". Even in the year
of the Society's formation it was recognized that there were two
distinct tasks before it; first, the reaching out after new facts and
new principles in illuminating engineering; and second, the dissemination of these and older truths to the industry and to the public.

At the present time, in fact, at least once *every* year, it seems
fitting that we should ask ourselves two questions: first, what at
present is our greatest need, *technically*, in the practice of our
profession; and second, wherein lies our greatest opportunity for
the dissemination of illuminating engineering information to the
public. No doubt, if a poll were taken there would be almost as
many answers to these questions as there are members of the
Society, but taking advantage of the opportunity which has been
afforded me I will attempt to state what my own answers would be.

PART I

In my opinion our greatest technical need today is for a better
working knowledge of *brightness* and *glare*, from both a quantitative and a qualitative standpoint. As the result of the general
adoption during the past ten years of such expressions as "foot-
candle", "lumen", and "coefficient of utilization", it is safe to say
that there are now several thousand persons in the United States
who have a very definite conception of the relation existing between the flux generated by a light source and the resultant foot-

*An Address presented before the Annual Convention of the Illuminating Engineering
Society, Lake George, N. Y., September 24-28, 1923.

The Illuminating Engineering Society is not responsible for the statements or opinions
advanced by contributors.

candles of illumination in a room. In other words, we think in lumens fairly well. But, how many are there who can say the same thing when the subject is that of brightness, the progeny of lumens and the factor with which finally we are all most vitally concerned; concerned with it on the working plane because there it is the thing which enables us to see; concerned with it in the case of the light sources for there it is the thing which in excess may readily defeat vision and even impair eyesight.

How little thought is really given to questions of brightness by practicing engineers is well exemplified by the general prescription of "about twice as much light for work on dark textiles as for work on light fabrics", whereas actual brightness measurements would indicate that 10 or 15 to 1 as nearer the proper ratio. Again, how many of us, if asked to criticize a lighting specification for any given interior can feel certain as to whether the luminaires are going to prove comfortable or glaring, unless, of course, we have had experience with an almost exactly similar installation before. Right at the present time we seem further than ever from any simple mathematical formula by which such questions can be answered.

It is fortunate that all of our more precise instruments for measuring illumination may be calibrated to read brightness as well, and this circumstance should be of substantial assistance in increasing our store of "installation data" on the subject. At the same time it would also prove very helpful if an instrument comparable in simplicity and cost to the foot-candle meter, or possibly a modification of this instrument itself, were developed with which brightness could be read directly with accuracy sufficient for practical purposes.

Strange as it may seem, glare is a subject to which but little attention has been given by those of our members who are primarily interested in scientific research. For example, there is at present apparently no data available from this group to answer even such a simple question as the following:—Which is the preferable of two light sources, (a) which has a uniform brightness throughout, and (b) which has a low brightness at the periphery, increasing to a higher value near the center, assuming that the candlepower of the units and their *average* brightness is the same in both cases?

Now your President lays no claim to originality in offering the suggestion that questions of brightness are entitled to greater consideration than is usually accorded them. This subject has been brought up for discussion on numerous occasions by members of this Society but unfortunately we have always stopped there; we have made little or no progress. We are prone to look at the whole subject as exceedingly complex, particularly so in contrast with the more usual calculations involving lumens and foot-candles. May it not be that in the face of a really serious effort this difficulty would prove to be only a fancied one arising simply from ignorance on our part?

One of the greatest helps in the solution of any problem is an exchange of thought between individuals and an exchange of thought presupposes a language—a nomenclature. I believe that our progress in the present matter is considerably impeded by the lack of a simple term in which to express brightness, one which will apply equally well to luminaires and to surfaces illuminated, and finally, one which will be quite readily comprehended by the man on the street; one encounters real difficulty in trying to explain millilambert to him.

It is generally accepted that whenever new fundamental units are established involving distance, weight or time, the C. G. S. System should be followed, and with this in view our Committee on Nomenclature and Standards supplied us some years ago with our fundamental definitions. Of these the lumen and the candle might be termed truly international units, for in their practical application at least, they involve neither a unit of length nor of weight; the C. G. S. unit of brightness then established was the lambert and likewise the lux, or meter-candle, was defined as our fundamental unit of illumination intensity. In practice, however, most Americans think in feet and inches and the only generally used unit of illumination intensity in this country today is the "Foot-Candle". What we need also for general use in this country is a unit of brightness based upon our accepted unit of intensity, the "Candle", and our common unit of length, the "Foot". A logical unit would be the brightness of a spherical source of one foot radius having a uniform intensity of 1 candle in all directions; or if you prefer, the brightness of a perfectly translucent sphere of one foot radius surrounding a source of 1 candle. Such a unit renders the computation of brightness values for many units a comparatively easy matter. For example, a spherical diffusing globe 8″ in diameter having an output of 80 per cent would, when equipped with a lamp of 150 spherical candlepower, have a mean brightness of 1080 units computed as follows:

$$\frac{150 \times 24^2 \times 0.8}{8^2} = 1080$$

This unit would also have the very practical advantage that it would represent the mean brightness of a surface of 100 per cent reflection factor when illuminated to an intensity of one foot-candle; that is, one lumen per square foot.

There is still one thing more to be said in favor of a unit of brightness of the dimensions stated above; namely, that it is of about the right size for practical use. With it the brightness of working surfaces, at least in interior lighting, will usually be found in the range of from 0.1 to 10, numbers that are easy to comprehend and when we speak of the brightness of light sources we will usually find ourselves up in the hundreds or thousands, figures which are far more certain to cause the layman to pause and consider, than would the same brightness expressed as a decimal fraction.

In the opinion of the writer there is much in the choice of the size of a unit, there is also much in the choice of the name. In the term "foot-candle" the precedent has been established of prefixing our non-metric units involving length with the term "foot" which at once shows to what system it belongs. The same plan followed in the case of the unit of brightness would forever eliminate any possibility of it being confused with lambert. If the term "foot" could be coupled, not with some abstract term, but rather with a short Anglo-Saxon word which conveyed to the listener some inkling of the meaning of the whole word, the resulting combination should prove most desirable. An expression which seems to come very near to fulfilling these requirements is "foot-bright", and it is my wish to submit this term for the consideration of the Committee on Nomenclature and Standards.

In weighing the relative advantages and disadvantages of this and other terms the fact must not be lost sight of that we are an Illuminating Engineering Society, not a Society of Illuminating Engineers; that to be of value our findings must be capable of dissemination; that success in influencing the design of any considerable proportion of the lighting installations in this country depends upon making our technique seem extremely simple, not upon making it complex. My strongest recommendation for "foot-bright" is that it sounds almost common-place.

PART II

The second division of this address has reference to our obligation and our opportunity for the dissemination of knowledge of the theory and practice of illuminating engineering and how we may best go about this task. Since our last convention the number of chapters of this Society has increased from three to seven; in other words, the number of cities in which local branches are established is now eleven. Do not these facts suggest an answer to our problem? Is it not true that the most effective work that can be undertaken by the present local branches and by additional ones as they are organized, is to concentrate upon the dissemination of lighting information?

Plans for monthly meetings and for lectures should be arranged with a view to interesting successively engineers, architects, fixture manufacturers, and similar groups. The co-operation of city and suburban school authorities we should seek perhaps, most of all, for through them can be secured, not only lighting in the schools which will cease to be a menace to the eyesight of our own and our neighbors' children, but also instruction in the principles of lighting, which will better fit those of the next generation to know good lighting for themselves. It may be recalled I spoke briefly on this phase of the subject at the last convention. This week we have for your consideration the draft of a chapter on lighting which has been prepared at the request of the Council as a suggestion to writers of text books on High School physics and also as a guide to instructors on this subject. The monograph will be open for discussion at the Friday meeting, and the thoughtful criticism of the membership of this Society is sincerely requested. When this pamphlet is printed in final form, it will become the very fitting task of our sections and chapters to secure its general adoption in their individual territories.

One may say that there is nothing new in these proposals for section and chapter activity; that some of our branches have for years arranged joint meetings and programs which would interest those outside of their own membership. This is true, but for the Society as a whole, the effort has not been organized. One section has not had the benefit of the experience of others, and never has a complete program for the year's work been outlined for them by this Society. Likewise, they have never been asked to show the tangible results of a year's effort along such lines. To those who are fortunate enough to be able to attend our annual conventions these four-day periods represent perhaps the high point of the Society's achievement; in any event, it is a most enjoyable one. But we must not lose sight of the fact that the real strength of this Society lies in its sections and chapters, and that is why I believe that our greatest potentiality for progress lies in the plans which will be developed in the Section and Chapter Conference which is scheduled for Friday morning next. On behalf of the Council I bespeak your attendance at this Conference and your support.

*　*　*　*　*　*　*　*　*　*　*　*

REPORT OF COMMITTEE ON PRESIDENTIAL ADDRESS

Your Committee on Presidential Address feels that President Harrison has rounded out a very successful administration with an address which constitutes a valuable contribution to the art. He points out the necessity of further elucidation of the subject of brightness and glare. With reference to brightness, he indicates the need for the adoption of some simple and comprehensible unit and terminology and makes a definite suggestion to this end.

Your Committee is in hearty sympathy with the President's view and without assuming to pass on the merit of his specific suggestions recommends that the Society's Committee, within whose scope this question lies, give early and serious consideration to the problem to which the President has so pertinently directed our attention.

P. S. Millar
G. H. Stickney
L. B. Marks, Chairman

The report was unanimously accepted by the convention.

HOW CAN THE I. E. S. BE MADE MORE TRULY
A NATIONAL BODY*

BY D. McFARLAN MOORE**

About 74 per cent of the membership of the Illuminating Engineering Society resides east of Chicago. There are nine States that have no members at all and twenty-seven States that have less than one dozen members and yet there are in the United States millions of people interested in, or associated in a business way with light. Various membership committees have carefully studied methods which would seem to follow the path of least resistance towards the goal of a very great increase in the total membership and also have it well distributed. However, the gain altho steady, has not been rapid. The deterrents to joining our Society are of many varieties but finance is always important. Council has approved of the formation of Chapters on such liberal terms that it seems to the Committee on New Sections and Chapters that the formation of New Chapters should be of much value to the National Society. It permits of our interesting an unlimited number of people in the objects of the Society, yet at a nominal expense to them. Meetings of Chapters not only develop interest in the minds of many who will later become full members of the National Society but also greatly stimulate the enthusiasm of those who are already members.

A number of New Chapters have recently been formed and seem to be well worth while and it is hoped that more assistance will be given towards the formation of such Chapters so that it can be said that each State in the union has at least one Chapter of the Illuminating Engineering Society. To assist those who are inclined to be helpful in this important matter the following information will prove of great assistance—

*A Paper presented before the Annual Convention of the Illuminating Engineering Society, Lake George, N. Y., September 24-28, 1923.

**Chairman, Committee on New Sections and Chapters, I. E. S.

The Illuminating Engineering Society is not responsible for the statements or opinions advanced by contributors.

If there are a large number of industries in a specific vicinity, that are either directly or indirectly interested in illuminating engineering, the formation of a local Chapter will be more easily accomplished, and also more good will result from its formation. The most essential requirement and the one most difficult of fulfillment is that ten regular members of the Illuminating Engineering Society, (they can be either full members or associate members) must file a petition approximately as follows—

"We the undersigned herewith petition the Council of the Illuminating Engineering Society to authorize the formation of a local Chapter of the Society in———— to cover the vicinity of————."

And send it to the Chairman of the Committee on New Sections and Chapters who in turn will endeavor to obtain the formal approval of Council and then inform the petitioners that all is ready for the organization meeting. The General Office in New York is requested to make a list of all the names of the members of the Society living in the vicinity where it is proposed to form a New Chapter. If less than ten are thus obtainable an effort should be made to interest a sufficient number of new members so that the petition will be in proper form for presentation to Council. Those interested in obtaining new members are reminded that the annual dues for full members are $15.00 and for associate members $7.50 and that an initiation fee of $2.50 is required for either full members or associate members. A proposed constitution for a new Chapter is also immediately forwarded to those interested but a statement is made that the Chapter dues can be of any amount those locally interested desire; one dollar for example. Therefore, members of such a Chapter only, can claim association with the Illuminating Engineering Society, yet as indicated pay only a very small amount per year. However, they will not receive copies of the Society Transactions or be eligible to its offices, etc. Such members of a Chapter often later become full members of the National organization. Our Society has now approved of seven Chapters as follows: Toronto, Cleveland, San Francisco Bay Cities, Columbus, Northern New Jersey, Michigan and Los Angeles and their increase during the last year together with the lively character of their meetings indicate that a definite effort should be made to very greatly increase this line of activity as already indicated. The unoccupied fields are so enormous that it is fitting to say that a start has only been made. Many of the definite localities where there seems to be no good reason why a Chapter should not be formed should be selected and an earnest effort made in every instance to promptly bring a Chapter into being and thereby not only greatly benefit our National Organization but also fulfill the primary object of our organization, viz., to further the interests of better lighting.

SECTION DEVELOPMENT CONFERENCE*

CHAIRMAN CLARENCE L. LAW: As you know, the object of the meeting this morning is to have an informal discussion on Section and Chapter development. We want to see what we can do to aid Sections, Chapters and Local Representatives throughout the country, in order to promote the work of the Society.

If there are any Section or Chapter Chairmen or Secretaries here will they please stand up and announce their names and Sections, so that we may all know each other?

The following Section and Chapter officers and Local Representatives were present:

W. V. Batson, Chairman, New England Section.

Julius Daniels, Secretary, New England Section.

W. S. Fitch, Board of Managers, New England Section.

L. J. Lewinson, Chairman-elect, New York Section.

S. K. Barrett, Chairman Papers Committee, New York Section.

C. L. Dows, Chairman, Cleveland Chapter.

J. M. Ketch, Chairman, Michigan Chapter.

W. H. Woods, Toronto Chapter.

George C. Cousins, Toronto Chapter.

Ellsworth Francisco, Northern New Jersey Chapter.

Charles Gallo, Northern New Jersey Chapter.

E. C. Crittenden, Local Representative, Washington, D. C.

F. L. Loomis, Pittsburgh, Pa.

J. D. Lee, Trenton, N. J.

CHAIRMAN LAW: We have not set a program this morning and I thought we could make the discussions very informal, therefore, I have not any definite plan with regard to this meeting. I would like to call on President Harrison to open the Conference.

PRESIDENT HARRISON: It seems to me that if we were to analyze this Society; thinking of it as divided into component parts, and were to try to rate the various parts, we would probably

*Held at Lake George, N. Y., during the Annual Convention of the Illuminating Engineering Society, September 28, 1923.

The Illuminating Engineering Society is not responsible for the statements or opinions advanced by contributors.

732

first think of our Conventions and on a scale of one hundred,—rate them very high. The enthusiasm is good, the attendance is good, papers are good, and we accomplish, I believe, a very considerable amount; that is judging from the opinions of many people with whom I talk.

Next we would probably consider the plan of operation; the Constitution, the arrangement of executive authority, etc., as well thought out. In other words, the business of the Society runs smoothly, and we would, therefore, give the plan of operation a high rating.

Likewise, I think you would give a high rating to the work of the General Committees; Committees like those on Lighting Legislation, Motor Vehicle Lighting, Nomenclature and Standards, and the other general committees.

In the minds of the majority of the members all of these things would probably be found to rate higher than most of the Section Activities. In fact, that is why the suggestion was made that this conference be called to exchange ideas among the Section Officers. What one has found to be successful in his locality the others can try with the result that it should be easier to plan a successful year's program. I hope some steps will be taken along that line today. Perhaps we can make an outline of an entire series of meetings to offer to a Section, or Chapter, and they can accept as much of it as they wish.

Another point that I believe we ought all to get our minds together on is that if we want this Society to grow, and to cover a wider territory, we must put it on a broader basis—and such basis, it seems to me, is the Section and the Chapter. It has been gratifying, indeed, to see the way new Chapters have sprung up during the past year. Now that we have these Chapters we should give them something very definite to do.

CHAIRMAN LAW: Thank you Mr. Harrison. Mr. Moore is Chairman of our Committee on New Sections and Chapters. Will you give us about five minutes on your work of last year, Mr. Moore?

D. McFARLAN MOORE: I don't know whether talking about the work of last year will be the thing of most interest. Our new President who has made a success of his administration already seems to be impressed with the importance of the development of

Chapters, largely, then trusts that the Sections will follow the Chapters.

For several years, I was Chairman of the Membership Committee, and it must be now three or four years ago that I began to realize, as Chairman of the Membership Committee, the utter lack of geographic area that was covered by the Society as it then existed.

The members of the Council know that I started in three or four years ago sort of ripping things up the back, as regards the claim that we were a National Society, when it isn't far from the truth to say we are about one-quarter of what we ought to be, and what I want to do is to see us rise to the point of four-quarters. That is, we only represent about a quarter of the country in a way that indicates that it is perfectly reasonable for us to be four times larger and do four times as much good. That is what I want to get at. Let us avoid this thing of traveling upward perhaps under false colors, or something of that kind.

When we investigated the ways in which this could be done, it looked as though the chief resistance was in the line of Chapters and because of my making some remarks of that kind, the job was thrown on me and we went ahead and organized some new Chapters.

The problem can be stated very specifically. Any man here from a managerial standpoint would say, if he was laying out a new business, "why that is an easy thing to do. All you need to do is just pick out the portions of the country where there ought to be a Chapter, where there are so many thousands of people who are interested in lighting, and yet don't know that the Illuminating Engineering Society exists." It's a fact, I think, as I pointed out the other day, there are a few states that haven't got a member—and there are 36 states that have less than a dozen members.

Now let us go out and pick out these areas and plant seed of the very best variety, not taking men merely because the general managers of big corporations say to a dozen men "Here, join this Society". That is mushroom growth. We don't want that. We want men who are deeply interested in the subject of lighting and will stick. Such men will gradually drift into our conventions, and there are many men here who are just going away from this convention inspired with a new feeling, and they will hereafter work for the good of the cause.

Mr. Moore presented his paper at this point. See page 729.

DISCUSSION

SECRETARY S. G. HIBBEN: Mr. President, it might be well to have a thorough understanding of the foundation of this discussion, because some of us are not acquainted with the Section and Chapter divisions.

You saw the map with the colored lamps indicating the positions of these Sections, Chapters and Local Representatives, and in order that you may refresh your minds on the subject I will read some statistics. Of the four Sections we have a membership as follows:

New York...............430
Philadelphia............216
Chicago................164
New England Section....143

In the Chapters formed this past year and including the others previously existing—in other words, in the seven Chapters we now have—we have a Cleveland membership of 74; Northern New Jersey, 103; Columbus, 25; Los Angeles, 20; Michigan, which is the Detroit Chapter, 34; San Francisco, 40; Toronto, 31. That is the registered membership, but does not represent the persons reached by the chapters.

It is rather interesting to notice the Northern New Jersey Chapter with a membership of 103, and the Chicago and New England Sections with memberships of around about 150. It is quite possible some of these Chapters may outstrip the Sections so far as actual numbers of members are concerned.

CHAIRMAN LAW: I think we may as well throw this open for further discussion. Before I do this, I want to offer what I think is a splendid suggestion; it was made by Mr. Millar. He suggested that if any of you have pictures or snapshots that you have taken of members of your different Sections around at this Convention, that you send them into Headquarters. Then we will ask the Committee on Editing and Publication whether they won't consider devoting a page or two in the TRANSACTIONS to these pictures. So if any of you have any such pictures, please send them into Mr.

Hibben, at Headquarters, and I am sure the Chairman of the Com-mitttee on Editing and Publication will do his best to get them in.

This meeting is now thrown open for general discussion. I should like to hear particularly from some Chapter Chairmen as to the experiences they have had with the development of their Chapters and how it is going along.

G. G. Cousins: Mr. Chairman, relative to Mr. Hibben's remarks on the membership of the Chapters, there is a condition that exists in our Toronto Chapter that possibly exists in some others. I believe Mr. Hibben's tabulation refers to Society members in the Chapters.

Secretary Hibben: Yes, not to others that are strictly Chapter members.

G. G. Cousins: That is the point I wish to ask about. We have a number of Chapter members whom we classify as "affili-ates" and we are going to recruit our membership from those.

We are entering now on the third year of our Toronto Chapter, and we have learned a lot since we started. In discussing our field for the development of the Chapters, we ought to get a lot of men who are very intimately connected with illumination and know practically nothing about illuminating equipment or lighting principles. Those men are very largely in the fixture business, and contractors. There are a number of contractors who are progressive and study lighting principles.

Our plan was to make the meetings such that those men could get some of the fundamental principles so that they could talk more intelligently when they were speaking of the various materials that they are dealing in. We found that apparently they didn't want to know. It was very hard to get them interested.

We had a very good year, our first year, I think. The second year was practically as good but we found it harder work to keep up interest. Now we have come to the point where we have been talking over means of getting out along more definite lines and there are two schools of thought amongst those who are very much concerned in it. It has been argued on one hand that if we charged five dollars a year for Chapter membership, our present rate is three dollars exclusive of Society fees, those who paid the money would feel they had to get something out of it and would come. On the other hand, there are several of us who think that a number who might be interested to the extent of paying two or three dollars, would simply balk at five dollars a year. With the type of subjects we have been handling, we have found it rather difficult to gauge the type of meeting and the papers presented. A good many think of "Engineering Society" as some highbrow stuff and balk at it from the start.

We want to go ahead. We don't want to stand still. There is a big field for it. It takes a little tact and some knowledge and experience, I believe, to develop that field, and we would appreciate any suggestion that you can give us.

WILLIAM H. WOODS: Mr. Chairman and Gentlemen. Mr. Cousins has practically stated the conditions existing in Toronto in connection with the Toronto Chapter. We are however, not discouraged by any means with the little difficulties encountered in increasing our membership.

Having been Secretary under Mr. Cousins in the organizations of the Toronto Chapter, I have gotten the essential experience in following up prospects for our Chapter. I find that the only way to get members for the chapter is to get out and do personal work amongst the contractors and manufacturing interests, and at the same time try and keep out what we might call deadwood from the chapter. Our policy is to get those vitally interested in our society and present to them simple papers from which good information may be obtained.

I think too, that a good strong point in our chapter meetings is when we have speakers from the large companies over here. We have had some very excellent addresses from Mr. Ward Harrison, Mr. Samuel G. Hibben, Mr. Davis Tuck and Mr. A. J. Sweet, and we are indebted to these gentlemen for the benefit our chapter has derived from their visits. It seems to have consolidated it and shown the membership at large, that there is something really of an international spirit in our Society, and that the chapter has a lot of good information in store for them.

This year our program will be even better than we have presented in the past. We propose going out to our large manufacturing centres and get them to subscribe to our fund, in order that we will be financially sound, I think in that way the committee will not have to worry then, about the necessary expenses involved in promoting the objects of our organization, and eliminate the undesirable necessity of asking our membership for a subscription, besides our set fees.

Our object this year is to increase the membership of Toronto Chapter to between fifty and sixty members, there are prospects of this additional increase to our roll already lined up by our present secretary, Mr. J. Scott of the Sunbeam Lamp Co. I am sure that when we get working, we will find a good many more prospects about Toronto.

I expect also that Mr. Kintner of the Westinghouse Lamp Co. at Hamilton, who is a member of the Society, with Mr. Stuart, also of the same Company, would like to form a Chapter in Hamilton. Of course they have not the membership in Hamilton to constitutionally form a Chapter at present. These Gentlemen are at present members of the Toronto Chapter, and when the time is opportune, the Toronto Chapter will assist in the organization of a Hamilton Chapter, as that city presents a good field to work in.

G. G. COUSINS: May I say just one more word, please? Toronto is reputed to have one of the best Sections of the A. I. E. E. in America, and in comparing their situation with our own, I think the answer lies in the fact that a larger percentage of the A. I. E. E. men are technical school and college men. There are comparatively few men who are educated technically in illumination, in the same way that the electrical men are educated along the lines of the activity of the A. I. E. E.

JAMES M. KETCH: I will have to admit that it was with some hesitation that we broached the plan of organizing a Chapter in Michigan, simply because several of the men had heard from other Chapters that, well, to put it bluntly, the class of stuff that was presented at the Chapters and Sections went over the heads of most of the people who came. And so, we have taken a rather different attitude to this thing.

In order to diversify, we have selected directors from various interests. One man being from the University of Michigan, who maintains the research and the scientific end of that institution; one man from the local architectural society and one of the big architects who maintains our contact with the architects; one man from the central station who maintains that contact; a research man from Dodge Brothers Corporation who is interested in headlighting and in headlighting legislation, and also in factory illumination; one man from the National X-Ray Reflector Company who maintains the interest of the sales end of the game; and one man representing fixture and equipment manufacturers.

We have also been fortunate in the past in having a number of these same men belonging to the local Electrical League which they call the Electrical Extension Bureau there, and it is proposed now that in the large technical high school there, they put up a lighting demonstration, which combines the industrial and commercial. It seems now as though this will go across, that there will be put on for the benefit of the technical high school students, a course by the members of the local Electrical Extension Bureau and the Chapter of the I. E. S.

There will be a course put on probably for the local contractors association. One of the directors is also secretary of the local Electrical Contractors Association and he is very much interested that we link up the contractors and Illuminating Engineering Society to iron out some of these old stone age ideas held by some of the so-called contractors.

I see one great difficulty and a thing that we are frankly worried about, and that is in getting the papers. I throw this out as a suggestion: That the national office of the Illuminating Engineering Society keep the secretaries of the Sections and Chapters somewhat informed as to what might be available. Here is a bit of research work going on here, or a large job there; materials that form the subject of papers at the National Society might also well form the subject of papers for the local Chapters and Sections. It might be that the Chapter secretary would welcome once a month, or as they come in, a letter saying that "Here is so-and-so near you who I think would be glad to submit a paper on something that would be of direct local interest to your people."

It is rather hard I will have to admit for a secretary who knows very few people to start right out in a blank field and find papers that are of the calibre that the people who attend the Chapter and Section meetings will understand and something that will be also of news value.

We have also adopted the plan of having one member of the local Chapter attend each of the important, or belong to each of the important associations in the city: the Detroit Engineering Society, the local section of the American Institute of Electrical Engineers, the S. A. E.; and there is a glorified Electricians' Society. We are represented in all those organizations to maintain the contact for the Chapter and try to bring about cooperative meetings or combination meetings with these various Chapters. We also have an architect's representative belong to our Society as one of our directors.

We can't speak from a whole lot of experience yet simply because we organized last spring, and we hope to have our first paper this month, which we are going to get, if possible, from our retiring President of the I. E. S., on something of a very fundamental and work-a-day kind, something that will start us off in the proper way.

CHAIRMAN LAW: Is there some one here who can speak for the Northern New Jersey chapter that has just been organized?

CHARLES GALLO: I think you will find that the electricians, plant engineers, some school authorities, store managers, and people who are selling lighting equipment—all those people who have more or less to do with lighting—are all interested; and if you explain to them in a proper way, just what the Illuminating Engineering Society is trying to do, and how it can help them, I think those people will all join because there are no reasons why they shouldn't. You can convince prospects by properly approaching them, and by constantly keeping at them, emphasizing the fact that the Illuminating Engineering Society can be of great benefit to them, and also that they can be of benefit to the Society, which is unquestionable.

I believe that is about all, but I think Mr. Moore can give a few more details on the organization of the Northern New Jersey Chapter.

CHESTER L. DOWS: I am the outgoing Chairman of the Cleveland Chapter but I am looking forward this coming year to some real work in Cleveland. We have as our new Chairman, Prof. H. B. Dates, Head of the Electrical Engineering Department, Case School of Applied Science. We have as Secretary, Mr. R. A. Fulton, from the Cleveland Illuminating Company, and we have as Chairman of the Papers Committee, Mr. E. W. Commery, National Lamp Works. We hope to increase our membership this year and to diversify it.

The large proportion of our membership comes from the National Lamp Works, where so many are interested in illumination in its various phases. Our problem is one of increasing and diversifying our membership to spread interest in the field of lighting and illumination. We might use profitably any service that brought to our attention suitable speakers and papers from outside our own immediate circle.

During the past year, the Secretary of the A. I. E. E. at Headquarters wrote to all the Secretaries of the Sections, asking for one or two of the most popular papers of the year. Each A. I. E. E. Section has eight or nine meetings through the season and as Secretary of the Cleveland Section, I had no difficulty in picking out two or three outstanding papers to send in to Headquarters. You may expect that a very fine collection of subjects was compiled in this way. A list of these subjects was turned over to the Chairman of the Papers Committee for the year 1923-24 for their consideration.

At Mr. Harrison's suggestion, we started a survey last year of those that might be interested in church lighting. One hundred questionnaires were sent out on which three questions were asked as follows:

1. Would you personally plan to visit a special I. E. S. Meeting at Nela Park, Cleveland, on improving artificial lighting in churches? Yes——— No———

2. Do you believe a sufficient number would be interested in church lighting to make this Meeting a success? Yes——— No———

3. Please give below the names and addresses of men interested in church improvements.

SUMMARY OF TWENTY-EIGHT RETURNS

Question	Yes	No	Doubtful	Total
1	14	13	1	28
2	13	12	3	28
3	Fifty-three names and addresses obtained.			

The results of this questionnaire lead us to believe that there was enough interest in this subject to hold a Meeting.

CHAIRMAN LAW: Mr. Dows, to whom did you send the questionnires?

MR. DOWS: To pastors or their assistants of the larger churches in the city.

CHAIRMAN LAW: Did you hold such a Meeting.

MR. DOWS: No. This questionnaire was sent out late in the season but we expect to hold this Meeting next year.

CHAIRMAN LAW: May I ask what you charge for affiliates?

MR. DOWS: We have none.

CHAIRMAN LAW: Who maintains the Chapter? Who takes care of the expenses?

MR. DOWS: Mr. Commery might answer this question better than I, but my impression is that we have had no expense except for postage, for which we have been reimbursed by Headquarters.

CHAIRMAN LAW: By the Headquarters of the Society in New York?

MR. DOWS: Yes. Our Meetings have been held either jointly with the A. I. E. E. or in the lecture room of the National Lamp Works, so we have had no rent to pay. Our only expense has been postage.

CHAIRMAN LAW: I am wondering whether it is not a good plan to establish some set amount for affiliates in Chapters. Of course the Chapter can prepare their by-laws and constitutions, elect their own officers, and really work independently of the Society. But in listening to this discussion, I find that one Chapter charges dues of $1.00, another $2.00, and I can readily see that some might charge $5.00. It seems to me that it would be a great idea to establish some set amount, although, of couse, the I. E. S. Headquarters really has no jurisdiction over this amount.

PRESIDENT HARRISON: If you read the Constitution, and particularly the section on Chapters, it will look to you as though the Society created the Chapters and then coldly turned them out to shift for themselves. It says they may arrange their financial affairs just as they see fit, so long as they do not obligate the Society in any way.

To get down to facts, the Sections are supported entirely by the General Office; they ask for a budget at the beginning of the year and it is assigned to them; they send their bills in and they are paid. This means that relatively speaking the Society makes money out of any associate or any member of the Society who happens to reside in a chapter district, because they take his $7.50 or $15 and instead of having to return $2.00 of it—which is the average section expense—they return nothing to the local organization. Of course, the members gets the Transactions, but so do the other members. When it seemed evident that the expenses of the Society as a whole for the past year were going to be substantially less than its income, I told the chairmen of a number of chapters that while it was true the Society was not obligated to pay their bills, nevertheless the Society, I knew wanted to deal fairly with them, and if they sent in to the General Office bills not exceeding an average of $2.00 per member I had rather definite hopes that the Council would approve them; I believe the Council has approved them to date. Furthermore, as I recall the Toronto Chapter and the Cleveland Chapter, which were the first ones organized, have a stated appropriation on the books of the Society. I feel that every Chapter should have some appropriation. Of course, the Sections are tied down more closely than the Chapters. They have to hold meetings at stated intervals and do various things, whereas the chapters have more privileges and are more on their own responsibility.

It seems to me that $5.00 is quite a sum to charge for membership in a chapter as long as an affiliate does not get the Transactions. As I recall, for many years the A. I. E. E. chapter dues were $1.00. It seems to me, too, that if a chapter takes in members at a dollar, these same people should not be solicited immediately by the General Membership Committee who would try to advance them to a $7.50 grade; the prospect may just drop the whole thing.

G. BERTRAM REGAR: Mr. Chairman: Speaking from the standpoint of the Membership Committee, the Society has elected 289 members this year. That growth is larger than that of any other year with the possible exception of the first two years of the organization when it reached its peak. Then it grew up as a mushroom, many of those men coming in because of the desire of their companies for co-operation, and they have since dropped out. The slates are absolutely clean today and the report of the Secretary as to the membership shows the Society to be in a very healthy condition.

The Society is for the dissemination of knowledge. I was very glad last night to hear that the lighting man was coming down the hill pretty fast and catching up with the scientist, but if we do not get that information out where it is usable, then the fellow still in back of the lighting man is not catching up with us but is losing ground.

N. D. MACDONALD: Mr. Chairman: I want to say that this
meeting has certainly been well worth while. Messrs. Ketch and
Dows came here to tell us how badly the General Office was functioning in getting information to them and I stayed over especially
to take a "fall" out of some of these section and chapter secretaries
because they were not reporting anything to the General Office.

I am talking from the point of view of the Society Transactions.
The Committee on Editing and Publications a year ago tried to add
to the TRANSACTIONS by putting in some new sections. We felt,
considering the figures which Mr. Hibben has given you of 953
persons connected with sections, 337 with chapters, and 181 people
not directly connected with any body within the society, that
it was time the society had some news-disseminating organ,
and therefore, we have tried to put in the TRANSACTIONS such
news of the society and of the industry as we could get.

We find a good deal of trouble in getting out the TRANSACTIONS
promptly to you because the machinery seems to need a little
lubrication. The Section Secretaries are a little bit slow at times in
sending in papers; authors do not get them back to us corrected
just as promptly as they might, and on that account papers which
might be available for your use promptly, are not published for a
month or two.

It seems to me that the TRANSACTIONS might be made of great
use to you section and chapter men if you would only send us some
information about what you are doing in your own chapters. After
all, we are not able in the General Office to dig up any more original
ideas than you are but if you will send us news of what you are
doing and put it in such shape that the other secretaries can use it
at later meetings you will have the problem settled.

I just wish to say one other word on that: Won't you please
try to send us something that tells us what you really do? Do not
write us that "At the Blank Section we held a meeting on such a
date. A paper was presented by John Jones. 153 people were
present. The paper was enjoyed by all."

Tell us something about what the man said and if there was a
fight on the floor, tell us about it. That is interesting and it is
news.

Send us some photographs from time to time. Mr. Millar's suggestion is a very excellent one, and if you find items in the daily press from time to time commenting on your meetings or on any new installations, send them in. We shall be very glad to publish them.

J. R. COLVILLE: Mr. Chairman: Mr. Harrison mentioned a few minutes ago that after a chapter is organized it is more or less left out on a limb as far as the parent organization is concerned. I should like to bring up for the consideration of the proper committee the fact that additional instructions in the By-laws and Constitution of this organization would be of very material assistance to a chapter not only in organizing but in carrying on its work. I may be speaking beside the point because perhaps this has been taken care of in the past two years. I organized the Cleveland chapter two years ago and I know at the time, it almost took a lawyer to get out of the constitution the facts as to what we could do and what we could not do. I should like to make the definite, concrete suggestion that the proper committee give consideration to expanding a little bit in the Constitution and By-laws on the matter of chapters so that a person who is really interested can tell without a great deal of digging what the chapter can do and what it can not do.

I am sure, speaking for the men out over the country who do not have the benefit of the Council meetings that a good many of the men who live close by have, that this would be of material help in promoting the chapters. In other words, it strikes me now as though the national body says "If you want to organize a chapter why go ahead; we have no serious objections."

Mr. Moore's paper may clear the whole matter up but I think something should be put in the constitution itself, regardless of whether everything is explained in Mr. Moore's paper or not.

E. F. NICHOLS: Is there time at this meeting for a moment's digression? It is a matter which, to my mind, applies to the Sections and Chapters and the general society alike. I want to make a comment on the question of the manner of presentation of papers.

There has been one allusion to "highbrow" paper here this morning, and I want to say that the "highbrow" papers at this convention haven't all been given by scientific men. We have had some "highbrow" commercial and "highbrow" art papers too.

The fact of the matter is that the man who gives a paper is usually a specialist and his tendency is to give a presentation which would be suitable to other specialists in his groove. He gives it in a way in which only other specialists in his field can understand it. I think that every man who presents a paper before such a society as ours is under obligation to use his imagination concerning the things in his work which will interest and be within the understanding of his audience.

Papers that are pre-printed have all of the details and all of the facts and all of the technicalities that are necessary. It seems to me that certain very simple rules should be followed in the giving of a paper: The speaker should plainly answer his hearers obvious questions. First, what is he driving at? That he should give in plain, simple English so that a man in the street or in the alley, even, can get some notion about it. How is it connected up with other things that are nearest it? What are its surroundings? Second, why is it worth driving at? Third, how is he driving at it? Fourth, what did he find at the end of the road? And let it go at that. If any member is interested in any particular or further details, such matters can be brought out in the discussion by the members who have read the pre-printed paper. But I think in giving the paper orally, it ought to be given in such a simple, straight-forward way that the audience will not only get the benefit of the paper itself but will know what its closest relatives are and why the work was worth doing and worth reporting.

PRESTON S. MILLAR: At our annual conventions advancement of the science and art of illumination is discussed by the leaders in the illuminating engineering field. It is very rare for enough of these leaders to gather at a meeting of a chapter to make such discussions practicable. The Northern New Jersey Chapter of the Society is organized with the idea not so much of furthering discussions for the advancement of the science and art as for the purpose of taking to those who can influence the practice of the public such ideas as have been worked out in the Society's technical meetings and promoting their application in practice. Thus the Chapter seeks to hold meetings with civic bodies, trade organizations, church organizations, etc., for the purpose of pointing out how illumination in their respective provinces can be improved to their advantage.

It seems to me that additional chapters organized for scientific discussion of the lighting problem will not succeed. Chapters organized for the purpose to which the Northern New Jersey Chapter is committed can perform a valuable service in extending the influence of the Society for better lighting by the public.

In the conduct of Sections, Chapters and Local Representatives' work there is a very evident need for a co-ordinating influence from headquarters. It seems to me that there ought to be an individual or a committee charged with following the work of each and communicating to the others information of a promotive character.

RAILWAY CAR LIGHTING*

BY GEORGE E. HULSE**

SYNOPSIS: Limitations encountered in the problem of supplying illumination to cars. Amount of energy available limited, due to car being on the move. Position of lighting fixtures determined by car construction, preventing flexibility in placing units.

Maintenance of reflecting and transmitting surfaces more difficult than in most other situations.

MEANS OF LIGHTING—
    Gas—incandescent mantle.
    Electricity—axle driven generator with storage battery.

STANDARDIZATION OF CAR ILLUMINATION—
    The postal car lighting tests of 1912.
    Determined and standardized.
    The amount of illumination necessary for postal clerks to properly handle mail.
    The types of reflectors best suited for use.
    Based on these test results, the Railway Mail Service issued specifications for lighting of postal cars, giving definite values for their illumination and other details, such as mounting height of lamps, and angle of cut-off.
    These specifications can be applied without the necessity of further investigation, in the case of the change in interior design of postal cars, the type of reflector available, or the type of light source available.

COACH LIGHTING TESTS OF 1913—
    Determined the amount of illumination obtained with the possible arrangements of fixtures, and the available types of reflectors, bowls and lamps.
    The results of these tests are still in use as the basis for designing lighting installations in practically all classes of cars.

ARRANGEMENT OF FIXTURES FOR VARIOUS TYPES OF CARS, AND THE RESULTING ILLUMINATION—
    Coaches, Dining cars, Sleeping cars, Postal cars, Business cars, Baggage cars, Parlor smoking cars.

TYPES OF GLASSWARE USED AND THE EFFICIENCY OF INSTALLATION WITH THIS GLASSWARE—
    Illumination values obtained. The illumination obtained runs lower than illumination values used in office or factory installations, but seems to be ample for the conditions under which it is used.

The special conditions surrounding the generation of light, and its application for obtaining correct illumination in car lighting, differentiate it considerably from other forms of lighting and illumination.

The amount of energy which can be used for lighting is limited. This energy can be obtained from a stored supply, or it may be

*A Paper presented before the Annual Convention of the Illuminating Engineering Society, Lake George N. Y. September 24-28, 1923.

**Chief Engineer, The Safety Car Heating and Lighting Co., New Haven, Conn.

The Illuminating Engineering Society is not responsible for the statements or opinions advanced by contributors.

generated on the car itself, which also entails obtaining it at times from a stored supply. This limitation in the amount of energy available makes its most efficient utilization necessary.

The lighting fixtures must be permanently secured in position, and out of the way of seats, baggage racks, and sleeping car berths, and their position is also determined by the actual conditions of car construction. This limits the flexibility possible in placing the units.

The problem is further somewhat complicated by the fact that the fixtures must be maintained with the minimum of labor expended in their upkeep, and as the conditions on a railroad car are particularly bad as regards the collection of dust and dirt on the fixtures, this item is one which has had careful study in the designing and placing of the lighting units.

METHODS OF LIGHTING

A large number of cars are at present, and will continue for some time to be lighted by gas. Practically all gas lighted cars in this country use oil gas as the illuminant. This gas is made by 'cracking' petroleum oil in generators. After it is put through the usual processes of cleaning and purification, it is compressed to a pressure of about 150 pounds per square inch, and is distributed under this pressure to the car yards, where it is transferred to suitable holders carried underneath the cars.

This gas has a high heating and illuminating value, which is not materially decreased by the compression to which it is subjected. It is this feature of the gas which makes it suitable for car lighting, as it provides the maximum amount of lighting value.

Most of the cars lighted by gas use the incandescent mantle, and the flat flame type of lamps which were installed in the older cars are being converted to use the incandescent mantle. The largest proportion of the lamps use a mantle giving 90 cp. and consuming 2 cu. ft. of gas per hour. Where the lamps were originally installed at considerable distances apart, when the lamps are remodelled it is not feasible to respace the lamps, a large mantle giving 125 cp. with a consumption of 2.5 cu. ft. of gas per hour is used. A smaller mantle, giving 25 cp. is available for use where smaller light sources are necessary.

The mantle, being of the inverted type, gives a candle power distribution which is well suited for the requirements of car lighting.

As the gas is of constant quality, and the pressure maintained is very uniform, the light output of the mantles is constant, and is maintained without any necessity of adjusting the air or gas supply.

Electric Lighting

A large proportion of the new cars now being built are fitted with electric light. In some cases, electric light is also being applied to older cars, displacing the former systems.

The Axle Generator System is almost exclusively used for generating the electricity for lighting cars.

The system known as the 'Straight Storage System' was formerly used to a considerable extent, but it has almost entirely disappeared. With this system, the light was obtained from storage batteries which were charged at a stationary plant, necessarily while the car was not in operation. This system has gone out of use, because the time required for charging the batteries lessened too much the availability of the cars for actual service, as it required an actual charging period of eight hours to properly charge the batteries.

Some trains are still lighted by the 'Head End System,' in which the current is obtained from a steam driven generator in the baggage car of the train. This generator supplies the current for the whole train. A few of the cars in the train are supplied with storage batteries to take care of the lighting during such times as the engine is detached. This system, however, is going out of use. In the winter time the steam used by the turbine is too much of a drain on the locomotive, which, at certain times, needs all the steam supply which it can generate to pull the train. This system is also quite inflexible, as it provides no means of lighting cars which may be detached from the train and sent to branch lines.

In the Axle Generator System, the current is generated by a dynamo driven from the axle of the car. This dynamo provides current for the lights, and charges a storage battery which provides current for lighting when the car is not in motion. With this system, each car is its own power plant, which is in operation whenever and wherever the car may be in service.

Such attention as is necessary for the proper operation of the system may be given to the car during the time of its regular layover periods.

The essentials of such a system are as follows:—(1) A generator mounted on either the car body or the truck with some form of driving system between the car axle and the generator. (2) A storage battery to furnish current when the car speed is not sufficient to drive the generator at a speed to properly charge the batteries and light the lamps. (3) A regulator to govern the output of the generator at varying speeds, and to properly charge the storage battery and protect it from overcharging. (4) A regulator to maintain proper voltage at the lamps. (5) Some means of keeping the polarity of the generator constant, when the direction of movement of the car changes.

These conditions have been successfully met in various ways, the majority of equipments in use now embodying the following features:—(1) A generator mounted on the car underframe, driven by a belt. (2) The generator is controlled for output at varying speeds by a carbon pile rheostat in its field circuit. (3) The battery is protected from overcharging by the voltage which is supplied to it being limited to a voltage which will properly charge the battery but will not overcharge it. (4) The lamp voltage is held constant by an automatic carbon pile rheostat.

At present, there is some demand for a drive which will be more positive in its action than a belt. Belts do give some trouble from slipping, and from being lost, especially in winter weather. Various types of gears or other positive drives are being investigated, but it is doubtful whether when all the conditions of application to the car, first cost, and the cost of maintenance are considered, the increase in service will be justified.

CAR ILLUMINATION

I have already said that the mounting positions for the lighting fixtures in railroad cars are limited—in fact, there are practically only two positions in which the fixtures can be placed. These are:—(1) On the upper deck, that is, dirtly under the roof of the car, or (2) on the side deck. As the ceiling height is only nine feet, it is impossible to keep the light sources out of the range of vision, and makes the screening of all light sources absolutely necessary.

While these limitations may have led to some inconvenience, they have also accomplished the adoption of certain standards for lighting cars, which must be very closely adhered to.

There are standards in use for the illumination of almost every type of car, and these standards are based on data obtained from tests made in cars with all possible combinations of fixture positions, light sources, and fixture accessories.

Postal Car Lighting Tests

In 1912, the Post Office Department, represented by the Railway Mail Service, together with a number of Railroads and manufacturers, conducted a series of tests at Washington to determine what was the proper illumination for postal cars, and how it could best be obtained. These tests included the measurement of illumination with all available arrangements of fixtures, light sources, reflectors, and glassware. They also included tests of the amount of illumination necessary for the postal clerks to do their work properly and without eye fatigue. These latter tests were made with a large number of operators, and with all possible combinations of fixture positions, light sources, reflectors and glassware.

As a result of these tests, specifications were issued by the Post Office Department covering completely the lighting of postal cars, as to the general position of the fixtures, and the amount of illumination necessary in different parts of the car, the latter being given a maximum and a minimum value. Differentiation was also made between the different types of reflectors, and where the tests showed that the illumination necessary differed with one type of reflector and another, proper allowances were made in the specifications.

The application of these specifications has resulted in a very satisfactory illumination of postal cars, and although since these tests were made the types of light sources have changed, the interior construction of the car has been modified, and new types of reflectors have been introduced, the information obtained, and the standards adopted as a result of these tests, have made it possible to rearrange the illumination and give satisfactory results without the necessity of any further investigation.

Following the postal car tests, a number of Railroads and manufacturers interested, conducted a series of tests on the lighting of railroad coaches at Cleveland. In these tests, the amount of illumination with all possible arrangements of fixtures and available types of reflectors and bowls and light sources, was determined. The results of these tests were sufficiently comprehensive to cover all the features which have been introduced in car lighting since that time, and are still in use as a basis for designing the lighting installations in practically all classes of cars, except postal cars.

TYPE C LAMP

The introduction of the type C lamp with its larger output of light per unit of energy has, of course, had some influence on car lighting where the question of efficiency in the light source is particularly important.

Up to the time of the introduction of the type C lamp, the use of open reflectors was universal for almost all of the different classes of cars. This was used on account of its increased efficiency over any enclosing bowl, and also on account of the fact that it was considerably easier to keep clean than enclosing glassware. Indirect lighting has never been used to any extent in car lighting due to the difficulty in keeping the reflecting surfaces in proper condition. The appearance of the fixture with open reflectors, is, of course, not so pleasing as fixtures with enclosing bowls, and the proper screening of the light source in considerably more difficult.

Since the introduction of the type C lamp, however, the use of enclosing glassware is increasing, as about the same output per unit of energy used can be obtained by using enclosing glassware with the type C lamp as can be obtained with the open reflector and the type B lamp, and the appearance of the former type of fixture is considerably better than the open reflector.

Type C lamps are being used to some extent in open reflectors, but this is not good practice, unless the lamp is white glass, or bowl frosted, as, even if the light source itself is screened, the bright reflection spots on the reflector are very trying to the eyes.

The following is an outline of the methods of lighting used in the various types of railway cars—

Coaches

Coach lighting is, of course, of the greatest importance as, coaches constitute the largest number of cars in use, and carry more people than any other class of car.

A certain amount of general illumination is necessary, but the principal illumination is required for reading. The lighting system should be designed for best illumination on the reading plane, which is 45° to the horizontal, and at right angles to the centre line of the car. The lighting tests which I have referred to, demonstrated that equally good illumination, as far as intensity, distribution, and uniformity go, could be obtained either with the fixtures mounted on the upper deck, or on the side deck.

Centre lighting is more largely used, however, on account of the fact that it requires less fixtures, and there are fewer lamps and reflectors to maintain. Centre lighting also gives less trouble from shadows cast by the passenger's head on his own or another passenger's reading. The standard spacing for lamps is 6-ft. both for those mounted on the centre deck and on the side deck. The mounting height for the centre lamps is about 8-ft. and for the side lamps 6-ft. 6-ins.

The open-mouth type of reflector, in either heavy density opal glass, or clear prismatic, is employed for coach lighting, preferably with type B lamp. For centre lighting, a 50-watt lamp is used, having an output of 560 lumens, and for the side lighting a 25-watt lamp, having an output of 270 lumens.

Using these lamps, the following illumination values are obtained with the open- mouth reflectors as in Table I.

TABLE I

FOOT-CANDLES

| | 45° PLANE | | | HORIZONTAL | EFFICIENCY PER CENT |
	Window Seats	Aisle Seats	Mean	Average	Horizontal Plane
Heavy Density Opal	2.87	3.85	3.39	4.95	49.7
Clear Prismatic	2.81	3.67	3.25	4.67	46.6

When type C lamps are used, they should be used exclusively for centre lighting, in fixtures placed close against the ceiling with an enclosing bowl, the mounting height of the fixtures being about 8-ft. 6-ins.

The bowls used are of the best type of transmitting glassware, and the lamps 50-watt type C, with an output of 740 lumens. With the grade of glass now available, the following illumination values are obtained—See Table II.

TABLE II
FOOT-CANDLES

| 45° PLANE | | | HORIZONTAL PLANE | EFFICIENCY PER CENT |
Window Seats	Aisle Seats	Mean	Average	Horizontal Plane
2.59	2.85	2.72	3.35	22

Using the 75 watt type C lamp (1215 lumens, or 202 lumens per running foot of car) these values would be as follows—See Table III.

TABLE III
FOOT-CANDLES

| 45° PLANE | | | HORIZONTAL PLANE | EFFICIENCY |
Window Seats	Aisle Seats	Mean	Average	Horizontal Plane PER CENT
3.58	3.44	3.75	4.65	22

It will be noted that using the same wattage the values obtained with the type C lamp and enclosing bowls are about 20 per cent lower than with the open-mouth reflectors. With the enclosing bowl, to obtain illumination equal in intensity to that obtained with a type B lamp and open-mouth reflectors, it is necessary to increase the wattage 50 per cent and to use a type C lamp. With the enclosing bowl, there is no danger of glare, and the appearance of the car is considerably improved.

With gas lighting, the lamps are generally spaced 9-ft. apart. The best arrangement uses an enclosing bowl around the mantle, as open-mouth reflectors are not suitable for use with gas burners. The type of bowl most generally used is one that is spherical in shape, the upper half being of heavy density opal glass, and the lower portion of clear glass optic ribbed. As the gas mantle has a rather low intrinsic brilliancy, these ribs give sufficient diffusion to prevent any bad effects from glare. The illumination values for a coach with such an equipment, using the 90-cp. mantle with 9-ft. spacing, 87 lumens per running foot of car, are as follows as in Table IV.

TABLE IV
FOOT-CANDLES

45° PLANE			HORIZONTAL PLANE	EFFICIENCY PER CENT
Window Seats	Aisle Seats	Average	Average	
1.75	2.04	1.89	2.42	25

With the 125 c.p. mantle, giving 120 lumens per running foot of car, these values would be as follows—See Table V.

TABLE V
FOOT-CANDLES

45° PLANE			HORIZONTAL PLANE	EFFICIENCY PER CENT
Window Seats	Aisle Seats	Average	Average	
2.44	2.84	2.64	3.37	25

Dining cars

The evolution of dining car lighting is an example of the fact that the type of lighting which may be satisfactory in one place may not be satisfactory under what appear to be very similar conditions. It is generally conceded that the requirements for dining room lighting are that the table itself should be made a place of high illumination, and that the remainder of the room should be at a considerably lower level.

Some years ago a system of lighting dining cars was designed, based on these requirements, and this system met these requirements fully. However, it did not give satisfaction to the patrons of these cars. The lighting fixtures were placed directly over the tables, with concentrating reflectors above the lamps, and directing plates under the lamps, placing a high illumination directly on the table. The general illumination was quite low, and was obtained by a few light sources in the upper deck of the car. While the tables were very well lighted, the rest of the car, of course, was not, and patrons of dining cars complained that the cars were gloomy. This has led to an installation, in which there is a considerable degree of general illumination. The lamps over the tables have been retained, but the general illumination is now of a higher value, and is obtained either by semi-indirect lamps, or by enclosing bowls, placed directly against the ceiling. The lamps for general illumination are mounted 6-ft. apart, and use 50-watt type C lamps. The illumination on the table top is from eight to twelve foot candles, and is obtained by the use of 50-watt type C lamps, mounted in the special fixture referred to.

The kitchens and pantries in the dining cars, of course, require a high degree of general illumination, as they are so small that the work of preparing meals is carried on in all parts of these rooms. The present practice is to light these rooms by ceiling fixtures, using a 50-watt type C lamp with a reflector of the RLM type.

As the work is carried on by the occupants of these rooms in a standing position, and this work is all below the level of the eye, this type of fixture can be used without any bad effects from glare.

Sleeping cars

Sleeping cars require lighting for general illumination, for reading, or for working at a table in the sections, for the illumination of the berths after they are made up, and for the illumination of the aisle for passengers who go through after the berths are made up.

The fixtures for general illumination must be placed close to the ceiling to prevent interference with the fixture from the upper berth. A shallow enclosing bowl is used with these fixtures.

Small units, with individual control, placed in the corner of each section, provide additional local illumination for reading in the seats, and for the necessary light after the berths have been made up.

The state rooms in a sleeping car and compartment car are lighted by the same type of ceiling lamp as is used in the berth section of the sleeper.

Smoking rooms are lighted by centre lamps in the ceiling, and also may have bracket lamps at the back of the fixed seats to ensure proper lighting for reading.

The passageways are lighted by ceiling fixtures, either with open-mouth reflectors, or a fixture having a reflector set into the ceiling, and a directing plate under the lamp set flush with the ceiling.

It was formerly the custom to leave one or more of the centre lights lit after the berths had been made up, but this, of course, was a nuisance to the occupants of the upper berths. Fixtures are now being used which are secured to the frame of the seat. These fixtures throw the light on the floor, making it easy to find the way along the aisle after the berths have been made up, and the ceiling lamps have been extinguished.

For the centre lamps, type C bulbs, 100-watt, having an output of 1700 lumens, are used. The general illumination on the reading plane with this fixture and 100 watt bulb, averages five foot candles. For the berth lamps, 25-watt type B bulbs are used.

Parlor cars

Parlor cars require illumination mainly for reading purposes, together, of course, with sufficient general illumination. The same type of installation is used in parlor cars as in sleeping cars, with, of course, the omission of the particular units having directly to do with the sleeping car berths.

Parlor-smoking cars

These cars require general illumination, and also good illumination on the reading plane. The general illumination is obtained by the same type of centre fixture as is used in sleeping cars, but as the seats are on the side of the car facing the centre, it is necessary to add to this illumination for reading purposes. The additional illumination is obtained either by pendants attached to the side deck, or by brackets on the side walls of the car.

Postal cars

The proper illumination of a postal car is, of course, very important as postal clerks are employed constantly while in the car on work which demands the best possible illumination.

Postal cars are divided into three sections—

(1) The section where the letters are taken from the sacks and distributed to the pigeon holes according to the route which they are to follow.

(2) The bag distributing rack, where the papers are distributed into boxes, and where packages or letters are placed in the mail bags.

(3) The storage end, where the filled sacks are stored.

The letter distributing cases require high illumination on the vertical plane for the box labels, and on the horizontal plane for the reading of the addresses on the mail.

The bag distributing racks require high illumination on the horizontal plane, for reading the addresses on the packages, and also for the labels on the bag rack.

The storage end of the car requires a fair general illumination. The following are the specifications of the Railway Mail Service for the initial illumination values for the various portions of the postal car as in Table VI

TABLE VI

INITIAL VALUES OF ILLUMINATION REQUIRED

| | FOOT-CANDLES | |
Bag Rack Portion	Minimum	Maximum
Centre of car—horizontal......................	3.75	12.00
Mouth of bags, measured 18 inches from side of car—		
horizontal.................................	2.00	12.00

Letter-cases		
Over table—horizontal........................	3.75	19.00
Face of case—vertical........................	1.66	19.00
Storage portion..............................	2.00	12.00

These initial values are so set so as to give adequate illumination in case there is a depreciation of 40 per cent in the efficiency of the installation.

If reflectors with a glazed reflecting surface, such as porcelain enamelled metal, are used, these values must be increased 25 per cent.

The reflector most largely used in postal car lighting is one made from aluminum, with a special matted surface, which gives a high reflecting efficiency, and which may be easily cleaned, and its original efficiency restored.

These specifications are fulfilled in the 60-ft. postal car of the present type, by using 15 type B 50-watt 560-lumen lamps. These lamps are used with an aluminum open-mouth reflector, the reflector at the lettercase being of the intensive type, and those in the bag rack and storage portion, giving the maximum outputat an angle of 45°.

If type C 50-watt lamps are used, eleven lamps are required. These lamps have an output of 740 lumens. The same type of reflector is employed.

The Railway Mail Service require that if the type C lamp is used, it must be bowl frosted, or enamelled.

For the 60-ft. apartment, with Pintsch gas, eleven lamps are required, using a 90-cp. mantle, with aluminum reflectors.

Private or business cars

The private or business cars generally consists of what may be termed the living room, or office, in the observation end, several state rooms, or bedrooms, dining room, and kitchen.

The living room or office, is provided with general illumination from centre lights. It may have chairs placed along the side facing the centre, and lighting for reading at these chairs is provided by brackets on the side wall, or pendants on the lower deck. Adjustable local lighting units are provided for the desks and typewriters with which these rooms are generally fitted. There are also local lights provided for lighting the gauges and speed indicators which are installed in this room.

The bedrooms may be fitted either with the ordinary form of sleeping car berth, or with a portable bed. These rooms generally have one centre lamp. If there are berths, the berths are provided with berth lights, and the beds have reading lights attached to the bed posts. The mirrors in these rooms are provided with local lighting brackets set on either side of the mirror. The centre lights used are generally of the enclosing bowl type, mounted close to the ceiling.

The dining room is ordinarily lighted by one centre light placed directly over the table. In some cases, the dining room may be also used for a sitting room. In this case, additional lighting is provided by brackets or pendants on the deck rail.

Baggage Cars

These cars must be provided with good illumination both on the horizontal and vertical planes.

The illumination on the vertical plane must be of a good value up to a height of seven feet. Trunks, and other articles, which are carried in baggage cars, are often piled very high in the car, and the illumination must be sufficient so that the labels, or markings, on all baggage can be easily read.

Glass reflectors, or enclosing bowls, are not desirable in this class of car on account of the danger of breakage.

The best practice at present is the use of metal reflectors of the RLM type, placed as close to the ceiling as possible. With such reflectors, and 50-watt, 740 lumen type C lamps, spaced 10-ft. apart, an illumination of about four foot candles will be obtained on the horizontal plane, and an illumination varying from one to five foot candles on the vertical plane.

Baggage cars should also be provided with a fixture over each door, having a metal reflector designed to light the floor of the car directly at the door, and also to reflect the light through the doorway of the car so that it may illuminate the floor of a baggage trunk placed at the doorway.

ILLUMINATION INTENSITIES

I have heard at times some criticism of the intensities of illumination which are obtained in car lighting, with the suggestion that these should be considerably increased in line with the increase in intensities in other lines of work.

With the best installations, in coaches the illumination on the horizontal plane will be about four foot candles, and on the 45° reading plane, will average about three foot candles.

In sleeping cars, with the present arrangement of lamps, the light on the 45° reading plane will average about five foot candles.

These values are, of course, lower than those recommended at present for the lighting of offices and similar rooms. It must be considered, however, that cars, with the exception of the baggage car, postal car, and kitchen of dining or business cars, are not used to any extent for work. The illumination is used principally for reading for short spaces of time. In view of this fact, these intensities of illumination seem to be adequate. The intensity obtained in Pullman cars at present is certainly sufficient for any use to which it is put.

In the case of postal cars, the horizontal illumination obtained with the new installation is about six foot candles, although the specifications based on actual tests and study of the problem, allow for a considerably less intensity than this.

As I have already said, the amount of energy available for use on the car is limited—therefore, the illumination provided must of necessity be as low as possible and still be adequate. The illumination obtained, however, from the best practice at the present time appears to be quite adequate.

ABSTRACTS

In this section of the TRANSACTIONS there will be used (1) ABSTRACTS of papers of general interest pertaining to the field of illumination appearing in technical journals, (2) ABSTRACTS of papers presented before the Illuminating Engineering Society, and (3) NOTES on research problems now in progress.

THE FIRST LOW VOLTAGE GAS FILAMENT LAMP*

BY D. MC FARLAN MOORE**

SYNOPSIS: For many years all luminosity due to the action of electricity in gases was associated with high voltages. The light between the two terminal electrodes of a small Geissler tube, or of a tube lamp such as used in the Moore long tube system, is positive column light, which in a measure accounts for its much greater intensity and efficiency. Several years ago it was found that luminosity in gases could be produced solely by voltages as low as 110, but it was a negative glow light of low intensity that surrounded the metal electrodes in a bulb of rarified neon gas. Although the ice on the problem was broken by the negative glow lamps, there remained the far more difficult problem of obtaining positive column light on low voltages. This paper announces such lamps, and contains, also, a brief resumé of the history of gaseous conduction and its many contacts with electrochemistry

*   *   *   *   *   *   *   *   *   *   *   *

It has been questionable from a scientific standpoint whether or not it would ever be possible to obtain a positive column lamp which will start and run on potentials less than 220 volts, but such a lamp is described in this paper.

The basic principles upon which it operates are such that hopes are entertained that it can be developed into a new and very useful light source. The steps leading up to this result have been interesting, and I have been requested to touch upon some of the high points in gaseous conduction lighting and keep in mind the various electrochemical problems involved.

*   *   *   *   *   *   *   *   *   *   *   *

Large tube lighting units have their commercial limitations, however, whether built *in situ* or in portable form, so that once

---

*Abstracted from a paper presented at the Forty-fourth General Meeting of the American Electrochemical Society, Dayton, Ohio, September 27, 28 and 29, 1923.

**Moore Light Dept., Edison Lamps Works of G. E. Co., Harrison, N. J.

The Illuminating Engineering Society is not responsible for the statements or opinions advanced by contributors.

more the desire for a simple bulb form of electric gas lamp became acute, and the first low voltage gas lamp already referred to resulted. However, the possible candle-power on 110 volt from such neon corona lamps was meager, and, therefore, the commercial possibilities restricted.

Higher candle-powers were reached in another new type of low voltage lamp*, in which the light emanated simultaneously from three sources: (1) the neon gas, (2) the tungsten arc, and (3) the more or less highly heated tungsten electrodes.

It was realized that if a positive column could be made by 220 or 110 volts it would probably have long life due to the small tendency of a positive column to blacken, and also that it could be made bright and efficient. But with ordinary electrodes, of iron for example, 220 volts even in neon gas would not produce any positive column light, and therefore special expedients needed to be devised that were simple, that is, that did not involve the use of a high potential transformer or other auxiliary apparatus. Figure 1 is a photograph of one of these new low voltage positive column lamps, that is immediately applicable to any 220-volt a. c. socket. Figure 2 indicates its construction as well as one form of circuits that involve the fundamental principle of first generating in a short gap the electrons needed by the long positive column.

The line potential on 1 and 2, which are attached to the cylindrical electrodes 4 and 5, can not normally produce any current or any light through the long positive column 3; but there is provided the electrode 6, that is less than 1/64 inch from 4 and it is connected to a similar electrode 7, so that the 220-volts immediately bridges the gaps 6-4 and 7-5, thereby generating electrons which then cause the 220-volts on 4 and 5 to produce a bright discharge through the positive column 3. One of the definitions of the word filament is "something continued in a long course" and since, when viewed from a short distance such a positive column discharge does not look unlike the carbon filament of a former type of series lamp, the term "gas filament lamp" has evolved in contra-distinction to a solid filament lamp. It will be noted that the gas filament is confined within a simple curved glass tube which has an opening or vent at its center. Since the bulb is supplied with neon gas at a

*A low voltage self-starting Neon Tungsten arc incandescent lamp. Trans. Am. Inst. Elec. Eng. Sept. 26, 1921.

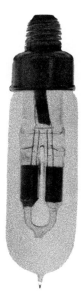

Fig 1—Moore low voltage positive column gaseous conductor lamp, 220 v.

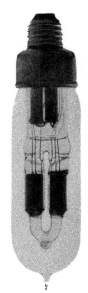

Fig. 3—Moore low voltage positive column gaseous conductor lamp, 110 v.

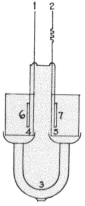

1 2

6 7
4 5

3

Fig. 2 —Low voltage (220 v.) positive column lamp.

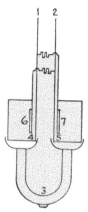

1 2

6 7
4 5

3

Fig. 4 Low voltage (110 v.) positive column lamp.

pressure of about 20 mm., this vent not only tends to regulate automatically the gas pressure, within the filament tube, when it is warmed by the discharge, but also prevents electrode impurities, though small, from seriously affecting the gas filament conductivity. It might be called a self-healing or self-repairing conductor.

Figure 3 is a photograph of such an experimental gas filament lamp suitable for 110 volts and Figure 4 is a diagram of its construction and circuits. It will be noted that in the 220-volt lamp there are two ionizing gaps in series, and in the 110-volt lamp the gaps of each electrode are in multiple, which dispenses with all wasteful positive column resistance. A large number of modifications of both circuits and electrodes have been tried, with wattages varying from 1 to 100.

Life tests have proven that some of these lamps have, without appreciable change, run for over 400 hours. But it is not claimed in any sense that this new type of lamp has reached a commercial stage. This paper is written merely as a means of recording scientific advance in what has been claimed to be an important direction. Theory seems to indicate that to increase materially the luminous efficiency of light sources in general, resort should be made to radiation from gases electrically agitated.

Perhaps the exact and complete theoretical explanation of the action of these lamps is not now known, but it may suffice to say that at a definite line wattage consumption, a given lamp may satisfactorily operate without apparent change for several hundred hours, yet if operated at about 30 per cent lower watts its positive column will cease, due to lower gas pressure, and if it is operated at about 30 per cent higher watts its positive column will also cease, but due to too high gas pressure. These actions suggest the query whether or not at the medium wattage there is going on a simple consumption of the electrode material, or is there a definite electrochemical cycle comparable to that of the mercury tube light.

*   *   *   *   *   *   *   *   *   *   *   *

SOCIETY AFFAIRS

THE LAKE GEORGE CONVENTION

"All work and no play make Jack a dull boy" apparently was one of the principles followed by the committee in charge of the Seventeenth Annual Convention of the Illuminating Engineering Society at Lake George, September 24th to 28th. The Fort William Henry Hotel served as headquarters to the largest assemblage its staff has ever known. Many nearby cottages and smaller hotels were called upon to house the throng.

An elaborate entertainment program alternated with the many business sessions which were held from day to day. Especial pains were taken to provide social events for the women guests. The Glens Falls Country Club allowed free use of its privileges to the delegates.

A business session was held Monday afternoon and the usual procedure of getting acquainted was followed, the day culminating in an informal dance in the hotel at night.

THE NIGHT OF LIGHT AND COLOR

One of the outstanding features of the session was a lecture, "Light", given by M. Luckiesh, of the National Lamp Works of the General Electric Company, Cleveland, at which time he introduced a special lantern for illustrating his talk. This device throws a kaleidoscopic color pattern on a screen, constantly changing to new designs, and proving of invaluable assistance in developing new color schemes. Various color cylinders are inserted in the machine and the resulting designs on the screen will not duplicate themselves if the apparatus is run constantly for 24 hours.

The lighting and fireworks spectacle the night of September 25th was the grand climax of the convention. Visitors from the surrounding country up to 75 miles away poured into the village during the latter part of the afternoon and early evening and took up positions on the shore of the lake for miles to view the display. It was asserted by state troopers who handled the traffic during the general exodus following, that only with the greatest difficulty was the string of automobiles kept moving, hours being consumed in emptying the region of visitors.

THE OUTDOOR DISPLAY

A multicolored lighting and fireworks spectacle designed by W. D'Arcy Ryan and unequalled since the Panama-Pacific International Exposition was the outstanding feature of the seventeenth annual convention of the Illuminating Engineering Society at the Fort William Henry Hotel, Lake George, N. Y., September 24th to 28th, inclusive. Over 300 members and guests of the Society were present at the gathering, and it is estimated over 50,000 viewed the lighting and fireworks display the night of the 25th.

Excursion Boat—Tuesday Afternoon Session

Apparatus on Pier for Fireworks and Steam Effect

Searchlight Battery

Fusing Sky-Rocket Battery

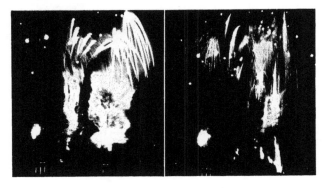

Fireworks—Night of Light and Color

For a number of days previous to the opening of the Convention, workmen and engineers were busy arranging the spectacular light and color effects which were introduced. This work was carried on under the general direction of W. D'Arcy Ryan, Chairman of the Convention Committee, and his assistants. A well-drilled corps of about 50 men was required to operate the many searchlights, colored torches, lanterns and fireworks which were utilized. Set pieces, mortars for discharging bombs, rockets and red fire were included in the pyrotechnic display.

The trees and shrubbery about the hotel and grounds were gorgeously illuminated by colored searchlights placed in advantageous positions, various shades and hues enchancing the natural beauty and form of the growth. One tree, bathed in rainbow tints by searchlights placed directly beneath, was an example of unusual chromatic beauty. It was in this work that Mr. Ryan excelled himself and more than sustained his reputation gained in the past for splendid effects of this nature. Torches were placed along the paths and colored lanterns hung in the groves. A jeweled emblem of the society, bedecked with glass pendants and brilliantly illuminated by a single powerful beam from one of the largest searchlights in use, formed a point of splendor in the trees.

This light and color combination on the grounds of the hotel formed the introduction to the fireworks display which followed. It was during the exhibition that there was called into use the famous Ryan color scintillator which was so successful at the San Francisco exposition. A dozen beams from as many searchlights, varied and changing in hue, formed designs of unusual beauty both on the clouds of smoke formed from the bursting bombs and fireworks, and with the aid of steam especially generated for the purpose. To this end, a Delaware and Hudson R. R. locomotive was utilized. Some of the effects gained in this manner were "Aurora Borealis," "Fighting Serpents and Chromatic Wheel," "Beam Drill," "Plumes of Paradise," "Spooks' Parade" and "Devil's Fan."

Another feature of the pyrotechnic program was the use of twelve-inch Ryan electric shells, which are among the largest ever used in such displays. The release of these and others ranging down to four inches in diameter was made from a total of 150 mortars arranged upon the steamship dock. Floating figures of animals, parachutes and flags were released from some of the bursting bombs. These were picked up by the many-colored searchlights and their descent illuminated.

The Convention Banquet was held on Thursday evening at which General George H. Harries officiated as toastmaster. The main address was given by Mr. Laurence A. Hawkins of the Research Laboratory of the General Electric Company.

One of the most pleasant surprises at the banquet was the presentation of the Past-President Badge to those Past-Presidents of the Society in attendance at the Convention. This badge is the regular design of the member badge and is made in gold with no enamel.

CLEVELAND AGAIN WINS ATTENDANCE PRIZE

The customary feature of the annual banquet was observed in the presentation of the traveling gavel to the Cleveland Chapter by Mr. D. McFarlan Moore for the largest representative attendance at the Convention. Mr. E. W. Commery accepted the award in a fitting manner.

SECTION ACTIVITIES

LOS ANGELES CHAPTER

At a meeting of the members of the Illuminating Engineering Society held at the Alexandria Hotel, Los Angeles, on September 24, the following officers were elected for the year 1923-24 for the new Los Angeles Chapter of this Society: *Chairman:* David C. Pence, Manager of the Illuminating Dept. Illinois Electric Company, Los Angeles; *Secretary-Treasurer:* J. F. Anderson, Illuminating Engineer, Southern California Edison Company, Los Angeles; *Directors:* F. S. Mills, Western Representative of the National X-Ray Reflector Company, Los Angeles; Herbert J. Mayo, Sales Engineer, Benjamin Electrical Manufacturing Company, Los Angeles; Walter A. Alden, Interior Lighting Engineer, Westinghouse Electric & Manufacturing Company, Los Angeles.

COUNCIL NOTES

ITEMS OF INTEREST

At the meeting of the Council, September 27, 1923, the following were elected to membership:

Four Members

FALES, THOMAS C., President, Dedham & Hyde Park Gas & Electric Light Co., 60 Congress Street, Boston, Mass.

LOOMIS, FRANKLIN WELLS, Illuminating Engineer, Duquesne Light Co., 435 Sixth Avenue, Pittsburgh, Pa.

TOENSFELDT, RALF THOMAS, Chief Electrical Engineer, Dept. Public Utilities, City of St. Louis, Mo.

WILLCOX, FRANCIS WALLACE, Manager, International General Electric Co. 120 Broadway, New York, N. Y.

Eighteen Associate Members

BISHOFF, ALEXANDER, Assistant Examiner in U. S. Patent Office, Art of Illumination, 1607 East Capitol Street, Washington, D. C.

CIAMPI, GUIDO, Electric Engineer, Compagnia Generale di Elettricita, Milano, Italy, International General Electric Co., Schenectady, N. Y.

DOANE, LEROY C., Engineer, Ivanhoe Regent Works of General Electric Co., 5716 Euclid Avenue, Cleveland, Ohio.

GRAVES, ROY MARTIN, Testing Engineer, Commonwealth Edison Co., 28 North Market Street, Chicago, Ill.

HAIL, JOSEPH C., Deputy Commissioner, Dept. of Gas and Electricity, City Hall, Chicago, Ill.

HERR, R. D., Interior Lighting Engineer, Westinghouse Electric & Mfg. Co., 1704 Union Bank Bldg., Pittsburgh, Pa.

HILLIS, ARTHUR W., Lighting Specialist, Western Electric Co., Kirby & Dequindre Sts., Detroit, Mich.

HOLLAND, E. J., Sales Manager, Consolidated Lamp and Glass Co., Coraopolis, Pa.

HOLMES, PERCY R., Engineer, National Lamp Works of G. E. Co., Nela Park. Cleveland, Ohio.

OLSON, MELVIN E., Salesman, National Lamp Works of G. E. Co., Buckeye Division, 1811 East 45th St., Cleveland, Ohio.

PLUMB, HYLON T., Intermountain Engineer, General Electric Co., 720 Newhouse Bldg., Salt Lake City, Utah.

RAMBUSCH, HAROLD, Rambusch Decorating Co., 2 West 45th St., New York, N. Y.

RANNENBERG, K. S., Deputy Supt., Street Lighting Dept. of Streets & Engineering, Room 410, Municipal Bldg., Springfield, Mass.

SCHUMACHER, RUDOLPH, JR., Lighting Specialist, 960 Anderson Ave., Bronx, New York City.

SWACKHAMER, ROBERT J., Assistant Illuminating Engineer, General Electric Co., Schenectady, N. Y.

WEAVER, K. S., Experimental Illuminating Engineer, Westinghouse Lamp Co., Clearfield Avenue, Bloomfield, N. J.

WILLIAMS, HARRY E., Illuminating Dept., Empire Gas & Electric Co., 103 Castle Street, Geneva, N. Y.

YOUTZ, J. PAUL, Electric Engineer, General Electric Co., Schenectady, N. Y.

Other Changes in Membership
One Associate Member Reinstated

Gleason, Homer E., Cascade Fixture Co., 1720 Yale Avenue, Seattle, Wash

---

The General Secretary reported the death, on August 13, 1923, of one associate member, George V. Strahan, Mitchell Vance Co., Inc., 507 West 24th St., New York, N. Y.

CONFIRMATION OF APPOINTMENTS

As Chairman of the Committee on Finance—Adolph Hertz.

As Chairman of the Committee on Papers—A. L. Powell.

As Chairman of the Committee on Editing and Publication—Norman D. Macdonald.

As Chairman of the Committee on Lighting Legislation—Louis B. Marks.

As Chairman of the General Board of Examiners—Leonard J. Lewinson.

As Chairman of the Committee on Membership—G. Bertram Regar.

As Chairman of the Committee to Co-operate with Fixture Manufacturers—
. M. Luckiesh.

As Chairman of the Committee on Research—Ernest Fox Nichols.

As Chairman of the Committee on Progress—Francis E. Cady.

As Chairman of the Committee on Nomenclature and Standards—E. C. Crittenden.

As Chairman of the Committee on Motor Vehicle Lighting—Clayton H. Sharp.

At the meeting of the Council, October 11, 1923, the following were elected to membership:

One Member

PIERCE, ELMER S., Principal, Seneca Vocational School, 787 Seneca St., Buffalo, N. Y.

Five Associate Members

BERNHARD, LESTER S., Student, Ohio State University, 92 Chittenden Ave., Columbus, Ohio.

BOYD, GEORGE M., Westinghouse Elec. & Mfg. Co., George Cutter Works, South Bend, Indiana.

BULLER, C. S., Westinghouse Elec. & Mfg. Co., 426 Marietta St., Atlanta, Ga.

CHAPMAN, A. J., Electrical Development & Equipment Co., 180 West Lake St., Chicago, Illinois.

HOLT, THOMAS T., President, The Skybryte Co., 1100 Keith Bldg., Cleveland, Ohio.

CONFIRMATION OF APPOINTMENTS

As members of the Council Executive Committee:
Clarence L. Law, Chairman; Louis B. Marks, Samuel G. Hibben, Preston .
S. Millar and H. F. Wallace.

As members of the Committee on Finance:
H. F. Wallace and Frank R. Barnitz.

As members of the Committee on Papers:
R. W. Shenton, Vice-Chairman, Julius Daniels, George C. Cousins, N. D. Macdonald, H. H. Higbie, E. C. Crittenden, H. P. Gage, F. C. Taylor, J. L. Stair and R. H. Maurer.

As members of the General Board of Examiners:
G. Bertram Regar and H. V. Bozell.

As members of the Committee on Membership:
J. F. Anderson, C. A. Atherton, W. T. Blackwell, Frederick W. Bliss, R. I. Brown, James E. Buckley, W. E. Clement, George G. Cousins, E. C. Crittenden, J. Daniels, Henry D'Olier, Jr., J. Carl Fisher, H. W. Fuller,

H. E. Hobson, M. C. Huse, Karl E. Kilby, James D. Lee, Jr., F. W. Loomis, H. E. Mahan, Charles C. Munroe, John A. O'Rourke, S. L. E. Rose, H. H. Smith, J. L. Stair, L. A. S. Wood.

As members of the Committee on Lighting Legislation:
W. F. Little, W. T. Blackwell, F. C. Caldwell, George S. Crampton, H. B. Dates, E. Y. Davidson, Jr., J. E. Hannum, Ward Harrison, S. G. Hibben, J. E. Hoeveler, James E. Ives, M. G. Lloyd, M. Luckiesh, R. H. Maurer, A. S. McAllister, E. G. Perrot, W. J. Sherrill, R. E. Simpson, George H. Stickney, R. C. Taggart.

As Chairman of the Committee on Sustaining Members: Herbert F. Wallace.

As members of the Committee on Nomemclature and Standards:
Howard Lyon, Secretary, G. A. Hoadley, E. P. Hyde, A. E. Kennelly, M. Luckiesh, C. O. Mallioux, A. S. McAllister, E. C. McKinnie, C. H. Sharp, W. J. Serrill, G. H. Stickney.

As Chairman of the Committee on New Sections and Chapters: D. McFarlan Moore.

As Vice-Chairman of the Committee on Motor Vehicle Lighting: G. H. Stickney.

As Secretary of the Committee on Motor Vehicle Lighting: W. F. Little.

As Local Representatives:
G. O. Hodgson, Denver, Colo., Charles F. Scott, New Haven, Conn., J. Arnold Norcross, New Haven, Conn., R. E. Simpson, Hartford, Conn., R. B. Patterson, Washington, D. C., E. C. Crittenden, Washington, D. C., Charles A. Collier, Atlanta, Ga., W. R. Putnam, Boise, Idaho, H. B. Heyburn, Louisville, Ky., J. F. Murray, Springfield, Mass., Carl D. Knight, Worcester, Mass., Charles C. Monroe, Detroit, Mich., G. D. Shepardson, Minneapolis, Minn., Louis D. Moore, St. Louis, Mo., Frank C. Taylor, Rochester, N. Y., Harwood E. Mahan, Schenectady, N. Y., A. M. Wilson, Cincinnati, Ohio, W. E. Richards, Toledo, Ohio, F. H. Murphy, Portland, Oregon, F. A. Gallagher, Providence, R. I., J. M. Bryant, Austin, Texas, W. S. Rodman, Charlottesville, Va., Fred A. Osborn, Seattle, Wash., F. A. Vaughn, Milwaukee, Wis., L. V. Webber, Montreal, Canada.

As Official Representatives to other Organizations:

The United States National Committee of the International Commission on Illumination—Louis B. Marks and Preston S. Miller.

American Institute of Electrical Engineers Standards Committee—Clayton H. Sharp.

Governing Board of the American Association for the Advancement of Science—Ernest F. Nichols and Clayton H. Sharp.

Advisory Committee, Engineering Division, National Research Council—Dugald C. Jackson.

A. E. S. C. Sectional Committee, Building Exits Code—R. E. Simpson.

New York State Committee on Places of Public Assembly—George H. Stickney.

CHARLES PROTEUS STEINMETZ

By E. W. Rice, Jr.

The sudden death of Doctor Steinmetz comes as a great shock to his friends in the General Electric organization, including the directors, officers and every employe. He joined our ranks some thirty years ago and during all this time has rendered services of the most conspicuous character and extraordinary value. He had a world wide reputation as a scientist, electrical engineer, author and teacher. He was as well known in Europe, South America, Africa, Australia and the Orient as here in his adopted America. Universally acknowledged as one of the world's greatest scientists, he was, if possible, a greater teacher. He was the author of many scientific papers and of a large number of electrical books which have long been the accepted standard text books in colleges, laboratories and workshops everywhere. He possessed marvelous insight into all scientific phenomena and unequalled ability to explain in simple language the most difficult and abstruse problems. Countless electrical engineers now occupying positions of great importance in our Company and elsewhere in the world, will gladly give testimony of their debt to him. All those who knew him mourn the loss not only of a great teacher and an inspiring personality, but of a cheerful and ever helpful friend.

THE DEATH OF DR. STEINMETZ

Charles P. Steinmetz, A. M. Ph. D., chief consulting engineer of the General Electric Company, and one of the world's most famous engineers, died suddenly at his home in Schenectady, N. Y., October 26, from acute dilatation of the heart following chronic myocarditis, a weakening of the heart muscles, of many years standing.

Dr. Steinmetz left Schenectady the first of September intending to take a rest. He wanted to see the Pacific Coast, which he had never visited. Word of his trip sped ahead of him, and requests that he speak at various places immediately resulted. His speaking arrangements were carefully made, but so many wished to hear him that in trying to comply he overtaxed himself. That this was the cause of his death is not certain, however, owing to the nature of his malady.

He returned home September 13th and, an examination by his physician disclosing the fact that his heart action was unsteady, a complete rest was advised. During the first part of the week in which he died, his condition seemed to be improving. He had been about the house and on Thursday night declared he was beginning to feel like himself again. The night before he died he was dressed and passed the evening in his library, discussing with his adopted son, J. LeRoy Hayden, and some friends, a scientific book, "The Physics of the Air." He was bright and cheerful and remarked that he would soon be out again and at work. He retired about 10:30, and awoke the following morning in the best of spirits. He walked up and down the hall and announced to the nurse that he was ready for breakfast. Mr. Hayden was with him for a few minutes, chatting, before he went down to his own breakfast. As Mr. Hayden

Charles P. Steinmetz
1865-1923

went down stairs, he passed his son, "Billy" Hayden, who was on his way up-stairs with Dr. Steinmetz's breakfast. Calling to the latter, young Hayden received no response, and immediately summoned his father. Physicians hurried to the Steinmetz home, but he was dead.

Word was immediately sent to the General Electric Works and the news spread throughout the city, causing great sorrow, for, in addition to their admiration for his scientific attainments, his fellow citizens held him in high personal esteem. Flags at the General Electric Works and in the city were half-masted, the latter by order of the acting mayor.

Charles Proteus Steinmetz was born April 9, 1865, at Breslau, Germany.

Educated at the gymnasium (high school) and then at the University of Breslau, where he studied mathematics and astronomy, then physics and chemistry, and finally, for a short time, medicine and national economy.

Involved in the social democratic agitation against the government, he escaped to Switzerland in 1888, and there studied mechanical engineering at the Polytechnisum.

In 1889 he immigrated to America, and found a position with the Osterheld & Eichemeyer Manufacturing Company, first as draftsman, then as electrical engineer and designer, and finally on research work in charge of the Eichemeyer Laboratory, in New York.

With the absorption of the Eickemeyer-Field interest by the General Electric Company, Dr. Steinmetz joined the latter, was attached to Mr. H. F. Parshall's calculating department in Lynn, Mass. With the transfer of the company's headquarters to Schenectady in the spring of 1894, Dr. Steinmetz organized and took charge of the calculation and design of the Company's apparatus and of the research and development work.

For a number of years Dr. Steinmetz was Professor of Electrical Engineering at Union College, and at the present time is Professor of Electro-Physics at Union University, at the same time retaining his connection with the General Electric Company as chief consulting engineer, and about the year 1910 again entered into closer relation with this company by organizing a consulting engineering department under his charge.

Some of the more important of his publications and articles are:

A series of mathematical papers on Polydimentional involutory correspondence.

A series of investigations on the magnetic circuit and the law of hysteresis.

A series of investigations on dielectric and electrostatic phenomena.

A series of papers on "The Design and Performance of Electrical Apparatus," such as transformers, induction machines, synchronous machines, commutating machines, etc."

A series of papers on "High Frequency Oscillations and Surges in Electric Circuits."

A series of papers on "Radiation, Light and Illumination,"

A series of papers on "Mechanical Thormodynamics and Steam Turbines."

Most of his papers on electrical subjects are published in the TRANSACTIONS of the A. E. E. E.

The following books have been published by Dr. Steinmetz: A popular work on "Astronomy and Meteorology," in the German Language, 1st edition 1889.

"Theory and Calculation of Alternating Current Phenomena," 1st edition 1897, 5th edition 1916.

"Theoretical Elements of Electrical Engineering," 1st, edition 1901, 4th edition 1916.

"General Lectures on Electrical Engineering," 1st edition 1908, 5th edition 1911.

"Theory and Calculation of Transient Electric Phenomena and Oscillations," 1909, 3rd edition 1919.

"Radiation, Light and Illumination," 1st edition 1911, 2nd edition 1915.

"Electric Discharges, Waves and Impulses, "1st edition 1911, end edition 1914.

"Electrical Engineering Mathematics," 1st edition 1911, 2nd edition 1915.

"Theory and Calculation of Electric Circuits," 1917.

"Theory and Calculation of Electrical Apparatus," 1917.

In 1902, Dr. Steinmetz received the honorary A.M. degree from Harvard University, 1903 the honorary Ph.D. degree from Union College.

Dr. Steinmetz is Past-President of the National Association of Corporation Schools, Vice-President of the International Association of Municipal Electricians, Past-President of the American Institute of Electrical Engineers, honorary member of the National Electric Light Association, Past-President of the Illuminating Engineering Society, fellow of the American Association for the Advancement of Science, member of the (British) Institution of Electrical Engineers, member of the American Mathematical Society, the Quaternion Society, the Society of Mechanical Engineers, the Electrochemical Society, the Illuminating Engineering Society, and the Physical Society.

He was a President of the Common Council and twice President of the Board of Education of the City of Schenectady.

NEWS ITEMS

SECTIONAL COMMITTEE ON CODE OF LIGHTING FOR SCHOOL BUILDINGS

Sponsors: Illuminating Engineering Society and American Institute of Architects. Chairman: L. B. Marks, 103 Park Avenue, New York, N. Y. Secretary: W. F. Little, Electrical Testing Laboratories, 80th Street and East End Avenue, New York, N. Y. American Gas Assn.—R. H. Maurer; American Institute of Architects—James O. Betelle, C. E. Dobbin, and S. W. Jones; American Institue of Electrical Engineers—C. E. Clewell, American Medical Assn. (Ophthalmological Section) Dr. Geo. S. Crampton; American Public Health Association—Dr. R. W. Elliott; American School Hygiene Assn— Dr. T. D. Wood; American Society of Safety Engineers—Wm. Darlington;

Eye Sight Conservation Council of America—J. E. Hannum; Illuminating Engineering Society—W. F. Little, L. B. Marks, and M. Luckiesh; National Association Public School Business Officials—James J. Maher; National Bureau of Casualty & Surety Underwriters—Thomas M. Nial; National Committee for the Prevention of Blindness—Mrs. Winifred Hathaway; National Council of Schoolhouse Construction—Hubert Eicher; National Education Association—Frank I. Cooper; National Electric Light Association—Ward Harrison; National Safety Council—G. H. Stickney; U. S. Bureau of Education—Dr. Wm. T. Bawden; U. S. Bureau of Standards—Dr. M. G. Lloyd; U. S. Public Health Service—Dr. Taliaferro Clark; Women's Bureau of the U. S. Department of Labor—Miss Mary N. Winslow.

ILLUMINATION INDEX

PREPARED BY THE COMMITTEE ON PROGRESS

AN INDEX OF REFERENCES to books, papers, editorials, news and abstracts on illuminating engineering and allied subjects. This index is arranged alphabetically according to the names of the reference publications. The references are then given in order of the date of publication. Important references not appearing herein should be called to the attention of the Illuminating Engineering Society, 29 W. 39th St., New York, N. Y.

*Not previously reviewed.

TRANSACTIONS
OF THE
ILLUMINATING ENGINEERING SOCIETY

| Vol. XVIII | November, 1923 | No. 9 |

Are Present Standards of Industrial Lighting Efficient?

IT IS TRUE that any illuminating engineer can point out several lighting installations in factories where the illumination can be fairly said to be not only well planned but adequate, according to our present judgment. Nevertheless taking plant by plant and judging each plant as a whole rather than by the best in each plant, it becomes apparent that there is much to be done before lighting can be called satisfactory. In the same factory or in adjoining buildings there usually can be found one or several miserable examples for ready contrast.

The physical laws governing light and illumination were quite adequately set down many years ago. It must be confessed, however, that in spite of dozens of volumes and hundreds of articles in the technical press, the practical application of the principles of vision, such as the effect of different qualities and quantities of illumination upon sight and upon the comfort and health and sustained efficiency in production have not been so clearly understood or emphasized even by those engaged in the sale of lamps and lighting equipment.

In recent years the scientific group have seemed to have taken an increased interest in the practical problems of lighting, and we have seen as a result papers from them in the TRANSACTIONS of the Society with a special usefulness from the view point of practice. Then the practical experimenters in shops have been applying a greater care in their studies of the effects of different systems of lighting under conditions of actual production. There is becoming available, therefore, an increasing fund of solid information as against speculation and snap judgment to guide practice.

And it would seem that the most important facts that have been developed are of apparently a rather simple and obvious character.

For instance, it is not hard to believe and seems to be substantiated, that for all around purposes, industrial illumination roughly approximating natural daylight in color combines the largest advantages in quickness and accuracy of sight and sustained comfortable vision, while conducing to a cheerful atmosphere. Other colors have advantages for special purposes no doubt, but an error seems to have been often made in applying the results of special experiments to the broadest run of conditions.

Next, light to man is a great deal like air—there can hardly be too much of it unless applied in a violent or unnatural manner. There has been so much of confusion of *glare* with *quantity* of illumination, many have observed bare glaring lamps and have concluded that the quantity of light must be kept low to avoid discomfort or even injury. Practically it appears that there is no limit other than cost to the quantity of light that is useful in industry.

It is to be noted that with higher levels a certain amount of care to diffuse the light reasonably becomes the more essential. But also where the higher levels can be obtained less need is felt for special arrangements of the lighting equipment.

Predictions are venturesome but there seems good reason to believe that in the next decade installations providing well diffused illumination of the order of 25 or 30 foot-candles will become as common as 10 foot-candles are today. No illumination production efficiency tests yet conducted have been carried to a level so high that the added cost of light has overbalanced the gain in production efficiency. Is this not something for thought?

EARL A. ANDERSON

REFLECTIONS

When Daylight Costs More Than Artificial Light

IN IMAGINATION one needs to look backward only a century
to reach a period of inadequate artificial light from candles and
other feeble flames, when light cost fifty times more than it does
now. During that century development gradually gained momentum until within the last score of years the possibilities of artificial
light have increased so rapidly that the general public has not been
able to keep pace with them. Proper and adequate lighting is just
as essential to human progress as clean and sufficient air. The
public has a pretty full appreciation of the hygienic value of fresh
air. The airtight bedchambers of a century ago have given way to
ventilated rooms, but the old attitude toward lighting— a most
natural inheritance of centuries of costly and inadequate artificial
light—persists far too tenaciously. This attitude cannot be changed
too soon in the interests of greater production, greater safety,
greater efficiency and greater progress. Human eyes are the product of countless centuries of adaptation to daylight intensities which
are more than a thousand times greater than artificial lighting
intensities generally but erroneously considered sufficient. The
psycho-physiological laws of vision are proof of this, and they are
well supported by the results from installations of good lighting.

There is another interesting argument to be derived from recently published investigations (by M. Luckiesh and L. L. Holladay)
on the cost of daylight. Daylight is free outdoors when it is available at all. By coming indoors mankind created conditions which
not only decreased the intensity available for indoor activities but
removed daylight from the free list. These investigations have
shown that by charging to daylight the additional cost of window
and skylights, their maintenance, the additional heat losses, etc., natural lighting costs as much as and in many cases more than good
artificial lighting. Owing to the growing complexity of our econom-

ic life such investigations cannot be ignored. The results are bound to have their influence on building in congested districts and perhaps on others.

However, without entering into a discussion of the economic phases, the central-station company and others interested in lighting progress should find much use for the knowledge that daylight indoors costs as much as artificial lighting. In order that the people shall enjoy proper and adequate lighting they must be relieved of the inhibiting attitude that artificial light is costly. That it is a very inexpensive necessity, or that its cost is insignificant as compared with its benefits, can be proved by actual data. But oftentimes an indirect argument is more effective. The simple statement that daylight in homes, offices and factories costs as much as proper and adequate artificial lighting, and sometimes more, is startling, true and convincing—three excellent characteristics for statements made for an educational purpose—*Editorial, Electrical World, Nov. 3, 1923.*

Incandescent Lamps Forty-four Years Old

THE incandescent electric lamp has just celebrated its 44th birthday. On October 21, 1879, the first carbon filament lamp was perfected by Edison in his little Menlo Park laboratory, but three years later, when the first electric central station in the world was started in New York City, there were not enough electric lamps in the world to light a modern office building.

In 1881, when the commercial manufacture of lamps was begun, 30,000 were produced. Last year the total, for the United States alone, was more than 500 million.

Research, experiment and improved manufacturing methods have worked, with decreasing prices of current, to bring the cost of electric light to the consumer to a point lower than it has ever been before.

With the use of tungsten filaments in 1907, there began an improvement in electric lamps that has continued, until the 40-watt lamp of to-day is more than eight times as efficient as the 40-watt lamp of 1907. In 1922 approximately $500,000,000 were spent for electric light. To produce the same amount of light in 1880 would have cost $3,500,000,000.

PAPERS

THE RELATION OF ILLUMINATION TO PRODUCTION*

BY D. P. HESS** and WARD HARRISON***

SYNOPSIS: This paper is a report of extensive tests on the time required for the inspection of parts of roller bearings under various levels of illumination from 5 to 20 foot-candles. Over 7,000,000 separate pieces of material were inspected during the test period. The types of lighting employed as well as the illumination levels were found to have an important bearing on the output of the department. Cost data on the lighting and the value of increased production are included in the paper.

OBJECT

This test was conducted in an effort to establish what relationship, if any, exists between illumination and production in a roller bearing plant.

DESCRIPTION OF TEST DEPARTMENT

About fifteen per cent of all those employed in the plant are inspectors and the section chosen for the test was known as the Green Inspection Department. The nature of the work in this Department consisted of inspecting the material in the green, i.e. just as it was turned out by the automatic screw machines and before heat treating. The material inspected as illustrated in Figure 1 consisted of various sizes of cups, cones and threaded cones, which are separate parts of a roller bearing.

The work is carried on in three stages, the first group of inspectors gauge the material for diameter and depth, the second inspect for defects such as chatter, tool marks, ingot breaks, thin ribs, and bad champfer, and the third group for imperfections in the thread on the inside of the cones, bad mill, bad champfer, and inadequate burnishing. Some of the work such as the inspection of threaded

---

*A Paper presented before the Annual Convention of the Illuminating Engineering Society, Lake George, N. Y., September 24-28, 1923.

**Manager of Columbus Plant of The Timken Roller Bearing Company, Columbus, Ohio.

***Illuminating Engineer, National Lamp Works of G. E. Co., Cleveland, Ohio.

The Illuminating Engineering Society is not responsible for the statements or opinions advanced by contributors.

cones, ingot breaks and chatter marks require close visual inspection, while in some of the gauging, which is done by means of indicating and limit gauges, as illustrated in Figure 2 and 3, relatively little is required of the eyes.

The personnel of this Department at the time the test was begun consisted of a foreman, clerk, and 38 inspectors. This number varied each week, the average number of inspectors during the test being forty-four. Table I shows the weekly change in personnel of the Department during the entire test.

TABLE I.

	Feb. 15–21	Feb. 22–28	Mar. 1–7	Mar. 8–14	Mar. 15–21	Mar. 22–28	Mar. 29–Apr. 4	Apr. 5–11	Apr. 12–18	Apr. 19–25
Total Inspectors	38	48	43	43	41	45	45	46	48	43
New Inspectors Hired	8	12	7	7	1	6	5	4	6	1
Inspectors' Services Discontinued	2	12	7	3	2	5	·3	4	6	6

LIGHTING SYSTEM

The area occupied by the Green Inspection Department has dimensions of approximately 30 by 60 feet and is located near one corner of a large one story building covering about 6 acres. Figure 4.

The lighting system in this Department at the beginning of the test consisted of six outlets, four of which were equipped with 200-watt clear lamps, and two with 150-watt clear lamps. The six lamps were equipped with enameled-steel reflectors and gave an average illumination of about 2 foot-candles. The distribution of light, however, due to wide and irregular spacing of the units, was uneven and caused bad shadows. Figure 5 shows the lighting effect produced by this system.

The Department is so located that it receives daylight from windows at a distance on one side and from skylights located in the saw tooth roof construction. During the greater portion of the time during the first two weeks of the test, the above mentioned lighting system was used as a supplement to the natural daylight. The resulting average illumination in the test section for this period was approximately 5 foot-candles.

Fig. 1.—Various Types of Material Inspected

Fig. 2.—Indicating Gauge

Fig. 3.—Taper Gauges

Fig. 4.—Sectional View of Green Inspection Department

Fig. 5.—Original Lighting System in Green Inspection Department

The new lighting system, Figure 6, consisted of 28 Glassteel Diffusers located on 8 by 10 foot-centers and mounted 12 feet from the floor. This type of lighting unit, Figure 7, has an enclosing globe that entirely surrounds the lamp producing an even distribution of light with soft shadows, and a noticeable absence of objectionable specular reflection. The lighting effect obtained by the above mentioned system is shown in Figure 8.

In an effort to maintain the levels of illumination as uniform as possible under this system, the skylights in the saw-tooth roof were blackened.

PROCEDURE

The test was planned to include investigations of the old system, a 6 foot-candle system, a 13 foot-candle system, and a 20 foot-candle system. The test weeks in each case were begun on a Thursday and ended on a Wednesday, continuing for a period of ten weeks. Due to the advancing season and much stronger daylight it was found impossible to return to the original 5 foot-candle system obtained from the unobstructed skylight and 6 incandescent lamps. Table II gives the order of tests.

TABLE II.

Week	Old Lighting 5.0 Foot-Candles	New Lighting 6.0 Foot-Candles	New Lighting 13.0 Foot-Candles	New Lighting 20.0 Foot-Candles
Feb 15-21	x			
Feb. 22-28	x			
Mar. 1-7				x
Mar. 8-14				
Mar. 15-21		x		
Mar. 22-28				
Mar. 29-Apr. 4				
Apr. 5-11				
Apr. 12-18				
Apr. 19-25				

Records were available in the Green Inspection Department which showed the number of pieces inspected per day with the total number of actual inspection hours. Often, especially in the case of overtime, an inspector is called upon to do general work in the department, such as moving containers and cleaning up. In that event, his time is not shown on the inspectors' records, but only on the time cards. If an experienced inspector is requested to instruct a new worker, the experienced inspector's time does not appear in the inspection records. All employees in this Department are paid on an hourly basis.

The question has often been raised as to whether atmospheric conditions materially affect the output of factory workers. In an effort to obtain data on this point wet and dry bulb thermometer readings were taken four times a day at the test area and from these the relative humidity calculated. The daily weather bureau record showing the per cent sunshine, exterior temperature, and amount of precipitation was also tabulated.

RESULTS

Table III and Figure 9 show the per cent of sunshine, average weekly relative humidity and average weekly interior temperature. It will be noted that the tests were run at a season of the year when artificial heat was necessary and as a result both the interior temperature and the humidity were fairly constant. There was a considerable variation in per cent of sunshine for the several weeks but the variation apparently did not influence the production.

TABLE III.

Week	Percent Sunshine	Relative Humidity	Interior Temperature
Feb. 15-21	63.0	—	—
Feb. 22-28	22.0	—	—
Mar. 1-7	61.0	66.4	74.8
Mar. 8-14	69.0	63.8	72.0
Mar. 15-21	58.0	66.5	71.0
Mar. 22-28	47.0	68.6	72.3
Mar. 29-Apr. 4	62.0	71.0	71.6
Apr. 5-11	76.0	70.3	72.0
Apr. 12-18	52.0	68.3	72.4
Apr. 19-25	90.0	65.5	72.7

Table IV shows the total number of pieces inspected and the number of actual inspection hours for the Department during each of the ten weeks of the test. The total number of pieces inspected was 7,313,323.

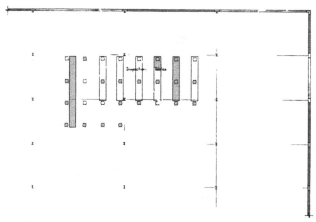

Fig. 6.—— Outlet for a Mazda C lamp fitted with a Glassteel Diffuser located
12 feet above the floor.

Fig. 7.—Glassteel Diffuser

FIG. 8.—New Lighting System in Green Inspection Department

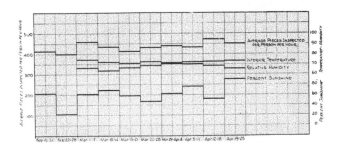

FIG. 9.

TABLE IV.

Foot-Candles		Total Pieces Inspected	Total Inspection Hours	Average Pieces Inspected per Person per Hour
5.0	(Old System)	684,164	1,644	415
5.0	"	581,709	1,449	400
20.0	(New System)	681,621	1,476	462
12.8	"	708,559	1,620	437
5.7	"	739,627	1,778	415
11.9		735,316	1,698	432
20.2		763,762	1,737	440
6.2		809,631	1,866	434
20.2		842,138	1,783	472
13.5		766,796	1,700	451

Table V shows the average of all weeks under the same lighting system with the percentage increase in production.

TABLE V.

Foot-Candles		Average Pieces Inspected per Person per Hour	Increase
20.0		458	12.5%
13.0		440	8.0%
6.0		424	4.0%
5.0	(Old System)	407	0.0%

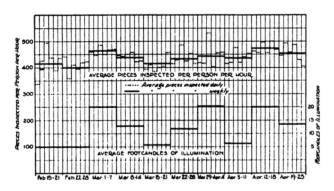

FIG. 10.

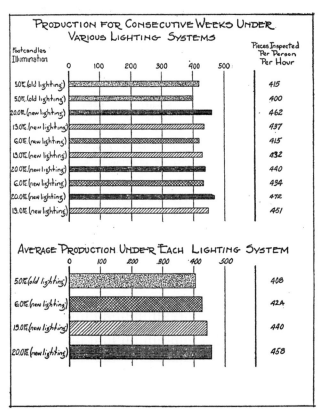

FIG. 11.

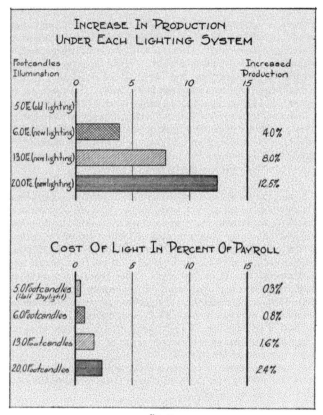

FIG. 12.

These results are also shown graphically on Figures 10, 11 and 12. Figure 10 also shows the daily rate of production for the entire period. The tests furnish apparently conclusive proof that for the class of work carried on in the Green Inspection Department the production is materially affected by the character of illumination supplied. It will be noted that a well designed system of illumination, giving approximately 6 foot-candles with a minimum of glare and objectionable specular reflection, results in an increase of 4 per cent in production over that obtained under a faulty system which gave about the same average foot-candles. Likewise, an increase in illumination from 6 foot-candles to 13 foot-candles with a well designed system results in a further 4 per cent increase in production. Increasing the illumination from 13 foot-candles to 20 foot-candles results in an additional 4.5 per cent increase, or comparing the 20 foot-candle system with the one originally in use a 12.5 per cent increase in production is found.

ECONOMICS

The current consumed with the 20 foot-candle system amounts to 8.4 kilowatts which at a rate of $0.03 per kilowatt hour results in a cost of $0.25 per hour for current to which should be added approximately $0.07 for lamp renewals and other charges making a total cost of $0.32 per hour. With the old 5 foot-candle lighting system about half of which was daylight, the cost for current and lamp renewals was $0.04 per hour. The inspectors receive an average wage rate of approximately $0.30 per hour or for 44 people a total of $13.20 per hour. A 12.5 per cent in production means, therefore, a saving of $1.47 per hour in labor which is five times the added cost of the lighting. To put the matter in another way, the new lighting increased the production 12.5 per cent at a cost of less than 2.5 per cent of the payroll.

The authors desire to express their appreciation of the valuable services rendered by Mr. C. M. Snyder in the conduct of the tests which have been described.

DISCUSSION

GEORGE AINSWORTH: Might I ask if the inspectors knew at about which level the illumination was maintained during the different periods of inspection, if they knew whether they were working under, say, a quarter or a half or twice as much light? Or were they kept in ignorance of these changes?

C. M. SNYDER: The inspectors did not know a test was being conducted. They, of course, could tell the illumination was lower, but they did not know that the lamp sizes were actually being changed. Some of them were under the impression that the lighting equipment was being tested, and a few thought that the lamps were being dimmed and made brighter by means of rheostats.

JULIUS DANIELS: I would like to ask the speaker which system they decided to install.

C. M. SNYDER: The 20 foot-candle system is installed at present.

CHARLES GALLO: I would like to ask when inspecting tool-marks, thin ribs, etc., whether there was any advantage in using that steel-glass diffuser over bowl enameled Mazda C lamps.

It would seem that too much diffused light, for examining such tool-marks, would tend to obliterate the tool-mark instead of making it more prominent, and in that way hinder fast inspection; and, that if there were more directed light, that is, not so great a diffusion, more rapid inspection could possibly be obtained.

C. M. SNYDER: We found that the Glassteel Diffuser gave a very satisfactory illumination for this type of work. If the light is so bright that specular reflection results, it is impossible for the inspectors to see the tool marks or other imperfections due to the amount of light that is thrown back from the material into their eyes. The Glassteel Diffuser gives a very soft, diffused light and still shows up these imperfections very plainly.

THOMAS G. WARD: I would like to ask Mr. Snyder if the worth of the production—the increase in production—was figured in comparison, instead of the wage per hour. The $1.47 per hour saving is a very minor consideration in comparison to the increase in production—the worth of the increase in production.

C. M. SNYDER: I do not believe I understand your question.

THOMAS G. WARD: You get above a standard cost on production. Is the increase in production worth more than the comparison between the saving of labor and the cost of light?

C. M. SNYDER: This is figured on the average wage rate of the inspectors per hour.

C. A. ATHERTON: I assume from the figures that the employees in the portion of the factory tested are all day or hour workers. I wonder if similar tests have ever been conducted in factories in which piece-workers are employed. In this case, there would be the added incentive to work as fast as possible independent of the illumination. I wonder whether the same general order of increase in work accomplished might be expected from increased illumination under conditions in which the workers already had an incentive to work as fast as possible."

C. M. SNYDER: There have been various tests of that description including piece workers and the regular scale. Although there was some overtime taken into consideration here, the results are calculated on an hourly basis.

DAVIS H. TUCK: Mr. Gallo brought up a point a minute ago that I had in mind to bring up and that is the specular from such pieces as this one (indicating) and this one (indicating). Under diffused lighting, you can't pick out imperfections on such specular surfaces as you could if you had some specular reflection, and I believe that in Classes 2 and 3 perhaps the increase in production could be further raised if they had specular reflection.

Furthermore, I think in Table IV on Page 791 if the amount of rejects had been tabulated, it would tend to increase the production shown, because if you pick out some bad pieces you look at them a little longer before you throw them away, and under good illumination there would be more rejects. In such a case, there would be more rejects and I think the increase in production would be a good deal higher than twelve and a half per cent.

G. H. STICKNEY: In connection with this question of lighting specular surfaces for inspection, my experience indicates strongly in favor of diffusing light sources. It is, of course, impracticable to illuminate such a surface. If, however, a light source of low, uniform brilliancy is reflected by the surface, scratches, cracks and distortions show up quite conspicuously.

It is a well established practice, for inspectors use the reflection of the sky in this way, and a lamp with proper diffusing equipment can easily be made to act as a miniature sky. If a lamp filament or other source of uneven brilliancy is so reflected, the contrasts conceal the irregularities of the surface.

I have verified this experience in the inspection of glossy discs for electric meters, rollers for bearings, glass, glossy rubber, etc. In the roller bearing factory, they had experimented for several months with exposed filament equipments, without any success. The diffused light scheme at one-fourth the wattage, worked from the start.

It is, of course, necessary to have the position of the lamp with regard to the operator and work such as to facilitate reflection to the eyes of the operator, that is, the lamp should be above and beyond the work.

J. C. FISHER: If I understood Mr. Snyder correctly, the skylight was blackened during the test and artificial light used entirely. Whether or not that is true I would like to know what the estimated increase in production would be if the daylight were used as far as possible and simply supplemented with the artificial lighting. In other words, I presume the artificial lighting would be used only a part of the total time and therefore there would be quite a fall-off in that percentage increase in production due solely to the artificial light.

I would like to know approximately what that would amount to, because that is a question often raised by people who have a so-called daylight factory in which they object to spending a great deal for an increased lighting installation on that ground.

C. M. SNYDER: Figure 4 shows that a considerable amount of actual daylight was present on the inspection tables from the wall windows, although the skylights were blackened during the test.

I think you will find that the season of the year makes a great deal of difference in the amount of daylight present. This test was run during the winter months, therefore, the percent of sunshine was appreciably lower.

I regret to say that I do not have any exact figures available at the present time as to the percentage increase or decrease that there would be if daylight were used in conjunction with the artificial light.

WARD HARRISON: There are one or two sidelights on this test which are also quite interesting. First, we tried to determine, if possible, if outside conditions such as sunshine, or the lack of it, high temperatures or low temperatures, had anything to do with how much the operator felt like working, or how much they accom-

plished.   It seems to be definitely established, as far as this test is
concerned, that such outside conditions had no appreciable effect
upon production.

One reason that the enclosing fixtures were chosen as a standard for this test was that the size of lamp could be changed without
the operators necessarily being aware of it.   They knew that some
test on fixtures was being conducted, but few, if any of them, sus-
pected that the amount of work which they turned out was the
factor to be determined.   Incidentally they did complain bitterly
when the illumination was reduced to 6 foot-candles, although they
had been able to get along fairly well before under the old, 5 foot-
candle system.

Mr Ward brought out the point that direct labor is only one of
the items in factory production which may be affected by the lighting; another is the quality of the product turned out.   We could
scarcely get any figures on this, however, from the department under
test, since their job is looking for defects in the work of others;   if
they reject 2.4 per cent one day and 1.7 per cent the next, it is
impossible to tell whether the high rejections in the first case is due
to their ability to see better or that a more defective product was
being inspected.

As to the amount of daylight; in February every bit of available light was utilized, that is, all the old artificial lighting was
turned on and all the daylight they could get was allowed to enter,
yet the sum of the two averaged throughout the day was only 5
foot-candles.   Accordingly the Timken Roller Bearing Co., was
interested particularly in securing a lighting system which would
carry them through the winter months of December, January, and
February,—when the daylight itself does not average more than
three or four foot-candles.

JAMES E. IVES: (Communicated)   This paper is an important contribution to the problem of the relation of production to
illumination.   Although this subject is of the greatest interest to
business men and engineers, very little definite information on it
exists at the present time.

The office of the Industrial Hygiene and Sanitation of the U.
S. Public Health Service is particularly interesting in this paper on
account of the investigation recently carried on by it on the same
problem, in the New York City Hall Post Office.   We found a

marked increase in speed of working, among the letter separators, when the illumination was raised from 2.8, successively, to 3.6, 8.0 and 14.0 foot-candles. The workers used in the test were divided into three groups, according as their vision *without glasses* was good, medium or poor. The percentage increase in speed of working varied somewhat with the group, being greater for the poor vision groups than for the good vision group. The intensity of illumination giving maximum speed, also differed for the different groups, the poor vision group reaching its greatest speed under the highest illumination, and the other two groups under 8.0 foot-candles. On reducing the illumination, however, to its original value of 2.8 foot-candles we found that the production did not return to its original value but to a value considerably above it. The same phenomenon was noted recently in an experiment in England on the relation of illumination to coal production. When a coal miner used an ordinary lamp the production was 2.47 tons; with a lamp six times as bright, it was 2.83 tons, an increase of 14.6 per cent. When the miner returned to the use of the ordinary lamp, the production did not return to its original value but still remained 5.4 per cent above it.

This failure to return to the original speed of working, or of production, when the illumination is returned to its original value makes the interpretation of the result rather difficult, and it is felt by us that before the results obtained by us can be completely understood it will be necessary to find an explanation of this peculiar phenomenon. Another question that naturally presents itself is: "Will the increased speed obtained under increased illumination be maintained, or disappear, in part or in whole, as the time goes on?" It is natural to expect that some of it at least will be maintained, but we feel that it will be necessary to make further tests to establish the facts. The tests recently carried on by the Detroit Piston Ring Company would seem to indicate that the increased speed is maintained. In connection with the tests made by Mr. Hess and Mr. Harrison the same question presents itself, namely, "Would the increase in production observed by them be maintained over a long period of time if the increased illumination is maintained?"

WARD HARRISON (Communicated): Dr. Ives stated that in the New York Post Office tests the speed of work increased with the illumination but did not drop down all the way to its original value when the illumination was again reduced to 2.8 foot-candles. I have heard indirectly of other instances where the same phenomenon has been observed. The best explanation that I have heard was that suggested by Dr. E. F. Nichols, namely that the ease of seeing under the high illumination establishes the habit of faster work and that this is not immediately shaken off when the illumination is reduced.to the former low value. Likewise, the ultimate high speed of work obtainable under a high level of illumination is quite possibly not secured during the first few days of work under such illumination.

SALIENT FEATURES IN POWER STATION LIGHTING*

BY RAYMOND A. HOPKINS**

SYNOPSIS: Unusual problems are met in the lighting of power stations on account of individual arrangement of equipment, severe service conditions and exacting requirements. The successful lighting system must be reliable, economical, easy to maintain and adequately suited to the specific local requirements which requirements are found to differ throughout the station. The most reliable and economical source of energy is usually the station auxiliary bus. The distribution wiring should be of the particular quality best suited to meet power station conditions and should be designed to give the best possible voltage regulation consistent with economy. An emergency lighting system should be provided and of several possible arrangements the one giving greatest dependability should be selected. All equipment such as cabinets, switches, receptacles, lamps, globes, shades and reflectors should be carefully selected to give maximum operating convenience, long life and high efficiency. A thorough survey of a large number of existing first-class power stations gives data for the solution of a number of typical station illumination problems so selected that the designing engineer may extend the data and conclusions given to meet the requirements of any ordinary station.

GENERAL CONSIDERATIONS

Steam and hydroelectric power generating stations and substations present unusual lighting problems seldom met elsewhere in the industrial field. In the coal bunker, the firing aisle, the ash room, the switch room, and many other parts of the station, the conditions are so severe and the requirements so exacting that the design deserves very careful study. It is the intention of this paper to discuss a limited number of specific problems and to submit solutions for them, based on actual data from a large number of successful installations to the end that the designing engineer may readily extend the data and conclusions to completely cover the requirements of any ordinary station.

Reliability of the lighting system is of first importance since light is depended upon for the reading of meters and gauges, the operation of controls, the inspection of equipment and the supervision of all working parts of the station which themselves are

*A Paper presented before the Annual Convention of the Illuminating Engineering Society, Lake George, N. Y., September 24-28, 1923.

**Engineer, Stone & Webster, Inc., Boston, Mass.

The Illuminating Engineering Society is not responsible for the statements or opinions advanced by contributors.

essential to the continuous production of electric energy. The complete lighting system should, therefore, be of rugged construction and of ample capacity and should include spare transformers and other essential parts. Moreover, in order to provide an immediate relay against even a momentary failure of the normal lighting system, an auxiliary or emergency system should be arranged to serve all important parts of the station.

As an example to the public who support the power station by purchasing its output, the station lighting should represent the latest achievements in correct illumination. The use of wasteful and antiquated lighting equipment should be avoided and the installation should be designed to permanently secure proper illumination intensities and satisfactory operating conditions.

Maintenance, with every lighting system, is essential to successful performance. Particularly in the power station, on account of the unusual ceiling heights, the hazardous locations and the excessive dust, the problem of maintenance deserves earnest attention. Equipment should be selected which endures severe service and which is readily disassembled for cleaning and relamping, and it should be located so as to be easily accessible.

Less illumination is generally required for normal operation of the station than is required for maintenance or repair work. During normal operation there are instruments, gauges, bearings and other details to be watched and certain routine operations to be performed, but many parts of the station can be left relatively unattended except for regular scheduled inspections. These periodical inspections, however, require ample lighting throughout the station and repair work and overhauling of machines, which is often done on a rush basis, deserve the best possible lighting. It is, therefore, important that the lighting system be made flexible by providing plenty of local switching and of plug receptacles.

SOURCES OF ENERGY

A choice is generally to be made between several available sources of energy for the lighting system such as the main station bus, the station auxiliary bus, the exciter bus, the storage battery, or an outside source. The chief considerations involved in making the choice are reliability, economy and voltage regulation.

The main station bus in a direct current station is generally the most reliable and economical source for the lighting and is usually chosen on this basis, although the voltage regulation may not be all that is desired. Lamps of the 110 to 125-volt or 220 to 250-volt class are generally used and when the bus voltage is 550 or 600 volts, the lamps are grouped in series.

The main station bus in an alternating current station is in itself a reliable and economical source. When the voltage is over 2300 volts, however, the transforming and switching equipment are expensive and also constitute a possible hazard to the lighting system. Moreover, the voltage on the main bus is often varied as much as 10 per cent above and below normal at different times of the day to suit the load requirements, this condition necessitating the use of a feeder voltage regulator on the lighting circuit. For these reasons, the main bus when of over 2300 volts, is seldom used as a source for the station lighting.

The station auxiliary bus from which all motor driven auxiliaries are fed is usually considered the best source of energy for the lighting system. It is generally operated at 2300 volts and proper transformers and circuit breakers are used to supply energy to the lighting system. Since the continuous operation of the station auxiliaries is very important even under conditions of power interruption or voltage surge on the main bus, the station auxiliary bus is usually supported by at least two sources of energy, at least one of which is a prime mover. This assures a reliable, economical and usually well regulated source. A feeder voltage regulator on the lighting feeder is necessary only when the auxiliary bus for some reason does not carry a uniform voltage.

The exciter bus is not considered a suitable source for the normal lighting. It is reliable but not particularly economical and when automatic voltage regulators are used on the exciters, the voltage regulation of the exciter bus is often as great as 60 per cent above and below normal. Furthermore, the lighting system, if fed from the exciter bus, is somewhat of a hazard to the excitation system. The exciter bus is sometimes used as a source for the emergency lighting system but even this practice is not to be recommended.

The storage battery is a most reliable secondary source of energy for the lighting system and with proper charging connec-

COMPARISON OF DISTRIBUTION SYSTEMS
(BASED ON EQUAL FEEDER LENGTHS AND EQUAL LOADS, AND USING SYSTEM "A" AS A BASE FOR COMPARISON)

SYSTEM	DIAGRAM	DESCRIPTION	CURRENT	COPPER SIZE BASED ON CURRENT DENSITY (SHORT FEEDERS)				COPPER SIZE BASED ON PERCENT VOLTAGE DROP (LONG FEEDERS)			
				COPPER SIZE	TOTAL COPPER WEIGHT	PERCENT VOLTAGE DROP	POWER LOSS	COPPER SIZE	TOTAL COPPER WEIGHT	PERCENT VOLTAGE DROP	POWER LOSS
A		1 Ph or d c 2 Wire 110 Volt	100	100	100	1.00	100	100	100	1.00	100
B	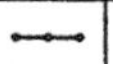	1 Ph or d c 3 Wire 220-110 Volt	Outside 50 Neutral 0	Outside 50 Neutral 50	75	.50	50	Outside 25 Neutral 25	37.5	1.00	100
C		2 Ph 3 Wire 110 Volt	Outside 50 Common 70.7	Outside 50 Common 70.7	85.3	1.00	85	Outside 50 Common 70.7	85.3	1.00	85
D		2 Ph 4 Wire 110 Volt	50	50	100	1.00	100	50	100	1.00	100
E	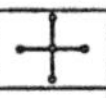	2 Ph 5 Wire 220-110 Volt	Outside 25 Neutral 0	Outside 25 Neutral 25	62.5	.50	50	Outside 12.5 Neutral 12.5	31.25	1.00	100
F		3 Ph 3 Wire 110 Volt	57.7	57.7	86.5	.865	86.5	50	75	1.00	100
G		3 Ph 4 Wire 191-110 Volt	Outside 33.3 Neutral 0	Outside 33.3 Neutral 33.3	66.7	.50	50	Outside 16.7 Neutral 16.7	33.3	1.00	100

FIG. 1.—Comparison of Distribution Systems

tions the voltage regulation is satisfactory. The cost of the battery and its poor efficiency at once prohibit its use for normal lighting but it is very commonly used as a source for emergency lighting. When so arranged the battery size is determined by the demand of the emergency lighting system plus the demand of the control system. The latter is likely to be heavy during a period of emergency since at such a time considerable switching is done. A maximum allowance of one hour for demand on the emergency lighting system at any one time is considered ample to meet the most extreme conditions. Since the control circuits constitute a very sensitive and important part of the station wiring, it is important that the emergency lighting system if fed from the control battery, be kept free from grounds or other faults which may cause trouble with the controls.

An outside service is sometimes used as a source of energy for the lighting system where the station voltage or frequency are not suitable for lighting circuits. Especially in small stations of this nature it is often found more economical to purchase energy for lighting than to install necessary transforming or converting equipment.

DISTRIBUTION SYSTEM

The materials and workmanship constituting the complete lighting system from the source of energy to and including the lamps should be of the best quality in order to insure dependable operation comparable to that of the major station equipment.

Data on a number of the most commonly used distribution systems are given in Figure 1. It will be seen that where the feeder is so short that the copper size is based on current density or carrying capacity, the use of one of the higher voltage systems as B, E or G results in a considerable saving in copper and in less voltage drop and power loss. Also, when the feeder is so long that the copper size is based on allowable voltage drop, the higher voltage systems result in even more saving in copper.

System B is probably the most commonly used. Being a single phase system it requires only one transformer while a polyphase system requires more than one. With a single phase system the lighting is all carried on one phase of the station, but this is seldom found objectionable.

The neutrals of systems B, E and G should always be grounded and the neutral conductor unfused. The size of the neutral should be determined by the size of the fuses in the outside conductors where the circuit is protected by fuses, since the blowing of a single fuse is liable to result in the neutral wire carrying the full load of one of the outside wires. Where a circuit breaker is used, all poles of which trip at the same time, the neutral wire need be only large enough to carry the unbalanced current. For system C the common conductor must have 1.41 times the capacity of each outside conductor. If the system is convertible the common conductor must have double the capacity of each outside. An example of this connection is found where the emergency lighting circuits are normally operated as system B, but in emergency are operated as system A by connecting the two outside conductors to the positive and the neutral to the negative.

The grouping of several cabinets on one feeder is not desirable from the standpoint of flexibility in switching, testing and maintenance but is often done for economic reasons. By referring to Figure 2 it will be noted that when two or more cabinets are located close to each other and at some distance from the bus as in A and B, the most economical arrangement is to group the

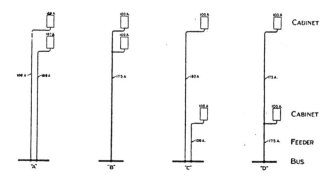

GROUPING OF DISTRIBUTION CABINETS SHOWING COMPARATIVE ECONOMIES
A & D   UNECONOMICAL
B & C   ECONOMICAL
FIG. 2.—Comparisons of Cabinet Groupings.

cabinets as shown in B.   This not only takes account of the fact
that one feeder costs less than two, even when equal weights
of copper are involved, but also of the fact that the single feeder
will contain less copper on account of diversity between cabinet
loads.   On the other hand when the cabinets are scattered as in
C and D, the most economical arrangement is separate feeders
as shown in C.   The common feeder as shown in D could be used
to advantage by fusing the feeder just beyond the first cabinet
but fuses in the feeders at other points than at the main
switchboard are not to be recommended.

Distribution cabinets of various styles are available and the
choice depends on a great many local conditions.   Main fuses are
necessary unless the cabinet buses are protected by the feeder main
fuse.   The main switch in the cabinet is optional.   It is convenient
for changing of branch fuses, for making changes in the cabinet
wiring and for controlling the entire cabinet in cases where control
of this kind is desirable.   The main switch is sometimes electrically
operated so that the entire cabinet can be controlled from a
distance.   Such a control is convenient for the distribution
cabinets supplying large turbine rooms, since the cabinets can be
located up near the lamps and the control switches can be located
at a convenient place on the turbine room floor.   Magnetic
switches are available in two types, one of which is operated by a

momentary contact control switch and the other by a single pole, single throw control switch. In laying out the lighting system it will be found that each type has distinct advantages over the other to meet certain conditions.

Branch fuses are necessary in all cases. At present two-pole fuses are required by code but it is expected that in the near future single-pole branch fuses will be allowed under specified conditions. Branch switches are optional. They are unnecessary where the cabinets are located with respect to the centers of distribution and are used only as fuse boxes, the control being entirely by local switches. On the other hand, branch switches are needed where the cabinets are located not necessarily in the centers of distribution but at convenient control points, and are used to control the lighting without local switches. The latter arrangement has the disadvantage that lights may be turned off by mistake in rooms where they are urgently needed.

Flush type and surface type, dead front and open front cabinets are available and the choice depends upon their locations. Directory frames are always convenient when replacing fuses. Wiring gutters are necessary in the majority of cases and it is convenient to provide wider gutters at the tops and bottoms than at the sides, since most of the wiring enters the cabinet at the top and the bottom.

Coding of the branch wiring where the neutral is grounded is becoming standard practice. It is convenient to use a white or light colored braid for the grounded wire and the usual black braid for the other wires. The shell sides of all sockets and receptacles are connected to the white wire, and all single-pole switches are placed in the black wire.

Special insulation is very desirable in boiler rooms and in condenser rooms where the wire is subject to severe heat and moisture conditions. In these locations the ambient temperature is often above the safe temperature for rubber insulation, and escaping steam sometimes produces condensation which is liable to saturate the braid. Several types of insulation and covering have been tried including asbestos compounds, fibre compounds, waterproof impregnations and varnished cambric with lead covering. It is very difficult to find an insulation and braid which will permanently resist both the moisture and the heat.

Generally if it is heat-resisting, it quickly absorbs moisture and on the other hand if it is impregnated to resist moisture the heat quickly dries out the impregnating compound. Varnished cambric insulation with lead covering is quite satisfactory but very expensive. Varnished cambric insulation with braid covering gives very good results. None of the usual rubber compounds are suitable, but certain vulcanized compounds are available which are satisfactory for boiler room use. Much trouble from excessive temperature can be avoided by carefully selecting the locations of conduit runs.

EMERGENCY LIGHTING

The emergency lighting system has been mentioned a number of times in the preceding paragraphs. The most carefully designed and maintained normal lighting system is not infallible. In case of system disturbance or loss of a lighting transformer or other accident that extinguishes the normal lighting system it is of utmost importance to station operation that the emergency lighting system be automatically put into operation at once. The important parts of the station to be covered by emergency lighting are water columns, boiler gauges and meters, stoker and fan controls, boiler feed pumps, turbine gauges and controls, turbine auxiliaries, valve controls, oil circuit breakers and disconnecting switches, lighting transformer and switch rooms, switchboard rooms, load dispatcher's office and all stairways and passages.

Three commonly used systems of emergency lighting connections are shown diagrammatically in Figure 3. In each system the emergency wiring is segregated from the normal wiring by the use of separate conduits, cabinets and outlets. It is often found convenient, however, and perfectly satisfactory, to use combination cabinets with barriers separating the normal from the emergency circuits. One of the three systems shown is generally selected although it is sometimes found in working out the details that combinations of all three are necessary.

System A consists of a number of small auxiliary lamps arranged to burn normally and in emergency on the emergency source. Since the lamps are kept burning continuously and are always fed from the emergency source, the number and size of the lamps must be kept small so as to conserve energy. The lamps therefore cannot be used to supply any portion of the normal

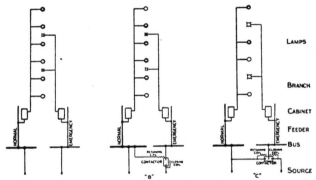

DIAGRAMS OF EMERGENCY DISTRIBUTION SYSTEMS

A - EMERGENCY LAMPS ALWAYS BURN ON EMERGENCY SOURCE
B - EMERGENCY LAMPS NORMALLY DARK, BUT IN EMERGENCY BURN ON EMERGENCY SOURCE
C - LAMPS NORMALLY BURN ON NORMAL SOURCE, BUT IN EMERGENCY BURN ON EMERGENCY SOURCE

FIG. 3—Diagrams of Emergency Distribution Systems

illumination but must be considered as auxiliary or supplementary
to the normal lamps. Thus each gauge, for instance, must have
two lamps, a normal and an emergency. Local switching is not
provided for fear the lamps will be turned off locally at the time
an emergency occurs. The feeders are sometimes switched at the
main switchboard and the operator is instructed to turn off the
lights during the day in those portions of the station which do
not require lighting in the daytime. This emergency system being
supplementary to the normal system is apt to be robbed of lamps
and fuses and to otherwise become deteriorated, since under
normal conditions its loss is not felt. The system therefore,
requires rigid routine testing and maintenance.

System B consists of a number of small auxiliary lamps
arranged to be normally dark but in emergency to burn on the
emergency source. Somewhat larger lamps can be used than in
system A, since they burn only in case of emergency, but as in A,
all gauges and other vital points must have duplicate lamps.
Local switching is not desirable. The main switch for the entire
system consists of a contactor, normally held open by a potential
coil energized from the normal source, but automatically closed
by energy from the emergency source, whenever the normal

potential fails. This being a supplementary system, requires the same rigid routine testing and maintenance as system A.

System C consists of a carefully selected portion of the normal lamps to be designated as emergency lamps and arranged to burn normally on the normal source and in emergency on the emergency source. The size and number of lamps, as with system B, are limited only by the energy available at the emergency source for the short time that the emergency exists. Moreover, since the lamps burn normally on the normal source, no duplication of lamps is required at gauges and other similar places. Local switching can be provided with perfect safety because the emergency lights are required for normal operation and will always be switched on when needed. It is important, however, to arrange the switching so that where an emergency light and a normal light occur in the same room for instance, both lights will be switched at the same operation. This has been successfully done by using a two-pole switch, wiring one pole in the emergency circuit, the other in the normal circuit. Likewise, in a large interior, such as the turbine room, a single emergency circuit can be controlled simultaneously with a number of normal circuits by the use of a double pole switch, one pole of which controls the emergency circuit directly, the other the normal circuits collectively through a contactor cabinet. The main switch for system C consists of two contactors interlocked so as to form a double throw switch. Normally the switch feeds the emergency lights from the normal source but in the event that the normal source fails, the contactor is automatically thrown over by energy from the emergency source so as to feed the lamps from the emergency source. The action is reversed when the normal potential is restored. Since with this system the lamps form a part of the regular lighting, their loss is immediately felt and therefore this system is more likely to be properly maintained than either of the other two systems.

EQUIPMENT

All equipment including cabinets, switches, receptacles, lamps and reflectors should be of especially rugged design consistent with the reliability required of the lighting system. Rugged construction can be obtained by the use of heavy conduit and outlet fittings, cast iron outlet boxes, cast iron hoods for supporting reflectors, porcelain enameled steel reflectors, heavy glassware, wire guards, rubber and steam gaskets, substantial sockets, positive locking devices on all screws supporting reflectors and other similar devices. Long supporting stems should be made flexible by the use of double strip flexible steel conduit or of flexible metallic steam hose. Aligners or flexible joints with rigid stems also give some degree of flexibility. Heavy equipment should not be supported from outlet box covers but should be attached directly to the boxes by substantial threaded fittings.

All equipment should be easy to clean, relamp and maintain This is of especial importance in the electrical bay where the lights are sometimes located in hazardous places. The ceiling lights in the turbine room are reached from the crane and it is very convenient to be able to unhook a complete unit and lower it to the floor for cleaning.

Steelware is generally used in the boiler house and condenser rooms and glassware is generally used in the turbine room, electrical bay and offices. Glare should be avoided by using bowl enameled lamps in all open reflectors and by using good quality diffusing glassware of large size for closed units.

Receptacles of the marine type are desirable for the ash room, boiler room and condenser room and are often used throughout the station for purposes of standardization. The screwed cap excludes dirt and moisture when the receptacle is not in use. Heater receptacles are often required in hydroelectric stations and substations where steam heat is not available and in electrical bays of steam stations where steam heat is objectionable. It is desirable to provide three-wire receptacles so that the heater frames may be grounded.

DESIGN DATA

In some parts of the station, such as the turbine room and the offices, the usual flux of light methods of design can be applied to advantage. For the majority of the station, however, the chief problem is to avoid shadows and glare and to locate the individual lights effectively with respect to the equipment and it is found that the efficiency of utilization and the watts per sq. ft. are themselves of very little assistance.

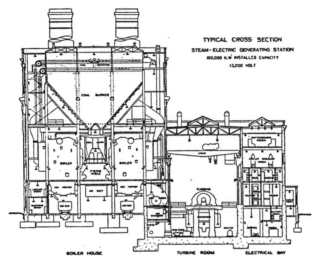

Fig. 4—Typical Cross Section of Steam Electric Generating System.

Figure 4 shows a cross sectional view of a typical coal burning steam-electric generating station of 100,000 kw. installed capacity, generating and delivering energy at 13,200 volts. The majority of the specified problems discussed below are illustrated in this cross section.

The design data given are typical of the actual conditions found in a recent survey of a number of first class power stations. The illumination intensities are the direct results of tests made in these stations with a Macbeth illuminometer. In making the tests no attempt was made to clean or renovate the lighting equipment before test, and the intensity values recommended are net, depreciated values. It is expected that in a new installation they would be from 20 to 40 per cent higher in order to allow for depreciation.

Fundamental design data compiled from a number of representative steam stations comparable to the one illustrated in Figure 4 are as follows:

	Boiler House	Turbine Room	Electrical Bay	Entire Station
Lighted Floor Area, sq. ft.	55,000	45,000	35,000	135,000
Number Normal Lights	300	250	400	950
Number Emergency Lights	120	50	140	310
Total Number Lights	420	300	540	1,260
Connected Normal Watts	31,000	49,000	33,000	113,000
Connected Emergency Watts	8,500	10,500	11,000	30,000
Total Connected Watts	39,500	59,500	44,000	143,000
Total Watts per sq. ft.	0.71	1.32	1.26	1.06

COAL PILE AND YARD

The coal pile and yard are not unlike any industrial plant yard. Ordinary operation of the station does not require night work in the yard but conditions often arise which make night work necessary and the lighting system should adequately meet all such demands. Emergency lighting is seldom used in the yard. Either flood lighting from projectors or local lighting from pole tops can be successfully applied. The former is generally less expensive to install and more easily maintained. The boiler house roof and the coal conveyor house offer convenient locations for projectors. Switching should be located in the boiler house near the door used by the workmen. It is convenient to use standard lamps in the projectors and to select them of the same size as those used elsewhere in the station.

For unloading coal and conveying coal to the crusher, an average intensity of 0.7 fc. is desirable. With 1000 watt projectors this requires about 0.1 watts per sq. ft. For roadways, an intensity of 0.2 fc. is sufficient and with 200 watt projectors this requires about 0.03 watts per sq. ft.

COAL BUNKER AND CONVEYOR

The coal bunker and conveyer illustrated in Figure 5 is generally the dustiest part of the station. The coal is usually crushed in the yard and conveyed by belt or bucket up over the bunker where it is dropped into the bunker. The crusher house, the conveyor housing and the bunker are usually provided with natural light during the day when most of the work is performed but must have sufficient electric light for operation at night in cases of necessity. Emergency lighting is required only for safe passage and should constitute 15 per cent of the total wattage.

Bare lamps are often used but better results are obtained by adding porcelain enameled steel reflectors. Outer globes and wire guards should be used where there is danger of mechanical injury. Lamps should be located within easy reach for cleaning and this is an important consideration, especially over the bunker, where a fall by the cleaner might be very serious. Three-way switching is usually necessary. An intensity of 2 fc. is desirable and with 75- or 100-watt lamps this requires 0.5 watts per sq. ft.

FIRING AISLE

The firing aisle illustrated in Figure 6 is the most important part of the boiler house. Here are located the water columns, steam gauges, draft gauges and the controls for the larry, stokers, fans and auxiliaries. In many stations the firing aisle receives no natural light although the present tendency is to dispose of the coal bunker at the end or middle of the building so as to allow a skylight to be located over nearly the entire length of the firing aisle. The normal lighting should be adequate for operation of the boilers both night and day. The emergency lighting should be sufficient to allow full operation without normal lighting which means that 25 per cent of the general lights and all of the local lights should be on the emergency system. If general overhead lighting is used the lamps should be carefully located and properly shielded to avoid glare in the eyes of the operator when looking upward at the water columns and steam gauges. These overhead lights must also be hung above the larry and this, of course, puts them in a difficult position to maintain. On account of these limitations, it is sometimes considered satisfactory to leave the upper part of the firing aisle generally dark with strong local lights on the water columns and gauges.

The water column being located some 30 ft. above the floor but at the same time requiring constant watching by the operator, deserves very careful consideration. The light should preferably be directed from below so as to avoid specular reflection from the gauge glass. Moreover, the lighting equipment must be very rugged in order to withstand the severe force of an exploding water glass. Being arranged to face upward it must have some efficient means of removing the dust which quickly accumulates. The water column illuminator shown in Figure 7 has been found

FIG. 5—Coal Conveyor and Bunker.

FIG. 6.—Boiler Room Firing Aisle.

Fig. 8—Ash Room.

Fig. 7.—Water Column Illuminator.

well qualified for this severe duty. The two beams of light from the two 40-watt lamps are focused by the lenses to impinge on the meniscus of the water column and cause it to glow so that it can be readily seen from any part of the firing aisle. A series of small holes on the casing around each lens are arranged for connection to the station compressed air system or to the forced air duct and the streams of air effectively prevent the accumulation of dust. The device has always proved its ability to withstand water glass explosions due to its general rugged construction and to the special heat-resisting quality of its lense.

The steam gauge which is some 20 ft. above the floor, the clock, and the load indicator should also be lighted from below to avoid reflection. Excellent results are obtained by using a single 40-watt unit of the water column illuminator just described.

The gauge board generally contains a group of indicators and gauges and is successfully lighted by one or two 75-watt angular steel reflector units.

The stokers and drive mechanisms are generally located under the hoppers. These are adequately lighted by 75-watt angular reflectors hung from the hoppers. The walkway over the hoppers and the tops of the hoppers themselves should be well lighted since this walkway is provided for the use of the operator while inspecting the contents of the hoppers and breaking up masses of coal that may become clogged. Deep bowl or RLM units of 100-watt size are satisfactory. They should, of course, be switched locally so as to be dark always except when actually required for the above mentioned operations.

The larry requires one or more 75-watt deep bowl or angular reflectors in the cab to light the controls and weighing scales. It is also desirable to supply one or two 100-watt angle reflector units for each chute to light the end of the chute and the hopper during the dumping process. The lights on the larry are generally wired in series circuits so as to be fed from the larry motor circuit.

At the throat of the bunker one or more 75-watt reflector units may be used advantageously to assist the operator in opening the gates and discharging coal into the larry.

BETWEEN AND BACK OF BOILERS

Between the boilers and back of the boilers the passages require sufficient lighting to allow the operators perfect freedom about the station. The emergency lighting should constitute 20 per cent of the total wattage. Stoker drives which may be between boilers and important valves which are often behind the boilers require particularly good emergency lighting. These passages are satisfactorily lighted with 100- or 150-watt RLM units. An average intensity of 3 fc. is desired and this requires 0.6 watts per sq. ft.

OVER BOILERS

Over the boilers are walkways for inspecting the boiler tops, safety valves and water columns and for operating non-return and other valves. The emergency lighting should provide for safe passage over the walkways and for operation of the non-return valves. Generally 50 per cent of the walkway lighting should be on the emergency system. Lights located over the boilers should be well back from the firing aisle so as to avoid glare in the eyes of the operator below. They should also be switched locally since they are only needed occasionally. Generally a 100-watt RLM unit over each boiler is sufficient. For the walkway over the front of the boilers and also for the walkway at the back of the boilers 50-watt marine brackets may be used with outer globes and wire guards. Those for the front walkways should be located close against the columns on the side away from the firing aisle so as not to interfere with the reading of the water gauges.

ASH ROOM

The ash room, Figure 8, presents a peculiar problem because most of the equipment to be operated is on the ceiling. The railroad cars are pushed under the ash hoppers, the hopper gates are opened and the ashes drop into the cars. The ashes are generally mixed with water so that the lighting equipment is likely to be injured by steam and acid as well as by dust. Emergency lighting is required only for passage and should constitute 15 per cent of the total wattage. Since daylight is generally excluded from the ash room, some of the lamps should burn continuously. Enameled steel angle reflector units of 100- to 150-watt size or smaller units of the totally enclosed and wire guarded marine type are found satisfactory depending upon conditions. They should be located so as to light the controls without causing glare in the eyes of the workmen. The forced draft duct often occupies the center bay and under this should be located 75- or 100-watt RLM units for general illumination. Receptacles should be used to facilitate repair work and the ash room is one of the places where marine type receptacles are necessary. The ash room lights should be switched locally at the columns since work is done only in portions of the room at a time.

TURBINE ROOM

The turbine room, Figure 9, presents the large interior problem. The crane spans the entire room and ceiling lights must be located above the crane trolley. Bracket lights if used must be close to the walls to avoid the crane hook. The horizontal component of illumination is important in order to read the vertical gauges. During normal operation an average horizontal intensity of 4 fc. is sufficient but when a generating unit is down for repairs an intensity of 8 fc. is desirable. Consequently a flexible switching arrangement is needed. A very convenient switching arrangement consists of one or more cabinets with branch switches and with magnetically operated main switches. The lighting for the entire room is controlled by one or two pilot switches and the amount of illumination is varied by adjusting the branch switches. Emergency lighting should be provided to give an average of 2 fc. which under stress of necessity will permit full operation. With prismatic reflector units of 750-watt size and with light buff colored brick walls it is found that an intensity of 4 fc. requires 0.5 watts per sq. ft.

CONDENSER FLOOR

The condenser floor often presents some confusion in the mind of the lighting designer. The turbine usually rests on a platform isolated from the rest of the building and standing on its own foundation. Under and around the turbine platform are assembled the various station auxiliaries, such as pumps, fans, valves and controls. The lights in general must be hung from the turbine platform or supported from the condenser and must be located below the mass of large and small pipes that often occupy most of the space overhead. Parts of the condenser floor are sometimes lighted naturally but often the entire room receives no daylight. Constant supervision of the auxiliaries is a vital part of the station operation. Emergency lighting should constitute 20 per cent of the total wattage. Local switching should be provided so that lights may be turned out around machines which are shut down. Receptacles are needed frequently for inspection and repair work and should be of the waterproof type. With RLM enameled steel reflectors and 100- or 150-watt lamps it is found that an intensity of 4.0 fc. requires 0.75 watts per sq. ft. of working space.

BATTERY ROOM

The battery room requires acidproof equipment. It is customary to use a totally enclosing prismatic reflector which screws into a cast-iron box embedded within the ceiling concrete so that only the glass itself is exposed to the action of the acid-laden fumes. The switch is placed outside and no receptacles are provided. Emergency lighting should constitute 25 per cent of the total wattage. Battery rooms have sometimes been painted black but this is unnecessary since light-colored acid-resisting paints are now available. With light gray walls and ceiling and with 75-watt lamps in totally enclosing prismatic reflectors, it is found that a density of 1.0 watt per sq. ft. gives an intensity of 4.5 fc. which is considered satisfactory.

BUS ROOM

The bus room which is purely a bus room without switches or other operative equipment does not present a serious problem. The lighting should provide for reading the captions over the cells and for inspecting and cleaning the insulators and copper work. In order to effectively light the spaces between the horizontal barriers, it is advisable to hang the lamps on stems, to use diffusing glassware and to provide light colored walls. Where stems can not be used, receptacles for portable lamps should be provided. Local switching is desirable and also three-way switching when the room is entered from both ends. Emergency lighting is needed only for safe passage and should constitute 20 per cent of the total wattage. An illumination of 3 fc. on a horizontal 30″ plane and of 1.5 fc. on the face of the bus structure is desirable. With white walls and ceiling and with 75-watt

FIG. 11.—Switchboard Room Lighted by Concealed Sources with Prismatic Glass Ceiling.

FIG. 12.—Night View of Outdoor Switching Station.

Fig. 10—Circuit Breaker and Disconnection Switch Room.

Fig. 9.—Turbine Room.

prismatic units, this requires 1.25 watts per sq. ft. With red brick or concrete walls and bus structure the wattage per sq. ft. would need to be increased to 1.75.

SWITCH ROOM

The switch room, Figure 10, generally contains oil circuit breakers and disconnecting switches. This equipment requires very close supervision and careful inspection.   The disconnecting switches are often opened and closed manually by means of a wooden pole with a hook on one end which is inserted into a ring or eye on the switch blade. For normal operation the most exacting requirements of the lighting system are to provide for clear reading of the captions over the compartments and for operation of the disconnecting switches. For maintenance the lighting should cover the breaker mechanisms and all insulators and copper work. Receptacles should also be provided for extension lamps for inspecting the inside of the breaker cells. Local three-way switching is desirable. Emergency lighting should be sufficient for operation and should constitute 30 per cent of the total wattage. It is very important that the operator shall not become confused as to the particular conductors and equipment constituting the circuit he wishes to find, and as an assistance to him the barriers separating circuits are generally striped in a pronounced color. The compartments are from 5 ft. to 7 ft. wide. It greatly assists in the marking of the compartments if the lamps are placed regularly with respect to the compartments; that is, on the beams between the compartments or in the centers of compartments. Good diffusion should be obtained by the use of diffusing glassware and light walls.

Among the many arrangements of equipment found in the switch room, two typical cases deserve special consideration, namely, subcell disconnecting switches and ceiling disconnecting switches. Generally the room is so narrow that ceiling lights do not adequately illuminate the subcell disconnecting switches and wall lights are necessary. The wall lights should be located either on center lines of compartments or between compartments. With concrete walls and barriers, and with 100-watt prismatic reflector units on the ceiling spaced 14 ft. apart and 50-watt totally enclosed guarded units on the walls, spaced 7 ft. apart, the intensity on a horizontal 30-inch plane is 4 fc. on the vertical front edge of the cell top 2 fc. and on the disconnecting switches 1.5 fc., and these valves are very satisfactory.

With disconnecting switches on the ceiling the only logical location for lights is on the beams between compartments. Needless to say, a very mild well diffused light should be provided by the use of small lamps, good quality closed bottom diffusing glassware, and white walls. The best results are obtained by using one unit under each beam which brings the spacing 5 ft. to 7 ft. If the barriers are painted white, 50-watt units are satisfactory. If the barriers are of gray concrete or red brick, 75-watt units are needed.

CABLE ROOM

The problems presented by the cable room are somewhat similar to those of the bus and switch rooms. Generally considerable work is done in the cable room in connection with testing and altering cables and the lighting should be designed with this in mind. Local switching is desirable. Emergency lighting is needed only for passage.

SWITCHBOARD ROOM

The switchboard room is the control point for the entire electrical portion of the station and should be perfectly lighted at all times. Emergency lighting is probably more important in this room than in any other part of the station. It should constitute 30 per cent of the total wattage and in all cases should cover the entire switchboard. The meters and instruments are arranged on vertical boards and the controls are arranged on either vertical boards or benchboards. The chief requirement of the lighting is to provide uniform, soft illumination on these instruments and controls and to avoid reflection and glare from the instrument cases. Two general schemes of switchboard room lighting are in successful use and deserve mention: one from visible sources, the other from concealed sources.

Switchboard room illumination is from visible sources. The prismatic reflectors and bowl enameled lamps are mounted in front of and above the board, the position being designed to avoid direct reflection from the meters. With a light colored ceiling and using 100-watt lamps a density of 0.75 watts per sq. ft. gives an intensity of 3.5 fc. on a horizontal 30-inch plane, and 2.5 fc. on a vertical plane through the instrument scales.

Switchboard room lighting from concealed sources has been accomplished by two methods, first by the use of a diffusing glass ceiling and second by the use of a prismatic glass ceiling. In either case the ceiling is lighted during the daytime by a skylight located directly over the ceiling and at night by prismatic reflectors and clear lamps hung between the ceiling of the skylight. In the first method the lamps are so placed as to direct the light obliquely toward the switchboard and the glass is simply of the diffusing or ripple type and serves to hide the sources. In the second method, which is illustrated in Figure 11, the light from the skylight and from the lamps is considered to reach the glass ceiling in a generally vertical direction and the glass is designed to redirect the light obliquely against the switchboards. The glass is smooth on top and prismatic on the bottom. It is realized, of course, that better results can be obtained by placing such glass with the prisms upward or probably still better results by using glass with prisms on both sides. The difficulty of keeping the upper surface of the glass clean if other than smooth, however, is thought to outweigh the advantages of having prisms on the upper side. The wattage required is practically the same for either kind of glass. With prismatic reflectors and 75-watt clear lamps a density of 3.5 watts per sq. ft. of glass ceiling will give under the glass ceiling an intensity of 8 fc. on a horizontal 30-inch plane and of 3 fc. on a vertical plane through the instrument scales and these values are very satisfactory.

OFFICES

The general offices require no particular departure from commonly accepted good office lighting practice. The load dispatcher's office, however, requires sufficient emergency lighting for satisfactory operation without the normal lighting. The load dispatcher's office usually contains a mimic switchboard or system diagram which requires special lighting either by means of general lighting on its front surface or by means of miniature lamps used to indicate switching operations.

OUTDOOR STATION

The lighting of the outdoor station must usually provide for the manual operation of disconnecting switches. Emergency lighting is seldom required.

Being an open structure the outdoor station can be lighted by projectors or flood lights when suitable supports are available for the light sources. Care must be exercised, of course, to locate the lights so that they will not blind the operator. It is also desirable to have the lights scattered so as to avoid shadows.

A second method which has proved successful uses totally enclosed diffusing glassware in weatherproof fittings. An example is shown in Figure 12. As with the indoor switch room, it is very important to have the lamps small in proportion to the glassware and to provide glassware of good quality so as to avoid glare.

DISCUSSION

F. C. CALDWELL: Mr. Chairman, in line with what the President has just said, the Dayton Power & Light Company have added one more feature to the lighting of their power station. As an example of how it should be done they wanted the public to see a well lighted plant. Consequently, they had in mind in the arrangement of their lighting the effect produced at night, from the outside, and lighted up the rooms and windows so that passers-by could see a brilliantly lighted building.

JULIUS DANIELS: There is one thing I want to mention: I hope that in the designing of the installation of a central station power plant, sufficient provision is made for a sign on the station buses. A sign load for a station of 100,000 K. V. should be approximately 75 kilowatts and that should go on the station lighting service.

Another point that came to my mind was that sufficient lighting should be furnished in back of the switchboard so that the secondary wiring can be tested and checked.

H. T. PLUMB: In connection with the last speaker's remarks, I have in mind a station where we flood-lighted the outside of the building so that people could see it for miles in every direction.

In the pictures shown, it seems to me there is a great deal of glare present in some positions. For example, when the man was looking at the switchboard instruments, there were some glaring fixtures hanging back of the switchboard.

All such things can be avoided by suspending the fixtures behind beams, or by dropping boards between the operator and the fixtures.

Davis H. Tuck: (Communicated) For illuminating coal crushing and pulverising equipment, where there is an excessive amount of black dust in the air, there has been developed a gas tight prismatic refractor unit having a symmetrical or asymmetrical distribution. The refractor has a closed bottom and is held in a gasketed chamber so that the only exposed surface is smooth glass. The lamp bulb and inner surface of the refractor is thus protected from the coal dust by a gas tight chamber.

For lighting turbine rooms where the lights must be placed above the craneways and where there is a switch room balcony on an approximate level with the craneway, prismatic intensive type reflector units similar to those described by Mr. Hopkins, have been built with an aluminum cover spun over the reflector, so that the transmitted light in the direction of the switch room is cut off but the transmitted light to the ceiling is allowed to pass through. Such an installation is in operation in the New Colfax Station of the Duquesne Light Company, Pittsburgh. This installation was designed by the D. P. Robinson Company to eliminate the glare from the switch balcony and at the same time maintain the daylight effect given by the indirect component of light to the ceiling.

A method of installing lighting units has been used which makes maintenance easy. Type T condulets are used and the downward leg terminates in a hook. The stem of the unit terminates in a loop and a piece of reinforced cord is used to plug in the unit on the condulet cover. The unit is unhooked, the plug pulled and the entire lighting fixture is lowered to the floor for cleaning.

ARTIFICIAL ILLUMINATION IN THE IRON & STEEL INDUSTRY*

BY W. H. RADEMACHER**

SYNOPSIS: During the last decade the application of artificial light in the Iron and Steel Industry has undergone marked changes with a distinct trend toward betterment. The modern incandescent lamp has rapidly displaced other forms of illuminants and in conjunction with modern reflecting equipment is today recognized standard. Altho much of the work involved in this industry is of a rough nature and does not necessitate lighting intensities of a relatively high magnitude, the requirements are nevertheless far from unimportant. Chief among the credits to the account of modern lighting are safety insurance and the twenty-four hour day, attended by the successful coping with keen competition and the affection of economies in production.

The selection and application of equipment for the various areas embraced in plant structure are exceptionally important problems, dictating as they do the success or failure of the resultant illuminating effect.

In this paper the requirements of the various sections and operations are treated in detail, recommendations being offered as to the best practice. Photographs illustrating the application of the modern principles discussed accompany the text.

INTRODUCTORY

The information embodied in this paper is based on an investigation and analysis of existing conditions in twelve of the leading steel mills located in the Pittsburgh, Youngstown and Chicago Districts. In view of the number of mills surveyed, their diversified location and their standing in this field of manufacture it is believed that the data presented may be considered as fair criteria of the practice in the industry as a whole.

The basic objectives of the investigation were to study building structure, manufacturing operations and general working conditions from the standpoint of lighting requirements; to determine the type of lighting equipment in present usage, to ascertain the lighting intensities employed and to compare these values with those reported in the previous surveys. It was further desired to analyze the suitability of the lighting equipment used and intensities in vogue with a view toward confirming their application or suggesting revisions if such procedure seemed advantageous.

*A Paper presented before the Annual Convention of the Illuminating Engineering Society, Lake George, N. Y., September 24-28, 1923.

**Edison Lamp Works of G. E. Co., Harrison, N. J.

The Illuminating Engineering Society is not responsible for the statements or opinions advanced by contributors.

In establishing a view point for the consideration of this discussion it should be noted that steel manufacture as now practised is essentially a twenty-four hour process. This is necessarily so for economic reasons. Very hot, even to the point of incandescence, and moulten metal of great bulk enters into most processes, and only thru the maintenance of temperatures and elimination of appreciable reheating is fuel consumption and production time—in a word, cost—kept within present low limits.

STRUCTURAL CONDITIONS

To fully understand the lighting requirements of any industry it is necessary to become familiar with the structural arrangements enclosing or affecting the areas to be lighted.

The working areas of the typical mill may be divided into two broad classes, as (1) Exterior,

(2) Interior,

The former consists of the throughfares and yards and the latter all enclosed or covered operations. These two groups, for the purpose of discussion and further analysis, may be subdivided. The exterior, with regard to the activities within the areas involved, as

(1) Throughfares,

(2) Active Working Areas,

(3) Storage Areas.

The interior, with regard to the grade of work carried on as

(1) Rough,

(2) Medium rough,

(3) Medium,

(4) Fine. See Table I

All exterior work may be included under the first classification.

The thoroughfares about the steel mill are of three types; foot, vehicle and railroad, the former are usually beaten earth or cinder paths leading from building to building and from one point in the yards to another. These paths cross railroad tracks, wind between buildings, pass storage piles, dumps, etc., and there exists an ever present possibility of obstructions being cast in the route of the unsuspecting pedestrian. The vehicle roads are of much the same character as the foot paths, though of course wider. The railroad tracks are connecting arteries between various yard points and buildings, providing a line of travel for locomotive cranes, switching engines, ingot, ladle, stock and freight cars etc. Considerable switching is involved in the handling of traffic, the burdens carried are often of a treacherous nature such as brimful ladles and red hot ingots, and danger lurks at every crossing and curve.

The active working areas of the typical mill yard include such sections as the ore dumps, stock piles, skull cracker yards, mould yards and in some cases, cooling tables and loading and shipping yards. The actual work in these areas is of a rough but usually hazardous nature as stumbling obstructions are common and cranes are constantly moving about with loads of material.

The activities on the thoroughfares and within the active working areas are practically as intense during the hours of dark, ness as during daylight.

The storage areas are those sections set aside for the dead storage or surplus stocks of coal, scrap iron, etc.

Interior conditions in steel mills are far different than those encountered in any other industry. The buildings housing what have been termed the rough and medium-rough operations are usually very large, covering considerable ground area—the widths commonly range from 30 to 100 ft. the lengths from 100 to 600 ft,, and the heights from floor line to roof trusses from 25 to 60 ft. The roofs are invariably of the steel truss supported monitor type. Heavy duty overhead cranes traverse most of the buildings and are in almost constant operation carrying heavy machine parts, stock, and incandescent and hot metal in bars and ladles. The atmosphere within these buildings is almost constantly charged with steam, created by water in contact with the hot rolls and metal and rising from the water jackets of furnaces, together with graphite particles from the converters and much smoke, ore, and fuel dust. Also by virtue of these same conditions equipment and surroundings—floors, walls and ceiling—are dark and sooty making them decidedly ineffective from the standpoint of light reflection.

The machinery and equipment is usually widely spread out covering a considerable part of the floor area with various parts projecting here and there about the areas traversed by the workers. The transfer rolls, cooling tables and many of the active machine parts are located at a waist high position. The footing is usually rough, the floor area being broken up by machine parts and material either permanently located or strewn about during the processes involved in manufacture. Narrow bridges with steep abrupt approaching steps or tunnels entered by steep narrow stairways afford the common means of crossing from one side of the buildings to another. In furnace buildings the equipment is usually lined along the sides, the central floor area being occupied by charging materials, charging machinery and the various equipment used in processing.

Those buildings housing operations of what are classed as a medium nature are much the same in their general construction and arrangement as those described above but the surroundings are usually somewhat lighter and the atmosphere less fogged, the work of this class usually involving the handling and working of cold metal.

Those operations classed as fine are rather distributed in location, some of them such as certain kinds of inspection work being carried on directly adjacent to the manufacturing processes while others, such as tin plate sorting, are most frequently handled in typical medium sized factory buildings.

CLASSIFICATION OF PROCESSES

The following tabulation (Table I) gives the author's conception of the proper classification of areas and operations under the headings previously mentioned.

Rough operations may be defined as those requiring no discrimination of detail and involving the handling of only bulky materials.

TABLE I

CLASSIFICATION OF IRON AND STEEL MILL WORK WITH REGARD TO VISUAL
REQUIREMENTS

Rough	Medium Rough	Medium	Fine
Yards	Blast Furnaces	Chipping	Tin Plate
Thoroughfares	Cast Houses	Cold Rolling	Sorting
Stock Houses	Mixing Houses	Close Shearing	and
Open Hearths	Bessemer Sheds	General Inspection	Inspecting
Soaking Pits	Stripping Sheds	Wire Drawing	
Reheating Furnaces	Blooming Mills	Pipe Threading	
Puddling Furnaces	Structural & Rail Mills	Nail Making	
Annealing Furnaces	Rough Shearing	Pickling	
	Rod & Tube Mills	Tinning	
	Hot Sheet Mills	Machine Shops	
	Cooling Tables	Power Houses	
	Warehouses		

Work classed as Medium rough is that involving some discrimination of detail but not of a close or accurate nature and applying in general to the manipulation of hot metals.

Medium operations are those requiring comparatively close discrimination of detail and in general the processing of cold metal.

Fine work is that involving close application and accurate visual perception as is required for inspection.

ESSENTIAL FEATURES OF OPERATION

The following brief sketch of the essential points of the manufacturing processes and their visual requirements is given so that knowledge will be had of all conditions bearing upon the problem of light application. Attempt has not been made to give a complete explanation of the theory and processes involving as such is not within the scope of this paper.

The function of the thoroughfares has already been fully covered. Workers must be able to find their way about with celerity and safety, stored material, tracks, crossings and obstruction should be easily visible. Vehicle roadways should be clearly defined to permit the rapid movement of automobiles and tracks should be lighted so as to facilitate the identification and switching of cars.

Ore dumps or as they are sometimes termed yards or stocks are in reality stock piles to which ore is brought (from the mines) by boat or rail unloaded by various mechanical means, conveyors, grab buckets or car dumpers, and from which the ore is taken to the stock houses by ore bridges or skip hoists. Sufficient light must be provided for the operation of ore handling mechanism and the movement of cars.

In the skull cracker yards the scrap steel used in the furnace charges is broken up to suitable sizes by the dropping of a large heavy steel ball, this being handled by a magnetic lift crane controlled by an operator located in the crane cab. The scrap is also picked up and loaded in cars for transit to the furnaces and stock houses. The crane operator must be able to clearly discern the materials he is demolishing or picking up, the location of cars and the movements of ground workers.

In the Mould Yards—moulds are cleaned, stored and prepared, crane operators must be able to easily locate materials and ground workers must have sufficient illumination for the performing of rough work.

Stock Houses are temporary storage buildings for limestone, ore, prepared scrap and materials entering into the basic manufacturing processes. Artificial light is required sufficient for car switching, the safe movement of switchmen and loaders, and the identification of materials.

In the Blast Furnaces the raw materials are reduced to pig iron. These furnaces are huge towering structures at the top of which is the charging platform reached by a steep narrow stairway. Around the lower section of the structure is located the tuyer mechanism and water jacket cooling system which are under the observation of attendants. Adjacent to the Blast Furnaces are the stoves which preheat the air blast, the blowing engine houses etc. Light should be provided to enable examination of the charging bells, safe movement along the stairways and to check the operation of the water cooling and tuyer mechanism. The areas about the stairs should be clearly visible so workers may move about promptly and safely.

At the base of the furnaces is the Cast House at which point the metal is run from the furnaces into huge ladles mounted on cars. This structure also sometimes shelters the pig casting equipment which in modern mills is of an automatic nature and receives the moulten metal, forms the pigs and conveys them to waiting cars. Workers require sufficient light for the preparation of the runways used in drawing off the metal, and operation of mud guns for sealing the furnaces after tapping etc. During actual pouring a high intensity of light is created by the moulten metal passing through the troughs to the ladle cars and casting machines.

The Open Hearth furnaces (Figure 1) receive and reduce moulten iron from the blast furnaces, scrap stock and various other materials used in manufacturing open hearth steel. They are charged both by machinery and hand. The charging floor occupies the position in front of the furnaces while the tapping pit, at which point the moulten steel is drawn off into ladles, is located at the rear. Workers are in constant attendance. Artificial light sufficient to facilitate the safe movement of workers about the constantly obstructed floors, to enable crane operators to readily locate workers and materials, and to enable the efficient handling of ladle cars etc. is essential. The furnaces when in operation cast great flares of light over the working areas adjacent to them.

Stripping consists of the removal of the ingot moulds by means of a huge hydraulic or electric arrangement. Artificial light is required here to enable the strippers to readily engage their mechanism on the moulds and to discern the location of cars and workers. The ingots when stripped present red hot surfaces which tend to build up the illumination about the area occupied.

Mixing houses shelter huge vessels which hold, maintain at a high temperature, and mix with manganese, charges of moulten metal from the Blast Furnaces preparatory to be charged in the Bessemer Converters. Sufficient light for maintaining safety, for manipulating valves and making repairs should be provided.

The Bessemer converters are located in huge structures and are usually controlled by operators located on a shielded or enclosed platform running along the side of the building on a level with the converters. When the converters are in operation no artificial light is required adjacent to the area they occupy as they send off a terrific flare which illuminates the surroundings to intensities ranging in the hundreds of foot candles. Light should be provided however, to permit safe movement about the building when the converters are down and are undergoing repairs. Good illumination is essential on the blowers platform from which point the operation is controlled as scales and gauges must be accurately read.

The Soaking Pits are gas-heated furnaces occupying practically the entire floor area of a building set aside for them. Their covers, which are mechanically operated, occupy the principal floor surface. Their purpose is to hold the ingots at a high temperature until required by the blooming mills. Light provided here should be sufficient to enable the workers to move about with safety and to enable crane operators to easily discern the location of the floor men and pit covers. The covers when rolled back permit a huge flare to issue from the furnaces thus illuminating the surroundings to a relatively high intensity.

The Blooming Mills (Figure 2) break down the ingots to bars or slabs of a pre-determined size. The manipulation of the ingot during the process is usually controlled by skilled operators located in a pulpit above or to the side of the rolls. The white hot

FIG. 1.—A typical open hearth charging floor as it appears by night. The smoky atmosphere, the material strewn about the floor, the flare of light from the furnaces on the left, and the dark surroundings are characteristic of the conditions encountered in steel mill lighting. An intensity of two foot-candles is provided by a system of overhead units which are 500-watt Mazda C lamps in porcelain enameled bowl shaped steel reflectors. These are hung 40 feet above the floor on centers 45 x 30 feet.

FIG. 2—This modern blooming and 18 inch merchant mill is illuminated to an average intensity of 2.5 foot-candles by a well designed system of general lighting, employing 750-watt Mazda C lamps in bowl type steel reflectors on centers 40 x 50 feet, 40 feet high.

Fig. 3—Night view of a billet cleaning, chipping and inspection department. A well distributed illumination of approximately four foot candles is supplied here by the application of overhead (500-watt deep bowl units) and side (300-watt angle units) lighting. This provides working conditions well above the average encountered in this class of work.

Fig. 4—This night view of a typical hot mill building illustrates the importance of well diffused light. The installation pictured is of the better grade. Approximately 2.5 foot-candles is provided over the floor area by lamps placed 37 feet high on centers 50 x 40 feet. The Mazda C lamp and the RLM Standard dome reflector is used. The 750-watt size over the mills and the 500-watt size at the right. The rolls though appearing dark are well illuminated when the hot plates which are being reduced are passing through. The reheating furnaces will be seen at the left.

ingot after being placed on the transfer tables by a crane is caused to travel to the rolls where it is passed back and forth and turned at will by mechanical means under the control of the roll operators. The ingot being incandescent illuminates the surrounding areas during its travels causing a shifting illumination thru the mill. From the blooming mill the slab passes thru shears which trim off the rough ends and cut the slabs approximately to size. Rail mill operation is similiar to the foregoing. In Plate mills the slabs go to reheating furnaces and are then rerolled on smaller mills. Rod, Skelp and Tube mill equipment is of various forms, the rods or tubes usually being passed thru the rolls under the guidance of an attendant who grasps the rod as it emerges and snakes it back thru another set of rolls. In what are known as continuous mills the operation is more nearly automatic. Sufficient general illumination for the safe movement of workers, for the easy examination of equipment, and to permit operators to see the movement of material and indicators at all times, is essential in these areas. Supplementary lighting on roll indicators is usually practised and desirable.

Rough shearing involves the cutting of bars, billets etc. to approximate size by crop or flying shears as the stock travels along the transfer tables from the rolls.

The Chipping and Cleaning operations (Figure 3) consist of chipping out cracks and flaws in the slabs, preparatory to their being further rolled. Chipping is done by pneumatic chippers and grinders are sometimes employed for further surfacing. From the standpoint of lighting this is exacting work as fine hair line cracks and flaws must be detected, chipped or ground out by the workmen and pickling of bars and billets is also sometimes undertaken in these areas as a preliminary operation. Such work is relatively coarse however, involving as it does the preparation of vats and immersion of the material.

Sheet Mills (Figure 4) reduce slabs to sheet stock. The slabs are "pulled out" folded and rerolled several times forming sheet steel. The sheets are then trimmed or sheared and pulled apart or as it is technically termed "opened". Hot mill operators must be able to move about quickly and safely. The hot sheets illuminate the rolls mechanism during passes and the reheating furnaces assist in lighting up the areas directly before them. Opening requires that the operator be able to easily discern the sheet edges and shearing demands sufficient illumination for accurate setting and trimming.

Cold rolling (Figure 5) and polishing is the next process in sheet making and consists of the passing of cold sheets thru high speed rolls. Cold rolling of strip is much the same and involves careful machine setting, micrometer reading and close watching of machines. In cold rolling no light is provided by the metal and good illumination is essential for the setting and operation of rolls, the reading of gauges and the promotion of safety. The sheet metal is treacherous, at best presenting exceedingly sharp cutting edges which frequently cause serious accidents when not readily visible.

Pickling consists in the running of the sheet in chemicals for the purpose of cleaning. Sufficient light for the safe movement of workers and handling of sheet is required in this operation.

The Annealing Process (Figure 6) consists of stacking the pickled sheets in large retorts which are then rolled into gas heated annealing furnaces. This is comparatively coarse work and the principal demand for light is to facilitate the movements of cranes and workers—and the safe handling of the sheets.

Tinning, galvanizing and dusting are finishing treatments of sheet steel. The vats and machines are usually lined in rows along the sides of the buildings and are under the constant attention of operators. A fairly high intensity of well diffused light is necessary over the equipment, so that setting, operation and finish can be readily observed by the attendants.

Tin Plate sorting and inspection is usually done by women workers on long benches in buildings especially set aside for the operation. The stacks of finished sheets are received on the benches, and the operators lift each sheet, examine each side and grade it in one of three classes with respect to the nature of the finish. The sheets have highly polished surfaces and are likely to cause objectionable specular reflections if the lighting is not carefully planned. This is undoubtedly the most exacting work in sheet mill operation. Flaws, scratches and surface blemishes must be easily and rapidly detected.

Wire drawing, pipe threading and nail making are semi-automatic operations. Supervision and setting of machines necessitates comparatively close work and one operator usually cares

Fig 5—In the cold rolling department there is no luminous metal. Artificial light must provide illumination on the rolls as well as around them. In this modern steel mill an average intensity of four foot-candles is furnished by the overhead 750-watt dome and side 300-watt angle type units.

Fig 6—The annealing department as typified in this picture offers no real difficulties from the lighting standpoint. The work is comparatively coarse and the light sand used adds materially to light reflection. The department illustrated is illuminated to an average intensity of over two foot-candles by 750-watt Mazda C lamps in RLM Standard dome reflectors. The building is 50 feet wide and a single row of outlets on 50-foot centers is to be noted 46 feet above the floor.

Fig. 7—A combination of general and local lighting is provided in this opening and shear building of one of the progressive sheet mills   300-watt RLM units, uniformly distributed, provide general illumination of 2.5 foot-candles, while 200-watt local lights provide a higher intensity of 12 5 foot-candles at the shears.

Fig. 8 —Night view of a well illuminated black sheet warehouse and shipping building. An average intensity of 3.5 foot-candles is provided here by 500-watt Mazda C lamps and RLM Standard dome reflectors on approximately 35-foot centers.   The illumination is quite satisfactory for the work involved.

for a battery of units. This work may be considered analogous to medium machine shop practice.

Steel mill machine shop practice usually involves a medium grade of work such as roll turning, equipment repairs etc. Visual requirements are the same as commonly encountered in this trade.

The power houses and substations represent the heart of the plant and such illumination as will facilitate inspection, cleanliness and safety is essential.

The Cooling Tables as the name implies are huge steel platforms on which the metal from various rolling processes travels for cooling. These tables are also quite frequently the point of inspection. Illumination must be adequate for observing the operation of the shifting fingers (frequently painted white to increase the visibility) which move the materials and for the inspection and making of the stock. Inspection is sometimes done by operators who draw a piece of wood along the hot stock, the examinations being made by the light of the resultant flare.

It is believed that the use of the visual demands within other areas is obvious and no further discussion regarding them is required here.

PRESENT PRACTICE

In every mill inspected, incandescent electric lamps exclusively were being used for the lighting of all areas. In all cases general illumination of some form was provided, either alone or in combination with localized lighting on particular operations.

Lamps employed for general illumination usually ranged from the 300- to 1000-watt size, the larger units predominating. Localized units commonly made use of lamps of sizes of 200 watts and below. Reflectors of some type were invariably found to be used with the lamps employed for general lighting.

Those observed were of various types, the RLM Standard Dome, flat enamel steel, shallow dome, deep bowl and angle type porcelain enameled steel being common. While the flat enameled steel and shallow dome reflectors were predominating in the older installation, the RLM Standard Dome and Deep Bowl Enameled Steel seem to represent modern practice.

The local lighting was usually found to be not so well provided for. A very nondescript collection of reflectors and bare lamps creating objectionable glare was common.

Spacings of units, in most instances, were found to be much wider than practicable for the provision of uniform light distribution.

While the importance of periodic cleaning of lighting equipment was found to be universally admitted, but few of the mills visited practiced regular maintenance schedules. This situation was usually explained by the difficulty experienced in obtaining suitable labor.

With but two exceptions, the plants investigated were found to be using 250 volt nominal lighting service with either 250 volt lamps or two 125 volt lamps in series. The lighting circuits in many cases were tapped direct from the power feeders supplying the auxiliary drive within the building to be lighted. The disadvantage of operating two lamps in series and superiority of 110 volt (or thereabout) lamps for general lighting service are too well known to require discussion here. The practice of operating lighting circuits from d-c. power feeders, particularly in steel mills, is extremely objectionable as severe voltage fluctuation is prevalent due to the constantly changing load on the generators. This results in bad flickering of the lights and of course affects the life and light output of the lamps. In one mill visited by the writer a variation of almost 2 to 1 in intensity was noticed at times. Figure 10 shows a voltage chart taken across the bus bars in a typical steel mill substation. From this chart it will be noted that a variation of 240 to 315 volts occurred over the period of observation which means that neglecting a slightly variable line drop the lamp voltage varied in the same proportion.

Separate lighting circuits, or in the case of a-c. drive mills, high voltage distribution with a step down thru transformers and buck and boost regulators where voltage variation is extreme are best adapted to the producing of satisfactory mill lighting. In many cases schemes of this kind are in the course of installation or under consideration.

THE TREND IN LIGHTING PRACTICE DURING

THE PAST TWELVE YEARS

Table II following, gives the prevailing intensities in certain areas of typical steel mills in 1911 as found by an investigation

FIG. 9.—Night view of a well illuminated bar storage warehouse. The intensity, 2.5 foot-candles, provided here is adequate for stacking and loading. A higher intensity is provided for the machines visible at the rear end by using 500—rather than 300-watt lamps.

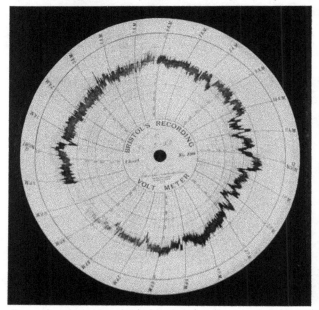

FIG. 10.— Reproduction of a typical d-c. bus voltage chart, illustrating the widely varying voltage conditions existing on d-c. power feeders.

FIG. 11.—Night view a of well lighted foundry. 300-watt Mazda C lamps in RLM Standard dome reflectors on centers 10 x 40 feet are mounted on the ceiling trusses, 35 feet above the floor. An intensity of over five foot-candles is provided.

FIG. 12.—This picture of a typical foundry gives one a good idea of the rough footing and generally hazardous conditions encountered in this industry. A number of 300-watt Mazda C lamps in RLM Standard dome reflectors on 10-foot centers in this 40-foot bay provide adequate light (five foot-candles) to make objects clearly visible and minimize accident hazards.

conducted at that time, and figures obtained by the author on the extremes good and bad constituting present practice.

Process or Area	Intensity—1911 Foot-Candle Average	Intensity—1923 Foot-Candle		Intensity Foot-Candle Recommended
		Average Low	Average High	
Thoroughfares	0.087	0.02	0.3	0.1 − 0.5
Ore Yards	0.19	0.2	0.3	
Loading Yards (No Inspection)	0.14	0.2	0.4	0.25− 1.0
Loading Yards (Inspection)	0.36	0.5	0.75	
Open Hearth Mould Yards	0.29	0.2	0.4	
Stock Houses	—	—	—	0.5 − 1.0
Open Hearth Charging Floor	0.14	0.3	1.5	1.0 − 2.0
Open Hearth Casting Floor	0.17	0.5	1.5	
Soaking Pits	—	0.15	1.5	1.0 − 2.0
Reheating Furnaces	0.46	0.3	3.	2.0 − 4.0
Puddling Furnaces	—	—	—	2.0 − 4.0
Annealing Furnaces	—	0.5	3.5	2.0 − 4.0
Blast Furnaces	0.25	0.4	0.7	1.0 − 2.0
Cast Houses	0.22	0.5	1.	1.0 − 2.0
Mixing Houses	—	—	—	2.0 − 4.0
Bessemer {Converter Houses} Bessemer {Blowers Platform}	—	2.5	4	3.0 − 6.0
Ingot Stripping	—	0.4	1	1.0 − 2.0
Blooming Mills	0.32	0.5	2	2.0 − 4.0
Cooling Tables	—	—	—	2.0 − 4.0
Rail & Structural Mills	—	0.5	2	2.0 − 4.0
Pipe & Tube Mills	0.3	0.5	1.5	2.0 − 4.0
Chipping	—	3.5	7	4.0 − 8.0
Hot Mills	0.34	0.75	3.5	2.0 − 4.0
Cold Rolling	0.65	1.5	5	4.0 − 8.0
Shearing (Close)	—	1.5	11	8.0 −12.0
Inspection (General)	—	—	—	4.0 − 8.0
Wire Drawing	0.87	1.	3	4.0 − 8.0
Pipe Threading	0.76	1.5	5	4.0 − 8.0
Nail Making	—	—	—	4.0 − 8.0
Pickling	—	0.25	1	2.0 − 4.0
Tinning		1.5	5	4.0 − 8.0
Tin Plate Sorting & Inspecting	—	5	12	10.0−15.0
Warehouses		1	2.5	2.0 − 4.0
Shipping	—	1.5	3	3.0 − 6.0
Machine Shops	1.37	2.5	4	4.0 − 8.0
Power Houses	1.13	2.5	6	4.0 − 8.0
Layout & Fabrication (Structural Steel)	—	—	—	4.0 − 8.0

Because of the wide variation in existing practice it was considered desirable to average the high and low readings separately citing each rather than quote a common average.

A careful study of manufacturing requirements and conditions has led the author to the conclusion that the intensities listed in the fourth column are economically desirable in mill lighting.

Consideration of these figures clearly illustrates that there has been a distinct trend toward higher intensities in the lighting of steel mills during the past twelve years, as with but few exceptions, the average low values of today are well above the average in 1911 while the average high values in all cases represent a marked increase.

In only three of the plants investigated however, were lighting values of the higher order found to exist generally, which fact combined with the radical departure between high and low averages, would seem to indicate that there has not been a full appreciation of the value of good lighting and there is still room for marked improvement in the lighting standards of the industry as a whole. It should here be noted that the mills having the higher intensities were all of relatively new construction having been built within the last six years whereas the others were considerably older. This in a measure explains the variation in values.

It is interesting to note the comparative attitude of the operating engineers responsible for lighting practice in mills having good and poor lighting. In the case of the better installations it was invariably expressed that good lighting had proven beneficial from the standpoints of safety, supervision and elevation of morale, in operations on hot metals and as an aid toward greater production and reduction in spoilage plus the aforementioned benefits in the case of processing of cold metal. In one mill an instance was cited of two buildings both housing hot sheet rolling operations one well lighted and the other poorly lighted. When operators were transferred from the well lighted to the poorly lighted building they objected so strenuously to the gloomy, forbidding and hazardous surroundings that the remodeling of the lighting equipment was practically forced.

On the other hand the operating engineers in the poorly lighted mills, while considering the necessity for artificial light in some form, usually looked upon it as being a consideration of secondary importance. Installation in most of these cases are a result of individual opinion and past practice within the mill. Often a guess is made as to how much light is needed, this is provided, and as long as no complaints are registered by the workers the illumination is considered satisfactory. It should be obvious that such practice is far from commendable and it is questionable whether the workers as individuals are capable of judging the suitability of lighting standards. Most workers are faced with a lack of knowledge of any conditions other than those beneath which they work and as a consequence are easily satisfied with the conditions at hand. That they would favor better lighting is indicated by the instance previously cited.

The opinion seems to prevail that in most areas it is necessary to simply furnish enough light to find one's way about and that the hot metals and furnaces will furnish enough illumination for closer work.

A careful consideration of the problem of steel mill lighting leads to the conclusion that while one of the prime functions of artificial light within and without the mills is to promote safety, this factor is somewhat more involved than would seem the case on hasty consideration and further there are other benefits to be derived from the proper application of light.

The visibility of objects in steel mills is inherently low because of the characteristically dark surroundings and dark nature of the objects themselves. Since we see objects by the light they reflect to the eye it follows that to easily distinguish dark objects requires more illumination than if they were light in color. Largely for these reasons the steel industry requires more illumination from the safety standpoint than do most other trades.

The hot metals worked, while unquestionably instrumental in building up the lighting at the working points, are present only at intermittent periods and are constantly moved from place to place. This condition creates a demand for more rather than less light. The incandescent metal represents sources of high brilliancy within the worker's field of view. If the adjoining areas are dark the eye does not see clearly as it moves to the dark surroundings from the bright work and vice versa. The wider the contrast in intensities the greater the degree of readjusting and the longer the time required by the eye to adapt itself as it moves about.

Mention has already been made of the abnormal amount of foreign material present in the atmosphere, hence to overcome the resultant absorption a relatively high wattage for a given area is necessary to produce the desired intensity on the work.

Moreover, even with frequent cleaning, depreciation of reflecting surfaces is a factor which must always be given careful consideration and adequate provision made to insure a satisfactory average value of illumination.

Production increases through speedier movements on the part of the workers, cleaner surroundings and easier supervision of labor, derivitives of good lighting often discussed in the past, are also made possible in many areas of the steel mill.

FOUNDRY LIGHTING

In view of the close relation between the modern foundry and the Iron & Steel Industry as herein discussed it is felt that a brief consideration of Foundry Lighting practice is appropriate at this point.

FLOOR MOULDING AND POURING

Inasmuch as both of these operations are carried on within the same area though at different times of the day the lighting system will be the same.   Moulding calls for the greatest accuracy and is usually done during the early hours of the day.   For pouring and stripping the lighting demands are less exacting.   This work is carried on during the latter part of the day when natural light is poor and the air in the foundry is full of dust and gases.   However, if provision is made for proper light for moulding, a lower though adequate intensity will automatically prevail for pouring and stripping.   The intensity employed should range from 4 to 12 foot-candles depending upon the character of the product.   For the manufacturing of heavy castings where the finish of the mould is not important intensities of from 4 to 6 foot-candles are sufficient.   In foundries casting small parts requiring careful finish and consequently closer discrimination of detail from 6 to 12 foot-candles should be provided.

BENCH MOULDING, MACHINE MOULDING AND CORE MAKING

The lighting of this type of work, which calls for some discrimination of detail since the pieces turned out are usually small and must be accurately moulded, can best be accomplished by what is termed a "localized general" system. A high intensity is required along the row of machines or benches while a lower illumination is sufficient for the remainder of the room. This effect of varied intensity is obtained by placing the lighting units relative to the machines. The maximum intensity is thus delivered from the correct direction and the spread of light takes care of the surrounding areas. Such a system has practically all the advantages of the drop light in getting the light where it is needed and yet the lighting units are not within reach.

CHARGING TUMBLING AND CLEANING

The work coming under the above classification is of a relatively rough nature and requires little discrimination of detail. Intensities of from 3 to 6 foot-candles are adequate. Distribution should be such that harsh shadows and dark corners will not result and hatchways and pits be readily visible.

In lighting equipment as is true of practically all other devices used in steel mills permanence, simplicity, durability, efficiency and ease of maintenance are essential requirements.

Porcelain enameled steel reflectors of the RLM Standard dome variety and, in the case of buildings which are narrow compared to their height, porcelain enameled steel reflectors of the angle type mounted below the crane rail represent the best types of the reflectors available for interior lighting. For exterior lighting the RLM standard dome, the radial wave, enclosed forms of street lighting units and floodlighting units are admirably adapted.

LIGHTING COSTS

From the standpoint of energy cost the operating expense of the average steel mill lighting installation is exceedingly small in comparison with the total production cost. Although conditions vary widely in different mills the following citation serves to illustrate the relative magnitude of the lighting expense. In a typical mill the cost of energy was found to be six-tenths of one cent per kw-hr. This figure is typical, the cost usually ranging from five-tenths to seven-tenths cents. The total average energy consumption per ton of Open Hearth steel produced in the

particular mill reckoned over a period of twenty-four hours
including all drives, cranes and lighting was approximately
45 kw-hr. It may be conservatively estimated in a mill of this
type that the lighting load represented about 2 per cent of the
total load. Since the energy consumption per ton figure is based
on a typical twenty-four hour period and since in this as well as
the average mill there are about 12 hours of daylight work and
12 hours of work under artificial light, the actual energy consumed
per ton output under artificial light was 2 x 0.02 x 45 = 1.8 kw-hr.
At a cost of six-tenths cents per kw-hr. this represents a cost of
light of 0.6 x 1.8 x 1.08 cents. In a mill rolling sheet this figure
would of course be somewhat larger but as a maximum would
probably not exceed 10 cents per ton.

In the face of these figures and considering the influence
of properly applied illumination in promoting safety, increasing
production, reducing spoilage, elevating the morals of the workers
and facilitating supervision it is difficult to excuse the low intensities so commonly encountered and easy to justify the occasional
installations of good lighting.

BIBLIOGRAPHY

Special Lighting for Steel Mills, H. M. Gassman, Proceedings, A.I. &
S.E.E. 1911.

Illumination of Iron & Steel Works, C. J. Mundo, Proceedings, A.I. &
S.E.E. 1911.

Some Features of Illumination of Iron & Steel Works, Ward Harrison,
Proceedings A.I. & S.E.E. 1911.

Modern Illumination in the Iron & Steel Industry, C. E. Clewell, Proceedings, A.I. & S.E.E. 1912.

Incandescent Lamps in the Steel Industry, Ward Harrison, Proceedings,
A.I. & S.E.E.

Steel Mill Lighting, B. G. Beck, Electrical World, Oct. 5, 1912.

A Progress Report on Illumination, Ward Harrison & B. H. Magsdick,
Proceedings, A.I. & S.E.E. 1913.

The Flaming Arc Lamp in the Iron & Steel Industry, A. T. Baldwin, Proceedings, A.I. & S.E.E. 1914.

High Candle Power Mazda Lamps for Steel Mill Lighting, G. H. Stickney,
Proceedings, A.I. & S.E.E. 1914.

Lighting of Steel Mills with Quartz Lamps, G. C. Keech, Electrical Review, Sept., 1916.

Illumination Iron & Steel Mills, C. E. Clewell, Electrical World, July 21, 1917.

Better Lighting of Iron & Steel Mills & Fabricating Plants, F. H. Bernhard, Electrical Review, Nov. 30, 1918.

Practical Methods in Lighting of Steel Mills, D. E. Atwater and J. J. McLaughlin, Electrical Journal, Sept., 1922.

Lighting a Large Iron Foundry, G. H. Stickney, American Machinist, Dec. 1, 1910.

The Efficient Lighting of Foundries, B. G. Worth, Castings, April, 1912.

Lighting for Core Room of Foundry, R. S. Rich, Electrical World, April 16, 1921.

Foundry Lighting, J. M. Shute, Industrial Engineer, April, 1922.

DISCUSSION

F. C. CALDWELL: In drawing up the Industrial Lighting Code for Ohio the only concerted opposition we experienced was from the steel industries of the state. The explanation for this I think is found in two conditions; one has been mentioned in the paper, that is, that the steel manufactures have always been used to doing their work by the light of the hot metal, and it is hard for them to feel that any more light is needed than that which has been used for generations.

Again, the steel industry is one of the oldest industries and in general we find that the older an industry, the more conservative it is. For instance, the rubber industry is one of the youngest, at least in its present state, and that was the industry from which we received the most cooperation. They looked most favorably upon the movement for better lighting in industry.

These two characteristics of the steel manufacturer must be overcome before a general improvement in steel mill lighting can be attained.

J. A. HOEVELER: When I began my work for the Industrial Commission of Wisconsin, I came across the argument several times that Professor Caldwell has mentioned; namely: that the hot metal in the foundry or mill would provide sufficient illumination, with some small amount of supplementary artificial lighting. But it was not until some time ago that I heard the prize argument against higher illumination, when, in talking to the night superintendent of a factory, after we had measured the illumination and found it to be of very low order, the man replied, "Well, you know, I wouldn't want more illumination in this factory; I am afraid if it were too bright it would put my men to sleep."

H. W. DESAIX: I believe that in the last few years there has been a marked difference among the steel people, but I note that the paper omits one phase of the steel industry, that of the concentrating mill.

Iron ore today is not sold in the crude form that it used to be. It is concentrated to a high value and the premium on the ore is based upon the value of it; that is, the percentage of iron.

You perhaps all know that when a man is buying raw material he endeavors to get it at as low a cost as possible, and that creates a condition of high competition among the concentrating mills.

There has been within the last few years considerable development in the concentrating mills. They have raised their amount of production and decreased their cost of production by automatic machinery, use of magnetic separators, magnetic clutches and the like. I have in mind a particular concentrating mill that used to turn out 1500 to 2,000 tons of ore a day, using 18 men within the mill; due to the use of automatic machinery they have reduced the number of men employed within the mill to 5—that means number of men per shift.

As soon as they did this however they came across a condition that caused them a considerable amount of trouble—that of tramp iron. Tramp iron would get into escalators and crushers and sometimes cause damage. Lighting intensities were therefore considerably increased so as to allow these few men in the mill to pick up the tramp iron and push the buttons which are used to release the magnetic clutches and stop the operation long enough to pick out the tramp iron.

I know that in northern New Jersey there is a considerable field right now for higher intensities of illumination in the steel industry; of course, there we get the mining end of it and not the production end.

F. W. LOOMIS: Have had no specific experiences with the iron and steel industry since I have been located in Pittsburgh but can cite something of a parallel nature in Canada; it is with regard to another metal—aluminum.

I had the privilege of visiting a plant in Quebec Province about eight months ago and I do not believe they even attempted to guess at intensities because when I went around with the superintendent to check the intensities as found, in one process that might be called medium fine at least—that is, changing rods to tubes—I checked about a quarter to one-half of one foot-candle. That was about as high as anything that was discovered in the plant. From this investigation, the superintendent decided to take definite steps to build up his intensities. He was sold on the idea of what higher standard illumination really means from the standpoint of production.

The reflectors found there were of an old type aluminum finish, and the fumes from the plant had so deteriorated them that they were worthless. I recommended something that could be more easily cleaned, and I believe that recommendation had been acted upon.

This would imply that the iron and steel industry and the metal industry as a whole are receptive to our suggestions if same are brought to their attention. It is perhaps sometimes quite a job to make them accept higher intensities but the field undoubtedly is awaiting us.

In the Pittsburgh district I understand something has been done toward increasing or attempting to increase the illumination standards but there is still a wonderful field awaiting us.

THOMAS G. WARD: I would like to ask on the matter of cost, if the hourly demand was taken into consideration on that in figuring the lighting and diversity factor, which would materially lower that cost factor.

A. S. TURNER, JR.: Mr. Rademacher's calculations of the lighting cost have all been figured with a considerable factor of safety, so that no steel mill engineer might state that his figures were lower than was actually the case. This would cover any discrepancies which might arise in the various processes of manufacture.

G. H. STICKNEY: My recent experience in steel mill lighting comes second-handed through Mr. Rademacher, so that I have not a great deal to add regarding present practice. I have been over Mr. Rademacher's work quite carefully and have considerable confidence in the conclusions which he has reached.

My most careful study in steel mill lighting was made about 1909 or 1910. At that time we succeeded in pointing out to the mill engineers the importance of better lighting. Our investigations showed that some mills were using something like ten times as much light for certain processes as were others. These comparisons called to the attention of the mill people, resulted in a considerable advance in the average practice, as referred to by Mr. Rademacher.

There has been a certain amount of advance since 1910, but I believe the practice is still far below what economic conditions would warrant. Steel making is a night and day process, requiring men to work for long hours by artificial light. Hence the advantages to be gained through enhanced safety and increased production are relatively large. Furthermore, fortunate circumstances provide steel mills with remarkably low price power, so that the cost of good illumination is comparatively small.

The failure to provide better lighting does not appear to be due to its cost, but rather the fact that those responsible for the lighting have many other complicated engineering problems, involving large expenditures, and this has a tendency to divert their attention so that lighting does not have the consideration which it really needs.

Due to the liability of mechanical accidents, many of the processes of steel making are quite dangerous, so that it is important that the steel industry should recognize that good illumination is one of the most important agencies for preventing accidents, and furthermore, that it is the one guard which speeds production rather than retards it.

O. F. HAAS: I would like to ask Mr. Turner if in the plants visited the exterior lighting in the main consisted of flood-lighting or general overhead lighting using enameled steel reflectors or refractor units.

A. S. TURNER, JR.: The equipment now being used in steel mill yards consists, in a small degree, of that employed for street lighting, such as prismatic refractors. The majority of this lighting is merely the incandescent lamp with radial wave reflectors or plain porcelain enamel steel reflectors.

WALTER STURROCK: As shown by this paper, the levels of illumination in steel plants in general are found to be quite low. About two years ago, one of the steel plants that I visited had an intensity of illumination of only about a half foot-candle. I talked with the electrician about the value of better lighting in steel mills and soon afterwards he began studying lighting from the standpoint of accident prevention. He was particularly interested to find out just what he could do to improve the conditions so that his plant would be a safe and better place to work.

As a result of his study as to the value of better lighting he planned the lighting system for a new building which they were adding to their plant. In this new building he installed 1,000-watt lamps on 20 x 20 foot centers, producing a level of illumination of about 18 foot-candles. He later reported that this level of illumination was very highly appraised by the executives of his plant.

I think this illustrates the favorable attitudes which many electricians would take toward better illumination in steel mills, if they would more carefully study the value of higher levels.

I would like to ask Mr. Turner in connection with this paper if he knows how many plants were visited.

A. S. TURNER, JR.: Twelve, I believe, in all.

HOSPITAL OPERATING ROOM LIGHTING WITH RE-PRODUCED DAYLIGHT*

BY NORMAN MACBETH**

SYNOPSIS: The hospital operating room is probably one of the most important fields for artificial illumination, a human life may depend upon what the surgeon does not see. Daylight color for the proper identification of the various tissues is of extreme importance. During the operating period—throughout the year, with a minimum in the heated summer months—the natural daylight is frequently inadequate due to dark days, storm clouds and the shortened period between dawn and sunset.

A description is here given of an artificial daylighting equipment furnishing a light of daylight color with the mechanical arrangements in two operating rooms where the demands of the surgeons have been satisfactorily met.

The intensities throughout the operating table zone generally run from 50 to over 65 foot-candles with current densitites of 12 and 16 KW per room. The light is multidirectional towards the center of the room; a scheme of localized general illumination. The shadows are softened and are illuminated to an extent just less than the high lights.

None of this equipment is below the ceiling and none of it is directly over the table thus removing the dust and radiant heat objections present with the usual table operating luminaire.

In a recently published article by an experienced investigator, the statement was made that in an inspection tour through over one hundred hospitals located in the East only two had lighting equipment of comparatively modern type—artificial illumination having received practically no attention—the fixtures and methods of lighting being those prevailing in the early days of electric lighting; while the latest in instruments, sterilizing apparatus—operating and antiseptic devices were in use; the lighting equipment almost without exception was on a par with the type of sterilizing apparatus which would be represented by a wash boiler and a gas burner.

In no other class of buildings, particularly in the operating rooms, are the lighting requirements so exacting. Probably in no other profession is the ability to see quickly and truthfully as important as in surgery where a human life is at stake.

The location of operating rooms on upper floors with liberal window and skylight exposures results in a high standard of illumination with which the customary artificial lighting intensities suffer by comparison. Artificial light as a consequence has been largely used for emergencies and for cleaning up purposes, and because of the almost universal belief that artificial light is expensive and daylight cheap, but slight effort has, in the past, been made to build up the illumination to the necessary high intensities.

*A paper presented before the New York Section of the Illuminating Engineering Society, April 19, 1923.

**Illuminating Engineer, Macbeth Daylighting Co., New York, N. Y.

The Illuminating Engineering Society is not responsible for the statements or opinions advanced by contributors.

Photometers have been but comparatively recently used by electrical engineers and hence it is only within a few years that it has been realized that the intensity of artificial light generally supplied was so far below that available in the well designed operating room on clear days. There has also been a failure to appreciate the considerable importance of color of light in its effect on the color of objects viewed under it.

Low hung fixtures directly over the operating table have also caused difficulty because of radiant heat. One surgeon recently described the equipment in his operating room as "an inverted bath tub full of hot glaring incandescent filament bulbs", which during an operation necessitated the constant attention of one nurse to wipe the perspiration from the faces of the operating surgeon and his assistants.

These typical bath-tub fixtures were not always sanitary, they collect dust, and the light distribution, when highly concentrated, was generally productive of dense shadows.

At all times the color of artificial light has differed so widely from that of natural daylight that tissues, blue veins, arteries, bile ducts and so forth were seen with extreme difficulty, if at all. Artificial light as most frequently used has failed on many counts but particularly in color, intensity and distribution. Shadows should be softened—they should not be in absolute darkness nor yet eliminated through perfect diffusion of light, with as high intensities of illumination on those surfaces where shadows are natural as on the highly lighted parts. An absolutely uniform illumination of all surfaces whether horizontal, inclined or vertical would be nearly as unsatisfactory as the other more usual extreme of high lights and dense shadows.

Nor has natural daylight been wholly ideal—it has only been comparatively better. It is not uniform in intensity throughout the day nor throughout the year. Even on equally clear days the intensities between summer and winter vary tremendously.

With cloudy days to contend with even natural daylight from day to day is none too good—at its worst it is quite as inadequate as has been the average artificial lighting equipment.

Because of the necessity for quick accurate vision natural daylight therefore has only partially met the situation—met it only on clear bright days and for a limited number of hours per day. This quality of daylight is exceedingly limited during many months of the year and in some localities, in cities and towns adjacent to the Great Lakes, the percentage of cloudy or not clear days is well over 50 per cent throughout the year—by far the greater number of clear days coming in the summer when the operating rooms are in less constant use because of vacation periods and the putting over of many operations to evade hot weather.

To adequately illuminate operating rooms with an electric daylight merely required a photometric study of operating rooms under satisfactory conditions of natural daylight illumination, and a working back to designs in lamps adapted to the reproduction of daylight, to enable plans to be worked out which would result in the proper illumination of an operating room. Dependence for the final approval of the result was left to the judgment of the operating surgeons.

Operating room lighting with the equipment used in these rooms is therefore not an experiment—nor is it particularly new—it is just an engineering adaptation, fitted to the operating room, of the equipment used in other fields during the past seven years. It has done more, however, than to just illuminate an operating room for emergencies, for late afternoons and exceptionally cloudy days. It is represented as an improvement over all but the few best natural daylight hours of the year in operating rooms equipped with more than the usual glass window and skylight areas. It has proven to be a system of lighting so adequate that operating rooms may be located with a view only to their convenience—no longer is it necessary to go to the top floor of the building.

An operating room with this system of lighting may be located in the basement just as satisfactorily if from the standpoint of convenience that location would be desirable.

So far as the eye can judge, and the operating surgeons' opinions taken, it is natural daylight duplicated or reproduced as to color, distribution, and the necessary intensity, and in addition is available at any hour from any electric service on which standard incandescent electric lamps can be used; direct or alternating current and voltages—110 to 125, and 220 to 250 volts.

A light permitting exact, accurate color discrimination and freedom from the usual dust collecting intense heat-radiating lighting fixture over the operating table—a lighting effect similar

Fig. 1.—Exterior view of Macbeth Daylight Lamp No. WD1co showing provision for ventilation, the suspension rods and the method of attachment to the ceiling.

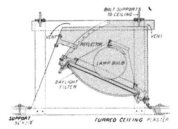

Fig. 2a.—Sectional view of Lamp. Lamp and reflector unit is adjustable as to maximum light distribution from vertically downward to 35° from the vertical, towards center of room. The final adjustment is dependent upon the size of room and the height of the ceiling. The procelain lamp socket used is also readily adjustable to properly center the filaments of the 500 watt or 1000 watt flood lighting bulbs for which this equipment is designed.

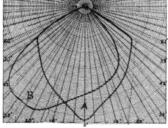

Fig. 2b.—Characteristic Distribution of Light. Fom Daylight Ceiling Lamp (Fig. 1). "A" Parallel with bulb axis. "B" Normal to bulb axis and directed towards center of room.

FIG. 3.—Natural Daylight. The admitted ideal on days when daylight of good quality is available. Operating Room, Women's Clinic Building, Johns Hopkins Hospital, Baltimore, Md.

North Window 10 ft. high by 8 ft. wide, "hammered" glass. Floor, green tile. Sidewall, to 60 in. height—green tile. Room 24 ft. by 25 ft. Ceiling height, 16 ft.

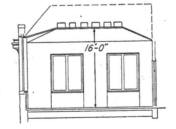

to that of an all glass ceiling with a clear sky continuously overhead.

In planning the installations here illustrated the hospital authorities emphasized, first,—the necessity of furnishing a high intensity of illumination; second,—a diffusion of light which would result in softened shadows—that is, illumination in the shadows; third,—the importance of keeping all lighting equipment away from the region over the operating table, thus eliminating the dust and heat hazards; and fourth,—an arrangement of lamps, screened in such a manner as to eliminate the glare factor.

Glare is objectionable because of those retinal burns resulting in more or less persistent after images which for an interval, long or short depending upon the exposure and the adaptation of the eye, blur the vision. This is an almost instantaneous result of exposing the normal eye, even for a flash view, to the over bright filaments of our present high efficiency, high temperature concentrated filament light sources. The glare possibilities lie in that zone above the critical angle of 60° from the vertical. The type of lamps used in these installations has a sharp cut off at 55° from the vertical,—a more than 5° safety factor.

This system of daylighting equipment has been developed during the past seven years for many specific uses in industrial, textile and mercantile fields where a demand existed for seeing colored objects and materials as they appear in good natural daylight. To enable exacting inspection and critical examination processes to be carried through on dark days and at night with results equal to the best natural daylight hours, and without the customary eye fatigue associated with artificial light.

The reproduced daylight from these lamps or lighting equipment is from ordinary clear glass Mazda or tungsten filament bulbs the light from which is corrected by passing through a special light filter so that it is modified in quality to be the same as natural daylight in its effect on colors. It is not loosely described as an "approximate daylight" nor as a light approximating the quality of sunlight nor of any particular daylight. So far as the eye and color are concerned it is daylight. That this effect has been secured has been vouched for during over seven years of practical every day use by hundreds of color experts checking against good natural daylight with thousands of colors, shades and hues in all kinds of fabrics and materials.

This system should not be confused with the various arrangements of ordinary blue glass lamp blubs. While these blue bulbs have been called "daylight lamps" by the lamp manufacturers it is generally known that this is regarded by them as a convenient commercial designation and is not a claim for technical accuracy. The fact is that the color of light with these blue-green bulbs lies on a light-color scale about midway between that from the ordinary clear glass lamp and the direct light of the sun at noon in the summer. This light is therefore, owing to its midway position, as close an approximation to ordinary artificial light as it is to sunlight.

All workers in color know that observations in direct sunlight are misleading and of little value in color identification, and they appreciate also that there is a wider difference between the color of light from a north sky exposure and direct sunlight than there is between sunlight and ordinary artificial light.

The filters used are solid glass with the color in the glass appearing dark blue by daylight and clear and colorless by transmitted artificial light. They need only be kept clean by an occasional washing and used with standard clear glass incandescent electric bulbs of the proper size and voltage rating.

If taken in time, when planning new buildings and credited with window and skylight construction costs which may be saved the total cost of this lighting equipment is exceedingly moderate.

Much depends upon the size of the rooms and particularly upon the height of the ceilings, which largely determine the number of lamps to use.

The maintenance costs—for bulb replacement and electrical energy are of course greater than for the present admittedly inadequate luminaries. These costs depend largely on the costs of the electric service and in terms of other equally important costs have been shown to be comparatively insignificant. They may run as high as fifty cents to a dollar per hour's use—a total cost most nearly comparable with that for gauze and bandages or a small percentage of the operating room charge or a lesser percentage of the standby charges for surgeons and attendants during an operation.

The relatively higher intensity of light on the operating table shown in Figure 4 resulted in an over-exposure of the photographic negative at this point. The intensity of light on a horizontal plane

FIG. 4.—Electric Daylight, Johns
Hopkins Hospital.

Fourteen Macbeth Daylight
Ceiling Lamps (Fig. 1), with
1000-watt clear glass bulbs and
Southern overcast-sky equivalent daylight filters, arranged
in a rectangle on 9 ft. by 12-
ft. centers.

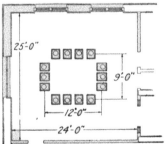

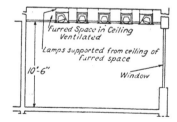

FIG. 5. Natural Daylight. Operating Room. Boston Lying-In Hospital, Boston, Mass.

Window 8 ft. wide extending to ceiling. Room, 14 ft. by 16 ft. Ceiling height 10 ft. 6 in. "Shadowless daylight" as shown here is a phrase that cannot be applied to the light distribution received even from this large window with northern exposure.

36 inches high and over the area extending radially more than 5 feet from the center of the room, that is over a 10 foot circle, ran above 55 foot-candles. This intensity was appreciably greater than that secured at noon on a clear day with natural daylight.

The artificial light has the additional advantage in being directed from four sides, equally illuminating vertical and oblique surfaces where with natural daylight, owing to the window locations, the surfaces facing the windows were highly illuminated with somewhat dense shadows on the other two sides.

The negatives of both operating rooms are without retouching and the prints are free of art work. Just plain photographic comparisons by local disinterested photographers, using exclusively of course the sources of light shown in each illustration.

In the illustrations of the Boston installation, Figures 5 and 6, the photographer used particular care to give both exposures equal time, with the same lens stop, same kind of plates, equal development, etc.

The curtains shown on the windows were fitted shortly after the electric daylighting was used. Visibility conditions were found to be much better when the view of the sky and the varying intensity of the natural daylight due to the alternating clear sky, light and dark clouds, were shut out.

At night the curtains reflect into the room that light which would otherwise be lost by transmission through the window glass.

Photometric measurements in the Boston room, Figure 6, showed an average of over 65 foot-candles on the 36-inch horizontal plane, uniform over a ten foot circle. Compared to the results in Figure 5 with natural daylight, the artificial light intensities were higher at all points beyond two feet from the window and from the center of the room to the wall opposite the window, the intensities were from three to seven times those of the natural daylight. The artificial light was appreciably more uniform while retaining, of course, all the color advantage of natural daylight.

Of course there have been objections raised to this system, relatively few by those who have seen either of these two installations; our chief difficulty is with those who are color blind either wholly or partially and with the greater number who are actually unaware of the enormous difference in color between ordinary

artificial light and daylight. They apparently have no appreciation of the unseen detail under artificial light and the visual
possibilities of daylight; and particularly of the advantage of intensities of over fifty foot-candles even with objects or materials
having high reflection factors. It has long been my conclusion
that much of the going to the window or the store door to see
what a fabric looked like was not so much because of color but
for the detail revealing high intensity daylight, which would help
out the sense of touch with that of sight—to see whether a material
was pressed hair, woven wool or matted cotton fiber.

The effective filtered lumens per watt on a photometric basis
from this equipment will probably not exceed five and may in
some instances be as low as two. This is not due to an inefficient
filter any more than a wire screen which would only pass sand and
gravel not exceeding the size desired would be inefficient if it
passed practically all of the size for which it was designed and none
larger.

This subtraction method of correcting the light from Madaz
lamps necessitates considerable absorption in the red, orange and
yellow to even up this part of the spectrum with the violet and
blue proportions in the light of the tungsten lamp wherein it
differs from visual daylight. To return to the screen and sand,
what would you do if desiring sand and gravel not exceeding that
size which would pass a one quarter inch mesh if you found the
laborer after shovelling ten cubic yards of gravel against the
screen had secured only one or two cubic yards of the desired size—
a three-eights or half inch screen wouldn't be considered more
efficient as it would probably pass too much of sizes above that
desired—you wouldn't blame the screen but the sand pit and you
would probably advise your laborer to select a part of the pit
further removed from the boulder size and ring more into the
desired sand and gravel size. The tungsten lamp is our gravel pit—
if we have to sift or filter out a considerable quantity of the total
flux it is because of the large proportion of long wave lengths compared to the much lesser proportion of shorter wave lengths in
the visible spectrum of that lamp—but don't charge the filter
with being inefficient if the resultant color of light has been properly corrected. An accurate reproduction of daylight is the sole
basis on which filters can be rated as efficient or not. The re-

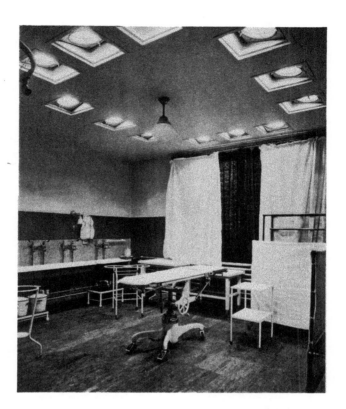

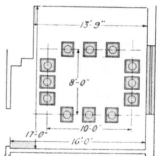

FIG. 6. Electric Daylight, Boston Lying-In Hospital.

Twelve Macbeth daylight ceiling lamps (Fig. 1) with 1000-watt clear glass bulbs and overcast-sky equivalent filters are here arranged in a rectangle on 8 ft. by 10 ft. centers.

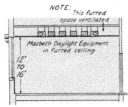

NOTE: This furred space ventilated

Macbeth Daylight Equipment in furred ceiling

12' TO 16'

Elevation

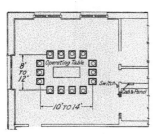

8' TO 12'

Operating Table

Switch

Cab'k Panel

—10' TO 14'—

P an

FIG. 7.—Plan and Elevation.

A suggested typical arrangement of daylight ceiling lamps. The number of lamps and size of rectangle depending upon the size of the room, the height of the ceiling and the intensity of light desired.

EMERGENCY LIGHT

FIG. 8.—Operating Room Emergency Lamp as installed in Boston Lying-In Hospital.

Electric hand lamp with ample cord, connected to the storage battery circuit supplying electric clocks, annunciators, etc. In the event of a breakdown of the regular electric service, Light is here immediately available at the turn of the socket key. This is, without doubt, the most reliable always ready for use adequate light distributing combination yet developed for operating room emergency lighting.

Extra 50-watt, 32-volt bulb also stored in metal, glass door cabinet conveniently located on side wall. Cabinet 22 in. high, 12 in. wide and 4 in. deep.

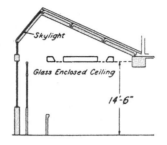

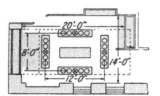

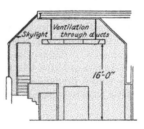

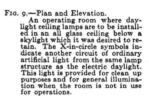

FIG. 9.—Plan and Elevation.
An operating room where daylight ceiling lamps are to be installed in an all glass ceiling below a skylight which it was desired to retain. The X-in-circle symbols indicate another circuit of ordinary artificial light from the same lamp structure as the electric daylight. This light is provided for clean up purposes and for general illumination when the room is not in use for operations.

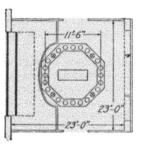

Fig. 10.—Plan and Elevation.
An amphitheater suggestion where in a comparatively small room with a 20-foot ceiling height it is planned to supply an octagonal fixture with lamps on practically 12-ft. centers, the light to be directed towards the operating table from eight sides and from 20 individual sources. Ample light from this equipment is also distributed throughout the room for its adequate general illumination. Four lamps in the same fixture on a separate circuit provide general illumination of ordinary artificial light quality. This circuit is to be electrically controlled in such a manner however, that it cannot be used along with the daylighting equipment.

sultant ratio of effective to total flux is most largely up to the distribution of luminous flux from the light source—its proportion of visible short vs. long wave lengths. And on that point the tungsten lamps we all admit as having a high efficiency are only relatively high as compared for instance to the carbon filament lamps, as in their conversion of energy into luminous flux they are still more than 90 per cent inefficient.

What about glass breakage—suppose a filter should break? That is possible but not probable. The glass is a laboratory product, is thoroughly annealed, and each filter is cut into quadrants, which are asbestos packed on the edges, supported with metal straps and bound together with a heavy channel band. In our several years' experience with thousands of these filters in use in all kinds of places we have not known of a single piece of glass, falling from a lamp. In a very small percentage of cases a filter may crack but we have never had a crack which released a piece of glass in this size filter.

About heat, with twelve to twenty kilowatts per room, twenty to forty watts per square foot, the heat generated must be considerable? It is. It is so uniformly distributed, however, and the lamp housing designed as it is to pass considerable quantities of air removes by convection what might be described as the excess heat, with the final result of a most uniformily ventilated room, with comparatively slow moving air of relatively large volume and an increased temperature due to the lamps of three to five degrees Fahrenheit and that without any forced draught or fan exhaust—simply provision above the ceiling for the escape of the heated air.

DISCUSSION

W. J. WEGLENER: Mr. Macbeth's talk has been enlightening literally as well as figureatively speaking and the information he has given us is particularly valuable, as it is based upon a life long study of this subject.

The Kny-Scheerer Constant Operating Room Light, is a practical light produced after many years of experience in operating rooms with surgeons. It not only meets but exceeds actuarequirements. The function of the operating room light is to produce a constant, clear and concentrated light on the field of operal tion, and one that will show the true pathological colors of the tissues. Briefly described, the Kny-Scheerer Light is a 24" parabolic reflector with six 75-watt blue daylight lamps, mounted about seven (7) feet from the floor.

In an operation there exists the possible need of moving the patient from one position to another. In the case of most concentrated lights, the area of intense illumination is so limited as to make it difficult to move the patient without interferring with the density of illumination on the field of operation. To overcome this, our Operating Room Light has been designed so that the light is sufficiently defused to permit the moving of the patient on the table without in any way interferring with the density of illumination.

Of greatest importance to the surgeons and hospital authorities, to say nothing of the patient is the auxilliary gas lighting attachment. At first consideration this seems of minor importance, but let us go into it further, and let me tell you of an actual fact, the hospitals in Roanoke, Virginia, rely entirely on electric power as a means of illuminating their operating rooms. Recently, the power was shut off for a considerable period, and but one hospital was able to continue operating without any delay whatever. This was because they had installed our light, and when the electricity was cut off, it required but the fraction of a second for the nurse to reach up and turn on the gas light and full illumination was resumed.

In instances such as this, not only is the surgeon's reputation, as well as the reputation of the hospital injured, but of far greater importance is the fact that the patient's life may be lost through even a minutes delay in the operation.

The fact that this has happened once, simply proves that it may happen again, and our light is but a form of insurance against tragedy.

In the construction of this light, we have been able to eliminate the disagreeable feature of intense heat. The temperature existing in the operating room is 90° F. Most lights give out so great aquantity of heat, as to make the temperature so high, that it retards the action of the surgeon. Our light is well ventilated, and it has been proven that a surgeon could work continuously for over two hours directly under this light. So far as actual construction of the light is concerned, the dome outside is constructed of brass and copper with baked white enamel finish. This means that it can be easily washed and kept perfectly clean. The bottom is of wired plate glass, which eliminates the possibility of broken glass falling on the field of operation.

CHARLES CRANE: This question of lighting operating rooms is a very vital question. We have indeed learned a great deal from this demonstration today.

The Kny-Scheerer lamp, with its instantaneous gas connection making it possible to get the same candle power in case of an emergency from the breaking down of your own plant or the current contracted for, is a wonderful thing.

The other lighting system demonstrated is also quite a wonderful improvement on account of the perfect daylight and the reproduction of the human flesh that is given to the human flesh present here today.

I think that both of these equipments should be sent to the American Hospital Association so that they may have the necessary data to furnish Hospital Supts. and others who are vitally interested in the lighting of operating rooms and accident rooms.

Both of these gentlemen should see that their firms give this data to the Liberty Bureau of the American Hospital Association.

S. S. FRANK: I have nothing particular to say other than on the subject of emergency lighting for operating rooms, just brought up by Mr. W. J. Weglener. In a hospital now in course of construction we have taken care of this in a very satisfactory way, but unfortunately it is not possible in all hospitals. There will be two sources of current supply—the New York Edison d.c. street mains for elevators and other motors; the United Electric Light & Power Company a.c. street mains for x-ray work and general illumination. Each of seven operating rooms has a circuit from each of the sources, controlled by an automatic switch that instantaneously throws in the second circuit on failure of the normal one.

At present we are contemplating installing the Artificial Daylighting system. It is quite expensive, but reports from two hospitals where it has been in use are so favorable that we have about come to the conclusion that we can not afford to omit it.

G. B. NICHOLS: The subject under consideration deals with two questions. There are two types of hospitals, two types of operating-rooms. In some of the hospitals, take our state hospitals for insane, with which I am familiar, there might not be more than one operation a day performed in the operating-room, and perhaps not as many as that. For that type of operating-room, it strikes me it would be rather impractical to go into Mr. Macbeth's system. In the private hospital, where a physician has his own private operating-room, he probably would go into the simpler system. But in a hospital where they were carrying on continuous operations that would require continuous service of the room, the hospital could afford the more elaborate system, and probably would, with more refinements.

A. L. POWELL: My experience in dealing with surgeons and hospital superintendents has indicated that the cost of operating the lighting is practically negligible. Not only are the conditions of use very critical but a relatively high rental per hour for the room is obtained. The money which is expended for electric power is indeed a very minor item and the hospital can well afford to put in the best system that is available. A special storage battery has been used in a number of cases as an auxilliary source of power. This needs very careful supervision, otherwise the house electrician is very likely to neglect it, if it has not been called upon for a period of several months and the very time when one wishes to use the battery it may be out of commission. A much better scheme is to have the auxiliary lighting connected to the line serving bells, clocks and indicators, which system being in constant use, will be properly maintained.

W. T. BLACKWELL: I remember some years ago when I was connected with the City Government in New York City that we found the lighting in hospitals antiquated and no provisions made for duplicate illumination. We hit upon an expedient of using duplicate sources of illumination,—that is both gas and electricity. While it did not prove altogether satisfactory, yet it was the only means that we had of solving the problem.

One of the difficulties experienced in lamp illumination was the heat affected the silvering on the mirrored reflectors used on hospital units. Another difficulty was keeping the gas mantle lights in proper working order. Almost invariably upon inspection of the gas mantel lighting it was found that the equipment had been neglected and was not in working order.

It would seem to me that the plan of using electricity solely with separate sources of supply is a much more dependable plan. It does away entirely with the difficult maintenance of the gas system.

C. C. COLBY, JR.: It appears to me that in a good many cases there would be cause to perform an operation without a general anesthetic, that is, cases in which the patient would be conscious. I wonder if Mr. Macbeth has any installations in mind which have been made with a view to eliminating the glare in the patient's eyes, which might be a source of annoyance.

NORMAN MACBETH: Mr. Powell raised the point with reference to storage batteries. Undoubtedly, the storage battery makes an excellent emergency lighting system but all emergency lighting systems should be used every day, the battery should have clock or annunciator duty to insure its regular maintenance.

I recall a recent breakdown of hospital operating room lighting service from an electric plant that had not been off before in twenty years. Can you imagine an emergency service, maintained solely as such for twenty-one years which would be one hundred per cent available unless the service was also used for some every day purpose?

Many of the members recall that I had some gas experience. It was found in department stores and other places where gas mantle lamps were used for emergency, that they were not there with the desirable light output when the emergency arose. A mantle is highly hydroscopic; it will take up a great deal of moisture and dust. The gas should be used frequently. You can use your electric with it, or use your electric for an emergency, but you can't have an emergency gas lamp unless it is practically in constant use. That was found to be the case in a city I recall where they had a good many electrical breakdowns and as a consequence all the department stores largely dublicated their installations. But they weren't used as duplicate installations; they were used as supplement installations. Gas and electric lamps were regularly in use throughout each floor.

In talking to a surgeon from Winnipeg a short time ago he told me that at one time, just in the midst of an operation a thunderstorm struck them, and since their plant was of the hydro-electric type, the lines went down. It happened that somebody had a flashlight; you can get remarkably good service from such a small lamp over a small area. The flashlight saved the situation for him.

I recall a discussion on hospital lighting in England a short time ago, and their conclusion was also that a good storage battery outfit with two or three lamps made the best emergency, provided they were used frequently.

Mr. Nichols speaks about an ordinary operating-room where they have an operation a day. I have been informed that there are days when they cannot operate, as natural daylight is not sufficient at any hour of the day.

While this system here may seem exceedingly elaborate from the standpoint of lighting, it isn't unduly expensive when compared with other hospital equipment. I was talking to Dr. Winford Smith director of John Hopkins. He was very proud of his new sterilizing outfit. That outfit cost $100 more than the lighting equipment. When our demonstration was made down there, one of the surgeons asked, "What have we got in the old operating-room?" I said, "You have a 300-watt lamp in the metal reflector. We have 16,000 watts here. He said, "We can't go from 300 to 16,000." But I explained we have to consider such a question relatively. If you want this kind of lighting you can have it. Let us consider its relative cost. Don't consider the value of the patient's life, or how important that patient may be, but merely the standby charges in the operating-room, even the steam heating element, don't neglect the minor cost of the cotton gauze that they sop up and throw away (they don't waste it any more, they reclaim it), and the cost of reclaiming.

After you have this total, I am not inclined to believe that this system will be considered relatively expensive with a cost of approximately 50 cents an hour to run. What is 50 cents an hour if you can practically lift the roof off your operating room and have day-light from all angles and at a desirable intensity 24 hours of the day?

That brings up another point. I receive inquiries, "Send us particulars about your lamp," To which we reply, "We haven't got a lamp. Send us particulars of your operating-room." In other words, it is a lighting system that has to be fitted into the place in which it is to be used.

CHAIRMAN POWELL: You don't answer Mr. Colby's question about glare.

NORMAN MACBETH: That question hasn't been raised in any of our installations. We can take care of any reasonable demand as far as glare is concerned for an observer or worker in any position in the room, of course, with a patient not under an anesthetic, lying flat on his back looking up at the ceiling, it is a different matter, but so far as any one else in the room is concerned, all the other out-off angles in this equipment are below 60 degrees from the vertical; some of the light is out-off at 45 degrees, so it is only by looking directly up at the lamps you can see the bright filaments which of course, are tremendously toned down, when viewed through the filters.

I saw an article, a short time ago, about some operation they performed with a local anesthetic, and they clamped a radio receiving set on the patient's head so he could get a little diversion. Perhaps he can get a little side diversion also from a view of these lamps.

ABSTRACTS

In this section of the TRANSACTIONS there will be used (1) ABSTRACTS of papers of general interest pertaining to the field of illumination appearing in technical journals, (2) ABSTRACTS of papers presented before the Illuminating Engineering Society, and (3) NOTES on research problems now in progress.

THE COLORIMETRY AND PHOTOMETRY OF DAYLIGHT AND INCANDESCENT ILLUMINANTS BY THE METHOD OF ROTARY DISPERSION*

BY IRWIN C. PRIEST

SYNOPSIS: Further studies, both theoretical and experimental have been made on toe methods of photometry and colorimetry previously proposed by the author. (Phy. Rev. (2) 9, p. 341; 1917. Phy. Rev. (2) 10, p. 208; 1917. J. Op. Soc. Am. 5, p. 178; 1921. J. Op. Soc. Am. 6, p. 27; 1922. B.S. Sci. Pap. 443; 1922).

In the light of this work, the method is now proposed as a complete and satisfactory solution of the practical problem of the visual photometry and colorimetry of the illuminants (including the important phases of daylight) whose spectral distribution approximates the Planckian formula closely enough to give a color match. This solution is based upon the principle of the additivity of homogenous luminosities and the assumption of a standard visibility function.

The method falls in the general class of substitution "equality of brightness" methods. All brightness matches are made at a color match. This color match is obtained by modifying the color of a constant comparison source by allowing its light to pass through a train of nicol prisms and quartz plates which form, in effect, a blue or yellow filter of continuously adjustable spectral transmission.

Tables and graphs have been prepared by which color temperature and candlepower or brightness may be readily obtained from the instrument readings on the basis of any visibility which it is desired to assume as standard.

* * * * * * * * * * * *

(Mr. Priest exhibited a model of the "Rotatory Dispersion Colorimetric Photometer" as constructed by the Bureau of Standards for use in the colorimetry and photometry of daylight and incandescent illuminants.)

This instrument represents the chief tangible result of the development of an idea with which I have been working more or less intermittently for more than eight years. During that period

*Stenographic report of a paper delivered by invitation before the Seventeenth Annual Convention of the Illuminating Engineering Society at Lake George, N. Y. Sept. 26, 1923, as revised and amended by the speaker. Published by permission of the Director of the Bureau of Standards, U. S. Department of Commerce, and the Director of the Munsell Research Laboratory. The instrument described was developed in part while the author was an employee of the Bureau of Standards and in part while a Research Associate of the Munsell Color Co., stationed at the Bureau of Standards.

The Illuminating Engineering Society is not responsible for the statements or opinions advanced by contributors.

I have published a number of somewhat fragmentary and inadequate papers on various phases of this subject, but, up-to date, there has been no adequate publication and no unified treatment of the subject. However, a paper which I hope will be an adequate presentation of the whole subject will appear in the December number of the Journal of the Optical Society.

My chief purpose in accepting the invitation of your Papers Committee to speak at this meeting was to present the subject personally in a more concrete manner by exhibiting this instrument and demonstrating it to you. With this point of view, I do not intend to weary you with a detailed exposition of the theory of the instrument which I think you can follow much more to your own satisfaction and convenience in the printed paper which will appear later. If you do this, you will find some rather formidable mathematical formulas; but I assure you they are perfectly harmless and they do not militate in any degree against the usefulness or practical value of the method. In the paper they serve the purpose of brevity of expression—and I believe that it is commonly understood that such decorations add to the dignity of a scientific paper—but their detailed consideration here might have a depressing effect on what might otherwise be an enjoyable occasion.

To get to the meat of this matter, this instrument serves two purposes which are somewhat distinct but at the same time are very closely related.

The first purpose is the color grading of all the ordinary phases of daylight and all the incandescent illuminants on a simple, intelligible scale color, that is on the scale of color temperature as first developed largely by Hyde, his colleagues, and others, and notably extended by the use of this instrument. The extension consists of extending the experimental range of the method from about 3,200 degrees absolute, where Hyde's work left it, up to the quality of the color of daylight and the blue sky, that is, up to color temperatures in the neighborhood of 20,000 or 24,000 degrees absolute.

The second purpose of the instrument is to afford a means of comparing the relative intensities of illuminants in spite of their difference of quality, and in doing this it covers the whole field of the various phases of daylight, and all the incandescent illuminants, down to the yellow Hefner lamp. That is, it circumvents the difficulties of heterochromatic photometry. It does this, of course, on the basis of an assumed standard for the visibility function which Dr. Gibson was discussing a few minutes ago.

I believe I can present this subject most clearly by describing first the instrument and its use, getting right down to brass and glass in its application, rather than by following the usual procedure of such a presentation and elaborating on the theory first and then showing its incorporation into the instrument.

* * * * * * * * * * * *

At this point the speaker described the instrument and its use, pointing out its principal parts both on the instrument itself and on diagrams (by lantern slides) showing its construction. He also exhibited, by a lantern slide diagram, representative standard data on the color temperature of various phases of daylight and various incandescent illuminants. Since this portion of the discourse can not be understood without reference to the diagrams, reference must be made to a paper which will appear under the same title in the Journal of the Optical Society of America, December, 1923.)

* * * * * * * * * * * *

This method and instrument provide means for the specification of color quality and the measurement of relative intensity for all incandescent illuminants and the important phases of daylight, including, on one extreme, the Hefner lamp, and, on the other, the blue sky. The practical solution of the problem of the photometry of illuminants of different quality here proposed is given explicit in the most simple and fundamental terms possible in the nature of the problem. The method is free of non-reproducible terms and temporary expedients in specification. At the same time it appears to be convenient and well adapted to ordinary routine work in the laboratory or in the field. This method and instrument are therefore recommended to the consideration of the Illuminating Engineering Society as a fundamental and convenient solution of the problems of the colorimetry and photometry of incandescent sources and daylight.

There is obviously a great deal of detail which I can not discuss within the time now avaliable. (Among other items is the use of the "spectral centroid of light" as a color index in some respects preferable to "color temperature".) I have planned to set up this instrument in the lobby upstairs with some lamps which I have brought along. I will exhibit it in operation and let those who may be interested stop there this afternoon and look into it. I have here the complete manuscript of the paper including the figures, and will have it upstairs with the instrument. Those who are interested may come and talk with me about it; those who are not may play golf. (The instrument was exhibited in operation in the hotel lobby during the afternoon)

DISCUSSION

M. LUCKIESH: As one interested in color measurement for a great many years I want to take this opportunity of expressing my appreciation to Dr. Priest for the culmination of a really fine job.

I confess that at times I was rather doubtful as to the outcome of an instrument using nicol prisms, not on account of Dr. Priest's ability at all but on account of the fact that I have abhorred nicol prisms in certain work.

I am sorry Dr. Priest did not have time to go into the details of this. To me, it is most fascinating because it puts color measurement on a mathematical basis. I have often used mathematical relations and spectrophotometric curves for getting many data results without having to make measurements with an instrument.

Mr. Priest's instrument, being based on mathematical curves and being developed to this high state, I think is going to provide a means for a good deal of valuable data. I hope that he can extend it so we can get over into the measurement of hue and even over into that great field beyond black body radiation. But even for that alone, it seems to me we have something now that we have never had before—an instrument from which we can build various scales.

A. H. TAYLOR: I would like to ask a few questions about this. One is whether there are any of the individual parts of this instrument which would require an experimental calibration; or, in other words, whether the curves which Mr. Priest has worked out apply alike to all instruments made up as nearly as possible identical with this one; second, whether it is possible with this instrument to match colors of sources which do not have a black body distribution of energy—for example, the gas mantle, or in a more extreme case, colored glass which doesn't have a systematic or uniform spectral distribution, or transmission; and third, as to the degree of accuracy which you can obtain with the instrument; that is, I don't mean so much in measurements of candlepower (although I would like to know that) but color temperature; for example, just how accurately you can measure color temperature.

MAX POSER: I would like to ask Mr. Priest whether he found any method of standardizing the effectivity of the nicol prisms when made in quantities.

NORMAN MACBETH: It may be of passing interest in this connection to state that we have developed an instrument for the measurement of daylight in units of color and intensity. This instrument enables us to measure also any color change from the ordinary Mazda lamp or even from a carbon filament lamp through the color changes to the extreme of blue sky daylight. The color temperature equivalents may also be noted to the point where we reach the maximum of direct sunlight at noon.

We have used a certain blue glass that we call daylight glass. Two wedges were made, these are about 15 centimeters long, 3 millimeters thick on one end and 1 millimeter at the other end. These wedges intercept a beam of light from a lamp house on a photometer track. Varying the distance between the lamp and Lummer-Brodhun cube permits an intensity match. The outside field of the comparison prism may be turned towards a white glass test plate or towards the sky or even a particular cloud in the sky. The color match depends upon the density of these daylight glass wedges interposed in the light path. Then as this particular blue glass is a definite mixture of three distinct colors of glass; a signal green, a full amethyst and a very pale blue; three additional sets of wedges of these glasses may also be used, so if at any time we should find a daylight color that varied from that which could ordinarily be matched with the daylight glass wedges we can add any or all of these colors, any combination of which can be reproduced in a single glass to be used as a filter with a Mazda lamp as the source, for an exact reproduction of any color of light matched in this daylight colorimeter.

We also have another daylight glass, the composition of which was worked out by Ives and Brady. This is a glass which Mr. Priest tells us perfectly possesses the property of raising the apparent color temperature of one black body to that of another, consequently with various densities of this glass and a Mazda filament source any black body color can be matched. The Gage-Corning glass, of which the first mentioned wedges were made, is not a "sunlight" glass. That is to say, it will not result in an exact sunlight or black body radiation color match. It is my understanding that natural daylight from which direct sunlight is absent is not an exact visual match for any black body temperature color and the Gage glass possesses this daylight color characteristic. This colorimeter is not on the market nor do I believe it is likely to be as it was brought out in the usual course of our daylighting development to maintain a check on our own work. This instrument of Mr. Priest's probably permits a great many determinations that we cannot make. At the same time, I believe that we can make many measurements of considerable value in our particular work that Mr. Priest cannot make so exactly, but as our work is highly specialized I do not believe that this in any manner detracts from Mr. Priest's achievement.

MR. I. G. PRIEST: Beginning with Mr. Macbeth's question, I should say that probably the chief utility of this instrument for his purpose would be the calibration of his wedges in terms of color temperature. Probably a much cheaper and simpler instrument can be made for a great deal of field work, although personally I should not prefer it. I would prefer to take this instrument into the field.

Coming to Mr. Taylor's and Dr. Poser's remarks, I should have answered Mr. Taylor's first question before it was asked. I intended to make it a part of my presentation, but did not do so because of the limited time. The instrument is completely reproducible from specifications. It is not dependent upon an emperic calibration. However, the computation from given constants of the instrument—which any one might be excused for questioning —has been checked experimentally so that over considerable ranges we have both the computed calibration curve, (*i.e.* the theoretical calibration) and the emperic calibration curve. Some data on the comparison of the theoretical and empiric calibrations were given in a paper which I presented at the Optical Society nearly two years ago and have been published in Bureau of

Standards Scientific Paper, Number 443. In that paper it is shown that color temperatures measured by means of the theoretical calibration of the instrument checked with color temperatures obtained by other means. For example, values reported independently by Dr. Forsythe of the Nela Research Laboratory were checked with surprising accuracy.

In regard to Mr. Taylor's question about measurement of the color filters of course it should be stated there are several cases in which you can make a color match with this instrument. In the first place, you have identity or what may for practical purposes be called identity of spectral distribution between the test lamp and the distribution which you manufacture by means of the instrument. That, of course, gives a perfect color match. Then you have the case where you make a color match with the test lamp, the test lamp having a different spectral distribution than that produced to color match it. A good example, perhaps the best example of that case is the case of the artificial daylight units such as have been proposed by Ives, Luckiesh and others. There we can measure the apparent color temperature of the artificial daylight unit with this instrument, but the daylight unit does not have the same spectral distribution because the glass has little kinks in its spectral transmission curve. The Ives-Brady glass, with the sources which are used with it, gives a perfect color matching. The conditions of the two cases are somewhat different.

There is a still further case where we get only an approximate color match, where it isn't possible with this instrument to match the source because it has not the spectral distribution characteristic of a black body. Such a case is the Moore carbon dioxide lamp. If we make observations on a Moore lamp (as we have done) our experimental data represent the temperature at which the color of a black body would most closely approximate to the Moore lamp, and the color difference then existent in the field is not much more than the least perceptible difference. But the Moore lamp is not a perfect color match to a black body and its spectral distribution of course as many of you know is widely different from a black body.

With regard to photometric accuracy, I published a great many years ago some preliminary photometric results which showed I think very good agreement, within a per cent or so, with values obtained otherwise.

I had some more slides to present dealing with the selection of standard visibility data. I will sum that all up without presenting the slides by saying that of course we assume a standard visibility function as a fundamental standard in this method. It makes absolutely no difference, however, for the determination of the relative intensities of illuminants, whether you choose the I. E. S. curve, the curve which I used, or the curve by Gibson and Tyndall, or any other reasonable average; the results will be the same within the ultimate possible precision of photometric work.

SOCIETY AFFAIRS
SECTION ACTIVITIES

NEW YORK Meeting—October 18, 1923

At the first meeting of the New York Section held on October 18, a paper, "Lighting and Signal System of the Leviathan", was presented by Mr. W. M. Zippler, Electrical Engineer, Gibbs Brothers, Inc., New York. Prior to the meeting, the Dinner Committee successfully arranged a table d'hote dinner at Shanley's Restaurant, 117 West Forty-Second Street. After the meeting the members and guests visited the S. S. Leviathan where an inspection trip was made. The time allotted for the members and guests of the I. E. S. for visiting the Leviathan was from 8:00 to 10:00 P. M. It was estimated that about 400 people visited the Leviathan, and there were 150 members and guests at the dinner.

PHILADELPHIA Meeting—October 22, 1923

The Philadelphia Section met at the Engineers' Club on October 22 to discuss a number of papers presented as follows:

"The relationship of the Scientist to Illuminating Engineering," By Laurence A. Hawkins, Research Laboratory of General Electric Co.; "The relation of Illuminating Engineering to the Central Station," By C. J. Russell, Vice-President, Philadelphia Electric Co.; "Proper lighting as an aid in the prevention of Accidents," By R. E. Simpson, Engineer, Travelers Insurance Co.; "Illuminating Engineering as an aid to the Architect," By Emile G. Perrot, Architect and Engineer, Philadelphia; "Artificial Light in the Home," By M. Luckiesh, Director, Laboratory of Applied Science, National Electric Lamp Works of G. E. Co.; "Illumination in Industry," By Jos. G. Crosby, Vice-President & General Manager, Whalen Crosby Electric Company, Philadelphia; "The Effect of Modern Illuminants on the Eye" By Dr. George S. Crampton, Opthalmologist, Philadelphia.

Preceding the meeting, members and guests to the number of twenty-five were served with dinner at the Engineer's Club. This being the opening meeting of the year, the Chairman opened the meeting with a brief address to the members and guests, stating the scope and purpose of the Society, the object of the Philadelphia Section and the aim of the Officers and Papers Committee during the coming Section Year.

Mr. G. B. Regar, Chairman of the I. E. S. Committee on Membership spoke of the increase in membership of the Society, calling particular attention to the results of last year. He urged that the Philadelphia Section be particularly active in this line.

After the presentation of the Papers for the evening an interesting discussion followed, which was participated in by Mr. Jos. D. Israel, Dr. Geo. A. Hoadley, Mr. Frank Lewis, of Baltimore, and Mr. M. E. Arnold.

CHICAGO Meeting—October 31, 1923

The section met at the Benjamin Electric Mfg. Co. Chicago Office at 12:30 P. M. and drove via automobile to the Copper Kettle at Oak Park, Illinois where lunch was served and then proceeded to the Benjamin Electric Mfg. Co. Factory, DesPlaines, Illinois.

At 3:00 P. M. the members and guests made an inspection trip through the factory accompanied by guides who explained the various processes in the manufacture of porcelain steel lighting equipment.

At 5 o'clock the various groups returned to the meeting place, Messrs. A. E. Clark and W. E. Quivey, of the Benjamin Electric Mfg. Company, gave a short talk on illumination as applied to their products. Mr. H. S. Thayer of the Atlas Electric Devices Company gave a short talk on the Violet Arc Fading Machine. Mr. W. J. Guntz of the Paul E. Johnson Co. gave a short talk followed by a discussion of the quartz tube mercury vapor unit used for light therapy.

NEW ENGLAND Meeting—October 26, 1923

The subject for discussion at the New England Section meeting on October 26 was a series of solutions of a lighting problem. Messrs. R. B. Brown, James A. Toohey, L. S. Purdy and W. D. McCabe presented solutions which were discussed by Messrs. R. B. Burnham, Philip Drinker and A. H. Hirons.

The meeting was held at the Engineers' Club and there was an attendance of fifty members and guests. Resolutions were adopted expressing regret of the death of Dr. Charles P. Steinmetz, Past-President, I. E. S.

MICHIGAN Meeting—November 2, 1923

The Michigan Chapter opened the *Year of Cooperation* with a joint meeting of the Industrial Engineers Society of Detroit and the Electrical Extension Bureau of Detroit. In the Detroit Edison Company's Auditorium, Mr. Ward Harrison Illuminating Engineer, National Lamp Works of G. E. Co., Cleveland, Ohio, gave a lecture and demonstration of "Control of Light".

There were 353 members and guests present at this lecture. Architects, factory and store executives safety engineers, factory maintenance engineers, and representatives of all branches of the electrical industry were present. This attendance was attained through extensive advertising, letters to a selected mailing list, 500 printed posters, and splendid cooperation from the Detroit Safety Council, Detroit Retail Merchants Association electrical jobbers, electrical Extension Bureau and the Detroit Society of Industrial Engineers.

The Chapter is planning a series of four or five more joint meetings for the *Year of Cooperation*, some of general interest and some of special interest to various groups which will be open to the public.

NORTHERN NEW JERSEY Meeting—October 31, 1923

A joint meeting with the "Ourselves Club" was held at Hackensack, N. J., October 31, 1923. The "Ourselves Club" is an organization of electrical contractors and dealers and other people actively engaged in the electrical industry. Mr. Samuel G. Hibben gave a very interesting talk on "Facts on Lighting and Use of Light".

Using some ingenious portable apparatus, Mr. Hibben accompanied his talk by giving very interesting demonstrations of light control, and showed what astonishing changes can be obtained from a single display by using light of various colors. Using a "shadow box" which permitted light to be thrown from different directions upon small statues and plaster masks, the effects of shadows on lighted displays were shown. Both the lecture and the demonstrations were not only interesting but of practical value to those present, as many of the lighting effects demonstrated can with advantage be applied in lighting of show windows or other displays, and may be directly applied by the electrical contractors in their work along these lines.

An audience of sixty-five was present including Mr. Clarence L. Law, President of the Illuminating Engineering Society. Mr. Law spoke briefly of the satisfaction which he felt on seeing the meeting so well attended and congratulated both the officials of the "Ourselves Club" and of the Northern New Jersey Chapter on the excellent work they were doing in furthering better lighting practice and the idea of cooperation between electrical contractors and dealers and the illuminating engineers.

TORONTO Meeting—October 15, 1923

The first meeting of the Toronto Chapter was held at the School of Science, University of Toronto on October 15, Messrs. George G. Cousins and W. H. Woods presented a resumé of the principal papers discussed at the Lake George Convention of the I. E. S.

At this meeting committees were appointed to draw up a tentative "Code of Industrial Illumination" for consideration of the Provincial Government at the request of the Canadian Electrical Association, and to consider "Automobile Number Plate Illumination" and report to the Provincial Government.

PITTSBURGH Meeting—October 25, 1923

The organization meeting of the Pittsburgh Chapter was held on October 25, in attendance there were forty members and guests. The following officers were elected: Chairman, Franklin W. Loomis; Secretary-Treasurer, Sanford C. Lovett; Executive Committee, L. J. Kiefer, G. W. Ward, J. P. Warner; Committee Chairmen: Papers, F. Y. Davidson, Jr., Membership, Douglas Wood; Attendance and Publicity, J. J. Husson; Entertainment, J. H. Van Aernam.

COUNCIL NOTES

ITEMS OF INTEREST

At the meeting of the Council, November 8, 1923 the following were elected to membership:

Eleven Associate Members

BYRNE, THOMAS W., Electrical Engineer, 710 Little Bldg., 80 Boylston St., Boston.

DOLBEARE, WALTER IRVING, Commercial Lighting Salesman, Blackstone Valley Gas & Electric Co. 231 Main St., Pawtucket, R. I.

FLANAGAN, E. C., Salesman, Illuminating Glassware, The Phoenix Glass Co. P. O. Box 757, Pittsburgh, Pa.

HUNTER, T. A., Manager, Lighting Dept. Pacific States Electric Co., 236 S. Los Angeles St., Los Angeles, Calif.

KARAPETOFF, VLADIMIR, Professor of Electrical Engineering, Cornell University, 607 East State St., Ithaca, N. Y.

KING, FLOYD E., Salesman, Wheeler Reflector Co., 156 Pearl St., Boston, Mass.

LOVETT, SANFORD C., Edison Lamp Manager General Electric Co., 1318 Oliver Bldg., Pittsburgh, Pa.

PIERSOL, ROBERT JAMES, Illumination Research Engineer, Westinghouse Elec. & Mfg. Company, East Pittsburgh, Pa.

SHRYOCK, EDWIN W., Illuminating Specialist, West Penn Power Co., 14 Wood St., Pittsburgh, Pa.

VAN AERNAM, J. H., Electrical Jobber, Iron City Electric Co., 436 Seventh Avenue, Pittsburgh, Pa.

WESTERVELT, A. E., Manager, Lighting Inspection Bureau, The New York Edison Co. 130 East 15th St., New York, N. Y.

OTHER CHANGES IN MEMBERSHIP

One Transfer to Member

MAYO, HERBERT J., Sales Engineer, Benjamin Electric Mfg. Co., 582 Howard St., San Francisco, Calif.

One Member Reinstated

OWENS, H. THURSTON, 51 East 42nd St., New York, N. Y.

The General Secretary reported the death, on October 27, 1923 of one member, L. E. Voyer; Illuminating Engineer, San Francisco Office, General Electric Co., San Francisco, Cal., and four associate members, in the recent Japanese earthquake, as follows: Tetsuya Fujii; Engineer in charge of Laboratory, Tokyo Electric Co., Kawasaki-machi, Kanagawa Ken, Tokyo, Japan. Kenjiro Kato; Chief of Engineering Section, Tokyo Electric Co., Kawasaki-machi, Kanagawa Ken, Japan.   K. Khikata; Tokyo Electric Co., Kawaski, Kanagawa-Ken, Japan.   Hannosuke Murota; Electrical Engineer, Tokyo Electric Co., Kawasaki-Kanagawa, Ken, Japan.

Upon the recommendation of the Council Executive Committee, Mr. Norman D. Macdonald, was appointed General Secretary to fill the unexpired term of Mr. Samuel G. Hibben. The resignation of Mr. Hibben as General Secretary was presented to the Council on October 11, 1923 to take effect November 8, 1923.

CONFIRMATION OF APPOINTMENTS

As members of the Committee on Motor Vehicle Lighting—F. C. Caldwell, E. C. Crittenden, A. W. Devine, C. E. Godley, C. A. B. Halvorson, Jr., John A. Hoeveler, W. A. McKay, A. L. McMurtry, H. H. Magdsick, L. C. Porter.

As members of Committee on Sky Brightness—H. H. Kimball, Chairman, E. C. Crittenden, E. H. Hobbie, James E. Ives, Bassett Jones, W. F. Little, M. Luckiesh, L. B. Marks, I. G. Priest.

As members of Committee on Membership—C. E. Addie and F. H. Murphy.

As members of Committee on Sustaining Members—W. T. Blackwell and E. J. Teberg.

As Official Representative on the U. S. National Committee of the International Commission on Illumination—G. H. Stickeny.

As Local Representative at Salt Lake City—L. B. Gawan.

At the meeting of the Executive Committee of the Council held on October 25, 1923, the petition for a charter presented by the Pittsburgh Chapter was granted.

NEWS ITEMS

Professor Vladimir Karapetoff of the School of Electrical Engineering, Cornell University, has been awarded a prize of four thousand francs by the Montefiore Foundation of the University of Liege, Belgium. The award was made for his kinematic computing devices of electrical machinery, described in the technical press during the last three years. A committee of five Belgian and five foreign members, which makes these awards, has characterized this work as an expression of a "new idea which may lead to important developments in the domain of electricity".

Mr. L. E. Voyer, assistant local sales manager of the Edison Lamp Works of the General Electric Company at San Francisco, died October 27, 1923 of double pneumonia after a five days' illness. He was well known as an illuminating engineer, especially throughout the Pacific Coast section, where he will be remembered for his notable work in connection with the headlight laws of the state motor vehicle act in California, his assistance in the Panama-Pacific Industrial Exposition and his work in advancing lighting practice in the state of California. He took an active and guiding part in the establishment of various lighting regulations there, including the Code of Lighting for industrial plants. One of his first activities was in connection with the headlights used on railway locomotives, when he secured a more logical interpretation of those laws both in California and Nevada.

Mr. Voyer was born September 10, 1887, at Junction City, Wisconsin. He entered the employ of the General Electric Company at Harrison, N. J., in 1911 as a student engineer. In 1912 he was transferred to the illuminating engineering department at Harrison and in 1913 he was transferred to the San Francisco office as a special illuminating expert. He was an alumnus of the University of Wisconsin, class of 1911. He has been constantly in communication with illuminating engineering work in the east, being active in the work of the Illuminating Engineering Society. As western representative of this organization on the Committee on Motor Vehicle Lighting he assisted in the co-ordination of regulations adopted in various states. He established the Bay Cities chapter of the Illuminating Engineering Society, being its first chairman.

The Japanese earthquake exacted a toll of lives among Illuminating Engineers. A report just received from Mr. M. Uchisaka, Tokyo Electric Company, Limited, gives the following members of the society as among those who lost their lives in that catastrophe—Tetsuya Fujii, Kenjiro Kato, Hamnosuke Murota, and Kimo Shikata. In addition, Mr. T. H. Amrine, who was a member of the Society when connected with the Edison Lamp Works of General Electric Company, at Harrison, N. J., was lost with his family.

Although several of the buildings of the Tokyo Electric Company's lamp factory at Kawasaki were razed by the earthquake, the Japanese have, with their customary energy, been rapidly reconstructing and are already manufacturing in quantity to meet the demand.

The negative of the group picture taken at the Lake George Convention last September has been sent to the General Office. Orders for the group photograph may be sent to the General Office up to December 31, 1923, the cost of the photograph will be $1.50 delivered.

ILLUMINATION INDEX

PREPARED BY THE COMMITTEE ON PROGRESS.

An INDEX OF REFERENCES to books, papers, editorials, news and abstracts on illuminating engineering and allied subjects. This index is arranged alphabetically according to the names of the reference publications. The references are then given in order of the date of publication. Important references not appearing herein should be called to the attention of the Illuminating Engineering Society, 29 W. 39th St., New York, N. Y.

OFFICERS & COUNCIL

1923-24

President

CLARENCE L. LAW,
Irving Place and 15th Street,
New York, N. Y.
Term expires Sept. 30, 1924

Junior Past-Presidents

GEO. S. CRAMPTON,
1819 Walnut Street,
Philadelphia, Pa.
Term expires Sept. 30, 1924

WARD HARRISON,
Nela Park,
Cleveland, Ohio
Term expires Sept 30, 1925

Vice-Presidents

WM. J. DRISKO,
Mass. Inst. of Technology,
Cambridge, Mass.
Term expires Sept. 30, 1924

D. McFARLAN MOORE,
Edison Lamp Works,
Harrison, N. J.
Term expires Sept. 30, 1925

O. L. JOHNSON,
230 South Clark Street,
Chicago, Ill.
Term expires Sept. 30, 1924

G. BERTRAM REGAR,
1000 Chestnut Street,
Philadelphia, Pa.
Term expires Sept. 30, 1924

General Secretary

NORMAN D. MACDONALD
80th St. and East End Ave.
New York, N. Y.
Term expires Sept. 30, 1923

Treasurer

L. B. MARKS,
103 Park Avenue,
New York, N. Y.
Term expires Sept. 30, 1923

Directors

Terms expire Sept. 30, 1924	*Terms expire Sept. 30, 1925*	*Terms expire Sept. 30, 1926*
F. C. CALDWELL, Ohio State University, Columbus, Ohio	FRANK R. BARNITZ, 130 East 15th St., New York, N. Y.	JAMES P. HANLAN, 80 Park Place, Newark, N. J.
AUGUSTUS D. CURTIS, 235 W. Jackson Blvd., Chicago, Ill.	A. L. POWELL, Fifth and Sussex Sts., Harrison, N. J.	HOWARD LYON, Welsbach Company, Gloucester, N. J.
PRESTON S. MILLAR, 80th St. and East End Ave. New York, N. Y.	W. D'A. RYAN, GeneralElectric Company, Schenectady, N. Y.	H. F. WALLACE, 84 State Street, Boston, Mass.

TRANSACTIONS
OF THE
ILLUMINATING ENGINEERING SOCIETY

Vol. XVIII   December, 1923   No. 10

Who Needs the Society?

WITH every new member there comes to the Society an increased revenue which, transformed into such items as educational work, publications and acquaintanceship represents a higher dividend to all of the member-stockholders in the organization. Yes, the Society needs members, and its members are each one benefited by the addition of more members,—in truth, the Society consequently needs within its membership everyone interested in upbuilding civilization with light.

The Society welcomes all such, needs all such, but who needs the Society? The Illuminating Engineer? Yes, although he should eventually constitute but a small minority in members. The Architect? Yes, but he is only one of many who may use light as a master Sculptor or as a means of making their structures more livable. The Physician? Yes, but here again is a class mighty in power though small in numbers, who spread the gospel of better living, through optical hygiene. The Electrical Contractor? Surely, for he holds the key to many a storehouse of illumination treasurers. Those and other special groups of professional men or artisans all need the Society, as an aid to business, but isn't the usefulness of the Society broader than that?

After all, does not everyone who wishes to better enjoy the business of living, need knowledge of the business of lighting, and does there not come to even the most ordinary one among us a sense of higher pleasure and a better appreciation of the beauties of things we see when we, through reading or by word of mouth, know something of the subject?

"I always have supper for him when he gets home, and lights, lights, plenty of lights—he always likes lights." So Lady Beaconsfield explained in her acute way how it was that Disraeli, Lord

Beaconsfield, was welcomed at home when he stepped out of the character of the Prime Minister of England into the character of the relaxed common man. Plenty of light, if such fitted the mood, represented ethereal medicine to that haunted and weary mind. But a Prime Minister cannot command a greater comfort than the poorest servant, for light is no slave of royalty or caste, and anyone who chooses may through it add to his pleasures if he will. He who wants the comfort that pleasureable lighting may bring to his leisure hours, needs and society?

Who needs the Society? Every man who reads may find in the Society's literature a wealth of knowledge and a pleasure in the reading thereof that, might it not increase his earning power, certainly increases his enjoyment. How slowly the world has discovered, is yet discovering the stubborn secrets of artificial light,— how much that protracted discovery has contributed, will yet contribute to the ease and luxury of life! And yet the reading of the story of lighting progress is the perusal of a vital chapter in civilization history. And it is a romance. He who reads, then, needs the Society's contributions to his store-house of pleasant learning. From the first to the last, no person of consequence dares cease to read authentic sources of knowledge. "That's good, go on, read some more" were President Harding's last words.

He needs the Society who needs friends. He who lives alone, who needs not the pleasant companionship of educated minds, who is a person whom what he *is* becomes a habit and whose brain atrophies in isolation, does not need the Society, nor any Society!

All of these things,—the appreciation of living, the enjoyment of reading, the growth from contact with friends, all come in fuller measure to him who finds he needs the Society and then fulfills the need. Mental ignorance is darkness of a kind, and with the wealth of the Society's offerings at hand, none may now be afraid of such darkness. Only those who never discover that they need the Society may feel during their life as did O. Henry at the end of his, when he said, "Turn the light a little higher. I'm afraid to go home in the dark."

SAMUEL G. HIBBEN

REFLECTIONS
Good Lighting Vital to Industrial Safety

"ALMOST every measure put forth to safeguard industrial employes against accidental injury requires, as a primary and fundamental condition, proper lighting," states A. C. Carruthers in *Safety Engineering*, the official publication of the American Society of Safety Engineers.

"It presupposes that at all times the operators will be able to see clearly and distinctly in performing their duties and moving about the plant, but in a great many cases the proper lighting has not been furnished and the beneficial results which could and should be secured through the installation of safety devices and other protective equipment and apparatus, safety rules and safety instructions imparted to the employes, have not yielded the expected results. Proper lighting is a very important safety measure and it removes a very definite and now well- recognized hazard of industry.

"It is safe to say that a large number of accidents both fatal and non-fatal have occurred because of bad lighting conditions, but industry has paid the price, both employer and employes having suffered severely through this insidious hazard. Few, if any, industrial hazards can be removed so completely from the midst of a busy industrial plant as the bad lighting hazard.

"It is due largely to the efforts of the Illuminating Engineering Society that seven states have thus far adopted State Lighting Laws, based upon standards formulated by the Society to protect workers in industrial plants from accidents due to bad lighting and from unnecessary eyestrain. The states that have passed laws making obligatory definite industrial lighting standards are New York, New Jersey, Pennsylvania, Wisconsin, Oregon, Ohio and California."

Modern Street Lights a Development of Last Decade

"A GROWING appreciation of the value of street lighting, new traffic problems, and the improvements in electric lamps, have produced such rapid development in the art of street illumination that the best practices of ten years ago have already become obsolete," says the *American City Magazine*.

"Ten or a dozen years ago illuminating engineers were seriously and emphatically debating whether street lighting intensities on residence streets should be equal to that of a half-moon or quarter-moon, or about those relative values. While there was some inkling of the possibility of developing a scientific illumination of the average street at relatively high intensities, and while there had been several examples of 'white ways'—largely for advertising purposes, however—rash indeed, would have been the man who dared to predict street lighting practices of the kinds that are already obsolescent.

"Street lighting has assumed a different character recently. In early days the principal object was to provide enough light to enable pedestrians to walk without colliding with obstructions or with other persons, and to find their homes, and to permit horse-drawn vehicles safely to navigate the more or less quiet thoroughfares and side streets.

"Today the problem of handling with safety the ever-increasing automobile traffic, the value of high illumination as a police safety precaution, the importance of well-lighted streets in forming public opinion on the desirability of certain districts, all serve to focus the attention of others than illuminating engineers and central station men on the problem of adequate street illumination."

ABSTRACTS

In this section of the Transactions there will be used (1) ABSTRACTS of papers of general interest pertaining to the field of illumination appearing in technical journals, (2) ABSTRACTS of papers presented before the Illuminating Engineering Society, and (3) NOTES on research problems now in progress.

EXCERPT FROM REPORT OF EDUCATIONAL COMMITTEE OF AMERICAN PHYSICAL SOCIETY ON "PHYSICS IN RELATION TO MEDICINE," 1923

Light. The effect of light in modifying the course of chemical reactions is almost too well known to need comment. The relation between the growth of green plants and sunlight has long been known. A more recent development is knowledge of a direct relation between exposure to light of certain wavelength and the prevention of rickets, a disease of childhood. One of our most important sense organs is the eye. A knowledge of geometrical optics and of various optical instruments is absolutely indispensable to the ophthalmologist. Optical instruments of many kinds are in use by biologists. Photography plays a part of increasing importance in scientific investigation. The spectroscope has long been used in chemical investigations. The microscope is preeminently the biologist's own instrument. The spectrophotometer promises to become a powerful instrument of research in biological chemistry.

The problems of physiological optics have been but little cultivated by physiologists since the time of Helmholtz, and many of these should be made subjects of research, notably color vision, vision itself, the part played by the ocular pigment and the visual purple, etc. In recent years physicists concerned with optics have been forced to make studies of some of these physiological problems bearing on the use of the eye in physical measurements, for example, studies of visual luminosity curves in connection with photometry of light of different wave-lengths. The physiologist who undertakes such problems as those mentioned above must of necessity be well informed on these investigations and on physical optics in general. There is great need of a course in ophthalmology of a comprehensive character as shall really fit the graduate for all the various exigencies of practice. It should naturally include the pathological and surgical aspects of the subject, and also thorough and adequate instruction in geometrical optics and physiological optics. There are few men at the present time competent to give really adequate instruction in the last, and, so far as the Committee are aware, no medical school offers to a graduate in medicine a comprehensive course comparable with that outlined above.

DEVELOPMENT OF THE ELECTRIC INCANDESCENT LAMP

BY B. P. DUDDING AND J. C. SMITHELLS

The first of a series of articles dealing with the development of incandescent lamps. The first article outlines the history of this type of lamp from the early platinum filaments of Grove and other experimenters to the modern gas-filled tungsten-filament lamp. Early metal filaments proved unsuitable, and the carbon filament took their place about 1880. The metallized carbon-filament lamps of 1907 had an efficiency of 3.5 watts to 4 watts per spherical candle. Experiments upon increased selective radiation led to the manufacture of carbon-silicon lamps and the Nernst lamp, the latter employing rare-earth filaments, with a degree of success which was overshadowed by the introduction of metallic-filament lamps. The latest development of gas-filled lamps has improved the efficiency of electric incandescent lamps from 9 watts per candle of the ordinary carbon lamp to about 0.7 watt per mean spherical candle.—*Beama*, October, 1923.

PROJECTION OF LIGHT

BY W. J. JONES AND E. A. MARX, JR.

Experiences with the use of gas-filled electric lamps in optical lanterns are summarized. It is shown that under the best conditions the percentage of the original light reaching the screen does not exceed 5 to 10, although it is evident that there is much difference between various forms of lanterns in this respect. In the motion-picture projector the proportion of light reaching the screen is even less.—*Illuminating Engineer (England)*, Vol. XVI, No. 3.

PAPERS

ELEVEN SOLUTIONS OF A STREET LIGHTING PROBLEM*

SYNOPSIS: Approximately three months ago the following letter and questionnaire, together with a blue print of the plan shown in Figure 1, and photograph, Figure 2, were sent to a selected list of representative street lighting specialists throughout the country. Eleven solutions of the problem were received which are incorporated in this symposium and should serve as a basis for discussion. These answers were coordinated by the Papers Committee in order that a uniform presentation might be made. The essential data, however, is given exactly as submitted.

At this year's convention the Papers Committee is planning somewhat of an innovation in the type of a symposium for the subject of Street Lighting. To facilitate discussion a thoroughfare lighting problem has been prepared and is being mailed to a selected list of engineers interested in this subject.

A typical thoroughfare in a city of 100,000 population has been chosen and those items necessary for the design of the system are given on the blueprint of the sketch of this street which is a part of this questionnaire. A photograph of the street is also attached and a table which is to be filled in giving your solution of this problem. The actual layout of units should be shown to scale on the blueprint. The discussion of the solution must be restricted to one page, double spaced typewritten, in addition to the spaces given in the table. This discussion, together with the photographs or sketches requested, should be bound in the folder and all be of uniform size.

It is the plan to treat other divisions of the street lighting art in the same manner, possibly also commercial and industrial lighting, at future conventions.

NOTE: Send completed design data to J. L. Stair, Chairman, Papers Committee, National X-Ray Reflector Company, Chicago, Illinois. Please do not delay—all designs should be in by *August 1* to give time for tabulation and arrangement of reports, by the Papers Committee, for printing.

*A Paper presented before the Annual Convention of the Illuminating Engineering Society, Lake George, N. Y., September 24-28, 1923.

The Illuminating Engineering Society is not responsible for the statements or opinions advanced by contributors.

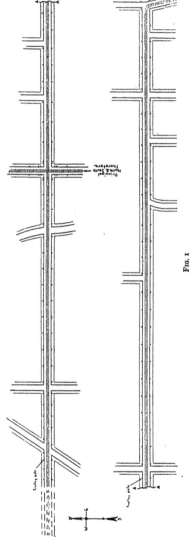

Fig. 1

Plan Submitted to Contributors

DESCRIPTIVE LEGENDS

The thoroughfare to be lighted is a section one mile in length of a brick-paved street in a city of 100,000 population and is one of the main arteries of travel in and out of the city. The width of pavement from curb to curb is 40 feet, and from property line to property line 70 feet.

A double street car track, as can be seen from the attached photograph, lies along the center of the street.

The houses on this thoroughfare are of the better grade and there is a sprinkling of neighborhood stores especially near important corners.

The section to be lighted begins one mile from the center of the city. At present there are wooden poles on the rear lot lines (a distance of 150 feet back from the curb) which carry the telephone wires, house lighting 115 volts distribution, and 2300 volt, 3-phase, primary feeders.

The estimated annual cost to the city is to include fixed charges on the equipment as well as current and maintenance.

The choice of equipment is not limited to types of equipment now in use in the city.

SOLUTION TO
A STREET LIGHTING PROBLEM

(Fill in under proper headings the details of
your suggested lighting system)

I. Please state what in your opinion a city should be prepared to
expend per running mile for the proper lighting of a street of
the class described on the blueprint in a city of 100,000 population...

What average and minimum horizontal foot-candles values of
illumination in service should be supplied on the street;
average...................., minimum.....................

II. Design your system in accordance with such an expenditure
and illumination level.

1. Show layout to scale on attached blueprint.

2. Type of Lamp (cross out two)
Incandescent filament   Electric Arc   Gas

3. Size of Lamp (fill in one)
Series Incandescent: Amperes........................
Lumen Output (10 times nominal horizontal c.p.)..........
Multiple Incandescent: Watts...............Volts.................
Lumen Output....................
Arc: Type............................Amperes...................................
Type of Electrodes.......................................
Gas: Type............C.P............Cubic Feet per Hour.........

4. Type of circuit (Cross out one)
Underground   Overhead

5. Arrangement of Lamps (Cross out two)
One side of street
Center span
Both sides of street (cross out one)
(a) Opposite   (b) Staggered

6. Give average distance between lamps measured along
center line of street...

III. Give description of fixture and post

    1. Describe fixture._____

    _____

    (Include (1) photo of sketch and (2) candlepower distribution curve showing average c.p. in vertical plane; also per cent lumens 0°—90°............and 0°—180°............)

    2. Describe Lamp Standard_____

    _____

    (Include photograph or sketch)

    3. Give height of light source (center) above ground............

    _____

    4. Give type and length of bracket, if used............................

IV. Describe circuit adopted

    1. Type of circuit (cross out two)
       Series   Multiple   Series-multiple

    2. If series circuit used specify (cross out two)
       Insulating transformers   Auto-transformers
                   Film Cutouts
       3. If multiple circuit is used state how lamps are to be switched on and off..

    _____

    4. If series-multiple circuit is used, discuss:

V. If you wish to give additional information or discuss any of the above more in detail do so here:

FIG. 2
Photo of street supplied to contributors

FIG. 3
Type of luminaire proposed by Mr. Dickerson

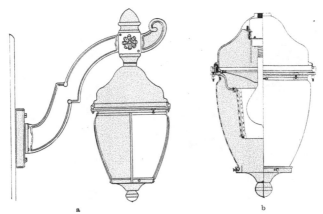

a   b

FIG. 4—Type of Luminaire proposed by Mr. Wood

Fig. 5—Type of Luminaire
proposed by Mr. Oday

FIG. 6—Type of Luminaire
proposed by Mr. Blackwell

(1) *Contributor.* A. F. DICKERSON.

Business Connection. Chief Engineer, Illuminating Engineering Laboratory, General Electric Company, Schenectady, N. Y.

I. Desirable Expenditure per Mile—$4000 to $5000.
Illumination Values—50 to 75 lumens per linear foot of street (based on lumens of light source)

II. Layout—Alternate trolley poles, two units on all intersecting corners and four units at principal intersection.
Type of Lamp—Incandescent filament.
Size of lamp—15- or 20-ampere—4,000 or 6,000 lumens.
Type of circuit—Underground
Arrangement of Lamps—Both sides of street staggered.
Spacing—81 feet.

III. Fixture—Lantern type with reflector or refractor in order to divert most of light downward. Upward light should be ¼ to ⅛ (Type L Figure 3 preferred).
Standard—Present trolley pole with bracket.
Height—18 to 20 feet.

IV. Series circuit with insulating transformers.

V. COMMENTS—The solution of the problem as submitted should be largely regarded as theoretical. Local conditions, particularly the character of lighting on other streets and existing electric service might tend to change the recommendations. In general, such a street requires the expenditure of from 50 to 75 lumens per linear foot of street from suitable luminaires.

The effectiveness of the illumination is of more importance than the actual foot-candles recorded by the photometer. For this reason I purposely omit reference to foot-candles in making recommendations. In surveying a city for street lighting, the first problem is to classify the streets, then assign lumen expenditures and lamps which will expend the available light to best advantage.

A certain amount of upward light even on residential streets is of material value in reducing glare by lessening the contrast between the light source and the background. This upward illumination is also necessary to preserve the night integrity of the street.

(2) *Contributor.* F. MISTERSKY.

Business Connection. Superintendent, Public Lighting Plant, Detroit, Michigan.

I. Desirable Expenditure per Mile—$2200. to $2400.

Illumination Values—Average 0.155—Minimum 0.082.

II. Layout—Alternate trolley poles, two units on all intersecting corners and four units at principal intersection.

Type of lamp—Incandescent filament.

Size of lamp—6.6 ampere—2500 lumens.

Type of circuit—Overhead.

Arrangement of Lamps—Both sides of street staggered.

Spacing—Approximately 87 feet.

III. Fixture—Street hood with porcelain head. 18-inch bowl reflector and prismatic band refractor giving maximum candle power at ten degrees below the horizontal.

Standard—Present trolley pole.

Height—18 feet.

IV. Series circuit with film cut outs.

V. COMMENTS. The street to be lighted being one of the main arteries of travel, in and out of the city, and carrying a double street car track, requires a fairly high average illumination. The houses on this thoroughfare, with a few exceptions, are of the better grade of residences. High candle power units are, therefore, objectionable. With such units, there is too much light upon some of the front porches and in front bedrooms. On the plan, trolley poles are shown, spaced approximately 87 feet between poles. The most ornate lamp pole or post that could be provided for a lighting fixture, if installed in addition to the trolley poles would result in a cluttered appearance of the street. Lighting fixtures should for this reason be mounted upon the trolley poles. An ornamental bracket fixture would not be as efficient as a pendent light with reflector, which may be mounted at a greater height, thus raising the near units out of the direct field of vision. A fixture with a band refractor is recommended. The band will give a better distribution to the light flux; and by leaving off the skirt a greater brilliancy is secured at the lamps, thus providing the effect of a better lighted street. Circuit to be operated by means of a Time Switch, from a Pole Type Moving Coil Regulator, connected to one phase of the three phase mains in the alley.

(3) *Contributor.* L. A. S. WOOD.

Business Connection. Manager of Illuminating Section, Westinghouse Electric & Mfg. Co., South Bend, Indiana.

   I. Desirable Expenditure Per Mile (Initial) $25,000.
      Operating Cost Per Mile—$5961.71.
      Illumination Values—Average 0.35—Minimum 0.25.

   II. Layout—Each trolley pole, with eight units at principal intersection.
      Type of lamp—Incandescent filament.
      Size of lamp—15-ampere, 4,000 lumens.
      Type of circuit—Underground.
      Arrangement of lamps—Both sides of street, opposite.
      Spacing—90 feet.

   III. Fixture—Ornamental Cast Iron Bracket supporting Egyptian Junior Reflectolux Pendant with Opalescent Rectilinear Globe, Two-Way Asymmetric Refractor and parabolic enameled reflector above and below the lamp. The globe encloses this equipment and is so assembled as to be dust proof. (Figure 4).
      Standard—Present trolley pole.
      Height—22 feet.
      Length of Bracket—24 inches.

   IV. Series circuit with insulating transformers.

   V. COMMENTS. In arriving at the solution of this problem, several different designs were worked out, including both Series and Multiple Circuits and in both cases on the assumption that the addition of the street lighting load would necessitate additional capacity in existing feeders, feeder regulators, distributing circuits, transformers and other items affected under the different systems considered. This procedure seems logical since the problem deals with a residential street where the street lighting load would overlap the peak of the existing residential load.

The series system recommended herein is divided into four groups each controlled by a 10 KVA Pole Type Regulator and oil break automatic time switch connected by means of 2500-volt, lead encased, steel taped, parkway cable to individual insulating transformers placed in gum filled concrete compartments at the base of each post.

An isolux curve indicates that a very uniform distribution of light is obtained by the use of aysmmetric refractors which build up the intensities toward the street and eliminate objectionable high intensity, otherwise directed toward porches and residence fronts. The design of the refractor is such as to produce moderate intensities towards the sidewalk and residences so as to afford ample illumination for protection against crime and to enable passing pedestrians to recognize each other.

(4) *Contributor.* A. B. ODAY.

Business Connection. Lighting Service Department, Edison Lamp Works, Harrison, New Jersey.

   I. Desirable Expenditure—See Comments.

      Illumination Values—Average 0.25 Minimum 0.05.

  II. Layout—Alternate trolley poles, two units on intersecting corners.

      Type of lamp—Incandescent filament.

      Size of lamp—15-ampere, 4000 lumens.

      Type of circuit—Underground.

      Arrangement of lamps—Both sides of street staggered.

      Spacing—85 feet.

 III. Fixture—Pendant with clear rippled glass globe and prismatic dome refractor giving maximum candle power at ten degrees below the horizontal. 62 per cent of the lumens, 0 to 90 degrees; 70 per cent lumens, 0 to 180 degrees. (Figure 5).

      Standard—Present trolley pole with ornamental base supplied.

      Height—18 feet if not obstructed by trolley wire support.

      Bracket—Ornamental Bishop's Crook with 2 ft. extension.

 IV. Series circuit with insulating transformer installed at base of pole.

V. COMMENTS—It is felt that the Illuminating Engineering Society is primarily interested in the engineering considerations of such a problem and that cost figures being of a commercial nature should not be given too much prominence. Further, it is rather difficult to determine from the question, just what cost figures are desired. It is also felt that local conditions have such a bearing on construction and operation costs that different estimates would not be comparable. For the above reasons cost figures have been omitted from this questionnaire.

In reference to the spacing, good practice would undoubtedly call for the use of four lighting units located at the intersection of this thoroughfare and the principal north and south thoroughfare. It is assumed that the north and south thoroughfare is lighted and that two lamps for this intersection would properly be chargeable to this street.

(5) *Contributor.* W. T. BLACKWELL.

Business Connection. General Lighting Representative, Public Service Electric Co., Newark, New Jersey.

I. Desirable Expenditure Per Mile—$2250.
Illumination Values—Average 0.13—Minimum 0.03.

II. Layout—Every fourth trolley pole with two units at principal intersection.
Type of lamp—Incandescent filament.
Size of lamp—20-ampere, 6000 lumens.
Type of circuit—Underground.
Arrangement of lamps—Both sides of street staggered.
Spacing—180 feet.

III. Fixture—General Electric Form 6 Novalux Pendant or Westinghouse CI Luxolite, giving 66 per cent lumens 0 to 90 degrees and 82 per cent lumens 0 to 180 degrees. Diffusing globe.
Standard—Present trolley pole.
Height—19 feet 6 inches.
Bracket—Type shown Figure 6, 3 ft. 6 inches extension.

IV. Series circuit with auto-transformers.

V. COMMENTS—The cost of lighting per mile has been arrived at by considering that thirty 600-candle-power lamps will be used per mile in all roadways. If the minimum number of lamps is 750, these lamps will be charged for at the rate of about $75.00 per lamp per year which will make the cost of lighting $2250. per mile.

In those sections of the streets where there are many stores, it would be advisable to use 1000-candle-power lamps in place of the 600-candle-power. For those sections of the street similar to that illustrated in photograph, 600-candle-power lamps are to be preferred.

The installation cost of a system as herein described will be approximately $12,700 per mile, when the existing trolley poles are used and equipped with brackets as herein illustrated. The wiring is to consist of armored park cable. This figure ($12,700 per mile) does not include any substation or other generating equipment.

(6) *Contributor.*   W. A. PORTERFIELD.

Business Connection.   Engineer, Union Metal Manufacturing Co.,
   Canton, Ohio.

   I. Desirable Expenditure Per Mile—$4,500.
      Illumination Values—Average 0.5—Minimum 0.3 foot-
      candle.

  II. Layout—See Figure 7.
      Type of lamp—Incandescent filament.
      Size of lamp—20-ampere, 10,000 lumens.
      Type of circuit—Underground.
      Arrangement of lamps—Both sides of street staggered.
      Spacing—125 feet.

 III. Fixture—General Electric Co.   Form 25 Large East
      Cleveland type fixture, Figure 8.   Clear rippled glass globe
      and refractor giving maximum candlepower 17 degrees be-
      low horizontal.   Lumens 0 to 90 degrees, 56 per cent; 0 to
      180 degrees, 62 per cent.
      Standard—Union Metal Manufacturing Co. No. 1260, con-
      sisting of two sections tubular steel pole 4″ and 3″ section
      diameter, with ornamental base 17″ diameter and 33″ high.
      Height—20 to 22 feet.
      Bracket—½″ pipe arm extending 6′ from pole supported by
      ornamental scroll.

  IV. Series circuit with insulating transformers.

   V. COMMENTS—The series circuit with individual insulating
      transformers gives:

   1. "Safety first" from electrical standpoint.

   2. "Safety first" from mechanical standpoint.

   3. Low voltage above ground—avoids injury and deaths in
      case of accidents or when workman on pole.
      Bury insulating transformers in manhole or in ground sur-
      rounded by insulating compound.

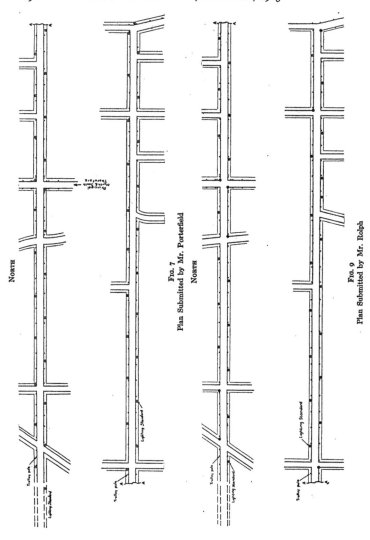

Fig. 7
Plan Submitted by Mr. Porterfield

Fig. 9
Plan Submitted by Mr. Rolph

(7) *Contributor.* T. W. ROLPH.

Business Connection. Managing Engineer, Street Lighting Dept., Holophane Glass Co., New York City.

I. Desirable Expenditure Per Mile—Installation cost $10,000. if necessary. Under usual conditions less is required.
Illumination Values—Average 0.35—Minimum 0.15 foot-candles.

II. Layout—See Figure 9.
Type of lamp—Incandescent filament.
Size of lamp—500-watt, 115-volt, 9,000 lumens.
Type of circuit—Underground.
Arrangement of lamps—Both sides of street staggered.
Spacing—132 feet.

III. Fixture—Ornamental fixture carrying 11¾″ two piece bowl refractor of latest type, lightly etched; maximum candlepower at 20 degrees below horizontal. (Holophane No. 4433 Velvet finish). Figure 10. Estimated lumens 0 to 90 degrees, 63 per cent; 0 to 180 degrees, 75 per cent.
Standard—Ornamental standard of concrete iron or tubular steel. See Figure 8a.
Height—20 feet.
Bracket—6′ ornamental.

IV. Multiple circuit with time switches.

V. COMMENTS—The illumination recommended—0.35 foot-candles is considered adequate according to the best standards of today. A higher illumination is desirable but is not justified until the general standard of street illumination for the entire city has been brought up to the best of modern practice.

The illumination recommended is initial illumination. This is subject to depreciation in service due to collection of dirt and to the slight depreciation in candle-power which the multiple Mazda lamp suffers during life. An adequate maintenance system is essential to preserve continuous good service.

A mounting height of at least 20 ft. is desirable in order to keep the blinding effect of glare from the luminaires as low as possible. This blinding effect is also kept down by setting the lamp in the refractor at the position which gives maximum candle-power at 70 degrees from the nadir, or 20 degrees below the horizontal. The maximum candle-power entering the eye from the luminaire will be at least 20 degrees from the usual line of vision.

This photometric distribution will produce approximately uniform illumination at a spacing of 6 times the mounting height. The spacing adopted averages 6.6 times the mounting height. The variation in illumination over the street area is 4 to 1.

Interference of foliage is eliminated by using a bracket arm 6 ft. in length.

The lighting standard adopted will give a certain degree of ornamental effect which a street of this character is entitled to. It would be desirable to have the trolley wire supported in part by the lighting standards in order to decrease the number of poles on the street. The appearance problem involved has not been worked out satisfactorily, however, for 20 ft. mounting height with 6 ft. bracket arm.

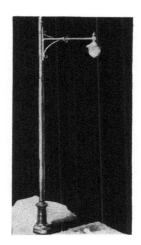

Fig. 8—Type of Luminaire
proposed by Mr. Porterfield

Fig. 8a—Type of Luminaire
proposed by Mr. Rolph

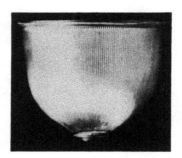

Fig. 10—Type of Refractor proposed by Mr. Rolph

Fig. 1:—Type of Luminaire proposed by Mr. Haas

Fig. 16—Type of Luminaire
proposed by Mr Manwaring

Fig. 18—Type of Luminaire
proposed by Mr. Sweet

(8) *Contributor*—O. F. HAAS.

Business Connection. Engineering Department, National Lamp Works, Cleveland, Ohio.

I. Desirable Expenditure Per Mile—$3000 to $4000.
   Illumination Values—Average from 0.25 to 0.50 in service.
   Minimum 3 to 5 times full moonlight in service 0.10 to 0.15 foot candles.

II. Layout—See Figure 11.
   Type of lamp—Incandescent filament.
   Size of lamp—20-ampere 10,000 lumens.
   Type of circuit—Underground.
   Arrangement of lamps—Both sides of street staggered.
   Spacing—153 feet.

III. Fixture—General Electric Form 25 Novalux Large Basket Unit with Clear Rippled Globe and Dome Refractor, giving maximum candlepower 20° below the horizontal. Figure 12. Without basket, lumens 0 to 90 degrees, 65.1 per cent; lumens 0 to 180 degrees 72 per cent; with basket lumens, 0 to 90 degrees, 56.1 per cent; 0 to 180 degrees, 62 per cent. (See comments).
   Standard—Tubular steel pole with cast iron base.
   Height—20 feet.
   Bracket—Tubular steel 6 ft. long.

IV. Series circuit with insulating transformers.

V. COMMENTS—Streets of the class described on the blue print carry a great amount of high speed traffic and a large proportion of the drivers are not familiar with the street. The lighting requirements are, therefore, very rigid. The layout submitted must be considered as providing the minimum amount of light which will satisfactorily meet the requirements. In the design of the system care was taken to locate a unit at the head of each side street which "dead-ends" into the thoroughfare. Such a lamp location provides an effective spread of light down the side street and creates an especially bright area clearly marking the intersection.

The fixture recommended employs a one-piece rippled glass globe supported by a basket formed of copper bands. The top of

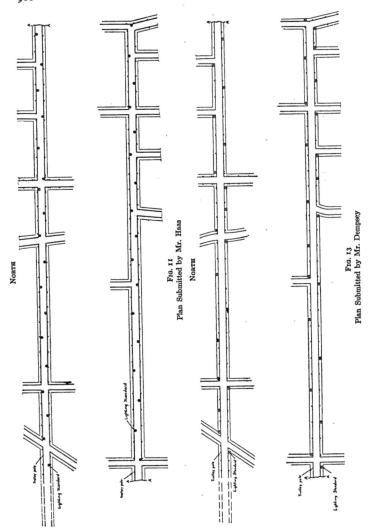

FIG. 11
Plan Submitted by Mr. Haas

FIG. 13
Plan Submitted by Mr. Dempsey

the unit is a single casting to which the basket is hinged. A fired porcelain-enameled reflector is bolted to the casting and serves the additional purpose of supporting the dome prismatic refractor. The dimensions of the lantern are such that the heat is safely dissipated by radiation. Consequently, no ventilation whatever is provided, and as the globe seat is heavily felted, all light-reflecting surfaces are enclosed in a bug-proof and virtually dust-tight chamber.

In practice one of the vertical bands should be placed to face the lamp post. With the position in which the open panels then fall, the effective illumination on the street is materially greater than the output of the fixture with basket would indicate.

The mounting height of 20 feet was chosen as a height which would remove the light sources from the ordinary line of vision of motor vehicle drivers and also allow a wide spread of light.

The bracket post construction was recommended as the bracket brings the fixture out over the street and thus increases the utilization of light on the pavement surface; tends to clear tree foliage; and by virtue of the position over the street each unit illuminates the faces of both curbs.

FIG. 14

(9) *Contributor.*   W. T. DEMPSEY.

Business Connection.   Superintendent,   Service
Maintenance Division, New York Edison Co.

I. Desirable expenditure per mile (see comments).

II. Layout (see figure 13).
   Type of lamp—Incandescent Filament.
   Size of Lamp—One 200-watt on curb side
   and three 100-watt on other sockets, 110-
   volt multiple.   (6,900 Total Lumens)
   Type of Circuit—Underground.
   Arrangement of Lamps—both sides of street
   staggered.
   Spacing—300 ft. maximum, 200 ft. minimum.

III. Fixture—Four pendent spherical diffusing
   globes with 15 per cent absorption.
   Standard—present trolley pole.
   Height 12 ft.
   Bracket—ornamental (see figure 14).

IV. Multiple circuit with automatic time clock.

V. COMMENTS—In submitting the following system of street
lighting in the proposition submitted, it is noted that a trolley railroad occupies the center of the roadway with, of course, necessary
trolley poles for the wire support. On the basis that this railroad
will remain on the street, the installation suggested in this instance
is a four light fixture on trolley poles, a drawing of which is submitted herewith. This suggestion is made to avoid any further
congestion of the thoroughfare by the installation of additional
posts along the curb. It will be noted in figure 14 that trolley poles
have been utilized wherever possible where the spacing between
lamps did not exceed 200 feet, however, in several instances, the
lamp location is other than a trolley pole. In such cases, however,
the trolley pole might be shifted slightly to the post location shown.
Furthermore, two lighting posts have been located at intersections

considered dangerous or where there may be heavy cross traffic, while at other intersections, only a single lighting unit post has been installed—assuming that the congestion of traffic here is not directly across the main thoroughfare.

The cost of a street lighting installation is dependent entirely upon the type of installation that may be selected by the City Officials. It is also governed by local conditions, price of labor and material. In view of this, the question is left open.

It is planned for all four lamps to burn from sun-down to midnight. After midnight, however, the three 100 watt lamps are extinguished, leaving the 200 watt lamp burning.

(10) *Contributor.*   A. H. MANWARING.

Business Connection.   Engineer of Transmission and Distribution,
Philadelphia Electric Co.

I. Desirable Expenditure per mile—$3618.
Illumination Values—0.23 Average—0.21 Minimum foot
candles.

II. Layout—(See figure 15).
Type of Lamp—Incandescent Filament.
Size of Lamp—6.6 ampere 4000 lumens.
Type of circuit—Underground.
Arrangement of Lamps—both sides of street, staggered.
Spacing—100 ft.

III. Fixture—Ornamental upright light alabaster rippled globe
and canopy with dome refractor (Figure 16) giving maximum
candle-power 10 degrees below horizontal.
Standard—Pressed steel shaft with ornamental cast iron
base.
Height—14 ft. 6 in.

IV. Series circuit with film cutouts.

V. COMMENTS—The system suggested has been chosen after
studying the physical aspect of the street as shown on photograph.
Tree conditions; width of street and relative importance of thoroughfare (which apparently is mostly residential), have been taken
into consideration; also the size of the city and its ability to pay
for good street lighting, as indicated by the population of 100,000.

The amount specified on question No. 1 ($3618) represents
the annual sum which the city would pay to the central station
company for one mile of street lighting, viz.—54-400 c.p. lamps,
spaced approximately 100 feet apart at $67.00 each; the company
owning and maintaining all equipment. This is the rate for 400
c.p. lamps on the company underground system in Philadelphia.

The straight 6.6 ampere system was chosen in preference to 15
or 20 amperes with compensators, as we have found from past experience that there is considerable disturbance on adjacent telephone lines, caused by inductive coordination, which occurs when
there are lamp outages.

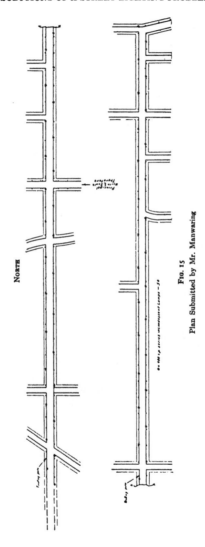

Fig. 15

Plan Submitted by Mr. Manwaring

(11) *Contributor.* ARTHUR J. SWEET.

Business Connection.   Consulting Engineer, Milwaukee, Wisconsin.

I. Desirable Expenditure—For installation $12,000 to $18,000, per mile. For operating cost including fixed (investment) charges, $2,700 to $4,000. per mile.

Illumination Values—Average 0.2 to 0.3 foot candles. Uniformity not poorer than 5 to 1.

(NOTE.) Estimated installation cost of system recommended $16,000. Estimated annual operation cost not including investment charges, $2,261. Estimated annual operation cost including investment charges, $3,541.

II. Layout—See Figure 17.

Type of lamp—Incandescent filament.

Size of lamp—20-ampere—6,000 and 10,000 lumens. See plan.

Type of circuit—Underground.

Arrangement of lamps—Center span.

Spacing—266 feet.

III. Fixture—It is proposed to employ an asymmetric prismatic refracting globe. At street intersection units, the globe employed is one designed to concentrate the light up and down both intersecting streets (four-way). For units not at street intersections, the globe employed is designed to concentrate the light up and down the street (two-way). Refractors produce maximum candle-power 10° below the horizontal.

The body of the fixture is porcelain. The body, metal supporting ring, and refracting globe are designed to constitute a dust-tight fixture. No reflector is used above the refracting globe. See Figure 18.

The fixture is supported from and fed through an absolute cutout, permitting the fixture to be lowered by rope or chain with its terminals dead.

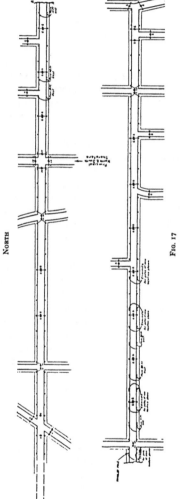

NORTH

FIG. 17

Plan Submitted by Mr. Sweet

LEGEND AND EXPLANATION

o New suspension pole installed at present trolley pole location.

• New suspension pole installed at new location.

• Single light unit.

•• Double light unit.

Numerals adjacent to unit indicate lamp size expressed in thousands of lumens.

Light units mounted above the center lines of two intersecting streets employ a prismatic refractor having 4 way asymetric distribution.   Other light units employ a prismatic refractor having a 2 way asymetric distribution.

The double fixture units indicated on the accompanying plan consist of two such fixtures as described above, respectively supported at the ends of a 2½ ft. asphaltum-treated wooden bar supported parallel to and directly above the axis of the main street. This bar is supported by three 5/16 in. extra high strength, extra galvanized steel cables of plow steel from each of the two supporting poles. One cable extends from the center of the bar and attaches directly to its suspension pole. The other two cables extend from either end of the bar to a short cross arm attached to the suspension pole.

In the event of breakage of any one of the cable supports, or of one in each direction, the remaining cables are amply strong to support the double fixture unit.

Where the suspension crosses perpendicular to the street axis, the drop between the suspension pole end and the fixture end of the cable is 5 ft. Where the suspension crosses the street axis diagonally, the similar drop is 9 ft.

At street intersections, the double unit is supported by four cables from four suspension poles.

Standard—The concrete lamp standard (suspension post) is a tapered reinforced concrete shaft with hollow duct through the center through which the circuit wiring is carried. Poles are provided under specifications to withstand without injury a stress of 3,000 lbs. applied at right angles to the pole axis at a distance of 22 ft. above the ground, the pole being firmly set in the earth to a distance of 6 ft., together with a stress, simultaneously applied, of 1,500 lbs. perpendicular to the axis of the pole and applied at a distance of 32 ft. Also to show a deflection not exceeding 3 in. when half the above stresses are simultaneously applied at each of the two points specified.

Height—25 feet.

IV. The use of a series circuit and film cutout, the circuit voltage being carried directly to the lamp terminals, is contingent upon the possibility of securing a special type of constant current transformer designed for maintaining the circuit current at 20 amperes, the apparatus being identical in other

respects with existing well-known types of current transformers. In the event that it is impossible to secure this
slightly special apparatus, the circuit would be fed through
multitap constant potential transformers, reactance type
insulating transformers being used at the individual light
units.

V. COMMENTS—A controlling consideration in the character of
the solution submitted herewith is the presence of the street car
line. As a general rule to which there are but few exceptions, it is
considered ill-advised to install additional street lighting posts on
a street where there are already a large number of poles for supporting the trolley of a street car system. Such additional posts, even
if individually ornamental, tend to detract from rather than add
to the attractiveness of the street.

It is proposed that the municipality provide and install all
poles used as suspension poles for street lighting units. It is
further proposed that the street railway use said municipally-
owned poles for supporting the trolley. No rental or other compensation should be required in view of the fact that the street
railway now has a satisfactory, "going" installation and that it
cannot make any saving under the said use for many years to come.
The street railway should, however, make without charge the new
attachments for trolley supports and should without charge move
or rearrange its other suspension poles as may be required to suitably fit in with the location of lighting unit suspension poles.

The character of the street makes it economically as well as
aesthetically desirable to provide underground circuit and concrete
suspension poles. Permanence, reliability of service, and low
maintenance cost, as well as the avoidance of any detraction from
the appearance of the street, are represented by the installation
recommended.

It is considered that the relatively bright street lighting recommended has a aesthetic value as well as a service value in providing
that the street may be used with increased safety and convenience.

A feature of the recommended installation is the double unit
system which substantially insures continuity of service in the

TABULATION OF ESSENTIAL FEATURES OF VARIOUS SOLUTIONS

Contributor	Desirable Expenditure	Illumination Horizontal Foot-Candles		Lamp	Arrangement	Spacing	Fixture	Standard	Suspension		Distribution System	Average Lumens per foot
		Ave.	Min.						Type	Height		
Manufacturer's Representatives												
A. F. Dickerson	$4000–$5000	50–75 lum. per ft.		15 or 20 amp. 4000–6000 lumens	Both sides staggered	81 ft.	Ornamental pendant lantern with reflector or refractor	Present trolley poles	Bracket	18–20 ft.	Series—underground. Insulating transformers	62.
L. A. S. Wood	$5962	0.35	0.25	15 amp. 4000 lumens	Both sides opposite	90 ft.	Ornamental pendant Double reflector and two-way refractor	Present trolley poles	24 in.	22 ft.	Series underground. Insulating transformers	90.
A. B. Oday	—	0.25	0.05	15 amp. 4000 lumens	Both sides staggered	85 ft.	Pendant—rippled globe-dome refractor	Present trolley poles	24 in.	18 ft.	Series underground. Insulating transformers	47.
W. A. Porterfield	$4500	0.5	0.3	20 amp. 10,000 lumens	Both sides staggered	125 ft.	Ornamental Pendant Lantern rippled globe dome refractor	Ornamental tubular steel	6 ft.	20–22 ft.	Series underground. Insulating transformers	80.
T. W. Rolph	$4000	0.35	0.15	500 watt 9000 lumens	Both sides staggered	132 ft.	Ornamental pendant 11¾" bowl refractor	Ornamental concrete Iron or steel	6 ft.	20 ft.	Multiple underground time switches	68.
O. F. Haas	$3000–$4000	0.25 0.50	0.10 0.15	20 amp. 10,000 lumens	Both sides staggered	153 ft.	Ornamental pendant lantern-rippled globe dome refractor	Ornamental tubular steel	6 ft.	20 ft.	Series underground. Insulating transformer	66.
Average Values	$4492	0.36	—	7000 lumens	—	111 ft.	—	—	—	20 ft.	—	69.

CENTRAL STATION REPRESENTATIVES

	Cost				Span	Fixture	Pole	Bracket	Height	System	
W. T. Blackwell	$2250	0.13	0.03	20 amp. 6000 lumens	Both sides staggered	180 ft.	Pendant—diffusing globe	Present trolley poles	3 ft. 6 in.	19 ft. 6 in.	Series underground auto transformers	33.
A. H. Manwaring	$3618	0.23	0.21	6.6 amp. 4000 lumens	Both sides staggered	100 ft.	Ornamental upright rippled alabaster globe and canopy dome refractor	Pressed steel shaft cast iron base	—	14 ft. 6 in.	Series ground film cutout	40.
W. T. Dempsey	—	—	—	100 & 200 watt 6,000 lumens	Both sides staggered	200 ft. min. 300 ft. max.	Pendant spherical diffusing globe	Present trolley poles	1 ft. ornamental four arm bracket	12 ft.	Multiple underground time clock	36.
Average Values	$2935	0.18	—	5400 lumens	—	177 ft.	—	—	—	15 ft.	—	33.

CONSULTING ENGINEER

A. J. Sweet	$2700 $3000	0.2–0.3	—	20 amp. 6000 & 10,000 lumens	Center span	266 ft.	Pendant two and four way refracting globes	Hollow concrete	cable	25 ft.	Series underground film cutout	55.

MUNICIPAL REPRESENTATIVE

F. Misteraly	$2200– $2400	0.155–0.082	—	6.6 amp. 2500 lumens	Both sides staggered	87 ft.	Pendant reflector and band refractor	Present trolley poles	Bracket	18 ft.	Series overhead film cutout	29.
Grand Average	$3748	0.28	—	6300 lumens	—	141 ft.	—	—	—	19 ft.	—	55.

event of any individual lamp burnout. At four of the street inter-sections, local conditions result in the substitution of two or four separate single units for the double fixture units.

It is estimated that the operating cost, including thereunder that of the units provided for lighting side streets near their junction with the main street, will be $0.43 per lineal foot of street, this figure covering all elements of energy and maintenance but not covering fixed charges on the investment; and $0.67 per lineal foot of street per year when fixed charges on the investment are included. The average illumination is estimated at slightly greater than 0.25 fc.

DISCUSSION

J. L. STAIR: Speaking for the Papers Committee, The idea back of the Street Lighting Symposium, briefly, is this: In preparing the program, it was considered very desirable to have something on street lighting at this Convention. Also, in looking around for possible authors of papers we found that we could probably pick out twenty to thirty members of the Society who could present papers. This, of course, brought about a difficulty. Another thing that puzzled the Committee was the selection of proper subjects.

It was thought that we might have two or three good papers on street lighting but at one of our meetings some one made the happy suggestion, "Why not assume a problem in street lighting and send the problem in the form of a questionnaire to those that we had in mind as possible authors?" That was done and we got a very happy response to the questionnaire. You will find a summary of the answers in the printed paper.

If you have had a chance to read the first paragraph of the paper, you will have seen that we assumed a thoroughfare lighting problem in a town of 100,000 people. In discussing the problem we will have the various contributors to this questionnaire to speak for themselves, calling on each contributor in succession to discuss his solution to the particular problem and enlarge on it in any way he chooses. If any of the contributors have apparatus they wish to show, they will be permitted to make demonstrations. After hearing from all the contributors to the questionnaire, we will throw the paper open for general discussion.

L. A. S. Wood: I will not read the solution in detail, because you have it in front of you, but, in considering this problem, I did the same thing Mr. Dickerson has done—I took advantage of the trolley poles to support the lighting unit.

This is always desirable because it is very unwise to clutter up the streets with additional obstructions. The street under consideration is a residential street, and, in preparing the solutions, I have used a type of lighting unit which will prevent too much light from reaching the front porches of the residences. There has always been considerable cirticism in residential districts when the light falls on the porches, and, in many cases, globes have been blackened at the back to prevent the light from falling in the direction of the front porches. I will, however, read the comments.

. . . Mr. Wood then read the comments contained in his contribution to "Solutions of a Street Lighting Problem". —

There are two points in connection with this solution that I particularly wish to call to your attention. The first is the estimated cost of the installation of the equipment which I have recommended. The Illuminating Engineering Society is a scientific body; but, in order to reach the interests which commercially apply the results of our research and development work, we must look at the commercial aspects.

I believe that one of the factors which has tended to retard the universal adoption of improved street-lighting systems throughout the country is the fact that we are apt to overlook these commercial aspects. Street lighting hitherto has not been a profitable load to the central stations. It has been based on rates which have in turn been based on concessions to obtain franchises and other considerations, and it is time for us to urge that street-lighting rates be such as to return a fair interest on the investment.

I have taken this point into consideration in the estimate of cost, and I have allowed adequate overhead and adequate profit in the estimate that I have given; because, unless street lighting is profitable to the interests involved and to the contractors who have to install it, we shall always have resistance when we are trying to recommend these improved systems. In other words, we must make the improved system profitable as well as advantageous from a scientific standpoint.

The other point I wish to draw to your attention is that in my solution I have adopted an old principle of light distribution, but a new one as applied to ornamental street lighting, viz., the asymmetric principle of light distribution. This (indicating) is the asymmetric refractor referred to in the comments. It was designed by our engineers in co-operation with the Holophane Company, and, as explained in the comments, is designed to build the light longitudinally up and down the streets, slightly tilted toward the road, and to take away the excess light on the roadway and on the building side. Hitherto street-lighting units have generally been designed to deliver the same amount of light in all directions around the unit; but the distance the light has to be thrown across the street or across the pavement is not so great as the distance up and down the street; therefore the excess light across the street is wasted.

In this unit we have taken away the excess light in both directions across the street and built it up into two beams up and down the street. If you will put the shades down, you will be able to see at once how those beams operate. You, gentlemen, there (indicating) are sitting in the road, and you will notice that from the roadside it is perfectly possible to look at the refractor with comfort. These gentlemen here (indicating towards the other side) are sitting in the gutter (laughter) and, as a result, they feel the full flush of the lumens. Here are some more sitting in the gutter. Those gentlemen (indicating) are brilliantly illuminated. (Laughter) This demonstration will bring home to you just exactly what the unit was designed to accomplish.

Now, then, if I turn the refractor around, you (indicating) will be sitting on the pavement and Mr. Harrison, quite properly, will be in the limelight, also our friend, the Chairman of the Papers Committee. But you gentlemen (indicating) are all sitting in the dark; you are on the front porches, with your friends. (Laughter)

I don't think any other demonstration than this is necessary to bring home to you why this refractor was designed. It can be used in ornamental post tops or as in this solution in pendant units installed adjacent to the curb-line.

I have some distribution curves which will bring the characteristics of the refractor home to you. Figure 4 shows the pendant unit mounted on the trolley pole with the asymmetric refractor mounted on the inside. Figure 19 shows the vertical distribution through an angle 22½ degrees from the curb in the direction of the roadway. You must take this lying down, because the slide was made sideways, but the point is very clearly shown. Imagine the curve turned up the other way and you have the distribution curve along the maximum beam toward the gutter. It will be noted that the 300-watt lamp used in this test gives 2200 candlepower up and down the street. Figure 20 shows the distribution across the roadway at an angle of 90 degrees from the curb line. The 300-watt lamp in this direction gives 800 candlepower delivered at right angles to the curb across the road.

The third curve, Figure 21, shows the illumination across the pavement. This is only 350 candlepower. In other words, the tests prove that the asymmetric refractor does exactly what you see in the example.

In figure 22 there are some isolux curves which also are interesting. These curves are made with the units spaced 90 feet apart and with 22 feet mounting height. With this spacing, curiously enough, the maximum intensity is at a point in the center of the road midway between the four posts, and the minimum is in the center of the roadway between two posts located at opposite sides of the street. It will be noted that the maximum is fifty-four one hundredths of a foot candle in the center of the road between four posts, and the minimum twenty-five one hundredths at a point in the center of the road between two posts on opposite sides of the street.

I also have an isolux curve with 150 foot spacing and the same mounting height, which is interesting. In figure 23 this Isolux Curve shows the units with 22 feet mounting height and 150 foot spacing, and, under these conditions, with 300 watt lamps, the minimum intensity is seventeen hundredths of foot candle in the center of the road, at a point midway between four posts, and the maximum is four tenths of a foot candle between two posts. The condition in this case is the reverse of that obtained with the spacing shown in the previous curve, and the maximum is between two opposite posts, while the minimum is in the center of the road between four posts. You will notice that the maximum and minimum show comparatively small variation.

A similar Isolux Curve, which I will not show at this time, has been prepared, with an ordinary refractor under the same conditions, and this curve shows a maximum illumination (fifty-four one hundreths foot candle) at a point in the center of the road, midway between two posts on opposite sides of the street, and a minimum illumination (one-tenth foot candle) at a point in the road midway between four posts. In other words, the Bi-Lux Refractor is able to give approximately 70 per cent more light at a point in the center of the roadway, midway between posts, than is obtained with the most efficient type of refractor hitherto used.

CHARLES GALLO: Mr. Blackwell could not be here, but I will speak for him. When you look at the last page of this report, you will probably be surprised at the differences in cost that you find there. Some of the contributors have treated the problem as a theoretical one. Mr. Blackwell, in his solution treated it as an actual problem, in the way the central station would meet it.

As illuminating engineers, we probably would like to spend $25,000 per mile but the question is: will the city pay for it? There is a great deal of work to be done in selling the idea of better street lighting to the different city governments and I think you will find that the solution sumbitted in Number 5 represents a condition which is considerably better than that generally found today in the majority of cities of that size. While it does not represent an enormous improvement compared to some of the other solutions, it does bring the standard of street lighting up and it is practical, in that such a solution would be found acceptable to the city government, whereas, perhaps, some of the others would not be found acceptable.

Discussing the distribution of light—while this street is a thoroughfare it is also a residential street, judging by the photograph given, and we have found in practice that people living on such streets object to an upward distribution of light, as Mr. Wood has said, because it goes into second story windows, and so on. So that it is more suitable to install a fixture which will give perhaps not quite so uniform a distribution of light but which will give more light directly under the fixture and less light away from it, and which will cut down the flux above 90 degrees.

The solution presented is one that can actually be carried out and that probably will prove acceptable to the government of the city to which it is submitted.

T. W. ROLPH: I took the question "Desirable Expenditure Per Mile" to mean the expenditure for installation cost which I estimate not to exceed $10,000. The conditions will vary greatly in different cities and often that can be put in for less than $10,000 per mile.

I used multiple incandescent lamps because in figuring out the cost of installation with the multiple circuit on the lot lines back of the street, it figured out considerably lower using multiple lamps and running underground cable out to the street and using time switches to control those lamps in small groups. The multiple lamp would be somewhat less efficient than the series lamp but that would be more than compensated for by the lower investment charge.

Nothing was said about the availability of series circuits and it might be that with a more complete study of the conditions in the actual city represented here, my choice would be changed. —Mr. Rolph then read his comments, contained in his contribution—.

It is desirable not to have any more posts on the street than necessary. However, the use of the trolley poles for street lights involves something of an appearance problem, to which I have no solution at hand. The posts recommended are inconspicuous in character, and judging from installations of a similar type which have been put in, they do not detract from the appearance of the street but rather add to it.

I would like to say in general discussion of this problem that the times through which we are passing in street lighting are not ordinary times. The use to which our streets are being put has undergone a revolutionary change in the past fifteen years. In that period of time the number of automobiles in service in this country has increased from a negligible quantity up to ten or twelve million today; and you are all familiar yourselves with the great changes that have taken place in the character of street traffic.

The most serious thing which has accomplished those changes is the increased hazard to life and limb. We are killing in the neighborhood of ten to twelve thousand people annually by automobiles. In addition to that there are innumerable accidents of a minor character. A large number of accidents take place at night, and statistics presented before this Society at the 1921 Convention showed, as a result of actual investigation, that approximately 18 per cent of these night accidents were due to inadequate lighting.

Therefore, I would place as the primary requirement of street lighting today that we provide, in so far as we as illuminating engineers can provide, for the elimination of those traffic accidents which are due to inadequate illumination.

Consequently, it is necessary to obtain the highest efficiency in the street lighting system that we can possibly obtain. We must deliver the maximum amount of light on the street, but that is not the only requirement. We must also make the illumination as effective as possible. In other words, the illumination must be such that vision will be as easy as possible, and the measure of the total light on the street will not be the only measure of that. Seeing ability is what is needed.

Seeing ability requires not only that the street itself should be lighted but that objects along the street should also be adequately lighted. For example, if a child steps from the curbing out into the street, the only way in which that child can be adequately lighted is by an illumination on the child itself, because the child has a background, which is essentially dark. That, I conceive, means obtaining an approximately uniform horizontal illumination, and I have incorporated that in the problem.

It also means that the glare from the lights themselves must be reduced to the minimum and to do that we should raise the lights above the usual line of vision as high as it is practical to raise them. With foliage and with side mountings, I believe that 20 feet is the highest we are able to raise them.

The design of the post and the idea for a long bracket type of mounting which enables us to obtain a 20 foot height were first worked out by the engineers of the National Lamp Works and I would like to give them a great deal of credit for it. Without the use of a long bracket of that character, foliage would make necessary a mounting height in the neighborhood of 15 feet. Long brackets have been used previously at the side of the street for non-ornamental lighting, but never before have they been incorporated in the form of an ornamental standard, and I think that is an important contribution to street lighting practice.

W. T. DEMPSEY: In working out this problem, I considered this street as a residential thoroughfare and therefore believed that the lighting installation should be of a somewhat ornamental design. In recommending the four light bracket on the trolley pole, I had in mind the same as the other gentlemen who have utilized this same form of support, to avoid as much as possible, any further congestion of the street surface. I selected a mutilple 200-watt lamp on the curb side and three 100-watt lamps in the other sockets on the house side. The multiple installation was adopted to avoid the duplication in investment which would be necessary if the series system was utilized. I believe the questionnaire stated that a 60 cycle secondary system was in the rear of the houses fronting on this street, supplying service for commercial purposes, and in utilizing this existing system of distribution, we avoid a duplicate installation and the extension of a series circuit from the distant Central Station to this particular street.

I did not quote the cost of installation because that figure is dependent entirely on the amount of money the City Fathers wish to spend. I did not know just where the town was located but that question would have to be settled entirely on the basis of the cost of labor and material in the localty adjacent to the town. To install a series circuit to light this particular street would mean a considerable initial cost and a circuit of a type that would only be working and bringing in revenue during the burning period. I did not think that should be done in view of the fact that service supply is so near at hand, in fact within possibly one hundred feet of the curb.

I also recommend that this multiple circuit be controlled by a time clock which would turn off the three 100-watt Lamps at midnight, leaving the 200-watt lamp burning until sunrise.

I did not recommend a very great intensity in illumination which is the tendency today on main street lighting. It is not my understanding that this is the main street of the town. This

particular street intersected the main thoroughfare or business
street and I think higher intensities should be reserved for the
business thoroughfares entirely. However, at the business intersections and all other intersections on this street, I did recommend
two four light brackets in some cases and four in the more important cross sections.

G. H. STICKNEY: When all is considered, it seems to me that
we have a surprising uniformity in these several solutions. We
must bear in mind that there are many things that an engineer
looking at the street would see, which it is not practicable to detail
in presenting the problem. The local psychological attitude is
likely to have quite a bearing. Omitting the figures which are
obviously investment, the operation cost figures range from $2,000
to $5,000 and average about $3500.

Mr. Manwaring's figures seem to be higher than the average in
proportion to the light supplied. I am wondering if this may not
be on the basis of a one year contract, which, I believe, is the legal
limit in his City. I mention this to bring out the point that such
local factors as contract conditions, type and cost of construction,
may materially effect the cost, so that uniformity cannot necessarily
be expected in different cities.

The trolley poles offer considerable opportunity for difference
of opinion. The desire to minimize the number of poles in the
street, favors the utilization of trolley poles. On the other hand,
their use imposes some limitation on ornamental appearance of
system and location of lamps. In this particular case I am inclined to go with the six who favor the use of the present trolley
poles. It seems to me that, by ingeniously selecting the poles,
suitable spacing can be secured and suitable relations of locations
to the street corners. The use of the trolley poles makes certain
heights of light centers—say 18 to 22 feet—unavailable, but a very
good illumination may be obtained with a height of 18 feet.

While some solutions of the problem appeal to me more than
others, I do not see any which could be classed as bad. I think
they are nearly all better than average practice, as we find it.

The recommendation of time switches by several, raised some
question as to whether such devices are proving entirely satisfactory in large installations, especially in a cold climate.

Another question which seems to require further study, is the method of expressing the quantity of illumination. I am somewhat in sympathy with Mr. Dickerson's idea of giving the volume of light rather than intensity on the horizontal surface, yet realize that it does not tell the whole story. Average illumination on the street surface does not mean as much as some people seem to think. The distribution and direction of light appear to have a distinct bearing on the seeing value, as was indicated by the tests run several years ago in the Bronx. At the levels used, uniform illumination did not seem to be particularly advantageous, while illumination on surfaces approaching the vertical assisted in seeing irregularities and obstructions. No satisfactory method of assigning a value to these elements has yet been found.

In general, it seems to me that a wider variation between maximum and minimum, than is indicated by most contributions, is likely to be realized and is likely to prove advantageous.

Only one contributor has called for the opposite arrangement, all but one of the others have called for the staggered arrangement. This presumably is on account of the street being comparatively narrow. While the opposite arrangement unquestionably presents a better appearance for the light units themselves, the staggered arrangement is likely to give a more effective distribution for seeing purposes.

R. A. Hopkins: The Committee, I believe, should, be congratulated on having obtained such fine replies to the questionnaire. They put the solution of the problem in concrete form which is valuable data.

I would like to suggest that the distribution system enters into the total cost in such a large proportion that it should be given some consideration, and I wonder if some of the contributors will be kind enough to give us their view points on such questions as carrying a trench on each side of the street versus staggering, and the use of single conductor cable versus twin cable.

Another point was suggested by Mr. Wood's paper mentioning asymetrical lighting. In our subways where direction of traffic is absolutely standardized we are all accustomed to seeing 100 per cent asymetrical lighting so that the motorman never receives glare from the lamps. In our United States streets our direction of traffic is practically as thoroughly standardized so that it seems some thought should be given to throwing the light principally in the direction of traffic, or in other words, away from the driver.

W. H. WOODS: I see the multiple system has been pretty well over-ridden by the series system. In Toronto we use entirely the multiple system for our street lighting service.

We have now under construction the installation of street lighting on trolley poles. You will find that the tubular pole of 28 feet is set about 5 feet in the ground and you cannot get an average height of about 20 feet for the hangers. Our practice there is to make a uniform height of 14 feet. We try to keep it 16, if possible. Of course, the contour of the street enters in. We have lights on the top of the grade probably a little low; which is about 13 feet in a few cases. But we have found the use of the trolley pole very practicable by using a 22-inch bracket with a 300-watt lamp, and at present we are using a Moonstone globe. But our fixture have been designed in such a manner that it is universal, that is our unit may be upright or pendant. We can also adapt any light control that we may wish to put in at a later date.

Naturally, in improving the city street lighting we are governed by the amount of money that the commissioner can allow. A street that we are particularly interested in at present is an 80 foot street, semi-business in character, where they have an iron trolley pole on one side of the street. On the other side of the street the distribution primary lines, which means a 50 foot cedar lead. We use the cedar pole for one of the trolley suspensions poles.

There is another point of great advantage to the city in the multiple system that probably may be overlooked by the average power station, and that is the usefulness of the street lighting circuit to control signs. The sign business in Toronto has become quite a revenue-producer and we are encouraging that line of business provided it does not interfere with the voltage regulation of a transformer section.

Another advantage in the mulitple system is that two trans-formers on one section can be used. The idea of this is continuity of service; if a transformer happens to drop out, there would then be only alternate lights on the street, out. This has proven very, satisfactory in case of storms. You may criticize this probably from an engineering standpoint on the lower efficiency obtained in using two transformers, in stead of one but when you consider that a large system has a number of small transformers on hand, due to increased power demand, you will realize it is a good way to use up small transformers which are obsolete stock with increase in service continuity. (Laughter)

Another point in favor of the multiple system is that a 4,000 volt circuit out of stations can be used and one side of the transformer grounded to the neutral. That gives about 2200 volts across the primary of its transformer. For instance, in Toronto there are over 55,000 lamps controlled with 21 pricircuits, which is not usual practice with the series system.

I don't know that I wish to make any other comments, only I am glad that the question of street lighting has come up in this practical way to start with, and I believe in our future conventions we will get something good out of the discussion.

An extreme example on this point occurred during the war in London, where owing to air raids it was necessary to darken the streets and dim the street lighting. The means employed for doing this was to send men out with brushes to paint the arc lamp globes black. The result was, the street lighting was definitely cut down to a permanent condition of almost complete darkness irrespective of whether the period was one of air raids or not. With the facilities given by series lighting on independent circuits it would have been possible to have kept the streets lighted normally and the lights need only have been decreased or extinguished during the hours when air raid attacks were made—as the authorities were always apprised in advance, in ample time, of any impending air raids. In this way the whole of London could have had the benefit of the fullest street lighting except for the few minutes or hours of the active air raids.

I am also interested to note the prevalence of recommendations for the *pendant* type of lantern as against the *upright* type. Only one recommendation is for the upright form of lamp,—The upright globe is ornamental type.

One thing that it seems to me we ought to try to avoid is the use of any form of equipment which increases the difficulty of cleaning or which aggravates the accumulation of dust and dirt on the lamp globe. This is the objectionable feature of the design of fixture shown in Figure 12, where you have metal strips surrounding the outside of the globe. The advantage of a fixture such as is shown in Figure 18 is in providing a clear globe without any outside strips or bands. The same thing applies to lanterns shown in Figure 3; these are all panel lamps with strips or bands. Such a design was necessary in the earlier days when the only form of lamp glass obtainable was sheet glass but now with the glass industry developed as it is, with all forms of glass obtainable, it seems a throwback to adopt lanterns with panelings in them.

F. W. WILLCOX: I think the industry is to be congratulated upon the standardization that appears to be indicated by the uniformity of the various recommendations contained in this report.

With the exception of 1919, this is the first Convention of the Society I have had the pleasure of attending in some fourteen years. Coming back now to find such a consistancy in street lighting work impresses me most favorably, and, I think the industry is to be felicitated. If you will recall the varieties of street lighting systems which have been promoted and tried during the last fifteen years, from arc lighting through all forms of incandescent lighting, I think you will agree it is remarkable and a great credit to illuminating engineering and this Society that such a uniformity of recommendation is to be found.

Several points are worthy of mention. It is interesting to note that every recommendation employs the incandescent lamp. I suppose for the particular type of street considered the arc lamp was not suitable. I am also gratified to note the predominance of recommendations for *series* lighting. Only two proposals recommend multiple lighting.

There is a good deal to say for the independant operation of street lights on series circuits, because you will have fogs and dark days or periods, and there are also periods of brilliant moonlight effects, to call for modulation of this lighting.

T. W. ROLPH: In Mr. Sweet's solution to the problem he used a four-way refractor and inasmuch as this refractor is a rather radical departure in design it might be of interest to show it. This is the four-way refractor (demonstrating), and it is made in two pieces. The inside piece carries horizontal prisms which give, by vertical re-direction or up-and-down re-direction, the characteristic vertical distribution of light for street lighting. The outside piece is divided into four equal prismatic constructions, as you see. Opposite the clear space, the light goes directly through without re-distribution laterally. These prisms at the side of the clear space turn the light in a lateral direction without affecting the up- -and-down distribution; the prisms half way between clear spaces turn the light in both directions, one side of the prism acting to turn it into one street and the other side acting to turn it into the other street. In that way you obtain a lateral distribution which is about like a four-leaf clover and you double the intensity of light at the maximum, as compared with a similar symmetrical refractor.

The increase in illumination which you can obtain by distributing the light more efficiently over the street with such a refractor is around 40 per cent over what is usually obtained with symmetrical refractors.

I would like to say that I think it is an excellent scheme for presenting a street lighting paper, to have a symposium of this character, so that many different minds all work on the same problem.

It is interesting to note that the mounting height generally recommended here is about as high as could be used with the tree conditions which obtain. Eighteen feet to twenty-five feet was generally recommended for mounting height. One mounting height was 14 feet 6 inches and one 12 feet. In connection with that 12-foot mounting height, there would seem to be little reason why it shouldn't be 15 feet because trees would surely not interfere at that mounting height.

Mr. Sweet in his solution used a spacing of 10½ times the mounting height and obtained an illumination variation of 5 to 1. That, for all practical purposes, is uniform illumination. In my solution to the problem I used a spacing of 6½ times the mounting height and my justification for this closer spacing is that it enables me to place the angle of maximum candlepower 10 degrees lower than in the other case and obtain as good a uniformity.

I believe that is an important thing in reducing glare.   However, I can't see any justification for spacings closer than 6 times the mounting height.   Some were of that character.

I don't believe that the illumination recommended here by any of us is high enough.   We must absolutely prevent accidents which are due to inadequate street lighting and I don't believe we can thoroughly do that with illuminating intensities such as we have recommended here.   I haven't any doubt that within the next few years we will see illuminations of one foot-candle at least on streets of this character.   However, we have this practical condition now: that it is a very difficult matter indeed to get cities to spend the amount of money they ought to spend to obtain complete safety.   Therefore I, along with the others, have recommended a lower intensity of illumination because I cannot justify the high intensity that really ought to be put on the street until the rest of the city is also brought up to the same high character of illumination, and I know very well, since this is a typical city, that that cannot be done right now.   It will be done before very many years pass by, I am sure.

H. T. PLUMB: Nature's way of street lighting is hard to beat. It consists of a very high source and very thoroughly diffused light from that source, and the point source of light is usually hidden from the driver's eye, if you please.

Long rows of brilliant street lights stuck on trolley poles or otherwise are undoubtedly artistic and beautiful to look at but they are naturally wrong because they are contrary to Nature's way of doing business.   Street lights should give light to see by, and should not be seen.   Therefore, some day you will go even beyond the schemes that are presented here and devise means of street lighting which will consist of a lamp that is totally invisible.

I like, therefore, suggestions such as those made by Mr. Wood and by another speaker who mentioned asymmetrical distribution in the direction of travel.

E. Y. DAVIDSON, JR.: I notice that in the second part of the questionnaire the question is asked: "What average and minimum horizontal foot-candles values of illumination in service should be supplied on the street?"

In reading that, I wondered if it would not be possible to express that question in terms with which we are more directly concerned. Should we not ask ourselves the question: "What is a city buying? What does it desire from a street lighting installation? Is it foot-candle intensity, average or maximum?" For my part, I think it is "ability to see" on the part of the pedestrian or motor driver; if one is able to see, one derives from the street lighting condition the thing for which the street lighting installation is made.

I hesitate to make this suggestion because I do not know in what terms those values can be clothed, but I feel that the amount of illumination, either average, maximum or minimum, expressed in foot-candles is only the means toward the end and not the end.

We are all familiar with the old distribution curve of an enclosing globe of one shape or another, seemingly throwing its light energy to the four winds. We are prone to think that it is unscientific in its distribution of light, and yet I am of the personal conviction that such a unit requires a lower foot-candle intensity of illumination on the street to produce a given condition of visibility than another systems, say the directional lighting system, required to produce the equivalent "ability to see."

I hope that in the future some work can be done in that direction: to determine just what is the relation between those factors, and how they contribute to the final result.

My second thought on this Symposium is in connection with depreciation. I think if figures could be given in each one of these installations to show in a given city with a given atmospheric condition the depreciation through a period of time, it would be also helpful. I know that in some installations in certain localities certain equipment can be used, and in other localities that equipment is disastrous.

PRESTON S. MILLAR: Having declined to participate in this Symposium, with a few uncomplimentary allusions to the project, I would like to take this occasion to compliment the Committee on having brought forward a very interesting Symposium.

One speaker remarked that the problem of lighting a street of this character is distinctly a problem of promoting visibility, and it is indeed so. When you consider the problem of promoting visibility on a street to be lighted with a relatively small amount of light at night you come flatly up against the silhouette question.

One speaker has suggested that the light be shot down the street in the direction in which the vehicles proceed. That is on the assumption that we are going to see people and irregularities in the street surface and objects on the street solely by the light reflected by them, whereas in a street of this character it is not so. We see most things on the street because they are silhouetted against the street background which is in the long run brighter than the objects on the street, and it is by availing of that contrast offered by the silhouette that we most readily perceive the things that must be avoided on the street.

Mr. Plumb has stated that Nature's way of street lighting is the best. It perhaps serves no useful purpose but I can't refrain from commending to him the words of Juliette,

"Oh swear not by the moon, the inconstant moon,
That monthly changes in her circled orb"—(Laughter)

If Nature lighted the streets properly we wouldn't be having this Symposium today. (Laughter)

In the interesting summary on the last page I observe that eight of the eleven contributors suggest that the refractor be used as an auxiliary to some larger outer glass envelope. That is the way I think the refractor should be used. That is the way the refractor can be used most advantageously. So used it is a very useful element of the street luminarie.

Mr. Sweet in his contribution has advocated a center span mounting for lamps on such a street as this. Some of the other contributors have advocated rather long brackets of the mast arm or some other type. I am distinctly in favor of that sort of lighting for this class of street because although in this street the pavement is brick, and the benefit on a brick pavement is perhaps less than on an asphalt pavement, yet if you can get the lamp over the driveway—that is, between the curb and the trolley track—you get the benefit of the streaks of light reflected from the street surface, those streaks proceeding perhaps from lamps a quarter or a half mile away, which greatly assist the nearby lamps in revealing irregularities in the street surface and objects on the street.

In concluding I would like to suggest to the Committee or to its successor that they carry on in this effort, and if they can, choose a street which is to be re-lighted which some municipal engineer perhaps has the problem of re-lighting, instead of naming a hypothetical street which is insufficiently described, as this one had of necessity to be. Let them name a particular street—North Eleventh Street in Squeedunck—and give definite information about that street and about the policy in that city with regard to street lighting expenditures and the character of street lighting in the other streets of the city, and including an expenditure limitation, if that is necessary. I believe that a further symposium of this kind would be even more successful than the present one and would go far toward bringing the best talent of the country to apply itself to this problem.

Further, if they could classify the answers so as to show whether the men who are offering the solutions are connected with the manufacturing companies which supply the equipment or with the operating companies or with the municipalities, or whether they are independent of both operating and manufacturing connections, I believe that would add to the interest of their summary. And if they find any man who is committeed commercially to a particular kind of system through selling it or recommending it, who offers a solution which is not in keeping with the system to which he is committed, I would suggest he be made an honorary member of the Society.

PRESIDENT HARRISON: Mr. Dempsey, you have been called upon twice to answer questions and explain things to us, and I think every one would be interested in what I do not know of in detail but what I have heard of in general: a new means of controlling lamps that you are putting in operation.

W. T. DEMPSEY: Yes, we have such a system.

Four or five questions, Mr. Harrison, have been asked, and if I may take the liberty, I shall answer them in their order.

Mr. Stickney asked about the installation of a multiple circuit on this street—

G. H. STICKNEY: Of the time switch.

W. T. DEMPSEY: Yes. My thought was that on either side of the street we would have a multiple duplex wire, looped up

into each post. It would be, I think, about Number 6 B. & S., two
covers, rubber insulated and lead covered, laid in a narrow trench
and that circuit would control six or ten lamps on one side of the
street. It would be fed in the center by a three-wire main from
the transformer. The other side of the street directly opposite
would be supplied by another transformer. In that way, the
liability of total outage in that section of the street is reduced to
a minimum.

The whole thoroughfare over a distance of a mile being so
lighted from many taps on different commercial primaries would,
I think, reduce the possibility of total outage over the entire
street area, as mentioned by Mr. Willcox; it would be reduced to
a minimum.

With regard to Mr. Stickney's question on time clocks, we
have been operating the 1500 lamps in Central Park for ten years
on time clock on the multiple system. About the first of November, we put a small heating coil in each clock, and I can report that
these clocks do give very excellent service.

On the remote control of multiple street lighting from the
central station, mentioned by Mr. Millar, we have 7,000 lamps on
the multiple system in New York that are controlled by a pilot
wire. This system was installed after the lamps were installed.
Previous to the installation of this system, to put it in another
way, each lamp was controlled individually. Well, that was quite
a job and quite a problem. Through the co-operation of two of
the large manufacturers, a very small type relay was designed and
we have a pilot wire leading from the sub-station, running along
a particular avenue, we will say for a mile, and entering another
sub-station. A circuit of this type can readily be opened up for
testing purposes along its entire length, including both ends. The
control wire is looped into a post on the corner of each intersecting
street, and in this particular post is a relay that closes the multiple
circuit by gravity, so that we do not depend upon potential to
turn the street lighting on. In other words, if anything should
happen to the control circuit, the worst that could happen would
be that the street lamps would go on and not off. That system
has been in operation now for five years and has given very excellent service.

O. F. HAAS: In designing the system submitted by the writer, mounting the lighting units on the existing trolley poles was considered. There were several reasons why this plan was abandoned. (1) From the print of the street it was evident that the concrete poles had a marked rake as they have in practically all installations, and therefore a rigid bracket fitted to them would not have a good appearance. (2) It was also evident from the print that the poles were located at varying distances back from the curb. (3) It was considered very desirable in locating the units to place one at the head of each "dead-end" street and thus provide an effective spread of light down the side street as well as an especially bright area clearly marking the intersection. This, of course, could not be accomplished by utilizing existing trolley poles. (4) As we were specifying underground distribution it was not thought that any material saving in total installation cost would result from adapting the system to trolley poles already in place.

Without doubt, additional poles add to the cluttered appearance of the street. However, with the design as submitted it does not mean duplicating the existing poles, in fact, it only means adding about one more pole in four.

Mr. Stickney remarked that the various solutions are quite similar. In general the annual operating costs and amount of illumination on the street recommended are similar. However, a study of the essential features of the various solutions show spacings ranging from 90 feet opposite to 266 feet center span, light generated per outlet from 2500 lumens to 10,000 lumens and mounting heights from 12 to 25 feet.

Before it will be possible to convince city and central station officials generally that their particular cities' lighting problems can be solved through the application of sound engineering principles, there must first exist among street lighting engineers themselves a reasonable degree of unanimity of opinion on the fundamental principles of the art. It has been generally felt that the difference of opinion on the part of engineers is primarily due to the incompleteness of fundamental data on the visibility of objects under various systems of street lighting. Investigations by various individuals and organizations have contributed substantially to this knowledge, but even with the information resulting from these tests there are still many important factors which have not as yet been definitely evaluated.

The Central Station, City, and lamp company of Cleveland have co-operated in installing a full-scale experimental set-up for investigating and demonstrating the principles underlying thoroughfare and residence district lighting. The installation covers a 2000-foot section of a typical Cleveland street, East 152nd Street. It consists of some 40 separate circuits—virtually forty streets in one—and each circuit with the lamps it controls illustrates a definite principle. Lamps ranging from 100 candlepower (1000 lumen) to 3000 candlepower (30,000 lumen) in size, spacings from 75 to 900 feet, mounting heights from 11 feet to 26 feet, and location of units from 2 feet back of the curb line in four successive steps (by the use of bracket arms) to the center of the street, are employed.

This installation affords by far the most comprehensive demonstration of street lighting principles and effects ever installed and to the best of my knowledge is the only place in the world where a comparison of a wide variety of lighting effects upon the street can be instantly made. It is planned to conduct a number of investigations on this street under the various lighting effects for a wide variety of conditions, with the view of properly evaluating the different factors involved and obtaining definite information regarding the particular characteristics of each system.

This installation will also serve another distinct purpose, apart from its research function as the selection of a suitable street lighting system by a town or city has always been rendered difficult on account of the lack of an opportunity to view a number of different street lighting systems at one time and to compare their characteristics with respect to the specific requirements of that town or city.

A cordial invitation is extended to all members of the Illuminating Engineering Society to view and study this demonstration.

L. A. S. WOOD: Mr. Haas in his discussion spoke of the rake of trolley poles. The rake on steel trolley poles is something that has been handed down to us from the wooden poles and there is no reason at all why a steel trolley pole should have any rake at all. But if there is a rake on a steel trolley pole it can be easily removed.

In Salt Lake City all the trolley poles, when they put in their ornamental trolley poles, were straightened, I believe by simply pulling them over.

Mr. Blackwell's and Mr. Dempsey's contributions are very interesting because they reflect the central station's attitude toward street lighting and I think that in their solutions they have been influenced by the question of cost and the availability of funds. I don't think that is the right way to approach this question of improved street lighting.

The gentleman from Toronto also spoke of the fact that he was governed by the amount that the commissioner had available.

Why consider that question? If we are sold on the idea of improved street lighting, let us go out and convince the cities and the municipal authorities and the central stations and the commissioners that it is desirable to spend such money to get an adequate system that will not only be efficient but that will also be ornamental and add beauty and charm to the city. That, I think, is what we have to do.

Mr. Hopkins asked the question whether we had considered uni-directional asymmetrical distribution. We did, but we realized that pedestrians would be walking, on one side of the street, in both directions and therefore to direct the light in one direction, only down the street, for the benefit of the vehicles, would work a hardship on pedestrians.

He also mentioned underground cable. Prior to my coming to the United States in 1911, I was connected with street lighting in England and had a drilling in underground cable construction. There underground cable construction is almost universally used, and I was sold on the steel-armoured, lead-covered cable. I am glad to say that in the United States steel-armoured, lead-covered cable is now becoming very common, and I believe that it is the solution to these underground street lighting systems.

I do not think there is any advantage in using twin cable. It is true that the cost of twin cable is a little less than the cost of two separate cables but the difficulty in slicing and making joints and connections in the twin cable run up the cost of that work, and in the long run you are very little better off than if you had laid two single conductor cables.

Mr. Willcox brought out a very important point when he said that it was desirable to have the street lighting system independent from the commercial system. I heartily agree with him in that.

Mr. Millar made an interesting confession. He said when he received this questionnaire he was dubious as to the desirability of it and I am very glad to say that he came across like a man and told the truth about what he found out afterwards.

I feel that the Committee is to be commended upon the work they have done in this Symposium, and I would like to ask the Chairman whether it would be in order to entertain a motion commending the Committee on the work that they have done.

PRESIDENT HARRISON: I think that would be quite in order.

L. A. S. WOOD: Then I move that the Committee on Papers be especially commended for this Symposium because it has given us an opportunity to discuss street lighting problems in a manner that I believe could not have been afforded in any other way. (Applause)

PRESIDENT HARRIS: You have heard the motion; is it seconded?

. . . The motion was seconded and unanimously carried . . .

PRESIDENT HARRISON: Mr. Stair, the meeting extends you a vote of thanks.

J. L. STAIR: I don't know that there is anything else the Committee has to offer.

This symposium idea was approached with a little bit of fear and trembling; it was an experimental sort of proposition as far as the Committee was concerned, but I think we are convinced now that it has worked out successfully. We will be glad to pass on to the next Committee the symposium idea and see if Mr. Millar's suggestion regarding another street lighting problem can be put into effect, or if the symposium idea can be carried into other fields of lighting.

CHARLES GALLO: In this particular problem, regarding the question as to whether there should have been series or multiple used, it would probably have been better to have the multiple if use could have been made of the rear lock line. However, in using that line, one question comes up and that is getting the authority to lay your line between the houses, which are private property, and I think perhaps sometimes it will be found that the trouble encountered in securing this right overbalances the advantage obtained by using that rear lock line.

PRESIDENT HARRISON: Mr. Halvorson, I have been asked to call on you for discussion.

C. A. B. HALVORSON, JR.: I am very much interested in this symposium and believe that some of the systems recommended will work very well. (Laughter)

The question of asymmetric distributions in general is a very interesting one and I might say a very delicate subject to handle; that is, in street lighting.

Mr. Wood made the statement that it was a new thing to use an asymmetric distribution or at least to suggest an asymmetric distribution. I will simply say that there have been in use for several years fixtures which give an asymmetric distribution, and very successful lighting. But their use thus far has been confined to the lighting of highways, where the conditions of road surface and so on are very advantageous to that type of distribution. In other words, the light reflected toward the users of the highway is of importance, in that the brightness of the road surface is affected, and that of course is due in turn to the character of the road surface.

In this particular case, the road surface is not one that lends itself especially to an asymmetric distribution, in my opinion. I feel that the total amount of light in the street is of far greater importance and that the light ought to be bunched, so to speak, and placed on the street in such a way that the silhouette effect can be obtained.

In other words, the question of visibility is the whole thing, and the brightness of the unit and the brightness of the street surface, and those factors, are of the greatest importance. In that connection, I believe Mr. Millar has contributed more of value to this discussion than any man who has talked on this floor.

I would like to hear Dr. Sharp talk on asymmetric distributions. He was one of the earliest, as far as I know, to propose distribution of that kind, and I am quite sure that Dr. Sharp can add a great deal of interest to this discussion. (Applause)

PRESIDENT HARRISON: Dr. Sharp, will you discuss the subject?

CLAYTON H. SHARP: It is quite a good many years ago now since I presented to this Society a form of reflector which was intended to conserve the light flux and to throw it up and down the street rather than to the side. And it did that. It did it pretty well. The street lighting art, however, had not reached the point where that sort of a system was demanded. There seemed to be no place for it in the then existing scheme of things, and after long experimentation with such reflectors, the matter was given up largely on account of the difficulties of manufacturing such an appliance and as I say because of the fact that the art had not reached the point where the need for such a contrivance was recognized.

However, I want to say that I have been very much pleased to see this idea, which really is quite an old one, revived and given new life by the production of asymmetrically distributing units for street lighting, not only of the reflecting but also of the refracting type. I believe that the underlying idea is sound; that is, in street lighting at best we have to deal with a very limited amount of luminous flux and it is very important to put that where it will do the most good. Where it will do the most good depends to a considerable extent upon the character of the street. In some cases, it may do the most good by smearing it generally over the landscape; in other cases, it undoubtedly will be most effective by concentrating it in the direction of the street at the proper angle to give the best appearance of brightness to the surface and to minimize the effect of glare on users of the street.

At the present time, even though the production of luminous flux is on a scale which was not approached twelve or thirteen years ago when I was working on this thing, I believe that the question of the economical use of our luminous flux is still of very great importance, and I believe that the systems which are now before the public offer at least an approximate solution for many conditions of street lighting.

PRESIDENT HARRISON: Mr. Chairman, on behalf of the Society I want to thank you for the very interesting symposium this afternoon. I feel that from those eleven papers, there is a great deal that one can learn. The points on which the contributors agree are most interesting and I think two of the points on which they disagree, such as those Dr. Sharp, Mr. Halvorson and Mr. Wood have brought out, are also very interesting.

On that question of cost, as I read the question, it seems to ask what the engineers—these people I believe who contributed are all engineers—think the city ought to spend for street lighting. In other words, they must establish first what they think is proper before they can expect the city fathers to come up to it. In that, I agree 100 per cent with Mr. Wood.

PAGEANT STREET LIGHTING*

BY SAMUEL G. HIBBEN**

SYNOPSIS: The advent of the new spray-colored or diffusing bulb Mazda lamps has made possible some remarkable decorative effects in the spectacular lighting of streets for pageants and festivals.

One of the latest and most striking of such engineering accomplishments is described by the author, being the decorations of Washington, D. C., during the National Shriner's Convention of June, 1923.

The more decorative lighting of prominent streets, employing permanent equipment, is becoming evident as the quality and harmonious shapes of the ornamental standards increase and as our communities devote more attention to city planning and civic beauty. This type of street lighting, however, involves underground wiring, lasting materials, and the fixed placing of lighting accessories. Its decorative features are usually considered secondary to the efficiency of street surface and building front illumination.

The other form of decorative street lighting using temporarily placed equipment and popularly known as pageant lighting, should consider the decorative and spectacular features preeminently, for usually the matter of operating efficiency is not a factor. Pageant lighting is not new, but in numerous recent instances it has assumed much larger proportions and forms a striking part of any special convention or celebration. The outdoor illumination becomes the much advertised feature, the chief drawing card, the "piece de resistance" of the pageant.

The recent advent of spray coated or diffusing bulb incandescent lamps has greatly aided such lighting displays because thereby

*A Paper presented before the Annual Convention of the Illuminating Engineering Society, Lake George, N. Y., September 24-28, 1923.

**Manager, Illumination Bureau, Westinghouse Lamp Co., New York, N. Y.

The Illuminating Engineering Society is not responsible for the statements or opinions advanced by contributors.

938

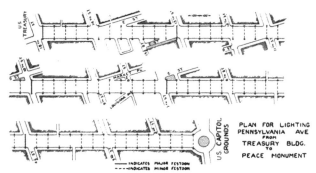

FIG. 1.—Plan of Pennsylvania Avenue showing location of Festoons

FIG. 2.—Entrance to "The Garden of Allah."

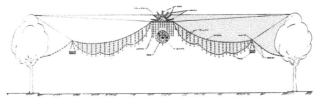

FIG. 3.—Design of the Major Festoon.

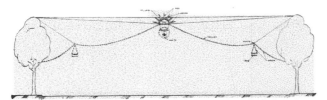

FIG. 4.—Design of the Minor Festoon.

FIG. 5 —The Center Crown.

it is possible to secure more satisfactory and permanent colors and particularly because the lamp bulbs so colored appear attractive in the daytime as well as when burning.

Many varieties of pageant street lighting have been known to the illuminating engineer, from the large banners or translucent boxes placed about the existing street luminaires to the elaborate arches of lattice-work or jeweled frame-work spanning streets one hundred feet or more in width. It is possible, however, to gain an idea of the construction and methods of arranging for such a special lighting installation by a study of one of the most recent and perhaps the largest pageant lighting display, namely the illumination of Washington, D. C., during the Shriners' Convention of June 1923. This installation demonstrates some of the possibilities with temporary equipment and, while involving only the ordinary materials that might be assembled by any good wiring contractor, it nevertheless deserves study in order to formulate a clear conception of the processes of erection, maintenance, and operating costs, and the general effects produced.

This installation was originally designed to provide a decorative canopy over Pennsylvania Avenue, Washington, D. C., for 6500 feet from the U. S. Capitol grounds to the Treasury Building in accordance with the street plan of Figure 1. The canopy was later extended about half a mile on Pennsylvania Avenue past the White House, and involved also some festoons and other equipment in the so-called "Garden of Allah" in front of the Treasury Building. See Figure 2. The plan called for a street illumination of between one and two foot-candles to enable dancing on the street surface, and the general idea was followed providing for the minimum number of obstructions, it being necessary to keep the avenue entirely clear of scaffolding and supports both during erection and after completion.

THE FESTOONS, AND LAMPS

Two styles of festoons were installed, shown in Figures 3 and 4. The spacing distance between festoons was approximately 100 feet, resulting in 58 minor and 5 major pieces between the Capitol and the Treasury Building. On the less prominent streets the

minor festoons were used spaced at 300 foot intervals. Each major
festoon consisted of the following lamp equipment:

146	15-watt	S-14	115 volt	Mazda B	Yellow Spray
208	10-watt	S-14	115 volt	Mazda B	Green Spray
220	10-watt	S-14	115 volt	Mazda B	Red Spray
30	10-watt	S-14	115 volt	Mazda B	White Spray
2	50-watt	PS-20	115 volt	Mazda C	White Spray

606 lamps, total for each festoon

Each minor festoon, consisting of a central crown and pendant
but using only single streamers of lamps, was made up of the
following:

116	15-watt	S-14	115 volt	Mazda B	Yellow Spray
104	10-watt	S-14	115 volt	Mazda B	Green Spray
82	10-watt	S-14	115 volt	Mazda B	Red Spray
30	10-watt	S-14	115 volt	Mazda B	White Spray
2	50-watt	PS-20	115 volt	Mazda C	White Spray

334 lamps, total for each festoon

Considering only the portion of Pennsylvania Avenue shown
in Figure 1, this resulted in 22, 402 lamps initially installed, while
the lamps in the Garden of Allah and in the minor street decorations a little more than doubled this number. The lamp bulbs
were supplied tipless, naturally reducing the handling breakage.

The connected load of each major festoon was 6.87 k.w., of
each minor festoon, 4.00 k.w., or a total for the Avenue between
the Capitol and the Treasury of 266.35 k.w.   This provided about
42 watts per running foot of street.

At voltage, the lumens of the various lamps were taken as:

Size of Lamps	Approximate Lumens
15-watt yellow	65
10-watt green	7
10-watt red	13
10-watt white	77
50-watt white	470

Consequently the installation provided roughly 1.2 generated
lumens per square foot of street area.

In this installation the Shrine colors of red, green, and yellow
were made predominant, while the intersections of the various
drapes and the outlines of the pendant circular plaques were

Fig. 6.—Pennsylvania Avenue looking toward the Capitol.

Fig. 7.—Portable Towers used for Lighting

Fig. 8.—More than a Mile of Canopy of Light

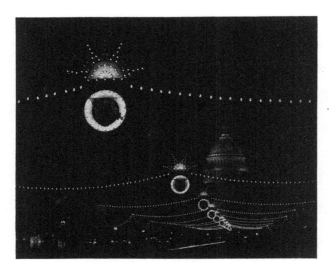

Fig. 9.—Looking toward the Capitol at Night

marked by white diffusing bulb lamps. Where the festoons ended in the foliage, the green lamps were carried to the supporting poles, and on the sunburst center pieces, the rays were alternate red and green, the lamp colors matching the color of the background. Golden yellow was the predominating color, simulating the yellow sands of the "road to Mecca".

The center pieces or crowns were built up of sheet metal wired with receptacles on both faces, the completed unit being about 6 feet high and 10 feet wide. Two sections of conduit were clamped to each sunburst, the upper conduit being slightly above the center of gravity, designed to carry the weight of the festoons. The lower conduit carried a guy cable that acted as a safety precaution as well as preventing the swing of the center pieces in the wind. Figure 5 illustrates one of these units before erection.

SUPPORTING AND PLACING THE FESTOONS

Wooden poles were set about 5 feet into the sidewalk inside the curb line, it being found necessary to remove a concrete block or to chip a hole thru the paved surface, being careful to avoid any damage to cellars and tunnels under the paving. Forty foot poles thus projected 35 feet above the sidewalk and were guyed where possible to the building fronts by 3/8 inch or 1/2 inch steel messenger cable. These poles were painted green and red and were not conspicuous among the foliage.

After the poles were in place, a 3/8 inch messenger cable was strung thru the upper conduit on the sunburst and the festoons were attached to the bottom portion of each sunburst, the center pieces being pulled out to their final position by sliding along the cable, feeding out the festoons from one side and pulling in on the other. In this manner the erection would not obstruct street car or vehicular traffic and the center pieces could be easily lined up by adjusting tension on the bottom cable. Calculating with a large safety factor, it was decided to have a maximum droop of the cable so that the center portion would be 4 feet below the ends. This placed the lowermost lamp of the major festoon about 17 feet above the car tracks. The average span of each festoon was 110 feet, curb to curb.

After the festoons were in place it was found necessary to reduce the windage on the pendant circular plaques by cutting away an area surrounding the star and within the arms of the crescent monogram that was painted thereon. These monograms were of various design as illustrated in the day view of Figure 6.

The lamping of the weatherproof sockets called for the erection of special wooden towers shown in Figure 7, which were rolled along the street surface and did not interfere with traffic. These towers carried ballbearing wheels at their bases and afforded easy access to the lamps for inspection and renewals. One tower was kept in operation after midnight each evening to replace burnt out lamps, and adjust sockets.

SERVICE, AND ERECTION COSTS

The electric service was carried on cables attached to the wooden poles along one side of the street, with 20 manually operated control switches that were in reach of the ground. One hundred and ten—two hundred and twenty volt direct current was available, the total Avenue canopy installation representing about 300 kilowatts. The magnitude of this complete installation may be better understood when noting that in all, including the Avenue extensions, slightly over 50,000 lamps were used, representing an average kilowatt hour consumption per day of 2900. The current was turned on at twilight of each day and during the convention was turned off at sunrise.

The total erection required about four weeks utilizing an average of thirty men and the maintenance required a crew of usually three men from midnight to dawn. The small number of lamps failing during service was very gratifying, indicating that the moderate swaying in the wind did not seriously effect their performance even tho one rather severe storm occurred during the pageant.

The idea in mind from the beginning was to provide what from a distance would appear to be a continuous canopy of golden light beneath which parades and other traffic would pass, and this effect was secured in a very magnificent and pleasing manner as may be gathered from Figures 8 and 9. One feature, adding to the beauty, was the symmetrical and uniform droop of the festoons, and the alignment of all units in the long vista.

The total installation cost in this case represented between $35,000 and $40,000, not counting the value of the salvaged lamp sockets, wire and poles.

Anyone fortunate enough to observe the final installation was impressed by the beauty of the setting and the symmetry and striking color of the jewel-like canopy. No similar installation on such a grand scale has heretofore been made in American cities and the success of this particular one encourages other pageant decorations of a like nature for Fourth of July celebrations, historical conventions, and other gatherings where careful attention to color arrangements and artistic or graceful grouping of colored lamps will give to the business streets a festive atmosphere that could not be equalled in any other manner. It is hoped that future displays of this character will be carefully planned and will be elaborated upon to the extent that such pageant lighting will be looked upon as a safe and beautiful display more permanent than fireworks and more striking than the haphazard stringing of incandescent lamps arranged without careful consideration of symmetry and color harmony.

DISCUSSION

D. W. ATWATER: I would like to add a few points to what Mr. Hibben has just said—some sidelights on the installation of these decorations in Washington.

The feature which impressed me most was the accurate alignment and the uniform spacing of the festoons along the Avenue. When the preliminary survey was made of Pennsylvania Avenue we felt that the decorations would have to be comparatively simple as there was apparently nothing to which festoons or decorations could be attached. When asked what could be done the chairman of the Shriners' Decoration Committee replied, "You can do anything you please; we have had an act of Congress passed permitting us to do anything we want".

It was decided to place the festoons 100 feet apart, which according to the map we were using represented 1 inch. The draftsman in indicating the positions for the festoons located them at exactly 1 inch intervals regardless of local conditions. As a result some of the positions happened to fall in the middle of intersecting streets. This apparently made no difference as the pavement was dug up and forty foot poles erected along the Avenue at exactly 100-foot spacings as indicated on the plan.

Preliminary tests were made to determine the relative value of dipped and sprayed coated lamps for decorative purposes. Festoons of each were hung up and observed from various positions. When observers were about 10 or 15 feet away there seemed to be very little difference as to the desirability of using one in preference to the other. When 50 feet away there was a favorable twinkle or sparkle to the dipped lamp. Upon going further from the festoons in positions approximating that of an observer looking at the actual installation over Pennsylvania Avenue, it was decided to use the spray coated lamp for with the same wattage the light source seemed considerably larger and brighter. As Mr. Hibben mentioned, the day-time appearance of the spray coated lamp was far superior to that of the dipped lamp. After approximately two weeks outdoors service and several hundred hours of burning the spray coating was found to be unaffected with no visible change in the lighted appearance of the lamp.

CHAIRMAN F. M. FEIKER: Would any one else like to discuss this paper?

Those of us who saw this installation realize that there aren't many samples of this sort in the country, but it occurs to me that as time goes on there is going to be more and more pageant street lighting to take the place of fire works in municipal celebrations, and so on.

A few weeks ago the fall festival in Cincinnati opened and I believe there they worked out a tower of jewels idea. The idea of the municipality seemed to be that electricity offered a certain opportunity for decoration at the time of fall festivals and similar local celebrations, in lieu of the ordinary arrangements of flags and so on.

It seems to me that if this paper does nothing more, it points out a very remarkable installation which is interesting many communities in the use of electricity for decorative lighting of this sort. I would like to have it discussed more, if there are other people who care to contribute to the discussion.

JOHN B. TAYLOR: What was done with the ordinary street lights during this special illumination?

S. G. Hibben: Mr. Taylor, the permanent lights in Washington are on rather low, ornamental posts, set into the foliage. They were left burning during this celebration, since the diffusing globes were not troublesome. For safety's sake, and chiefly because there were no bad effect from them we left them burning. There usually exists a street lighting contract that necessitates the maintenance of the regular service.

John B. Taylor: Was there no change in globes?

S. G. Hibben: No, sir. It was thought best at first to color those globes, but that was a rather difficult job. They were hot, since most of them are arcs and the only thing we might have done would have been to put some sort of banner or enclosed, diffusing box around them, but in among the foliage that wasn't considered advisable. They had no deleterious effect on the overhead light.

John B. Taylor: The difference in spacing did not break up the symmetry of the whole arrangement then?

S. G. Hibben: No. Standing at one end of the avenue you could see only a few of those ornamental posts at one time, unless you got in the middle of the street. There was a line of trees almost all the way down the avenue, especially at the eastern end.

H. C. Doughty: What was done with the lamps at the close of the convention? Where is the market wherein they were disposed of?

S. G. Hibben: There are several markets in which these lamps might have been disposed of. If you ask me what was the particular one in this case, I would say that the Shrine Committee chose to purchase lamps themselves; some of them were kept for their own use, for various meetings and decorations of their own, and quite a number, I believe over half, were sold to be used in other decorations elsewhere. Of course, each time the lamps are sold as "seconds" they probably decrease in price. There is, naturally, an amortization every time they are put up and taken down, but at any rate they are passed on for use in other meetings, and used in minor schemes.

C. A. Atherton: I would like to ask Mr. Hibben why he decided to use 15-watt lamps for the yellow and 10-watt lamps for the red and green portions of his display. This would of course not result in equal brightness in each of the colored portions. It would seem to me that the lamp wattages should if anything have been reversed, if equal or anything approaching equal brightness were desired.

S. G. Hibben: On the basis of brightness or lumen output, we had about this ratio; of the 15-watt yellow, 65 lumens; that is approximately; of the 10-watt green, 7 lumens; of the 10-watt red, 13 lumens; of the 10-watt white, 77 lumens. Of course, the yellow at the same wattage would have the larger lumen output; the green the least. We had from this overhead equipment somewhere in the neighborhood of one foot-candle on the street surface. We could have used 10-watt yellow, getting a little less illumination, without seriously affecting the decorative balance.

C. A. Atherton: I questioned whether or not equal brightness in the various colored parts of the display was not desirable. Would you not have gotten better results and a more pleasing display by having more or less the same color strength in the green and red that you had in the yellow. For example, if you had put say 25 and 50-watt lamps in the red and green portions, respectively and 10-watt lamps in the yellow, you would have gotten somewhere near equivalent color effects from the three sets of lamps.

S. G. Hibben: We wanted to get the yellow predominant, of course, because the parades and all the festivities were supposed to be over "The Golden Sands of the Road to Mecca" and it was necessary to have the gold color emphasized.

N. D. Macdonald: May I ask Mr. Hibben just what the proportion of salvaged lamps was. It would appear that in handling incandescent lamps out in the open with more or less unskilled labor the breakage must have been very high. Have you any figures as to how many were reclaimed?

S. G. Hibben: No, I can't give you exact figures on that. One cannot differentiate the breakage from—I won't call it theft, just natural depreciation of lamps. (Laughter) If I were to make a guess, I would say we got back about 40 to 50 percent of the lamps.

The handling of those lamps you know is unusually rough. The actual operation is to fill a peach crate with a number of them and pour them from one peach crate to another. (Laughter) If a peach basket happens to topple off the top of one of the towers in the street, the lamps do suffer sometimes. (Laughter)

We had a pretty high depreciation. In such cases as this, I would say that it is lucky if you got back more than 50 percent of good lamps after everything was down. Souvenirs, you know, are useful (Laughter). Small lamps, too, are not as sturdy as larger ones.

L. C. PORTER: I would like to ask Mr. Hibben if he had considered at all the use of skidoodle plugs in that major curtain. It would seem to me with all those various curtains in there blinking at different times, it would make a very beautiful and very effective stunt.

S. G. HIBBEN: That would have been a good idea, discounting cost. We had thought of trying not individual skidoodle plugs, but a master control system with relay operation so that we could flash sections of streamers down the street, or better yet, take any one color—say yellow—and run all the yellows through, then fade those and bring up the greens and reds. But the cost in connection with that sort of thing is prohibitive.

In smaller installations, I presume skidoodle plugs could handle it very well. It would give a rather novel effect certainly.

CHAIRMAN F. M. FEIKER: Are there any other questions?

H. C. DOUGHTY: On thinking this over, I haven't felt entirely satisfied with Mr. Hibben's reply as to the disposal of the lamps and possibly a word along that line might not be amiss.

Our company has conducted pageant installations for a number of years at one plant that I have in mind, and our practice has always been to use the dipped lamps, which could be washed afterwards and re-sold if not used too long. (Laughter) We did experiment with the sprayed lamps. It is not my purpose to discuss the merits of the sprayed lamp as against the dipped lamp, although I have my own opinion as to the light and its efficiency, but our experience did show us that for this particular purpose, dipped lamps were preferable to the sprayed lamps for temporary installations. And before any one went into that, it seemed to me that would be a good thought to leave for consideration.

In regard to putting them up, we found the outfit that suited our purpose best was a tower erected on a Ford, and we got considerable speed out of that.

PRESTON S. MILLAR: Mr. Chairman, may I ask the last speaker at what price those used lamps were sold and what is the name of the company that followed that practice? (Laughter)

H. C. DOUGHTY: Professional etiquette makes me refrain from answering that question, Mr. Chairman. (Laughter)

MOTOR VEHICLE LIGHTING REGULATIONS

1923 REPORT OF COMMITTEE ON MOTOR VEHICLE LIGHTING

BY CLAYTON H. SHARP, CHAIRMAN

SYNOPSIS: The Committee on Motor Vehicle Lighting in the Report for the present year summarize their activities. Specifications for rear lamps have been drawn up and adopted, receiving the endorsement of the Society of Automotive Engineers. The Headlighting Specifications have become a tentative American Standard. The Conference of Motor Vehicle Administrators, consisting of the New England, Middle Atlantic States and Ohio, are approving headlighting devices under the Standard Specifications. The California Headlight law is the standard specifications with but slight modification. International relationship has been established through the Committee Chairman with the International Commission on Illumination, and work is in progress with England, France, and Switzerland. No further change can be made in the Specifications except through the Sectional Committee of the A. E. S. C.

SPECIFICATIONS FOR REAR LAMPS

As noted in last year's report, the question of specifications for tests of rear lamps had been taken up and put in the hands of a subcommittee for report to a joint meeting of this Committee and of the Lighting Division of the Society of Automotive Engineers. This subcommittee made its report in December of last year and its proposals for specifications governing the acceptability of electric tail lamps on motor vehicles were unanimously adopted at a joint meeting of the Committee on Motor Vehicle Lighting of the Society with the Lighting Division of the Standards Committee of the Society of Automotive Engineers. These specifications were approved by the I. E. S. Council on January 18th, 1923. They will be found on page 208, Volume XVIII, TRANSACTIONS, I. E. S., February, 1923.

The considerations which led to the adoption of these specifications have been well set forth by Mr. G. H. Stickney, who, as Vice-Chairman of the Committee on Motor Vehicle Lighting, very ably

---

*A Report presented before the Annual Convention of the Illuminating Engineering Society, Lake George, N. Y., September 24-28, 1923.

The Illuminating Engineering Society is not responsible for the statements or opinions advanced by contributors.

conducted the operations which led up to the framing and adoption of the specifications. Mr. Stickney's exposition of the matter is in the form of an editorial.*

ADOPTION BY AMERICAN ENGINEERING STANDARDS COMMITTEE

At the time of the last report, the matter of the adoption of the rules governing the approval of headlighting devices for motor vehicles as an American Standard by the American Engineering Standards Committee was still pending. During the past year the American Engineering Standards Committee has adopted these rules insofar as they apply to tests for purpose of approval. The Illuminating Engineering Society and the Society of Automotive Engineers have been appointed joint sponsors in the matter of the revision and extension of these rules. The first duty of the joint sponsors is to select a Sectional Committee of the A.E.S.C. for this purpose, and this matter, which is one requiring considerable time on account of the large number of organizations which must be communicated with, has not as yet been completed, but is under way. The headlight testing rules therefore have now the added authority and prestige which accrues from their adoption by the American Engineering Standards Committee as a tentative American Engineering Standard.

ACTION OF CONFERENCE OF MOTOR VEHICLES ADMINISTRATORS

The report of last year noted that the headlight specifications had been adopted by the Conference of Motor Vehicle Administrators as their standard for the approval of headlighting devices. This Conference is now composed of motor vehicle administrators. of the States of Maine, New Hampshire, Vermont, Massachusetts, Connecticut, New York, New Jersey, Pennsylvania, Maryland and Ohio. The administrators have examined the list of approved devices for each state and have gone over the reports of tests of each of these devices. By a process of elimination, and by the aid of auxiliary tests, they have been able to ascertain that only twenty-two of all the devices which have formerly been approved would meet with approval under the standard testing specifications. They have therefore put out this list of twenty-two devices which are approved in all the states represented at the Conference. In general, putting out this list has not involved the immediate revoca-

*Stickney, G. H., "Tail Lights and License Plates" TRANS., I. E. S., 1923, XVIII, pp. 111-113.

tion of approvals previously granted in the individual states, so that it has not been necessary for motorists at once to change from a headlighting equipment which had previously been approved to one of the restricted list of the administrators. In most of the states the old approval list stands for a while. In the State of New Jersey, however, all devices not approved under the Conference list have already been ordered discontinued in use. The following table shows something of the status year by year of the number of devices which had been approved by five states as a result of tests made by one testing organization.

State	Number 1918 Specifications		Number 1920 Specifications		Number 1922 Specifications
	Tested	Approved	Tested	Approved	Approved
New York	86	55	55	92	22
Pennsylvania	107	80	8	98	22
Connecticut	68	52	9	47	22
Maryland	—	—	66	57	22
Massachusetts	{ Massachusetts Specifications }		58	32	22

It is apparent from the table that the reduction in number of approved devices is a very radical one. It should be said, however, that a large percentage of the devices which have failed to receive approval under the standard specifications are no longer commercial. Indeed many of them never have been manufactured. In a number of important cases older designs of standard makes of devices have been discarded in favor of new designs which will conform with the standard specifications. Reduction of the list to twenty-two represents a very thorough housecleaning on the part of the state authorities and it is safe to say that no device is found on this latest approval list which is not a practical and meritorious one from the standpoint of road illumination and the reduction of glare.

CALIFORNIA LAW

The State of California has recently revised its motor vehicle legislation and has changed its requirements for headlights, so as to bring them in general conformity with the Standard specifications. It is to be noted, however, that California has made in two respects variations from the standard which the Committee cannot view with anything but regret. The first of these variations is changing the limits at the C point; namely, one degree of arc above the center of the lamps and directly ahead, from a maximum of 2400 cp. and a minimum of 800 cp. to a maximum of 1500 cp. and a minimum of 500 cp. Thus the glare limit at this point is lowered and the graduation of intensity from the B point to the C point is made more abrupt. This change may not be a disadvantageous one in itself. However, it constitutes a variation from the Standard specifications, the utility of which is not obvious and the results of which, from the point of view of interstate uniformity, cannot well fail to be unfortunate. The second variation is with respect to the size of incandescent lamp used by the testing agency in making the laboratory test. The California law reads:—

"(d) The testing agency shall conduct an exact scientific and laboratory test of every device submitted to it as herein provided using twenty-one standard candlepower lamp or bulb or thirty-two standard candlepower lamp or bulb, or any standard candlepower lamp or bulb between these two limits, and determine whether or not the device will conform with the requirements of this act when used in accordance with instructions of the testing agency, stating the candlepower lamp or bulb and any particular adjustments to be used in connection with such device.

This procedure, as compared with the simple and definite procedure laid down in the Standard specifications, calling for a test with a 21-cp. lamp only, is manifestly cumbersome, and seems quite likely to lead to confusion. The leading lamp manufacturers of this country do not at the present time list any standard sizes of headlighting lamps of candlepower intermediate between 21 and 32. Even the 32-cp. lamp is not considered as "standard." What this provision therefore says, is that if a device can be approved with a 21-cp. lamp and some adjustment which the ingenuity of the testing authority may be able to find, it may be approved for use with a 21-cp. lamp. If, on the other hand, the device will meet the specifications with a 32-cp. lamp and some particular adjustment found by experiment, it may be used with a 32-cp. lamp and with that particular adjustment. It seems quite probable that devices which have been designed to meet the specifications with a 21-cp. lamp will in many cases not meet them with a 32-cp. lamp on account of the slightly different dimensions of the filament. The obverse of this may also be true. At any rate, it is conceivable that a design which would bring the device within the specifications with both candlepowers, would be a compromise which would not be so good as a design made for either one candlepower or the other alone. Furthermore, the adjustment required with a lamp of one size would very likely not be the same as that required with the other size. Evidently the complexities introduced are many, and it becomes a question again whether the advantages which will accrue from this arrangement will compensate for such complications and for the disadvantages arising from a variation from uniformity.

INTERNATIONAL RELATIONSHIPS

The matter of the control of automobile headlighting devices is assuming no little degree of international importance. The problem which we have in this country exists in other lands as well, though probably not to quite the same extent as here, because of the smaller density of motor vehicle traffic on the highways. In recognition of this condition, the International Commission on Illumination, as its meeting in Paris, in 1921, appointed an International Committee to consider this question. The membership of this Committee has been filled out, and is now as follows: R. Bossu, France; K. Edgcumbe, England; M. Payot, Switzerland; and C. H. Sharp, United States (Chairman).

The Chairman of your Committee was able to discuss this problem last winter with M. Tur of the Department of Public Works of France, and with Sir Henry Maybury of the Department of Transport of Great Britain. The exchange of views with these officials was most helpful, and an opportunity was afforded to explain to them what has been done under the auspices of the Illuminating Engineering Society in this country. The Chairman also had a conference with Major Edgcumbe of the International Committee on these matters. These various gentlemen expressed interest in what was going on in this country, and as a result of these conferences, samples of American front glasses for headlights, such as are approved here, together with test data under the I.E.S. specifications for these glasses have been forwarded to the members of the International Committee of the I.C.I. These front glasses are now in the hands of these gentlemen in England, France and Switzerland, all of whom have promised to examine them carefully, and to use them on cars in an endeavor to obtain a concrete idea of the kind of motor vehicle headlighting which we in America consider to be best adapted to public convenience and safety, in the present state of the art. It is to be hoped that further conferences between members of the Committee of the I.C.I. may prove fruitful toward bringing about a more complete understanding of the problem, for it must be remembered that there are differences in conditions and differences in practice in automobile driving in different countries which may have the result that what is best here may not be best elsewhere. At any rate, an earnest attempt is being made toward securing an international understanding on this important question.

FUTURE WORK

With regard to the work of the Committee for the following year, it is to be observed that the standards have now reached a point where they cannot be revised except by action of the Sectional Committee of the A.E.S.C. Evidently no revision can or should be attempted unless there is a very well marked reason for it. The working out of the problems of application of these rules and the enforcement of the laws is the main thing for the future At the same time there is a necessity for the existence of a technical committee of the Society to cover this matter. The art is an advancing one and new ideas and new devices are constantly being put forward and no one can say when a more or less radical alteration of our notions on the matter may be called for. Furthermore, the movement toward the general illumination of the highways by fixed lamps may well lead to conditions which will require the co-ordination of motor vehicle lighting and street lighting, and this may become a profitable field for the activities of a technical committee.

Therefore, while it may be said that the Committee on Motor Vehicle Lighting has for the present completed one stage of its work, yet its opportunities for usefulness are by no means closed.

Respectfully submitted,

Committee on Motor Vehicle Lighting,

Clayton H. Sharp, *Chairman*

G. H. Stickney, *Vice-Chairman*

F. C. Caldwell	W. F. Little
A. W. Devine	W. A. McKay
C. E. Godley	A. L. McMurtry
C. A. B. Halvorson	H. H. Magdsick
J. A. Hoeveler	L. C. Porter

L. E. Voyer

DISCUSSION

CHAIRMAN WM. J. DRISKO: I know there are a great many present who have very positive ideas about this matter of heading and it seems to me we ought to have a very free discussion. A lot of the questions which have arisen are suggested in Dr. Sharp's report: the question of what relation between street lighting and automobile headlighting should prevail in our larger cities. The question of the uniformity of standards as set forth seems to be pretty well agreed upon; that is, the various Societies are all in practical agreement. I may say, I think, they are all in legal agreement; in minor details, they differ.

There are other questions, however, than the questions of simply laboratory testing. My attention was recently called to a difference of something like 50 per cent between the test of one particular point in a particular lens by two different authorities. It sounded to the individual as though it were a terrible discrepancy. The public needs educating. That is, perhaps, the real question.

E. L. ELLIOTT: The Chairman of the Committee I think very rightly said that there is a lot yet to be done before the subject is finished. So far as the menace of an automobile headlight is concerned, all attempts to remove it with a fixed lens are of little avail. A lens can do two things: it can throw the light down, which can be done by tilting the headlight; and it can widen the beam sidewise. Distributing a light sidewise is useful when you have the road to yourself; but when you meet a passing automobile, which is the occasion of most of the accidents, just at the point where you want to see the clearest the side light from the approaching car is in your eyes. A worse thing could hardly be done.

Legislation that goes no further than giving a list of approved lenses or reflectors which may be used on fixed headlights will never remove, nor very materially reduce the dangers of night driving. The mechanical and physical difficulties of the problem are by no means unsuperable; the most serious obstacle at present is the education of legislative bodies to an appreciation of the fact that the menace of the headlight can be removed only by putting it under the constant control of the driver. So I say again, that Dr. Sharp's point that there is much more work to be done is very well taken.

MAX POSER: For a good many years I have been studying dangerous glaring of automobile headlight lamps. Mr. Elliott is quite right in saying that one meets on the road frequently, dangerous glaring headlights, even reflecting glaring rays sideways and it seems that this sort of thing is going on unchallenged, which must make the automobile headlight lamp law appear as a farce.

Such glaring rays of headlight lamps however, would not be met with if the entire lamp adjustments were carried out strictly within the boundary of the headlight lamp law and I defy Mr. Elliott to show me dangerous glare with an automobile headlight lamp adjusted in the proper manner as defined by the law.

According to my observation, this whole thing is largely a question of enforcing the law and demand correct construction and adjustment of the entire automobile headlight lamp.

It is a well-known fact that if direct emanating light is not permitted to enter our eyes, there is no dangerous glare. The law of course, cannot prevent dangerous glare, if the people using headlight lenses on their lamps indiscriminately do not pay attention to the law and this is the very trouble we are up against.

I experimented with a hundred candlepower lamp with the rays optically guided in the proper manner, in order to meet law conditions and no dangerous glare whatsoever was experienced. We have a headlight lamp law, but chiefly scrutinize headlight lamp lenses, and since the lens being only a part of the entire headlight lamp, cannot give entire satisfaction in general, if the reflector and entire lamp construction is faulty. It seems evident that the construction of the entire headlight lamp should be considered. By this I mean to say that an automobile headlight lamp lens may furnish excellent results with a properly constructed reflector and lamp casing, while that same lens may be quite inferior with a poorly constructed reflector and lamp casing.

It stands to reason that the construction of the entire automobile headlight lamp should be subject to the automobile headlight law; otherwise I do not think that the results aimed at by the law, will be achieved.

What is the use of mounting a carefully designed automobile headlight lens to a shaky lamp body, flimsy in construction, owing to the price of such lamps being forced down for commercial competitive reasons, to a value of about $2.60? No reliable maker could produce automobile headlight lamps, efficiently constructed and provided with the proper focusing and centering adjustment of the lamp bulbs, with first-class reflector and optical device, essential to meet the condition of the law, at such a low price?

I have seen automobile headlight lamps fitted with the best optical device, properly adjusted, after having gone only a short distance over a bumpy road, getting completely out of adjustment, owing to the loosely-fitted reflector and lamp socket, and which, through the slightest jerk, get out of alignment.

If this sort of thing is permitted to pass, no careful setting and adjustment of the lamps can be considered secure and consequently, the Law is bound to be infringed. For that same reason, it is very inadvisable to issue certificates for headlight lamp adjustments for lamps constructed in such poor manner.

Imagine a person having obtained a certificate for his automobile headlight lamps, but the lamp bodies being of a poor construction as mentioned, got out of adjustment the very moment he passed over a rough road. He is afterwards questioned by an inspector as to his glaring lights and of course, will present his certificate, very likely feeling most indignant at being stopped. The inspector also will be puzzled and both parties, to say the least, may have an argument.

To sum up the matter in brief, it amounts to this: To establish a non-glare light emanating from an automobile headlight lamp for road illumination at night, it is essential to have the entire automobile headlight lamp construction specified in detail and to permit only such designs which warrant the possibility of furnishing the desired road illumination.

Everybody involved in this question should agree to such specifications and the cost involved of a headlight lamp of this kind should be a secondary consideration. A specification of the entire automobile headlight lamp should be controlled by the automobile headlight lamp Law when we will get a great step further towards ideal road illumination than we can boast of to-day.

It is to be regretted that sound and efficient headlight lamp constructions are often rejected on account of the cost, since headlight lamps are merely considered as an ordinary part of the car. Right here a great deal of education is required, so that motorists will observe and consider that it is more important to have efficient automobile headlight lamps than all kinds of fancy and expensive accessories on their cars of much less importance. Nothing is too expensive to save human lives and that is what the whole affair really amounts to.

It seems to me that the authorities should insist upon the automobile headlight lamp in its entirety being subject to the Law, so that the tests could be made for the entire equipment and not only the lens alone.

Flimsy and unreliable constructions as we meet them in numerous cases today, will then be a thing of the past and a great obstacle to establish safe travelling at night overcome.

L. C. PORTER: I have been studying this headlight problem for a number of years, and have had the pleasure and privilege of working with Dr. Sharp on the Headlight Committee. It seems to me the work has about reached the point where it should take a little different line. What I mean is this: a few years ago we didn't know very much about headlights. We didn't know how to control the light; we didn't know what was good light. We didn't know where to put the light to secure good driving conditions and to reduce glare.

A great deal of study has been put into that end of the problem and the technical men have found out what constitutes good lighting. Furthermore, they have found out how to secure and use it and a number of technical papers have been written about it. Data on this subject has been presented to technical societies, and a number of instruction sheets have been gotten up to go with headlight equipment, telling how to adjust it and so forth and so on.

But now that we have all that information I think we have reached the point where what we really need is more popular publicity on the thing. The public today does not know how to handle the equipment that is available. These technical instruction sheets are pretty hard to follow and they haven't been given general distribution.

I have been making a little investigation among our own men at the factory and none of them ever found out how to focus a headlight. Some of them after they had been told how to do it and tried to follow the technical instruction sheets came back to me and said, "Maybe my headlight has a focusing device on it but I can't find it," or they told me, "I had to get a big Stillson wrench and bend the brackets."

The motoring public can't do that and they are not going to do it if they could. Even should they get it done, as Dr. Poser pointed out, the present equipment is not going to stay in adjustment. It is not good equipment.

I think the two things which are most needed are improvement in the mechanics of present headlight equipment, not only to make better equipment, more rigid equipment and more accurate equipment, but also equipment which is easy to handle and easy to get at. We ought to have headlights on which the focusing device can be operated by hand without having to resort to the use of tools, and we ought to have devices on which we can easily open the headlights to change the lamps and clean the reflectors, etc.

Today you cannot do that, and I think it is largely a question of cost competition. Instead of having a nice hinge on the door to open it, they punch a little bayonet joint on the thing and snap it in, and maybe it stays there, and maybe it doesn't. Probably it gets rusted and then requires a mechanic and tools to get it off.

The reflectors which are in use in a great many cases are not accurate and where that is the case it is absolutely impossible to get good focusing on your headlights. If you follow your technical instructions, you cannot get the results because the equipment isn't right.

So, I say, it seems to me we have passed the investigating stage—you might say the technical stage—and what we really need now is to get after the manufacturers to make better equipment then teach the public how to use it.

The incandescent lamp manufacturers are making very rapid progress along that line. For a considerable time they were passing the buck back and forth with the brass lamp manufacturers. The brass lamp fellows would say, "What is the use of our making accurate equipment when your lamps vary in light center length and axial alignment.

The incandescent lamp fellows would come back and say, "Well, your reflectors aren't true, your sockets wobble, and it is no use making any more accurate lamps."

But the lamp manufacturers have taken the initiative and they are putting out lamps today which are really good. They are putting out lamps in which the filaments come within plus or minus three sixty-fourths of an inch, and that is close enough if you get good equipment to go with it. When you start talking about one sixty-fourth of an inch or a half of a sixty-fourth of an inch, you are getting down to a point where the refinement is not practical. I have had some of my fellows trying to find out what one sixty-fourth of an inch variation in light center length would do to a headlight beam or even a half of a sixty-fourth and they have had to use micrometer equipment to get those adjustments and to measure the effects.

I have seen a number of demonstrations recently of what these variations mean, and to demonstrate it the fellows have had to make up really very expensive special apparatus, and they have had to have three or four fellows working with it weeks before they could get it into condition to demonstrate.

If the technical men—the engineers—have that trouble, what is the poor public going to do who haven't any tools and don't know how to do it? So I say again, I think the next step should be to get the manufacturers to put out more accurate, simple and more rugged equipment and then to give popular publicity to the methods of using that equipment. I think we ought to have some motion picture films on the subject, perhaps a little comedy or a little tragedy, or something that will put it before the general public. We ought to get a story into the Saturday Evening Post, or something of that sort, because the technical literature does not reach the general motoring public. Thank you. (Applause)

E. C. CRITTENDEN: There are a number of points I would like to mention, with particular reference to an attempt which the Bureau of Standards is making to meet some of the needs which have been well set forth in this report and its discussion.

In the first place, regulation of headlights by law has been tried long enough to show that it can not give satisfactory results and we ought now, as Mr. Porter says, to follow a different tack. With reference to the suggestion which has been made that there ought to be a federal law, I am afraid there is a tendency to think that any problem which is too stiff for local officers or states to handle can be settled by having Congress pass a law. That certainly isn't the way to get at this question. As Dr. Sharp has said in the report, the working out of the application of these specifications is the thing which is needed now. The problem is how are you going to apply them in actual practice.

Perhaps most of you are aware that the Bureau of Standards has no legal authority in practically any field and doesn't seek such authority, but in many fields we have occasion to serve as a means of getting together the officers of various states and of transmitting information from one state to another, because requests come to us from state officials for help in meeting their problems. There have been so many such inquiries, so many urgent calls for help in this lighting problem, that we have felt compelled to divert to it some resources we have needed elsewhere.

We have tried to assist for some years past in the work of this Committee and have felt that its experimental way of going at the problem was the very best thing that could be done. Perhaps Mr. Elliott in his remarks has overlooked the early history of the work of the Committee, which may well be recalled. These numbers in the specifications were not made up by anybody sitting down at a desk and figuring them out, but by actually getting out on the road and trying different lighting conditions. They are the result of thousands of observations by many men including not merely illuminating engineers but also many other experienced drivers. The requirements of the specifications were drawn up with a full realization of the difficulties to be met and also with considerable knowledge of the possibilities in the material that was available. At the Bureau we have felt that this experimental way of determining what you ought to do and what you can do was the best plan that any organization could follow and we have felt that there was no use of duplicating this work.

It has seemed to us that the best thing to do is to take these specifications as they now exist, especially since they have furnished the basis on which such an excellent beginning has been made in cooperation among the different states. We should now try to get the intent of the specifications actually into effect by following up the use of the approved devices which we know are capable of giving good results.

As a beginning on that, we set up a simple testing station with canvas sheets for observing the beams given by headlights, and invited those drivers in our own city—Washington—who chose to stop in, to do so during a certain week. Slightly over 400 cars came in for inspection. A considerable proportion of these belonged to members of the Bureau staff men, who are probably above the average not merely in their interest in the subject, but in their technical education, men who ought to have known better than the general public how to adjust their lamps.

I will not trouble you with the statistics showing the results of these tests, since they have been published in various articles. The outstanding fact is that of the 400 cars about 22 had their lights in really good condition. Of course many others were passable, but certainly 75 per cent of the cars had their headlamps in very poor adjustment. A large proportion of them had lenses which were on the approved list, but having good lenses doesn't give you approved adjustment—we will all grant that. This test was run, incidentally, to show our own Police Department and the motor associations that the use of the approved devices was only a first step, that something more had to be done.

With this small collection of cars as an example we took the matter up with several groups of people: the manufacturers of cars, the dealers—in our own locality at first—and with the two national associations, the National Motorists and the American Automobile Association. Fortunately, from those two organizations we have gotten very active cooperation. As a result we got through them the thing that I think is really need that is, some incentive for the service stations and the dealers to see that cars are put out in good condition and kept in good condition so far as the lighting is concerned.

Undoubtedly, a first step in that direction ought to be for car makers to put on better equipment, to put more money into lighting equipment. We haven't felt we could say they should do that, but we have urged them to put out the equipment that they do use in reasonably good condition, and to provide adequate instructions for its maintenance in good adjustment.

Arrangements for maintenance and inspection service can probably best be made through motorists organizations. In our case the local organization of the National Motorists Association arranged with about 20 garages to put in these simple testing stations, and to give such service to any of their members without charge. Furthermore, this has worked out so well in Washington that the Association is suggesting to all their other local clubs the operation of a similar plan. Likewise, the American Automobile Association is getting out a booklet which they are sending to all their branches, urging that they establish these adjusting stations.

We hope that at least a fair percentage of car owners will be able by such means to keep more of a check on their own cars, and especially that service stations and sales stations will be made to realize the importance of giving such service. We believe that this matter of impressing upon the people who furnish cars and who service them the necessity for giving attention to head-lamps is the first step, and that the other one of education of drivers themselves to the need for such adjustment is supple-mentary.

In this work we have felt that there is an opportunity for our organization to put into actual application the very valuable results gained by this Society. We realize it isn't going to be a perfect solution of the problem, but it ought to help some, and we hope to do our bit in keeping the movement under way.

D. H. TUCK (communicated): I believe that the practical application of the committee's work now consists of teaching the motorist how to drive at night. In the first place, it is against the law (except in Ohio), to dim the headlights when approaching another motorist (or at any time while driving). It is at this critical moment that the maximum road illumination is required to prevent running down a pedestrian walking along the side of the road and to prevent accidents due to running off the road.

In the second place, it is not necessary for a driver to watch the head lamps of the approaching motorist. Watch the right edge of the road and don't bother about the approaching motorist.

I believe that if effective publicity were given these two points, —"Don't dim",—"Don't watch the head lamps", that night driving would automatically become more safe and pleasant to all.

THE GENERAL SECRETARY'S REPORT
FOR THE FISCAL YEAR 1922-23

By Samuel G. Hibben

The seventeenth fiscal year of the Society terminating October 1, 1923 constituted a very important period, inasmuch as a decided upward trend has been developed and the items of finance, membership, sectional development, publications, and authoritative recognition have all advanced to a notably higher plane. Coincident with the growth of the electrical industry, the status and solidarity of the Society ought to improve, following the same upward trend until many additional activities and much more valuable work can be looked forward to when, one or two years hence, the financial surplus and the enthusiastic members' support make possible the enlargement of the general activities. The encouragement and support of all Society officers directed towards the general executive office has been exceedingly gratifying and the successful furthering of the Society's established policies has indicated the wisdom of continuing the general operations with perhaps only minor changes, along the lines laid down in previous years.

THE COUNCIL

The regular monthly meetings of the Council have been held with the exception of July and August when the Executive Committee, being so empowered, enacted the necessary routine business. The members of the Council have individually expressed a keen interest in the Society's affairs and, both socially and in a business way, have profited by these Council meetings just as the membership at large have profited by the careful consideration and the sacrifice of personal interest that Council members have manifested in order to advance the Society's affairs. The Society's president has personally supervised the year's activities and attended all Council meetings, while many other Council members have likewise praiseworthy attendance records. The Assistant

Secretary and the central office staff are responsible for much of the progress in simplification of routine and the reduction of expenses. Changes of personnel but not in numbers of persons have been made.

MEMBERSHIP

Considering the four divisions of the Society's membership exclusive of honorary members, there remain the divisions of members, associate members, sustaining members, and student members. No activities have been undertaken in connection with the last classification, but in regard to the other divisions, the following facts are of interest:

Sustaining Members

Organizations and business concerns to the number of 113 now contribute sums ranging from $25 to $300 towards the general support of the Society's work. As of September 30th, 1923, revenue from this class was $6680.00 paid in, with a probable further amount of unpaid dues of about $350.00. The corresponding figures for last year show 80 sustaining members with collections slightly in excess of $6,580. The increase in revenue from this source during the past fiscal year is $450, or about 7 per cent increase. This sustaining membership represents about 30 per cent of the total revenue accruing to the Society. The increase, notable in numbers of Sustaining Member Companies more than in total revenue, is gratifying particularly in the New England territory, and if extended to other territories, will firmly establish the Society on a financial basis wherefrom it can and should give more concrete help in the way of disseminating useful information to these sustaining members.

Members

The number of members at the end of this fiscal year was 395 as compared with 384 of last year. The revenue derived this year was about $5,795 as against $5,732 of last year, with a probability of a slight increase from a few delinquent dues. The net gain in this income was $63. The percentage net increase in members was 3%. From this source is, therefore, derived about 25 percent of the total revenue and it can be assumed that these figures will not radically change within the next year.

Associate Members

From this source is derived about 28 per cent of our revenue. The number of associate members this year is 931 as compared with 810 of last year, and 799 of the year previous. There was a net gain of 121 members representing about a 15 per cent increase. The revenue derived this year was approximately $6,648 as against $5,700 of last year, or a net gain in revenue of more than 16 per cent. It will be evident that the tendency is towards a growth in associate membership and that this growth is quite healthy and quite appreciable. The net increase mentioned does not express the entire facts inasmuch as a rather thorough elimination of delinquent members was necessary and the membership list put on a much better basis by the resignation of disinterested persons.

Further details of membership classification can be obtained from the printed list of Society membership issued during the year, and from the accompanying charts.

LOCAL REPRESENTATIVES

The local representatives of the Society now number 34, practically the same as last year, and are distributed among the various states chiefly where sections and chapters are not located, as well as in Canada and South Africa. These representatives are to be encouraged in their activities, as being the nuclei of chapters and the lieutenants of the chairman of the Membership Committee. It is urged that such local representation be developed next year and that by correspondence and otherwise, these representatives be connected more closely with the executive office and with the various committee chairmen.

TABLE I

Class of memberships	Honorary members	Members	Associates	Total	Sustaining members
Membership Oct. 1, 1921	2	384	799	1183	87
Additions					
Transferred		3	4		
New members qualified		34	137		
Re-elected		3	1		
Re-instated		5	8		
Sub-total		45	150		
Deductions					
Deceased		3	5		
Resigned		20	47		11
Dropped		18	84		
Transferred		4	3		
Sub-total		45	139		
Increase		0	11	11	
Membership Oct. 1, 1922	2	384	810	1194	80
Net Increase During Year in Membership		0	11	14	

TABLE II

Class of memberships	Honorary members	Members	Associates	Total	Sustaining members
Membership Oct. 1, 1922	2	384	810	1194	80
Additions					
Transferred		7	6		
New members		38	240		34
Re-elected		0	1		
Re-instated		1	8		
Sub-total		46	255		
Deductions					
Deceased		3	12		
Resigned		12	44		
Dropped		9	53		
Transferred		6	7		
Sub-total		30	116		
Increase		16	139	155	33
Membership Oct. 1, 1923		400	949	1349	113
Net Increase in Membership During Year		16	139		33

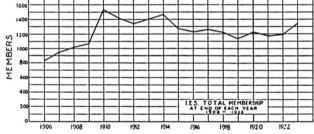

I.E.S. TOTAL MEMBERSHIP AT END OF EACH YEAR 1906 - 1923

SECTIONS AND CHAPTERS

The following table gives the statistics on section and chapter meetings indicating about the average activity when compared with the preceding year.

TABLE III

Chapter or Section	No. of Meetings	No. Papers Presented	Average Attendance
Chicago	6	17	40
New York	8	13	105
New England	4	4	40
Philadelphia	8	8	56
Cleveland	2	—	—
Columbus	2	—	—
Los Angeles	1	—	—
Michigan	1	—	22
Northern New Jersey	3	3	50
Pittsburgh	1	—	—
San Francisco	—	—	—
Toronto	6	6	38

Certain sections and chapters feel the need of more frequent meetings even though these be luncheon gatherings or inspection trips. Versatility of the program is to be sought for and attendance at sectional meetings increased through demonstrations and if necessary by means of some social diversions that have been tried with success in several instances.

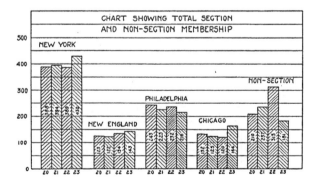

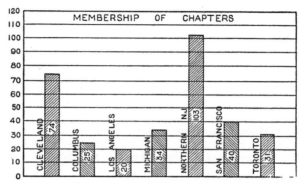

The same sections functioned this year at last, but there has been a notable increase in the number of chapters for which a regular charter is now granted. In addition to the Cleveland and Toronto chapters of last year, we now have chapters as follows:

San Francisco Bay Cities Chapter, San Francisco, California.
Northern New Jersey Chapter, Newark, New Jersey.
Columbus Chapter, Columbus, Ohio.
Michigan Chapter, Detroit, Michigan.
Los Angeles Chapter, Los Angeles, California.
Pittsburgh Chapter, Pittsburgh, Pennsylvania.

This last Chapter was organized in September, but charter not granted until after the close of last fiscal year.

The policy of holding joint meetings with other organizations has generally been successful. Papers discussed at the meetings have been of the usual high character and the majority have been accepted for publication in the TRANSACTIONS. The policy is to be encouraged of having an exchange of papers and speakers among neighboring chapters and sections, and of inspections of prominent installations or more concrete discussions of lighting problems.

COMMITTEES

Committee on Editing and Publication

This committee has done especially good work in the face of numerous petty difficulties and has provided better and more prompt issues of the TRANSACTIONS as well as being successful in keeping printing costs to a minimum. During the year the printing of the TRANSACTIONS has been transferred from eastern Pennsylvania to Ithaca, New York, in order to effect a considerable saving, this being a very large item in the expense budget.

A very remarkable sale of leaflets and pamphlets has been secured during the year due to the activities of the executive office staff, amounting to about $3,300. Of this amount some $1,530 has been secured from TRANSACTION advertising, $1,358 from TRANSACTION sales, and somewhat more than $400 from miscellaneous sale of pamphlets, etc.

It should not be overlooked that a wider distribution of the TRANSACTIONS is being made with a growing membership and that the average size of each issue is greater than heretofore.

Committee on Membership

The activities of the Membership Committee have been exceedingly gratifying as indicated by the former statistics and by the fact that as of September 1st approximately 275 new members and associates had been secured for this past fiscal year. The districts most fruitful have been New York, New Jersey, New England, and eastern Pennsylvania, Ohio, Michigan, Illinois and California. The work of this committee has been greater than the net membership indicates, by virtue of the elimination of delinquent and unsatisfactory members carried over from previous years as before mentioned.

Committee on Nomenclature and Standards

Detailed activities of this committee will be presented in its own report, the dominant feature of the year's work being the effort to co-ordinate with other organizations both abroad and at home in order to facilitate the establishment of a convenient and effective system of nomenclature throughout the world. Its work has resulted in a definition of lighting as distinct from light and various explanatory definitions of brightness.

The word "Luminaire" has been continued, with the word "Lamp" as applying to the light producing part of the luminaire.

Committee on Lighting Legislation

Details of this committee's activities will be found in their own report but it may be well to summarize the progress, noting that the code of lighting of factories, etc., in its final revised form has been distributed and quite generally adopted, and that work is well in hand on simplified explanations of the code and suggestions on its application suitable for state inspectors.

A revision of the school lighting code has been completed after much careful study and conference, and appears elsewhere.

Members of this committee have frequently given talks and explanations to various states adopting the code or contemplating it, notably in Massachusetts, Ohio and Pennsylvania.

Committee on Motor Vehicle Lighting

A summary of the activities of this committee have been presented during the recent Convention as well as of the Committee on the Tentative Code of Luminaire Design and the Committee of Progress. The good work of other committees is evident from the general condition of the Society, and, building upon the foundation of good work during the past year it is hoped that much encouraging progress may be reported at the eighteenth annual convention.

FINANCES

The details of the financial standing of the Society will be found in an addendum to this report. In general, the income as of September 30th, 1923 was $23,360.98 as compared to an income for the twelve months of the previous fiscal year of $22,885.80. The gross operating expenses of this year, $20,542.85 have been materially reduced over last, leaving a net surplus in the Society's treasury of $5,769.61 as compared to the 1922 surplus of $2,532.77.

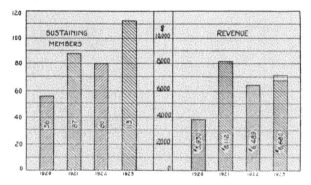

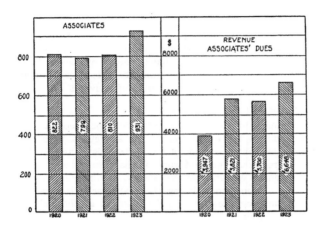

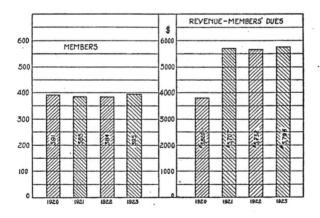

ADDENDUM

New York, October, 22, 1923.

Illuminating Engineering Society,
29 West 39th Street,
New York.

Dear Sirs:

Pursuant to engagement, we have made an audit of your books and accounts for the year ended September 30, 1923, and submit herewith the following described exhibits:

EXHIBIT "A"—Balance Sheet, September 30, 1923.

EXHIBIT "B"—Statement of Income and Profit & Loss for the Year Ended September 30, 1923.

The cash on deposit was verified by certification obtained from the depositary, and the petty cash fund at the New York office was counted. We did not verify the petty cash fund at Chicago.

The inventories, with the exception of identification bars, are as estimated by the Assistant Secretary.

The United States liberty loan bonds were not inspected by us.

According to the budget for 1922-1923, prepared by the Committee on Finance, there is a possibility of a charge of $300.00 for the support of the International Commission on Illumination for the year 1923.

Yours truly,

HASKINS & SELLS.

EXHIBIT "A"

BALANCE SHEET—SEPTEMBER 30, 1923

ASSETS

CASH:		
On Deposit	$3,213.55	
Petty cash—New York	100.00	
Petty cash—Chicago	146.62	
Total cash		$3,460.17
ACCOUNTS RECEIVABLE:		
Members' dues	$ 10.00	
Associate Members' dues	11.25	
Sustaining members' dues	350.00	
Initiation fees	7.50	
"Transactions"	18.68	
Advertising	226.00	
Miscellaneous	4.62	
Total accounts receivable		628.05
UNITED STATES LIBERTY LOAN BONDS		3000.00
INTEREST RECEIVABLE ON BONDS OWNED		35.51
INVENTORIES:		
Badges	$ 70.20	
Reprints	120.00	
"Transactions"	300.00	
Illustration cuts	50.00	
Identification bars	117.25	
Total inventories		657.45
OFFICE FURNITURE AND FIXTURES	$1,033.30	
Less reserve for depreciation	155.00	878.30
RENT PAID IN ADVANCE		130.00
TOTAL		$8,789.48

LIABILITIES

ACCOUNTS PAYABLE	$2,590.05
DUES RECEIVED IN ADVANCE	423.07
ADVANCE COLLECTION FOR "TRANSACTIONS"	6.75
SURPLUS PER EXHIBIT "B"	5,769.61
TOTAL	$8,789.48

EXHIBIT "B"

STATEMENT OF INCOME AND PROFIT & LOSS FOR THE YEAR
ENDED SEPTEMBER 30, 1923

INCOME

Dues:

Members		$5,794.63	
Associate members		6,647.97	
Sustaining members		6,680.00	
Student members		2.00	$19,124.60
Initiation fees			712.50
"Transactions" sales			1,357.81
Advertising sales			1,529.40
Miscellaneous sales			393.82
Royalties			23.40
Interest on bonds			130.00
Interest on bank balances			89.45
Total income			$23,360.98

EXPENSES:

"Transactions"		$ 7,565.54	
General office expenses:			
Salaries	$6,567.30		
Rent	1,430.00		
Stationery, printing, and supplies	544.63		
Postage	421.33		
Telegraph, telephone, and expressage	234.52		
Depreciation of office furniture and fixtures	155.00		
Miscellaneous	632.83		
Total			9,985.61
Committee expenses		138.18	
1923 convention expenses		1,379.44	
New York Section	$ 491.39		
Philadelphia section	504.73		
Chicago section	299.95		
New England section	172.51		
Toronto chapter	5.50		
Total		1,474.08	
Total expenses			20,542.85
GROSS INCOME			$2,818.13
INCOME CHARGE—ADJUSTMENT OF "TRANSACTIONS" ACCOUNTS RECEIVABLE			8.98
NET INCOME FOR THE YEAR			$2,809.15
SURPLUS, OCTOBER 1, 1922			2,532.77
OTHER PROFIT & LOSS CREDITS:			
Chicago petty cash fund, charged to expense in prior year		$146.62	
Interest on bonds received, applicable to prior years		105.57	
1922 convention—Recovery of expenses previously written off		175.50	
Total			427.69
SURPLUS, SEPTEMBER 30, 1923			$5,769.61

ELECTRIC AFFAIRS

PROGRESS REPORT ON CODE OF LIGHTING SCHOOL BUILDINGS

A preliminary draft of the proposed revised Code of Lighting School Buildings was presented for discussion at the Annual Convention of the Society last September. The suggestions then made received the careful consideration of the Committee in a later revision of the draft. Numerous criticisms received subsequently from various sources indicated the need of substantial revisions of the text and illustrations.

The introduction has been rewritten. Rule 1 (Illumination required), Rule 4 (Color and finish of interior), Rule 5 (Switching and controlling apparatus), Rule 6 (Exit and emergency lighting), and Rule 8 (Blackboards), have been revised. Part II (Why the fulfillment of the rules is important) has been amplified considerably. Some of the material in Part III (How to comply with the rules) has been transferred to Part II.

A revised drawing of the plan and elevation of a typical classroom has been completed to take the place of Fig. 3; two new illustrations are in hand to take the place of Figs. 10 and 16; two characteristic illustrations on daylighting of school building interiors have been added. New captions for many of the illustrations and cuts have been prepared. Larger type will be used for all captions and footnotes and for Table I (Illumination required).

An index has been prepared and will be added.

The Rule relating to exit and emergency lighting has been submitted for approval to the I. E. S. representative on the Building Code Committee (A. E. S. C.).

The definitions of technical terms used in the code have been submitted to the Chairman of the I. E. S. Committee on Nomenclature and Standards.

Immediately after the revised draft has been received from the printer (probably early in the month of January) it will be submitted to the Sectional Committee (A. E. S. C.) for approval and then to the Illuminating Engineering Society and the American Institute of Architects (joint sponsors) and finally to the American Engineering Standards Committee.

Recommendations made recently to the American Engineering Standards Committee indicate that the Sectional Committee may be urged to include in the code a treatment of fire-hazard and consequent life hazard of compressed gas, acetylene generators, and kerosene lamps used in country school houses. This brings up a question as to the scope of the lighting code, which question is now being considered by a special committee of the A. E. S. C. in cooperation with the sponsors.

A COURSE IN ILLUMINATING ENGINEERING

The School of Industrial Arts, Trenton, New Jersey, offers a course in the theory and practice of Illuminating Engineering, supplemented by popular lectures, Mr. James D. Lee, Jr., Illuminating Engineer, during the season November 21-23 to May 14, 1924.

This course will be especially valuable to all interested in the planning, installation and use of electric lighting, in residences, stores, theatres, manufacturing plants, etc.

As the School of Industrial Arts has regularly offered courses by non-resident lecturers as part of its system of instruction, an arrangement has been effected by which the regular course will be supplemented by popular lectures given by some of the foremost engineers in the profession.

This course has three objects:—(1) to indicate those arts and sciences which constitute illuminating engineering; (2) to furnish a condensed outline of study of the subject; and (3) to give to practicing engineers and architects an opportunity to obtain a conception of the art and science of illumination as a whole.

The course will give to the student a viewpoint of this particular branch of engineering and make him familiar with the sources of information. The progress and developments in the art of illumination have been so rapid, that special emphasis will be given this phase of the work in the many splendid examples of modern installations to be shown and a thorough study made of the results obtained.

The School of Industrial Arts, recognizing the fact that the facilities available for the specialized instruction required are inadequate, determined to encourage the establishment of a course of instruction on illuminating engineering.

The subjects and scope of the lectures have been carefully selected to supplement the instruction. Those regularly enrolled in the classes now forming, will be given an opportunity to study the many problems presented in the lectures, with the assistance of a competent instructor who will be in charge of the work.

An outline of the course follows:—

Theory—(1) Fundamental Concepts, (2) Units and Terms, (3) Commercial Illuminants, (4) Photometry, (5) Calculation of Illumination, (6) Systems of Illumination, (7) Engineering Data.

Practice:—(a) *Commercial Lighting*, (b) *Industrial Lighting*, (c) Residence, (d) Street, (e) Church, (f) School, (g) Special Problems, (h) Aesthetic problems.

ILLUMINATION INDEX

PREPARED BY THE COMMITTEE ON PROGRESS.

An INDEX OF REFERENCES to books, papers, editorials, news and abstracts on illuminating engineering and allied subjects. This index is arranged alphabetically according to the names of the reference publications. The references are then given in order of the date of publication. Important references not appearing herein should be called to the attention of the Illuminating Engineering Society, 29 W. 39th St., New York, N. Y.

OFFICERS & COUNCIL

1923-24

President

CLARENCE L. LAW,
Irving Place and 15th Street,
New York, N. Y.
Term expires Sept. 30, 1924

Junior Past-Presidents

GEO. S. CRAMPTON,
1819 Walnut Street,
Philadelphia, Pa.
Term expires Sept. 30, 1924

WARD HARRISON,
Nela Park,
Cleveland, Ohio
Term expires Sept 30, 1925

Vice-Presidents

WM. J. DRISKO,
Mass. Inst. of Technology,
Cambridge, Mass.
Term expires Sept. 30, 1924

D. MCFARLAN MOORE,
Edison Lamp Works,
Harrison, N. J.
Term expires Sept. 30, 1925

O. L. JOHNSON,
230 South Clark Street,
Chicago, Ill.
Term expires Sept. 30, 1924

G. BERTRAM REGAR,
1000 Chestnut Street,
Philadelphia, Pa.
Term expires Sept. 30, 1924

General Secretary

NORMAN D. MACDONALD
80th St. and East End Ave.
New York, N. Y.
Term expires Sept. 30, 1925

Treasurer

L. B. MARKS,
103 Park Avenue,
New York, N. Y.
Term expires Sept. 30, 1925

Directors

Terms expire Sept. 30, 1924	*Terms expire Sept. 30, 1925*	*Terms expire Sept. 30, 1926*
F. C. CALDWELL, Ohio State University, Columbus, Ohio	FRANK R. BARNITZ, 130 East 15th St., New York, N. Y.	JAMES P. HANLAN, 80 Park Place, Newark, N. J.
AUGUSTUS D. CURTIS, 235 W. Jackson Blvd., Chicago, Ill.	A. L. POWELL, Fifth and Sussex Sts., Harrison, N. J.	HOWARD LYON, Welsbach Company, Gloucester, N. J.
PRESTON S. MILLAR, 80th St. and East End Ave. New York, N. Y.	W. D'A. RYAN, General Electric Company, Schenectady, N. Y.	H. F. WALLACE, 84 State Street, Boston, Mass

SECTION & CHAPTER OFFICERS

Chicago Section

CHAIRMAN.....F. A. Rogers, Lewis Institute, Madison and Robey Sts., Chicago, Ill.
SECRETARY.........................E. J. Teberg, 72 West Adams Street, Chicago, Ill.

MANAGERS

Albert L. Arenberg.........................316 S. Wells Street, Chicago, Ill.
W. S. Hamm....................319 West Ontario Street, Chicago, Ill.
Norman B. Hickox...............235 West Jackson Boulevard, Chicago, Ill.
Wyllis E. Quivey..................847 West Jackson Boulevard, Chicago, Ill.
E. D. Tillson...........................72 West Adams Street, Chicago, Ill.

New England Section

CHAIRMAN.........................W. V. Baston, 210 South Street, Boston, Mass.
SECRETARY.........................J. Daniels, 39 Boylston Street, Boston, Mass.

MANAGERS

Cyrus Barnes.........................200 Devonshire Street, Boston, Mass.
A. W. Devine...............Room 11, Commonwealth Pier, Boston, 9, Mass.
W. S. Fitch...............Dennison Manufacturing Co., Framingham, Mass.
R. A. Hopkins...............................147 Milk St., Boston, Mass.
J. A. Toohey......................................Box 3396, Boston, Mass.

New York Section

CHAIRMAN............Leonard J. Lewinson, 80th St. & East End Ave., New York, N. Y.
SECRETARY...............J. E. Buckley, Irving Place and 15th St., New York, N. Y.

MANAGERS

S. K. Barrett.........New York University, University Heights, Bronx, N. Y.
H. W. Desaix.........................214 Straight Street, Paterson, N. J.
E. E. Dorting.....................600 West 59th Street, New York, N. Y.
J. R. Fenniman........................130 E. 15th Street, New York, N. Y.
E. H. Hobbie...........................220 Fifth Avenue, New York, N. Y.

Philadelphia Section

CHAIRMAN.........................H. Calvert, 2114 Samsom Street, Philadelphia, Pa.
SECRETARY.........................J. J. Reilly, 1000 Chestnut Street, Philadelphia, Pa.

MANAGERS

H. B. Andersen.........................20 Maplewood Ave., Philadelphia, Pa.
M. C. Huse...........................1000 Chestnut Street, Philadelphia, Pa.
Howard Lyon.....................Welsbach Company, Gloucester, N. J.
G. A. Hoadley............................518 Walnut Lane, Swarthmore, Pa.
E. L. Sholl...................................Woodside Ave., Berwyn, Pa.

Cleveland Chapter

CHAIRMAN...............H. B. Dates, Case School of Applied Science, Cleveland, Ohio
SECRETARY.......................Robert A. Fulton, 75 Public Square, Cleveland, Ohio

San Francisco Bay Cities Chapter

TREASURER.........................J. A. Vandegrift, 1648 16th Street, Oakland, Calif.

Toronto Chapter

CHAIRMAN..................W. H. Woods, 226 Yonge Street, Toronto, Ontario, Canada
SECRETARY-TREASURER....J. T. Scott, 221 Dufferin Street, Toronto, Ontario, Canada

Northern New Jersey Chapter

CHAIRMAN.........................W. T. Blackwell, 80 Park Place, Newark, N. J.
SECRETARY ....H. C. Calahan, 52 Lafayette St., Newark, N. J.
TREASURER.........................George E. Davis, 15 Central Ave., Newark, N. J.

Columbus Chapter

CHAIRMAN.........................G. F. Evans, 162 West Long Street, Columbus, Ohio
SECRETARY.....................R. C. Moore, 146 North Third Street, Columbus, Ohio

Michigan Chapter

CHAIRMAN.........................James M. Ketch, 642 Beaubien St., Detroit, Mich.
SECRETARY-TREASURER.................A. L. Lent, 2000 Second Ave., Detroit, Mich.

Los Angeles Chapter

CHAIRMAN.....................David C. Pence, Illinois Electric Co., Los Angeles, Calif.
SECRETARY-TREASURER, J. F. Anderson, Southern Calif. Edison Co., Los Angeles, Calif.

Pittsburgh Chapter

CHAIRMAN.....................Franklin, W. Loomis, 435 Sixth Avenue, Pittsburgh, Pa.
SECREARY-TREASURER.......Sanford C. Lovett, 1318 Oliver Building, Pittsburgh, Pa.

LOCAL REPRESENTATIVES

COLORADO:	Denver........G. O. Hodgson, Edison Lamp Works.
CONNECTICUT:	New Haven....Charles F. Scott, Yale University.
	J. Arnold Norcross, 80 Crown Street.
	Hartford......R. E. Simpson, Travelers Insurance Co.
DISTRICT OF COLUMBIA:	Washington....R. B. Patterson, 213 14th Street.
	E. C. Crittenden Bureau of Standards.
GEORGIA:	Atlanta........Charles A. Collier, Electric and Gas Building.
IDAHO:	Boise..........W. R. Putnam, Idaho Power Co.
KENTUCKY:	Louisville......H. B. Heyburn, Second and Washington Sts.
LOUISIANA:	New Orleans...Douglas Anderson,
	Tulane University of Louisiana.
MASSACHUSETTS:	Springfield.....J. F. Murray, United Electric Light Co.
	Worcester......Carl D. Knight,
	Worcester Polytechnic Institute.
MINNESOTA:	Minneapolis....G. D. Shepardson, University of Minnesota.
MISSOURI:	St. Louis......Louis D. Moore, 1130 Railway Exchange Bldg.
NEW YORK:	Rochester.....Frank C. Taylor, 166 Shepard St.
	Schenectady...H. E. Mahan, General Electric Co.
OHIO:	Cincinnati.....A. M. Wilson, University of Cincinnati.
	Toledo........W. E. Richards, Toledo Railway and Light Co.
OREGON:	Portland.......F. H. Murphy,
	Portland Railway Light and Power Co.
	Electric Building.
RHODE ISLAND:	Providence.....F. A. Gallagher
	Narragansett Elec. Lighting Co.
TEXAS:	Austin.........J. M. Bryant, University of Texas.
UTAH:	Salt Lake City. L. B. Gawan, 134 Main Street.
VIRGINIA:	Charlottesville W. S. Rodman, University of Virginia.
WASHINGTON:	Seattle........Fred A. Osborn, University of Washington.
WISCONSIN:	Milwaukee.....F. A. Vaughn, Metropolitan Block, Third and
	State Sts.
CANADA:	Montreal......L. V. Webber, 285 Beaver Hall Hill.

OFFICIAL REPRESENTATIVES TO OTHER ORGANIZATIONS

On the United States National Committee of the International
Commission on Illumination................................. { L. B. Marks
Preston S. Millar
G. H. Stinkney

On Governing Board of the American Association for the Advancement
of Science... { Ernest F. Nichols
Clayton H. Sharp

On the Advisory Committee, Engineering Division, National Research
Council... Dugald C. Jackson

On A. E. S. C. Sectional Committee, Building Exits Code................R. E. Simpson

On New York State Committee on Places of Public Assembly.............G. H. Stickney

COMMITTEES

1923-1924

Except as noted below, all committees are appointed by the President, subject to the approval of the Council, and terminate at the time of the first Council meeting of each new administration, in the month of October. The duties of each committee are indicated.

CLARENCE L. LAW, President, Ex-officio member of all Committees

(1) STANDING COMMITTEES AUTHORIZED BY THE CONSTITUTION AND BY-LAWS

COUNCIL EXECUTIVE—*(Consisting of the President, General Secretary, Treasurer and two members of the Council.) Act for the Council between sessions of the latter.*

Clarence L. Law, Chairman
Irving Place and 15th Street
Norman D. Macdonald Preston S. Millar
L. B. Marks. H. F. Wallace

FINANCE.—*(Of three members: to continue until successor is appointed.) Prepare a budget; approve expenditures; manage the finances; and keep the Council informed on the financial condition.*

Adolph Hertz, Chairman.
Irving Place & 15th St., New York, N. Y.
F. R. Barnitz H. F. Wallace

PAPERS.—*(Of at least five members.) Provide the program for the annual convention; pass on papers and communications for publication; and provide papers and speakers for joint sessions with other societies.*

A. L. Powell, Chairman.
Fifth and Sussex Sts., Harrison, N. J.
R. W. Shenton, Vice-Chairman
Nela Park, Cleveland, Ohio

E. C. Crittenden, H. H. Higbie,
Geo. G. Cousins, J. L. Stair,
Julius Daniels, F. C. Taylor,
H. P. Gage, Norman D. Macdonald,
R. H. Maurer,

Chairman of Section and Chapter Papers Committees, Ex-officio Members.

EDITING AND PUBLICATION.—*(Of three members.) Edit papers and discussions; and publish the Transactions.*

Norman D. Macdonald, Chairman.
80th St. and E. End Ave., New York, N. Y.

GENERAL BOARD OF EXAMINERS.—*(Appointed by the President.) Pass upon the eligibility of applicants for membership or for changes in grade of membership.*

L. J. Lewinson. Chairman.
80th St. and East End Ave.,
G. Bertam Regar, H. V. Bozell.

(2) COMMITTEES THAT ARE CUSTOMARILY CONTINUED FROM YEAR TO YEAR.

LIGHTING LEGISLATION.—*Prepare a digest of laws on Illumination; cooperate with other bodies in promoting wise legislation on illumination; and prepare codes of lighting in certain special fields. To function also as a Technical Committee on Industrial Lighting.*

L. B. Marks, Chairman.
103 Park Avenue, New York, N. Y.
W. F. Little, Secretary.
80th St. and E. End Ave., New York, N. Y.

W. T. Blackwell, Clarence L. Law,
F. C. Caldwell, M. G. Llyod,
Geo. S. Crampton, M. Luckiesh,
H. B. Dates, E. G. Perrot,
E. Y. Davidson, Jr., R. H. Maurer,
J. E. Hannum, A. S. McAllister,
Ward Harrison, W. J. Serrill,
S. G. Hibben, R. E. Simpson,
J. A. Hoeveler, G. H. Stickney,
James E. Ives, R. C. Taggart.

MEMBERSHIP.—*To obtain additional individual memberships.*

G. Bertram Regar, Chairman.
1000 Chestnut St., Philadelphia, Pa.,
C. E. Addie F. W. Bliss,
J. F. Anderson, R. I. Brown,
C. A. Atherton, J. E. Buckley
W. T. Blackwell, W. E. Clement.

IV

George G. Cousins,
E. C. Crittenden,
J. Daniels,
Henry D'Oliver, Jr.
J. Carl Fisher,
H. W. Fuller,
H. E. Hobson,
Merritt C. Huse,
Karl E. Kilby,
James D. Lee, Jr.,
F. W. Loomis,
H. E. Mahan,
C. C. Munroe,
F. H. Murphy,
J. A. O'Rourke,
S. L. E. Rose,
H. H. Smith,
J. L. Stair,
L. A. S. Wood.

MOTOR VEHICLE LIGHTING.

C. H. Sharp, Chairman.
80th St. and E. End Ave., New York, N. Y.

G. H. Stickney, Vice-Chairman.
Fifth and Sussex Sts., Harrison, N. J.

W. F. Little, Secretary
80th St. and East End Ave., New York, N. Y.

F. C. Caldwell,
A. W. Devine,
C. E. Godley,
C. A. B. Halvorson,
J. A. Hoeveler,
W. A. McKay,
A. L. McMurtry,
H. H. Magdsick,
L. C. Porter,

NOMENCLATURE AND STANDARDS—

Define the terms and standards of Illumination; and endeavor to obtain uniformity in nomenclature.

E. C. Crittenden, Chairman.
Bureau of Standards, Washington, D. C.

Howard Lyon, Secretary.
Welsbach Co., Gloucester, N. J.

G. A. Hoadley,
E. P. Hyde,
A. E. Kennelly,
M. Luckiesh,
C. O. Mailloux,
A. S. McAllister,
E. C. McKinnie,*
C. H. Sharp,
W. J. Serrill,
G. H. Stickney.

PROGRESS.—*Submit to the annual convention a report on the progress of the year in the science and art of Illumination.*

F. E. Cady, Chairman.
Nela Park, Cleveland, O.

RESEARCH.—*Stimulate research in the field of illumination; and keep informed of the progress of research.*

E. F. Nichols, Chairman.
Nela Park, Cleveland, O.

(3) TEMPORARY COMMITTEES FOR SPECIAL PURPOSES.

SKY BRIGHTNESS.
H. H. Kimball, Chairman.
U. S. Weather Bureau, Washington, D. C.

E. C. Crittenden,
E. H. Hobbie,
James E. Ives,
Bassett Jones,
W. F. Little,
M. Luckiesh,
L. B. Marks,
I. G. Priest.*

COMMITTEE TO COOPERATE WITH FIXTURE MANUFACTURERS

To Prepare Code of Luminaire Design
M. Luckiesh, Chairman,
Nela Park, Cleveland, O.

COMMITTEE ON NEW SECTIONS AND CHAPTERS

D. McFarlan Moore, Chairman.
Edison Lamp Works, Harrison, N. J.

COMMITTEE ON SUSTAINING MEMBERS

H. F. Wallace, Chairman.
84 State St., Boston, Mass.

W. T. Blackwell   E. J. Teberg.

*Not members of the Society but cooperating in the work of these committees.

SUSTAINING MEMBERS

Alexalite Company, New York, N. Y.
  Official Representative, Herman Plaut.
American Gas & Electric Company, New York, N. Y.
  Official Representative, George N. Tidd.
American Optical Company, Southbridge, Mass.
  Official Representative, E. D. Tillyer.
Amherst Gas Company, Amherst, Mass.
  Official Representative, J. J. O'Connell.
Bangor Railway & Electric Company, Bangor, Maine.
  Official Representative, W. E. Hooper.
Bausch & Lomb Optical Company, Rochester, N. Y.
  Official Representative, Max Poser.
Beardslee Chandelier Mfg. Company, Chicago, Ill.
  Official Representative, F. Lee Farmer.
Benjamin Electric Mfg. Company, Chicago, Ill.
  Official Representative, Wyllis E. Quivey.
Boss Electrical Supply Company, Providence, R. I.
  Official Representative, F. A. Boss.
Boston Consolidated Gas Company, Boston, Mass.
  Official Representative, E. N. Wrightington.
Brooklyn Edison Company, Inc., Brooklyn, N. Y.
  Official Representative, W. F. Wells.
Brooklyn Union Gas Company, Brooklyn, N. Y.
  Official Representative, Arthur F. Staniford.
Cambridge Electric Light Company, Cambridge, Mass.
  Official Representative, R. M. Miller.
Central Electric Company, Chicago, Ill.
  Official Representative, A. L. Arenberg.
Central Hudson Gas & Electric Co., Poughkeepsie, N. Y.
  Official Representative, T. M. Barr.

Commonwealth Edison Company, Chicago, Ill.
Official Representative, Oliver R. Hogue.

Consolidated Electric Lamp Company, Danvers, Mass.
Official Representative, Frank W. Marsh.

Consolidated Gas E'ectric Light & Power Company of Baltimore.,
Baltimore, Md.
Official Representative, Wm. Schmidt, Jr., Secy.

Consolidated Gas Company of N. Y., New York, N. Y.
Official Representative, R. H. Maurer.

Consumers Power Company, New York, N. Y.
Official Representative, B. C. Cobb.

Cooper-Hewitt Electric Company, Hoboken, N. J.
Official Representative, W. A. D. Evans.

Corning Glass Works, Corning, N. Y.
Official Representative, E. C. Sullivan.

Cox, Nostrand & Gunnison, Brooklyn, N. Y.
Official Representative, Edward L. Cox.

Dawes Brothers, Chicago, Ill.
Official Representative, Rufus C. Dawes.

Detroit Edison Company, Detroit, Mich.
Official Representative, Charles C. Munroe.

Duquesne Light Company, Pittsburgh, Penn.
Official Representative, Joseph McKinley.

Edison Electric Illuminating Company of Boston, Boston, Mass.
Official Representative, C. L. Edgar.

Edison Lamp Works of Gen. Elec. Co., Harrison, N. J.
Official Representative, A. D. Page.

Electrical Testing Laboratories, New York, N. Y.
Official Representative, Preston S. Millar.

Fall River Electric Light Company, Fall River, Mass.
Official Representative, R. F. Whitney.

Fitchburg Gas & Electric Light Co., Fitchburg, Mass.
Official Representative, Frank S. Clifford.

Franklin Electric Light Company, Turners Falls, Mass.
  Official Representative, C. E. Bankwitz.

I. P. Frink, Inc., New York, N. Y.
  Official Representative, Wm. H. Spencer.

Gas & Electric Improvement Co., Boston, Mass.
  Official Representative, Philip B. Jameson.

Georgia Railway & Power Company, Atlanta, Ga.
  Official Representative, Charles A. Collier.

Gillinder Brothers, Port Jervis, N. Y.
  Official Representative, Edwin B. Gillinder.

Gillinder & Sons, Inc., Philadelphia, Penn.
  Official Representative, Edgar A. Gillinder.

Gleason-Tiebout Glass Company, Brooklyn, N. Y.
  Official Representative, Marshal Tiebout Gleason.

Gloucester Electric Company, Gloucester, Mass.
  Official Representative, Walter L. Brown, Jr.

Greenfield Electric Light & Power Co., Greenfield, Mass.
  Official Representative, H. E. Duren.

Edwin F. Guth Company, St. Louis, Mo.
  Official Representative, Walter Wippern.

Hartford Electric Light Company, Hartford, Conn.
  Official Representative, W. D. Gorman.

Haverhill Electric Company, Haverhill, Mass.
  Official Representative, G. W. Hurn.

Holophane Glass Company, Inc., New York, N. Y.
  Official Representative, Charles Franck.

Horn & Brannen Mfg. Co., Inc., Philadelphia, Penn.
  Official Representative, William Horn.

Hygrade Lamp Company, Salem, Mass.
  Official Representative, E. J. Poor.

Hydro-Electric Power Commission of Ontario, Toronto, Ontario, Canada.
  Official Representative, F. A. Gaby.

Irving & Casson—A. H. Davenport Co., Boston, Mass.
  Official Representative, David Crownfield.

Kristiania Elektrecitetsvork, Kristiania, Norway.
  Official Representative, Ragnvald Steen.

E. B. Latham & Company, New York, N. Y.
  Official Representative, Harry Heckert.

Lawrence Gas Company, Lawrence, Mass.
  Official Representative, Fred H. Sargent.

Leeds & Northrup Company, Philadelphia, Penn.
  Official Representative, I. M. Stein.

Lightolier Company, New York, N. Y.
  Official Representative, Moses D. Blitzer.

Lundin Electric & Machine Company, Boston, Mass.
  Official Representative, Emil O. Lundin.

Lynn Gas & Electric Company, Lynn, Mass.
  Official Representative, H. K. Morrison.

Macbeth-Evans Glass Company, Pittsburgh, Penn.
  Official Representative, Howard S. Evans.

Malden Electric Company, Malden, Mass.
  Official Representative, J. V. Day.

Malden & Melrose Gas Light Co., Malden, Mass.
  Official Representative, Harry Walton.

McGraw-Hill Publishing Company, New York, N. Y.
  Official Representative, F. M. Feiker.

Millville Electric Light Company, Millville, N. J.
  Official Representative, Wm. C. Buell.

Montpelier & Barre Light & Power Co., Montpelier, Vt.
  Official Representative, C. J. Cookson.

Narragansett Electric Lighting Co., Providence, R. I.
  Official Representative, Francis A. Gallagher, Jr.

National Lamp Works of General Electric Co., Cleveland, Ohio.
  Official Representative, W. M. Skiff.

National X-Ray Reflector Company, Chicago, Ill.
 Official Representative, Augustus D. Curtis.

New Bedord Gas & Edison Light Co., New Bedford, Mass.
 Official Representative, Charles R. Price.

Newburyport Gas & Electric Company, Newburyport, Mass.
 Official Representative, J. Lee Potter.

New England Power Company, Worcester, Mass.
 Official Representative, Hugo Rocktaschel.

New Haven Gas Light Company, New Haven, Conn.
 Official Representative, J. Arnold Norcross.

New York Edison Company, New York, N. Y.
 Official Representative, Clarence L. Law.

N. Y. & Queens Elec. Lt. & Power Co., Long Island City, N. Y.
 Official Representative, Ray Palmer.

Pennsylvania Power & Light Co., Allentown, Penn.
 Official Representative, J. S. Wise.

Peoples Gas Light & Coke Company, Chicago, Ill.
 Official Representative, Charles A. Luther.

Pettingell-Andrews Company, Boston, Mass.
 Official Representative, Frank S. Price.

Philadelphia Electric Company, Philadelphia, Penn.
 Official Representative, Joseph B. McCall.

Philadelphia Electrical & Mfg. Co., Philadelphia, Penn.
 Official Representative, C. L. Bundy.

Pittsfield Electric Company, Pittsfield, Mass.
 Official Representative, E. P. Dittman.

L. Plaut & Company, New York, N. Y.
 Official Representative, Herman Plaut.

Portland Railway Light & Power Co., Portland, Oregon.
 Official Representative, Francis H. Murphy.

Potomac Electric Power Company, Washington, D. C.
 Official Representative, H. A. Brooks.

Providence Gas Company, Providence, R. I.
  Official Representative, Charles H. Manchester.
Public Service Corp. of N. J., Newark, N. J.
  Official Representative, P. S. Young.
Public Service Co. of Northern Ill., Chicago, Ill.
  Official Representative, Frank J. Baker.
Rochester Gas & Electric Corp., Rochester, N. Y.
  Official Representative, Frank C. Taylor.
Salem Electric Lighting Company, Salem, Mass.
  Official Representative, S. Fred Smith.
Max Schaffer Company, New York, N. Y.
  Official Representative, Max Schaffer.
Shapiro & Aronson, Inc., New York, N. Y.
  Official Representative, N. W. Belmuth.
Slocum & Kilburn, New Bedford, Mass.
  Official Representative, A. H. Smith.
Societa Idroelettrica Piemonte, Turin, Italy.
  Official Representative, Gian Giacomo Ponti.
Southern Electric Company, Baltimore, Md.
  Official Representative, J. G. Johannesen.
Sterling Bronze Company, New York, N. Y.
  Official Representative, Charles F. Kinsman.
Stone & Webster, Boston, Mass.
  Official Representative, Wm. H. Blood.
Suburban Gas & Electric Company, Revere, Mass.
  Official Representative, C. F. Chisholm.
Charles H. Tenney & Company, Boston, Mass.
  Official Representative, Cyrus Barnes.
Union Electrical Supply Company, Providence, R. I.
  Official Representative, Frank L. Falk.
Union Gas & Electric Co., Cincinnati, Ohio.
  Official Representative, W. W. Freeman.

Union Light & Power Company, Franklin, Mass.
 Official Representative, E. S. Hamblem.

United Electric Light Company, Springfield, Mass.
 Official Representative, J. Frank Murray.

United Electric Light & Power Co., New York, N. Y.
 Official Representative, Frank W. Smith.

The United Gas Improvement Co., Philadelphia, Penn.
 Official Representative, Wm. J. Serrill.

Alfred Vester & Sons, Inc., Providence, R. I.
 Official Representative, J. Wm. Schulze.

Welsbach Company, Gloucester City, N. J.
 Official Representative, Townsend Stites.

Welsbach Street Lighting Co. of America, Philadelphia, Penn.
 Official Representative, Arthur E. Shaw.

Western Electric Co., New York, N. Y.
 Official Representative, E. J. Dailey, Jr.

Westinghouse Electric & Mfg. Co., East Pittsburgh, Penn.
 Official Representative, George W. Sawin.

Westinghouse Lamp Company, Bloomfield, N. J.
 Official Representative, Samuel G. Hibben.

Wetmore-Savage Company, Boston, Mass.
 Official Representative, R. M. Topham.

Worcester Electric Light Company, Worcester, Mass.
 Official Representative, F. H. Smith.

Worcester Suburban Electric Co., Palmer, Mass.
 Official Representative, H. M. Parsons.

OLD TRANSACTIONS WANTED

Vol. XV, 1920—Nos. 1, 4, 5, 7, 8 and 9
Vol. XVI, 1921—Nos. 2, 3, 7, 8 and 9
Vol. XVII, 1922—Nos. 5, 8, 9 and 10
Vol. XVIII, 1923—Nos. 1, 3, and 4

Advise the General Office of Issues for sale and prices will be quoted.

OFFPRINT ORDERS

In order to minimize the expense of holding type, the printer has been instructed to throw the pages of the TRANSACTIONS shortly after publication. Authors wishing to secure offprints at the minimum rate should notify the General Office in advance. For offprints ordered after publication in the TRANSACTIONS a a higher fee must be charged.

COMMITTEE ON EDITING AND PUBLICATION

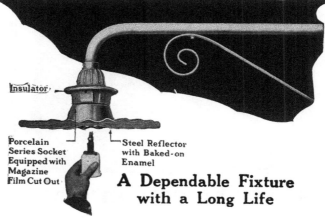

Insulator

Porcelain
Series Socket
Equipped with
Magazine
Film Cut Out

Steel Reflector
with Baked-on
Enamel

A Dependable Fixture with a Long Life

All G-E Novalux fixtures are adapted for bracket or center span mounting and are equipped with the famous G-E 25,000-volt porcelain insulator.

This insulator protects the equipment against grounding which extinguishes lamps (increasing maintenance expense) and against jumping of current on high voltage circuits in case of surge. Necessary openings in the insulator are so covered as to prevent moisture [or dirt from interfering with the circuit.

Another feature of G-E Novalux fixtures is the G-E Series Socket which has a substantial center contact of sturdy phosphor bronze springs. A locking wire effectually prevents lamps from loosening. All G-E Series Sockets are adapted to the use of the convenient G-E Magazine Film Cutout—15 fresh dielectric surfaces in a single compact cartridge.

General Electric
Company
General Office
Schenectady
New York

Street lighting specialists at all G-E district offices have detailed information on various types of G-E street lighting equipment.

35C-119D

GENERAL ELECTRIC

TRANSACTIONS
OF THE
ILLUMINATING ENGINEERING SOCIETY
OFFICE OF PUBLICATION: 125 W. STATE ST., ITHACA, N. Y.

Published monthly except June and August under the direction of the Committee on Editing and Publication.

DATES OF PUBLICATION

No. 1, January   No. 2, February   No. 3, March
No. 4, April   No. 5, May   No. 6, July
No. 7, September   No 8, October   No. 9, November
                     No. 10, December

SUBSCRIPTION, $7.50 PER ANNUM   SINGLE COPIES $1.00
FOREIGN SUBSCRIPTION $8.00 PER ANNUM

ADVERTISING RATES MAY BE HAD UPON APPLICATION
TO THE GENERAL OFFICES OF THE SOCIETY
29 WEST 39TH STREET, NEW YORK, N.Y.

SUPPLEMENT TO THE 1923 VOLUME

TRANSACTIONS
OF THE
ILLUMINATING ENGINEERING SOCIETY

LIST
of
MEMBERS

Corrected to JULY 31, 1923

GEOGRAPHICAL DISTRICTS

Geographical districts of the United States showing Section and Chapter boundaries. New England Section: Maine, New Hampshire, Vermont, Massachusetts and Rhode Island; Headquarters Boston, Mass. New York Section: Connecticut, New York and Northern New Jersey north of latitude 40° 20′; Headquarters, New York City. Philadelphia Section: Southern New Jersey, Delaware, Maryland, District of Columbia and Eastern Pennsylvania; Headquarters, Philadelphia, Pa. Chicago Section: Illinois, Indiana, Michigan and Wisconsin; Headquarters, Chicago, Ill. Cleveland Chapter: Cleveland, Ohio, and vicinity. Columbus Chapter: Columbus, Ohio, and vicinity. Michigan Chapter: Detroit, Mich., and vicinity. Northern New Jersey Chapter: New Jersey north of Trenton, N. J.; Headquarters, Newark, N. J. Toronto Chapter: City of Toronto and vicinity. Los Angeles Chapter: Southern California; Headquarters, Los Angeles Calif. San Francisco Bay Cities Chapter: Northern California, Oregon and Washington; Headquarters, San Francisco, Calif. Non-Section territory as indicated.

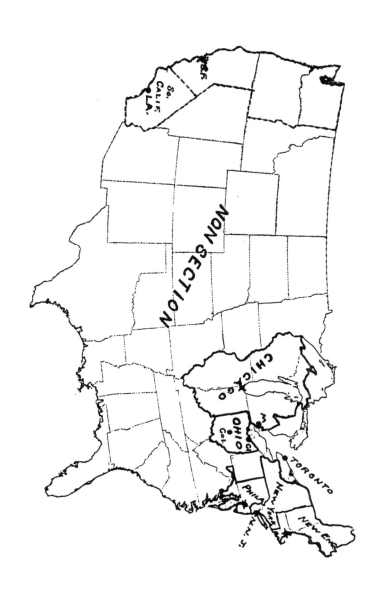

EXPLANATORY NOTE

The list of Members of the Illuminating Engineering Society contains the following groups, arranged alphabetically:

1. Sustaining Members, with name of Official Representative.

2. Honorary Members.

3. Members of all Grades.

4. Members of Sections and Chapters.

The Member's name, business title and affiliation, and address is given.

The Grade of Membership is indicated by (M) for Member, (A) for Associate Member and (S) for Student Member.

Members who have served the Society as officers of the Council, have the official position and duration of term of office indicated in italics.

Charter Members of the Society are indicated by a star (*) following their names.

SUSTAINING MEMBERS AND THEIR
OFFICIAL REPRESENTATIVES

Alexalite Company
432 East 23d Street
New York, N. Y.
  Official Representative
  Herman Plaut, Secy.

American Gas & Electric Company
30 Church Street
New York, N. Y.
  Official Representative
  George N. Tidd

American Optical Company
Southbridge, Mass.
  Official Representative
  E. D. Tillyer

Amherst Gas Company
11 Pleasant Street
Amherst, Mass.
  Official Representative
  J. J. O'Connell

Bangor Railway & Electric Company
Bangor, Maine.
  Official Representative
  W. E. Stooper

Bausch & Lomb Optical Company
635 St. Paul Street
Rochester, N. Y.
  Official Representative
  Max Poser

Beardslee Chandelier Mfg. Company
216-220 S. Jefferson Street
Chicago, Ill.
  Official Representative
  F. Lee Farmer

Benjamin Electric Mfg. Company
120 S. Sangamon Street
Chicago, Ill.
  Official Representative
  Wyllis E. Quivey

Boss Electrical Supply Company
11 Peck Street
Providence, R. I.
  Official Representative
  F. A. Boss

Boston Consolidated Gas Company
24 West Street
Boston, Mass.
  Official Representative
  E. N. Wrightington

Brooklyn Edison Company, Inc.
360 Pearl Street
Brooklyn, N. Y.
  Official Representative
  W. F. Wells

Brooklyn Union Gas Company
176 Remsen Street
Brooklyn, N. Y.
  Official Representative
  Arthur F. Staniford

Cambridge Electric Light Company
46 Blackstone Street
Cambridge, Mass.
  Official Representative
  R. M. Miller

Central Electric Company
316 S. Wells Street
Chicago, Ill.
  Official Representative
  A. L. Arenberg

Central Hudson Gas & Eelectric Co.
50 Market St.
Poughkeepsie N. Y.
  Official Representative
  T. M. Barr

Commonwealth Edison Company
28 N. Market Street
Chicago, Ill.
  Official Representative
  Oliver R. Hogue

Consolidated Electric Lamp Company
128 Maple Street
Danvers, Mass.
  Official Representative
  Frank W. Marsh

Consolidated Gas Electric Light &
Power Company of Baltimore
100 W. Lexington Street
Baltimore, Md.
  Official Representative
  Wm. Schmidt, Jr., Secy.

Consolidated Gas Company of N. Y.
130 East 15th Street
New York, N. Y.
  Official Representative
  R. H. Maurer

Consumers Power Company
14 Wall Street
New York, N. Y.
  Official Representative
  B. C. Cobb

Cooper-Hewitt Electric Company
95 River Street
Hoboken, N. J.
Official Representative
W. A. D. Evans

Corning Glass Works
Corning, N. Y.
Official Representative
E. C. Sullivan

Cox, Nostrand & Gunnison
335-337 Adams Street
Brooklyn, N. Y.
Official Representative
Edward L. Cox, Pres.

Dawes Brothers
1615 Harris Trust Bldg.
111 West Monroe Street
Chicago, Ill.
Official Representative
Rufus C. Dawes

Detroit Edison Company
Detroit, Mich.
Official Representative
Charles C. Munroe

Duquesne Light Company
435 Sixth Avenue
Pittsburgh, Penn.
Official Representative
Joseph McKinley

East Side Metal Spinning Co.
451 Greenwich Street
New York, N. Y.
Official Representative
Arthur O. Burger

Edison Electric Illuminating Company
of Boston
39 Boylston Street
Boston, Mass.
Official Representative
C. L. Edgar

Edison Lamp Works of Gen. Elec. Co.
Harrison, N. J.
Official Representative
A. D. Page

Electric Outlet Company, Inc.
8 West 40th Street
New York, N. Y.
Official Representative
E. Cantelo White

Electrical Testing Laboratories, Inc.
80th & East End Avenue
New York, N. Y.
Official Representative
Preston S. Millar

Fall River Electric Light Company
85 N. Main Street
Fall River, Mass.
Official Representative
R. F. Whitney

Fitchburg Gas & Electric Light Co.
537 Main Street
Fitchburg, Mass.
Official Representative
Frank S. Clifford

Franklin Electric Light Company
Avenue A
Turners Falls, Mass.
Official Representative
C. E. Bankwitz

Oscar O. Friedlaender, Inc.
40 Murray Street
New York, N. Y.
Official Representative
William M. Friedlaender

I. P. Frink, Inc.
24th St. & 10th Ave.
New Yo'k, N. Y.
Official Representative
Wm. H. Spencer

Gas & Electric Improvement Co.
77 Franklin Street
Boston, Mass.
Official Representative
Philip B. Jameson

Georgia Railway & Power Company
Electric & Gas Building
Atlanta, Ga.
Official Representative
Charles A. Collier

Gillinder Brothers
Port Jervis, N. Y.
Official Representative
Edwin B. Gillinder

Gillinder & Sons, Inc.
Tacony & Devereaux Streets
Philadelphia, Penn.
Official Representative
Edgar A. Gillinder

Gleason-Tiebout Glass Company
99 Commercial Street
Brooklyn, N. Y.
Official Representative
Marshall Tiebout Gleason

Gloucester Electric Company
26 Vincent Street
Gloucester, Mass.
Official Representative
Walter L. Brown, Jr.

Greenfield Electric Light & Power Co.
41 Federal Street
Greenfield, Mass.
Official Representative
H. E. Duren

Edwin F. Guth Company
2615 Washington Avenue
St. Louis, Mo.
Official Representative
Walter Wippern

Hartford Electric Light Company
266 Pearl Street
Hartford, Conn.
Official Representative
W. D. Gorman

Haverhill Electric Company
121 Merrimack Street
Haverhill, Mass.
Official Representative
G. W. Hurn

Holophane Glass Company, Inc.
340 Madison Avenue
New York, N. Y.
Official Representative
Charles Franck

Horn & Brannen Mfg. Co., Inc.
427 N. Broad Street
Philadelphia, Penn.
Official Representative
William Horn, Pres.

Hygrade Lamp Company
60 Boston Street
Salem, Mass.
Official Representative
E. J. Poor

Hydro-Electric Power Commission of Ontario
190 University Avenue
Toronto, Ontario, Canada
Official Representative
F. A. Gaby

Irving & Casson-A. H. Davenport Co.
573 Boylston Street
Boston, Mass.
Official Representative
David Crownfield

Krich Light & Electric Company
306 Market Street
Newark, N. J.
Official Representative
Max H. Krich, Manager

Kristiania Elektrecitetsvork
Kristiania, Norway
Official Representative
Ragnvald Steen

E. B. Latham & Company
550 Pearl Street
New York, N. Y.
Official Representative
Harry Heckert

Lawrence Gas Company
370 Essex Street
Lawrence, Mass.
Official Representative
Fred H. Sargent

Leeds & Northrup Company
4901 Stenton Avenue
Philadelphia, Penn.
Official Representative
I. M. Stein

Lightolier Company
569 Broadway
New York, N. Y.
Official Representative
Moses D. Blitzer, Vice-Pres.

Lundin Electric & Machine Company
10 Thacher Street
Boston, Mass.
Official Representative
Emil O. Lundin

Lynn Gas & Electric Company
90 Exchange Street
Lynn, Mass.
Official Representative
H. K. Morrison

Macbeth-Evans Glass Company
Wabash Building
Pittsburgh, Penn.
Official Representative
Howard S. Evans

Malden Electric Company
139 Pleasant Street
Malden, Mass.
Official Representative
J. V. Day

Malden & Melrose Gas Light Co.
137 Pleasant Street
Malden, Mass.
Official Representative
Harry Walton

McGraw-Hill Publishing Company
36th Street & 10th Avenue
New York, N. Y.
Official Representative
F. M. Feiker, Editor

Millville Electric Light Company
233 High Street
Millville, N. J.
Official Representative
Wm. C. Buell

Montpelier & Barre Light & Power Co.
20 Langdon Street
Montpelier, Vt.
Official Representative
C. J. Cookson

Narragansett Electric Lighting Co.
Turks Head Bldg.
Providence, R. I.
Official Representative
Francis A. Gallagher, Jr.

National Lamp Works of Gen. Elec. Co.
Nela Park
Cleveland, Ohio
Official Representative
W. M. Skiff

National X-Ray Reflector Company
235 West Jackson Blvd.
Chicago, Ill.
Official Representative
Augustus D. Curtis

New Bedford Gas & Edison Light Co.
New Bedford, Mass.
Official Representative
Charles R. Price

Newburyport Gas & Electric Company
Newburyport, Mass.
Official Representative
J. Lee Potter

New England Power Company
35 Harvard Street
Worcester, Mass.
Official Representative
Hugo Rocktaschel

New Haven Gas Light Company
80 Crown Street
New Haven, Conn.
Official Representative
J. Arnold Norcross

New York Edison Company
Irving Place & 15th Street
New York, N. Y.
Official Representative
Clarence L. Law

N. Y. & Queens Elec. Lt. & Power Co.
444 Jackson Avenue
Long Island City, N. Y.
Official Representative
Ray Palmer

Pennsylvania Power & Light Co.
Allentown, Penn.
Official Representative
J. S. Wise

Peoples Gas Light & Coke Company
Peoples Gas Building
Chicago, Ill.
Official Representative
Charles A. Luther

Pettingell-Andrews Company
160 Pearl Street
Boston, Mass.
Official Representative
Frank S. Price

Philadelphia Electric Company
1000 Chestnut Street
Philadelphia, Penn.
Official Representative
Joseph B. McCall

Philadelphia Electrical & Mfg. Co.
1228 North 31st Street
Philadelphia, Penn.
Official Representative
C. L. Bundy

Pittsfield Electric Company
Pittsfield, Mass.
Official Representative
E. P. Dittman

L. Plaut & Company
432 East 23d Street
New York, N. Y.
Official Representative
Herman Plaut

Portland Railway Light & Power Co.
Electrical Bldg.
Portland, Oregon
Official Representative
Francis H. Murphy

Potomac Electric Power Company
14th & C Streets N. W.
Washington, D. C.
Official Representative
H. A. Brooks

Providence Gas Company
Providence, R. I.
Official Representative
Charles H. Manchester

Public Service Corp. of N. J.
Broad & Banks Streets
Newark, N. J.
Official Representative
P. S. Young

Public Service Co. of Northern Ill.
72 W. Adams Street
Chicago, Ill.
Official Representative
Frank J. Baker

Rochester Gas & Electric Corp.
34 Clinton Avenue North
Rochester, N. Y.
Official Representative
Frank C. Taylor, Elec. Engr.

Salem Electric Lighting Company
205 Washington Street
Salem, Mass.
Official Representative
S. Fred Smith

Max Schaffer Comapny
31-35 West 15th Street
New York, N. Y.
Official Representative
Max Schaffer

Shapiro & Aronson, Inc.
20 Warren Street
New York, N. Y.
Official Representative
N. W. Belmuth

Slocum & Kilburn
23 N. Water Street
New Bedford, Mass.
Official Representative
A. H. Smith

Societa Idroelettrica Piemonte
Turin, Italy
Official Representative
Gian Giacomo Ponti, Mgr. Dir.

Southern Electric Company
Baltimore, Md.
Official Representative
J. G. Johannesen, Manager

Sterling Bronze Company
201 East 12th Street
New York, N. Y.
Official Representative
Charles F. Kinsman

Stone & Webster
147 Milk Street
Boston, Mass.
Official Representative
Wm. H. Blood

Suburban Gas & Electric Company
150 Beach Street
Revere, Mass.
Official Representative
C. F. Chisholm

Charles H. Tenney & Company
200 Devonshire Street
Boston, Mass.
Official Representative
Cyrus Barnes

Tungsten Wire Works, A. B.
98 Ringvagen
Stockholm, Sweden
Official Representative
J. Gustaf V. Lang

Union Electrical Supply Company
Providence, R. I.
Official Representative
Frank L. Falk

Union Gas & Electric Co.
4th & Plum Streets
Cincinnati, Ohio
Official Representative
W. W. Freeman

Union Light & Power Company
Franklin, Mass.
Official Representative
E. S. Hamblem

United Electric Light Company
73 State Street
Springfield, Mass.
Official Representative
J. Frank Murray

United Electric Light & Power Co.
130 East 15th Street
New York, N. Y.
Official Representative
Frank W. Smith

The United Gas Improvement Co.
Broad & Arch Streets
Philadelphia, Penn.
Official Representative
Wm. J. Serrill

Alfred Vester & Sons, Inc.
5 Mason Street
Providence, R. I.
Official Representative
J. Wm. Schulze

Welsbach Company
Gloucester, N. J.
   Official Representative
   Townsend Stites

Welsbach Street Lighting Co. of America
1934 Market Street
Philadelphia, Penn.
   Official Representative
   Arthur E. Shaw

Western Electric Co.
195 Broadway
New York, N. Y.
   Official Representative
   E. J. Dailey, Jr.

Westinghouse Electric & Mfg. Co.
East Pittsburgh, Penn.
   Official Representative
   George W. Sawin

Westinghouse Lamp Company
Bloomfield, N. J.
   Official Representative
   Samuel G. Hibben

Wetmore-Savage Company
78 Pearl Street
Boston, Mass.
   Official Representative
   R. M. Topham

Worcester Electric Light Company
11 Foster Street
Worcester, Mass.
   Official Representative
   F. H. Smith

Worcester Suburban Electric Co.
Palmer, Mass.
   Official Representative
   H. M. Parsons

HONORARY MEMBERS

Edison, Thomas A.
   Inventor, Orange, N. J.

Nichols, Edward L.*
   Professor of Physics, Emeritus, Cornell University, 5 South Avenue, Ithaca, N. Y.

MEMBERS OF ALL GRADES

Abbott, Arthur L.   **M**
Commonwealth Electric Co., 182 E. 6th St., St. Paul, Minn.

Abell, H. C.   **M**
Electric Bond and Share Co., 71 Broadway, New York, N. Y.

Acheson, Albert R.   **M**
Consulting Engineer, Bureau of Gas & Electricity, Syracuse, N. Y.

Addicks, W. R.*   **M**
Vice-President, Consolidated Gas Co. of N. Y., 130 E. 15th St., New York, N. Y.

Addie, Charles E.   **A**
Supt., Commercial Lighting Division, Denver Gas & Elec. Light Co., 900 15th St., Denver, Colo.

Africa, Walter G.   **A**
Treasurer & Manager, Manchester Gas Co., 47-49 Manchester St., Manchester, N. H.

Ainsworth, George   **A**
Mfg. & Designer of Luminaires, 101 Park Ave., New York, N. Y.

Alden, Walter A.   **A**
Interior Lighting Engineer, Westinghouse Elec. & Mfg. Co., 420 S. San Pedro St., Los Angeles, Cal.

Alexander, George L.   **A**
Outdoor Illumination Specialist, International General Electric Co., Schenectady, N. Y.

Alexander, Lowell M.   **A**
Asst. Prof. of Physics, University of Cincinnati, Burnet Woods, Cincinnati, Ohio.

Alger, Ellice M., M. D.   **A**
Director, 1914-17.
Ophthalmologist, New York Post Graduate Medical School & Hospital, 303 E. 20th St., New York, N. Y.

Allen, Asa E.   **A**
Engineer, Southwest General Electric Co., 516 Interurban Bldg., Dallas, Tex.

Allen, H. V.*   **M**
Sales Engineer, Ridgewood Dynamo & Engine Co., 350 Madison Ave., New York, N. Y.

Allen, J. H.   **A**
General Electric Co., Monadnock Bldg., Chicago, Ill.

Allen, Walter C.   **M**
Executive Secretary, Public Utilities Commission, D. C., District Bldg., Washington, D. C.

Allis, C. L.   **A**
Wooster Electric Co., Public Square, Wooster, Ohio.

Alhouse, Daniel M.   **A**
Chief Electrician, Strawbridge & Clothier, 8th & Market Sts., Philadelphia, Pa.

Andersen, H. B.   **A**
United Gas Improvement Co., 1401 Arch St., Philadelphia, Pa.

Andersen, L. W.   **A**
President, Waterbury Metal Wares Co., P. O. Box 228, Waterbury, Conn.

Anderson, Douglas S.   **A**
Professor of Electrical Engineering, Tulane University of Louisiana, New Orleans, La.

Anderson, Earl A.   **M**
Illuminating Engineer, Engineering Dept., National Lamp Works of G. E. Co., Cleveland, Ohio.

Anderson, George R.   **A**
Professor of Engineering Physics, University of Toronto, Toronto, Ont., Canada.

Anderson J. F.   **M**
Illuminating Engineer, Southern Calif. Edison Co., 3d & Broadway, Los Angeles, Calif.

Anderson, Oscar P.   **A**
Commercial Engineer, Edison Lamp Works, G. E. Co., Harrison, N. J.

Andrews, Edgar M.   **A**
Commercial & Industrial Lighting, 15 N. 12th St., Richmond, Va.

Andrews, William S.   **M**
Consulting Engineer, Engineering Dept., General Electric Co., Schenectady, N. Y.

Antrim, William D.   **A**
Assistant Engineer, Welsbach Co., Gloucester, N. J.

Arbuckle, Samuel F.   **M**
Vice-President, General Manager Monogram Lens Corp., 2467 Cass Blvd., Detroit, Mich.

Arenberg, Albert L.   **M**
Sales Engineer, Central Electric Co., 316 S. Wells St., Chicago, Ill.

Arnold, M. Edwin   A
Electrical Contractor, M. E. Arnold
& Co., 1019 Cherry St., Philadel-
phia, Pa.

Aschner, L.   A
United Incandescent Lamps & Electrical Co., Ltd., Ujpest, near Budapest, Hungary.

Ashinger, H. H.   A
Sales Engineer, Westinghouse Elec.
& Mfg. Co., Notre Dame &
Division Sts., South Bend, Ind.

Ashley, Edward E., Jr.   M
Mechanical & Electrical Engineer,
Starrett & Van Vleck, 8 W. 40th
St., New York, N. Y.

Ashworth, James   A
Supt. of Distribution, Public Service
Gas Co., St. Pauls & James Aves.,
Jersey City, N. J.

Ashworth, R. H.
Asst. to General Manager, Utah
Power & Light Co., 519 Kearns
Bldg., Salt Lake City, Utah.

Atherton, Carlyle A.   A
Engineering Dept., National Lamp
Works of G. E. Co., Nela Park,
Cleveland, Ohio.

Atkins, David F.   M
Engineer, 5045 Grand Central Terminal New York, N. Y.

Atmore, A. L.
Chief Clerk, Engineering Dept.,
Philadelphia Electric Co., 1000
Chestnut St., Philadelphia, Pa.

Atwater, D. W.
Illuminating Engineer, Westinghouse
Lamp Co., 165 Broadway, New
York, N. Y.

Auler, Henry   A
Architect, F. R. A. Bldg., Washington Blvd., Oshkosh, Wis.

Axman, Daniel   A
Edison Lamp Works, of G. E. Co.
Room 945, Ill., Merchants Bank
Bldg., Chicago, Ill.

Babson, Thomas E.   A
Wiring Device Specialist, General
Electric Co., Bldg. 33B, Boston
Rd. & Bond St., Bridgeport,
Conn.

Baker, Clark   A
Asst. to Manager for Pacific Coast
National Mazda Lamp Div., G. E.
Co., 1648 16th St., Oakland, Cal.

Baker, S. A.
Macbeth-Evans Glass Co., 715-719
Call Bldg., San Francisco, Calif.

Balkam, Herbert H.   A
Illuminating Engineer, Consumers
Power Co., 252 W. Main St.
Jackson, Mich.

Ball, William J.   A
Secretary, Tri-City Electric Co.,
1529-31 Third Ave., Moline, Ill.

Barnes, Albert M.   A
President & General Mgr., Cambridge Gas Light Co., 719 Massachusetts Ave., Cambridge, Mass.

Barnes, Cyrus   M
General Sales Manager, Charles H.
Tenney & Co., 200 Devonshire
St., Boston, Mass.

Barnes, H. Alden   A
Lighting Salesman, Wheeler Re-
flector Co., 156 Pearl St., Boston,
Mass.

Barnes, H. Freeman   A
Dept. of Publicity, Edison Lamp
Works of G. E. Co., Harrison, N. J.

Barnitz, Frank R.   M
Director, 1922-25.
Asst. Secretary & Gen. Supt., Con-
solidated Gas Co., 130 E. 15th
St., New York, N. Y.

Barr, Taylor M.   A
Central Hudson Gas & Elec. Co., 50
Market St., Poughkeepsie, N. Y.

Barrett, Sampson K.   M
Asst. Prof. of Electrical Engineering,
New York University, University
Heights, Bronx, N. Y.

Barstow, William S.   A
President, W. S. Barstow & Co.,
Inc., 50 Pine St., New York, N. Y.

Bartlett, C. C.   A
1511 S. Wilton Place, Los Angeles,
Calif.

Bartlett, John S.   A
Public Service Corp., 84 Sip Ave.,
Jersey City, N. J.

Bartlett, P. H.   M
Supt. Meter & Installation Dept.,
Philadelphia Elec. Co., 10th &
Chestnut Sts., Philadelphia, Pa.

Barton, Charles A.   M
General Sales Mgr., N. Y. & Queens
Elec. Light & Power Co., Electric
Bldg., Bridge Plaza N., Long
Island City, N. Y.

Bates, Walter A.   M
Vice-President, Bates Expanded Steel
Truss Co., 232 S. Clark St.,
Chicago, Ill.

Batson, Walter V. M
Consulting Electrical Engineer, Hollis French & Allen Hubbard, 210 South St., Boston, Mass.

Battey, Paul L. A
Consulting Engineer, 123 W. Madison St., Chicago, Ill.

Battle, Robert T. A
Electrical Contractor, 220 W. 42nd St., New York, N. Y.

Bay, Charles H. A
Illuminating Engineer, Detroit Edison Co., 2000 Second Ave., Detroit, Mich.

Baylis, Roger V. M
Supt., American Gas Accumulator Co., 999 Newark Ave., Elizabeth, N. J.

Beach, R. L. A
Business Supervisor, Eng. Dept., Edison Lamp Works of G. E. Co., Harrison, N. J.

Beadenkopf, George A
Chief Engineer, Gas Dept., Consolidated Gas, Elec. Light & Power Co., Lexington & Liberty Sts., Baltimore, Md.

Beal, Charles A. A
General Electric Co., Harrison, N. J.

Beal, Thaddeus R.* A
President, Central Hudson Gas & Elec. Co., 50 Market St., Poughkeepsie, N. Y.

Beals, G. W. A
Ivanhoe-Regent Works of G. E. Co., 5716 Euclid Ave., Cleveland, Ohio.

Beam, J. W. A
Consulting Engineer, Ivanhoe-Regent Works of G. E. Co., 5716 Euclid Ave., Cleveland, Ohio.

Beardsley, Daniel H. A
Superintendent, Citizens' Electric Co., Battle Creek, Mich.

Beauchamp, Leon M
Electrical Engineer & Sales Manager, Solex Co., Ltd., 1202 St. Lawrence Blvd., Montreal, Canada

Bechtel, Ernest J. A
Consulting Electrical Engineer, Hodenpyl Hardy & Co., 14 Wall St., New York, N. Y.

Beedle, H. W. A
Electric Storage Battery Co., 718 Beacon St., Boston, Mass.

Beggs, Eugene W. A
Lamp Development, Westinghouse Lamp Co., Bloomfield, N. J.

Behan, Roscoe J. A
Assistant Field Engineer, Automobile Club of Southern California, 2601 S. Figueroa St., Los Angeles, Calif.

Beiswanger, John A
Vice-President, Gill Brothers Co., 7th & Franklin Sts., Steubenville, Ohio.

Belden, F. A. A
Vice-President & General Mgr. Rockingham County Light & Power Co., 29 Pleasant St., Portsmouth, N. H.

Bell, Archibald D. A
Edison Lamp Works of G. E. Co., Harrison, N. J.

Bell, Howard H. A
Illum. Eng. Laboratory, General Electric Co., Schenectady, N. Y.

Beman, Ralph A
Standardizing Dept., National Lamp Works of G. E. Co., Nela Park, Cleveland, Ohio.

Benford, F. A., Jr. M
Physicist, Illum. Eng. Lab. General Electric Co., Schenectady, N. Y.

Benjamin, R. B. M
President, Benjamin Elec. Mfg. Co., 128 Sangamon St., Chicago, Ill.

Bennett, Edward* A
Professor of Electrical Engineering, University of Wisconsin, Madison, Wis.

Bensi, R. J. A
South American General Electric Co., 560 Av. de Mayo, Buenos Aires, Argentina, S. A.

Berens, Conrad, M. D. M
Ophthalmologist, 9 E. 46th St., New York, N. Y.

Berger, Charles F. A
District Manager, Philadelphia Elec. Co., 18th & Columbia Ave., Philadelphia, Pa.

Berger, William W. M
Electrical Contractor, Wm. W. Berger Co., 4913 Capitol Ave., Omaha, Neb.

Bernhard, Albert H. A
Electrical & Illuminating Engineer, U. S. Navy, Bureau of Steam Engineering, 44 Court St., Brooklyn, N. Y.

Bernhard, F. H.   **M**
Editor, E. M. F. Electrical Year Book, 824 Monadnock Block, Chicago, Ill.

Berry, Clarence J.   **A**
Edison-Clerici Ital. Lamp Works, 4 rue d'Aguesseau, Paris, France.

Besinsky, Vaclav   **A**
Czechoslovak League of Electrotechnics, 1 Palackeho Ulice, Prague VII, Czecho-Slovak Republic.

Betts, Philander   **M**
Board of Public Utility Commissioners, 790 Broad St., Newark, N. J.

Beutell, Alfred W.   **A**
Cohen & Armstrong Disposal Corp., 2 Victoria St., London S. W. 1, England.

Bickerstaff, Ernest   **A**
Messrs. Templin & Toogood, P. O. Box 91, Wellington, New Zealand.

Bicknell, George W.   **A**
John Hancock Mutual Life Insurance Co., 197 Clarendon St., Boston, Mass.

Biddle, Robert   **A**
President, Biddle-Gaumer Co., 3846-56 Lancaster Ave., Philadelphia, Pa.

Birge, Nathan R.*   **A**
Street Lighting Dept., General Electric Co., Schenectady, N. Y.

Blackmore, Charles T.   **A**
Electrical Engineer, Stone & Webster, Inc., 147 Milk St., Boston, Mass.

Blackwell, W. T.   **M**
General Lighting Representative, Public Service Electric Co., 80 Park Place, Newark, N. J.

Blake, S. H.   **A**
General Electric Co., 829 Union St. Schenectady, N. Y.

Blakeslee, D. W.   **M**
Electrical Engineer, Jones & Laughlin Steel Corp., 26th & Carson Sts., Pittsburgh, Pa.

Bliss, Frederick W.   **A**
Local Sales Manager, Edison Lamp Works of G. E. Co., 1012 Turks Head Bldg., Providence, R. I.

Bliven, Joseph E.   **A**
Empire Theatre Co., 394 Bank St., New London, Conn.

Blondel, Andre*   **A**
41 Ave. de la Bourdonnais, Paris, France.

Blumenauer, C. H.   **A**
President & Treasurer, Jefferson Glass Co., Follansbee, W. Va.

Bodine, Samuel T.   **A**
President, The United Gas Improvement Co., Broad & Arch Sts. Philadelphia, Pa.

Boiler, W. F.   **A**
Professor of Ophthalmology, State Univ. of Iowa, College of Medicine, 12½ S. Clinton St., Iowa City, Ia.

Bonner, John J.   **A**
Technical Dept., Welsbach Co., Gloucester, N. J.

Boss, Frederic A.   **A**
Proprietor, Boss Electrical Supply Co., 11 Peck St., Providence, R. I.

Bossi, Santiago J.   **A**
Messrs. Bossi & Co., Calle 48 No. 644, La Plata, Argentina.

Bostock, Edgar H.   **A**
Bostock Rhodes Co., 9 N. Moore St., New York, N. Y.

Botsford, C. J.   **A**
Westinghouse Lamp Co., 165 Broadway, New York, N. Y.

Bowers, Corwin J.   **A**
Illumination Engineer, Columbus Ry. Power & Light Co., 102 N. 3d St., Columbus, Ohio.

Bowles, Richard H.   **A**
Asst. Chief Elec. Engineer, Sao Paulo Tramway, Light & Power Co., Sao Paulo, Brazil.

Boyce, Ernest W.   **M**
N. Y. Electric Lamp Co., 208 W. 17th St., New York, N. Y.

Bozell, H. V.   **M**
Editor, Electrical World, 36th St. & 10th Ave., New York, N. Y.

Bradbury, Edward L.   **A**
Asst. Sales Mgr., Holophane Glass Co., 340 Madison Ave., New York, N. Y.

Brady, Edward J.   **M**
Photometrist, United Gas Improvement Co., 3101 Passyunk Ave., Philadelphia, Pa.

Brady, N. F.   **M**
c/o A. H. Laidlaw, 80 Broadway, New York, N. Y.

Bramoso, Juan   **A**
Chief Electrician, Municipality of Villa Maria, 729 Entre Rios, Villa Maria, F. C. C. A., Argentina, S. A.

Braun, Eugene · A
Chief Electrician, Comstock & Gest,
104 W. 39th St., New York, N. Y.

Braun, W. C. M
Montgomery Ward & Co., 618 Chicago Ave., Chicago, Ill.

Briggs, Wallace W. M
Westinghouse Lamp Co., 165 Broadway, New York, N. Y.

Brock, Arthur, Jr. A
Arthur Brock Jr., Tool & Mfg.
Works, 533 N. 11th St., Philadel-
phia, Pa.

Brooks, Morgan M
Professor of Elec. Engineering, University of Illinois, Urbana, Ill.

Brown, David S. A
Engineer, New York Telephone Co.,
104 Broad St., New York, N. Y.

Brown, Harry A
Electrical Contractor, H. Brown
Electrical Co., 113 St. Marks FL,
New York, N. Y.

Brown, Harry W. A
Electrical Engineer & Sales Agent,
General Electric Co., 84 State St.,
Boston, Mass.

Brown, R. I. A
Commercial Manager, Little Rock
Ry. & Electric Co., Box 859, Little
Rock, Ark.

Brown, R. B. A
General Mgr., Milwaukee Gas Light
Co., 182 Wisconsin St., Milwaukee
Wis.

Brown, Richard B., Jr. A
Illuminating Engineer, Edison Elec.
Illuminating Co. of Boston, 39
Boylston St., Boston, Mass.

Brown, Sumner E. A
Engineer, 560 4th St., Boston, Mass.

Brown, Wendell S. A
Engineer, F. P. Sheldon & Son, 1009
Hospital Trust Bldg., Providence,
R. I.

Brown, Willard C. A
Electrical Engineer, National Lamp
Works of G. E. Co., Nela Park
Cleveland, Ohio.

Browne, William Hand, Jr.* A
Professor of Elec. Engineering, North
Carolina State College of Agri-
culture & Engineering, West
Raleigh, N. C.

Bruce, Howard A
First Vice-President & General Mgr.
Bartlett, Hayward & Co., Balti-
more, Md.

Brumbaugh, G. Edwin A
Architect, Real Estate Trust Bldg.,
Broad & Chestnut Sts., Philadel-
phia, Pa.

Brundage, H. M. A
Secretary, Consolidated Gas Co., of
N. Y., 130 E. 15th St., New York,
N. Y.

Brush, C. F. A
3725 Euclid Ave., Cleveland, Ohio.

Bryant, Alice G., M.D. M
Physician, Ear, Nose, Throat, 502
Beacon St., Boston, Mass.

Bryant, John M. M
Professor of Elec. Engineering, University of Texas, Austin, Tex.

Buckley, James E. A
Asst. Mgr., Bureau of Ltg. & Decorations, N. Y. Edison Co., 130 E.
15th St., New York, N. Y.

Bucknam, Paul C. A
Salesman, Pettingell-Andrews Co.,
511 Atlantic Ave., Boston, Mass.

Bull, John H. M
215 Ditmars Ave., Astoria, Long
Island, N. Y.

Burchard, A. W. A
General Electric Co., 120 Broadway,
New York, N. Y.

Burgess, A. E. C. A
Consulting Elec. & Mech. Engineer,
A. C. F. Webb & Burgess, 1105
Culwulla Chambers, Castlereagh
St., Sydney, Australia.

Burke, Thomas A
District Manager, Philadelphia Electric Co., 9 S. 40th St., Philadelphia, Pa.

Burnap, Robert S. A
Lamp Engineer, Engineering Dept.,
Edison Lamp Works of G. E. Co.,
Harrison, N. J.

Burnett, Douglass* M
Commercial Manager, Elec. Div.,
Consolidated Gas, Elect. Light &
Power Co., 100 W. Lexington St.,
Baltimore, Md.

Burnett, H. D. M
Manager, Canadian Tungsten Lamp
Co., Ltd., 428 Cannon St. E.,
Hamilton, Ont., Canada.

Burnham, Roy R.   A
Consulting Engineer, 119 Water St.,
Boston, Mass.

Burr, Alfred R.   A
Sales Agent, New Haven Gas Light
Co., 70 Crown St., New Haven,
Conn.

Burr, C. B.   A
Electrical Contractor, 100 Main St.,
Ansonia, Conn.

Burrill, Paul C.   A
Office Manager, Trester Service
Electric Co., 47 Oneida St., Mil-
waukee, Wis.

Burrows, W. R.   A
General Manager, Edison Lamp
Works of G. E. Co., Harrison, N. J.

Burton, Robert B.   A
Sales Manager, Duplex-a-Lite Dept.
of Edward Miller and Co., 68 Park
Place, New York, N. Y.

Bush, William E.   M
Illuminating Engineer, British Thom-
son Houston Co., Ltd., 77 Upper
Thames St., London, England.

Buss, Arthur F.   A
Electrical Engineer, 315 S. Broadway, Los Angeles, Calif.

Cadby, John N.   A
Executive Manager, Wisconsin Utili-
ties Association, 445 Washington
Bldg., Madison, Wis.

Cady, Francis E.   M
Director, 1919-22.
Manager, Research Dept., National
Lamp Works of G. E. Co., Nela
Park, Cleveland, Ohio.

Cairnie, H. W.   A
Western Electric Co., 385 Summer
St., Boston, Mass.

Calahan, Harry C.   A
Tri-City Electric Co., 52-56 Lafayette
St., Newark, N. J.

Caldwell, Francis C.   M
Director, 1921-24.
Professor of Electrical Engineering,
Ohio State University, Columbus,
Ohio.

Caldwell, O. H.   A
Editor, Electrical Merchandising,
36th St. & 10th Ave., New York,
N. Y.

Calhoun, Ervin   A
Supt. of Stores, United Gas Improvement Co., 19th St. & Allegheny Ave., Philadelphia, Pa.

Callender, Charles L.   A
The Varney Electrical Supply Co.,
121-123 South Meridian St., Indianapolis, Ind.

Callender, D. E.   A
General Manager, Wisconsin Gas &
Elec Co., 305 6th St., Racine, Wis.

Calvert, H.   M
Director, 1914-17
President, Baird-Osterhout Co., 2114
Sansom St., Philadelphia, Pa.

Cameron, David G.   A
Illuminating Engineer, Phoenix Glass
Co., 230 5th Ave., New York, N.Y.

Camp, H. M.   A
Lighting Sales Mgr., Tennessee Elec-
tric Power Co., 620 Market St.
Chattanooga, Tenn.

Campbell, Clarence J.   A
Illuminating Engineer, Westinghouse
Lamp Co., 165 Broadway, New
York, N. Y.

Campbell, Guy   A
Managing Director, Benjamin Elec-
tric Ltd., Brantwood Works, Tariff
Road, Tottenham, London, N. 17,
England.

Campbell, James P.   A
Illuminating Company, 75 Public
Square, Cleveland, Ohio.

Campbell, John   M
Director, 1908.
Edison Elec. Illuminating Co. of
Boston, 100 Boylston St., Boston,
Mass.

Carder, Frederick   A
Secretary & General Mgr., Steuben
Glass Works, Erie Ave., Corning,
N. Y.

Carl, Fred L.   A
District Illuminating Engineer,
Westinghouse Elec. & Mfg. Co.,
111 W. Washington St., Chicago,
Ill.

Carpenter, E. P.   A
Service & Lighting Engineer, Moore-
Handley Hardware Co., Birming-
ham, Ala.

Carson, Clarence   A
Experimental Engineer, Dodge
Brothers, 7900 Jos. Campau Ave.,
Detroit, Mich.

Carter, Robert A., Jr.   A
Experimental Engineer, Consolidated
Gas Co., 130 E. 15th St., New
York, N. Y.

Cary, Irving B.   A
Sales Dept., Corning Glass Co., 501 5th Ave., New York, N. Y.

Cary, Walter*   M
Westinghouse Lamp Co., 165 Broadway, New York, N. Y.

Castor, W. A.   A
Supt. Meters Div., United Gas Improvement Co., 19th & Allegheny Ave., Philadelphia, Pa.

Cattell, J. H.   A
Secretary & Treasurer, Warner-Patterson Co., 914 S. Michigan Ave., Chicago, Ill.

Chadbourn, Ralph W.   A
Edison Elec. Illuminating Co. of Boston, 1165 Massachusetts Ave., Roxbury, Mass.

Chamberlin, Guy N.   A
Chief Engineer, Street Ltg. Dept., General Electric Co., 928 Western Ave., West Lynn, Mass.

Chandler, Charles F.*   A
Emeritus Professor of Chemistry, Columbia University, New York, N. Y.

Channon, H. O.   A
Manager, Quincy Gas, Electric & Heating Co., 422 Maine St., Quincy, Ill.

Chapin, Edwin S.*   A
Vice-President & Manager, Welsbach Gas Lamp Co., 392 Canal St., New York, N. Y.

Chase, Philip H.   A
Asst. Engineer, Transmission & Distribution Dept., Philadelphia Elec. Co., 2301 Market St., Philadelphia, Pa.

Christman, H. S.   A
Asst. Commercial Agent, Phila. Gas Works, U. G. I. Co., 134-142 N. 13th St., Philadelphia, Pa.

Christy, Henry B.   A
Asst. Supt., United Gas Improvement Co., 20 Maplewood Ave., Philadelphia, Pa.

Church, Arthur C.   M
Illuminating Specialist, Pacific States Electric Co., 236-240 S. Los Angeles St., Los Angeles, Calif.

Church, Fermor S.   A
Los Alamos Ranch School, Otowi, N. Mex.

Claparols, Manuel   M
Electrical Engineer, College of Wooster, Wooster, Ohio.

Clark, John C. D.   M
Director, 1911–13 and 1918–21
Boston Consolidated Gas Co., 149 Tremont St., Boston, Mass.

Clark, John F.   A
Chief Engineer, Custom House, Room 139, New York, N. Y.

Clark, Walton   M
2d Vice-President, United Gas Improvement Co., Broad & Arch Sts., Philadelphia, Pa.

Clark, William E.   A
President, Clark & Mills Elec. Co., 75 Newbury St., Boston, Mass.

Clark, William J.*   M
Vice-President, 1919–1921.
Vice-President, Westchester Lighting Co., 1st Ave. & 1st St., Mt. Vernon, N. Y.

Clarke, Frank A.   A
Sales Engineer, George Cutter Works, Westinghouse Elec. & Mfg. Co., South Bend, Ind.

Clarke, George R.   A
Manager, Radio Dept., A. T. Knowlson Co., 415 E. Congress St., Detroit, Mich.

Clasen, Arthur J.   A
Chief Electrician, American Bank Note Co., Garrison Ave. & Tiffany St., Bronx, N. Y.

Cleaver, N. S.   A
Philadelphia Electric Co., 1000 Chestnut St., Philadelphia, Pa.

Clement, Wm. E.   A
Commercial Agent, N. O. Railway & Light Co., 201 Baronne St., New Orleans, La.

Clewell, C. E.   M
Professor of Elec. Engineering, Univ. of Pennsylvania, Philadelphia, Pa.

Clewell, George A.   A
Lighting Service Dept., Westinghouse Lamp Co., 165 Broadway, New York, N. Y.

Clifford, H. E.   M
Professor of Elec. Engineering, Pierce Hall, Harvard University, Cambridge, Mass.

Clinch, Edward S., Jr.   M
Consulting Engineer, 110 W. 39th St., New York, N. Y.

Close, R. C.   A
Central Electric Co., 316 S. Wells St., Chicago, Ill.

Clover, George R.   **M**
District Mgr., Cooper-Hewitt Elec.
Co., 604 Ford Bldg., Detroit,
Mich.

Coahran, J. M.   **M**
Engineer, Keystone Wood Products
Co., Colegrove, Pa.

Cobb, P. W.   **M**
Director, 1913–15
Nela Research Laboratory, National
Lamp Works of G. E. Co., Cleve-
land, Ohio.

Cobby, E. V.   **A**
Building Engineer, Pacific Telephone & Telegraph Co., 807 Sheldon Bldg., San Francisco, Calif.

Coffin, F. A.   **A**
Sales Manager, Milwaukee Elec. Ry.
& Light Co., Public Service Bldg.,
Milwaukee, Wis. .

Coghlin, Peter A.   **A**
Economy Electric Co., 22 Foster St.,
Worcester, Mass.

Cogan, D. E.   **M**
Salesman, General Electric Co., 84
State St., Boston, Mass.

Cohen, Carl   **A**
Commercial Engineer, Chas. Atkins
& Co., Ltd., 894 Hay St., Perth,
Western Australia.

Cohn, Charles M.   **M**
Vice-President, Consolidated Gas,
Elec. Light & Power Co., 100 W.
Lexington St., Baltimore, Md.

Cohn, Leo P.   **A**
Chief Inspector, Cooper-Hewitt Elec.
Co., 730 Grand St., Hoboken, N. J.

Colby, Charles C., Jr.   **A**
Westinghouse Lamp Co., Bloomfield, N. J.

Cole, Rex J.   **M**
President, Edward Miller & Co., 68
Park Place, New York, N. Y.

Coleman, Herman   **A**
Proprietor, Penn Gas & Elec. Sup-
ply Co., 26 N. 7th St., Allentown,
Pa.

Coleman, Noah   **A**
Manager, Coleman Elec. Co., 121 N.
7th St., Allentown, Pa.

Colket, James H.
New York Telephone Co., 15 Dey
St., New York, N. Y.

Collamore, Ralph   **M**
Mechanical Engineer, Smith, Hinch-
man & Grylls, 800 Marquette
Bldg., Detroit, Mich.

Collar, Olcott N.   **A**
Elec. Draftsman, Sargent & Lundy,
72 W. Adams St., Chicago, Ill.

Collier, Charles A.   **M**
General Sales Manager, Georgia Rail-
way & Power Co., 301 Elec. & Gas
Bldg., Atlanta, Ga.

Colville, J. R.   **A**
Illuminating Engineer, National
Lamp Works of G. E. Co., Nela
Park, Cleveland, Ohio.

Commery, Eugene W.   **A**
Engineer of Light Utilization, Laboratory of Applied Science, National
Lamp Works of G. E. Co., Cleveland, Ohio.

Conant, Lewis C.   **M**
Illuminating Engineer, Buffalo General Elec. Co., Electric Bldg.,
Washington & Genesee Sts., Buffalo, N. Y.

Conlisk, Raimon F.   **M**
Westinghouse Elec. & Mfg. Co.,
First National Bank Bldg., San
Francisco, Calif.

Conner, George C.   **A**
Frank H. Stewart Elec. Co., Philadelphia, Pa.

Cook, Henry A.   **A**
Edison Lamp Works of G. E. Co.,
Harrison, N. J.

Cook, W. H.   **A**
Secretary, The Owsley Co., 1301 Ma-
honing Bank Bldg., Youngstown,
Ohio.

Copley, I. C.   **A**
President, Western United Gas &
Elec. Co., Aurora, Ill.

Corbus, Frederick G.   **A**
Welsbach Company, Glouster, N. J.

Cordner, A. R.   **A**
Meter Tester, Philadelphia Elec. Co.,
23d & Market Sts., Philadelphia, Pa.

Corraz, Zoe N.   **A**
Technical Assistant, Electrical Test-
ing Laboratories, 80th St. & East
End Ave., New York, N. Y.

Cousins, George G.   **A**
Senior Asst. Laboratory Engineer,
Hydro-Electric Power Commis-
sion of Ontario, 8 Strachan Ave.,
Toronto, Ont., Canada.

Cowles, J. W.   **M**
Vice-President, 1913–14
Supt. of Installation, Edison Elec.
Illuminating Co. of Boston, 39
Boylston St., Boston, Mass.

Cowles, R. Roy   **A**
Elec. Engineer, Pacific Gas & Elec. Co., 445 Sutter St., San Francisco, Calif.

Crampton, George S., M.D.   **M**
Vice-President, 1916–18 and 1920–21
President, 1921–22
Junior Past-President, 1922–24
Ophthalmologist, 1819 Walnut St., Philadelphia, Pa.

Cravath, J. R.*   **M**
President, Pioneer Elec. Co., 1113 Macdonald Ave., Richmond, Calif.

Crawford, David F.   **M**
President, Westinghouse Union Battery Co., Swissvale, Pa.

Cressman, Russell B.   **M**
Sales & Adv. Mgr., Gleason-Tiebout Glass Co., 200 5th Ave., New York, N. Y.

Crittenden, E. C.   **M**
Director, 1919–21
Physicist, Chief of Electrical Div., Bureau of Standards, Washington, D. C.

Crockett, Charles H.   **A**
Engineer, C. W. Crockett & Sons, Stow Ave., Troy, N. Y.

Crofoot, Clarence E.   **A**
Head of Elec. Dept., Utica Free Academy, Kemble St., Utica, N.Y.

Croft, Terrell   **M**
Directing Engineer, Terrell Croft Engineering Co., 6600 Delmar Blvd., University City, St. Louis, Mo.

Crosby, Halsey E.   **M**
Building Engineer, Western Electric Co., Inc., 463 West St., New York, N. Y.

Crosby, Joseph G.   **M**
Vice-President & Gen. Mgr., Whalen Crosby Elec. Co., 140 N. 11th St., Philadelphia, Pa.

Crownfield, David   **M**
Designer & Illuminating Engineer, Irving & Casson-A. H. Davenport Co., 573 Boylston St., Boston, Mass.

Currier, Burleigh   **M**
Philadelphia Electric Co., 23d & Market Sts., Philadelphia, Pa.

Curtis, Augustus D.   **M**
Director, 1921–24
President, National X-Ray Reflector Co., 235 W. Jackson Blvd., Chicago, Ill.

Curtis, Darwin   **M**
Sales Engineer, X-Ray Reflector Co. of N. Y., 31 W. 46th St., New York, N. Y.

Curtis, Kenneth   **M**
Chairman Board of Control, National X-Ray Reflector Co., 235 W. Jackson Blvd., Chicago, Ill.

Cutler, Coleman W., M.D.   **A**
Physician, 24 E. 48th St., New York, N. Y.

Dahl, Thure   **A**
Designer, Lightolier Co., 569 Broadway, New York, N. Y.

Dailey, E. J., Jr.   **A**
Sales Engineer, Western Electric Co., 195 Broadway, New York, N.Y.

Daniels, Julius   **M**
Illuminating Engineer, Edison Electric Illuminating Co. of Boston, 39 Boylston St., Boston, Mass.

Danley, P. Y.   **M**
Illuminating Engineer, Westinghouse Electric & Mfg. Co., 717 S. 12th St., St. Louis, Mo.

Dansboe, W. F.   **A**
Lighting Specialist, Southwest Gen. Electric Co., 522 Interurban Bldg., Lamar St., Dallas, Tex.

Dates, Henry B.,   **M**
Professor of Electrical Engineering, Case School of Applied Science, Euclid Ave. at Wade Park, Cleveland, Ohio.

Davidson, Edward Young, Jr.   **A**
Illuminating Engineer, Macbeth-Evans Glass Co., 157 N. Craig St., Pittsburgh, Pa.

Davidson, W. F.   **M**
Director, Engineering Investigation, Brooklyn Edison Co., 360 Pearl St., Brooklyn, N. Y.

Davies, H. W.   **A**
Representative, The Safety Car Heating & Lighting Co., 1738 Commercial Trust Bldg., Philadelphia, Pa.

Davis, Charles S.   **A**
District Manager, Westinghouse Lamp Co., 60 India St., Boston, Mass.

Davis, Ernest H.   **A**
President, Lycoming Edison Co., West & Willow Sts., Williamsport, Pa.

Davis, L. Wood   A
Branch Manager, Westinghouse Lamp Co., 420 S. San Pedro St., Los Angeles, Calif.

Davy, Humphrey   A
Consulting Elec. Engineer, Causbey Engineering Co., 1011 Chestnut St., Philadelphia, Pa.

Day, B. Frank   A
Pay Master & Statistician, Philadelphia Electric Co., 1000 Chestnut St., Philadelphia, Pa.

deChatelain, Michael A.   A
Professor of Electrical Engineering, Petrograd Polytechnic Institute, Lesnoy St., Petrograd, Russia.

deGraaff, Antonius   A
Chemical Engineer, Library Dept., Philips' Glowlampworks, Eindhoven, Holland.

deKosenko, Stepan   M
Vice-President & Art Director, Sterling Bronze Co., 18 East 40th St., New York, N. Y.

DeLacy, Thomas   A
Commercial Inspector, Philadelphia Electric Co., 1000 Chestnut St., Philadelphia, Pa.

DeLay, Frederic A.   A
Head Lecturer, Chicago Central Station Institute, 72 West Adams St., Chicago, Ill.

Dempsey, William T.   A
Supt. Service Maintenance Division, New York Edison Co., 38 West 17th St., New York, N. Y.

Dennington, Arthur R.   A
Electrical Engineer, Westinghouse Lamp Co., Bloomfield, N. J.

Desaix, Herbert W.   M
Manager, Watson-Flagg Engineering Co., 214 Straight St., Paterson, N. J.

Deshler, Charles   A
Consulting Engineer, Engineering Dept., Edison Lamp Works of G. E. Co., Harrison, N. J.

Deskins, Hiram T.   A
Construction Foreman, Kilgore Electric Co., Box 371, Williamson, W. Va.

Detweiler, Paul George
Treasurer, The Detweiler-Bell Co., 152 Temple St., New Haven, Conn.

Devereux, Washington   A
Asst. Chief, Electrical Dept., Phila. Fire Underwriters Association, Bullitt Bldg., Philadelphia, Pa.

Devine, Alfred W.   M
Illuminating Engineer, Registry of Motor Vehicles, Room 11, Commonwealth Pier, Boston 9, Mass.

Dibelius, Ernest F.   A
Sales Agent & Illuminating Engineer, Edison Lamp Works of G. E. Co., 120 Broadway, New York, N. Y.

Dick, A. C.   A
Study of Light & Illumination, Electrical Testing Laboratóries, 80th St. & East End Ave., New York, N. Y.

Dickerson, A. F.   M
Chief Engineer, Illuminating Engineering Laboratories, General Electric Co., Bldg. 31, Schenectady, N. Y.

Dickey, Charles H.   A
Manufacturer, Maryland Meter Works, 229 Guilford Avenue., Baltimore, Md.

Dickey, Edmund S.   A
Manager, Maryland Meter Works, Guilford Ave. & Saratoga St., Baltimore, Md.

Diggles, George L.   A
Inspector in Charge, Harrison Bureau of Electrical Testing Laboratories, 80th St. & East End Ave., New York, N. Y.

Diggs, D. M.   M
Charge of Lighting, International General Electric Co., Schenectady, N. Y.

Dion, A. A.   M
General Manager & Engineer, Ottawa Elec. Co. & Ottawa Gas Co., 35 Sparks St., Ottawa, Ontario, Canada.

Dixon, Fred R., Jr.   M
Supervisor, Commercial Lighting Dept., Consolidated Gas, Elec. Lt. Power Co., Lexington & Liberty Sts., Baltimore, Md.

Doane, Francis H.   A
Principal, School of Electrical Engineering, International Correspondence Schools, Scranton, Pa.

Doane, Samuel Everett* M
Director, 1907
President, 1919–20
Junior Past-President, 1920–22
Chief Engineer, National Lamp
Works of G. E. Co., Nela Park,
Cleveland, Ohio.

Dobson, W. P. M
Laboratory Engineer, Hydro-Elec.
Power Commission of Ontario, 8
Strachan Ave., Toronto, Canada.

Dodds, George A
Salesman, Industrial Machinery, 611
Ferguson Bldg., Pittsburgh, Pa.

Dodge, Kern A
Consulting Engineer, Morris Bldg.,
1421 Chestnut St., Philadelphia,
Pa.

Doherty, Henry L.* A
60 Wall St., New York, N. Y.

Dolbier, F. Van Buren A
Supply Dept., Philadelphia Electric
Co., 132 S. 11th St., Philadelphia,
Pa.

D'Olier, Henry, Jr. A
Research Work, Standardization
Service, Watson Bldg., Bridgeport,
Conn.

D'Olive, Eugene R. A
Testing Engineer, Photometric
Laboratory, Commonwealth Edi-
son Co., 28 N. Market St., Chicago,
Ill.

Dolkart, Leo* A
Managing Engineer, Moline Electric
Co., 1307 Fifth Ave., Moline, Ill.

Dolman, Percival, M. D. M
Oculist, Flood Bldg. Room 1167, San
Francisco, Calif.

Domoney, Earl R. A
Illuminating Engineer, Consumers
Power Co., 134 S. Washington St.,
Saginaw, Mich.

Donahue, E. J. A
Engineer of Distribution, Hudson
Division, Public Service Gas Co.,
St. Pauls & James Ave., Jersey
City, N. J.

Donahue, Rev. Joseph N. A
Professor of Physics, Columbia University, Portland, Oregon.

Donley, H. B. M
Chief Engineer, The John W. Brown
Mfg. Co., Marion Road, South
Columbus, Ohio.

Donley, W. H. A
Assistant to Personal Supervisor,
Philadelphia Electric Co., 1000
Chestnut St., Philadelphia, Pa.

Doolittle, C. Melvin A
Illuminating Engineer, Westing-
house Lamp Co., 165 Broadway,
New York, N. Y.

Dorey, W. A. A
Designing Engineer, Holophane
Glass Co., Inc., Newark, Ohio.

Dorting, E. E. M
Lighting Engineer, Interborough
Rapid Transit Co., 600 West 59th
St., New York, N. Y.

Dougall, A. C. A
Sales Agent, Edison Lamp Works
of G. E. Co., Room 2304, 120
Broadway, New York, N. Y.

Dougherty, John E. A
Salesman, Philadelphia Electric Co.,
1000 Chestnut St., Philadelphia,
Pa.

Doughty, Herbert C. M
Manager, Division 2, Atlantic City
Electric Co., 3300 Atlantic Ave.,
Wildwood, N. J.

Douthirt, W. F. A
United Gas Improvement Co., 1401
Arch St., Philadelphia, Pa.

Dow, John Stewart A
Assistant Editor, The Illuminating
Engineer, 32 Victoria Street, Lon-
don, S. W., England.

Downs, William S. A
Public Service Electric Co., 695
Bloomfield Ave., Montclair, N. J.

Dows, Chester L. M
Electrical Engineer, Engineering
Dept., National Lamp Works of
G. E. Co., Nela Park, Cleveland, O.

Doyle, Walter H. A
Lighting Salesman, Central Hudson
Gas & Electric Co., 129 Broadway,
Newburgh, N. Y.

Drinker, Philip M
Instructor, Ventilation & Illumination, Harvard School of Public
Health, 240 Longwood Ave., Boston, Mass.

Drisko, William J. M
Vice-President, 1922–24
Associate Professor of Physics, Mass.
Institute of Technology, Cam-
bridge, Mass.

Duane, Alexander, M. D.   A
139 East 37th St., New York, N. Y.

Duncan, F. B.   A
Manager, Advertising and Sales
Promotion, Duplex-a-lite Dept.,
Edward Miller & Co., Meriden,
Conn.

Duncan, Ronald Bruce   A
Supt. Spring Garden District, The
United Gas Improvement Co., 1615
N. 9th St., Philadelphia, Pa.

Dunlap, Knight   A
Professor of Experimental Psychology, The Johns Hopkins University, Baltimore, Md.

Dunn, Gano   A
President, J. G. White Engineering
Corporation, 43 Exchange Place,
New York, N. Y.

Dunn, J. M.   A
General Manager, Radio Appliance
Co., 123 Pleasant St., Morgan-
town, W. Va.

Dyre, Walter T.   A
Salesman, Philadelphia Electric Co.,
1000 Chestnut St., Philadelphia,
Pa.

Eames, Paul H.   A
Salesman, Lamp & Lighting Dept.,
George H. Wahn Co., 69 High St.,
Boston, Mass.

Eckstein, Herman   A
Electrical Contractor, 1742 Sansom
St., Philadelphia, Pa.

Edgar, Charles L.   M
President & General Manager, The
Edison Electric Illuminating Co.
of Boston, 70 State St., Boston,
Mass.

Edwards, Evan J.   M
Director, 1918-21
Electrical Engineer, National Lamp
Works of G. E. Co., Nela Park,
Cleveland, Ohio.

Egeler, C. E.   A
Engineering Dept., National Lamp
Works of G. E. Co., Nela Park,
Cleveland, Ohio.

Eglin, W. C. L.   M
Vice-President & Chief Engineer,
Philadelphia Electric Co., 1000
Chestnut St., Philadelphia, Pa.

Eichelberger, C. G.   A
Commercial Engineer, Union Gas &
Electric Co., 25 West Fourth St.,
Cincinnati, Ohio.

Eichengreen, L. B.   M
Assistant to General Superintendent,
United Gas Improvement Co.,
1401 Arch St., Philadelphia, Pa.

Elliott, Leavenworth E.   M
Consulting Engineer, Cooper-Hewitt
Electric Co., 161 Summer St.,
Boston, Mass.

Ely, Robert B.   M
Director, 1919-22
Illuminating Engineer, Westinghouse
Lamp Co., 165 Broadway, New
York, N. Y.

Emerich, LeRoy E.   A
Sales Engineer, Leeds & Northrup
Co., 4901 Stenton Ave., Philadel-
phia, Pa.

Emerson, Guy C.   A
Consulting Engineer for Municipal
Work, Boston Finance Commis-
sion, 73 Tremont St., Boston,
Mass.

Emerson, Harrington   A
Consulting Engineer, The Emerson
Co., 30 Church St., New York,
N. Y.

Engelfried, Henry O.   A
Engineer, Consolidated Electric
Lamp Co., 88 Holton St., Dan-
vers, Mass.

Ericksen, Thomas   A
Electrician, Detroit Stove Works,
6900 Jefferson Ave., East, Detroit, Mich.

Eteson, Franklin C.   A
Sales Dept., Blackstone Valley Gas
& Electric Co., 231 Main St., Paw-
tucket, R. I.

Evans, Dean P.   A
Columbus Railway Power & Light
Co., 102 North Third St., Colum-
bus, Ohio.

Evans, Edward A.   A
Salesman, Westinghouse Lamp Com-
pany, 1205 Walker Bank Bldg.,
Salt Lake City, Utah.

Evans, Gladden F.   A
Federal Manufacturing Co., 162
West Long St., Columbus, Ohio.

Evans, H. S.   A
1452 Beechwood Ave., Pittsburgh,
Pa.

Evans, William A. D.   M
President & Treasurer, Cooper-
Hewitt Electric Co., 95 River St.,
Hoboken, N. J.

Evans, William H. A
Publicity Dept., Philadelphia Electric Co., 1000 Chestnut St., Philadelphia, Pa.

Everson, R. W. A
District Manager, Westinghouse Lamp Co., 1321 Candler Bldg., Atlanta, Georgia.

Ewart, Frank R. A
ConsultingElectricalEngineer,Ewart, Jacob, Armer & Byam, Ltd., 207 Excelsior Life Bldg., Toronto, Canada.

Ewing, R. E. A
Designer, Lamp Factory, Welsbach Co., Gloucester, N. J.

Fabry, Charles A
Professeur a l'Université de Paris, Laboratoire de Physique 1 rue Victor Cousin, Paris, France.

Falge, R. N. A
Illuminating Engineer, National Lamp Works of G. E. Co., Nela Park, Cleveland, Ohio.

Farmer, Clifford M. A
Power Engineer, New Bedford Gas & Edison Light Co., 693 Purchase St., New Bedford, Mass.

Farmer, F. Lee M
Sales Manager, Beardslee Chandelier Mfg. Co., 216 S. Jefferson St., Chicago, Ill.

Farrand, Dudley M
Vice-President, Public Service Corporation of N. J., 80 Park Place, Newark, N. J.

Fee, C. Edward A
Eastern Manager, Sunbeam Incandescent Lamp Division, National Lamp Works of G. E. Co., 401 Hudson St., New York, N. Y.

Feiker, Frederick Morris M
Vice-President, 1921–23
The Society for Electrical Development, 522 Fifth Ave., New York, N. Y.

Feingold, Marcus, M.D. A
Ophthalmologist, 4206 St. Charles Ave., New Orleans, La.

Fenniman, John R. A
Assistant Treasurer, Consolidated Gas Co. of N. Y., 130 East 15th St., New York, N. Y.

Ferguson, Louis A. A
Vice-President, Commonwealth Edison Co., 72 West Adams St., Chicago, Ill.

Ferguson, Lynn W. A
Sales Engineer, Benjamin Electric Manufacturing Co., 247 West 17th St., New York, N. Y.

Ferguson, Olin Jerome M
Professor of Electrical Engineering, University of Nebraska, Station A, Lincoln, Neb.

Ferree, C. E. M
Associate Professor of Psychology & Director of Psychological Laboratory, Bryn Mawr College, Bryn Mawr, Pa.

Ferris, C. H. A
Assistant Business Manager, Illuminating Glassware Guild, 19 West 44th St., New York, N. Y.

Field, Oscar S. M
Chief Engineer, Hall Switch & Signal Co., Garwood, N. J.

Firehock, Milton B. A
Illuminating Engineer, Cooper-Hewitt Electric Co., Hoboken, N. J.

Firth, C. J. A
Philadelphia Electric Co., N. W. Corner 23d & Market Sts., Philadelphia, Pa.

Fisher, B. F., Jr. M
Consulting Engineer of Mazda Service, General Electric Co., Research Laboratory, Schenectady, N. Y.

Fisher, J. Carl A
Industrial Lighting Salesman, Consolidated Gas, Elec. Light & Power Co., Lexington Bldg., Baltimore, Md.

Fisk, Eugene, M. A
Engineer, Western Union Telegraph Co., 195 Broadway, New York, N. Y.

Fiske, John M. A
Sub-Station Operator, Chicago, Milwaukee & St. Paul Railway, Box 434, Three Forks, Montana.

Fitch, Walter S. A
Construction Engineer, Dennison Mfg. Co., Framingham, Mass.

Fitzpatrick, K., Jr. A
Lighting Sales Agent, Dayton Power & Light Co., 20 S. Jefferson St., Dayton, Ohio.

Fitzsimons, Thomas E. A
Westinghouse Lamp Co., 1052 Gas & Electric Bldg., Denver, Colo.

Flaherty, H. J.   A
Assistant Engineer, Street Lighting Dept., General Electric Co., River Works, West Lynn, Mass.

Fleisher, Walter A.   A
Sales Service, S. B. & B. W. Fleisher, Inc., 25th & Reed Streets, Philadelphia, Pa.

Fleming, E. F.   A
Sales Engineer, Central Electric Co., 316 S. Wells St., Chicago, Ill.

Fletcher, James Y.   M
Director, General Electric Co., Ltd., Magnet House, Kingsway, London W. C. 2, England.

Flowers, Dean W.   A
General Superintendent & Engineer, St. Paul Gas Light Co., 51 East Sixth St., St. Paul, Minn.

Fobes, George Shaw   A
Manager, Boston Office, Macbeth-Evans Glass Co., 60 High St., Boston, Mass.

Foeller, Henry A.   A
Architect, Foeller, Schober & Stephenson, Nicolet Bldg., Green Bay, Wis.

Foersterling, H.   A
Research Chemist, The Abor Farm, Jamesburg, N. J.

Fogg, Oscar H.   A
Secretary & Manager, American Gas Association, 342 Madison Ave., New York, N. Y.

Fogler, William A.   A
Superintendent Laboratories, Philadelphia Electric Co., 2301 Market St., Philadelphia, Pa.

Folsom, Charles E.   A
Optometrist, 7 Alabama St., Atlanta, Georgia.

Foltz, Leroy S.   A
Associate Professor of Electrical Engineering, Michigan Agricultural College, East Lansing, Michigan.

Ford, Arthur H.   A
352 Sixth Ave. S., St. Petersburg, Fla.

Ford, Frederick Howe   M
Manager, Headlight Lens Dept., C. A. Shaler Co., Waupun, Wis.

Forstall, Alfred   A
Consulting Engineer, Forstall, Robison & Luqueer, 15 Park Row, New York, N. Y.

Forstall, Walton   M
Director, 1920–23
Engineer of Distribution, United Gas Improvement Co., 1401 Arch St., Philadelphia, Pa.

Foster, Spottswood C.   M
Mechanical & Electrical Engineer, Bedford Pulp & Paper Co., Coleman's Falls, Va.

Foster, William   A
3218 Turner St., Station C., Philadelphia, Pa.

Fowle, Frank F.   M
Consulting Engineer, Frank F. Fowle & Co., 1201 Monadnock Bldg., Chicago, Ill.

Fox, Clifford S.   A
Superintendent of Distribution, The East River Gas Co. of Long Island City, Webster Ave. & East River, Long Island City, N. Y.

Fox, Edward G. B.   A
Commercial Engineer, Edison Lamp Works of G. E. Co., Harrison, N. J.

Fox, William A.   A
Vice-President, Commonwealth Edison Co., 72 West Adams St., Chicago, Ill.

Francisco, Ellsworth   A
Supervisor of Gas Street Lighting, City Hall, Newark, N. J.

Franck, Charles   M
Manager, Holophane Glass Co., Inc., 340 Madison Ave., New York, N. Y.

Francklyn, Gilbert   A
Engineer of Distribution, Consolidated Gas Co. of N. Y., 130 East 15th St., New York, N. Y.

Frechie, Joseph E.   A
Lighting Fixture Manufacturer, Joseph E. Frechie & Co., Inc., 27 N. Seventh St., Philadelphia, Pa.

Freeman, C. K.   M
Vice-President & Treasurer, Armspear Manufacturing Co., 447 West 53d St., New York, N. Y.

Freeman, E. H.   A
Professor of Electrical Engineering, Armour Institute of Technology, 33d & Federal Sts., Chicago, Ill.

Freeman, Frederick Charles   A
Providence Gas Co., Providence, Rhode Island.

Freeman, W. W.   M
President, Union Gas & Electric Co., Fourth & Plum Sts., Cincinnati, O.

Frei, Fred J. A
Lighting Representative, Public Service Electric Co., 188 Ellison St., Paterson, N. J.

Friedrich, Ernest G. A
Westinghouse Electric & Manufacturing Co., Third & Elm Sts., Cincinnati, Ohio.

Frisbie, Walter L. A
Division Superintendent, Edison Electric Illuminating Co. of Boston, 39 Boylston St. Boston, Mass.

Froget, Andre A
Holaphane Company, 156 Boulevard Haussmann, Paris (8) France.

Fuchs, Theodore, Jr. A
Asst. in Illumination Service, Edison Lamp Works of G. E. Co., Harrison, N. J.

Fuerst, G. P. A
Harrington Electric Company, 413 Caxton Bldg., Cleveland, Ohio.

Fujii, Tetsuya A
Engineer in charge of Laboratory, Tokyo Electric Co., Kawasaki machi, Kanagawa Ken, Tokyo, Japan.

Fukuda, Yutaka A
Electrical Engineer, Japan Hydraulic Power Co., Ltd., Tokyo, Japan.

Fuller, Carl T. A
Incandescent Lamp Engineer, Edison Lamp Works of G. E. Co., Harrison, N. J.

Fuller, H. W.
Vice-President in charge of Operation, Northern States Power Co., 15 S. Fifth St., Minneapolis, Minn.

Fulton, Robert A.
Power Salesman, Cleveland Electric Illuminating Co., 75 Public Sq., Cleveland, Ohio.

Fulweiler, W. Herbert A
Engineer with Dept. of Tests, United Gas Improvement Co., 319 Arch St., Philadelphia, Pa.

Fussell, Lewis A
Professor of Electrical Engineering, Swarthmore College, Swarthmore, Pa.

Gaffey, John J. M
Assistant Engineering Manager, Harry M. Hope Engineering Co., 230 Boylston St., Boston, Mass.

Gage, Henry Phelps M
Physicist, Corning Glass Co., Corning, N. Y.

Gage, Otis Amsden A
Sales Department, Corning Glass Works, Corning, N. Y.

Gallagher, Francis A., Jr. A
Manager of Lighting Division, Narragansett Electric Lighting Co., 814 Turks Head Building, Providence, R. I.

Gallo, Charles A
Illuminating Engineer, Public Service Electric Co., 80 Park Place, Newark, N. J.

Ganser, Herbert H. A
Manager, Counties Gas & Electric Co., 212 DeKalb St., Norristown, Pa.

Gardner, George Norman A
Lighting Specialist, Canadian Westinghouse Co., Ltd., 602 Hastings St., W., Vancouver, British Columbia, Canada.

Garms, Peter A
In charge of Service Department, Electrical Testing Laboratories, 540 East 80th St., New York, N. Y.

Garrison, A. C. A
General Manager, Columbia Lamp Division, 1201 Title Guaranty Bldg., 706 Chestnut St., St. Louis, Mo.

Gartley, W. H. M
Director, 1908
President, 1909.
Junior Past-President, 1910–11
Vice-President, Equitable Illuminating Gas Light Company of Phila., 1401 Arch Street, Philadelphia, Pa.

Gast, Fred W. M
Electrical Engineer, U. S. Treasury Department, Washington, D. C.

Gawan, Louis B. A
Commercial Ltg. Salesman, Utah Power & Light Co., 132 Main St., Salt Lake City, Utah.

Gawtry, Lewis B. A
Vice-President, Bank of Savings, 280 Fourth Avenue, New York, N. Y.

Gens, Morris H. A
Superintendent, Peerless Electric Co., 37 Exchange St., Boston, Mass.

Getz, A. M. A
Delaware County Electric Co., 29 West State St., Media, Pa.

Gibbs, L. D.   A
Superintendent, Advertising Department, Edison Electric Illuminating Co. of Boston, 39 Boylston St., Boston 10, Mass.

Gibson, Fairley J.   A
District Supervisor, Philadelphia Electric Company, Ogontz, Pa.

Gifford, N. W.   A
Vice-President, Consolidated Gas Co., 38 Central Sq., East Boston, Mass.

Gilbert, William M., M. D.   M
Eye Specialist, Room 208, Franklin Trust Bldg., 20 S. 15th St., Philadelphia, Pa.

Gilchrist, James M.   A
Secretary & General Manager, Federal Electric Co., 8700 S. State St., Chicago, Ill.

Gilliam, H. R.   M
Superintendent, Lancaster Lens Co., 126 W. Main St., Lancaster, Ohio.

Gillinder, Edwin B.   M
Manufacturer, Gillinder Brothers, 23 Sullivan Ave., Port Jervis, N. Y.

Glameyer, William, Jr.   A
Meter Tester, United Electric Light & Power Co., 514 West 147th St., New York, N. Y.

Glasgow, A. G.   M
Chairman, Humphreys & Glasgow, Ltd., 38 Victoria St., London, S. W. 1, England.

Gleason, Marshall T.   A
Sales Manager & Assistant Secretary, Gleason-Tiebout Glass Co., 99 Commercial St., Brooklyn, N. Y.

Gleeson, T. P.   A
Manager, Illuminating Engineering Dept., Commercial Electric Supply Co., 320 S. Broadway, St. Louis, Mo.

Godley, Charles E.   M
Engineer & Designer, Edmunds & Jones Corporation, 4440 Lawton Ave., Detroit, Mich.

Goepel, C. P.   A
Patent Counsel, Goepel & Goepel, 165 Broadway, New York, N. Y.

Golden, John W.   A
Vice-President, Savannah Gas Co., 114 Barnard St., Savannah, Ga.

Goldmark, C. J.   M
Consulting Engineer, 103 Park Ave., New York, N. Y.

Goodale, Samuel Perley   A
Electrical Engineer, Thorndike Co., Thorndike, Mass.

Goodchild, Albert E.   M
Goodchild Electrical Co., 128 St. Peter St., Room 5, Montreal, Canada.

Goodenough, Francis W.   M
Controller of Gas Sales, The Gas Light & Coke Co., Horseferry Road, Westminster, London, S.W., England.

Goodrich, William M.   M
Manager, Lighting Department, Western Electric Co., 500 S. Clinton St., Chicago, Ill.

Goodwin, Harold Jr.   A
Sanderson & Porter, 52 William St., New York, N. Y.

Goodwin, William L.   M
Assistant to President, Society for Electrical Development, 522 Fifth Ave., New York, N. Y.

Gorge, S. V.   A
Electrical Contractor-Dealer, 8416 Eighteenth Avenue, Brooklyn, N. Y.

Gorton, Robert E.   A
General Manager, Packard Lamp Division of G. E. Co., Warren, Ohio.

Gosling, Edward P.   A
General Manager, Newport Electric Corporation, 449 Thames St., Newport, R. I.

Gottsche, A. Lynd   A
Interior Ltg. Section, Westinghouse Elec. & Mfg. Co., George Cutter Works, South Bend, Ind.

Gould, Herman P.   M
General Manager, Eastman Optical Shop, Inc., 12 Maiden Lane, New York, N. Y.

Goward, A. T.   A
Manager, B. C. Electric Railway
Co., P. O. Drawer 100, Victoria,
B. C., Canada.

Graf, Carl H.   A
Vice-President & General Manager,
Municipal Gas Co., 124 State St.,
Albany, N. Y.

Graff, Wesley M.   A
Consulting Engineer, Graff Engi-
neering Corporation, 150 Nassau
St., New York, N. Y.

Graham, E. C.   A
President & General Manager, Na-
tional Electrical Supply Co., 1330
New York Ave., Washington,
D. C.

Graham, Malcolm M.   A
Secretary & Treasurer, New Amster-
dam Gas Co., 130 East 15th St.,
New York, N. Y.

Grant, Albert W., Jr.   A
Superintendent, Minnesota By-
Products Coke Co., 1000 Ham-
line Ave., St. Paul, Minn.

Graves, L. H.   M
Vice-President, X-Ray Reflector Co.
of N. Y., Inc., 31 West 46th St.,
New York, N. Y.

Gray, G. Robin   A
Supervisor of Buildings, Bell Tele-
phone Co. of Canada, Montreal,
Canada.

Gray, Samuel McK.   A
Technical Assistant, Electrical Test-
ing Laboratories, 80th St. & East
End Ave., New York, N. Y.

Green, George Ross   A
Director, 1909-1911.
Engineer, Philadelphia Electric Co.,
1000 Chestnut St., Philadelphia, Pa.

Greene, Joseph E.   A
Joseph E. Greene Co., Inc., 85
Pearl St., Boston, Mass.

Greenewalt, Mary Hallock   A
1424 Master St., Philadelphia, Pa.

Greiner, Robert E.   A
Illuminating Engineer, Edison Lamp
Works of G. E. Co., Harrison, N. J.

Gribbel, John   A
Vice-President, Brooklyn Borough
Gas Co., 1513 Race St., Philadel-
phia, Pa.

Griffin, G. Brewer*   M
Vice-President, 1907.
82 Worthington St., Springfield,
Mass.

Griffin, J. H., Jr.   A
Special Inspector, Edison Electric
Illuminating Co. of Boston, 39
Boylston St., Boston, Mass.

Griswold, Clarence F.   A
Lighting Salesman, Narragansett
Electric Lighting Co., 814 Turks
Head Bldg., Providence, R. I.

Gritzan, L. LeRoy   M
Claude B. Hellman Co., 403 N.
Charles St., Baltimore, Md.

Grondahl, L. O.   M
Director of Research, Union Switch
& Signal Co., Swissvale, Pa.

Gross, E. L.   M
President, Gross Chandelier Co.,
2036 Morgan St., St. Louis, Mo.

Grossberg, Arthur S.   A
Mechanical Engineer, Albert Kahn,
1000 Marquette Bldg., Detroit,
Mich.

Gunnison, Foster   M
Secretary, Cox, Nostrand & Gunnison, Inc., 337 Adams St., Brooklyn, N. Y.

Guth, Edwin F.   M
President, The Edwin F. Guth Co.,
2615 Washington Ave., St. Louis,
Mo.

Haas, O. F.   A
Illuminating Engineer, National
Lamp Works of G. E. Co., Nela
Park, Cleveland, Ohio.

Hahn, Harold W.   M
Sales Manager, Cox, Nostrand &
Gunnison, Inc., 337 Adams St.,
Brooklyn, N. Y.

Haines, E. L.   A
Secretary & General Sales Manager,
National X-Ray Reflector Co.,
235 West Jackson Blvd., Chicago, Ill.

Hake, Harry G.   A
Assistant Professor of Electrical
Engineering, Washington Univer-
sity, St. Louis, Mo.

Hake, William S.   A
Manager, Fixture Dept., Brown &
Pierce Co., Inc., Main & Franklin
Sts., Rochester, N. Y.

Halbertsma, Nicolaas Adolf   A
Illuminating Engineer, Philips Glowlampworks, Ltd., Eindhoven, Ko-
ekoeklaan 8, Netherlands, Holland.

Hale, H. S. A
Interior Lighting Engineer, Westinghouse Electric & Manufacturing Co., 717 S. 12th St., St. Louis, Mo.

Hale, Robert S. M
Superintendent, Special Research, Edison Electric Illuminating Co. of Boston, 39 Boylston St., Boston, Mass.

Hall, Arthur H. A
Assistant Treasurer & Superintendent of Distribution, Central Union Gas Co., 519 Courtland Ave., Bronx, New York, N. Y.

Hall, James D. A
Commercial Engineering Dept., Westinghouse Lamp Company, Bloomfield, N. J.

Hall, Lee D. A
Commercial Agent, Alabama Power Co., Huntsville, Ala.

Hall, William Parker A
Salesman, Macbeth-Evans Glass Co., 1722 Ludlow St., Philadelphia, Pa.

Halstead, Charles W. · M
Sales Superintendent, Lowell Electric Light Corporation, 31 Market St., Lowell, Mass.

Halvorson, C. A. B., Jr. M
Vice-President, 1914–16
Designing Engineer, General Electric Co., 928 Western Ave., West Lynn, Mass.

Hamblen, Ephraim S. M
Manager, Union Light & Power Co., 25 East Central St., Franklin, Mass.

Hamm, William S. A
The Adams & Westlake Co., 319 W. Ontario St., Chicago, Ill.

Hanlon, James P. M
Assistant to New Business Agent, Public Service Gas Co., 80 Park Place, Newark, N. J.

Hannum, Joshua E. A
Research Engineer, Eye Sight Conservation Council of America, 1206 Times Bldg., New York, N. Y.

Hanscom, William W. M
Consulting Engineer, 848 Clayton St., San Francisco, Calif.

Harman, George H. A
Thomas Day Co., 725 Mission St., San Francisco, Calif.

Harries, George H. M
President, 1920–21.
Junior Past-President, 1921–23.
H. M. Byllesby & Co., Continental & Commercial Bank Bldg., Chicago, Ill.

Harris, Arthur C. A
District Sales Agent, Milwaukee Electric Railway & Light Co., Racine, Wis.

Harris, Henry A
President, United Light Co., Wilmerding, Pa.

Harrison, Haydn T. A
11 Victoria St., London S. W., England.

Harrison, Ward M
Vice-President, 1913–15.
President, 1922–23.
Illuminating Engineer, National Lamp Works of G. E. Co., Nela Park, Cleveland, Ohio.

Hartman, Harris V. A
Manager of Bureau of Lighting & Decorations, New York Edison Co., 130 East 15th Street, New York, N. Y.

Hartman, Leon Wilson A
Professor of Physics, University of Nevada, 215 Maple St., Reno, Nevada.

Harty, Edgar A. A
Electrical Tester, River Works of G. E. Co., Lynn, Mass.

Hastings, Milton B. A
Sales Engineer, A. H. Winter Joyner, Ltd., 62 Front St., W., Toronto, Ontario, Canada.

Hatch, B. E. M
Interior Lighting Engineer, Westinghouse Elec. & Mfg. Co., 1442 Widener Bldg., Philadelphia, Pa.

Hatzel, J. C. M
Electrical Engineer & Contractor, Hatzel & Buehler, 373 Fourth Ave., New York, N. Y.

Hausmann, George A. A
Electrical Engineer, Board of Education, Sixth & Rockwell Sts., Cleveland, Ohio.

Hawkins, Lawrence A. M
Engineer, Research Laboratory, General Electric Co., Schenectady, N. Y ·

Haynes, Pierre E.   A
881 Ellicott Square Bldg., Buffalo, N. Y.

Hecker, Louis M.   A
Street Illumination, Commercial Light Co., 127 N. Dearborn St., Chicago, Ill.

· Heckmann, L. F.
District Illuminating Engineer, Westinghouse Electric & Manufacturing Co., 814 Ellicott Sq., Buffalo, N. Y.

Heffron, Joseph F.   A
Macbeth-Evans Glass Co., Chamber of Commerce Bldg., Pittsburgh, Pa.

Helme, William E.   A
Helme & McIlhenny, 700 Bulletin Bldg., Philadelphia, Pa.

Heman, Stanley A.   A
Illuminating Engineer, George Cutter Works, Westinghouse Electric & Manufacturing Co., South Bend, Ind.

Hendry, W. Ferris   A
Chief Engineer, Manhattan Electrical Supply Co., 17 Park Place, New York, N. Y.

Henninger, John G.   M
District Representative, Fostoria Incandescent Lamp Works of G. E. Co., 4101 Fifth St., N. W., Washington, D. C.

Henry, J. R.   A
National X-Ray Reflector Co., 628 Witherspoon Bldg., Philadelphia, Pa.

Henry, Charles T.   A
Electrical Contractor, 13 West 42d St., New York, N. Y.

Herrmann, W. S.   A
Westinghouse Electric & Manufacturing Co., 1535 Sixth St., Detroit, Mich.

Herron, James C.   M
President & General Manager, Reflector & Illuminating Co., 565 West Washington Blvd., Chicago, Ill.

Hibben, Samuel G.   M
Director, 1914-16.
General Secretary, 1922-24.
Manager, Illumination Bureau Westinghouse Lamp Company, 165 Broadway, New York, N. Y.

Hibbs, Weston James   A
Treasurer, U. G. I. Contracting Co., 1401 Arch St., Philadelphia, Pa.

Hickox, Norman B.   A
Sales Manager, National X-Ray Reflector Co., 235 West Jackson Blvd., Chicago, Ill.

Higbie, H. Harold   A
Professor of Electrical Engineering, University of Michigan, Ann Arbor, Mich.

Higgins, A. W.   M
President, South Minnesota Gas & Electric Co., 110 S. Dearborn St., Chicago, Ill.

Higgins, C. M.   A
Standard Oil Co., 26 Broadway, New York, N. Y.

Higgins, W. S.   A
Professor, Southwestern Presbyterian University, Clarkesville, Tenn.

Hill, Charles F.   A
Electrician, W. F. Schrafft & Sons, 160 Washington St., N., Boston, Mass.

Hill, Emory, M. D.   A
Ophthalmologist, 501 East Franklin St., Richmond, Va.

Hill, George A.   A
Illuminating Engineer, National X-Ray Reflector Co., 235 W. Jackson Blvd., Chicago, Ill.

Hill, Marvin   A
Benton, Kansas.

Hinton, James W.   A
Salesman, Westinghouse Lamp Co., 1005 Market St., Philadelphia, Pa.

Hirons, Frank K.   A
Sales Engineer, Pettingell-Andrews Co., 511 Atlantic Ave., Boston, Mass.

Hitchcock, George G.   A
Pomona College, Claremont, Calif.

Hixson, Morris C.   A
Building Equipment Specialist, General Electric Co., Rialto Bldg., San Francisco, Calif.

Hoadley, George A.   M
Vice-President, 1914-16.
President, 1918-19.
Junior Past-President, 1919-21.
Professor of Physics & Electrical Engineering, Swarthmore College, 518 Walnut Lane, Swarthmore, Pa.

Hobbie, Edward H.   **M**
Publicity & Promotion Manager, Mississippi Wire Glass Co., 220 Fifth Ave., New York, N. Y.

Hobbs, Leonard A.   **M**
District Manager, The Edwin F. Guth Co., 1331 West Seventh St., Los Angeles, Calif.

Hobson, Henry Elmer   **A**
Sales Manager, Southwest General Electric Co., Interurban Bldg., Dallas, Texas.

Hockaday, Irving T.   **A**
Salesman, Southwest General Electric Co., 1109 City National Bank Bldg., San Antonio, Texas.

Hodge, Percy   **M**
Professor of Physics, Stevens Institute of Technology, Hoboken, N. J.

Hodgson, George Oscar   **A**
Edison Lamp Works of G. E. Co., 508 U. S. National Bank Bldg., Denver, Colo.

Hoeller, Otto A.   **A**
Division Sales Manager, Central Illinois Public Service Co., Beardstown, Ill.

Hoeveler, John A.   **M**
Electrical Engineer, Industrial Commission of Wisconsin, Capitol Annex, Madison, Wis.

Hofrichter, Charles H.   **M**
President, Crescent Brass Products Co., 233 Gordon Sq., Bldg., Cleveland, Ohio.

Holden, Alfred R.   **A**
Department of Design, The B. F. Goodrich Co., S. Main St., Akron, Ohio.

Holdrege, H. A.   **M**
1936 South 33d St., Omaha, Neb.

Holladay, Lewis L.   **A**
Research Physicist & Engineer, Applied Science Laboratory, National Lamp Works of G. E. Co., Nela Park, Cleveland, Ohio.

Holman, E. R.   **A**
Commercial Lighting Specialist, Solar Lighting Co., Inc., 219 East Fourth Street, Los Angeles, Calif.

Holmes, James Thomas   **A**
Electrical Engineer, Superintendent of Construction Dept., I. P. Frink, Inc., 239 Tenth Ave., New York, N. Y.

Hopkins, Raymond A.   **M**
Electrical Engineer, Stone & Webster, Inc., 147 Milk St., Boston, Mass.

Hopper, Charles H.   **A**
Salesman, Canadian Westinghouse Co., Ltd., Traders Bank Bldg., Toronto, Ontario, Canada.

Horn, Carrie E.   **A**
Technical Assistant in study of Light and Illumination, Electrical Testing Laboratories, 80th St. & East End Ave., New York, N. Y.

Horner, Merritt, Jr.   **A**
Vice-President, 1918-20.
Illuminating Engineer, Edison Lamp Works of G. E. Co., Harrison, N. J.

Howard, G. T.   **A**
Electric Contractor, Althoff-Howard Electric Co., 314 First Ave., Evansville, Ind.

Howard, Harold W.   **A**
Illum. Eng. Laboratory, General Electric Co., Schenectady, N. Y.

Howard, T. H.   **A**
President, Phoenix Glass Co., Ninth Street, Monaca, Pa.

Howe, Floyd W.   **A**
Lighting Salesman, Narragansett Electric Lighting Co., 814 Turks Head Bldg., Providence, R. I.

Howell, John W.   **M**
General Electric Co., Harrison, N. J.

Hower, Harry S.   **M**
Director of Research, Macbeth-Evans Glass Co., Chamber of Commerce Bldg., Pittsburgh, Pa.

Howland, Lewis A.   **A**
General Superintendent, Queens Borough Gas & Electric Co., Far Rockaway, N. Y.

Hoyt, W. Greeley   **M**
Vice-President, 1917-19.
President, Standard Gas Light Co., 130 East 15th St., New York, N. Y.

Hubbard, Arthur Sherwood   **A**
Cooper-Hewitt Electric Co., 161 Summer St., Boston, Mass.

Hubbard, William C.*   **A**
General Sales Agent, Cooper-Hewitt Electric Co., Eighth & Grand Sts., Hoboken, N. J.

Hubbert, Edward I.   **A**
Inspector, Philadelphia Electric Co., Milnor & Robbins Sts., Philadelphia, Pa.

Hubert, Conrad A
Glenalla Realty Corporation, 305 East 43d St., New York, N. Y.

Hudson, R. A. A
Consulting Engineer, Hunter & Hudson, 505 Rialto Bldg., San Francisco, Calif.

Huffnagle, Arthur A. A
Lighting Salesman, Commonwealth Edison Co., 72 West Adams St., Chicago, Ill.

Hughes, David M. A
Grosbard & Hughes, 802 Garland Bldg., Los Angeles, Calif.

Hulswit, Frank T. M
President, United Light & Railways Co., Michigan Trust Bldg., Grand Rapids, Mich.

Humez, J. F. A
Representative, Macbeth-Evans Glass Co., 5-134 General Motors Bldg., Detroit, Mich.

Humpfer, George Jr. A
Electrician, Strawbridge & Clothier, Eight & Market Sts., Philadelphia, Pa.

Humphrey, Arthur F. A
Salesman, Biddle-Gaumer Co., 3846 Lancaster Ave., Philadelphia, Pa.

Humphreys, Alexander C. M
President, Stevens Institute of Technology, Hoboken, N. J.

Hunter, George Leland A
122 East 82d St., New York, N. Y.

Hunter, James F. A
Assistant Engineer of Construction, Consolidated Gas Co., 130 East 15th St., New York, N. Y.

Hunter, Morris A
Electrical Engineer & Contractor, 113 North 8th St., Richmond, Virginia.

Huntley, William R. A
Vice-President, Buffalo General Electric Co., 1200 Electric Bldg., Buffalo, N. Y.

Hurley, Edward D., M. D. A
Oculist, 419 Boylston St., Boston, Mass.

Huse, Merritt C. A
Assistant to Commercial Manager, Philadelphia Electric Co., 1000 Chestnut St., Philadelphia, Pa.

Hussey, R. B.* A
Assistant Engineer, Street Lighting Dept., General Electric Co., 928 Western Ave., West Lynn, Mass.

Hutton, Donald J. A
Contract Agent, Mexican Light & Power Co., Ltd., Apartado Postal 124-Bis., Mexico, D. F. Mexico.

Hyde, E. N. M
E. N. Hyde Electric Co., 114 N. 11th St., Philadelphia, Pa.

Hyde, Edward P. M
President, 1910.
Junior Past-President, 1911-12.
Morgan-Harjes & Co., 14 Place Vendome, Paris, France.

Hyde, Edward B. A
Superintendent, Commercial Dept., New Amsterdam Gas Co., 140 East 15th St., New York, N. Y.

Hyldahl, Niels A
Contract Dept., Commonwealth Edison Co., 72 West Adams St., Chicago, Ill.

Hyodo, Masaru A
Chief of Eng. Dept., Osaka Electric Lamp Co., Daini, Osaka-fu, Japan.

Ilgner, Howard F. A
Engineer in charge of Illumination, City of Milwaukee, Bureau of Illumination Service, City Hall, Milwaukee, Wis.

Illing, I. L. A
Illuminating Engineer, Milwaukee Electric Railway & Light Co., Public Service Bldg., Milwaukee, Wis.

Ingalls, L. O. M
Electrical Engineer, United Electric Light Co., 210 Alden St., Springfield, Mass.

Insull, Samuel M
President, Commonwealth Edison Co., 72 ,West Adams St., Chicago, Ill.

Irwin, Beatrice A
Writer & Chromatologist, The Color Science Centre, 121 East 57th St., New York, N. Y.

Irwin, Walter E. L. A
Cadet Engineer, United Gas Improvement Co., 1931 S. Ninth St., Philadelphia, Pa.

Ishikawa, Yasuta A
Illuminating Engineering Dept., Tokyo Electric Co., Kawasaki-Kanagawa-ken, Japan.

Ishikawa, Yoshijiro A
Director & Manager, Commercial Dept., Kyoto Electric Light Co., Kyoto, Japan.

Israel, Joseph D.   **M**
Director, 1909–10.
General Secretary, 1913–14.
Director, 1914–16.
District Manager, Philadelphia Electric Co., 1000 Chestnut St., Philadelphia, Pa.

Ives, James E.   **M**
Physicist, U. S. Public Health Service, 16 Seventh St., S. W. 1, Washington, D. C.

Ives, John Nash   **M**
Stone & Webster, Inc., 147 Milk St., Boston, Mass.

Jackson, Dugald C.   **M**
Engineer, 387 Washington St., Boston, Mass.

Jackson, Thomas H.   **A**
General Agent, United Gas Improvement Co., 1401 Arch St., Philadelphia, Pa.

James, Leonard V.   **A**
1460 Maryland Ave., Milwaukee, Wis.

Jaquet, George E.   **A**
Editor, Electrical Directory of Canada, Gardenvale, Quebec, Canada.

Jarrard, O. O.   **A**
Vice-President, Mauch Chunk Heat, Power & Electric Light Co., Ltd. Lehigh Ave., Mauch Chunk, Pa.

Jenkins, David R.   **A**
Professor of Electrical Engineering, University of North Dakota, Grand Forks, N. D.

Johnson, Alfred   **A**
Mechanical & Electrical Engineer, Mandel Brothers, State, Madison & Wabash Avenues, Chicago, Ill.

Johnson, Charles E.   **A**
Lighting Specialist, Westinghouse Electric & Manufacturing Co., 420 S. San Pedro St., Los Angeles, Calif.

Johnson, L. B.   **A**
Sales Engineer, General Electric Co., 719 Newhouse Bldg., Salt Lake City, Utah.

Johnson, E. H.   **A**
Professor of Physics, Kenyon College, Gambier, Ohio.

Johnson, Otis L.   **M**
Vice-President, 1918–20 and 1922–24
Commercial Engineer, King Manufacturing Co., 230 S. Clark St., Chicago, Ill.

Johnson, W. H.   **M**
Senior Vice-President, Philadelphia Electric Co., 1000 Chestnut St., Philadelphia, Pa.

Johnston, Howard L.   **A**
Westinghouse Electric & Manufacturing Co., East Pittsburgh, Pa.

Johnston, J. K.   **A**
General Manager, Bryan-Marsh Division of National Lamp Works of G. E. Co., 4622 Grand Central Terminal, New York, N. Y.

Johnston, Richard J.   **A**
Engineer, George Cutter Works, Westinghouse Electric & Manufacturing Co., South Bend, Ind.

Jones, Bassett   **M**
Director, 1908–10.
Consulting Engineer, 101 Park Ave., New York, N. Y.

Jones, Lloyd A.   **M**
Physicist, Research Laboratory, Eastman Kodak Co., Bldg. 2, Kodak Park, Rochester, N. Y.

Jones, Philip C.   **M**
Electrical Engineer, The Goodyear Tire & Rubber Co., Akron, Ohio.

Jones, Theodore I.   **A**
General Sales Agent, Edison Electric Illuminating Co. of Brooklyn, 360 Pearl St., Brooklyn, N. Y.

Jones, William R.   **M**
Engineer of Plant, University of Pennsylvania, 3446 Walnut St., Philadelphia, Pa.

Jourdan, James H.   **A**
President, Brooklyn Union Gas Co., 176 Remsen St., Brooklyn, N. Y.

Kase, Daniel B.   **A**
Lighting Specialist, Rumsey Electric Co., 1007 Arch St., Philadelphia, Pa.

Kato, Kenjiro   **A**
Chief of Engineering Section, Tokyo Electric Co., Kawasaki-machi, Kanagawa Ken, Japan.

Kauffman, Paul F.   **A**
Salesman, Commonwealth Edison Co., 72 West Adams St., Chicago, Ill.

Kaulke, Johannes   **A**
Electrical Engineer, General Electric Co., Harrison, N. J.

Keenan, John R.   **A**
Philadelphia Electric Co., 18th & Columbia Ave., Philadelphia, Pa.

Kelcey, G. G.   M
Manager, Traffic Engineering Division, American Gas Accumulator Co., 999 Newark Ave., Elizabeth, N. J.

Keller, Phil C.   A
Commercial Engineer, Ivanhoe-Regent Works of G. E. Co., 5716 Euclid Ave., Cleveland, Ohio.

Kelley, J. Bayard   M
Illuminating Engineer, 100 Linden Ave., Collingswood, N. J.

Kellogg, Alfred S.   A
Consulting Engineer, 89 Franklin St., Buston, Mass.

Kellogg, Raymond C.   M
Consulting Engineer, Room 1112 Harris Bldg., 111 West Monroe St., Chicago, Ill.

Kelly, James B.   A
U. S. Veterans' Bureau, 1037 N. S. National Bank Bldg., Denver, Col.

Kennedy, James S.   A
Superintendent of Works, Standard Gas Light Co., East 115th St., New York, N. Y.

Kennedy, Jeremiah J.   A
Consulting Engineer, 52 Broadway, New York, N. Y.

Kennelly, Arthur Edwin   M
President, 1911.
Junior Post-President, 1912-13.
Professor of Electrical Engineering, Harvard University, Cambridge, Mass.

Kenslea, Daniel L.   A
Electrical Instructor, Mullane Engineering School, 17 Yarmouth St., Boston, Mass.

Kent, L. C.   A
Engineering Dept., National Lamp Works, of G. E. Co., Nela Park, Cleveland, Ohio.

Kerens, J. T.   A
Yale Electric Corporation, Tillary & Pearl Sts., Brooklyn, N. Y.

Kershaw, K. O.   A
Illuminating Specialist, Panama Lamp Co., 757 S. Los Angeles St., Los Angeles, Calif.

Ketch, James M.   A
Engineer, National Lamp Works, of G. E. Co., Nela Park, Cleveland, O.

Keuffel, Carl W.   A
Supervisor of Optical Dept., Keuffel & Esser Co., Third & Adams Sts., Hoboken, N. J.

Keyes, Sumner R.   A
Assistant Superintendent, Purchasing Dept., Edison Electric Illuminating Co., 39 Boylston St., Boston, Mass.

Kiefer, Lewis J.   M
Superintendent of Building, McCreery & Co., Wood St. & Sixth Ave., Pittsburgh, Pn.

Kilby, Karl E.   M
Advertising Manager, Coleman Lamp Co., 222 N. St. Francis Avenue., Wichita, Kansas.

Killeen, John F.   A
General Sales Manager, Mitchell-Vance Co., 505 W. 24th St., New York, N. Y.

Kimball, H. H.   A
Meteorologist, U. S. Weather Bureau, 24th & M Sts., Washington, D. C.

Kinney, Clarence W.   A
Consulting Engineer, P. O. Box 753, Worcester, Mass.

Kinney, Raymond C.   A
Asst. to Plant Engineer, Western Electric Co., Hawthorn Plant, Dept. 6723, Chicago, Ill.

Kinsey, Frederick S.   A
Manager, Syndicate Dept., Westinghouse Lamp Co., 165 Broadway, New York, N. Y.

Kintner, Watson   M
Factory Engineer, Canadian-Westinghouse Co., Ltd., Aberdeen Ave., Hamilton, Ontario, Canada.

Kirby, Daniel B., M.D.   A
Ophthalmologist, Bellevue Hospital, New York, N. Y.

Kirk, James J.   M
Director, 1918-20.
Vice-President, 1920-22.
Illuminating Engineer, Commonwealth Edison Co., 72 West Adams St., Chicago, Ill.

Kirlin, Ivan M.   A
Manager, Detroit Office, National X-Ray Reflector Co., 400 Penobscot Bldg., Detroit, Mich.

Kirschberg, Harold   M
President, The Lighting Specialties Co., 543 Fourth Ave., Pittsburgh, Pa.

Kittle, Robert G.   A
Salesman, Benjamin Electrical Man-
ufacturing Co., 243 W. 17th St.,
New York, N. Y.

Knight, A. R.   M
Professor, Department of Electrical
Engineering, University of Illinois,
Urbana, Ill.

Knight, Carl D.   M
Asst. Professor of Electrical Engi-
neering, Worcester Polytechnic
Institute, Worcester, Mass.

Knight, Charles   A
Master Mechanic, Eastern Malleable
Iron Co., Bridge St., Naugatuck,
Conn.

Knight, J. Harmer   A
Electrical & Mechanical Draftsman,
Room 532, City Hall, Philadel-
phia, Pa.

Knoedler, E. L.   A
General Superintendent, Welsbach
Co., Gloucester City, N. Y.

Koenes, Orman   A
Manager, Wisconsin Gas & Electric
Co., 205 Main St., Watertown,
Wis.

Koerber, Jerome A.   A
Display Manager, Strawbridge &
Clothier, Eighth & Market Sts.,
Philadelphia, Pa.

Koockogey, H. A.   A
Philadelphia Electric Co., 1000
Chestnut St., Philadelphia, Pa.

Kopecky, F. E.   A
P. O. Box 27, Birmingham, Ala.

Krapf, Edgar W.   A
Electrician, Huffer & Cuddy, 1208
1208 Liberty Ave., Brooklyn,
N. Y.

Kruse, O. J.   A
Superintendent, Lighting Sales, Milwaukee Electric Railway & Light
Co., 215 Sycamore St., Milwaukee, Wis.

Kurlander, John H.   A
Illuminating Engineer, Edison Lamp
Works of G. E. Co., Fifth & Sus-
sex Sts., Harrison, N. J.

Labelle, Philip R.   A
Power Sales Mgr., Shawinigan Water
& Power Co., 621 Power Bldg.,
Montreal, Que., Canada.

Lancaster, Walter B., M. D.   A
Ophthalmic Surgeon, 522 Com-
monwealth Ave., Boston, Mass.

Lang, J. Gustav V.   M
Electrical & Mechanical Engineer,
8 West 40th St., New York, N. Y.

Lansingh, Van Rensselaer*   M
Treasurer, 1906.
General Secretary, 1907–08.
Vice-President, 1910–11.
Treasurer, 1911.
President, 1912.
Junior Past-President, 1913–14.
President, York Metal & Alloys
Co., 100 East 42d St., New York,
N. Y.

Lathrop, Alanson P.   A
President, American Light & Trac-
tion Co., 120 Broadway, New
York, N. Y.

Laughton, A. Abbott   A
Local Manager, Athol Gas & Elec-
tric Co., 426 Main St., Athol,
Mass.

Law, Clarence L.   M
Vice-President, 1915–17.
General Secretary, 1917–22.
Director, 1922–25.
Assistant to General Commercial
Manager, New York Edison Co.,
130 East 15th St., New York, N.Y.

Lawrence, Ralph R.   A
Associate Professor of Electrical
Engineering, Massachusetts In-
stitute of Technology, Cambridge,
Mass.

LeClear, Gifford   A
Densmore & LeClear, 88 Broad St.,
Boston, Mass.

Lee, James D.   M
Lighting Representative, Public Ser-
vice Electric Co., 418 Federal St.,
Camden, N. J.

Lee, Claudius   A
Professor, Virginia Polytechnic Institute, Blacksburg, Va.

Lent, A. L.   A
Detroit Edison Co., 2000 Second
Ave., Detroit, Mich.

Le Page, Clifford B.   M
Assistant Secretary, A. S. M. E.
(Standards Research) 29 West
39th St., New York, N. Y.

Leurey, Louis F.   A
Consulting Electrical Engineer, 58
Sutter St., San Francisco, Calif.

Levan, Milton Bryan   A
Chief Clerk, Spring Garden District,
United Gas Improvement Co.,
1615 N. 9th St., Philadelphia, Pa.

Lever, Thomas S., Jr. A
Office Manager & Employment Supervisor, United Gas Improvement Co., 24 N. 22d St., Philadelphia, Pa.

Leveridge, Rowland H. M
Chief, Bureau of Electrical & Mechanical Equipment, Dept. of Labor, State of N. J., State House, Trenton, N. J.

Levy, Max J. A
President, Edwards Electrical Construction Co., 70 East 45th St., Room 4039, New York, N. Y.

Lewinson, Leonard J. M
Engineer of Lamp Tests, Electrical Testing Laboratories, Inc., 80th St., & East End Ave., New York, N. Y.

Lewis, Frank L. A
Salesman, Ivanhoe-Regent Works of G. E. Co., 108 S. 59th St., Philadelphia, Pa.

Lieb, John W. M
Vice-President, New York Edison Co., 130 East 15th St., New York, N. Y.

Lillie, Lewis A
Treasurer, United Gas Improvement Co., 1401 Arch St., Philadelphia, Pa.

Lindsay, Carl C. A
P. O. Box 1023, City Hall Station, New York, N. Y.

Lingard, Herbert Arthur A
Electrical Engineer, The British Thomson - Houston Co., Ltd., Crown House, Aldwych, London, W. C. 2, England.

Lisle, Arthur B. A
General Manager, Narrangansett Electric Lighting Co., Room 802 Turks Head Bldg., Providence, R. I.

Lissfelt, H. L. A
Macbeth-Evans Glass Co., Room 1012, 19 West 44th St., New York, N. Y.

Little, William F. M
Engineer in charge of Photometry, Electrical Testing Laboratories, Inc., 80th St. & East End Ave., New York, N. Y.

Liversidge, Horace P. A
Operating Engineer, Philadelphia Electric Co., 1000 Chestnut St., Philadelphia, Pa.

Livor, J. E. A
Pettingell-Andrews Co., 160 Pearl St., Boston, Mass.

Livingston, Robert E. A
Publisher & Advertising Manager, Consolidated Gas Co. of N. Y., 1 Madison Ave., New York, N. Y.

Lix-Klett, Ernesto A
Electrical Engineer & Manager Director, E. Lix-Kleet & Co., 1088 Libertad St., Buenos Aires, Argentine Republic, S. A.

Lloyd, Lorne K. A
Vice-President, Morris-Lloyd Lighting Studios, Ltd., 274 Union Ave., Montreal, Quebec, Canada.

Lloyd, M. G. M
Vice-President, 1916–18.
Technical Editor, Bureau of Standards, Washington, D. C.

Locker, Frank H. A
Lighting Engineer, Detroit Edison Co., 2000 Second Ave., Detroit, Mich.

Lockhart, W. J. A
District Manager, Philadelphia Electric Co., 7 West Chelten Ave., Germantown, Pa.

Logan, Henry M
Electrical Engineer, Holoplane Glass Co., 12808 Phillips Ave., Cleveland Ohio.

Long, Walter E. A
Philadelphia Electric Co., 1000 Chestnut St., Philadelphia, Pa.

Love, Robert M. A
Street Lighting Specialist, Canadian General Electric Co., 212 King Street W., Toronto, Canada.

Luckiesh, M. M
Director, 1916–18.
Director, Laboratory of Applied Science, National Lamp Works of G. E. Co., Cleveland, Ohio.

Luqueer, Robert O. A
Forstall, Robison & Luqueer, 15 Park Row, New York, N. Y.

Lyle, A. Ernest A
Assistant to Factory Manager, Canadian Westinghouse Co., Aberdeen Ave., Hamilton, Ontario, Canada.

Lyman, James A
Sargent & Lundy, 1412 Edison Bldg., Chicago, Ill.

Lynch, J. H. A
District Illuminating Sales Engineer, Westinghouse Electric & Manufacturing Co., 10 High St., Boston, Mass.

Lyon, Howard   M
Engineer, Welsbach Co., Gloucester, N. J.

Macbeth, Norman   M
Vice-President, 1912–13.
General Manager, Macbeth Daylighting Co., Inc., 227 West 17th St., New York, N. Y.

MacCreery, R. B.   A
Division Manager, Controller's Dept., Philadelphia Electric Co., 1000 Chestnut St., Philadelphia, Pa.

Macdonald, Norman D.   M
Assistant to General Manager, Electrical Testing Laboratories, Inc., 80th St., & East End Ave., New York, N. Y.

MacGuffin, William D.   A
Lighting Specialist, General Electric Co., 84 State St., Boston, Mass.

Mackall, Kenneth W.   A
Illumination Work, Crouse Hinds Co., Syracuse, N. Y.

MacMullen, Charles   A
President, Clark, MacMullen & Riley, 101 Park Ave., New York, N. Y.

Macy, Carleton   A
President, Queensborough Gas & Electric Co., Far Rockaway, Long Island, N. Y.

Magdsick, H. H.   M
Engineering Dept., National Lamp Works of G. E. Co., Nela Park Cleveland, Ohio.

Magee, Ralph R.   A
Lighting Specialist, Westinghouse Electric & Manufacturing Co., 165 Broadway, New York, N. Y.

Magnuson, A. H.   A
Chief Electrician, Graton & Knight Manufacturing Co., 344 Franklin St., Worcester, Mass.

Mahan, Harwood E.   M
Illuminating Engineering Laboratory, General Electric Co., Schenectady, N. Y.

Mailloux, C. O.   M
Consulting Engineer, 111 Fifth Ave., New York, N. Y.

Maisonneuve, Henry   A
Engineer, Compagnie des Lampes, 41 rue La Boetie, Paris 8, France.

Malone, P. S.   A
Chief Inspector, Incandescent Lamps, Edison Lamp Works of G. E. Co., Harrison, N. J.

Malone, James F.   S
Polytechnic Institute of Brooklyn, 99 Livingston St., Brooklyn, N. Y.

Mange, John I.   A
Vice-President, Associated Gas & Electric Co., New York, N. Y.

Manter, Everett W.   A
Westinghouse Electric & Manufacturing Co., 10 High St., Boston, Mass.

Marcou, Arthur W.   A
General Manager, Bryan-Marsh Division of G. E. Co., 99 Chauncey St., Boston, Mass.

Mark, Isaac, Jr.   A
New York Edison Co., 865 Broadway, New York, N. Y.

Markley, Ralph E.   A
Illuminating Engineer, Philadelphia Electric Co., 1000 Chestnut St., Philadelphia, Pa.

Marks, Louis B.*   M
President, 1906.
Junior Past-President, 1907–08.
Director, 1910–12.
Treasurer, 1913–24.
Consulting Illuminating Engineer, 103 Park Ave., New York, N. Y.

Marley, Robert M.   A
Engineer, Marley Electric Co., 116 N. Camac St., Philadelphia, Pa.

Marmon, Guy D.   A
General Manager, United Electric Supply Co., Marion, Ohio.

Marsh, L. W.   M
New England Manager, American Luxfer Prism Co., 49 Federal St., Boston, Mass.

Marshall, W. H.   A
Assistant to Third Vice-President, United Gas Improvement Co., 1401 Arch St., Philadelphia, Pa.

Marthai, William G.   A
Illuminating Engineer, The New York Edison Co., 130 East 15th St., New York, N. Y.

Martin, Allen J.   A
Engineering Research Dept., Room 272 Engineering Bldg., University of Michigan, Ann Arbor, Michigan.

Martin, Carl O.   A
Benjamin Electric Manufacturing Co., 580 Howard St., San Francisco, Calif.

Martin, Edwin R. A
Professor, Electrical Engineering Dept., University of Minnesota, Minneapolis, Minn.

Martin, Thomas Commerford* A
Secretary, National Electric Light Association, 29 West 39th St., New York, N. Y.

Martin, William A. M
General Manager, Jefferson Glass Co., Ltd., 388 Carlow St., Toronto, Canada.

Martinson, Arthur N. A
Ludwig Hommel & Co., 530 Fernando St., Pittsburgh, Pa.

Mason, Sidney A
President, Welsbach Co., Gloucester, City, N. J.

Masson, Charles M. M
Chief Illuminating Engineer, Southern California Edison Co., Edison Bldg., 306 West Third St., Los Angeles, Calif.

Matchett, Fred D. A
Victor Electric Supply Company, 131 Jefferson Avenue E., Detroit, Mich.

Mather, E. H. A
Public Service Corporation Expert, Hayden, Stone & Co., 87 Milk St., Boston, Mass.

Matsuda, C. A
Research Student, Electrical Engineering Dept., Kyoto Imperial University, Kyoto, Japan.

Maurer, R. H. M
Illuminating Engineer, Consolidated Gas Company, 130 East 15th St., New York, N. Y.

Mausk, Raymond E. A
Engineer, Laboratory of Applied Science, National Lamp Works of G. E. Co., Nela Park, Cleveland, O.

Maxwell, V. C. A
Sales Manager, Supply Dept., W. G. Nagel Electrical Co., 28 St. Clair St., Toledo, Ohio.

Mayo, Herbert J. A
Sales Engineer, Benjamin Electric Manufacturing Co., 582 Howard St., San Francisco, Calif.

Mayo, J. F. A
Illuminating Engineer, Consumers Power Co., 236 W. Main St., Jackson, Mich.

McAleer, William C. A
Philadelphia Electric Co., 1000 Chestnut St., Philadelphia, Pa.

McAllister, A. S. M
Director, 1909-11.
President, 1914-15.
Junior Past-President, 1915-16
Room 302 South Bldg., Bureau of Standards, Washington, D. C.

McArdle, Leon C. A
Lighting Service Dept., Philadelphia Electric Co., 1000 Chestnut St., Philadelphia, Pa.

McBan, Patrick Fred M
Technical Supervision of Motion Pictures, Trimble-Murfin Productions, Hollywood, Calif.

McCall, Joseph B. M
President, Philadelphia Electric Co., 1000 Chestnut St., Philadelphia, Pa.

McCallion, William James A
Electrical Inspector, Department of Public Works, Sydney, N. S. Wales, Australia.

McDonald, Donald A
General Superintendent, American Meter Co., 105 West 40th St., New York, N. Y.

McDonough, Thomas F. A
District Representative, Benjamin Electric Manufacturing Co., 1225 L. C. Smith Bldg., Seattle, Wash.

McDougall, George K. A
Consulting Engineer, 85 Osborne St., Montreal, Quebec, Canada.

McGowan, Robert H. A
Lighting Specialist, Philadelphia Electric Co., 1000 Chestnut St., Philadelphia, Pa.

McGraw, James H. A
Publisher, McGraw-Hill Co., Inc., Tenth Ave. & 36th St., New York, N. Y.

McGregor, William A
Pawtucket Gas & Electric Co., 231 Main St., Pawtucket, R. I.

McGuire, Frederick J. M
Supervising General Inspector, Dept. of Water Supply, Gas & Electricity, New York, N. Y.

McIlhenny, John D. A
Helme & McIlhenny, 17th & Clearfield Sts., Philadelphia, Pa.

McKay, William A.   M
Commercial Engineer, Westinghouse
Lamp Co., Box 100, Bloomfield,
N. J.

McKinlock, George A.   A
President, Central Electric Co., 326
S. Fifth Ave., Chicago, Ill.

McLaughlin, Joseph J.   M
Illuminating Engineer, Westing-
house Lamp Co., 165 Broadway,
New York, N. Y.

McManis, Thomas J.   A
Manager, Dept. of Publicity, Edison
Lamp Works of G. E. Co., Harri-
son, N. J.

McNair, Grayson B.   A
Westinghouse Electric & Manufac-
turing Co., 1052 Gas & Electric
Bldg., Denver, Colo.

McMurtry, Alden L.   M
Consulting Engineer, Bacon Engi-
neering Co., Inc., Sound Beach,
Conn.

McParland, John J.   A
Sales Manager, Pittsburgh Reflector
& Illuminating Co., Room 214,
1452 Broadway, New York, N. Y.

M'Coy, W. R.   M
Sales Manager, Fixture Dept. Cas-
sidy Co., Inc., 15 Wilbur Ave.,
Long Island City, N. Y.

McQueston, James H.   A
Public Service Electric & Gas Com-
pany, 271 N. Broad St., Elizabeth,
N. J.

McQuiston, J. C.   A
Westinghouse Department of Publicity, East Pittsburgh, Pa.

Meara, William J.*   A
District Manager, New York.Edison
Co., 15 East 125th St., New York,
N. Y.

Megonigal, William   A
Manager, Power Sales Dept., Philadelphia Electric Co., 1000 Chestnut St., Philadelphia, Pa.

Merrill, G. S.   A
Assistant to Chief Engineer, Na-
tional Lamp Works of G. E. Co.,
Nela Park, Cleveland, Ohio.

Mervish, I. Joseph
Salesman, Philadelphia Electric Co.,
1000 Chestnut St., Philadelphia,
Pa.

Meyer, August H.   M
Bryan-Marsh Division of G. E. Co.,
623 S. Wabash Ave., Chicago, Ill.

Meyer, J. W.   M
Engineer, Commercial Dept., Philadelphia Electric Co., 1000 Chestnut St., Philadelphia, Pa.

Miley, Frank S.   M
National Carbon Co., Inc., 30 East
42d St., New York, N. Y.

Millar, Harold H.   M
Engineer, Electrical Testing Labora-
tories, Inc., 80th St., & East End
Ave., New York, N. Y.

Millar, Preston S.*   M
General Secretary, 1909-12.
President, 1913.
Junior Past-President, 1913-15.
Director, 1920-21 and 1921-24.
General Manager, Electrical Testing
Laboratories, Inc., 80th St. &
East End Ave., New York, N. Y.

Miller, Alten S.   M
The Bartlett Hayward Co., Baltimore, Md.

Miller, Arthur   A
Salesman, Ivanhoe Regent Works of
G. E. Co., 105 West 40th St.,
New York, N. Y.

Miller, Joseph C.   A
Office Manager, L. Plaut & Co., 434
East 23d St., New York, N. Y.

Mills, E. A.   A
Engineer, Electric Signs, New York
Edison Co., 10 Irving Place, New
York, N. Y.

Mills, F. S.   M
National X-Ray Reflector Co., 631
Pacific Finance Bldg., Los Ange-
les, Calif.

Minor, W. F.   A
Manager, Ivanhoe-Regent Works of
G. E. Co., 5716 Euclid Ave., Cleve-
land, Ohio.

Mistersky, Frank R.   A
General Superintendent, Public
Lighting Plant, 138 East Atwater
St., Detroit, Mich.

Mixer, Charles A.   M
Engineer, Rumford Falls Light &
Water Co., Rumford, Maine.

Mohr, H. K.   A
Vice-President, 1907-08.
Manager, Advertising Bureau, Philadelphia Electric Co., 1000 Chestnut St., Philadelphia, Pa.

Monville, Francis X.   A
Superintendent of Building, Philadelphia Electric Co., 1000 Chestnut, Philadelphia, Pa.

Moore, D. McFarlan M
Director 1916-19 and 1920-21.
Vice-President, 1921-22.
Moore Light Dept., General Electric Co., Harrison, N. J.

Moore, David H. M
Electrical Engineer, Day & Zimmermann, Inc., 611 Chestnut St., Philadelphia, Pa.

Moore, Louis D. M
Electrical Engineer, Missouri Pacific R. R. Co., 1130 Railway Exchange Bldg., St. Louis, Mo.

Moore, Richard C. A
The Coe Manufacturing Co., Painesville, Ohio.

Morgan, Lyman D. M
Illuminating Engineering, Perfeclite Mfg. Co., 119 Main St., Seattle, Wash.

Morris, W. Cullen M
Treasurer, 1910.
Director, 1913-15.
Engineer of Construction, Consolidated Gas Co. of N. Y., 130 E. 15th St., New York, N. Y.

Morrison, George F. M
Vice-President, General Electric Co., 120 Broadway, New York, N. Y.

Morrison, Harry K. M
Vice-President, 1918-20.
General Manager, Lynn Gas & Elec. Co., 90 Exchange St., Lynn, Mass.

Morrison, Montford A
Acme-International Co., 341 W. Chicago Ave., Chicago, Ill.

Morrison, William G. A
Illuminating Engineer, Edison Lamp Works of G. E. Co., Harrison, N. J.

Morrow, G. T. A
Resident Engineer, National X-Ray Reflector Co., 331 4th Ave., Pittsburgh, Pa.

Mossgrove, J. R. A
Salesman, Erner & Hopkins Elec. Co., 146 N. 3d St., Columbus, Ohio.

Motschenbacher, August A
Asst. Supt., L. Plant & Co., 432 E. 23d St., New York, N. Y.

Mott, William R. M
Chemical Engineer, 97 Homer Lee Ave., Jamaica, N. Y.

Mountcastle, H. W. A
Professor of Physics, Western Reserve University, Cleveland, Ohio.

Moxey, Louis W., Jr. M
President & Engineer, Keller-Pike Co., 1213 Race St., Philadelphia, Pa.

Moyer, Wayne R. A
Asst. Supt., Spring Garden District, United Gas Improvement Co., 4650 Market St., Philadelphia, Pa.

Mudgett, Guernsey F. A
Sales Engineer, Westinghouse Electric & Manufacturing Co., South Bend, Ind.

Muncy, Victor E. A
Professor of Mechanics & Applied Electricity, Ohio Mechanics Institute, Cincinnati, Ohio.

Munroe, Charles C. M
Chief of Lighting Sales Staff, Detroit Edison Co., 2000 Second Ave., Detroit, Mich.

Murota, Hannosuke A
Elec. Engineer, Tokyo Electric Co., Kawasaki-Kanagawa-ken, Japan.

Murphy, F. H.* M
Portland R. R., Lt. & Power Co., Electric Bldg., Broadway & Alder St., Portland, Ore.

Murphy, John M
Elec. Engineer, Dept. of Railways & Canals & Board of Railway Commissioners for Canada, West Block, Ottawa, Canada.

Murray, J. Frank A
Supt., Meter & Light Dept., United Electric Light Co., 73 State St., Springfield, Mass.

Murray, Thomas E. M
Vice-President, New York Edison Co., Irving Place & 15th St., New York, N. Y.

Musson, Chester A. A
United Elec. Light & Power Co., 146th St. & Broadway, New York, N. Y.

Myers, Joseph B. A
Commercial Agent, United Gas Improvement Co., 1401 Arch St., Philadelphia, Pa.

Myers, R. E. M
Chief Engineer, Westinghouse Lamp Co., Bloomfield, N. J.

Myers, Romaine W. M
Consulting Elec. & Illum. Engineer, 204 Bacon Bldg., Oakland, Calif.

Naland, C. W. A
Manager, New Business Dept., Union Electric Co., Brown Block, Abilene, Kansas.

Nast, Cyril   A
Advertising Manager, New York Edison Co., 130 East 15th St., New York, N. Y.

Nathanson, J. B.   A
Assistant Professor of Physics, Carnegie Institute of Technology, Pittsburgh, Pa.

Needham, Harry H.   A
Incandescent Electric Lamp Engineer, International General Electric Company, 4 rue d'Aguesseau, Paris 8, France.

Neel, William Trent   , A
Electrical Engineer, Philadelphia Electric Company, 1000 Chesnut St., Philadelphia, Pa.

Neiler, Samuel G.   A
President, Neiler, Rich & Company, 314 S. Dearborn St., Chicago, Ill.

Nelson, Albert F.   M
Manager, Edison Electric Illuminating Co., of Brockton, 42 Main St., Brockton, Mass.

Nelson, George E.   A
Lamp Salesman, National Lamp Works of G. E. Co., Byran-Marsh Division, 642 Beaubien St., Detroit, Mich.

Netting, C. J.   A
The C. J. Netting Co., Dealer in Lighting Fixtures, 1502 Randolph St., Detroit, Mich.

Nettleton, Charles H.   A
President, Derby Gas & Electric Co., 22 Elizabeth St., Derby, Conn.

Neumuller, Walter   M
Secretary, New York Edison Co., 130 Irving Place, New York, N. Y.

Newbold, E. S.   A
1934 Market St., Philadelphia, Pa.

Newhouse, Richard E.   A
Superintendent of Construction, Electric Supply Co., 815 East Third St., Tulsa, Oklahoma.

Newkirk, Elmer F.   A
Display Advertising, Edison Lamp Works of G. E. Co., Harrison, N. J.

Newton, Arthur H.   A
Technical Clerk, American Telephone & Telegraph Co., Room 836, 195 Broadway, New York, N. Y.

Nichols, Ernest Fox
Director of Pure Science, Nela Research Laboratories, National Lamp Works of G. E. Co., Cleveland, Ohio.

Nichols, G. B.   M
Designing & Consulting Engineer, 300 Madison Ave., New York, N. Y.

Nichols, H. G., Jr.   A
Illuminating Engineer, Macbeth-Evans Glass Co., Chamber of Commerce Bldg., Pittsburgh, Pa.

Nicolai, G. O.   M
Superintendent, Light & Power, Terre Haute, Indianapolis & Eastern Traction Co., 820 Wabash Avenue, Terre Haute, Ind.

Neider, Edward J.   A
Lighting Representative, Illuminating Company, Illuminating Bldg., Public Square, Cleveland, Ohio.

Nodell, W. L.   M
1405 8th Ave., Brooklyn, N. Y.

Norcross, J. Arnold   M
Director, 1913–16.
Secretary & Treasurer, New Haven Gas Light Co., 80 Crown St., New Haven, Conn.

Norcross, Josiah C.   A
Assistant Superintendent, Installation Dept., Edison Electric Illuminating Co. 39 Boylston Street, Boston, Mass.

Nord, Carl E.   A
The Nord Company, 535 Central Bldg., Seattle, Wash.

Norris, B. H.   A
Illum. Eng. Laboratory, General Electric Co., Schenectady, N. Y.

Norris, George T.   A
Electrical Engineer, Philadelphia Electric Co., 1000 Chestnut St., Philadelphia, Pa.

Norris, Rollin   A
General Superintendent, United Gas Improvement Co., 1401 Arch St., Philadelphia, Pa.

Norton, C. A.   A
Assistant to Sales Manager, Westinghouse Lamp Company, 165 Broadway, New York, N. Y.

Norton, Guy P.   A
Vice-President, Edward Miller & Co., Meridan, Conn.

Nutty, G. R.   A
Frick Block Annex, Pittsburgh, Pa.

Nye, Arthur W.   A
Professor of Physics, University of Southern California, 3351 University Avenue, Los Angeles, Calif.

Oday, A. B. A
Assistant in Illuminating Engineering, Edison Lamp Works of G. E. Co., Harrison, N. J.

Odenath, Harry E. M
Chief Engineer, Sears, Roebuck & Co., 4640 Roosevelt Blvd., Phila., Pa.

O'Donnell, William W. A
Lighting Salesman, Narragansett Elec. Lighting Co., Arctic, R. I.

Ogden, Kenneth C. A
Manager, New Brunswick Dept., Westchester Lighting Co., 1st 1st Ave. & 1st St., Mount Vernon, N. Y.

O'Leary, J. J. M
President, Buffalo Electric Contracting Co., 20 Broadway, Buffalo, N. Y.

O'Leary, Joseph J. A
Illuminating Engineer, Westinghouse Electric & Manufacturing Co., 111 W. Washington St., Chicago, Ill.

Olmstead, Henry C. A
Sales Engineer, Western Electric Co., 500 S. Clinton St., Chicago, Ill.

OMeara, Thomas J. A
Electrical Engineer, Hall Switch & Signal Co., Garwood, N. J.

Onken, William H., Jr. M
Editor, Electrical World, 36th St. & 10th Ave., New York, N. Y.

Orange, John A. M
Research Chemist, Usher Lodge, Bradford, England.

Orbin, Frank A
Carnegie Technical School, Schenley Park, Pittsburgh, Pa.

O'Rourke, John A. A
Columbus Ry. Power & Light Co., 102 N. 3d St., Columbus, Ohio.

Osborn, Frederick A. M
Professor of Physics, University of Washington, Seattle, Washington.

Osborne, George C. A
Assistant to Manager, Edison Lamp Works of G. E. Co., Harrison, N. J.

Osborne, L. A. M
Westinghouse Electric & Manufacturing Co., 165 Broadway, New York, N. Y.

O'Shea, James P. M
General Sales Agent, Cooper-Hewitt Electric Co. 95 River St., Hoboken, N. J.

Oshima, Hiroyoshi A
Chief Engineer, Osaka Electric Lamp Company, Ltd., 70 Daini Sagisu - machi, Nishinari - gori, Osaka-fu, Japan.

Otis, Harrison C. A
Sales Dept., Cooper-Hewitt Electric Co., 95 River St., Hoboken, N. J.

Outley, E. J. A
Detroit Edison Co., 2000 Second Ave., Detroit, Mich.

Owen, Raymond V. M
Lighting Fixture Designer, Edwin F. Guth Co., 2615 Washington Ave., St. Louis, Mo.

Owens, H. Thurston M
Manufacturer's Agent, 51 East 42nd New York, N. Y.

Page, A. D.* M
Sales Manager, Edison Lamp Works of G. E. Co., Harrison, N. J.

Page, Howard E. A
General Electric Edison Corp. of China, Shanghai, China.

Palmer, Harry C. M
Vice-President, Iroquois Gas Corp., 37 Church St., Buffalo, N. Y.

Palmer, Ray M
President, N. Y. & Queens Elec. Light & Power Co., Electric Bldg., Bridge Plaza, Long Island City, N. Y.

Parker, Charles H. A
Edison Electric Illuminating Co., 39 Boylston St., Boston, Mass.

Parker, Joseph Smith A
Superintendent, Spring Garden District, United Gas Improvement Co., 1615 N. 9th St., Philadelphia, Pa.

Parker, Edwin S. A
Manager, Commercial Lighting Dept. Pettingell Andrews Co., 160 Pearl St., Boston, Mass.

Parker, Karr A
Engineering Manager, McCarthy Bros. & Ford, 75 West Mohawk St., Buffalo, N. Y.

Parker, John C. A
Electrical Engineer, Brooklyn Edison Company, Brooklyn, N. Y.

Parker, Robert* A
Assistant to Manager, Incandescent Lamp Sales Dept., Edison Lamp Works of G. E. Co., Harrison, N. J.

Parkhurst, George W.   **M**
Vice-President, Northern Union Gas Co., 1815 Webster Ave., Bronx, N. Y.

Parnell, Eric   **A**
Illuminating Engineer, Edison Electric Illuminating Co., 39 Boylston St., Boston, Mass.

Pate, C. B.   **M**
Sales Engineer, National X-Ray Reflector Co., 16 State Street, Rochester, N. Y.

Patrick, James S.   **A**
New Eng. Representative, Benjamin Electric Manufacturing Co., P. O. Box 2848, Boston, Mass.

Patterson, George W.   **A**
Professor of Electrical Engineering, University of Michigan, 2101 Hill St., Ann Arbor, Mich.

Patterson, R. B.   **M**
Supt., Eng. Dept., Potomac Electric Power Co., 213 14th St., Washington, D. C.

Pearl, Allen S.   **A**
Secretary & Treasurer, Delta-Star Electric Co., 2433 Fulton St., Chicago, Ill.

Peaslee, W. D. A.   **M**
Chief Engineer, Belden Manufacturing Co., 2300 S. Western Ave., Chicago, Ill.

Peck, Eugene H.   **A**
Sales Manager, The Phoenix Glass Co., 230 Fifth Ave., New York, N. Y.

Peck, Edward L.   **A**
Inspector, Electrical Testing Laboratories, Inc., 80th St. & East End Ave., New York, N. Y.

Pembleton, F. D.   **M**
Public Service Electric Co., 80 Park Place, Newark, N. J.

Pence, David C.   **M**
Manager, Illuminating Dept., Illinois Electric Co., 313 S. San Pedro St., Los Angeles, Cal.

Percival, J. T., Jr.   **A**
Sales Engineer, Puget Sound Power and Light Co., Electric Bldg, 7th & Olive Sts., Seattle, Washington.

Perkins, Earl G.   **A**
Photogenic Machine Co., 512 Andrews Ave., Youngstown, Ohio.

Perkins, Oscar H.   **A**
Sales Manager, Stuart Howland Co., 234 Congress St., Boston, Mass.

Perrot, Emile G.   **M**
Architect & Engineer, Boyertown Bldg., 1211 Arch St., Philadelphia, Pa.

Perry, Allen M.   **A**
Engineering Editor, Electrical World, 36th & 10th Ave., New York, N. Y.

Peterson, Clarence A.   **A**
Electrical Engineer & Draftsman, Treasury Department, Supervising Architect's Office, Washington, D. C.

Pevear, Munroe R.   **A**
Color & Illuminating Engineer, 71 Brimmer St., Boston, Mass.

Pfeil, Hubert H.   **A**
Salesman, Macbeth-Evans Glass Co., Bourse Bldg. 4th & Market Sts., Philadelphia, Pa.

Pfenning, Phil. C.   **A**
Manager, Mdse. Div., Advertisin Dept., Westinghouse Lamp Co.' 165 Broadway, New York, N. Y.

Phelan, G. Stanley   **M**
Inspector Assistant to Transmission Engineer, Penna. Water & Power Co., 1161 Lexington Bldg., Baltimore, Md.

Phillips, Charles Travers   **A**
Consulting Engineer, Bank of Italy Bldg. San Francisco, Calif.

Piasecki, Harry A.   **A**
252 May St., Buffalo, N. Y.

Picado, Ramon Maria   **A**
City Engineer and Engineer to the Cartago Electric Light Co., Cartago, Costa Rica, C. A.

Piez, Karl A.   **A**
City Salesman, National X-Ray Reflector Co., 235 W. Jackson Blvd., Chicago, Ill.

Pim, Frank A.   **M**
Bethard Auto Co., 1117 Main St., Richmond, Ind.

Pinkerton, Andrew   **A**
Electrical Engineer, American Sheet & Tin Plate Co., 1008 Frick Bldg., Pittsburgh, Pa.

Pinkney, W. F. T.   **A**
Newcastle Electric Supply Co., Hood St., Newcastle-upon Tyne, England.

Piser, Theodore H.   **A**
Vice-President, 1911 and 1916–18.
Welsbach Company of New England, 100 Federal St., Boston, Mass.

Plaut, Herman M
L. Plaut & Co., 432 E. 23 Street,
New York, N. Y.

Plaut, Richard L. A
L. Plaut & Co., 432 E. 23 Street,
New York, N. Y.

Plumpton, Arthur G. A
Photometric Laboratory Assistant,
Hydro-Elec. Power Commission of
Ontario, 8 Strachan Ave., Toronto,
Ont., Canada.

Poey, Charles D. A
Illuminating Engineer, N. Y. &
Queens Elec. Light & Power Co.,
Elec. Bldg., Bridge Plaza Long
Island City, N. Y.

Polacheck, Phil. A
Secretary, Chas. Polacheck & Bro-
ther Co., 217 Third St., Milwaukee, Wis.

Pope, A. A.* A
Vice-President, 1906–07.
Assistant General Commercial Mana-
ger, New York Edison Co., Irving
Place & 15th St., New York,
N. Y.

Porter, Lawrence C. M
Edison Lamp Works of G. E. Co.,
Harrison, N. J.

Porter, Royal A. A
Professor of Physics, Syracuse University, Syracuse, New York.

Porterfield, William A. A
General Sales Manager, Union Metal
Mfg. Co., 1435 Maple Ave. N.E.,
Canton, Ohio.

Poser, Max M
Special Representative, Bausch and
Lomb Optical Co., Rochester.
N. Y.

Post, Rudin W. A
Superintendent of Street Lighting,
Municipal Bldg., Rochester, N. Y.

Potter, Herbert S. M
Electrical Engineer & Contractor,
240 State St., Boston, Mass.

Potter, J. Lee A
General Manager, Newburyport Gas
& Electric Co., 57 Pleasant St.,
Newburyport, Mass.

Powell, Alvin L. M
Director, 1922–25.
Manager, Lighting Service Dept.,
Edison Lamp Works of G. E. Co.,
Harrison, N. J.

Power, William R. M
Manager, Consolidated Light Heat
& Power Co., 1043 4th Ave.,
Huntington, W. Va.

Price, Frank Shreve M
Director, 1920–23.
President, Pettingell-Andrews Co.,
160 Pearl St., Boston, Mass.

Prichett, George P. A
National X-Ray Reflector Co., 628
Witherspoon Bldge., Philadelphia,
Pa.

Priest, Lucian Charles A
Electrician, 59 High St., Charles-
town, Mass.

Prigge, John J. A
Electrical Engineer, 441 44th St.,
Brooklyn, N. Y.

Prince, J. Lloyd M
Instructor, Commercial School, New
York Edison Co., 130 East 15th
St., New York, N. Y.

Prine, C. W. A
Assist. Professor of Physics, Carnegie Institute of Technology, Pittsburgh, Pa.

Probst, Louis M. A
Bailey & Sons, 105 Vandeveer St.,
Brooklyn, N. Y.

Proffatt, Charles P. A
Owner, Universal Electric Co., 701
Asbury Ave., Ocean City, N. J.

Prussia, Robert S. M
District Illuminating Engineer, West-
inghouse Lamp Co., 701 First Nat'l
Bank Bldg., San Francisco, Calif.

Pullen, M. W. M
School of Engineering, Johns Hopkins University, Homewood, Baltimore, Md.

Putnam, W. R. A
Vice-President and General Mana-
ger, Idaho Power Co., Boise,
Idaho.

Quirk, W. G. A
Chief Inspector in charge of Street
Lighting, Room 2307 Municipal
Bldg., New York, N. Y.

Quivey, Wyllis E. A
Benjamin Electric Manufacturing
Co., 847 W. Jackson Blvd., Chi-
cago, Ill.

Radcliff, John P. Jr.* M
Manager, Sales Dept., Yonkers Elec.
Light & Power Co., Manor House
Square, Yonkers, N. Y.

Rademacher, W. H. A
Illuminating Engineer, Edison Lamp Works of G. E. Co., Harrison, N. J.

Ralston, F. W. A
Street Lighting Dept., General Electric Co., 928 Western Ave., Lynn, Mass.

Ralston, P. C. A
Secretary to Senior Vice-President, Philadelphia Electric Co., 1000 Chestnut St., Philadelphia, Pa.

Ramsey, Harold E. A
Atlantic Refining Co., 3144 Passyunk Ave., Philadelphia, Pa.

Rasin, Unit A
District Manager, Westinghouse Lamp Co., 1005 Market St., Philadelphia, Pa.

Randall, John E. M
Consulting Engineer, National Lamp Works of G. E. Co., Nela Park, Cleveland, Ohio.

Reast, F. M. A
George Cutter Works, Westinghouse Electric & Manufacturing Co., South Bend, Ind.

Reeder, Charles L. M
Consulting Engineer, 916 N. Charles St., Baltimore, Md.

Regar, G. Bertram M
Director, 1921–22.
Vice-President, 1922–24.
Illuminating Engineer, Philadelphia Electric Co., 1000 Chestnut St., Philadelphia, Pa.

Reilly, J. J. A
Supt., Meter Tests, Philadelphia Electric Co. N.W. 23d & Market Sts.. Philadelphia, Pa.

Reinach, Hugo B. A
Assistant General Superintendent, Consolidated Gas Co., 130 East 15th St., New York, N. Y.

Renshaw, E. N. A
Telephone Engineer, Amer. Tel. & Tel. Co., 195 Broadway, New York, N. Y.

Rhea, C. J. A
General Manager, Dillion Lens & Mfg. Co., Bridgeport, Ohio.

Rhodes, S. G.* M
New York Edison Co., 124 E. 15th St., New York. N. Y.

Rice, Edwin Wilbur, Jr. M
General Electric Co., Schenectady, N. Y.

Rice, E. Y. A
Electrical Engineer, The Hartford Electric Light Co., 266 Pearl Street, Hartford, Conn.

Richards, Jesse A
Manager, Corn Exchange Bank, Sutphin Blvd., Jamaica, Queens, N. Y.

Richards, W. E. M
Supt., Electrical Dept., Toledo Edison Co., Toledo, Ohio.

Richman, H. H. A
Assistant Manager Philadelphia Dept., Welsbach Co., 1008 Filbert St., Philadelphia, Pa.

Rider, T. J., Jr. A
Assistant Manager, Sunbeam Lamp Dept., Western Electric Co., 500 S. Clinton St., Chicago, Ill.

Riley, Milton D. A
Pettingell-Andrews Co., 160 Pearl St., Boston, Mass.

Risdon, T. J. M
Commercial Inspector, Philadelphia Electric Co., 10th & Chestnut Sts. Philadelphia, Pa.

Rishell, George Le Bar A
Electrical Engineer, Novelty Incandescent Lamp Co., Emporium, Pa.

Robb, Wm. Lispenard M
Professor of Electrical Engineering, Rensselaer Polytechnic Inst., Box 592, Troy, N. Y.

Roberts, Edward S. A
Blackstone Valley Gas & Electric Co., Pawtucket, R. I.

Robertson, E. E. A
Meter Tester, Philadelphia Electric Co., N.W. cor 23d & Market Sts., Philadelphia, Pa.

Robertson, William L. A
Philadelphia Electric Co., N.W. cor 23rd & Market St., Philadelphia, Pa.

Robins, Orrin A. A
Manager, The Electric League of Columbus, 9 E. Long St., Columbus, Ohio.

Rodgers, Ralph C. A
Assistant Secretary, Illuminating Engineering Society, 29 West 39th St., New York, N. Y.

Rodman, Walter Sheldon M
Professor of Electrical Engineering, University of Virginia, P. O. Box 222 University, Virginia.

Roese, Harry E.   A
Westinghouse Lamp Co., 1909 Union
Bank Bldg., Pittsburgh, Pa.

Rogers, Fred A.   M
Professor of Physics & Electrical En-
gineering, Lewis Institute 1949 W.
Madison St., Chicago, Ill.

Rogers, Henry B.   A
Assistant to Manager, Edison Lamp
Works of G. E. Co., Harrison, N. J.

Rogers, Stephen C.   M
Director, 1917-20
Commercial Illuminating Engineer,
General Electric Co., Western Ave.
West Lynn, Mass.

Rolinson, W. H.   M
502 West 173d. St., New York, N. Y.

Rolph, T. W.   M
Managing Engineer, Street Lighting
Dept., Holophane Glass Co., Ne-
wark, Ohio.

Rolston, Robert J.   A
Manager, Welsbach Co., 1008 Fil-
bert St., Philadelphia, Pa.

Rose, S. L. E.   A
Street Lighting Specialist, Supply
Dept., General Electric Co., 112
N. 4th St., St. Louis, Mo.

Rosenquest, Eugene H.   A
President & General Manager, The
Bronx Gas & Electric Co., 43
Westchester Sq., New York, N. Y.

Ross, Leland S.   A
Great Western Power Co., 1700
Broadway, Oakland, Calif.

Ross, Robert G.   A
Public Service Co., 80 Park Place,
Newark, N. J.

Rucker, George   M
Pacific Coast Sales Manager, Holo-
phane Co., 1066 Mission St., San
Francisco, Calif.

Rusby, John M.   A
Engineer of Tests, United Gas Im-
provement Co., Broad & Arch
Sts., Philadelphia, Pa.

Russell, C. J.*   M
Director, 1912-14.
District Manager, Philadelphia
Electric Co., 1000 Chestnut St.,
Philadelphia, Pa.

Russell, S. P.   A
Vice-President, H. B. Squires Co.,
583 Howard St., San Francisco,
Calif.

Rutledge, Fred J.   M
Director, 1913-15.
Assistant Manager, New Business
Dept., the United Gas Improve-
ment Co., Broad & Arch Sts.
Philadelphia, Pa.

Ryan, Harris J.   A
Professor of Electrical Engineering,
Leland Stanford Junior Univer-
sity, Stanford University, Calif.

Ryan, John Julius   A
Electrical Inspector, Department of
Gas and Electricity, 606 City Hall,
Chicago, Ill.

Ryan, W. D'A.*   M
Director, 1906-07.
Vice-President, 1908.
Illuminating Engineer, General Elec-
tric Co., Schenectady, N. Y.

Sampson, A. T.   A
Sampson & Allen, 434 Union St.,
Lynn, Mass.

Sanderson, Frank N.   A
Manager, The Electric Light &
Power Co. of Abingdon and Rock
land, 64 Charles St., N. Abingdon,
Mass.

Sands, Howard T.   A
Vice-President, Charles H. Tenny &
Co., 200 Devonsire St., Mass.

Sandoval, H. E.   A
Electrical Engineer, Pacific & Elec-
tric Co., 445 Sutter St., San
Francisco, Calif.

Sargent, Fred H.   A
Assistant Agent, Lawrence Gas Co.
370 Essex St., Lawrence, Mass.

Sasse, Frederick C.   A
Sales Engineer, X-Ray Reflector Co.
of N. Y., 31 W. 46th St., New
York, N. Y.

Satter, Henry A.   A
Illuminating Engineer, New York
Edison Co., 130 E. 15th St., New
York, N. Y.

Saunders, William E.   A
Engineer, United Gas Improvement
Co., N.W. cor. Broad & Arch Sts.,
Philadelphia, Pa.

Savage, Arthur H.   M
A. H. Savage Co., 914 Pioneer Bldg.,
St. Paul, Minn.

Sawin, George A.   M
Westinghouse Electric & Manufacturing Co., E. Pittsburgh, Pa.

Sawyer, Leroy P.   A
National Lamp Works of G. E. Co.,
Nela Park, Clveland, Ohio.

Schachter, M. F.   A
Chatham Lighting Fixture Co., 143
Bowery, New York, N. Y.

Scheel, Alfred A.   A
Lee Electric Co., 605 Davis Ave.,
Corning, Iowa.

Scheible, Albert   M
Vice-President, 1907-08.
Patent Attorney and Research En-
gineer, 110 S. Dearborn St.,
Chicago, Ill.

Schildhauer, Edward   M
405 Commercial Bank Bldg., Washington, D. C.

Schaldt, G. J.   A
Picatinny Arsenal, Dover, N. J.

Schneider, G. A.   A
Salesman, Philadelphia Electric Co.,
Power Dept., 1000 Chestnut St.,
Philadelphia, Pa.

Schroeder, Henry   A
Assistant to Sales Manager, Edison
Lamp Works of G. E. Co., Harri-
son, N. J.

Schuchardt, R. F.   A
Electrical Engineer, Commonwealth
Edison Co., 72 W. Adams St.,
Chicago, Ill.

Schumacher, John H.   M
Schumacher Gray Co., Ltd, 187
Portage Ave., Winnipeg, Mani-
toba, Canada.

Schwartz, H. M.   A
President, Robert Findlay Manu-
facturing Co., 100 Lexington Ave.,
Brooklyn, N. Y.

Schweitzer, Edmund O.   A
Commonwealth Edison Co., 28 N.
Market St., Chicago, Ill.

Scott, Charles F.   A
Professor of Electrical Engineering,
Yale University, New Haven,
Conn.

Scott, John T.   M
Electrical Engineer, Sunbeam Lamp
Div.,Canadian G. E. Co., 221
Dufferin St., Toronto, Ontario,
Canada.

Seaman, Joseph B.   A
Philadelphia Electric Co., 4600
Frankford Ave., Philadelphia, Pa.

Searle, Robert M.*   A
Vice-President, Rochester Gas &
Electric Corp., 34 Clinton Ave.,
N. Rochester, N. Y.

Sellers, Joseph P.   A
Philadelphia Electric Co., 132 S.
11th St., Philadelphia, Pa.

Serrill, William J.   M
Treasurer, 1912.
Vice-President, 1913-14.
President, 1916-17.
Junior Past-President, 1917-19.
Assistant to General Manager,
United Gas Improvement Co.,
1401 Arch Street, Philadelphia, Pa ·

Seville, J. A.   A·
Electrical Construction, Seville Electric Co., 209 N. 11th St., Philadelphia, Pa.

Shaad, George C.*   M
Professor, of Electrical Engineering,
Engineering Bldg., University of
Kansas, Lawrence, Kansas.

Shackelford, Benj. E.   A
Chief Physicist, Westinghouse Lamp
Co., Bloomfield, N. J.

Sharp, Clayton H.*   M
Vice-President, 1906.
President, 1907.
Junior Past-President, 1908-09.
Technical Director, Electrical Test-
ing Laboratories, Inc., 80th St. &
East End Ave., New York, N. Y.

Shaw, Arthur E.   A
President, Welsbach Street Lighting
Co. of America, 1934 Market St.,
Philadelphia, Pa.

Shaw, Harold   A·
Secretary, Electrical Extension Bur-
eau, 613 Lincoln Bldg., Detroit,
Mich.

Shaw, H. J.   A
President, Electrical Specialities Co.,
325 State St., Detroit, Mich.

Shaw, Joseph E.   M
Manager, Lighting Fixture Dept.,
Frank H. Stewart Electric Co.,
37 N. Seventh St., Philadelphia,
Pa.

Shay, John P.   A
General Foreman, Installation & In-
spection Depts., United Electric
Light & Power Co., 130 E. 15th
St., New York, N. Y.

Sheibly, Frank D.   M
Assistant Engineer, Cons. Teleg. &
Electric Subway Co., 54 Lafayette
St., New York, N. Y.

Sheldrake, J. S. A
Philadelphia Electric Co., 1000 Chestnut St., Philadelphia, Pa.

Shenton, R. W. M
Engineering Dept., National Lamp Works of G. E. Co., Nela Park, Cleveland, Ohio.

Shepardson, G. D. A
Professor of Electrical Engineering Univ. of Minnesota, Minneapolis, Minn.

Sherwood, Edward L. M
Consulting Engineer, E. L. Sherwood and Associate Engineers, 149 Broadway, New York , N. Y.

Shikata K. A
The Tokyo Electric Co., Kawasaki, Kanagawa-ken, Japan.

Shimidzu, Yoshichiro A
Engineer, Tokyo Electric Co., Kawaski, Kanagawa-Ken, Japan.

Shohan, Abraham M
6 Dennison St., Boston-19-Mass.

Sholl, Edward L. A
Salesman, Holophane Glass Co., 342 Madison Ave., New York, N. Y.

Shute, James M. A
Illuminating Engineer, Central Illinois Light Co., Peoria, Ill.

Sias, Arthur G. A
Manager, Municipal Light Dept., Municipal Bldg., Lowell St., Reading, Mass.

Silbert, Richard H. A
Philadelphia Electric Co., 23d & Market St., Philadelphia, Pa.

Silverman, Philip A
L. Plaut & Co., 432 E. 23d St., New York, N. Y.

Simkins, H. M. A
Inspector, Philadelphia Electric Co., 1000 Chesnut St., Philadelphia, Pa.

Simms, Frederick F. A
Director of Trades School, Virginia Normal & Industrial Institute, Ettrick P. O. Station, Petersburg, Va.

Simons, S. A. A
Junior Executive, Save Elec. Corp., 254 36th St., Brooklyn, N. Y.

Simonson, G. M. A
Chief Electrical Engineer, State of California, Dept. of Public Works, 615 Forum Bldg., Sacramento, Calif.

Simpson, Richard E. M
Engineer, Travelers Insurance Co., 700 Main St., Hartford, Conn.

Sinclair, H. A. A
Tucker Electrical Construction Co., 526 W. 34th St., New York, N. Y.

Skiff, Warner M. M
Manager Engineering Dept., National Lamp Works of G. E. Co , Nela Park, Cleveland, Ohio.

Skogland, J. F. A
Associate Physicist, Bureau of Standards, Washington, D. C.

Slocum, Chester A. A
Douglas Sprague & Slocum, Consulting Engineers, 50 E. 41 St., New York, N. Y.

Smith, Alphonso Howes A
Salesman, Slocum & Kilburn, 23 N. Water St., New Bedford, Mass.

Smith, Arthur G. A
Illuminating Engineer, Central Electric Co., 320 S. Wells St., Chicago, Ill.

Smith, E. A. M
Electrical Engineer, Warren & Wetmore, 16 E. 47th St., New York, N. Y.

Smith, F. C.* A
Secretary-Treasurer, J. P. Hall-Smith Co., Inc., 320 5th Ave., New York, N. Y.

Smith, Frank W. M
Vice-President, United Electric Light & Power Co., Irving Place & 15th St., New York, N. Y.

Smith, Fred H. A
Vice-President, Worcester Electric Light Co., 11 Foster St., Worcester Mass.

Smith, H. R. A
Solicitor, Philadelphia Electric Co., 7 Chelton Ave., Philadelphia, Pa.

Smith, Harry H. A
Manager, Lighting Fixture Dept., W. B. Catlett Electric Co , Inc., 114 West Grace St., Richmond, Va.

Smith, Helen A. A
Engineer, Lighting Service Dept., Edison Lamp Works of G. E. Co., Harrison, N. J.

Smith, Raymond A. A
Sales Engineer, Crescent Electric Co., 150 Jefferson St., E., Detroit, Mich.

Smith, S. W.   A
Philadelphia Electric Co., S.W. cor.
6th & Diamond Sts., Philadelphia,
Pa.

Snyder, Augustus E.   A
Assistant to Commercial Engineer,
Westinghouse Lamp Co., Bloom-
field, N. J.

Snyder, Charles S.   A
Supt., West Phila. Distribution Dist.,
United Gas Improvement Co.,
4650 Market St., Philadelphia, Pa.

Snyder, H. L.   A
Treasurer, New York & Queens
Electric Light & Power Co., 444
Jackson Ave., Long Island City,
N. Y.

Snyder, Samuel   M
Illuminating Specialist, United Gas
Improvement Co., 134 N. 13th
St., Philadelphia, Pa.

Souza, Edgard Eygdio   A
Electrical Engineer, Sao Paulo Tram-
way Light & Power Co., Ltd. 162
Caixa Postal, Sao Paulo, Brazil.

Spaide, G. Alvin   A
Lighting Salesman, Walker & Kep-
ler, 531 Chestnut St., Phila,. Pa.

Spaulding, H. T.   A
Bryan-Marsh Division of G. E. Co.,
99 Chauncy St., Boston, Mass.

Spencer, Eugene Jaccard   M
Secretary-Manager, St. Louis Elec-
trical Board of Trade, 1298 Arcade
Bldg., St. Louis, Mo.

Spencer, Melvin   A
Treasurer, I. P. Frink, Inc., 239
10th Ave., New York, N. Y.

Spencer, Paul*   M
Electrical Engineer, United Gas
Improvement Co., 1401 Arch St.,
Philadelphia, Pa.

Spencer, W. H.   M
Engineer & Designer, I. P. Frink,
239 10th Ave., New York, N. Y.

Spindell, David   A
Electrical Contractor, 607 Marcy
Ave., Brooklyn, N. Y.

Spinney, L. B.   A
Professor of Physics, Iowa State
College, Ames, Station A, Iowa.

Spitz, Edward William   A
Sales Engineer, Duplex-a-lite Dept.
of Edward Miller & Co., Lexington Bldg., Baltimore, Md.

Springer, R. D. N.   A
Optometrist, Dr. Kline's Sanitarium,
Drawer 111, Anoka, Minn.

Sproule, Thomas   M
Assistant to Gen'l Supt. of Distribution, Public Service Electric Co.,
80 Park Place, Newark, N. J.

Stahl, Chas. J.   A
Commerical Engineer, Illuminating
Section, Westinghouse Electric &
Manufacturing Co., Notre Dame
& Division Sts., South Bend,
Indiana.

Stair, J. L.   A
Chief Engineer, National X-Ray Re-
flector Co., 235 W. Jackson Blvd.,
Chicago, Ill.

Standish, Myles, M. D.   M
256 Newbury St., Boston, Mass.

Stanley, Robert W.   M
Illuminating Specialist, Holophane
Glass Co., 340 Madison Ave.,
New York, N. Y.

Stannard, Clare N.   A
Vice-President & General Manager,
Denver Gas & Electric Light Co.,
900 15th St., Denver, Colorado.

Stansel, N. R.   A
Power & Mining Dept., General
Electric Co., Schenectady, N. Y.

Steel, Miles F.   A
Pacific Coast Manager, Benjamin
Manufacturing Co., 580 Howard
St., San Francisco, Calif.

Steel, Walter D.   A
Vice-President, Benjamin Electric
Mfg. Co., 847 Jackson Blvd.,
Chicago, Ill.

Steinmetz, Charles P
President, 1915-16.
Junior Past-President, 1916-18.
Consulting Engineer, General Elec-
tric Co., Wendell Ave., Schenec-
tady, N. Y.

Stephens, John Newton   A
Electrical Engineer, British Thom-
son Houston Co., Ltd., 77 Upper
Thomas St., London, England.

Sterling, W. T.*   A
Standard Oil Co., Room 520, 26
Broadway, New York, N. Y.

Stetser, Jesse R.   A
Technical Dept., Welsbach Com-
pany, Gloucester, N. J.

Stevens, E. E.   A
Commerical Manager, New Beford
Gas & Edison Light Co., 693 Pur-
chase St., New Bedford, Mass.

Stevick, C. H. A
New Amsterdam Gas Co., Ravenwood Works, Webster Avenue & East River, Long Island City, N. Y.

Stewart, Frank H. A
President, Frank H. Stewart Electric Co., 37 N. 7th St., Philadelphia, Pa.

Stewart, Jacob R. A
Lighting Specialist, Westinghouse Electric & Manufacturing Co., 3rd & Elm Sts., Concinnati, Ohio.

Stickney, George H. M
Vice-President, 1913-15.
General Secretary, 1916-17.
President, 1917-18.
Junior Past-President, 1918-20.
Illuminating Engineering Assistant to Sales Manager, Edison Lamp Works of G. E. Co., Harrison, N. J.

Stites, Townsend A
General Manager, Welsbach Co., Gloucester City, N. J.

Stockwell, W. L. A
Pacific Gas & Electric Co., 445 Sutter St., San Francisco, Calif.

Stoll, Albert F. M
Russell and Stoll Co., 17 Vandewater St., New York, N. Y.

Stolpe, Ernest F. A
Assistant Main & Service Foreman, United Gas Improvement Co., 20 W. Maplewood St., Philadelphia, Pa.

Stone, C. W. A
Manager, Lighting Dept., General Electric Co., Schenectady, N. Y.

Story, Ernest D.
Manager, Merchandising Bureau, Westinghouse Lamp Co., · 165 Broadway, New York, N. Y.

Stott, Edward B. A
President, Edward B. Stott & Co., 461 Eighth Ave., New York, N. Y.

Strahan, George V. A
Designer, Lighting Fixtures, Mitchell Vance Co., Inc. 507 W. 24th St., New York, N. Y.

Strait, E. N. A
Public Utility Expert, H. M. Byllesby & Co., 208 S. LaSalle St., Chicago, Ill.

Strauss, Lawrence L. M
Contracting Electrical Engineer, 422 W. 42d St., New York, N. Y.

Strebig, Charles F. A
Sales Manager, Cooper-Hewitt Electric Co., 95 River St., Hoboken N. J.

Strecker, Alex. H. A
Engineer of Construction, Public Service Gas Co., Box 590 Newark, N. J.

Streich, H. C. A
Supt. of Lighting, City of St. Paul, 25 E. 5th St., St. Paul, Minn.

Strong, James Remsen A
President, Tucker Electrical Construction Co., 114 30th St., New York, N. Y.

Stuart, Ralph C. M
Superintendent, Lamp Factory, Canadian Westinghouse Co. Ltd., Aberdeen Ave., Hamilton, Ontario, Canada.

Sturrock, Walter A
Illuminating Engineer, National Lamp Works of G. E. Co., Nela Park, Cleveland, Ohio.

Sue, Osake A
Senpakuba, Mitsui & Company, Kobe, Japan.

Sulzer, G. A., M.D. A
Ophthalmologist, 327 E. State St., Columbus, Ohio.

Summers, John A. A
Assistant Illuminating Engineer, Edison Lamp Works of G. E. Co., Harrison, N. J.

Swallow, Joseph G. M
Chief of Installation & Inspection Dept., United Electric Light & Power Co., Irving Place & 15th St., New York, N. Y.

Sweeney, George J. A
Printometer Inspector, United Elec. Light & Power Co., 514 W. 147th St., New York, N. Y.

Sweet, Arthur J. M
Consulting Engineer, Palace Theatre Bldg., Milwaukee, Wis.

Swindell, Harry A
Ivanhoe-Regent Works of G. E. Co., 5716 Euclid Ave., Cleveland, Ohio.

Sylvester, Elmer L. A
Lighting Service Dept., Edison Lamp Works of G. E. Co., Harrison, N. J.

Sylvester, J. Wilson A
Assistant Supt., Philadelphia Electric Co., 2301 Market St., Philadelphia, Pa.

Symmes, Whitman   **M**
President, Thomas Day Co., 715 Mission St., San Francisco, Calif.

Taggart, Ralph C.   **M**
Chief Engineer, Dept. of Architecture, The Capitol, Albany, N. Y.

Tait, Frank M.   **A**
80 Broadway, New York, N. Y.

Tallman, Vernon M. F.   **A**
Power Engineer, Charles H. Tenney & Co., 200 Devonshire St., Boston, Mass.

Tansey, Frank D.   **A**
Supervisor, Lighting Division, Brooklyn Union Gas Co., 180 Remsen St., Brooklyn, N. Y.

Tanzer, E. Dean   **A**
Engineer, Transmission and Distribution Dept., Philadelphia Electric Co., N.W. cor. 23d & Market Sts., Philadelphia, Pa.

Taylor, A. Hadley   **A**
Physicist, Nela Research Laboratory, National Lamp Works of G. E. Co., Cleveland, Ohio.

Taylor, Dent A.   **A**
Vice-President & Manager, Northwood Glass Co., 36th St., Wheeling, W. Va.

Taylor, Frank C.   **A**
Engineer, Industrial Sales Dept., Rochester Gas & Electric Corp., 34 Clinton Ave., N., Rochester, N. Y.

Taylor, John B.   **A**
Consulting Engineer, 23 Lowell Road, Schenectady, N. Y.

Taylor, Mabel H.   **A**
Technical Work in Connection with Photometry, Electrical Testing Laboratories, 80th St. and East End Ave., New York, N. Y.

Teberg, E. J.   **A**
Public Service Company of Northern Illinois, 72 W. Adams St., Chicago, Ill.

Teller, J. Paul   **A**
Westinghouse Electric & Manufacturing Co., 814 Ellicott Square, Buffalo, N. Y.

Terry, F. S.   **A**
Manager, National Lamp Works of G. E. Co., Nela Park, Cleveland, Ohio.

Thomas, Stephen A.   **M**
Chief, Electrical Division, Board of Education, Flatbush Ave. Extension and Concord St., Brooklyn, N. Y.

Thomas, C. G. M.*   **M**
Vice-President, Consolidated Gas Co., 130 East 15th St., New York, N. Y.

Thomas, N. Wiley   **M**
Chief of Bureau of Gas, 338 City Hall, Philadelphia, Pa.

Thompson, Allison J.   **A**
President, Thompson Electric Co., 226 St. Clair Ave., N.E., Cleveland, Ohio.

Thompson, Elihu   **M**
Consulting Engineer, General Electric Co., Lynn, Mass.

Thompson, E. L.   **A**
Illuminating Engineer, Westinghouse Electric & Manufacturing Co., George Cutter Works, South Notre Dame St., South Bend, Ind.

Thompson, Roy B.   **A**
Commercial Agent, Central Hudson Gas & Electric Co., 129 Broadway, Newburgh, N. Y.

Thompson, R. R.   **A**
Illuminating Specialist, Westinghouse Electric & Manufacturing Co., 515 Hanna Bldg., Cleveland, Ohio.

Thompson, Robert J.   **A**
Supt., Welsbach Co., 863 Mission St., San Francisco, Calif.

Thomson, George L. A.   **A**
Public Service Electric Co., 102 River St., Newark, N. J.

Tiffany, Edward L.   **A**
Electrical Engineer, United States Rubber Co., 393 E. Chapel, Box 606, New Haven, Conn.

Tilden, E. A.
Westinghouse Electric & Manufacturing Co., 40 Clinton St., Newark, N. J.

Tileston, Roland R.   **M**
Professor, Colorado College, Colorado Springs, Colo.

Tillson, Edwin D.   **A**
Testing Engineer, Commonwealth Edison Co., 72 W. Adams St., Chicago, Ill.

Timm, Edward W.   **A**
Western Electric Co., Inc., 458 Milwaukee, St., Milwaukee, Wis.

Timper, Walter R. A
Pettingell-Andrews Company, 511 Atlantic Ave., Boston, Mass.

Tingley, Louisa Paine, M.D. M
Ophthalmologist, 9 Massachusetts Ave., Boston, Mass. −17.

Tizley, A. J. M
Mechanical Engineer, 15 Chester Court, Brooklyn, N. Y.

Toohey, James A. A
Sales Engineer, Holophane Glass Co., Inc., 342 Madison Ave., New York, N. Y.

Toohey, Thomas J. A
Lighting Salesman, Narragansett Electric Lighting Co., 814 Turks Head Bldg., Providence, R. I.

Topham, Richard M. A
Salesman, Wetmore Savage Co., 96 Pearl St., Boston, Mass.

Torchio, Phillip A
Chief Electrical Engineer, New York Edison Co., Irving Place and 15th St., New York, N. Y

Torrens, Robert John M
Electrical Engineer, U. S. Naval Observatory, Washington, D. C.

Tousey, Sinclair, M.D. A
Physician, 850 Seventh Ave., New York, N. Y.

Tousley, Ben A
Trade Relations Representative, Electric Outlet Co., Inc., 8 West 40th St., New York, N. Y.

Trawick, Samuel W., Jr. A
New Orleans Public Service, Inc., 201 Baronne St., New Orleans, La.

Tremaine, B. G. A
Manager, National Lamp Works of G. E. Co., Nela Park, Cleveland, Ohio.

Trimming, Percy H. A
Chief Electrician, Dominion Flour Mills, Ltd., Montreal, Quebec, Canada.

Troland, Leonard T. A
Instructor in Psychology, Harvard University, Emerson Hall Cambridge, Mass.

Trott, W. J. A
Salesman, Fostoria Incandescent Lamp Div. of G. E. Co., 269 Charles Ave., Grand Rapids, Mich.

Trueax, Clyde P. M
Assistant Engineer, Electric Engineering Dept., C. T. I., Illinois Central R. R., Michigan Ave. & Roosevelt Road, Chicago, Ill.

Tuck, Davis H. M
Electrical Engineer, Holophane Glass Co., 340 Madison Ave., New York, N. Y.

Tucker, Edward A. A
Electric Salesman, F. D. Lawrence Electric Co., 217 W. 4th St., Cincinnati, Ohio.

Tudbury, John L. M
Salem Gas Light Co., 247 Essex St., Salem, Mass.

Tufts, Bowen M
Banker, Vice-President, Central Massachusetts Electric Co., 150 Congress St., Boston, Mass.

Turnbull, Thomas S. A
Manager, Electric Department, Tallman Brass & Metal, Ltd., Wilson St., Hamilton, Ontario, Canada.

Turner, Allen S., Jr. A
Illuminating Engineer, Lighting Service Department, Edison Lamp Works of G. E. Co., Harrison, N. J.

Tyler, Randolph E. A
Manager, Lighting Service Dept., Schimmel Electric Supply Co., 526 Arch St., Philadelphia, Pa.

Uchisaka, Motoo M
Engineer & Assistant Manager, Sales Dept. Tokyo Electric Co., Kawasaki-Machi, Kanagawa-Ken, Japan.

Umbach, Walter R. A
Assistant Manager of Works, Westinghouse Lamp Co., Bloomfield, N. J.

Upham, W. H. A
Sales Manager, General Electric Co., 18 14th St., Wheeling, W. Va.

Uptegraff, W. D. M
207 Westinghouse Bldg., Pittsburgh, Pa.

Van Bergen, Curtis E. A
Vice-President and General Manager, Duluth Edison Electric Co., 216 West First Street, Duluth, Minn.

Van Bloem, P. Schuyler M
President, Viking Products Corporation, 422 W. 42d St., New York, N. Y.

Vandergrift, J. A.   A
General Sales Manager for the Pacific Coast, National Lamp Works of G. E. Co., 1648 16th St., Oakland, Calif.

Van Derzee, G. W.   A
Assistant to Vice-President, Milwaukee Electric Railway & Light Co., Public Service Bldg., Milwaukee, Wis.

Van Dyck, William Van Bergen   A
Caixa do Correio 109, Rio de Janeiro, Brazil, S. A.

Van Gieson, C. J.   A
Assistant Agent, Public Service Electric Co., Orange, N. J.

van Gilluwe, Frank   M
Lighting Specialist, Westcon Electric Co., 301 E. 9th St., Los Angeles, Calif.

Van Rensselaer, S. W.   A
X-Ray Reflector Co. of N. Y., 31 West 46th St., New York, N. Y.

Van Volkenburg, Ray   A
407 Seventh St., Jackson, Mich.

Vanzwoll, Henry B.   A
General Manager, Sunbeam Incandescent Lamp Co., 500 So. Clinton St., Chicago, Ill.

Vassar, H. S.   A
Laboratory Engineer, Public Service Electric Co., 21st St. & Clinton Ave., Irvington, N. J.

Vaughn, Francis Arthur   M
Director, 1913–14.
Vice-president, 1914–16.
F. A. Vaughn, Inc., Consulting Engineers, Metropolitan Block, Third and State Sts., Milwaukee, Wis.

Vaughan, Hollis B.   A
Sales Manager, George H. Wahn Co., 69 High St., Boston, Mass.

Vincent, Elliott W.   A
Sales Engineer, Edison Lamp Works of G. E. Co., 120 Broadway, New York, N. Y.

Vincent, H. B.   M
Engineer, Day & Zimmermann Inc., 611 Chestnut St., Philadelphia, Pa.

Vogan, Frank C.   M
Consulting Engineer, 1213 Race St., Philadelphia, Pa.

Vogel, Joshua H.   M
Architect, Mission Architects' Bureau, Shanghai, China.

Voyer, Leonard E.   M
Illuminating Engineer, San Francisco Office, General Electric Co., Rialto Bldg., San Francisco, Calif.

Wager, A. J.   A
Supt., Power Sales, The Milwaukee Elec. Railway & Light Co., Public Service Bldg., Milwaukee, Wis.

Wagner, Herbert A.   A
President, Consolidated Gas Electric Light & Power Co., Lexington St. Building, Baltimore, Md.

Wagner, Walter C.   A
Chief Engineer's Department, The Philadelphia Electric Co. 1000 Chestnut St., Philadelphia, Pa.

Wagschal, George   A
Electrical Engineer, 8926 Clarendon Ave., Detroit, Mich.

Wakefield, Albert F.   M
The F. W. Wakefield Brass Co., Vermilion, Ohio.

Wakefield, F. W.   A
President, F. W. Wakefield Brass Co., Vermilion, Ohio.

Walker, Clarence C.   A
Salesman, Wetmore-Savage Co., 46 Hampden St., Springfield, Mass.

Wall, William L.   A
Specialist in Spectacle Optics, Wall & Ochs, Opticians, 1716 Chestnut St., Philadelphia, Pa.

Wallace, Herbert F.   M
Vice-President, 1920–22.
District Sales Manager, Edison Lamp Works of G. E. Co., 84 State St., Boston, Mass.

Wallace, Wyndham S.   A
McKenney & Waterbury Co., 181 Franklin St., Boston, Mass.

Wallis, Louis R.   M
Supt., Sales Dept., Edison Electric Illuminating Co., of Boston, 39 Boylston St., Boston, Mass.

Walmsley, W. N.   A
Vice-President, Alabama Power Co., Brown-Morx Building, Birmingham, Ala.

Walsh, Vincent   A
Electrician, Kev Unit Lighting System, 258 Broadway, Room 608, New York, N. Y.

Ward, Charles W.   A
Sales Engineer, Westinghouse Lamp Co., 1909 Union Bank Building, Pittsburgh, Pa.

Ward, Thomas G.   A
Detroit Edison Co., 2000 Second Ave., Detroit, Mich.

Wardell, C. W.   A
5021 Walton Ave., Philadelphia, Pa.

Warner, G. H.   A
Secretary-Treasurer, William M. Crane Co., 16 W. 32 St., New York, N. Y.

Warner, Hugh A.   A
District Illuminating Sales Engineer, Westinghouse Electric & Manufacturing Co., 1333 Candler Bldg., Atlanta, Ga.

Warner, James P.   A
Instructor, Carnegie Institute of Technology, 3122 Iowa St., Pittsburgh, Pa.

Waterbury, John H.   A
Engineering Department, National Lamp Works, of G. E. Co., Nela Park Cleveland, Ohio.

Waterman, Fred E., Jr.   A
Treasurer & Manager, Waterman Supply Co., Inc., 79 Purchase St. Fall River, Mass.

Watkins, Frederick A.   A
District Sales Agent, Pittsburgh Reflector & Illuminating Co., 565 W. Washington St., Chicago, Ill.

Webber, Lewis V.   M
Manager, Jefferson Glass Co., Ltd., 285 Beaver Hall Hill, Montreal, Canada.

Weiler, Edward W.   A
Edison Lamp Works of G. E. Co., Harrison, N. J.

Weiner, Paul S.   A
Weiner & Weiner, Food Consulting Chemists, Room 604, 505 Fifth Ave., New York, N. Y.

Wells, Walter F.*   A
Vice-President & General Manager, Brooklyn Edison Co., Inc., 360 Pearl St., Brooklyn, N. Y.

Welsh, Don M.   A
Assistant Engineer, Interborough Rapid Transit Co., Room 508, 165 Broadway, New York, N. Y.

Weniger, Willibald   A
Professor of Physics, Oregon State College, Corvallis, Oregon.

Wenzel, Herman W.   A
Assistant District Manager, Branch Office, Philadelphia Electric Co., 4382 Main St., Manayunk, Philadelphia, Pa.

Werner, Louis H.   A
Vice-President, Central Union Gas Co., 529 Courtland Ave., Bronx New York, N. Y.

Wernert, A. L.   A
Salesman, Commercial Dept., Texas Central Power Co., First Nat'l Bank Bldg., San Antonio, Tex.

Wessels, W. T.   A
Sales Engineer, John Y. Parke Co., 31 N. 7th St., Philadelphia, Pa.

West, James William   A
Plant Engineer, The Forbes Lithograph Manufacturing Co., Chelsea, Mass.

Westcott, John Townsend   A
Ladbroke, Chine Crescent, West Cliff, Bournemouth, England.

Westermaier, Francis Victor   A
Engineer, Welsbach Street Lighting Co. of America, 1934 Market St., Philadelphia, Pa.

Weston, Edward   M
President, Weston Electrical Instrument Co., Waverly Park, Newark, N. J.

Whaling, Thomas G.   M
General Manager, Westinghouse Lamp Co., 165 Broadway, New York, N. Y.

Wheat, Herbert C.   A
The British Thomson-Houston Co., Ltd., Rugby, England.

Wheeler, Harry E.   M
Counsellor in Industrial Organization & Business Administration, P. O. Box 247, General Post Office, New York, N. Y.

Wheeler, L. E.   A
District Illuminating Engineer & Street Lighting Specialist, Westinghouse Electric & Manufacturing Co., 1803 Union Bank Bldg., Pittsburgh, Pa.

White, Albert H.   A
Lighting Representative, Public Service Electric Co., 80 Park Place, Newark, N. J.

White, Bryant   M
General Supt., Electrical Engineer, Lynchburg Traction & Light Co., Box 725, Lynchburg, Va.

White, E. Cantelo   M
President, The Tork Co., 8 W. 40th St., New York, N. Y.

White, Wilfred F. **A**
Illuminating Section, Westinghouse
Electric & Manufacturing Co.,
Co., George Cutter Works, Notre
Dame St., South Bend, Ind.

Whitely, Benjamin **A**
Consolidated Gas Co., 130 E. 15th
St., New York, N. Y.

Whiting, Herbert S. **M**
Room 4839, Grand Central Terminal,
New York, N. Y.

Whitney, E. H. **M**
Westinghouse Electric & Manufac-
turing Co., George Cutter Works,
South Bend, Ind.

Whitney, R. F. **M**
President, Fall River Electric Light
Co., 85 North Main St., Fall
River, Mass.

Whitney, Willis R. **A**
Director of Research Laboratory,
General Electric Co., Schenectady
N. Y.

Whitton, W. H.* **A**
General Agent, New York Edison
Co., Irving Place & 15th St., New
York, N. Y.

Wilcox, Norman T. **A**
Mississippi River Power Co., Keo-
kuk, Iowa.

Wiley, Charles T. **M**
Illuminating Engineer, The National
Screw & Manufacturing Co., A.B.
Division, 2440 75th St., Cleveland,
Ohio.

Williams, Arthur* **M**
Vice-President, 1908–09.
Director, 1911–13.

General Commercial Manager, New
York Edison Co., Irving Place &
15th St., New York, N. Y.

Williams, L. M. **A**
New Business Man, Wisconsin Power
Light & Heat Co., Ripon, Wis.

Williams, R. O. **A**
Sales Manager, Harter Manufactur-
ing Co., 1850 Fulton St., Chicago,
Ill.

Willis, Ben S. **A**
Assistant Physicist, Bureau of Standards, Washington, D. C.

Wilson, Elmer D. **A**
Electrical Contractor, 209 Clinton
Ave., Newark, N. J.

Willy, John **A**
Editor & Publisher, The Hotel
Monthly, 443 S. Dearborn St.,
Chicago, Ill.

Wilson, Alexander **A**
Consulting & Supervising Engineer,
Room 614, New Birks Bldg. Mon-
treal, Quebec, Canada.

Wilson, Alexander Massey **M**
Professor of Electrical Engineering,
University of Cincinnati, Cincin-
nati, Ohio.

Wilson, Edwin S. **A**
District Manager, Philadelphia Elec-
tric Co., 3100 Kensington Ave.,
Philadelphia, Pa.

Wilson, Ray H. **A**
Sales Engineer, X-Ray Reflector Co.,
of New York, 31 W. 46th St., New
York, N. Y.

Wily, James H. **A**
Assistant Professor of Physics, Le-
high University, Dept. of Physics,
South Bethlehem, Pa.

Winetsky, Michael **A**
Meterman, Public Service Electric
Co., 71 Murray St., Elizabeth,
N. J.

Winkler, Clem R. **A**
Sales Engineer, Arkansas Central
Power Co., 115 W. 4th St., Little
Rock, Ark.

Winslow, William H. **A**
Vice-President, Superior Water Light
& Power Co., 1516 Tower Ave.,
Superior, Wis.

Wise, John E. **A**
Instructor, Electrical Engineering,
Electric Lab., University of Wis-
consin, Madison, Wis.

Wiske, Prescott B. **M**
Manager, New Business Department,
Brooklyn Union Gas Co., 176
Remsen St., Brooklyn, N. Y.

Wohlauer, Alfred A. **M**
Allied Engineering Co., Inc., 1400
Broadway, New York, N. Y.

Wolf, J. L. **A**
Secretary, Lighting Fixture Dealers
Society of America, Builders Ex-
change, Cleveland, Ohio.

Wolff, William B.　　　　　　　　A
Manager, Lighting Dept., Electrical
Construction Sales Co., 934 Pros-
pect Ave., Cleveland, Ohio.

Wood, Arch K.　　　　　　　　　A
General Supt., McKee Glass Co.,
Jeannette, Pa.

Wood, Douglass　　　　　　　　M
General Manager, Colonial Electric
Div. of G. E. Co., 404 Chamber
of Commerce Bldg., Pittsburgh,
Pa.

Wood, E. Ellsworth　　　　　　　A
Engineering, Minature Incandescent
Lamp Corp., 95 Eighth Ave.,
Newark, N. J.

Wood, Edwin J.　　　　　　　　A
Charge of Lighting Salesmen, Narragansett Electric Lighting Co.,•
814 Turks Head Bldg., Providence,
R. I.

Wood, L. A. S.　　　　　　　　M
Manager Illuminating Section, Westinghouse Electric & Manufacturing Co., George Cutter Works,
South Bend, Ind.

Wood, Howard I.　　　　　　　　A
Manufacturing Engineer, Edison
Lamp Works of G.E.Co., Harrison,
N. J.

Wood, William G.　　　　　　　M
Vice-President & Manager, Berke-
ley Light Corporation, Humboldt
Bank Bldg., San Francisco, Calif.

Woods, George E.
Assistant & Chief Engineer, Consolidated Gas Co. of New York,
130 E. 15th St., New York, N. Y.

Woods, William H.
Toronto Hydro-Electric System, 226
Yonge St., Toronto, Ontario, Can-
ada.

Woodwell, Julian Ernest*　　　　M
Director, 1907–09.
501 Fifth Ave., Rooms 2214-15, New
York, N. Y.

Woolfson, Monroe George　　　　M
Assistant to Engineer of Lamp
Tests, Electrical Testing Labora-
tories, 80th St., & East End Ave.,
New York, N. Y.

Wooster, L. F.　　　　　　　　A
Apperson Hall, Oregon Agricultural
College, Corvallis, Oregon.

Work, William M.　　　　　　　A
Sales Dept., George Cutter Works,
Westinghouse Electric & Manu-
facturing Co., South Bend, Ind.

Worssam, Frank H.　　　　　　A
Lamp Salesman, Pettingell-Andrews
Co., 511 Atlantic Ave., Boston,
Mass.

Worthing, A. G.　　　　　　　M
Physicist, Nela Research Laboratory,
National Lamp Works of G. E. Co.,
Nela Park, Cleveland, Ohio.

Wright, Howard L.　　　　　　A
National X-Ray Reflector Co., 750
Prospect Ave., Cleveland, Ohio.

Wylie, Arthur G.　　　　　　　A
157 Pleasant St., Holyoke, Mass.

Xylander, Peter　　　　　　　　A
New Business Dept., Public Service
Gas Co., 235 Main St., Hacken-
sack, N. J.

Young, Fred R.　　　　　　　　A
Salesman, Sieck-Boyington Electric
Co., 211 West Main St., Freeport,
Ill.

Young, James W.　　　　　　　M
P. O. Box 244, Wilkes Barre, Pa.

Young, P. S.　　　　　　　　　M
Director, 1917–20.
Vice-President, Public Service Corp.
of N. J., 80 Park Place, Newark,
N. J.

Young, Richard R.　　　　　　A
New Business Agent, Public Service
Gas Co. & Public Service Electric
Co., 80 Park Place, Newark, N. J.

Younglove, G. Wilson　　　　　A
Student, University of Michigan,
Ann Arbor, Mich. Home address:
1308 E. Ann Arbor, Mich.

Zieme, Harry　　　　　　　　A
Supt. New Business Dept., Penn
Public Service Corp., 222 Lever-
good St., Johnstown, Pa.

Zillessen, Clara H.　　　　　　A
Asst. Advertising Manager, Philadelphia Electric Co., 1000 Chestnut St., Philadelphia, Pa.

Zimmerman, J. Harold   A
  Illuminating Engineer, Illuminating
  Bureau, Westinghouse Lamp Co.,
  165 Broadway, New York, N. Y.

Zimmerman, P. B.   A
  Assistant   Advertising   Manager,
  National Lamp Works of G. E.
  Co., Nela Park, Cleveland, Ohio.

Zoller, F.   A
  Philadelphia Electric Co., 319 S.
  51st St., Philadelphia, Pa.

Zorger, William H., M.D.   M
  President, Zorger Lens Company, 5
  Main St., Champaign, Ill.

MEMBERS OF SECTIONS
AND CHAPTERS

Chicago Section

Allen, J. H.
Arbuckle, Samuel F.
Arenberg, A. L.
Ashinger, H. H.
Auler, Henry
Axman, Daniel
Balkam, Herbert H.
Ball, Wm. J.
Bates, Walter A.
Battey, Paul L.
Bay, Charles H.
Beardsley, Daniel H.
Benjamin, R. B.
Bennett, Edward
Bernhard, F. H.
Braun, W. C.
Brooks, Morgan
Brown, R. B.
Burrill, Paul C.
Cadby, John N.
Callender, C. L.
Callender, D. E.
Carl, Frederick
Carson, Clarence
Cattell, J. H.
Channon, H. O.
Clarke, Frank A.
Clarke, Goerge R.
Close, R. C.
Clover, George R.
Coffin, F. A.
Collamore, Ralph
Collar, Olcott N.
Copley, I. C.
Curtis, Augustus D.
Curtis, Kenneth
DeLay, Frederic A.
D'Olive, Eugene R.
Dolkart, Leo
Domoney, E. R.
Ericksen, Thomas
Farmer, F. Lee
Ferguson, L. A.
Fleming, E. F.
Foeller, Henry A.
Foltz, Leroy Stewart
Ford, Frederick Howe
Fowle, Frank F.
Fox, Wm. A.
Freeman, E. H.
Gilchrist, James M.
Godley, Charles E.
Goodrich, William M.
Gottsche, A. L.
Grossberg, A. S.
Haines, E. H.
Hamm, William S.
Harries, George H.
Harris, Arthur C.
Hecker, Louis M.
Heman, S. A.
Herrmann, W. S.
Herron, J. C.
Hickox, Norman B.
Higbie, H. H.
Higgins, A. W.
Hill, George A.
Hoeller, Otto A.
Hoeveler, John A.
Howard, G. T.
Huffnagle, Arthur A.
Hulswit, Frank

Humez, J. F.
Hyldahl, Niels
Ilgner, H. F.
Illing, I. L.
Insull, Samuel
James, L. V.
Johnson, Alfred
Johnson, O. L.
Johnston, Richard J.
Kauffman, Paul F.
Ketch, J. M.
Kellogg, Raymond C.
Kinney, Raymond C.
Kirk, James J.
Kirlin, Ivan
Knight, A. R.
Koenes, Orman
Kruse, O. J.
Lent, A. L.
Locker, Frank H.
Lyman, James
McKinlock, George A.
Martin, Allen J.
Matchett, Fred D.
Mayo, J. F.
Meyer, August H.
Mistersky, F. R.
Morrison, Montford
Mudgett, Guernsey F.
Munroe, Charles C.
Neiler, Samuel G.
Nelson, George E.
Netting, C. J.
Nicholai, G. O.
O'Leary, Joseph J.
Olmstead, Henry C.
Outley, E. J.
Patterson, G. W.
Pearl, Allen S.
Peaslee, W. D. A.
Piez, K. A.
Pim, Frank A.
Polacheck, Phil.
Quivey, Wyllis E.
Reast, F. M.
Rider, T. J., Jr.
Rogers, F. A.
Ryan, John J.
Scheible, Albert
Schuchardt, R. F.
Schweitzer, Edmund O.
Shaw, Harold
Shaw, H. J.
Shute, J. M.
Smith, Arthur G.
Smith, Raymond A.
Stahl, Charles J.
Stair, J. L.
Steele, Walter D.
Strait, E. N.
Sweet, A. J.
Teberg, E. J.
Thompson, E. L.
Tillson, Edwin D.
Timm, Edward W.
Trott, W. J.
Trueax, Clyde P.
Van Derzee, G. W.
Vanzwoll, H. B.
Vaughn, F. A.
Van Volkenburg, Ray
Wager, A. J.
Wagschal, George
Ward, Thomas G.

Watkins, Fred A.
White, Wilfred F.
Whitney, E. H.
Williams, L. M.
Wiley, Chas. T.
Williams, R. O.
Willy, John
Winslow, W. H.
Wise, J. E.
Wood, L. A. S.
Work, W. M.
Young, Fred R.
Younglove, G. Wilson
Zorger, Wm. H.

New England Section

Africa, Walter G.
Barnes, Albert M.
Barnes, Cyrus
Barnes, H. Alden
Batson, Walter V.
Beedle, H. W.
Beldon, F. A.
Bicknell, George W.
Blackmore, Charles T.
Bliss, Frederick W.
Boss, Frederic A.
Brown, Harry W.
Brown, Richard B., Jr.
Brown, Sumner C.
Brown, Wdnell S.
Bryant, Alice G.
Bucknam, Paul C.
Burnham, Roy R.
Cairnie, H. W.
Campbell, John
Chadbourn, Ralph W.
Chamberlin, Guy N.
Clark, Jonn C. D.
Clark, Wm. E.
Clifford, H. E.
Cogan, D. E.
Coghlin, Peter A.
Cowles, J. W.
Crownfield, David
Daniels, Julius
Davis, Charles S.
Devine, Alfred W.
Drinker, Philip
Drisko, William J.
Eames, Paul H.
Edgar, C. L.
Elliott, E. L.
Emerson, Guy C.
Engelfried, Henry O.
Eteson, Franklin C.
Farmer, C. E.
Fitch, Walter S.
Flaherty, H. J.
Fobes, George Shaw
Freeman, Frederick C.
Frisbie, Walter L.
Gaffey, John J.
Gallagher, F. A., Jr.
Gens, Morris H.
Gibbs, L. D.
Gifford, N. W.
Goodale, Samuel P.
Gosling, Edward P.
Greene, Joseph E.
Griffin, G. Brewer
Griffin, J. H., Jr.

Griswold, Clarence F.
Hale, Robert S.
Halstead, Charles W.
Halvorson, C. A. B., Jr.
Hamblen, Ephraim S.
Harty, Edgar A.
Hill, Charles F.
Hirons, Frank K.
Hopkins, Raymond A.
Howe, Floyd W.
Hubbard, Arthur S.
Hurley, E. D.
Hussey, R. B.
Ingalls, L. O.
Ives, J. Nash
Jackson, D. C.
Kellogg, Alfred S.
Kennelly, Arthur E.
Kenslea, Daniel L.
Kerens, J. T.
Keyes, S. R.
Kinney, Clarence W.
Knight, Carl D.
Lancaster, W. B.
Laughton, A. A.
Lawrence, R.
LeClear, Gifford
Lisle, Arthur B.
Livor, J. E.
Lynch, J. H.
MacGuffin, W. D.
Magnuson, A. H.
Manter, E. W.
Marcou, A. W.
Marsh, L. W.
Mather, E. H.
McGregor, Wm.
Mixer, Chas. A.
Morrison, Henry Kent
Murray, J. Frank
Nelson, Albert F.
Norcross, Josiah C.
O'Donnell, William W.
Parker, Charles H.
Parker, Edwin S.
Parnell, Eric
Patrick, James S.
Perkins, Oscar H.
Pevear, Munroe R.
Piser, Theo. H.
Potter, Herbert S.
Potter, J. Lee
Price, Frank S.
Priest, Lucian C.
Ralston, F. W.
Riley, Milton D.
Roberts, Edward S.
Rogers, Stephen C.
Sampson, A. T.
Sanderson, Frank N.
Sands, H. T.
Sargent, Fred H.
Shohan, Abraham
Sias, Arthur G.
Smith, Alphonso H.
Smith, Fred H.
Spaulding, H. T.
Standish, Myles
Stevens, E. E.
Tallman, Vernon M. F.
Thompson, Elihu
Timper, Walter R.
Tingley, Louisa P.
Toohey, Thomas J.

Topham, Richard M.
Troland, Leonard T.
Tudbury, John L.
Tufts, Bowen
Vaughan, Hollis B.
Walker, Clarence C.
Wallace, Herbert F.
Wallace, Wyndham S.
Wallis, Louis R.
Waterman, F. E., Jr.
West, James W.
Whitney, R. F.
Wood,Edwin J.
Worssam, Frank H.
Wylie, Arthur G.

New York Section

Abell, H. E.
Acheson, A. R.
Addicks, W. R.
Ainsworth, George
Alger, Ellice M.
Allen, H. V.
Alexander, George L.
Andersen, I. W.
Anderson, O. P.
Andrews, Wm. S.
Ashley, Edward E., Jr.
Ashworth, Jas.
Atkins, David F.
Atwater, De N. W.
Babson, Thomas E.
Barnes, H. F.
Barnitz, Frank R.
Barr, T. M.
Barrett, Sampson K.
Barstow, W. S.
Bartlett, John S.
Barton, Charles A.
Battle, Robert T.
Baylis, R. V.
Beach, R. L.
Beal, Charles A.
Beal, Thaddeus R.
Bechtel, E. J.
Beggs, E. W.
Bell, Archibald D.
Bell. Howard H.
Benford, F. A., Jr.
Berens, Conrad
Bernhard, Albert H.
Betts, Philander
Birge, Nathan R.
Blackwell, W. T.
Blake, S. H.
Bliven, Joseph E.
Bostock, Edgar H.
Botsford, C. J.
Boyce, Ernest W.
Bozell, Harold V.
Bradbury, Edward L.
Brady, N. F.
Braun, Eugene
Briggs, Wallace W.
Brown, David S.
Brown, Harry
Brundage, Henry M.
Buckley, James E.
Bull, John H.
Burchard, A. W.
Burnap, Robert S.
Burr, Alfred R.
Burr, C. B.
Burrows, W. R.
Burton, Robert B.
Calahan, H. C.
Caldwell, O. H.
Cameron, David G.

Campbell, C. J.
Carder, Frederick
Carter, Robert A., Jr.
Cary, I. B.
Cary, Walter
Chandler, C. F.
Chapin, F. S.
Clark, John F.
Clark, Wm. J.
Clasen, Arthur J.
Clewell, George A.
Clinch, Edward S.
Cohn, Leo
Colby, Charles C.
Cole, Rex J.
Colket, J. H.
Conant, Lewis C.
Cook, Henry A.
Corraz, Zoe N.
Cressman, Russell B.
Crockett, Chas. H.
Crofoot, Clarence E.
Crosby, Halsey E.
Curtis, Darwin
Cutler, C. W.
Dahl, Thure
Dailey, F. J., Jr.
Davidson, W. F.
deKosenko, Stepan
Dempsey, W. T.
Dennington, Arthur R.
Desaix, H. W.
Deshler, Charles
Detweiler, Paul George
Dibelius, Ernest F.
Dick, A. C.
Dickerson, A. F.
Diggles, G. L.
Diggs, D. M.
Doherty, Henry L.
D'Olier, Henry, Jr.
Donahue, E. J.
Doolittle, C. Melvin
Dorting, E. E.
Dougall, A. R.
Downs, W. S.
Doyle, Walter H.
Duane, Alexander
Duncan, F. B.
Dunn, Gano
Edison, Thomas A.
Ely, Robert B.
Emerson, Harrington
Evans, W. A. D.
Farrand, Dudley
Fee, C. Edward
Feiker, Frederick M.
Fenniman, J. R.
Ferguson, Lynn W.
Ferris, C. H.
Field, Oscar S.
Firehock, Milton B.
Fischer, B. F., Jr.
Fisk, Eugene W.
Foersterling, H.
Fogg, Oscar H.
Forstall, Alfred
Fox, Clifford S.
Fox, E. G.
Francisco, Ellsworth
Franck, Charles
Francklyn, Gilbert
Freeman, C. K.
Frei, Fred J.
Fuchs, Theodore, Jr.
Fuller, Carl T.
Gage, Henry P.
Gage, Otis Amsden
Gallo, Charles

Garms, Peter
Gawtry, Lewis B.
Gillinder, Edwin B.
Glameyer, William, Jr.
Gleason, M. T.
Goepel, C. P.
Goldmark, C. J.
Goodwin, H., Jr.
Goodwin, Wm. L.
Gould, Herman P.
Graf, Carl H.
Graff, Wesley M.
Graham, Malcom
Graves, L. H.
Gray, Samuel McK.
Greiner, Robert E.
Gunnison, Foster
Hahn, Harold W.
Hake, William S.
Hall, A. H.
Hall, James D.
Hanlan, James P.
Hannum, J. E.
Hartman, H. V.
Hatzel, J. C.
Hawkins, Lawrence A.
Haynes, Pierre E.
Heckmann, L. F., Jr.
Hendry, W. F.
Henry, Charles T.
Hertz, Adolph
Hibben, Samuel G.
Higgens, C. M.
Hobbie, Edward H.
Hodge, Percy
Holmes, James Thos.
Horn, Carrie E.
Horner, Merritt, Jr.
Howard, Harold W.
Howell, John W.
Howland, Lewis A.
Hoyt, W. Greeley
Hubbard, Wm. C.
Hubert, Conrad
Humphreys, Alex. C.
Hunter, George Leland
Hunter, James F.
Huntley, W. R.
Hyde, Edward B.
Irwin, Beatrice
Johnston, J. K.
Jones, Bassett
Jones, Loyd A.
Jones, T. I.
Jourdan, J. H.
Kaulke, Johannes
Kelcey, G. G.
Kennedy, J. J.
Kennedy, J. S.
Keuffel, Carl W.
Killeen, John F.
Kinsey, Frederick S.
Kirby, Daniel B.
Kittle, Robert G.
Knight, Charles
Krapf, Edgar W.
Kurlander, John H.
Lang, J. Gustaf
Lansingh, V.
Lathrop, A. P.
Law, Clarence L.
LePage, Clifford B.
Levy, Mav J.
Lewinson, L. J.
Lieb, John W.
Little, W. F.
Lindsay, Carl C.
Lissfelt, H. L.
Livingston, Robert E.

Luqueer, Robert O.
Macbeth, Norman
Macdonald Norman D.
Mackall, Kenneth W.
MacMullen, Charles
Macy, Carleton
Magee, Ralph R.
Mahan, Harwood E.
Mailloux, C. O.
Malone, P. S.
Mange, John I.
Mark, Isaac, Jr.
Marks, Louis B.
Marthai, W. G.
Martin, Thomas C.
Maurer, R. H.
M'Coy, W. R.
McDonald, Donald
McGraw, James H.
McGuire, Frederick J.
McKay, Wm. A.
McLaughlin, Jos. J.
McManis, Thos. J.
McMurtry, Alden L.
McParland, John J.
McQueston, J. H.
Meara, W. J.
Miley, Frank S.
Millar, Harold H.
Millar, Preston S.
Miller, Arthur
Miller, Joseph C.
Mills, E. A.
Moore, D. McFarlan
Morris, W. Cullen
Morrison, George F.
Morrison, William G.
Motschenbacher. A.
Mott, William Roy
Murray, Thos. E.
Musson, Chester A.
Myers, R. E.
Nast, Cyril
Nettleton, Chas. H.
Neumuller, Walter
Newkirk, E. F.
Newton, Arthur H.
Nichols, Edw. L.
Nichols, G. B.
Nodell, W. L.
Norcross, J. A.
Norris, B. H.
Norton, C. A.
Norton, Guy P.
Oday, A. B.
Ogden, Kenneth C.
OMeara, Thomas J.
Onken, Wm. H., Jr.
Osborne, G. C.
Osborne, L. A.
O'Shea, James P.
Otis, H. G.
Owens, H. Thurston
Page, A. D.
Palmer, H. C.
Palmer, Ray
Parker, John C.
Parker, Karr
Parker, R. B.
Parkhurst, George W.
Pate, C. B.
Peck, Edward L.
Peck, Eugene H.
Pembleton, F. D.
Perry, Allen M.
Pfenning, Phil. C.
Piasecki, H. A.
Plaut, Herman
Plaut, Richard L.

Pope, A. A.
Porter, L. C.
Porter, R. A.
Poser, Max
Post, R. W.
Powell, Alvin L.
Prigge, John, Jr.
Prince, J. L.
Probst, Louis M.
Radcliff, John P., Jr.
Rademacher, W. H.
Reinach, Hugo B.
Renshaw, E. N.
Rhodes, S. G.
Rice, E. W., Jr.
Rice, E. Y.
Richards, Jesse
Robb, William L.
Rodgers, Ralph C.
Rogers, Henry B.
Rolinson, W. H.
Rosenquest, E. H.
Ross, R. G.
Ryan, W. D'A.
Sasse, Frederick C.
Satter, H. A.
Schachter, M. F.
Schladt, G. J.
Schroeder, Henry
Schwartz, H. M.
Scott, Charles F.
Searle, R. M.
Shackelford, Benj. E.
Sharp, Clayton H.
Shay, John
Sheibley, Frank D.
Sherwood, Edward L.
Silverman, Philip
Simons, S. A.
Simpson, R. E.
Sinclair, H. A.
Slocum, C. A.
Smith, E. A.
Smith, F. C.
Smith, Frank W.
Smith, Helen A.
Snyder, A. E.
Snyder, H. L.
Spencer, Melvin
Spencer, W. H.
Spindell, David
Sproule, Thomas
Stanley, Robert W.
Stansel, N. R.
Steinmetz, C. P.
Sterling, W. T.
Stevick, C. H.
Stickney, George H.
Stoll, Albert F.
Stone, C. W.
Story, Ernest D.
Stott, Edward B.
Strahan, George V.
Strauss, Lawrence L.
Strebig, Charles F.
Strecker, A. H.
Strong, Jas. R.
Summers, John A.
Swallow, Joseph G.
Sweeney, George J.
Sylvester, Elmer L.
Taggart, Ralph C.
Tait, Frank M.
Tansey, Frank D.
Taylor, Frank C.
Taylor, Jno. B.
Taylor, Mabel H.
Teller, J. Paul
Thomas, C. G. M.

Thomas, Stephen A.
Thompson, Roy B.
Thomson, Geo. L. A.
Tiffany, Edward L.
Tilden, E. A.
Tizley, A. J.
Toohey, James A.
Torchio, Philip
Tousey, Sinclair
Tousley, Ben
Tuck, Davis H.
Turner, A. S.
Umbach, W. R.
Van Bloem, P. Schuyler
Van Gieson, C. J.
Vassar, H. S.
Vincent, Elliott W.
Walsh, Vincent
Warner, G. H.
Weiler, Edward W.
Wejner, Paul S.
Wells, Walter F.
Welsh, Don M.
Werner, Louis H.
Weston, Edward
Whaling, Thomas G.
Wheeler, Harry E.
White, Albert
White, E. Cantelo
Whiteley, Benj.
Whiting, Herbert S.
Whitney, Willis R.
Whitton, W. H.
Williams, Arthur
Wilson, Elmer D.
Wilson, Ray H.
Winetsky, Michael
Wiske, Prescott B.
Wohlauer, Alfred A.
Wood, E. Ellsworth
Woodwell, Julian
Wood, Howard I.
Woods, Geo E.
Woolfson, M. G.
Xylander, Peter
Young, P. S
Young, R. R
Zimmerman, J. Harold

Philadelphia Section

Allen, W. C.
Althouse, Daniel M.
Andersen, H. B.
Antrim, Wm. D.
Arnold, M. Edwin
Atmore, A. L.
Bartlett, P. H.
Beadenkopf, Geo.
Berger, Chas. F.
Biddle, Robert
Bodine, Saml. T.
Bonner, John
Brady, Edward, Jr.
Brock, Arthur, Jr.
Bruce, Howard
Brumbaugh, G. Edwin
Burke, Thos.
Burnett, Douglass
Calhoun, Ervin
Calvert, H.
Castor, W. A.
Chase, Philip H.
Christman, H. S.
Christy, Henry B.
Clark, Walton
Cleaver, N. S.
Clewell, C. E.

Coahran, Jesse M.
Cohn, Chas. M.
Coleman, Herman
Coleman, Noah
Conner, Geo. C.
Corbus, Frederick G.
Cordner, A. R.
Crampton, G. S.
Crittenden, E. C.
Crosby, Joseph
Currier, B.
Davies, H. W.
Davis, Ernest H.
Davy, Humphrey
Day, B. Frank
DeLacey, Thos.
Devereux, Washington
Dickey, Chas. H.
Dicky, E. S.
Dixon, Fred R., Jr.
Doane, Francis H.
Dodge, Kern
Dolbier, F. Van Buren
Donley, W. H.
Dougherty, J. E.
Doughty, H. C.
Douthirt, W. F.
Duncan, R. B.
Dunlap, Knight
Dyre, Walter T.
Eckstein, Herman
Eglin, W. C. L.
Eichengreen, L. B.
Emerich, L. C.
Evans, William H.
Ewing, R. E.
Ferree, C. E.
Firth, C. J.
Fisher, J. Carl
Fleisher, Walter A.
Fogler, William A.
Forstall, Walton
Foster, Wm.
Frechie, Joseph E.
Fulweiler, W. H.
Fussell, Lewis
Ganser, Herbert H.
Gartley, W. H.
Gast, Fred W.
Getz, A. M.
Gibson, F. J.
Gilbert, William M.
Graham, E. C.
Green, Geo. Ross
Greenewalt, Mary H.
Gribbel, John
Gritzan, L. Leroy
Hall, Wm. Parker
Hatch, B. E.
Helme, W. E.
Henninger, John
Henry, J. R.
Hibbs, W. J.
Hinton, James W.
Hoadley, Geo. A.
Hubbert, Edw. I.
Humpfer, George, Jr.
Humphrey, A. F.
Huse, Merritt C.
Hyde, E. N.
Irwin, Walter E. L.
Israel, J. D.
Ives, James E.
Jackson, Thomas H.
Jarrard, O. O.
Johnson, W. H.
Jones, W. R.
Kase, Daniel B.
Keenan, John R.

Kelley, Bayard J.
Kimball, H. H.
Knight, J. Harmer
Knoedler, E. L.
Koerber, Jerome A.
Koockogey, H. A.
Lee, James D.
LeVan, Milton B.
Leveridge, R. H.
Lever, Thomas S, Jr.
Lewis, Frank L.
Lillie, Lewis
Liversidge, H. P.
Lloyd, M. G.
Lochart, Wm. J.
Long, Walter E.
Lyon, Howard
McAleer, W. C.
McAllister, A. S.
McArdle, Leon C.
McCall, Joseph B.
McGowan, Robert H.
McIlhenny, John D.
MacCreery, R. B.
Markley, Ralph E.
Marley, Robert M.
Marshall, W. H
Mason, Sidney
Megonigal, Wm.
Mervish, I. J.
Meyer, John
Miller, Alten S.
Mohr, H. K.
Monville, Francis X.
Moore, David H.
Moxey, Louis W., Jr.
Moyer, Wayne Rettew
Myers, Joseph B
Neel, William Trent
Newbold, E. S.
Norris, George T.
Norris, Rollin
Odenath, H. E.
Parker, J. S.
Patterson, Robert B.
Perrot, Emile
Peterson, C. A.
Pfeil, Hubert H.
Phelan, G. Stanley
Prichett, George P.
Proffatt, C. P.
Pullen, M. W.
Ralston, P. C.
Ramsey, Harold E.
Rasin, Unit
Reeder, C. L.
Regar, G. Bertram
Reilly, J. J.
Richman, H. H.
Risdon, T. J, Jr
Robertson, E. E.
Robertson, W. L.
Rolston, R. J.
Rusby, J. M.
Russell, Chas. J.
Rutledge, F. J.
Saunders, William E.
Schildhauer, Edward
Schneider, G. A.
Seaman, Joseph B.
Sellers, Jas. P.
Serrill, Wm. J.
Seville, J. A.
Shaw, Arthur E.
Shaw, Joseph E.
Sheldrake, J. S.
Sholl, Edward L.
Silbert, Richard H.
Simkins, H. M.

Skogland, J. F.
Smith, H. R.
Smith, S. W.
Snyder, Charles S.
Snyder, Samuel
Spaide, G. A.
Spencer, Paul
Spitz, Edward W.
Stetser, Jesse R.
Stewart, Frank H.
Stites, Townsend
Stolpe, Ernest F.
Sylvester, J. Wilson
Tanzer, E. Dean
Thomas, N. Wiley
Torrens, Robert J.
Tyler, R. E.
Vincent, H. B.
Vogan, Frank C.
Wagner, Herbert A.
Wagner, Walter C.
Wall, William L.
Wardell, C. W.
Wenzel, H. W.
Wessells, W. T.
Westermaier, F. Victor
Willis, Ben S.
Wilson, Edwin S.
Wily, James H.
Young, James W.
Zillessen, Clara H.
Zoller, Fred

Cleveland Chapter

Allis, C. L.
Anderson, Earl A.
Atherton, Carlyle A.
Beals, G. W.
Beam, J. W.
Beiswanger, John
Beman, Ralph
Brown, W. C.
Brush, C. F.
Cady, Francis E.
Campbell, James P.
Claparols, Manuel
Cobb, Percy W.
Colville, J. R.
Commery, Eugene W.
Cook, W. H.
Dates, H. B.
Doane, Samuel E.
Dows, Chester L.
Edwards, Evan J.
Egeler, C. E.
Falge, R. N.
Fuerst, G. P.
Fulton, Robert A.
Gorton, Robert E.
Haas, O. F.
Harrison, Ward
Hausmann, George A.
Hofrichter, Charles H.
Holden, Alfred R.
Holladay, L. L.
Johnson, E. H.
Jones, Philip C.
Keller, Phil C.
Kent, L. C.
Ketch, James Henry
Logan, Henry
Luckiesh, M.
Magdsick, H. H.
Mausk, R. E.
Maxwell, V. C.
Merrill, G. S.
Minor, W. F.

Mountcastle, H. W.
Nichols, E. F.
Nieder, Edw.
Perkins, Earl G.
Porterfield, William A.
Randall, J. E.
Rhea, C. J.
Richards, W. E.
Sawyer, Leroy P.
Shenton, R. .
Skiff, Warner M.
Sturrock, Walter
Swindell, Harry
Taylor, A. Hadley
Terry, F. S.
Thompson, Allison J.
Thompson, H. R.
Tremaine, B. G.
Wakefield, Albert F.
Wakefiled, F. W.
Waterbury, John H.
Wolf, J. L.
Wolff, William B.
Worthing, A. G.
Wright, Howard L.
Zimmerman, P. B.

Columbus Chapter

Alexander, L. M.
Bowers, Corwin J.
Caldwell, F. C.
Donley, H. B.
Dorey, W. A.
Eichelberger, C. G.
Evans, Dean P.
Evans, Gladden F.
Fitzpatrick, K., Jr.
Freeman, W. W.
Friedrich, Ernest G.
Gilliam, H. R.
Marmon, Guy D.
Moore, R. C.
Mossgrove, J. R.
Muncy, V. E.
O'Rourke, John A.
Robins, Orrin A.
Rolph, T. W.
Robins, O. A.
Stewart, J. R.
Sulzer, G. A.
Tucker, Edward A.
Wilson, Alexander M.

Los Angeles Chapter

Alden, Walter A.
Anderson, J. F.
Bartlett, C. C.
Behan, R. J.
Buss, A. F.
Church, A. C.
Davis, L. W.
Hitchcock, G. G.
Hobbs, L. A.
Holman, E. R.
Hughes, D. M.
Johnson, C. E.
Kershaw, K. O.
Masson, C. M.
Mayo, H. J.
McBan, P. F.
Mills, F. S.
Nye, A. W.
Pence, D. C.
van Gilluwe, Frank

Michigan Chapter

Arbuckle, Samuel F.
Balkam, Herbert H.
Bay, Charles H.
Carson, Clarence
Clarke, George R.
Clover, George R.
Collamore, Ralph
Ericksen, Thomas
Foltz, Leroy Stewart
Godley, Charles E.
Grossberg, A. S.
Herrmann, W. S.
Higbie, H. H.
Humez, J. F.
Ketch, James M.
Kirlin, Ivan
Lent, A. L.
Locker, Frank H.
Martin, Allen J.
Matchett, Fred D.
Mayo, J. F.
Mistersky, F. R.
Munroe, Charles C.
Nelson, George E.
Netting, C. J.
Outley, E. J.
Patterson, G. W.
Shaw, H. J.
Shaw, Harold
Smith, Raymond A.
Van Volkenburg, Ray
Wagschal, George
Ward, Thomas G.
Younglove, G. Wilson

Northern New Jersey Chapter

Anderson, O. P.
Ashworth, James
Atwater, D. W.
Babson, T. E.
Barnes, H. F.
Bartlett, J. S.
Baylis, R. V.
Beach, R. L.
Beal, C. A.
Beggs, E. W.
Bell, A. D.
Betts, Philander
Blackwell, W. T.
Burnap, R. S.
Burrows, W. R.
Calahan, H. C.
Cohn, Leo
Colby, C. C., Jr.
Cook, H. A.
Dennington, A. R.
Desaix, H. W.
Deshler, Charles
Donahue, E. J.
Doolittle, C. M.
Doughty, H. C.
Downs, W. S.
Edison, Thomas A.
Evans, W. A. D.
Farrand, Dudley
Ferguson, L. W.
Field, O. S.
Firehock, M. B.
Foersterling, H.
Fox, Edw. G. B.
Francisco, Ellsworth
Fuchs, Theodore, Jr.
Fuller, C. T.
Gallo, Charles
Greiner, R. E.

Hahn, H. W.
Hall, J. D.
Hall, M. W.
Hanlan, J. P.
Hodge, Percy
Howell, J. W.
Hubbard, W. C.
Humphreys, A. C.
Kaulke, Johannes
Kelcey, G. G.
Keuffel, C. W.
Kurlander, J. H.
Law, Clarence L.
Leveridge, R. H.
Malone, P. S.
McManis, T. J.
McQueston, J. H.
Meara, W. J.
Moore, D. McF.
Morrison, W. G.
Myers, R. E.
Oday, A. B.
OMeara, T. J.
Osborne, G. C.
O'Shea, J. P.
Otis, H. G.
Page, A. D.
Parker, R. B.
Pembleton, F. D.
Pondelick, R. M.
Porter, L. C.
Powell, A. L.
Rademacher, W. H.
Rogers, H. B.
Ross, R. G.
Schladt, G. J.
Shackelford, B. E.
Sherwood, E. L.
Smith, Helen A.
Snyder, A. E.
Sproule, Thomas
Stanley, R. W. .
Stickney, G. H.
Strebig, C. F.
Strecker, A. H.
Summers, J. A.
Sylvester, E. L.
Thomson, G. L. A.
Tilden, E. A.
Turner, A. S., Jr.
Umbach, W. R.
Van Gieson, C. J.
Vassar, H. S.
Weiler, E. W.
Welsh, D. M.
Weston, Edward
White, Albert
Wilson, Elmer D.
Winetsky, Michael
Wood, E. E.
Wood, H. I.
Xylander, Peter
Young, P. S.
Young, R. R.

San Francisco Bay Cities Chapter

Baker, Clark
Baker, S. A.
Cobby, E. V.
Conlisk, Raimon F.
Cowles, R. Roy
Cravath, J. R.
Dolman, Percival
Donahue, J. N.
Hanscom, W. W.
Harman, George H.
Hixson, Morris C.

Hudson, R. A.
Leurey, Louis F.
Martin, Carl O.
Mayo, Herbert J.
McDonough, Thomas F.
Morgan, Lyman D.
Murphy, F. H.
Myers, Romaine W.
Nord, Carl C.
Osborn, Frederick A.
Percival, J. T., Jr.
Phillips, Charles T.
Prussia, Robert S.
Ross, Leland S.
Rucker, George
Russell, S. P.
Ryan, Harris J.
Sandoval, H. E.
Simonson, G. M.
Steel, Miles F.
Stockwell, W. L.
Symmes, Whitman
Thompson, Robert J.
Vandegrift, J. A.
Voyer, L. E.
Weniger, Prof. W.
Wood, William G.
Wooster, L. F.

Toronto Chapter

Anderson, G. R.
Beauchamp, Leon
Burnett, H. D.
Cousins, George G.
Dion, Alfred A.
Dobson, W. P.
Ewart, Frank R.
Goodchild, A. E.
Gardner, G.
Goward, A. T.
Gray, G. Robin
Hastings, M. B.
Hopper, C. H.
Jaquet, George E.
Kintner, Watson
Labelle, Philip R.
Lloyd, Lorne K.
Love, R. M.
Lyle, A. Ernest
Martin, William A.
McDougall, George K.
Murphy, John
Plumpton, Arthur G.
Schumacher, John H.
Scott, John T.
Stuart, Ralph C.
Turnbull, T. S.
Trimming, P. H.
Webber, Louis V.
Wilson, Alexander
Woods, William H.

Non-Section
(Domestic)

Abott, Arthur L.
Addie, Charles E.
Allen, Asa E.
Anderson, D. S.
Andrews, Edgar M.
Ashworth, R. H.
Berger, William W.
Blakeslee, D. W.
Blumenauer, C. H.
Boiler, W. F.
Brown, R. I.
Browne, W. H. Jr.
Bryant, John M.
Camp H. M.
Carpenter, E. P.
Church, Fermor S.
Clement, William E.
Collier, Charles A.
Crawford, D. F.
Croft, Terrell
Dandley, P. Y.
Dansboe, W. F.
Davidson, E. Y. Jr.
Deskins, Hiram T.
Dodds, George
Dunn, J. M.
Evans, Edward A.
Evans, H. S.
Everson, R. W.
Feingold, Marcus
Ferguson, Olin Jerome
Fiske, John M.
Fitzsimons, T. E.
Flowers, Dean W.
Folsom, C. E.
Ford, A. H.
Foster, Spottswood C.
Fuller, H. W.
Garrison, A. C.
Gawan, Louis B.
Gleeson, T. P.
Golden, Jno. W.
Grant, Albert W., Jr.
Grondahl, L. O.
Gross, E. L.
Guth, Edwin F.
Hake, Harry G.
Hale, H. S.
Hall, Lee D.
Harris, Henry
Hartman, L. W.
Heffron, J. F.
Heyburn, Henry B.
Higgins, W. S.
Hill, Emory
Hill, Marvin
Hobson, Henry Elmer
Hockaday, Irving T.

Hodgson, George Oscar
Holdrege, H. A.
Howard, T. H.
Hower, Harry S.
Hunter, Morris
Jenkins, D. R.
Johnson, L. B.
Johnston, H. L.
Kelly, James B.
Kiefer, Lewis J.
Kilby, Karl B.
Kirschberg, Harold
Kopecky, F. F.
Lee, Claudius
Martin, Edwin R.
Martinson, Arthur N.
McNair, G. B.
McQuiston, J. C.
Moore, Louis D.
Morrow, G. T.
Naland, C. W.
Nathanson, J. B.
Newhouse, R. E.
Nichol, H. G., Jr.
Nutty, G. R.
Orbin, Frank
Owen, Raymond V.
Pinkerton, Andrew
Power, William R.
Prine, C. W.
Putnam, W. R.
Rishell, George L.
Rodman, Walter S.
Roese, Harry E.
Rose, S. L. E.
Savage, Arthur H.
Sawin, George A.
Scheel, Alfred A.
Shaad, Geo. C.
Shepardson, G. D.
Simms, Frederick F.
Smith, Harry H.
Spencer, Eugene J.
Spinney, L B.
Springer, R. D. N.
Stannard, Clare N.
Streich, H C.
Taylor, Dent A.
Tileston, Roland R.
Trawick, S W., Jr.
Upham, W H.
Uptegraff, W. D
Van Bergen, Curtis E.
Walmsley, W. N.
Ward, Charles W.
Warner, Hugh A
Warner, James P
Wernert, A. L.
Wheeler, L. E
White, Bryant
Wilcox, Norman T.

Winkler, Clem. R.
Wood, Arch K.
Wood, Douglass
Zieme, Harry

Non-Section
(Foreign)

Aschner, L.
Bensi, R. J.
Berry, Clarence J.
Besinsky, Vaclav
Beuttell, Alfred W.
Bickerstaff, Ernest
Blondel, Andre
Bossi, Santiago J.
Bowles, Richard
Bramoso, J.
Burgess, A. E. C.
Bush, Wm. Edward
Campbell, Guy
Cohen, Carl
deChatelain, M. A.
deGraaff, A.
Dow, John S.
Fabry, Charles
Fletcher, J. Y.
Froget, Andre
Fujii, Tetsuya
Fukuda, M.
Glasgow, A. G.
Goodenough, F. W.
Halbertsma, N. A.
Harrison, Hayden T.
Hutton, Donald J.
Hyde, Edwin P.
Hyodo, M.
Ishikawa, Yosuta
Ishikawa, Yoshijiro
Kato, Kenjiro
Lingard, Herbert A.
Lix-Klett, Ernesto
McCallion, Wm. J.
Maisonneuve, Henry
Matsuda, C.
Murota, Hannosuke
Neednam, H. H.
Orange, John A.
Oshima, Hiroyoshi
Page, Howard E.
Picado, Ramon M.
Pinkney, W. F. T.
Shikata, K.
Shimidzu, Yoshichiro
Souza, Edgard E.
Stephens, John N.
Sue, Osaki
Uchisaka, Motoo
Ven Dyck, Wm. V. B.
Vogel, Joshua
Wescott, J. T.
Wheat, H. C.